Variational Techniques for Elliptic Partial Differential Equations

Theoretical Tools and Advanced Applications

Variational Techniques for Elliptic Partial Differential Equations

Theoretical Tools and Advanced Applications

Francisco-Javier Sayas
Thomas S. Brown
Matthew E. Hassell

University of Delaware, Rice University, Lockheed Martin

CRC Press
Taylor & Francis Group
Boca Raton London New York

CRC Press is an imprint of the
Taylor & Francis Group, an **informa** business

CRC Press
Taylor & Francis Group
6000 Broken Sound Parkway NW, Suite 300
Boca Raton, FL 33487-2742

First issued in paperback 2020

ISBN-13: 978-1-138-58088-6 (hbk)
ISBN-13: 978-0-367-65664-5 (pbk)

Library of Congress Cataloging-in-Publication Data

Names: Sayas, Francisco-Javier, author. | Brown, Thomas S. (Mathematician), author. | Hassell, Matthew E., author.
Title: Variational techniques for elliptic partial differential equations : theoretical tools and advanced applications / Francisco J. Sayas, Thomas S. Brown, Matthew E. Hassell.
Description: Boca Raton, Florida : CRC Press, [2019] | Includes bibliographical references and index.
Identifiers: LCCN 2018050821| ISBN 9781138580886 (hardback : alk. paper) | ISBN 9780429507069 (ebook)
Subjects: LCSH: Differential equations, Elliptic. | Differential equations, Partial.
Classification: LCC QA377 .S33925 2019 | DDC 515/.3533--dc23
LC record available at https://lccn.loc.gov/2018050821

To the sum of our biological and mathematical ancestries

Contents

Preface   **xiii**

Authors   **xxi**

I   Fundamentals   **1**

1   Distributions   **3**

   1.1   The test space   3
   1.2   Distributions   7
   1.3   Distributional differentiation   10
   1.4   Convergence of distributions   13
   1.5   A fundamental solution (*)   15
   1.6   Lattice partitions of unity   17
   1.7   When the gradient vanishes (*)   20
   1.8   Proof of the variational lemma (*)   22
   Final comments and literature   23
   Exercises   24

2   The homogeneous Dirichlet problem   **27**

   2.1   The Sobolev space $H^1(\Omega)$ 27
   2.2   Cutoff and mollification   29
   2.3   A guided tour of mollification (*)   31
   2.4   The space $H_0^1(\Omega)$ 34
   2.5   The Dirichlet problem   38
   2.6   Existence of solutions   41
   Final comments and literature   43
   Exercises   44

3   Lipschitz transformations and Lipschitz domains   **47**

   3.1   Lipschitz transformations of domains   47
   3.2   How Lipschitz maps preserve H^1 behavior (*)   49
   3.3   Lipschitz domains   52
   3.4   Localization and pullback   56
   3.5   Normal fields and integration on the boundary   59
   Final comments and literature   62
   Exercises   62

4 The nonhomogeneous Dirichlet problem — **65**

4.1 The extension theorem 66
4.2 The trace operator 68
4.3 The range and kernel of the trace operator 70
4.4 The nonhomogeneous Dirichlet problem 73
4.5 General right-hand sides 75
4.6 The Navier-Lamé equations (*) 79
Final comments and literature 83
Exercises 84

5 Nonsymmetric and complex problems — **89**

5.1 The Lax-Milgram lemma 89
5.2 Convection-diffusion equations 93
5.3 Complex and complexified spaces 95
5.4 The Laplace resolvent equations 98
5.5 The Ritz-Galerkin projection (*) 101
Final comments and literature 103
Exercises 103

6 Neumann boundary conditions — **107**

6.1 Duality on the boundary 107
6.2 Normal components of vector fields 108
6.3 Neumann boundary conditions 111
6.4 Impedance boundary conditions 114
6.5 Transmission problems (*) 116
6.6 Nonlocal boundary conditions (*) 118
6.7 Mixed boundary conditions (*) 120
Final comments and literature 122
Exercises 123

7 Poincaré inequalities and Neumann problems — **125**

7.1 Compactness 126
7.2 The Rellich-Kondrachov theorem 128
7.3 The Deny-Lions theorem 129
7.4 The Neumann problem for the Laplacian 132
7.5 Compact embedding in the unit cube 133
7.6 Korn's inequalities (*) 137
7.7 Traction problems in elasticity (*) 142
Final comments and literature 144
Exercises 145

8 Compact perturbations of coercive problems   **149**

    8.1   Self-adjoint Fredholm theorems   150
    8.2   The Helmholtz equation   152
    8.3   Compactness on the boundary   156
    8.4   Neumann and impedance problems revisited   157
    8.5   Kirchhoff plate problems (*)   159
    8.6   Fredholm theory: the general case   162
    8.7   Convection-diffusion revisited   165
    8.8   Impedance conditions for Helmholtz (*)   167
    8.9   Galerkin projections and compactness (*)   169
    Final comments and literature   173
    Exercises   173

9 Eigenvalues of elliptic operators   **177**

    9.1   Dirichlet and Neumann eigenvalues   177
    9.2   Eigenvalues of compact self-adjoint operators   180
    9.3   The Hilbert-Schmidt theorem   182
    9.4   Proof of the Hilbert-Schmidt theorem (*)   185
    9.5   Spectral characterization of Sobolev spaces   188
    9.6   Classical Fourier series   192
    9.7   Steklov eigenvalues (*)   195
    9.8   A glimpse of interpolation (*)   198
    Final comments and literature   200
    Exercises   201

II Extensions and Applications   **207**

10 Mixed problems   **209**

    10.1   Surjectivity   210
    10.2   Systems with mixed structure   213
    10.3   Weakly imposed Dirichlet conditions   217
    10.4   Saddle point problems   221
    10.5   The mixed Laplacian   223
    10.6   Darcy flow   226
    10.7   The divergence operator   228
    10.8   Stokes flow   234
    10.9   Stokes-Darcy flow   236
    10.10 Brinkman flow   242
    10.11 Reissner-Mindlin plates   245
    Final comments and literature   248
    Exercises   249

11 Advanced mixed problems   **253**

11.1 Mixed form of reaction-diffusion problems   253
11.2 More indefinite problems   255
11.3 Mixed form of convection-diffusion problems   259
11.4 Double restrictions   264
11.5 A partially uncoupled Stokes-Darcy formulation   266
11.6 Galerkin methods for mixed problems   273
Final comments and literature   275
Exercises   275

12 Nonlinear problems   **277**

12.1 Lipschitz strongly monotone operators   277
12.2 An embedding theorem   279
12.3 Laminar Navier-Stokes flow   282
12.4 A nonlinear diffusion problem   286
12.5 The Browder-Minty theorem   289
12.6 A nonlinear reaction-diffusion problem   292
Final comments and literature   293
Exercises   293

13 Fourier representation of Sobolev spaces   **295**

13.1 The Fourier transform in the Schwartz class   296
13.2 A first mix of Fourier and Sobolev   300
13.3 An introduction to H^2 regularity   302
13.4 Topology of the Schwartz class   307
13.5 Tempered distributions   311
13.6 Sobolev spaces by Fourier transforms   314
13.7 The trace space revisited   318
13.8 Interior regularity   321
Final comments and literature   323
Exercises   323

14 Layer potentials   **327**

14.1 Green's functions in free space   327
14.2 Single and double layer Yukawa potentials   330
14.3 Properties of the boundary integral operators   333
14.4 The Calderón calculus   338
14.5 Integral form of the layer potentials   341
14.6 A weighted Sobolev space   343

14.7   Coulomb potentials   347
14.8   Boundary-field formulations   351
Final comments and literature   357
Exercises   358

15 A collection of elliptic problems   363

15.1   T-coercivity in a dual Helmholtz equation   364
15.2   Diffusion with sign changing coefficient   370
15.3   Dependence with respect to coefficients   374
15.4   Obstacle problems   379
15.5   The Signorini contact problem   385
15.6   An optimal control problem   387
15.7   Friction boundary conditions   391
15.8   The Lions-Stampacchia theorem   395
15.9   Maximal dissipative operators   396
15.10 The evolution of elliptic operators   399
Final comments and literature   404
Exercises   405

16 Curl spaces and Maxwell's equations   409

16.1   Sobolev spaces for the curl   409
16.2   A first look at the tangential trace   412
16.3   Curl-curl equations   417
16.4   Time-harmonic Maxwell's equations   423
16.5   Two de Rham sequences   428
16.6   Maxwell eigenvalues   431
16.7   Normally oriented trace fields   432
16.8   Tangential trace spaces and their rotations   435
16.9   Tangential definition of the tangential traces   439
16.10 The curl-curl integration by parts formula   444
Final comments and literature   448
Exercises   449

17 Elliptic equations on boundaries   453

17.1 Surface gradient and Laplace-Beltrami operator   453
17.2 The Poincaré inequality on a surface   456
17.3 More on boundary spaces   459
Final comments and literature   461
Exercises   462

Appendix A Review material   **465**

    A.1 The divergence theorem 465
    A.2 Analysis 466
    A.3 Banach spaces 469
    A.4 Hilbert spaces 471

Appendix B Glossary   **477**

    B.1 Commonly used terms 477
    B.2 Some key spaces 478

Bibliography   **479**

Index   **489**

Preface

To summarize in a single sentence, this book offers a self-contained presentation of the basic variational Theory for elliptic partial differential equations in nonsmooth (Lipschitz) domains.

Novelties and treatment

First of all, our goal is to give the reader access to the analytical techniques for elliptic partial differential equations (PDE), namely distributions, Sobolev spaces, variational formulations, etc. Everything will be done keeping Galerkin methods in sight, which makes this textbook ideal for those who plan to continue with finite element and boundary element methods. As opposed to much of the literature, we introduce a theoretical topic (a new definition or a property of Sobolev spaces, and also a theorem of functional analysis) when it is going to be needed for a new PDE problem. That is why Sobolev space theory and Hilbert space functional analysis are developed in parallel with the PDE constructions. We hope that this will give the reader a strong feeling about what is needed at which stage and how it is used.

With very few theorems, we will not provide the proof, but we will tell the reader where to find it. In some cases, we will point out easy to understand proofs that do not require additional knowledge beyond what already appears in this book, but we have preferred not to copy-and-adapt them. In other cases though, the missing proofs require the introduction of more advanced tools in analysis, and we will point out a reference containing the result or from whose contents the result follows.

We include some topics and equations that do not appear in (many) standard introductory books to elliptic PDE. Here is a short list of some of them:

- A careful presentation of the Helmholtz-like equations and their relation with the Fredholm theory.

- Eigenvalues and a hint at the theory of Hilbert spaces defined by generalized Fourier series, as well as results of a special case of Hilbert space interpolation theory.

- A thorough and step-by-step introduction of mixed variational problems with a gallery of examples mainly extracted from fluid mechanics. This includes a careful treatment of the divergence operator (where we will focus on the problem until we identify a part of a proof which we will decide not to include) and the Stokes problem.

- Model problems for Reissner-Mindlin plates and Brinkman flow, exploring how weighted norms can be used to prove weak convergence to the Kirchhoff plate and Darcy flow models respectively.

- The study of mixed formulations for nonsymmetric problems and reaction-diffusion equations, as well as coupled models (like Stokes-Darcy flow), benefiting from an extension of the classical theory of mixed problems.

- A rigorous treatment of the single layer and double layer potentials and their associated boundary integral equations, starting from variational definitions and ending in integral forms.

- The projection and sign-flipping trick that brings out hidden coercivity, which we will call T-coercivity.

- A rigorous treatment of different formulations for steady-state problems for Maxwell equations, including the trace spaces associated to the curl operator.

- The Laplace-Beltrami equation on nonsmooth surfaces.

We will also offer quick glimpses on other topics of interest, like dependence of solution operators with respect to coefficients, or optimal control problems for elliptic PDE.

The first part of the book. The first nine chapters of the book can be used for a rigorous one semester course on elliptic PDE for students who have some fundamental knowledge of real and functional analysis. We list some of the needed key results in Appendix A.1 for easy reference, but it might not be the best idea to try to learn this material from it. In particular, the reader should be comfortable with the very basics of Lebesgue integration (sets with zero measure, the dominated convergence theorem, completeness of L^p spaces) and Banach space theory (the Banach isomorphism theorem, the uniform boundedness principle). We will use two results that are often not included in the basic analysis curriculum, namely the Lebesgue differentiation theorem and Rademacher's theorem on Lipschitz functions. These results are easy to state and understand and we will leave it to the reader to look for their proofs. We do cover in its entirety the basic theory of Hilbert spaces as we need it, including duality theory, spectral theory of compact self-adjoint operators, and the Fredholm alternative. We understand that readers acquainted with the theory of Banach spaces will have already been exposed to many of these concepts (Hilbert spaces are the friendly guys in functional analysis courses), but have made the decision to include everything about them[1].

We present topics slowly but progressively, trying to get to variational PDE as soon as possible, and keeping hypotheses quite general.

1. Introduction to distributions. Just the basics.

2. The homogeneous Dirichlet problem for the Laplace equation, with a review of the Riesz-Fréchet theorem. No hypotheses on the domain, just the Poincaré-Friedrichs inequality is needed. Mollification and cutoff techniques are presented in detail.

3. Lipschitz domains and pullbacks. This will be needed to prove extension theorems, with them density of smooth functions, and with them the trace operator.

4. The trace operator, the trace space, and the non-homogeneous Dirichlet problem.

5. Nonsymmetric problems and problems with complex data and coefficients, ergo, the Lax-Milgram lemma.

6. The normal component in the space $\mathbf{H}(\mathrm{div})$ and how to use it to prescribe Neumann boundary conditions. At this stage, we will still be unable to study the Neumann problem for the Laplacian.

7. Here we prove the Poincaré inequality and with it we treat the Neumann problem for the Laplace equation and some related problems. We will prove the Poincaré inequality as a consequence of a Deny-Lions theorem which will be a corollary of the Rellich-Kondrachov compactness theorem.

8. Problems that need Fredholm theory. First the easy cases, with self-adjoint operators, and then the general case.

9. Problems that need the Hilbert-Schmidt theory. Eigenvalues of elliptic operators.

What is in the second part. The goal of the second part of the book is to introduce the reader to more advanced topics and problems. Part of this will require spending some time in learning more about Sobolev spaces using techniques from harmonic analysis. We will also deal with some nonlinear problems.

---

[1]... with the exception of the few theorems of Hilbert spaces that are actually Banach space results.

10. Darcy flow, the mixed Laplacian, the Stokes equation, Brinkman flow (and its weak limit to Darcy), the Reissner-Mindlin plate model (and its weak limit to the Kirchhoff-Love biharmonic plate equation), the Stokes-Darcy coupled flow.

11. Mixed formulations of reaction-diffusion and convection-diffusion equations, Stokes-Darcy flow coupled through a side condition.

12. Navier-Stokes flow with large viscosity, nonlinear diffusion problems (Lipschitz strongly monotone nonlinear operators) and reaction-diffusion problems (Browder-Minty theory).

13. Equations in free space and local regularity theory for the Laplace equations. This is a fundamental chapter, used to introduce many new tools, among which tempered distributions are included.

14. Layer potentials, integral equations for diffusion problems, boundary-field formulations.

15. Sign-changing diffusion problems (through T-coercivity), dependence of solutions with respect to coefficients, the obstacle problem, the Signorini problem, and Tresca friction. We will also briefly introduce techniques to study the evolution of elliptic systems (heat and wave equations).

16. The steady-state Maxwell equations (eddy currents, zero-frequency problems, time-harmonic problems) slowly building the theory of tangential traces for $\mathbf{H}(\mathrm{curl})$.

17. The Laplace-Beltrami equation on the boundary of a Lipschitz domain.

Where it all started: an account in the first person

How and when did this book get started? Well, it was definitely long ago, but the galaxy was pretty close. It was 1990, and as a senior undergraduate in Applied Mathematics at the Universidad de Zaragoza (Spain), with a keen interest in classical mechanics and theoretical astronomy, I (yours truly, FJS) had to take a class in Partial Differential Equations. This was the class that changed my life[2]. The two semester course was taught by Francisco 'Paco' Lisbona, who had built the course around the variational theory of elliptic PDE and the finite element method.

---

[2]At the end of the semester I gave up on celestial mechanics, which was going to be my graduate topic, and decided to get into finite elements. Little did I know that I would end up doing a thesis on boundary integral equations, which are not 'that close.'

Textbooks were not common in Spanish universities, and so we just sat down in class and took notes, but we were made aware that the material was a mix of Haim Brezis's *Functional Analysis* (which was one of the few advanced books of its kind translated into Spanish) and Pierre-Arnaud Raviart and Jean-Marie Thomas's *Introduction to Finite Element Methods* (this one existed in French only).

Right after graduating, I did my doctoral work with Michel Crouzeix[3], and learned about layer potentials and integral equations (but also his little tricks to find simple direct proofs), from him, from his handwritten notes of a course taught by Martin Costabel at the Université de Rennes, from struggling with Jean-Claude Nédélec's old course-notes on numerics of integral equations, and from other sources that soon became part of the 'what I know but I cannot remember how I learned' area.[4] Mixed formulations and the Stokes-like equations were brought to me through my collaboration and friendship with Salim Meddahi from the Universidad de Oviedo.

In the late nineties, as a tenured member of the same department that had seen me grow, I started teaching a graduate class called 'Mathematical Models in Partial Differential Equations,' which developed first in a format for mathematicians only and then (under the title 'Mathematical Models in Mechanics') in a more digestible shape for a combined crowd of computational mathematicians and mechanical engineers. Albeit without the theoretical depth, that course contained the core of what this book's philosophy is: can we learn just what we need to approach a particular problem?[5]

Fast forward, I am in the US, working for the University of Delaware and I get the chance to teach a theoretical class on 'Elliptic Partial Differential Equations.' I go full throttle and work hard in devising a completely self-contained course that has become the first half of this book. Matt Hassell was one of the students in that class and he eventually earned his PhD under my supervision. We decided to write the notes together as a book. But life happens, and other things took precedence, so we left the book half finished. Then came Tom Brown, another one of my students at UD, and we decided to finish what we started, taking advantage of my teaching of the Elliptic PDE class again. I decided to add a second part including more advanced topics that are of interest for users of Elliptic PDE that are not that easy to find in

---

[3]Michel was my advisor in Rennes, although I never moved away from Zaragoza and Paco acted as my proxy-mentor adviser at home, helping me to get a doctoral degree in Spain, because in those years French degrees were not good enough in Spain (not a joke!) and that was my excuse for being lazy and not moving to France.

[4]My professional life has given me the chance to meet Jean-Claude and Martin in person, and they are spectacular people, bright, funny, and gentle.

[5]By the way, this doesn't mean that you should not learn more about the theoretical tools that you need in your business and just stick to the basics that you will use. This book places the emphasis on what you need at what time. From my point of view, this is reverse-Bourbaki becoming Bourbaki, and I have learned a lot about this view by reading Luc Tartar's lecture notes on Sobolev spaces and PDEs.

textbooks. This forced the introduction of much more theoretical material on the go, and this is where we are now.

I have already mentioned (and it should be understood that I am thanking them with that) some big influences in my learning about elliptic PDE, my two bosses and my colleagues. With time I have learned to appreciate the carefulness and love of details that were part of the four course (double semester each) sequence on Mathematical Analysis that was force-fed to us students in Zaragoza by our 'ruthless' analyst crowd. I might not have liked everything in those courses, I might have loved to learn some other materials, and in retrospect it would have been better (for me!) to learn some of the applications on the go, but the courses were splendidly designed for those of us who grasped them and got to use them as professional researchers. So I will thank all my analysis professors (José Luis Cuadra, Jesús Bastero, Oscar Blasco, José 'Pepe' Galé, and Francisco 'Pacho' Ruiz), because they planted the seeds or fertilized the ground, or some other cheesy metaphor to express how what people teach you and make you learn sticks to you and shapes your future.

Before you start reading

Starred sections in Part I can be skipped in a first reading. We also utilize two types of framed statements.

[Proof not provided]

Theorem 0.1. *This is a result that we will use but not prove. We will keep these to the minimum.*

Warning. This is an important notational or conceptual warning. There will not be many of these.

We hope this two-part book can help instructors and students learn the basic material of variational elliptic PDE in a way that is approachable. Concepts and ideas will often be repeated and we try to guide the reader through long technical arguments. While this text is competing with strong and well-established textbooks, we know what we can offer as well as what we are not trying to do. While the first part of the book can be considered standard, the second part can be used as a source to get acquainted with equations and formulations that are of interest to numerical analysts and computational scientists and are not usually a regular part of the curriculum for researchers in the theory of PDE. Part of this book was written while the first author was IBM Visiting Professor of Applied Mathematics at Brown University. Additionally, all of the authors would like to thank the Department of Mathematical Sciences at the University of Delaware for hosting their work.

Tom Brown (Rice University), Matt Hassell (Lockheed Martin), Francisco-Javier 'Pancho' Sayas (University of Delaware).

Authors

Francisco-Javier Sayas grew up in the central Pyrenees, close to Spanish–French border. He earned a BS in 1991 and a PhD in 1994, both in Applied Mathematics at the University of Zaragoza (Spain). His thesis adviser was Michel Crouzeix (University of Rennes, France). In 1997, he obtained a tenured position from the same university. His early research centered on deep asymptotics for numerical approximation of boundary integral equations. The analysis techniques used in that work allowed him to develop a new class of methods which evolved into the "finite difference looking" deltaBEM methods, for which there is a freely available entirely open source MATLAB package. In 2001, he was awarded the Spanish Society of Applied Mathematics Prize to Young Researchers.

After some time working on time-harmonic problems and the coupling of Boundary Element Methods with volume methods, his research moved into Time Domain Boundary Integral Equations and the Hybridizable Discontinuous Galerkin Method. In between came a breakthrough with the proof of convergence of the Johnson–Nédélec coupling of BEM and FEM, an open question for almost thirty years. The results on TDBIE ended up as the first monograph of the author.

In 2007, Sayas took a three-year leave to visit the University of Minnesota. He then joined the University of Delaware as an associate professor in 2010 and became a tenured full professor in 2013. In Delaware, he has developed a group-centered research unit (Team Pancho), combining numerical analysis with scientific computing. Prof. Sayas has graduated eleven PhD students so far. For his overall research trajectory in 2016, he was awarded the Centennial Prize for Mathematics by the Royal Academy of Sciences of Zaragoza. Currently his research interests have focused on wave propagation in elastic media.

Thomas S. Brown grew up in central Virginia. He received a BA in Music Performance/Mathematics and later an MEd in Science Education from Lynchburg College. After teaching high school mathematics for five years for Pittsylvania County Schools, he resumed his studies and earned an MS and a PhD in mathematics from the University of Delaware. His graduate research, conducted under the advisement and supervision of Francisco-Javier Sayas, involved the analysis and implementation of numerical methods for elastic waves propagating through solids, in particular piezoelectric solids. Since obtaining his PhD, Thomas has been working as a lecturer in the Computational and

Applied Mathematics department at Rice University. His continuing research focuses on PDE constrained optimization or so-called optimal control problems.

Matthew E. Hassell is originally from New York. He earned a BS in Mathematics from the State University of New York at Binghamton, followed by an MS and PhD from the University of Delaware in Applied Mathematics under the advisement of Francisco-Javier Sayas. While an undergraduate, Matthew completed a National Science Foundation Research Experience for Undergraduates program in the summer of 2010 at Wabash College in Crawfordsville, Indiana, where he completed research on a finite element method for elliptic partial differential equations with singular and degenerate coefficients. This experience started him on his path into numerical analysis, which he continued at the University of Delaware. His thesis focused on numerical methods for time domain boundary integral equations (TDBIEs), specifically in the modeling and analysis of fluid flow and acoustics. Matthew also contributed to a novel analysis method for the TDBIEs arising in transient acoustics, as well as analysis and implementation of the coupling of boundary and finite elements for transient acoustic problems. In addition to these pursuits, he is also a co-author of a collection of detailed course notes on the derivation and implementation of Convolution Quadrature, a key tool for time discretization of TDBIEs.

After completing his PhD, he worked as an applied mathematician in industrial automation specializing on automated forklifts and other industrial vehicles. Presently he is a systems engineer with Lockheed Martin.

Part I

Fundamentals

1

Distributions

1.1   The test space   3
1.2   Distributions   7
1.3   Distributional differentiation   10
1.4   Convergence of distributions   13
1.5   A fundamental solution (*)   15
1.6   Lattice partitions of unity   17
1.7   When the gradient vanishes (*)   20
1.8   Proof of the variational lemma (*)   22
Final comments and literature   23
Exercises   24

In this chapter we are going to ease our way into the theory of distributions. Distributions are linear functionals defined on a **test space** made up of smooth compactly supported functions. Distributions will allow us to define generalized derivatives of any order of a large class of functions, including those in the Lebesgue spaces L^p.

1.1   The test space

Let $\Omega \subset \mathbb{R}^d$ be an open subset and suppose that $\varphi : \Omega \to \mathbb{R}$ is a continuous function. We define the support of φ as

$$\operatorname{supp} \varphi := \overline{\{\mathbf{x} \in \Omega \ : \ \varphi(\mathbf{x}) \neq 0\}}.$$

The closure here is taken in $\mathbb{R}^d$ and can, therefore, include points that are not in Ω (see Figure 1.1). We may then define the test space $\mathcal{D}(\Omega)$ by

$$\mathcal{D}(\Omega) := \{\varphi \in \mathcal{C}^\infty(\Omega) : \operatorname{supp} \varphi \text{ is compact}, \operatorname{supp} \varphi \subset \Omega\}.$$

When we state that the support of the function φ is contained strictly within Ω, we mean that the support does not touch $\partial\Omega$, so the function must therefore be zero in some neighborhood of $\partial\Omega$ as well. Some authors denote this space $\mathcal{C}_{00}^\infty(\Omega)$, where the double zero indicates the elements are zero on $\partial\Omega$ and zero *near* $\partial\Omega$.

A bump. We now provide examples of functions in $\mathcal{D}(\mathbb{R}^d)$. We start with a smooth bump function in one dimension (see the left side of Figure 1.2):

$$g(x) := \begin{cases} \exp\left(\frac{1}{x^2-1}\right) & |x| < 1, \\ 0 & |x| \geq 1. \end{cases}$$

In Exercise 1.1 you are asked to prove that $g \in \mathcal{C}^\infty(\mathbb{R})$, which reduces to showing that all side derivatives of g at ± 1 vanish. We now note that $\operatorname{supp} g = [-1, 1]$ and we thus have an element of $\mathcal{D}(\mathbb{R})$. With translations and dilations of g we can construct a large collection of elements of $\mathcal{D}(\mathbb{R})$. It is then simple to show that

$$\varphi(\mathbf{x}) := g(|\mathbf{x}|) = \begin{cases} \exp\left(\frac{1}{|\mathbf{x}|^2-1}\right) & |\mathbf{x}| < 1, \\ 0 & \text{otherwise,} \end{cases}$$

is in $\mathcal{D}(\mathbb{R}^d)$.

A smoothened step function. In a second step, we build a smooth version of the Heaviside (step) function (see the right side of Figure 1.2). We first shift and scale g to be supported in $[0, 1]$ and then integrate from the left and scale:

$$h(x) := \frac{1}{\int_{-\infty}^\infty g(2r-1)\,\mathrm{d}r} \int_{-\infty}^x g(2r-1)\,\mathrm{d}r.$$

This function trivially has all the following properties: $h \in \mathcal{C}^\infty(\mathbb{R})$, $h \equiv 0$ in $(-\infty, 0]$, $h \equiv 1$ in $[1, \infty)$, $0 \leq h \leq 1$, and $h' \geq 0$.

Smoothened characteristic functions of Euclidean balls. By translating h and flipping it, we can create a smooth version of the characteristic function of any interval. Take $R_1 < R_2$, and consider the function

$$\varphi(x) := h\left(\frac{R_2 - |x|}{R_2 - R_1}\right).$$

It is quite obvious that $\varphi \in \mathcal{C}^\infty(\mathbb{R} \setminus \{0\})$, since we are composing the absolute value function $x \mapsto |x|$ with $h \in \mathcal{C}^\infty(\mathbb{R})$. However, $\varphi \equiv 1$ in $[-R_1, R_1]$, which means that $\varphi \in \mathcal{C}^\infty(\mathbb{R})$. Finally, since $\operatorname{supp} \varphi = [-R_2, R_2]$, it follows that $\varphi \in \mathcal{D}(\mathbb{R})$. We can use the same strategy in any dimension to build smoothened versions of the characteristic function of the ball

$$B(\mathbf{x}_0; R_1) := \{\mathbf{x} \in \mathbb{R}^d : |\mathbf{x} - \mathbf{x}_0| < R_1\},$$

by using radial coordinates around the center $\mathbf{x}_0 \in \mathbb{R}^d$:

$$\varphi(\mathbf{x}) := h\left(\frac{R_2 - |\mathbf{x} - \mathbf{x}_0|}{R_2 - R_1}\right).$$

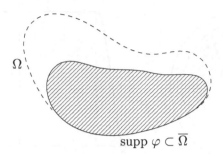

$$\text{supp}\,\varphi \subset \overline{\Omega}$$

Figure 1.1: The support of φ is a subset of the closure of Ω, and might not be contained in Ω.

Note that

$$\text{supp}\,\varphi = \overline{B(\mathbf{x}_0; R_2)} = \{\mathbf{x} \in \mathbb{R}^d \,:\, |\mathbf{x} - \mathbf{x}_0| \leq R_2\}.$$

The proof that $\varphi \in \mathcal{C}^\infty(\mathbb{R}^d)$ is very similar to that of the one-dimensional case.

In Figure 1.3, there is an example of such a function in two dimensions. Its graph resembles the shape of a mesa.

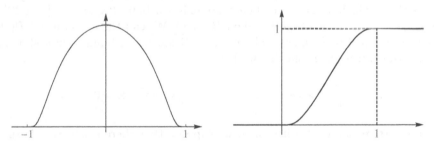

Figure 1.2: The 'bump' function, g (left), and smooth Heaviside function, h (right).

The functions being tested. We have already shown that the space $\mathcal{D}(\Omega)$ contains a large collection of functions: smoothened versions of the characteristic functions of Euclidean neighborhoods and their linear combinations. The variational lemma below will show that we actually have enough functions to be able to observe (test) locally integrable functions by weighted averaging with elements of $\mathcal{D}(\Omega)$. A function f is said to be **locally integrable** on a domain Ω if f is integrable on any compact set contained in Ω. The space of such functions is denoted by $L^1_{\text{loc}}(\Omega)$. Note that locally integrable functions in Ω can be extremely singular when we approach $\partial\Omega$, since integrability is only demanded on compact sets strictly contained in Ω.

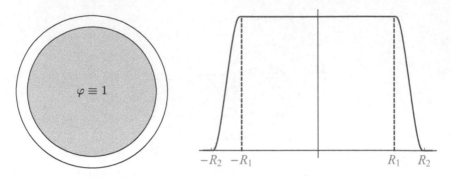

Figure 1.3: The support of a two-dimensional characteristic function (left) and a radial cross section (right).

Proposition 1.1 (Variational lemma). *If $f \in L^1_{\text{loc}}(\Omega)$ and*

$$\int_\Omega f(\mathbf{x})\varphi(\mathbf{x})\,\mathrm{d}\mathbf{x} = 0 \quad \forall \varphi \in \mathcal{D}(\Omega),$$

then $f = 0$ almost everywhere in Ω.

Proof. We first prove a simpler case for when $f \in C(\Omega)$. If there is a point $\mathbf{x}_0 \in \Omega$ for which $f(\mathbf{x}_0) > 0$, then there exists a ball $B(\mathbf{x}_0; r) := \{\mathbf{x} \in \mathbb{R}^d : |\mathbf{x} - \mathbf{x}_0| < r\} \subset \Omega$ such that $f > 0$ on $B(\mathbf{x}_0; r)$. We can then choose $\varphi \in \mathcal{D}(\Omega)$ such that $\varphi > 0$ in $B(\mathbf{x}_0; r)$ and supp $\varphi = \overline{B(\mathbf{x}_0; r)}$. (A bump centered at $\mathbf{x}_0$ serves this purpose.) From this we have

$$\int_\Omega f(\mathbf{x})\varphi(\mathbf{x})\,\mathrm{d}\mathbf{x} = \int_{B(\mathbf{x}_0;r)} f(\mathbf{x})\varphi(\mathbf{x})\,\mathrm{d}\mathbf{x} > 0,$$

and we arrive at a contradiction. If we suppose that there exists $\mathbf{x}_0 \in \Omega$ such that $f(\mathbf{x}_0) < 0$ we arrive at a similar contradiction. Thus $f \equiv 0$ on Ω. For the more general case of when $f \in L^1_{\text{loc}}(\Omega)$ see Section 1.8 at the end of the chapter. □

Notation for derivatives. For differentiation, we will make use of the following notation interchangeably:

$$\partial_{x_i}\phi = \frac{\partial \phi}{\partial x_i} = \phi_{x_i}.$$

All of these denote the partial derivative with respect to the x_i-th variable. We will also make use of multi-index notation for partial derivatives in $\mathbb{R}^d$. Consider a vector (a multi-index) $\alpha \in \mathbb{N}^d := \{(\alpha_1, \ldots, \alpha_d) : \alpha_i \in \mathbb{N}\}$, where

$$\mathbb{N} := \{0, 1, 2, \ldots\} = \{n \in \mathbb{Z} : n \geq 0\},$$

and denote $|\alpha| := \alpha_1 + \alpha_2 + \cdots + \alpha_d$. With this we define

$$\partial^\alpha := \frac{\partial^{|\alpha|}}{\partial_{x_1}^{\alpha_1} \partial_{x_2}^{\alpha_2} \cdots \partial_{x_d}^{\alpha_d}}.$$

We say that ∂^α for $|\alpha| = k$ is a k-th order derivative. At this stage we are only interested in differentiation of smooth functions and therefore the order of differentiation does not matter. It is also clear that $\partial^\alpha : \mathcal{D}(\Omega) \to \mathcal{D}(\Omega)$ is a well-defined linear operator for all $\alpha \in \mathbb{N}^d$.

Convergence in $\mathcal{D}(\Omega)$. Consider a sequence $\{\varphi_n\}$ in $\mathcal{D}(\Omega)$ and an element $\varphi \in \mathcal{D}(\Omega)$. We say $\varphi_n \to \varphi$ in $\mathcal{D}(\Omega)$ if:

(a) There exists a compact set $K \subset \Omega$ such that supp $\varphi_n \subset K$ for all n and supp $\varphi \subset K$.

(b) For all $\alpha \in \mathbb{N}^d$,

$$\max_{\mathbf{x} \in K} |\partial^\alpha \varphi_n(\mathbf{x}) - \partial^\alpha \varphi(\mathbf{x})| \to 0 \qquad \text{as } n \to \infty.$$

Before we move on to working with functionals acting on $\mathcal{D}(\Omega)$, let us observe that $\partial^\alpha : \mathcal{D}(\Omega) \to \mathcal{D}(\Omega)$ is a **sequentially continuous operator**, that is, if $\varphi_n \to \varphi$ in $\mathcal{D}(\Omega)$, then $\partial^\alpha \varphi_n \to \partial^\alpha \varphi$ in $\mathcal{D}(\Omega)$, which follows from the fact that differentiation cannot enlarge the support of a function.

1.2 Distributions

The definition of distribution. In short, a distribution is a sequentially continuous linear functional defined on $\mathcal{D}(\Omega)$. Let $T : \mathcal{D}(\Omega) \to \mathbb{R}$ be a linear map (a linear functional). For $\varphi \in \mathcal{D}(\Omega)$ we momentarily denote the action of T on φ by $\langle T, \varphi \rangle$. Any such T that is sequentially continuous (by which we mean $\langle T, \varphi_n \rangle \to \langle T, \varphi \rangle$ for any $\varphi_n \to \varphi$ in $\mathcal{D}(\Omega)$) is called a distribution. The vector space of all distributions will be denoted $\mathcal{D}'(\Omega)$ and the action of $T \in \mathcal{D}'(\Omega)$ on φ will be denoted $\langle T, \varphi \rangle_{\mathcal{D}'(\Omega) \times \mathcal{D}(\Omega)}$ whenever we want to display the open set Ω explicitly.

A topological remark. Before we show some examples, let us give here some additional information on the definition of this 'dual' space $\mathcal{D}'(\Omega)$. The space of test functions $\mathcal{D}(\Omega)$ can be endowed with a topology whose concept of convergence is the one we have given in the previous section. Moreover, a linear functional defined on this space is continuous with respect to that topology if and only if it is sequentially continuous, which allows us to take the shortcut of ignoring the definition of the topology in $\mathcal{D}(\Omega)$ and deal only with sequentially continuous functionals. The topology in $\mathcal{D}(\Omega)$ is actually an induced limit topology which is not so easy to introduce. This has become the standard reason to introduce distributions by using sequential continuity and 'ignoring' the topology in the background.

The Dirac delta distribution. Let $\mathbf{x}_0 \in \Omega$ and define

$$\langle \delta_{\mathbf{x}_0}, \varphi \rangle := \varphi(\mathbf{x}_0).$$

We claim that $\delta_{\mathbf{x}_0}$ is a distribution. The linearity is clear. For sequential continuity, consider a sequence $\{\varphi_n\}$ that converges to φ in $\mathcal{D}(\Omega)$. This implies that $\varphi_n \to \varphi$ uniformly in Ω, and so $\varphi_n(\mathbf{x}_0) \to \varphi(\mathbf{x}_0)$. Therefore

$$\langle \delta_{\mathbf{x}_0}, \varphi_n \rangle = \varphi_n(\mathbf{x}_0) \longrightarrow \varphi(\mathbf{x}_0) = \langle \delta_{\mathbf{x}_0}, \varphi \rangle,$$

which shows $\delta_{\mathbf{x}_0}$ is a distribution.

Characteristic functions. Let $\widetilde{\Omega} \subset \Omega$ be an open subset of Ω. Consider the action of the characteristic function of $\widetilde{\Omega}$ acting on an element of $\mathcal{D}(\Omega)$:

$$\langle \chi_{\widetilde{\Omega}}, \varphi \rangle = \int_{\widetilde{\Omega}} \varphi(\mathbf{x}) \, d\mathbf{x} = \int_{\Omega} \chi_{\widetilde{\Omega}}(\mathbf{x}) \varphi(\mathbf{x}) \, d\mathbf{x}.$$

This function defines a distribution. To prove it, suppose $\{\varphi_n\}$ is a sequence in $\mathcal{D}(\Omega)$ converging to φ. Therefore, there is a compact set $K \subset \Omega$ containing the supports of φ_n and φ, and

$$\int_{\widetilde{\Omega}} \varphi_n(\mathbf{x}) \, d\mathbf{x} = \int_{\widetilde{\Omega} \cap K} \varphi_n(\mathbf{x}) \, d\mathbf{x}.$$

Note that $\widetilde{\Omega} \cap K$ is a bounded set. By the uniform convergence of the sequence φ_n, we have

$$\int_{\widetilde{\Omega} \cap K} \varphi_n(\mathbf{x}) \, d\mathbf{x} \longrightarrow \int_{\widetilde{\Omega} \cap K} \varphi(\mathbf{x}) \, d\mathbf{x} = \int_{\widetilde{\Omega}} \varphi(\mathbf{x}) \, d\mathbf{x}.$$

Delta distributions on smooth bounded surfaces in free space. Let Γ be a closed and bounded smooth surface in $\mathbb{R}^3$. Consider the map

$$\langle \delta_\Gamma, \varphi \rangle = \int_\Gamma \varphi(\mathbf{x}) \, d\sigma_{\mathbf{x}},$$

where $d\sigma_{\mathbf{x}}$ is the area element on Γ. Using similar arguments to those given in the previous examples, it is simple to show that $\delta_\Gamma \in \mathcal{D}'(\mathbb{R}^3)$.

Locally integrable functions define distributions. Let $f \in L^1_{\text{loc}}(\Omega)$, and consider the action of f as a distribution, which is also equal to the integral of the product $f\varphi$ as a function:

$$\langle f, \varphi \rangle = \int_\Omega f(\mathbf{x})\varphi(\mathbf{x})\, \mathrm{d}\mathbf{x}.$$

Let us prove this. Let $\{\varphi_n\}$ be a sequence in $\mathcal{D}(\Omega)$ converging to φ. Therefore, there is a compact set K containing the supports of φ_n and φ such that φ_n converges to φ uniformly in K. Note that f is integrable in K, and moreover

$$\int_\Omega f(\mathbf{x})\varphi_n(\mathbf{x})\, \mathrm{d}\mathbf{x} = \int_K f(\mathbf{x})\varphi_n(\mathbf{x})\, \mathrm{d}\mathbf{x}.$$

We can bound

$$|\varphi_n(\mathbf{x})| \leq \sup_n \left(\max_{\mathbf{y} \in K} |\varphi_n(\mathbf{y})| \right) \leq C \quad \forall \mathbf{x} \in \Omega,$$

using the fact that the functions φ_n are continuous, compactly supported, and that they converge uniformly to φ. Since $f\varphi_n$ converges to $f\varphi$ pointwise, we can apply the dominated convergence theorem, and so

$$\int_K f(\mathbf{x})\varphi_n(\mathbf{x})\, \mathrm{d}\mathbf{x} \longrightarrow \int_K f(\mathbf{x})\varphi(\mathbf{x})\, \mathrm{d}\mathbf{x} = \int_\Omega f(\mathbf{x})\varphi(\mathbf{x})\, \mathrm{d}\mathbf{x}.$$

Note that the variational lemma guarantees that if $f, g \in L^1_{\text{loc}}(\Omega)$ satisfy $f = g$ as functions (that is, they are equal almost everywhere), then $f = g$ in the sense of distributions.

We say that $T \in \mathcal{D}'(\Omega)$ is a **regular distribution** if $T = f \in L^1_{\text{loc}}(\Omega)$. Because of this, we can make statements about distributions $T \in \mathcal{D}'(\Omega)$ such as $T \in L^2(\Omega)$ or $T \in \mathcal{C}^3(\Omega)$.

We cannot yet understand functions with strong non integrable singularities (in the domain) as distributions. For example, $1/x \notin \mathcal{D}'(\mathbb{R})$, but we will discuss this later. At this point, we only have the tools to show, for example, $1/x \in \mathcal{D}'((0, \infty))$ or $1/x \in \mathcal{D}'(\mathbb{R} \setminus \{0\})$.

A distribution that is not regular. Consider the Dirac delta distribution from earlier. We will next show that $\delta_{\mathbf{x}_0}$ is not a regular distribution. Suppose by way of contradiction that there is some $f \in L^1_{\text{loc}}(\Omega)$ such that

$$\int_\Omega f(\mathbf{x})\varphi(\mathbf{x})\, \mathrm{d}\mathbf{x} = \varphi(\mathbf{x}_0) \quad \forall \varphi \in \mathcal{D}(\Omega).$$

Define $\widetilde{\Omega} := \Omega \setminus \{\mathbf{x}_0\}$. If such an f exists, then $f \in L^1_{\text{loc}}(\widetilde{\Omega})$. On the other hand,

$$\int_\Omega f(\mathbf{x})\varphi(\mathbf{x})\, \mathrm{d}\mathbf{x} = \int_{\widetilde{\Omega}} f(\mathbf{x})\varphi(\mathbf{x})\, \mathrm{d}\mathbf{x} = 0 \quad \forall \varphi \in \mathcal{D}(\Omega).$$

Therefore $f \equiv 0$ in $\widetilde{\Omega}$, and so $f \equiv 0$ in Ω, which is impossible.

1.3   Distributional differentiation

Distributional derivatives of any order. Suppose T is a distribution and $\alpha \in \mathbb{N}^d$. We define the α-distributional derivative of T to be

$$\langle \partial^\alpha T, \varphi \rangle := (-1)^{|\alpha|} \langle T, \partial^\alpha \varphi \rangle,$$

where we recall that

$$\partial^\alpha = \frac{\partial^{|\alpha|}}{\partial_{x_1}^{\alpha_1} \partial_{x_2}^{\alpha_2} \cdots \partial_{x_d}^{\alpha_d}}.$$

The distributional derivative $\partial^\alpha T$ is itself another distribution. To see this, suppose $\varphi_n \to \varphi$ in $\mathcal{D}(\Omega)$. It then follows that

$$\langle \partial^\alpha T, \varphi_n \rangle = (-1)^{|\alpha|} \langle T, \partial^\alpha \varphi_n \rangle \longrightarrow (-1)^{|\alpha|} \langle T, \partial^\alpha \varphi \rangle = \langle \partial^\alpha T, \varphi \rangle,$$

so $\partial^\alpha T \in \mathcal{D}'(\Omega)$ as well.

Remark. Note that derivatives of distributions are defined all at once, not as derivatives of derivatives. However, they can be composed. Note first that

$$\partial^\alpha \partial^\beta \varphi = \partial^{\alpha+\beta} \varphi = \partial^\beta \partial^\alpha \varphi \qquad \forall \varphi \in \mathcal{D}(\Omega).$$

Therefore

$$\partial^\alpha \partial^\beta T = \partial^{\alpha+\beta} T = \partial^\beta \partial^\alpha T \qquad \forall T \in \mathcal{D}'(\Omega).$$

This shows that our multi-index notation for distributional differentiation is coherent.

How this concept is an extension. In the case of a continuously differentiable function, the distributional derivative coincides with the classical derivative. To prove this, consider a function $f \in \mathcal{C}^1(\Omega) \subset L^1_{\text{loc}}(\Omega)$, so that we have

$$\langle \partial_{x_i} f, \varphi \rangle = -\langle f, \partial_{x_i} \varphi \rangle = -\int_\Omega f(\mathbf{x}) \partial_{x_i} \varphi(\mathbf{x}) \mathrm{d}\mathbf{x}.$$

We now seek to apply integration by parts to the last term in the chain of equalities. However, we do not know if $\partial\Omega$ has a well-defined normal vector. Instead consider a polyhedral domain $\widetilde{\Omega} \subset \Omega$, containing supp φ. (This is easy to do. Cover supp φ with a finite number of open d-cubes contained in Ω.) We now have a well-defined normal vector almost everywhere on the boundary and we can use the divergence theorem (See Appendix A.1), a.k.a., integration by parts:

$$-\int_\Omega f(\mathbf{x}) \partial_{x_i} \varphi(\mathbf{x}) \mathrm{d}\mathbf{x} = -\int_{\widetilde{\Omega}} f(\mathbf{x}) \partial_{x_i} \varphi(\mathbf{x}) \mathrm{d}\mathbf{x} = \int_{\widetilde{\Omega}} \partial_{x_i} f(\mathbf{x}) \varphi(\mathbf{x}) \mathrm{d}\mathbf{x}.$$

In the last integral, the term $\partial_{x_i} f$ is the classical derivative of the function f, and this shows that the distribution associated to $\partial_{x_i} f \in C(\Omega) \subset L^1_{\text{loc}}(\Omega)$ is the distributional derivative of the distribution associated to f.

The Heaviside function. If we compute the derivative of the Heaviside function with the formal calculus that is so common to physicists and engineers, we find that its derivative is the delta function at the origin. We can show that this holds in the sense of distributions with a simple calculation:

$$\langle H', \varphi \rangle = -\langle H, \varphi' \rangle = -\int_{-\infty}^{\infty} H(t)\varphi'(t)\, dt = -\int_0^{\infty} \varphi'(t)\, dt = \varphi(0) = \langle \delta_0, \varphi \rangle.$$

Since our choice of $\varphi \in \mathcal{D}(\mathbb{R})$ was arbitrary, the variational lemma shows that $H' = \delta_0$.

Derivative of a tent function. Let f be the tent-shaped function defined by

$$f(x) = \max\{0, 1 - |x|\} = \begin{cases} x + 1 & 1 \leq x \leq 0, \\ -x + 1 & 0 < x \leq 1, \\ 0 & |x| > 1. \end{cases}$$

We compute its distributional derivative in the standard way:

$$\langle f', \varphi \rangle = -\langle f, \varphi' \rangle = -\int_{-1}^0 f(t)\varphi'(t)\, dt - \int_0^1 f(t)\varphi'(t)\, dt$$

$$= \int_{-1}^0 f'(t)\varphi(t)\, dt - f(0)\varphi(0) + f(-1)\varphi(-1)$$

$$+ \int_0^1 f'(t)\varphi(t)\, dt - f(1)\varphi(1) + f(0)\varphi(0).$$

Therefore

$$\langle f', \varphi \rangle = \int_{-\infty}^{\infty} f'(t)\varphi(t)\, dt.$$

The derivative of f thus corresponds to the classical derivative of a piecewise $\mathcal{C}^1$ function that is globally $\mathcal{C}^0$, see Figure 1.4. In a similar way we can show that $f'' = \delta_{-1} - 2\delta_0 + \delta_1$.

The derivative of the delta distribution. A simple calculation shows that $\langle \delta_0', \varphi \rangle = -\varphi'(0)$.

The derivative of the logarithm. We now study the distributional derivative of the logarithm, and see how this leads naturally to the inclusion of the non locally integrable function $1/x$ in the space of distributions using Cauchy principal values. Let $f(x) = \log|x|$. Clearly $f \in C^{\infty}(\mathbb{R} \setminus \{0\})$ with derivative $f'(x) = 1/x$. Moreover, $f \in L^1_{\text{loc}}(\mathbb{R})$, but $1/x \notin L^1_{\text{loc}}(\mathbb{R})$. The only difficulty

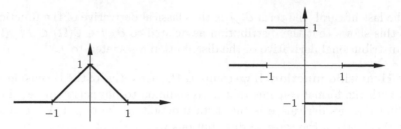

Figure 1.4: A tent function (left) and its distributional derivative (right).

in $1/x$ defining a regular distribution seems to occur around the point $x = 0$. Let us then compute the distributional derivative of f in $\mathbb{R}$:

$$\langle f', \varphi \rangle = -\langle f, \varphi' \rangle = -\int_{-\infty}^{\infty} \log|x| \varphi'(x) \mathrm{d}x.$$

To compute the integral, we introduce a parameter $\varepsilon > 0$ and split the integral as

$$
\begin{aligned}
-\int_{-\infty}^{\infty} \log|x|\varphi'(x)\mathrm{d}x &= \lim_{\varepsilon \to 0^+} -\left(\int_{-\infty}^{-\varepsilon} \log|x|\varphi'(x)\mathrm{d}x \right.\\
&\qquad\qquad \left. + \int_{\varepsilon}^{\infty} \log|x|\varphi'(x)\mathrm{d}x \right)\\
&= \lim_{\varepsilon \to 0^+} \left(\int_{-\infty}^{-\varepsilon} \frac{1}{x}\varphi(x)\mathrm{d}x - \varphi(-\varepsilon)\log(\varepsilon) \right.\\
&\qquad\qquad \left. + \int_{\varepsilon}^{\infty} \frac{1}{x}\varphi(x)\mathrm{d}x + \varphi(\varepsilon)\log(\varepsilon) \right)\\
&= \lim_{\varepsilon \to 0^+} \left(\int_{-\infty}^{-\varepsilon} \frac{1}{x}\varphi(x)\mathrm{d}x + \int_{\varepsilon}^{\infty} \frac{1}{x}\varphi(x)\mathrm{d}x \right)\\
&=: \mathrm{p.v.} \int_{-\infty}^{\infty} \frac{1}{x}\varphi(x)\mathrm{d}x,
\end{aligned}
$$

where p.v. denotes the Cauchy principal value of the otherwise divergent integral. Therefore we say that

$$(\log|x|)' = \mathrm{p.v.} \frac{1}{x}.$$

A very logical question. Suppose that the gradient of a distribution is zero, that is, all of its first partial derivatives are the zero distribution. It seems reasonable from our experience with the classical derivative that the distribution should be regular and equal to a constant. This turns out to be nontrivial to prove. We present the theorem and proof when Ω is an open interval here and save the proof of the general case for Section 1.7.

Theorem 1.1. *If $\Omega \subset \mathbb{R}^d$ is open and connected and $T \in \mathcal{D}'(\Omega)$ satisfies $\nabla T = 0$, then $T = c$ for some constant $c \in \mathbb{R}$, that is*

$$\langle T, \varphi \rangle = c \int_\Omega \varphi(\mathbf{x}) \, d\mathbf{x} \quad \forall \varphi \in \mathcal{D}(\Omega).$$

Proof. First we suppose that $\Omega = (a, b) \subset \mathbb{R}$ and $T \in \mathcal{D}'(a, b)$ such that $T' = 0$. The result in this case follows from a very well known argument, that can be found in any basic textbook. Take first

$$\varphi_0 \in \mathcal{D}(a, b) \qquad \int_a^b \varphi_0(x) \, dx = 1.$$

Given $\varphi \in \mathcal{D}(a, b)$, it then follows that

$$\varphi(x) = \varphi(x) - \left(\int_a^b \varphi(\tau) \, d\tau \right) \varphi_0(x) + \left(\int_a^b \varphi(\tau) \, d\tau \right) \varphi_0(x)$$

$$= \frac{d}{dx} \int_a^x \left(\varphi(\tau) - \left(\int_a^b \varphi(\rho) \, d\rho \right) \varphi_0(\tau) \right) d\tau + \left(\int_a^b \varphi(\tau) \, d\tau \right) \varphi_0(x)$$

$$= \psi'(x) + \left(\int_a^b \varphi(\tau) \, d\tau \right) \varphi_0(x), \qquad \psi \in \mathcal{D}(a, b).$$

Therefore

$$\langle T, \varphi \rangle = \langle T, \psi' \rangle + \int_a^b \varphi(x) \, dx \, \langle T, \varphi_0 \rangle = -\langle T', \psi \rangle + \langle T, \varphi_0 \rangle \int_a^b \varphi(x) \, dx.$$

Recalling that $\langle T', \psi \rangle = 0$, we obtain the result by taking $c := \langle T, \varphi_0 \rangle$. Note that the condition on the integral of φ_0 is needed so that

$$\langle T, \varphi_0 \rangle = c \int_a^b \varphi_0(x) \, dx = c.$$

For the proof of the more general case see Section 1.7. □

1.4   Convergence of distributions

The concept of convergence. Convergence of distributions is pointwise convergence, that is, the convergence you observe when you apply the distribution to a general element of the test space. Suppose $\{T_n\}$ is a sequence of elements of $\mathcal{D}'(\Omega)$, and that $T \in \mathcal{D}'(\Omega)$. We say that the sequence $\{T_n\}$ converges to T in $\mathcal{D}'(\Omega)$ if

$$\langle T_n, \varphi \rangle \longrightarrow \langle T, \varphi \rangle \qquad \forall \varphi \in \mathcal{D}(\Omega).$$

Moving Dirac deltas. Let $\{\mathbf{x}_n\}$ be a sequence in $\mathbb{R}^d$ such that $\mathbf{x}_n \to \mathbf{y}$ as $n \to \infty$. We then have that $\delta_{\mathbf{x}_n} \to \delta_{\mathbf{y}}$ in $\mathcal{D}'(\Omega)$, since

$$\langle \delta_{\mathbf{x}_n}, \varphi \rangle = \varphi(\mathbf{x}_n) \longrightarrow \varphi(\mathbf{y}) = \langle \delta_{\mathbf{y}}, \varphi \rangle \quad \forall \varphi \in \mathcal{D}(\Omega).$$

Digital signals. Consider the distribution given by the sum

$$T := \sum_{n=-\infty}^{\infty} c_n \delta_n,$$

where $\{c_n\}$ is a sequence of real numbers. Consider a sequence $\{\varphi_m\}$ in $\mathcal{D}(\mathbb{R})$ with a limit φ. There exists an $M > 0$ such that these functions have support in $[-M - 1/2, M + 1/2]$. Therefore

$$\langle T, \varphi_m \rangle = \sum_{n=-M}^{M} c_n \varphi_m(n) \longrightarrow \sum_{n=-M}^{M} c_n \varphi(n) = \langle T, \varphi \rangle,$$

and T is a distribution. Now define a sequence of distributions

$$T_m := \sum_{n=-m}^{m} c_n \delta_n,$$

and we see that as $m \to \infty$,

$$\langle T_m, \varphi \rangle = \sum_{n=-m}^{m} c_n \varphi(n) \longrightarrow \sum_{n=-\infty}^{\infty} c_n \varphi(n) = \langle T, \varphi \rangle \quad \forall \varphi \in \mathcal{D}(\mathbb{R}).$$

Therefore $T_m \to T$ in $\mathcal{D}'(\mathbb{R})$.

About $L^p(\Omega)$ convergence. For an open bounded domain Ω, we have the inclusion $L^p(\Omega) \subset L^1_{\text{loc}}(\Omega)$ for $p \in [1, \infty]$. Let $\{f_n\}$ be a sequence in $L^p(\Omega)$ such that $f_n \to f$ in $L^p(\Omega)$, for $p \in (1, \infty)$. Using Hölder's inequality we have

$$|\langle f_n, \varphi \rangle - \langle f, \varphi \rangle| \le \int_{\Omega} |(f(\mathbf{x}) - f_n(\mathbf{x}))\varphi(\mathbf{x})| \ d\mathbf{x}$$

$$\le \left(\int_{\Omega} |f(\mathbf{x}) - f_n(\mathbf{x})|^p \ d\mathbf{x} \right)^{1/p} \left(\int_{\Omega} |\varphi(\mathbf{x})|^{p^*} \ d\mathbf{x} \right)^{1/p^*},$$

where p^* is the conjugate exponent of p,

$$\frac{1}{p} + \frac{1}{p^*} = 1.$$

This shows that $f_n \to f$ as distributions. A similar argument can be used (do it) for the cases $p = 1$ and $p = \infty$.

1.5 A fundamental solution (*)

In this section we are going to solve a more or less complicated exercise to illustrate how the language of distributions gives a precise meaning to fundamental solutions (Green functions in free space). We are going to show that the distribution associated to the locally integrable function

$$\mathbb{R}^3 \ni \mathbf{x} \longmapsto \Phi(\mathbf{x}) := \frac{1}{4\pi|\mathbf{x}|}$$

is the fundamental solution of the negative Laplacian, that is,

$$-\Delta\Phi = \delta_{\mathbf{0}} \quad \text{in } \mathcal{D}'(\mathbb{R}^3).$$

We begin by showing that in a classical sense

$$\Delta\Phi = 0 \quad \text{in } \mathbb{R}^3 \setminus \{\mathbf{0}\}.$$

A simple computation shows that

$$\partial_{x_i}\Phi(\mathbf{x}) = -\frac{x_i}{4\pi|\mathbf{x}|^3} \qquad \forall \mathbf{x} \neq \mathbf{0},\ i = 1, 2, 3,$$

$$\partial_{x_i}^2\Phi(\mathbf{x}) = \frac{3x_i^2}{4\pi|\mathbf{x}|^5} - \frac{1}{4\pi|\mathbf{x}|^3} \qquad \forall \mathbf{x} \neq \mathbf{0},\ i = 1, 2, 3.$$

From this we obtain

$$\Delta\Phi(\mathbf{x}) = \sum_{i=1}^{3}\left(\frac{3x_i^2}{4\pi|\mathbf{x}|^5} - \frac{1}{4\pi|\mathbf{x}|^3}\right) = \frac{3|\mathbf{x}|^2}{4\pi|\mathbf{x}|^5} - \frac{3}{4\pi|\mathbf{x}|^3} = 0 \quad \forall \mathbf{x} \neq \mathbf{0}.$$

Now, consider the distributional Laplacian of Φ:

$$\langle \Delta\Phi, \varphi \rangle_{\mathcal{D}'(\mathbb{R}^3) \times \mathcal{D}(\mathbb{R}^3)} = \sum_{j=1}^{3}\langle \partial_{x_j}^2\Phi, \varphi \rangle = \sum_{j=1}^{3}\langle \Phi, \partial_{x_j}^2\varphi \rangle = \langle \Phi, \Delta\varphi \rangle.$$

Since $\Phi \in L^1_{\text{loc}}(\mathbb{R}^3)$, it is a regular distribution and we can write

$$\langle \Phi, \Delta\varphi \rangle = \int_{\mathbb{R}^3} \frac{1}{4\pi|\mathbf{x}|}\Delta\varphi(\mathbf{x})\, d\mathbf{x} = \lim_{\varepsilon \to 0^+} \int_{\mathbb{R}^3 \setminus B(\mathbf{0};\varepsilon)} \frac{1}{4\pi|\mathbf{x}|}\Delta\varphi(\mathbf{x})\, d\mathbf{x},$$

where we have the last equality from the dominated convergence theorem. Using Green's second identity we get that this limit is equal to

$$\lim_{\varepsilon \to 0^+}\left(\int_{\mathbb{R}^3 \setminus B(\mathbf{0};\varepsilon)} \Delta\Phi(\mathbf{x})\varphi(\mathbf{x})\, d\mathbf{x} + \int_{\partial B(\mathbf{0};\varepsilon)} (\Phi(\mathbf{x})\partial_n\varphi(\mathbf{x}) - \partial_n\Phi(\mathbf{x})\varphi(\mathbf{x}))\, d\sigma_{\mathbf{x}}\right), \quad (1.1)$$

where for an outward pointing unit normal vector $\mathbf{n}$ to a surface, we define the normal derivative
$$\partial_n f := \nabla f \cdot \mathbf{n}.$$
In this case, since the region we are integrating over is $\mathbb{R}^3 \setminus B(\mathbf{0}; \varepsilon)$, $\mathbf{n}$ is pointing into $B(\mathbf{0}; \varepsilon)$. The volume term in (1.1) is zero (since $\Delta\Phi = 0$ for $\mathbf{x} \neq 0$). The first boundary term also vanishes in the limit, since we can bound it as

$$\left| \int_{\partial B(\mathbf{0};\varepsilon)} \Phi(\mathbf{x}) \partial_n \varphi(\mathbf{x}) \, d\sigma_{\mathbf{x}} \right| \leq 4\pi\varepsilon^2 \frac{1}{4\pi\varepsilon} \max_{\mathbf{y} \in \partial B(\mathbf{0};\varepsilon)} |\nabla\varphi(\mathbf{y})| \longrightarrow 0 \quad \text{as} \quad \varepsilon \longrightarrow 0.$$

We now turn our attention to the only nonvanishing term

$$- \int_{\partial B(\mathbf{0};\varepsilon)} \partial_n \Phi(\mathbf{x}) \varphi(\mathbf{x}) \, d\sigma_{\mathbf{x}}.$$

To compute the value of this integral as $\varepsilon \to 0$, we begin by computing the normal derivative of the fundamental solution. Note that for z smooth enough the normal derivative on $\partial B(\mathbf{0}; \varepsilon)$ is

$$\partial_n z = -\nabla z \cdot \frac{1}{\varepsilon} \mathbf{x},$$

where $|\mathbf{x}| = \varepsilon$. Therefore

$$\partial_n \frac{1}{|\mathbf{x}|} = -\nabla \frac{1}{|\mathbf{x}|} \cdot \left(\frac{1}{\varepsilon} \mathbf{x} \right) = \frac{1}{|\mathbf{x}|^3} \mathbf{x} \cdot \left(\frac{1}{\varepsilon} \mathbf{x} \right) = \frac{1}{\varepsilon^2}.$$

From this computation, we can rewrite (1.1) as

$$- \lim_{\varepsilon \to 0^+} \int_{\partial B(\mathbf{0};\varepsilon)} \partial_n \Phi(\mathbf{x}) \varphi(\mathbf{x}) \, d\sigma_{\mathbf{x}} = - \lim_{\varepsilon \to 0^+} \frac{1}{4\pi\varepsilon^2} \int_{\partial B(\mathbf{0};\varepsilon)} \varphi(\mathbf{x}) \, d\sigma_{\mathbf{x}}.$$

Note now that

$$\frac{1}{4\pi\varepsilon^2} \int_{\partial B(\mathbf{0};\varepsilon)} \varphi(\mathbf{x}) d\sigma_{\mathbf{x}} = \frac{1}{4\pi\varepsilon^2} \int_{\partial B(\mathbf{0};\varepsilon)} \Big(\varphi(\mathbf{x}) - \varphi(\mathbf{0}) \Big) d\sigma_{\mathbf{x}} + \varphi(\mathbf{0}),$$

and that using the mean value theorem, we can bound

$$\left| \frac{1}{4\pi\varepsilon^2} \int_{\partial B(\mathbf{0};\varepsilon)} (\varphi(\mathbf{x}) - \varphi(\mathbf{0})) \, d\mathbf{x} \right| \leq \frac{1}{4\pi\varepsilon^2} \int_{\partial B(\mathbf{0};\varepsilon)} |\mathbf{x} - \mathbf{0}| \max_{\mathbf{y} \in B(\mathbf{0};\varepsilon)} |\nabla\varphi(\mathbf{y})| \, d\mathbf{x}$$

$$\leq \varepsilon \max_{\mathbf{y} \in B(\mathbf{0};\varepsilon)} |\nabla\varphi(\mathbf{y})|,$$

which vanishes when we take the limit $\varepsilon \to 0$.

Therefore, we have shown that

$$\langle \Delta\Phi, \varphi \rangle = -\varphi(\mathbf{0}) = -\langle \delta_{\mathbf{0}}, \varphi \rangle \quad \forall \varphi \in \mathcal{D}(\mathbb{R}^3),$$

which gives the result.

In two dimensions, the fundamental solution of the negative Laplacian is

$$-\frac{1}{2\pi}\log|\mathbf{x}|,$$

while for dimensions $d \geq 3$ it is given by

$$\frac{1}{\omega_d|\mathbf{x}|^{d-2}},$$

where ω_d is the area of the unit sphere in $\mathbb{R}^d$. You are asked to prove the case when $d = 2$ in Exercise 1.12.

1.6 Lattice partitions of unity

In this section we build new elements of $\mathcal{D}(\Omega)$ and learn to decompose given test functions into a finite sum of test functions with supports on 'little cubes.' We are going to use these ideas to prove Theorem 1.1 in general regions. The techniques we present here will be used in subsequent chapters. This section is somewhat technical, although the details are very simple. For the reader who is not interested in the details, we state the main results here.

Theorem 1.2 (Smooth separation from the boundary). *If Ω is a bounded open set and $\varepsilon > 0$, then there exists*

$$\varphi \in \mathcal{D}(\Omega), \qquad s.t. \qquad \varphi \equiv 1 \text{ in } \Omega_\varepsilon := \{\mathbf{x} \in \Omega : \text{dist}(\mathbf{x}, \partial\Omega) > \varepsilon\}.$$

Theorem 1.3 (Break down of test functions into cubic lattices). *Let Ω be any open set and $\phi \in \mathcal{D}(\Omega)$. There exist cubes Q_j and $\phi_j \in \mathcal{D}(\Omega)$ with*

$$\phi = \sum_{j=1}^{N} \phi_j, \qquad \text{supp } \phi_j \subset Q_j \subset \Omega.$$

Cutoff functions on cubes. We begin by returning to the idea of a smoothened characteristic function, but now in the context of cubes. Recall the definitions

$$g(r) := \begin{cases} \exp\left(\frac{1}{r^2-1}\right) & r < 1, \\ 0 & r \geq 1, \end{cases}$$

and

$$h(r) := \frac{1}{\int_{-\infty}^{\infty} g(2x-1)\,\mathrm{d}x} \int_{-\infty}^{r} g(2x-1)\,\mathrm{d}x.$$

Now consider two concentric closed parallelepipeds

$$Q := [-m_1, m_1] \times \ldots \times [-m_d, m_d], \quad Q^{\text{ext}} := [-M_1, M_1] \times \ldots \times [-M_d, M_d],$$

where

$$0 < m_j < M_j \quad \forall j.$$

The function

$$\varphi(\mathbf{x}) := \prod_{j=1}^d h\left(\frac{M_j - |x_j|}{M_j - m_j}\right)$$

satisfies

$$\varphi \in \mathcal{D}(\mathbb{R}^d), \qquad \text{supp}\,\varphi = Q^{\text{ext}}, \qquad \varphi \equiv 1 \text{ in } Q, \qquad 0 \le \varphi \le 1.$$

Once again, we can change the center of the set Q to get new functions centered at arbitrary points in $\mathbb{R}^d$.

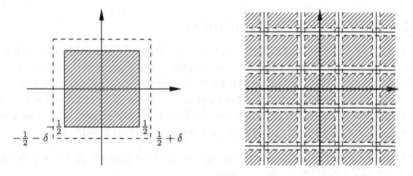

Figure 1.5: The support of φ_0 (left) and s (right) in $\mathbb{R}^2$. The functions are identically equal to 1 on all filled regions.

A smooth tiling of the space. Using the ideas above, we consider a basic configuration of two concentric d−cubes

$$Q_0 := (-\tfrac{1}{2}, \tfrac{1}{2})^d, \qquad Q_0^\delta := (-\tfrac{1}{2} - \delta, \tfrac{1}{2} + \delta)^d,$$

and a function

$$\varphi_0 \in \mathcal{D}(\mathbb{R}^d), \qquad \text{supp}\,\varphi_0 = \overline{Q_0^\delta}, \qquad \varphi_0 \equiv 1 \text{ in } \overline{Q_0}, \qquad 0 \le \varphi_0 \le 1.$$

Consider then the cubes

$$Q_{\mathbf{n}} := \mathbf{n} + Q_0 = \prod_{j=1}^d (n_j - \tfrac{1}{2}, n_j + \tfrac{1}{2}), \qquad Q_{\mathbf{n}}^\delta := \mathbf{n} + Q_0^\delta, \qquad \mathbf{n} \in \mathbb{Z}^d,$$

and the functions

$$\varphi_{\mathbf{n}} := \varphi_0(\cdot - \mathbf{n}).$$

The sum of all these functions

$$s := \sum_{\mathbf{n} \in \mathbb{Z}^d} \varphi_{\mathbf{n}}$$

satisfies (see Figure 1.5)

$$s \in \mathcal{C}^\infty(\mathbb{R}^d), \qquad s > 0, \qquad s \text{ is 1-periodic in all variables.}$$

We finally get to our goal functions:

$$\psi_{\mathbf{n}} := s^{-1}\varphi_{\mathbf{n}} \in \mathcal{D}(\mathbb{R}^d), \quad 0 \le \psi_{\mathbf{n}} \le 1, \quad \operatorname{supp}\psi_{\mathbf{n}} = \overline{Q_{\mathbf{n}}^\delta}, \quad \sum_{\mathbf{n} \in \mathbb{Z}^d} \psi_{\mathbf{n}} \equiv 1.$$

By taking δ small enough, we can enforce $\psi_{\mathbf{n}}$ to be identically equal to one in a neighborhood of $\mathbf{n}$. We are now ready to prove the main results which were stated at the beginning of this section.

Proof of Theorem 1.2. Let $\ell_\varepsilon := \varepsilon/(2\sqrt{d})$ and consider the cubes

$$Q_{\mathbf{n}} := \ell_\varepsilon \mathbf{n} + (-\tfrac{1}{2}\ell_\varepsilon, \tfrac{1}{2}\ell_\varepsilon)^d \subset Q_{\mathbf{n}}^{\mathrm{ext}} := \ell_\varepsilon \mathbf{n} + (-\tfrac{2}{3}\ell_\varepsilon, \tfrac{2}{3}\ell_\varepsilon)^d, \qquad \mathbf{n} \in \mathbb{Z}^d.$$

A simple change of scale in the previous construction allows us to get

$$\psi_{\mathbf{n}} \in \mathcal{D}(\mathbb{R}^d) \qquad \operatorname{supp}\psi_{\mathbf{n}} \subset Q_{\mathbf{n}}^{\mathrm{ext}}, \qquad \sum_{\mathbf{n} \in \mathbb{Z}^d} \psi_{\mathbf{n}} \equiv 1.$$

(Note that we are asking for $\psi_{\mathbf{n}}$ to be supported inside $Q_{\mathbf{n}}^{\mathrm{ext}}$, which can be easily accomplished by choosing δ small enough in the previous construction.) Now consider the finite set of indices

$$\mathcal{I} := \{\mathbf{n} \in \mathbb{Z}^d : Q_{\mathbf{n}}^{\mathrm{ext}} \cap \Omega_\varepsilon \ne \emptyset\},$$

and the function

$$\varphi := \sum_{\mathbf{n} \in \mathcal{I}} \psi_{\mathbf{n}} \in \mathcal{D}(\mathbb{R}^d).$$

What is left is just a collection of easy observations.

(a) If $\mathbf{x} \in \Omega_\varepsilon$, then $B(\mathbf{x}; \varepsilon) \subset \Omega$.

(b) The diameter of $Q_{\mathbf{n}}^{\mathrm{ext}}$ is $\sqrt{d}\,\tfrac{4}{3}\ell_\varepsilon = \tfrac{2}{3}\varepsilon$. Therefore, if $\mathbf{n} \in \mathcal{I}$, we can take $\mathbf{x} \in Q_{\mathbf{n}}^{\mathrm{ext}} \cap \Omega_\varepsilon$ and note that

$$\mathbf{y} \in Q_{\mathbf{n}}^{\mathrm{ext}} \implies |\mathbf{y} - \mathbf{x}| < \tfrac{2}{3}\varepsilon \implies \mathbf{y} \in B(\mathbf{x}; \varepsilon) \subset \Omega.$$

(c) As a consequence,

$$\operatorname{supp}\varphi \subset \bigcup_{\mathbf{n} \in \mathcal{I}} Q_{\mathbf{n}}^{\mathrm{ext}} \subset \Omega.$$

(d) Finally, if $\mathbf{x} \in \Omega_\varepsilon$, then $\psi_\mathbf{n}(\mathbf{x}) = 0$ for all $\mathbf{n} \notin \mathcal{I}$ and thus

$$\varphi(\mathbf{x}) = \sum_{\mathbf{n} \in \mathcal{I}} \psi_\mathbf{n}(\mathbf{x}) = \sum_{\mathbf{n} \in \mathbb{Z}^d} \psi_\mathbf{n}(\mathbf{x}) = 1.$$

The result is thus proved.   □

Proof of Theorem 1.3. Our argument will use the construction of the previous proof. Let us first prove the result when Ω is bounded. We start by defining $\varepsilon := \frac{1}{2}\text{dist}(\text{supp}\,\phi, \partial\Omega)$ so that $\text{supp}\,\phi \subset \Omega_\varepsilon$ (see Theorem 1.2). We can now have a finite set of indices $\mathcal{I}$, cubes $\{Q_\mathbf{n} : \mathbf{n} \in \mathcal{I}\}$, and smooth functions

$$\psi_\mathbf{n} \in \mathcal{D}(\mathbb{R}^d), \qquad \text{supp}\,\psi_\mathbf{n} \subset \Omega, \qquad \sum_{\mathbf{n} \in \mathcal{I}} \psi_\mathbf{n} \equiv 1 \text{ in } \Omega_\varepsilon.$$

Therefore

$$\phi = \phi \sum_{\mathbf{n} \in \mathcal{I}} \psi_\mathbf{n} = \sum_{\mathbf{n} \in \mathcal{I}} (\phi\,\psi_\mathbf{n}),$$

and the result follows by taking $\phi_j := \phi\,\psi_\mathbf{n}$ and renumbering the set $\mathcal{I}$. When Ω is unbounded, we can repeat this argument using $\Omega \cap B(\mathbf{0}; R)$ where $\text{supp}\,\varphi \subset B(\mathbf{0}; R)$.   □

1.7   When the gradient vanishes (*)

In this section we present the proof of Theorem 1.1 in full generality. The proof will be presented through the following two propositions. The first one proves the theorem when Ω is a cube, and the second uses a tiling argument to prove it for all domains.

Proposition 1.2. *Let* $Q := (a_1, b_1) \times \ldots \times (a_d, b_d)$. *If*

$$\langle \partial_{x_j} T, \varphi \rangle = 0 \qquad \forall \varphi \in \mathcal{D}(Q), \quad \forall j,$$

then there exists $c \in \mathbb{R}$ *such that*

$$\langle T, \varphi \rangle = c \int_Q \varphi(\mathbf{x})\,d\mathbf{x} \qquad \forall \varphi \in \mathcal{D}(Q).$$

Proof. The result is proved by induction on d. We have already proved it for $d = 1$ at the end of Section 1.3. Let $\widetilde{Q} := (a_1, b_1) \times \ldots \times (a_{d-1}, b_{d-1})$.

We start with

$$\varphi_0 \in \mathcal{D}(a_d, b_d), \qquad \int_{a_d}^{b_d} \varphi_0(t)\,dt = 1,$$

and consider $\widetilde{T} : \mathcal{D}(\widetilde{Q}) \to \mathbb{R}$ given by

$$\langle \widetilde{T}, \widetilde{\varphi} \rangle := \langle T, \widetilde{\varphi} \otimes \varphi_0 \rangle,$$

where

$$(\widetilde{\varphi} \otimes \varphi_0)(\mathbf{x}) := \widetilde{\varphi}(\widetilde{\mathbf{x}}) \, \varphi_0(x_d) \qquad \mathbf{x} = (\widetilde{\mathbf{x}}, x_d) \in Q = \widetilde{Q} \times (a_d, b_d).$$

It is very easy to prove that $\widetilde{T} \in \mathcal{D}'(\widetilde{Q})$ and that for $0 \le j \le d-1$

$$\begin{aligned} \langle \partial_{x_j} \widetilde{T}, \widetilde{\varphi} \rangle &= -\langle \widetilde{T}, \partial_{x_j} \widetilde{\varphi} \rangle \\ &= -\langle T, \partial_{x_j} \widetilde{\varphi} \otimes \varphi_0 \rangle \\ &= -\langle T, \partial_{x_j}(\widetilde{\varphi} \otimes \varphi_0) \rangle \\ &= \langle \partial_{x_j} T, \widetilde{\varphi} \otimes \varphi_0 \rangle = 0 \qquad \forall \widetilde{\varphi} \in \mathcal{D}(\widetilde{Q}), \end{aligned}$$

and therefore there exists a constant c such that $\widetilde{T} = c$. On the other hand, if we associate

$$\varphi \in \mathcal{D}(Q) \qquad \longmapsto \qquad \widetilde{\varphi} := \int_{a_d}^{b_d} \varphi(\cdot, t) \, \mathrm{d}t \in \mathcal{D}(\widetilde{Q}),$$

and decompose (see the proof of Theorem 1.1 in one dimension at the end of Section 1.3, where $\widetilde{\varphi}$ is just a scalar)

$$\varphi = \varphi - \widetilde{\varphi} \otimes \varphi_0 + \widetilde{\varphi} \otimes \varphi_0 = \partial_{x_d} \psi + \widetilde{\varphi} \otimes \varphi_0,$$

where

$$\psi(\mathbf{x}) = \psi(\widetilde{\mathbf{x}}, x_d) := \int_{a_d}^{x_d} \left(\varphi(\widetilde{\mathbf{x}}, t) - \widetilde{\varphi}(\widetilde{\mathbf{x}}) \varphi_0(t) \right) \mathrm{d}t, \qquad \psi \in \mathcal{D}(Q),$$

we show that

$$\begin{aligned} \langle T, \varphi \rangle &= \langle T, \partial_{x_d} \psi \rangle + \langle T, \widetilde{\varphi} \otimes \varphi_0 \rangle = -\langle \partial_{x_d} T, \psi \rangle + \langle T, \widetilde{\varphi} \otimes \varphi_0 \rangle = \langle \widetilde{T}, \widetilde{\varphi} \rangle \\ &= c \int_{\widetilde{Q}} \widetilde{\varphi}(\widetilde{\mathbf{x}}) \, \mathrm{d}\widetilde{\mathbf{x}} = c \int_Q \varphi(\mathbf{x}) \, \mathrm{d}\mathbf{x}, \end{aligned}$$

and the proof is finished. $\qquad \square$

Proposition 1.3. *If Ω is open and connected and $T \in \mathcal{D}'(\Omega)$ satisfies $\nabla T = 0$, then there exists $c \in \mathbb{R}$ such that*

$$\langle T, \varphi \rangle = c \int_\Omega \varphi(\mathbf{x}) \mathrm{d}\mathbf{x} \qquad \forall \varphi \in \mathcal{D}(\Omega), \qquad \mathrm{supp}\, \varphi \subset Q,$$

where Q is any cube contained in Ω and c does not depend on Q.

Proof. Proposition 1.2 shows that for all Q there exists c_Q such that

$$\langle T, \varphi \rangle = c_Q \int_\Omega \varphi(\mathbf{x}) d\mathbf{x} \qquad \forall \varphi \in \mathcal{D}(\Omega), \qquad \operatorname{supp} \varphi \subset Q.$$

If $Q_1 \cap Q_2 \neq \emptyset$, we choose

$$\phi \in \mathcal{D}(Q_1 \cap Q_2), \qquad \int_\Omega \phi(\mathbf{x}) d\mathbf{x} = 1,$$

and then note that

$$c_{Q_1} = c_{Q_1} \int_\Omega \phi(\mathbf{x}) d\mathbf{x} = \langle T, \phi \rangle = c_{Q_2}.$$

Now let Q_a and Q_b be any two nonintersecting cubes contained in Ω. We can find a finite sequence of cubes such that

$$Q_a = Q_1, Q_2, \dots, Q_N = Q_b, \qquad Q_j \cap Q_{j+1} \neq \emptyset, \qquad Q_j \subset \Omega.$$

(This is done by a connection and compactness argument. We join the center of the cubes Q_a and Q_b with a continuous arc, associate a cube to each point of the arc, and choose a finite subcover of the arc using compactness.) The previous argument shows then that $c_{Q_j} = c_{Q_{j+1}}$ and therefore $c_{Q_a} = c_{Q_b}$. $\quad\square$

Now we present the proof for a general open and connected Ω.

Proof of Theorem 1.1. This is an easy consequence of Proposition 1.3 and the decompositions of Theorem 1.3. Given $\phi \in \mathcal{D}(\Omega)$ we can write

$$\phi = \sum_{j=1}^{N} \phi_j \qquad \phi_j \in \mathcal{D}(Q_j), \qquad Q_j \subset \Omega,$$

where Q_j are cubes. We then use Proposition 1.3 to show that

$$\langle T, \phi \rangle = \sum_j \langle T, \phi_j \rangle = \sum_j c \int_\Omega \phi_j(\mathbf{x}) d\mathbf{x} = c \int_\Omega \phi(\mathbf{x}) d\mathbf{x},$$

which proves the result. $\quad\square$

1.8   Proof of the variational lemma (*)

In this section we prove Proposition 1.1 for a general locally integrable function. Let $f \in L^1_{\text{loc}}(\Omega)$ or, more properly speaking, let f be a particular element of the class of functions that are equal almost everywhere and locally integrable, so that we can take point values. We choose a point $\mathbf{x}_0 \in \Omega$ for which

$$\lim_{\varepsilon \to 0^+} \frac{1}{|B(\mathbf{x}_0; \varepsilon)|} \int_{B(\mathbf{x}_0; \varepsilon)} |f(\mathbf{x}) - f(\mathbf{x}_0)| \, d\mathbf{x} = 0. \tag{1.2}$$

The Lebesgue differentiation theorem guarantees that the set of points not satisfying (1.2) has zero measure. Now we choose $\varphi \in \mathcal{D}(\mathbb{R}^d)$ such that

$$\operatorname{supp} \varphi = \overline{B(0; 1)}, \qquad \int_{\mathbb{R}^d} \varphi(\mathbf{x}) \, d\mathbf{x} = 1, \qquad 0 \leq \varphi \leq 1.$$

The rescaled functions

$$\varphi_\varepsilon(\mathbf{x}) := \frac{1}{\varepsilon^d} \varphi\left(\frac{1}{\varepsilon}(\mathbf{x} - \mathbf{x}_0)\right)$$

are elements of $\mathcal{D}(\mathbb{R}^d)$ and they satisfy

$$\operatorname{supp} \varphi_\varepsilon = \overline{B(\mathbf{x}_0; \varepsilon)}, \qquad 0 \leq \varphi_\varepsilon \leq \frac{1}{\varepsilon^d}, \qquad \int_{B(\mathbf{x}_0; \varepsilon)} \varphi_\varepsilon(\mathbf{x}) d\mathbf{x} = 1.$$

Therefore for small enough ε, $\varphi_\varepsilon \in \mathcal{D}(\Omega)$ and

$$|f(\mathbf{x}_0)| = \left| \int_\Omega f(\mathbf{x}) \varphi_\varepsilon(\mathbf{x}) d\mathbf{x} - f(\mathbf{x}_0) \right| = \left| \int_{B(\mathbf{x}_0; \varepsilon)} (f(\mathbf{x}) - f(\mathbf{x}_0)) \varphi_\varepsilon(\mathbf{x}) d\mathbf{x} \right|$$

$$\leq \frac{1}{\varepsilon^d} \int_{B(\mathbf{x}_0; \varepsilon)} |f(\mathbf{x}) - f(\mathbf{x}_0)| d\mathbf{x}.$$

Taking the limit as $\varepsilon \to 0$ and using (1.2) (note that the volume of $B(\mathbf{x}_0; \varepsilon)$ is proportional to ε^d, it follows that $f(\mathbf{x}_0) = 0$. Since (1.2) holds almost everywhere, the result follows.

Note that there are many other possible proofs of the variational lemma that do not require the use of the Lebesgue differentiation theorem, but involve the use of convolutional (mollification) techniques. We will come back to this issue later on.

Final comments and literature

The theory of distributions was created by Laurent Schwartz in the 1940s. His first treatise on the subject (*Théorie des distributions*, published in 1950-51; reference [95] is the second edition of the two-volume set) is still a classic reference to learn about this theory. An easier learning material for this theory can be found in Schwartz's textbook [94]. One of the goals of Schwartz was putting together in a single structure all kinds of mathematical entities that could not be considered as functions, like Paul Dirac's delta distributions. His theory also gave a natural justification of Oliver Heaviside's functional calculus and could be used for a rigorous construction of Sergei Sobolev's spaces, which included a concept of weak differentiation that distributional differentiation generalizes.

The topological structure of the test space $\mathcal{D}(\Omega)$ fits into the class of induced limit topologies. This complicated topology was part of the catalyst that led to the theory of topological vector spaces, created by Schwartz's student Alexander Grothendieck in his doctoral dissertation.

Distributions are very general and rich tools in analysis and we will make use of them frequently in this textbook. One of the nice features of distributions comes from the fact that, because the test space is 'relatively small,' its dual contains many interesting mathematical entities. You can make the test space larger and you lose distributions: this is the origin of the class of tempered distributions that we will visit in Chapter 13. It does have to be understood that distributions are not the end of the story and that they bring along some problems: because they are functionals and not functions, distributions cannot be multiplied [93], which is clearly a problem when one wants to deal with nonlinear problems.

Exercises

1.1. Show that the function

$$g(x) := \begin{cases} \exp\left(\frac{1}{x^2-1}\right) & |x| < 1, \\ 0 & \text{otherwise}, \end{cases}$$

is in $\mathcal{D}(\mathbb{R})$. Show that $g(|\cdot|) \in \mathcal{D}(\mathbb{R}^d)$.

1.2. Let $\varphi \in \mathcal{D}(\mathbb{R}^d)$ be such that

$$\int_{\mathbb{R}^d} \varphi(\mathbf{x}) \, \mathrm{d}\mathbf{x} = 1,$$

and define

$$\varphi_\varepsilon(\mathbf{x}) := \frac{1}{\varepsilon^d} \varphi\left(\frac{1}{\varepsilon}\mathbf{x}\right).$$

Show that for all $f \in \mathcal{C}(\mathbb{R}^d)$,

$$\int_{\mathbb{R}^d} \varphi_\varepsilon(\mathbf{x}) f(\mathbf{y} - \mathbf{x}) \mathrm{d}\mathbf{x} = \int_{\mathbb{R}^d} \varphi_\varepsilon(\mathbf{y} - \mathbf{x}) f(\mathbf{x}) \mathrm{d}\mathbf{x} \xrightarrow{\varepsilon \to 0} f(\mathbf{y}) \qquad \forall \mathbf{y} \in \mathbb{R}^d.$$

1.3. Let $\lim_{n\to\infty} \mathbf{x}_n = \mathbf{x}$ in $\mathbb{R}^d$ and let $\varphi \in \mathcal{D}(\mathbb{R}^d)$. Show that

$$\varphi(\cdot - \mathbf{x}_n) \stackrel{n\to\infty}{\Longrightarrow} \varphi(\cdot - \mathbf{x}) \qquad \text{in } \mathcal{D}(\mathbb{R}^d).$$

1.4. Let $\mathbf{h} \in \mathbb{R}^d$, $\{c_n\}$ be a sequence of nonzero real numbers such that $\lim_{n\to\infty} c_n = 0$, and $\varphi \in \mathcal{D}(\mathbb{R}^d)$. Show that

$$\frac{1}{c_n}(\varphi(\cdot + c_n\mathbf{h}) - \varphi) \stackrel{n\to\infty}{\Longrightarrow} \mathbf{h} \cdot \nabla\varphi \qquad \text{in } \mathcal{D}(\mathbb{R}^d).$$

1.5. Let $\phi \in \mathcal{C}^\infty(\Omega)$. Show that

$$\varphi_n \stackrel{n\to\infty}{\Longrightarrow} \varphi \quad \text{in } \mathcal{D}(\Omega) \qquad \Longrightarrow \qquad \phi\varphi_n \stackrel{n\to\infty}{\Longrightarrow} \phi\varphi \quad \text{in } \mathcal{D}(\Omega),$$

that is, multiplication by a $\mathcal{C}^\infty(\Omega)$ function is a sequentially continuous operator in $\mathcal{D}(\Omega)$.

1.6. Show that in the sense of distributions $\partial^{\alpha+\beta} = \partial^\alpha\partial^\beta$ for every pair of multi-indices.

1.7. **Complex-valued distributions.** Prove that the following three definitions lead to the same concept:

(a) Given $T_1, T_2 \in \mathcal{D}'(\Omega)$, we define $T := T_1 + \imath T_2 : \mathcal{D}(\Omega) \to \mathbb{C}$ by

$$\langle T_1 + \imath T_2, \varphi \rangle := \langle T_1, \varphi \rangle + \imath \langle T_2, \varphi \rangle.$$

(b) We consider $T : \mathcal{D}(\Omega) \to \mathbb{C}$, $\mathbb{R}$-linear (that is, we only admit real scalars) and sequentially continuous.

(c) We consider the complex test space

$$\mathcal{D}(\Omega; \mathbb{C}) := \mathcal{D}(\Omega) + \imath\,\mathcal{D}(\Omega),$$

and consider $T : \mathcal{D}(\Omega; \mathbb{C}) \to \mathbb{C}$ linear and sequentially continuous.

Note that in the definition (c) we have a different test space. In this case you need to prove that there is a bijection relating the maps defined in (a) and (b) with those defined with (c).

1.8. **Fourier series.** Let $\{c_n\}$ be a sequence of complex numbers.

(a) Show that if the sequence is bounded, then the series of functions

$$\sum_{n=-\infty}^{\infty} c_n \exp(\imath n \cdot),$$

converges in $\mathcal{D}'(\mathbb{R})$. (**Hint.** Show that

$$\left| \int_{-\infty}^{\infty} \varphi(x)e^{\imath nx} \, dx \right| \le C_m \, n^{-m} \qquad \forall n \ge 0, \qquad \forall m,$$

for every $\varphi \in \mathcal{D}(\mathbb{R})$.)

(b) Extend the previous exercise to sequences $\{c_n\}$ satisfying

$$|c_n| \leq C|n|^M \qquad \forall n \in \mathbb{Z} \setminus \{0\}$$

for given $C > 0$ and $M > 0$.

1.9. Show that $\partial^\alpha : \mathcal{D}'(\Omega) \to \mathcal{D}'(\Omega)$ is sequentially continuous.

1.10. **Sequences converging to a Dirac delta.** Let $\xi \in L^1(\mathbb{R}^d)$ be such that

$$\int_{\mathbb{R}^d} \xi(\mathbf{x}) \, d\mathbf{x} = 1.$$

Consider then the functions

$$\xi_\varepsilon := \frac{1}{\varepsilon^d} \xi \left(\frac{1}{\varepsilon} \cdot \right).$$

Show that

$$\xi_\varepsilon \xrightarrow{\varepsilon \to 0} \delta_0 \qquad \text{in } \mathcal{D}'(\mathbb{R}^d).$$

1.11. **Multiplication of distributions by C^∞ functions.** Let $T \in \mathcal{D}'(\Omega)$ and $\phi \in C^\infty(\Omega)$. We define $\phi T : \mathcal{D}(\Omega) \to \mathbb{R}$ with the formula

$$\langle \phi T, \varphi \rangle_{\mathcal{D}'(\Omega) \times \mathcal{D}(\Omega)} := \langle T, \phi \varphi \rangle_{\mathcal{D}'(\Omega) \times \mathcal{D}(\Omega)}.$$

Show that $\phi T \in \mathcal{D}'(\Omega)$.

1.12. **Fundamental solution of the two-dimensional Laplacian.** Show that the fundamental solution to the Laplacian in $\mathbb{R}^2$ is

$$\Phi(\mathbf{x}) := -\frac{1}{2\pi} \log |\mathbf{x}|.$$

(**Hint.** We can follow the same argument as in Section 1.5: (1) Show $\Phi \in L^1_{\text{loc}}(\mathbb{R}^2)$, (2) note that $\Delta\Phi = 0$ except at $\mathbf{x} = \mathbf{0}$, then (3) study Φ carefully around $\mathbf{x} = \mathbf{0}$.)

1.13. **Fundamental solution for the Helmholtz equation.** Consider the function $\Phi : \mathbb{R}^3 \to \mathbb{C}$, given by

$$\Phi(\mathbf{x}) := \frac{e^{-\imath k|\mathbf{x}|}}{4\pi|\mathbf{x}|}.$$

Show that

$$\Delta\Phi + k^2\Phi = -\delta_0 \qquad \text{in } \mathcal{D}'(\mathbb{R}^3).$$

1.14. Consider the function

$$f(x, y) = \begin{cases} 1 & x + y > 0, \\ 0 & x + y < 0. \end{cases}$$

Compute the distributional derivatives $\partial_x f$ and $\partial_y f$. What equation does f satisfy?

1.15. Let $f(\mathbf{x}) = \chi_{|\mathbf{x}| \leq 1}$ in 2D. Compute $\partial_x f$ and $\partial_y f$.

2

The homogeneous Dirichlet problem

2.1   The Sobolev space $H^1(\Omega)$ 27
2.2   Cutoff and mollification 29
2.3   A guided tour of mollification (*) 31
2.4   The space $H_0^1(\Omega)$ 34
2.5   The Dirichlet problem 38
2.6   Existence of solutions 41
Final comments and literature 43
Exercises 44

In this chapter we begin solving PDEs in earnest. We will start with the homogeneous Dirichlet problem for the Laplacian, and introduce and develop theory for all of the necessary Sobolev spaces along the way. The Dirichlet problem will be presented in three equivalent formulations: as a distributional PDE, as a variational problem, and as a minimization problem. Inhomogeneous boundary conditions are introduced after a rigorous construction of the trace operator (restriction to the boundary) on Lipschitz domains in Chapter 4. We will consider the elliptic equation

$$-\Delta u = f \quad \text{in } \Omega, \qquad u = 0 \quad \text{on } \partial\Omega,$$

on an open subset $\Omega \subset \mathbb{R}^d$. At this point, we will not make assumptions on the regularity of the boundary of the domain. It may be considered nonconvex, fractal, or poorly behaved in any other number of ways. The partial differential equation will be understood in the sense of distributions, i.e., that u and f are distributions, and that the equation holds when tested by an element of the set of test functions $\mathcal{D}(\Omega)$.

2.1   The Sobolev space $H^1(\Omega)$

Let Ω be an open and possibly unbounded subset of $\mathbb{R}^d$. We define the Sobolev space
$$H^1(\Omega) := \{u \in L^2(\Omega) : \nabla u \in \mathbf{L}^2(\Omega)\},$$

where we are using the notation $\mathbf{L}^2(\Omega) := L^2(\Omega; \mathbb{R}^d) \equiv L^2(\Omega)^d$. We note that $L^2(\Omega) \subset L^1_{\text{loc}}(\Omega)$, and that the partial derivatives $\partial_{x_i} u$ are regular distributions in $L^2(\Omega)$. The space $H^1(\Omega)$ is equipped with the inner product

$$(u, v)_{1,\Omega} := (\nabla u, \nabla v)_\Omega + (u, v)_\Omega = \int_\Omega \nabla u(\mathbf{x}) \cdot \nabla v(\mathbf{x}) \, d\mathbf{x} + \int_\Omega u(\mathbf{x})v(\mathbf{x}) \, d\mathbf{x}.$$

This inner product is easily verified to be bilinear, symmetric, and positive definite. The leading term $(\nabla u, \nabla v)_\Omega$ will be of particular importance later, and is called the Dirichlet form. The associated norm will be denoted

$$\|u\|^2_{1,\Omega} := (u, u)_{1,\Omega} = \int_\Omega |\nabla u(\mathbf{x})|^2 d\mathbf{x} + \int_\Omega |u(\mathbf{x})|^2 d\mathbf{x}.$$

Remark. If we suppose that Ω is a bounded domain, then the set

$$\mathcal{C}^1(\overline{\Omega}) := \{U|_\Omega : U \in \mathcal{C}^1(\mathbb{R}^d)\}$$

is a subset of $H^1(\Omega)$. This is easy to verify, since $u \in \mathcal{C}(\overline{\Omega}) \subset L^2(\Omega)$. Moreover, the partial derivatives of u satisfy $\partial_{x_i} u \in \mathcal{C}(\overline{\Omega}) \subset L^2(\Omega)$ for all i. We state without ambiguity that *the* partial derivatives of u satisfy this, since the distributional and classical derivatives coincide in $\mathcal{C}^1(\Omega)$.

Warning. In the closure of an open set Ω we will always consider

$$\mathcal{C}^k(\overline{\Omega}) := \{U|_\Omega : U \in \mathcal{C}^k(\mathbb{R}^d)\}, \qquad k \geq 0 \quad \text{or} \quad k = \infty.$$

On the other hand, we will use the spaces

$$\mathcal{C}(B) := \{u : B \to \mathbb{R} : u \text{ continuous at every point}\}$$

for any subset of $\mathbb{R}^d$. If $u \in \mathcal{C}^0(\overline{\Omega})$, then $u = U|_\Omega$ where $U = \mathcal{C}(\mathbb{R}^d)$.

Theorem 2.1. *For any open set Ω, $H^1(\Omega)$ is a Hilbert space.*

Proof. We begin by noting that if $a_n \to a$ in $L^2(\Omega)$, then $a_n \to a$ in $\mathcal{D}'(\Omega)$. In addition, if $a_n \to a$ in $\mathcal{D}'(\Omega)$, then $\partial_{x_i} a_n \to \partial_{x_i} a$ in $\mathcal{D}'(\Omega)$.

Now, let $\{u_n\}$ be a Cauchy sequence in $H^1(\Omega)$, that is

$$\|\nabla u_m - \nabla u_n\|^2_\Omega + \|u_m - u_n\|^2_\Omega \longrightarrow 0 \quad \text{as} \quad m, n \longrightarrow \infty.$$

From this we can see that the sequence $\{u_n\}$ is a Cauchy sequence in $L^2(\Omega)$, hence there is a $u \in L^2(\Omega)$ such that $u_n \to u$ in $L^2(\Omega)$. Similarly, the sequence of gradients $\{\nabla u_n\}$ is Cauchy in $\mathbf{L}^2(\Omega)$, implying the existence of a $\mathbf{v} \in \mathbf{L}^2(\Omega)$ such that $\nabla u_n \to \mathbf{v}$ in $\mathbf{L}^2(\Omega)$. By our first observation, these sequences also converge in $\mathcal{D}'(\Omega)$ and $\mathcal{D}'(\Omega)^d$ respectively and we have that $\nabla u_n \to \nabla u$ in $\mathcal{D}'(\Omega)^d$. This shows that $\nabla u = \mathbf{v}$ as regular distributions. Thus $\nabla u \in \mathbf{L}^2(\Omega)$ and $u \in H^1(\Omega)$. Finally, from the above, we achieve the convergence $u_n \to u$ in $H^1(\Omega)$. $\qquad \square$

Leibniz's rule. We saw in Exercise 1.11 that if $u \in \mathcal{D}'(\Omega)$ and $\phi \in \mathcal{C}^\infty(\Omega)$ the definition

$$\langle \phi u, \varphi \rangle := \langle u, \phi \varphi \rangle \quad \forall \varphi \in \mathcal{D}(\Omega),$$

creates a new distribution ϕu. It is then a simple consequence of the following chain of equations

$$\begin{aligned}
\langle \partial_{x_i}(\phi u), \varphi \rangle &= -\langle \phi u, \partial_{x_i}\varphi \rangle = -\langle u, \phi \, \partial_{x_i}\varphi \rangle \\
&= -\langle u, \partial_{x_i}(\phi \, \varphi) - (\partial_{x_i}\phi)\varphi \rangle = -\langle u, \partial_{x_i}(\phi \, \varphi) \rangle + \langle u, \varphi \partial_{x_i}\phi \rangle \\
&= \langle \partial_{x_i} u, \phi \, \varphi \rangle + \langle u \partial_{x_i}\phi, \varphi \rangle = \langle \phi \, \partial_{x_i} u, \varphi \rangle + \langle u \, \partial_{x_i}\phi, \varphi \rangle,
\end{aligned}$$

that we arrive at Leibniz's formula

$$\partial_{x_i}(\phi \, u) = \phi \, \partial_{x_i} u + (\partial_{x_i}\phi) u,$$

for the product of a $\mathcal{C}^\infty(\Omega)$ function and a distribution. We can also write this as

$$\nabla(\phi \, u) = \phi \nabla u + u \nabla \phi. \tag{2.1}$$

2.2 Cutoff and mollification

In this section we give a very detailed account of the proof that $\mathcal{D}(\mathbb{R}^d)$ is dense in $H^1(\mathbb{R}^d)$. This proof uses two techniques (cutoff and mollification) that are of general interest, even if they are somewhat technical.

Cutting off from infinity. Recall the smoothened Heaviside function h defined in Chapter 1,

$$h(x) := \frac{1}{\int_{-\infty}^{\infty} g(2y-1)\,\mathrm{d}y} \int_{-\infty}^{x} g(2y-1)\,\mathrm{d}y,$$

where

$$g(x) := \begin{cases} \exp\left(\frac{1}{x^2-1}\right) & |x| < 1, \\ 0 & |x| \geq 1. \end{cases}$$

Now consider the family of functions

$$h_n(\mathbf{x}) := h(n+1-|\mathbf{x}|) \quad \mathbf{x} \in \mathbb{R}^d.$$

For all n, we have

$$h_n \equiv 1 \quad \text{in } \overline{B(0;n)}, \qquad \operatorname{supp} h_n = \overline{B(0;n+1)},$$

$$h_n \in \mathcal{D}(\mathbb{R}^d), \qquad \|\nabla h_n\|_{L^\infty} \leq C.$$

This function is identical to the smoothened characteristic function $\varphi = h\left(\frac{R_2-|\cdot|}{R_2-R_1}\right)$ introduced in Chapter 1 with $R_2 = n+1$ and $R_1 = n$.

Proposition 2.1. *For any $u \in L^2(\mathbb{R}^d)$ we have $h_n u \to u$ in $L^2(\mathbb{R}^d)$. Additionally, if $u \in H^1(\mathbb{R}^d)$, then $h_n u \to u$ in $H^1(\mathbb{R}^d)$.*

Proof. For $u \in L^2(\mathbb{R}^d)$, we notice that $|h_n u - u|^2 \leq 2|u|^2$ almost everywhere, and therefore we can apply the dominated convergence theorem to obtain

$$\lim_{n \to \infty} \int_{\mathbb{R}^d} |h_n(\mathbf{x})u(\mathbf{x}) - u(\mathbf{x})|^2 \, d\mathbf{x} \longrightarrow 0,$$

leading to the first result. To show that for $u \in H^1(\mathbb{R}^d)$ the sequence $h_n u \to u$ in $H^1(\mathbb{R}^d)$, we apply the Leibniz rule (see (2.1)):

$$\nabla(h_n u) = h_n \nabla u + u \nabla h_n,$$

and note that $h_n \nabla u \to \nabla u$ in $\mathbf{L}^2(\mathbb{R}^d)$ by our first result. The second term $u \nabla h_n \to 0$ in $L^2(\mathbb{R}^d)$. We can see this by observing that

$$|u \nabla h_n|^2 \leq C^2 |u|^2 \quad \text{a.e.,} \qquad \nabla h_n \longrightarrow 0 \quad \text{a.e.,}$$

and again appealing to the dominated convergence theorem. From this we obtain the convergence in $H^1(\mathbb{R}^d)$. □

Mollification. This is the name that is given to the process of smoothing a function while focusing around a point by convolving with a sequence obtained by scaling a fixed element of $\mathcal{D}(\mathbb{R}^d)$. Let us first introduce the concept of convolution: the **convolution** of $u \in L^1_{\text{loc}}(\mathbb{R}^d)$ with $\varphi \in \mathcal{D}(\mathbb{R}^d)$ is defined as

$$(u * \varphi)(\mathbf{x}) := \int_{\mathbb{R}^d} u(\mathbf{y})\varphi(\mathbf{x} - \mathbf{y}) \, d\mathbf{y} = \int_{B(\mathbf{x};M)} u(\mathbf{y})\varphi(\mathbf{x} - \mathbf{y}) \, d\mathbf{y},$$

assuming $\text{supp}\,\varphi \subset \overline{B(\mathbf{0}; M)}$. Now let $\varphi \geq 0$ be an element of the test space $\mathcal{D}(\mathbb{R}^d)$ such that $\text{supp}\,\varphi = \overline{B(\mathbf{0}; 1)}$ and $\int_{\mathbb{R}^d} \varphi \equiv 1$. Consider next the family of functions depending on a parameter $\varepsilon > 0$ given by $\varphi_\varepsilon := \varepsilon^{-d}\varphi(\cdot/\varepsilon)$. These functions satisfy

$$\text{supp}\,\varphi_\varepsilon = \overline{B(\mathbf{0}; \varepsilon)}, \qquad \varphi_\varepsilon \geq 0, \qquad \int_{\mathbb{R}^d} \varphi_\varepsilon(\mathbf{x}) \, d\mathbf{x} = 1.$$

We will use the convolution of the functions φ_ε with a function u to focus only locally on u. We will need two technical results for the main theorem of this section. Instead of proving them right away, we will use them now, and leave the technicalities for Section 2.3.

Proposition 2.2 (Smoothing by convolution)**.** *If $\varphi \in \mathcal{D}(\mathbb{R}^d)$ with $\text{supp}\,\varphi \subset B(\mathbf{0}; \varepsilon)$ and $u \in H^1(\mathbb{R}^d)$ vanishes outside $B(\mathbf{0}; R)$, then*

(a) $u * \varphi \in \mathcal{D}(\mathbb{R}^d)$.

(b) $\text{supp}\,(u * \varphi) \subset B(\mathbf{0}; R + \varepsilon)$.

(c) $\partial_{x_i}(u * \varphi) = \partial_{x_i} u * \varphi$.

Proposition 2.3 (Approximations of the identity). *Let*

$$\varphi \in \mathcal{D}(\mathbb{R}^d), \qquad 0 \le \varphi \le 1, \qquad \int_{\mathbb{R}^d} \varphi(\mathbf{x}) \, d\mathbf{x} = 1,$$

and consider the functions

$$\varphi_\varepsilon := \varepsilon^{-d} \varphi(\varepsilon^{-1} \cdot) \qquad \varepsilon > 0.$$

For all $u \in L^2(\mathbb{R}^d)$, as $\varepsilon \to 0$

$$u * \varphi_\varepsilon \longrightarrow u \qquad in \ L^2(\mathbb{R}^d).$$

We have now built up the necessary tools to prove the following theorem.

Theorem 2.2. *$\mathcal{D}(\mathbb{R}^d)$ is dense in $H^1(\mathbb{R}^d)$.*

Proof. Suppose that $u \in H^1(\mathbb{R}^d)$. Let h_n be a cutoff sequence as defined at the beginning of this section. We have shown that $h_n u \to u$ in $H^1(\mathbb{R}^d)$. Therefore the set

$$H^1_{\text{comp}}(\mathbb{R}^d) := \left\{ u \in H^1(\mathbb{R}^d) : u \equiv 0 \text{ in } \mathbb{R}^d \setminus B(\mathbf{0}; R) \text{ for some } R \right\}$$

is a dense subset of $H^1(\mathbb{R}^d)$, and we need only show $\mathcal{D}(\mathbb{R}^d)$ is dense in $H^1_{\text{comp}}(\mathbb{R}^d)$. To this end, consider $u \in H^1_{\text{comp}}(\mathbb{R}^d)$. Proposition 2.2 tells us that $u * \varphi_\varepsilon \in \mathcal{D}(\mathbb{R}^d)$ and Proposition 2.3 shows that $u * \varphi_\varepsilon \to u$ in $L^2(\mathbb{R}^d)$. Furthermore, using Propositions 2.2 and 2.3 again, it follows that

$$\partial_{x_i}(u * \varphi_\varepsilon) = \partial_{x_i} u * \varphi_\varepsilon \longrightarrow \partial_{x_i} u \qquad in \ L^2(\mathbb{R}^d).$$

Therefore $u * \varphi_\varepsilon \to u$ in $H^1(\mathbb{R}^d)$, which establishes the theorem. $\square$

2.3 A guided tour of mollification (*)

In this section we prove some of the technical results related to mollification that were left unproved in Section 2.2.

Proposition 2.4. *Let $\varphi \in \mathcal{D}(\mathbb{R}^d)$ and $u \in L^1_{\text{loc}}(\mathbb{R}^d)$. The convolution product $u * \varphi$ satisfies:*

$$u * \varphi \in \mathcal{C}(\mathbb{R}^d) \qquad and \qquad \partial_{x_i}(u * \varphi) = u * \partial_{x_i} \varphi.$$

Therefore $u * \varphi \in \mathcal{C}^\infty(\mathbb{R}^d)$. Additionally,

$$\begin{aligned} \operatorname{supp} \varphi \subset \overline{B(\mathbf{0}; \varepsilon)} \\ u \equiv 0 \ in \ B(\mathbf{x}; 2\varepsilon) \end{aligned} \quad \Longrightarrow \quad u * \varphi \equiv 0 \ in \ B(\mathbf{x}; \varepsilon).$$

Proof. Let $R_\varphi > 0$ be such that $\operatorname{supp}\varphi \subset B(\mathbf{0}; R_\varphi)$. For any $R > 0$ and every $\mathbf{x} \in B(\mathbf{0}; R)$ we have

$$(u * \varphi)(\mathbf{x}) = \int_{B(\mathbf{x};R_\varphi)} \varphi(\mathbf{x} - \mathbf{y})u(\mathbf{y})\,\mathrm{d}\mathbf{y} = \int_{B(\mathbf{0};R+R_\varphi)} \varphi(\mathbf{x} - \mathbf{y})u(\mathbf{y})\,\mathrm{d}\mathbf{y}.$$

Since $u \in L^1_{\mathrm{loc}}(\mathbb{R}^d)$, it follows that $u \in L^1(B(\mathbf{0}; R + R_\varphi))$. With this at hand, it is easy to prove that $u * \varphi$ is continuous at any point of the ball $B(\mathbf{0}; R + R_\varphi)$. Differentiability can also be shown to hold under integral sign. The last property is easy to verify. $\qquad\square$

Note that a simple consequence of Proposition 2.4 is the following statement: if $u \equiv 0$ outside $B(\mathbf{0}; R)$ and $\operatorname{supp}\varphi \subset \overline{B(\mathbf{0}; \varepsilon)}$, then $u * \varphi \equiv 0$ outside $B(\mathbf{0}; R + \varepsilon)$. This shows that the local smoothing by convolution with φ is made at the price of losing part of the area where u vanishes.

Proposition 2.5. *If $\varphi \in \mathcal{D}(\mathbb{R}^d)$ and $u \in H^1(\mathbb{R}^d)$, then*

$$\partial_{x_i}(u * \varphi) = (\partial_{x_i}u) * \varphi \qquad 1 \leq i \leq d.$$

Proof. By Proposition 2.4, in the classical way, $\partial_{x_i}(u * \varphi) = u * \partial_{x_i}\varphi$. Note that when $u \in H^1(\mathbb{R}^d)$, $u * \partial_{x_i}\varphi$ and $\partial_{x_i}u * \varphi$ are continuous functions by Proposition 2.4. We are next going to show that they are equal pointwise.

Fix $\mathbf{z} \in \mathbb{R}^d$, consider the function $\varphi_\mathbf{z} := \varphi(\mathbf{z} - \cdot) \in \mathcal{D}(\mathbb{R}^d)$, and note that

$$(\partial_{x_i}\varphi)(\mathbf{z} - \cdot) = -\partial_{x_i}\varphi_\mathbf{z}.$$

We now have

$$(u * \partial_{x_i}\varphi)(\mathbf{z}) = \int_{\mathbb{R}^d} u(\mathbf{y})(\partial_{x_i}\varphi)(\mathbf{z} - \mathbf{y})\,\mathrm{d}\mathbf{y} = -\int_{\mathbb{R}^d} u(\mathbf{y})\,\partial_{x_i}\varphi_\mathbf{z}(\mathbf{y})\,\mathrm{d}\mathbf{y}$$

$$= -\langle u, \partial_{x_i}\varphi_\mathbf{z}\rangle = \langle \partial_{x_i}u, \varphi_\mathbf{z}\rangle = \int_{\mathbb{R}^d} (\partial_{x_i}u)(\mathbf{y})\varphi_\mathbf{z}(\mathbf{y})\,\mathrm{d}\mathbf{y}$$

$$= \int_{\mathbb{R}^d} (\partial_{x_i}u)(\mathbf{y})\varphi(\mathbf{z} - \mathbf{y})\,\mathrm{d}\mathbf{y} = (\partial_{x_i}u * \varphi)(\mathbf{z}).$$

This finishes the proof. $\qquad\square$

Note that Propositions 2.4 and 2.5 prove Proposition 2.2. Before going on with the proof of Proposition 2.3 let us establish a simple consequence of Proposition 2.5, showing the effect of convolving $H^1(\mathbb{R}^d)$ functions with a particular subclass of test functions.

Proposition 2.6. *Let*

$$\varphi \in \mathcal{D}(\mathbb{R}^d), \qquad \varphi \geq 0, \qquad \int_{\mathbb{R}^d} \varphi(\mathbf{x})\,\mathrm{d}\mathbf{x} = 1.$$

If $u \in L^2(\mathbb{R}^d)$, then $u * \varphi \in L^2(\mathbb{R}^d)$ and

$$\|u * \varphi\|_{\mathbb{R}^d} \leq \|u\|_{\mathbb{R}^d}.$$

Therefore, if $u \in H^1(\mathbb{R}^d)$, *then* $u * \varphi \in H^1(\mathbb{R}^d)$ *and*

$$\|u * \varphi\|_{1, \mathbb{R}^d} \leq \|u\|_{1, \mathbb{R}^d}.$$

Proof. We begin by noting that proving the bound on the norm of $u * \varphi$ will prove that it is in $L^2(\mathbb{R}^d)$. To this end, we consider

$$\int_{\mathbb{R}^d} |(u * \varphi)(\mathbf{x})|^2 \, d\mathbf{x} = \int_{\mathbb{R}^d} \left| \int_{\mathbb{R}^d} \varphi(\mathbf{x} - \mathbf{y}) u(\mathbf{y}) \, d\mathbf{y} \right|^2 d\mathbf{x}$$

$$= \int_{\mathbb{R}^d} \left| \int_{\mathbb{R}^d} \varphi^{1/2}(\mathbf{x} - \mathbf{y}) \varphi^{1/2}(\mathbf{x} - \mathbf{y}) u(\mathbf{y}) \, d\mathbf{y} \right|^2 d\mathbf{x}$$

$$\leq \int_{\mathbb{R}^d} \left(\int_{\mathbb{R}^d} \varphi(\mathbf{x} - \mathbf{y}) \, d\mathbf{y} \right) \left(\int_{\mathbb{R}^d} \varphi(\mathbf{x} - \mathbf{y}) |u(\mathbf{y})|^2 \, d\mathbf{y} \right) d\mathbf{x},$$

where we have used the Cauchy-Schwarz inequality. Recalling that the integral of φ is one and using Fubini's theorem, we obtain

$$\int_{\mathbb{R}^d} \left(\int_{\mathbb{R}^d} \varphi(\mathbf{x} - \mathbf{y}) |u(\mathbf{y})|^2 \, d\mathbf{y} \right) d\mathbf{x} = \int_{\mathbb{R}^d} |u(\mathbf{y})|^2 \int_{\mathbb{R}^d} \varphi(\mathbf{x} - \mathbf{y}) \, d\mathbf{x} \, d\mathbf{y}$$

$$= \int_{\mathbb{R}^d} |u(\mathbf{y})|^2 \, d\mathbf{y}.$$

This proves the first part of the statement. The second part follows from this and Proposition 2.5. $\qquad\square$

Lemma 2.1 (Translations are continuous in L^2). *For all* $u \in L^2(\mathbb{R}^d)$

$$u(\cdot - \mathbf{h}) \stackrel{|\mathbf{h}| \to 0}{\longrightarrow} u \qquad in \ L^2(\mathbb{R}^d).$$

Proof. Let $\varphi \in \mathcal{D}(\mathbb{R}^d)$. It is then easy to show that $\varphi(\cdot - \mathbf{h}) \to \varphi$ uniformly and therefore in $L^2(\mathbb{R}^d)$. For general u, we bound

$$\|u(\cdot - \mathbf{h}) - u\|_{\mathbb{R}^d} \leq \|u(\cdot - \mathbf{h}) - \varphi(\cdot - \mathbf{h})\|_{\mathbb{R}^d} + \|\varphi(\cdot - \mathbf{h}) - \varphi\|_{\mathbb{R}^d} + \|\varphi - u\|_{\mathbb{R}^d}$$
$$= 2\|u - \varphi\|_{\mathbb{R}^d} + \|\varphi(\cdot - \mathbf{h}) - \varphi\|_{\mathbb{R}^d}.$$

We next notice that the variational lemma implies that

$$\mathcal{D}(\mathbb{R}^d)^\perp = \{u \in L^2(\mathbb{R}^d) : (u, \varphi)_{\mathbb{R}^d} = 0 \quad \forall \varphi \in \mathcal{D}(\mathbb{R}^d)\} = \{0\},$$

and therefore $\mathcal{D}(\mathbb{R}^d)$ is dense in $L^2(\mathbb{R}^d)$. (This is due to the fact that for every subspace M of a Hilbert space H, we have the orthogonal decomposition $H = \overline{M} \oplus M^\perp$.) To finish the proof, given $u \in L^2(\mathbb{R}^d)$ and $\varepsilon > 0$, we can find $\varphi \in \mathcal{D}(\mathbb{R}^d)$ such that $\|u - \varphi\|_{\mathbb{R}^d} < \varepsilon$ and then $\delta > 0$ such that if $|\mathbf{h}| < \delta$, then $\|\varphi - \varphi(\cdot - |\mathbf{h}|)\|_{\mathbb{R}^d} < \varepsilon$. Therefore, $\|u - u(\cdot - |\mathbf{h}|)\|_{\mathbb{R}^d} < 3\varepsilon$, which shows that continuity of the translation operators in $L^2(\mathbb{R}^d)$. $\qquad\square$

Proof of Proposition 2.3. Recall that we are dealing with the limits of convolutions $u * \varphi_\varepsilon$, where $\varphi_\varepsilon(\mathbf{x}) = \varepsilon^{-d} \varphi(\mathbf{x}/\varepsilon)$ and $\varphi \geq 0$ has unit integral. Consider the function $\omega(\cdot; u) : \mathbb{R}^d \to [0, \infty)$ defined by

$$\omega(\mathbf{h}; u) := \int_{\mathbb{R}^d} |u(\mathbf{x} - \mathbf{h}) - u(\mathbf{x})|^2 \, \mathrm{d}\mathbf{x} = \|u(\cdot - |\mathbf{h}|) - u\|_{\mathbb{R}^d}^2.$$

Note that $\omega(\cdot; u)$ is continuous at zero by Lemma 2.1. Also

$$\omega(\mathbf{h}; u) = \omega(\mathbf{h} - \widehat{\mathbf{h}}; u(\cdot + \widehat{\mathbf{h}})) \qquad \forall \mathbf{h}, \widehat{\mathbf{h}} \in \mathbb{R}^d,$$

and therefore $\omega(\cdot; u) \in \mathcal{C}(\mathbb{R}^d)$ and $\omega(\mathbf{0}; u) = 0$. In particular, $\omega(\cdot; u)$ is bounded on compact sets of $\mathbb{R}^d$.

With a simple change of variables, we can write

$$(u * \varphi_\varepsilon)(\mathbf{x}) - u(\mathbf{x}) = \int_{\mathbb{R}^d} \varphi(\mathbf{z})(u(\mathbf{x} - \varepsilon \mathbf{z}) - u(\mathbf{x})) \, \mathrm{d}\mathbf{z},$$

and therefore, we can easily bound

$$\|u * \varphi_\varepsilon - u\|_{\mathbb{R}^d}^2 \leq \int_{\mathbb{R}^d} \left(\int_{\mathbb{R}^d} \varphi(\mathbf{z}) \, \mathrm{d}\mathbf{z} \right) \left(\int_{\mathbb{R}^d} \varphi(\mathbf{z}) |u(\mathbf{x} - \varepsilon \mathbf{z}) - u(\mathbf{x})|^2 \, \mathrm{d}\mathbf{z} \right) \mathrm{d}\mathbf{x}$$

$$= \int_{\mathbb{R}^d} \varphi(\mathbf{z}) \omega(\varepsilon \mathbf{z}; u) \, \mathrm{d}\mathbf{z}.$$

By the continuity of $\omega(\cdot; u)$ and the compactness of the support of φ, we can use the dominated convergence theorem and take the limit above, which finishes the proof of the proposition. $\qquad \square$

2.4   The space $H_0^1(\Omega)$

In the last section we showed that $\mathcal{D}(\mathbb{R}^d)$ is a dense subset of $H^1(\mathbb{R}^d)$. This is not true, however, if we consider a bounded domain Ω. The failure of density in the bounded case leads us to define the following Sobolev space.

Definition of $H_0^1(\Omega)$. Let Ω be an open set in $\mathbb{R}^d$. We define the space

$$H_0^1(\Omega) := \{u \in H^1(\Omega) : \exists \{\varphi_n\} \text{ in } \mathcal{D}(\Omega) \text{ s.t. } \|\varphi_n - u\|_{1,\Omega} \to 0\}.$$

In other words, the space $H_0^1(\Omega)$ is the closure of $\mathcal{D}(\Omega)$ with respect to the $H^1(\Omega)$ norm. This is, by definition, a closed subspace of $H^1(\Omega)$. Note that $H_0^1(\mathbb{R}^d) = H^1(\mathbb{R}^d)$. For elements of $H_0^1(\Omega)$ on bounded domains, we have the following theorem.

Theorem 2.3 (Poincaré-Friedrichs inequality). *If $\Omega \subset \mathbb{R}^{d-1} \times (a,b)$ (or, more generally, if Ω is bounded in at least one direction), then*

$$\|u\|_\Omega \le \frac{b-a}{2}\|\nabla u\|_\Omega \quad \forall u \in H_0^1(\Omega).$$

Proof. Let us first show why it suffices to prove the theorem holds for elements $\varphi \in \mathcal{D}(\Omega)$. Given $u \in H_0^1(\Omega)$, we can consider a sequence $\{\varphi_n\}$ in $\mathcal{D}(\Omega)$ such that $\varphi_n \to u$ in $H^1(\Omega)$, and then $\|\varphi_n\|_\Omega \to \|u\|_\Omega$ and $\|\nabla\varphi_n\|_\Omega \to \|\nabla u\|_\Omega$. Therefore

$$\|\varphi_n\|_\Omega \le \frac{b-a}{2}\|\nabla\varphi_n\|_\Omega$$

$$\downarrow \qquad\qquad \downarrow$$

$$\|u\|_\Omega \le \frac{b-a}{2}\|\nabla u\|_\Omega.$$

We now prove the inequality in the one-dimensional case. For $\varphi \in \mathcal{D}(a,b)$, we write

$$\varphi(x) = \frac{1}{2}\int_a^x \varphi'(t)\,\mathrm{d}t - \frac{1}{2}\int_x^b \varphi'(t)\,\mathrm{d}t.$$

A simple computation yields the bound

$$|\varphi(x)| \le \frac{1}{2}\int_a^b |\varphi'(t)|\,\mathrm{d}t \quad \forall x \in (a,b),$$

which we can use to bound

$$\int_a^b |\varphi(x)|^2\,\mathrm{d}x \le (b-a)\max_{a \le t \le b}|\varphi(t)|^2 \le \left(\frac{b-a}{4}\right)\left|\int_a^b |\varphi'(t)|\,\mathrm{d}t\right|^2$$

$$\le \left(\frac{b-a}{4}\right)\left(\int_a^b 1\,\mathrm{d}x\right)\left(\int_a^b |\varphi'(t)|^2\,\mathrm{d}t\right)$$

$$= \frac{(b-a)^2}{4}\int_a^b |\varphi'(t)|^2\,\mathrm{d}t.$$

This establishes the result in one dimension. To prove the statement of the theorem, let $\varphi \in \mathcal{D}(\Omega)$ where $\Omega \subset \mathbb{R}^{d-1} \times (a,b)$. We can now apply the previous argument in the dimension in which Ω is bounded by writing $\mathbf{x} = (\tilde{\mathbf{x}}, x_d) \in$

$\mathbb{R}^{d-1} \times (a, b)$. Doing so, we obtain

$$\int_\Omega |\varphi(\mathbf{x})|^2 \, \mathrm{d}\mathbf{x} = \int_{\mathbb{R}^{d-1}} \left(\int_a^b |\varphi(\widetilde{\mathbf{x}}, x_d)|^2 \, \mathrm{d}x_d \right) \mathrm{d}\widetilde{\mathbf{x}}$$

$$\leq \int_{\mathbb{R}^{d-1}} \frac{(b-a)^2}{4} \left(\int_a^b |\partial_{x_d}\varphi(\widetilde{\mathbf{x}}, x_d)|^2 \, \mathrm{d}x_d \right) \mathrm{d}\widetilde{\mathbf{x}}$$

$$= \frac{(b-a)^2}{4} \int_{\mathbb{R}^{d-1} \times (a,b)} |\partial_{x_d}\varphi(\mathbf{x})|^2 \, \mathrm{d}\mathbf{x}$$

$$\leq \frac{(b-a)^2}{4} \|\nabla\varphi\|_\Omega^2.$$

This finishes the proof. □

Corollary 2.1. *If Ω is a bounded domain, then the Dirichlet form defines a norm in $H_0^1(\Omega)$ that is equivalent to the usual norm, namely there exists a $C > 0$ such that*

$$C\|u\|_{1,\Omega} \leq \|\nabla u\|_\Omega \leq \|u\|_{1,\Omega} \qquad \forall u \in H_0^1(\Omega).$$

A feeling for $H_0^1(\Omega)$. Using the Poincaré-Friedrichs inequality, it follows that if Ω is bounded, then constant functions are elements of $H^1(\Omega)$ that do not belong to $H_0^1(\Omega)$, so these two spaces are different. So far we have introduced $H_0^1(\Omega)$ and defined an alternative norm for this space, but we have not yet demonstrated its importance or usefulness. This will be partially rectified by way of the next Proposition and its consequences. The space of $H^1(\Omega)$ functions that vanish in a neighborhood of $\partial\Omega$ will be given a nonstandard name in Proposition 2.7. This space will play a very important role in some technical proofs in Chapter 3.

Proposition 2.7. *Let Ω be a bounded domain. For each $u \in H_0^1(\Omega)$, consider the function $\widetilde{u} : \mathbb{R}^d \to \mathbb{R}$ given by*

$$\widetilde{u} := \begin{cases} u & in \ \Omega, \\ 0 & in \ \mathbb{R}^d \setminus \Omega. \end{cases}$$

Also, define

$$H_{\equiv}^1(\Omega) := \{u \in H^1(\Omega) : u \equiv 0 \ in \ a \ neighborhood \ of \ \partial\Omega\}.$$

With these definitions, we have

$$\mathcal{D}(\Omega) \subset H_{\equiv}^1(\Omega) \subset H_0^1(\Omega) \subset \{u \in H^1(\Omega) : \widetilde{u} \in H^1(\mathbb{R}^d)\}.$$

Proof. The first inclusion is a straightforward consequence of the definition.

For the second inclusion, we detail three facts. First, if $u \in H^1(\Omega)$ and $u \equiv 0$ near $\partial\Omega$, then there exists an $\varepsilon > 0$ such that

$$u \equiv 0 \quad \text{in} \quad \{\mathbf{x} \in \Omega : \text{dist}(\mathbf{x}, \partial\Omega) < \varepsilon\}.$$

In this case $\tilde{u} \in L^2(\mathbb{R}^d)$, as it is a simple extension of u by zero. It is also simple to show that $\partial_{x_i}\tilde{u} = \widetilde{\partial_{x_i}u} \in L^2(\mathbb{R}^d)$ for each i (see Exercise 2.5). Therefore $\tilde{u} \in H^1(\mathbb{R}^d)$ and

$$\tilde{u} \equiv 0 \quad \text{in} \quad \{\mathbf{x} \in \Omega : \text{dist}(\mathbf{x}, \partial\Omega) < \varepsilon\}.$$

The second fact is that for any mollifying sequence $\{\varphi_\eta\}$ with $\eta \to 0$ we have $\tilde{u} * \varphi_\eta \in \mathcal{D}(\mathbb{R}^d)$ for each η and $\tilde{u} * \varphi_\eta \to \tilde{u}$ in $H^1(\mathbb{R}^d)$. Therefore

$$(\tilde{u} * \varphi_\eta)\big|_\Omega \longrightarrow u \quad \text{in} \quad H^1(\Omega).$$

The final fact which we shall use is that for η small enough, $\tilde{u} * \varphi_\eta \in \mathcal{D}(\Omega)$. This fact follows from the first fact that $\tilde{u} \equiv 0$ close to $\partial\Omega$ and the effect of convolving $\tilde{u}$ with φ_η for η small enough. See Figure 2.1 for a visual representation of this. From these three facts, we immediately achieve the inclusion of $H^1_{\equiv}(\Omega)$ in $H^1_0(\Omega)$.

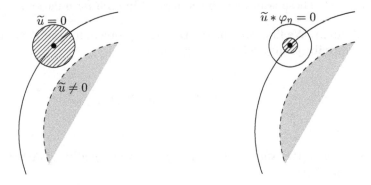

Figure 2.1: For small enough η, we have $(\tilde{u} * \varphi_\eta)\big|_\Omega \in \mathcal{D}(\Omega)$.

To prove the third inclusion, let $u \in H^1_0(\Omega)$ and $\{\varphi_n\}$ be a sequence in $\mathcal{D}(\Omega)$ converging to u in $H^1(\Omega)$. This means that

$$\varphi_n \longrightarrow u \quad \text{in } L^2(\Omega) \qquad \text{and} \qquad \nabla\varphi_n \longrightarrow \nabla u \quad \text{in } \mathbf{L}^2(\Omega).$$

The extensions $\widetilde{\varphi_n}$ are in $H^1(\mathbb{R}^d)$ and by smoothness we have $\nabla\widetilde{\varphi_n} = \widetilde{\nabla\varphi_n}$. Furthermore the extensions satisfy

$$\widetilde{\varphi_n} \longrightarrow \tilde{u} \quad \text{in } L^2(\mathbb{R}^d) \qquad \text{and} \qquad \widetilde{\nabla\varphi_n} \longrightarrow \widetilde{\nabla u} \quad \text{in } \mathbf{L}^2(\mathbb{R}^d).$$

To finish the proof we need only show that $\widetilde{\nabla u} = \nabla\tilde{u}$, since the above convergence would guarantee $\tilde{u} \in H^1(\mathbb{R}^d)$. Since $\widetilde{\varphi_n} \to \tilde{u}$ in $\mathcal{D}'(\Omega)$, it follows that $\nabla\widetilde{\varphi_n} \to \nabla\widetilde{u}$ in $\mathcal{D}'(\Omega)^d$. At the same time, $\nabla\widetilde{\varphi_n} = \widetilde{\nabla\varphi_n} \to \widetilde{\nabla u}$ in $\mathcal{D}'(\Omega)$. Equating both distributional limits, it follows that $\widetilde{\nabla u} = \nabla\widetilde{u}$, and therefore $\widetilde{u} \in H^1(\mathbb{R}^d)$. □

As a result of Proposition 2.7, we can think of a function $u \in H_0^1(\Omega)$ as a function in $H^1(\Omega)$ with a weak notion of $u = 0$ on the boundary of Ω.

2.5   The Dirichlet problem

Having developed the necessary tools for understanding the space $H_0^1(\Omega)$, we move to understand three equivalent representations of the Poisson problem with homogeneous Dirichlet boundary conditions:

$$-\Delta u = f \quad \text{in } \Omega, \qquad u = 0 \quad \text{on } \partial\Omega.$$

Note that, at least in this form, we are not making any statements about the regularity of the solution u, and in what way the boundary conditions are being imposed. These issues will be addressed in each formulation.

The first form. We first present the Dirichlet problem for the Poisson equation as a **distributional PDE**. The problem is

$$u \in H_0^1(\Omega),$$
$$-\Delta u = f \quad \text{in } \Omega \quad (\text{in } \mathcal{D}'(\Omega)), \tag{2.2}$$

which we read as

find u in $H_0^1(\Omega)$ such that $-\Delta u = f$ holds as equality of distributions in $\mathcal{D}'(\Omega)$.

In principle, we only seek $u \in H_0^1(\Omega)$, and therefore $-\Delta u$ may not be a regular distribution, but if we take $f \in L^2(\Omega)$, then it is. The problem imposes a weak form of zero boundary conditions within the solution space. The statement that an equation holds in $\mathcal{D}'(\Omega)$ means that the equation is satisfied when tested by an element of the test space $\mathcal{D}(\Omega)$. Therefore, the distributional PDE

$$-\Delta u = f \quad \text{in } \Omega \quad (\text{in } \mathcal{D}'(\Omega))$$

means

$$-\langle\Delta u, \varphi\rangle = -\sum_{j=1}^{d}\langle\partial_{x_j}^2 u, \varphi\rangle = \sum_{j=1}^{d}\langle\partial_{x_j} u, \partial_{x_j}\varphi\rangle = \langle f, \varphi\rangle \qquad \forall\varphi \in \mathcal{D}(\Omega).$$

The second form. Suppose that $u \in H^1_0(\Omega)$. Since $u, \partial_{x_i} u \in L^2(\Omega)$ are regular distributions, we can write the distributional PDE in a strong way as

$$\sum_{j=1}^{d} (\partial_{x_j} u, \partial_{x_j} \varphi)_\Omega = (f, \varphi)_\Omega \qquad \forall \varphi \in \mathcal{D}(\Omega),$$

which we claim is equivalent to

$$\begin{aligned} & u \in H^1_0(\Omega), \\ & (\nabla u, \nabla v)_\Omega = (f, v)_\Omega \quad \forall v \in H^1_0(\Omega). \end{aligned} \tag{2.3}$$

We will call this second form the **variational formulation**.

Proposition 2.8 (Equivalence of BVP and VF). *The variational formulation (2.3) is equivalent to the distributional PDE (2.2).*

Proof. Suppose that u satisfies the variational formulation. Recalling that $\mathcal{D}(\Omega) \subset H^1_0(\Omega)$, we see that we can take $v = \varphi \in \mathcal{D}(\Omega)$ and therefore

$$\langle \nabla u, \nabla \varphi \rangle = (\nabla u, \nabla \varphi)_\Omega = (f, \varphi)_\Omega \qquad \forall \varphi \in \mathcal{D}(\Omega), \tag{2.4}$$

or, in other words,
$$-\Delta u = f \qquad \text{in } \mathcal{D}'(\Omega).$$

For the converse, take $v \in H^1_0(\Omega)$ and a sequence $\{\varphi_n\}$ in $\mathcal{D}(\Omega)$ such that $\varphi_n \to v$ in $H^1_0(\Omega)$, hence $\varphi_n \to v$ in $L^2(\Omega)$ and $\nabla \varphi_n \to \nabla v$ in $\mathbf{L}^2(\Omega)$. We know such a sequence exists due to the density of $\mathcal{D}(\Omega)$ in $H^1_0(\Omega)$. Using the Cauchy-Schwarz inequality we can estimate

$$|(f, \varphi_n)_\Omega - (f, v)_\Omega| = |(f, \varphi_n - v)_\Omega| \le \|f\|_\Omega \|\varphi_n - v\|_\Omega \longrightarrow 0,$$

and similarly

$$|(\nabla u, \nabla \varphi_n - \nabla v)_\Omega| \le \|\nabla u\|_\Omega \|\nabla \varphi_n - \nabla v\|_\Omega \longrightarrow 0.$$

Therefore

$$(f, \varphi_n)_\Omega \longrightarrow (f, v)_\Omega \qquad \text{and} \qquad (\nabla u, \nabla \varphi_n)_\Omega \longrightarrow (\nabla u, \nabla v)_\Omega.$$

Since $-\Delta u = f$ as distributions, which means that (2.4) holds, this convergence implies the variational formulation. $\qquad \square$

Back to the variational form. Consider again the variational form (2.3). The equation $(\nabla u, \nabla v)_\Omega = (f, v)_\Omega$ only contains information about the PDE. We rely on the denseness of $\mathcal{D}(\Omega)$ in $H^1_0(\Omega)$ to impose the boundary conditions.

Corollary 2.2 (Uniqueness). *Problem (2.3) admits at most one solution.*

Proof. If u_1 and u_2 are both solutions to (2.3), or equivalently, the distributional PDE with the same right-hand side f, we can define $u = u_1 - u_2$, which satisfies

$$u \in H_0^1(\Omega), \qquad (\nabla u, \nabla v)_\Omega = 0 \quad \forall v \in H_0^1(\Omega).$$

Now take $v = u$ to derive that $\nabla u = 0$ (therefore u is constant). By the Poincaré-Friedrichs inequality (Theorem 2.3), it follows that $u = 0$. □

The third form. The third equivalent formulation for the Dirichlet problem is as a **minimization problem**. First we introduce some notation. We write

$$\phi(x) = \text{min!} \qquad x \in X,$$

to denote x as the unique minimizer of ϕ among elements of X, that is,

$$\phi(x) = \min_{y \in X} \phi(y).$$

With this definition, the minimization form of the Dirichlet problem is

$$\frac{1}{2} \int_\Omega |\nabla u|^2 - \int_\Omega fu = \text{min!} \qquad u \in H_0^1(\Omega). \tag{2.5}$$

The equivalence of the minimization problem to the variational form follows from the next lemma.

Lemma 2.2 (VF and minimization problem). *Let V be a real vector space, $a : V \times V \to \mathbb{R}$ a symmetric positive semidefinite bilinear form, and $\ell : V \to \mathbb{R}$ a linear functional on V. The minimization problem*

$$\tfrac{1}{2}a(u, u) - \ell(u) = \text{min!} \qquad u \in V,$$

is equivalent to the variational problem

$$u \in V,$$
$$a(u, v) = \ell(v) \quad \forall v \in V.$$

Proof. Let $u, v \in V$ and $t \in \mathbb{R}$. Note that a polynomial $\varphi(t) = C + Bt + At^2$ with $A \geq 0$ has a minimum at $t = 0$ if and only if $B = 0$ with the minimum value being C. We then proceed to compute

$$\frac{1}{2}a(u + tv, u + tv) - \ell(u + tv)$$

$$= \frac{1}{2}a(u, u) + \frac{t}{2}a(u, v) + \frac{t}{2}a(u, v) + \frac{t^2}{2}a(v, v) - \ell(u) - t\ell(v)$$

$$= \left(\frac{1}{2}a(u, u) - \ell(u) \right) + t\left(a(u, v) - \ell(v)\right) + \frac{t^2}{2}a(v, v).$$

Therefore

$$\tfrac{1}{2}a(u + tv, u + tv) - \ell(u + tv) \geq \tfrac{1}{2}a(u, u) - \ell(u) \quad \forall t \in \mathbb{R}, \; v \in V,$$

if and only if

$$a(u, v) - \ell(v) = 0 \quad \forall v \in V.$$

This shows the equivalence of the problems.   □

2.6   Existence of solutions

So far we have proved that problems (2.2), (2.3), and (2.5) are equivalent and that they have at most one solution. We still need to show that they do actually have a solution. It is the goal of this section to do this with the variational form of the problem (2.3), and by briefly *rephrasing the well-known Riesz-Fréchet representation theorem as an existence theorem*.

Review of functional analysis. Let V be a real normed space, and $\ell : V \to \mathbb{R}$ be a linear map. The functional ℓ is continuous if and only if it is sequentially continuous, or equivalently, bounded, meaning that there is a $C > 0$ such that

$$|\ell(x)| \leq C\|x\| \quad \forall x \in V.$$

For a linear map $\ell : V \to \mathbb{R}$ we define its dual norm by

$$\|\ell\|_{V'} := \sup_{0 \neq v \in V} \frac{|\ell(v)|}{\|v\|},$$

where V' is the dual space of V, consisting of all bounded linear functionals on V. The dual norm induces a norm in V', and with this norm V' is a complete normed (Banach) space. When we consider V to be an inner product space, we can study the maps from V to its dual given by

$$V \ni v \longmapsto (v, \cdot)_V : V \to \mathbb{R}.$$

These functionals satisfy

$$\| (v, \cdot) \|_{V'} = \sup_{0 \neq w \in V} \frac{|(v, w)_V|}{\|w\|_V} = \|v\|_V,$$

i.e., the map $V \to V'$ given by $v \mapsto (v, \cdot)_V$ is a linear isometry, and is therefore injective. The Riesz-Fréchet theorem shows that this map is in fact surjective.

Theorem 2.4 (Riesz-Fréchet). *If V is a real Hilbert space, the map $V \to V'$ given by $v \longmapsto (v, \cdot)_\Omega$ is surjective. Therefore:*

(a) *V' is isometrically isomorphic to V,*

(b) *for all $\ell \in V'$ there is a unique $v \in V$ such that $\ell(w) = (v, w)_V$ for all $w \in V$, and*

(c) *the element v found in (b) satisfies $\|v\|_V = \|\ell\|_{V'}$.*

Proof. We have already shown (a). To show (b), let $0 \neq \ell \in V'$ and $x \in (\ker \ell)^{\perp}$ with unit norm. Note that $\ell(\ell(y)x - \ell(x)y) = 0$, therefore

$$(x, \ell(y)x - \ell(x)y)_V = 0,$$

which shows

$$\ell(y) = \ell(x)(x, y)_V = (\ell(x)x, y)_V,$$

and therefore $\ell(x)x$ is the desired element in V. Now (c) follows from (a) and (b). □

Review. Recall our variational problem:

$$u \in H_0^1(\Omega),$$
$$(\nabla u, \nabla v)_\Omega = (f, v)_\Omega \quad \forall v \in H_0^1(\Omega).$$

We have shown that the space $H_0^1(\Omega)$ is a Hilbert space when endowed with the $H^1(\Omega)$ norm, and by way of the Poincaré-Friedrichs inequality, the Dirichlet form defines an equivalent norm on $H_0^1(\Omega)$. We can apply the Riesz-Fréchet theorem to the space $H_0^1(\Omega)$ with the Dirichlet form as an inner product. The map $\ell = (f, \cdot)_\Omega$ is linear and satisfies

$$|\ell(v)| = |(f, v)_\Omega| \leq \|f\|_\Omega \|v\|_\Omega \leq C_\Omega \|f\|_\Omega \|\nabla v\|_\Omega,$$

i.e., $\|\ell\|_{V'} \leq C_\Omega \|f\|_\Omega$. Therefore there is a unique $u \in H_0^1(\Omega)$ such that $(\nabla u, \nabla \cdot)_\Omega = (f, \cdot)_\Omega$ as elements of the dual space of $H_0^1(\Omega)$. This is equivalent to

$$(\nabla u, \nabla v)_\Omega = (f, v)_\Omega \quad \forall v \in H_0^1(\Omega).$$

In addition,

$$\|\nabla u\|_\Omega = \|f\|_{V'} \leq C_\Omega \|f\|_\Omega.$$

When a problem has a unique solution that depends continuously on the data, we say that the problem is well posed . We also say that a particular problem is **well posed** if it has a unique solution for all right-hand sides and the solution operator mapping the right-hand side to the solution is continuous (though not necessarily linear). This is illustrated in the following Proposition.

Proposition 2.9 (Symmetric coercive variational problems). *Let V be a real Hilbert space and $a : V \times V \to \mathbb{R}$ be a symmetric bilinear form that satisfies*

$$\alpha \|u\|_V^2 \leq a(u, u) \leq M \|u\|_V^2.$$

For every $\ell \in V'$, the variational problem

$$u \in V,$$
$$a(u, v) = \ell(v) \quad \forall v \in V,$$

has a unique solution u, which we can bound as

$$\|u\|_V \leq \frac{1}{\alpha}\|\ell\|_{V'},$$

and the solution operator mapping $\ell \mapsto u$ is linear and bounded.

Proof. The bilinear form $a(\cdot, \cdot)$ defines an inner product in V that is equivalent to the usual inner product, and so V is complete with this new inner product. By the Riesz-Fréchet theorem there is a unique solution such that

$$\alpha\|u\|_V^2 \leq a(u, u) = \ell(u) \leq \|\ell\|_{V'}\|u\|_V.$$

This last chain of inequalities requires some discussion. The dual norm of ℓ, $\|\ell\|_{V'}$ is the dual norm with respect to the original inner product on V, and not the inner product induced by a. The linearity of the solution operator follows from the superposition principle and the uniqueness of the solution. $\square$

When a bilinear form a satisfies the first inequality $\alpha\|u\|_V^2 \leq a(u, u)$, we say a is **coercive**. This is a concept that we will apply frequently in subsequent chapters.

Final comments and literature

Sobolev spaces are ubiquitous in the modern literature of partial differential equations. We have opted here to introduce only the simplest ones (there are more spaces defined in Exercises 2.3, 2.9, and 2.10), and to start using them right away in some basic boundary value problems. There are many more ways to define Sobolev spaces. One equivalent way to define them is by using weak derivatives instead of distributions: for instance, we say that $u \in L^2(\Omega)$ has weak first derivatives in $L^2(\Omega)$, when there are functions $v_i \in L^2(\Omega)$ such that

$$\int_\Omega v_i(\mathbf{x})\varphi(\mathbf{x})\mathrm{d}\mathbf{x} = -\int_\Omega u(\mathbf{x})\,\partial_{x_i}\varphi(\mathbf{x})\mathrm{d}\mathbf{x} \quad \forall \varphi \in \mathcal{D}(\Omega).$$

Note that, in our language, we would just say that $\partial_{x_i}u \in L^2(\Omega)$, because we are allowed to differentiate u in the sense of distributions. This way of introducing Sobolev spaces can be found in [23] for example. Both definitions are included (and reconciled) in Robert Adams's classic *Sobolev Spaces*, probably the most cited monograph on the subject (see [2] for the most recent edition). Sobolev spaces can also be defined by completion. For instance, we can consider the completion of

$$\{u \in \mathcal{C}^\infty(\Omega) \,:\, \|u\|_{1,\Omega} < \infty\}$$

with respect to the $\|\cdot\|_{1,\Omega}$ norm. The fact that this space is identical to $H^1(\Omega)$ is the object of Meyer and Serrin's two page article $H=W$ [80], a mathematical paper with a brilliantly short title. ($H^{1,2}(\Omega)$ would be the space defined by completion and $W^{1,2}(\Omega)$ the one defined with distributions. The index 2 refers to the use of $L^2(\Omega)$ in the definition.) A careful look at the history of Sobolev spaces can be found in [83].

Global definitions of Sobolev spaces in $\mathbb{R}^d$ can be given using Fourier transforms. We will deal with this definition in Chapter 13, since this definition helps in some other situations.

Exercises

2.1. Show that if Ω is an open bounded domain, $u \in H^1(\Omega)$, and $v \in \mathcal{C}^\infty(\Omega) \cap \mathcal{C}^1(\overline{\Omega})$, then $uv \in H^1(\Omega)$ and the product (Leibniz's) rule applies.

2.2. **Singularities in H^1 functions.** For $d \geq 2$, let $\Omega = B(0;1)$ and $u(\mathbf{x}) = (\log|\mathbf{x}|)^\mu$. Show that there is some $\mu > 0$ such that $u \in H^1(\Omega)$. (This example shows that there are discontinuous functions in $H^1(\Omega)$ for dimensions two and higher.)

2.3. **The space $H^2(\Omega)$.** Consider the space

$$
\begin{aligned}
H^2(\Omega) \;&:=\; \{u \in L^2(\Omega) \,:\, \partial^\alpha u \in L^2(\Omega) \quad \forall \alpha \in \mathbb{N}^d, |\alpha| \leq 2\} \\
&=\; \{u \in H^1(\Omega) \,:\, \nabla u \in \mathbf{H}^1(\Omega) := H^1(\Omega)^d\},
\end{aligned}
$$

endowed with the norm

$$\|u\|_{2,\Omega}^2 := \|u\|_{1,\Omega}^2 + \sum_{|\alpha|=2} \|\partial^\alpha u\|_\Omega^2 = \|u\|_\Omega^2 + \|\nabla u\|_\Omega^2 + \|\mathrm{D}^2 u\|_\Omega^2,$$

where $\mathrm{D}^2 u$ is a vector containing the $d(d+1)/2$ second partial derivatives of u.

(a) Show that $H^2(\Omega)$ is a Hilbert space. (Note that this includes finding an inner product whose associated norm is the one we have given.)

(b) Show that the inclusion $I : H^2(\Omega) \to H^1(\Omega)$ is a continuous operator.

(c) Show that $\partial_{x_i} : H^2(\Omega) \to H^1(\Omega)$ is a continuous operator.

2.4. Let Ω be a bounded open set such that the measure of $\partial\Omega$ is zero. Show that $\chi_\Omega \notin H^1(\mathbb{R}^d)$. (**Hint.** Show that the partial derivatives are not regular with the same technique we used for the Dirac delta in Chapter 1.)

2.5. Let $u \in H^1(\Omega)$ satisfy

$$u \equiv 0 \quad \text{in } \{\mathbf{x} \in \Omega : \text{dist}(\mathbf{x}, \partial\Omega) < \varepsilon\} =: \Omega_\varepsilon,$$

and let $\tilde{u} \in L^2(\mathbb{R}^d)$ be defined as

$$\tilde{u}(\mathbf{x}) = \begin{cases} u(\mathbf{x}) & \mathbf{x} \in \Omega, \\ 0 & \text{otherwise.} \end{cases}$$

Show that

$$\partial_{x_i}\tilde{u} = \widetilde{\partial_{x_i}u}.$$

(**Hint.** Find $\varphi_1, \varphi_2 \in \mathcal{C}^\infty(\mathbb{R}^d)$ with the following properties: $\varphi_1 + \varphi_2 \equiv 1$, $\text{supp}\,\varphi_1 \subset \Omega$ and $\varphi_1 \equiv 1$ in $\Omega \setminus \Omega_{\varepsilon/2}$. Next write $\varphi \in \mathcal{D}(\mathbb{R}^d)$ as $\varphi\varphi_1 + \varphi\varphi_2$, with $\varphi\varphi_1 \in \mathcal{D}(\Omega)$.)

2.6. Show that on a bounded set Ω,

$$\mathcal{C}^1_{00}(\Omega) := \{u \in \mathcal{C}^1(\Omega) : \text{supp}\,u \text{ is compact in } \Omega\}$$

is dense in $H^1_0(\Omega)$.

2.7. Assume that Ω is bounded and $v \in \mathcal{C}^1(\overline{\Omega})$. Show that $u \mapsto v\,u$ maps $H^1_0(\Omega)$ into itself. (**Hint.** Note that we are not demanding $v \in \mathcal{C}^\infty(\Omega)$, so in principle, it is not clear whether we can assert that $v\,u \in H^1(\Omega)$. You will need to use a density argument.)

2.8. **Reaction-diffusion problems.** Let Ω be a bounded domain, let $\kappa, c \in L^\infty(\Omega)$ be such that

$$\kappa(\mathbf{x}) \geq \kappa_0 > 0, \qquad c(\mathbf{x}) \geq 0, \qquad \text{almost everywhere,}$$

and let $f \in L^2(\Omega)$. Consider the problem

$$u \in H^1_0(\Omega) \qquad -\text{div}\,(\kappa\nabla u) + c\,u = f,$$

with all the differential operators understood in the sense of distributions.

(a) Find an equivalent variational formulation for this problem.

(b) Write the equivalent minimization problem.

(c) Show existence and uniqueness of the solution to the problem.

(d) Find a constant (independent of f) so that

$$\|u\|_{1,\Omega} \leq C_{\text{pb}}\|f\|_\Omega.$$

How does this constant depend on the coefficients of the equation (κ and c) and on the domain?

2.9. **The clamped Kirchhoff plate.** Consider the space (see Exercise 2.3)

$$H_0^2(\Omega) := \left\{ u \in H^2(\Omega) : \begin{array}{l} \exists \{\varphi_n\} \text{ sequence in } \mathcal{D}(\Omega) \\ \varphi_n \to u \quad \text{in } H^2(\Omega) \end{array} \right\}.$$

For all the following questions, assume that Ω is bounded.

(a) Show that if $u \in H_0^2(\Omega)$, then $u, \partial_{x_i} u \in H_0^1(\Omega)$.

(b) Use (a) to show that you can find $C_\Omega > 0$ such that

$$\|u\|_{2,\Omega} \le C_\Omega \|\mathrm{D}^2 u\|_\Omega.$$

(c) Using a density argument and differentiation in the sense of distributions, show that

$$(\partial_{x_i} \partial_{x_j} u, \partial_{x_i} \partial_{x_j} u)_\Omega = (\partial_{x_i}^2 u, \partial_{x_j}^2 u)_\Omega \qquad \forall u \in H_0^2(\Omega).$$

(d) Use (b) and (c) to show that $u \longmapsto \|\Delta u\|_\Omega$ defines a norm in $H_0^2(\Omega)$ that is equivalent to the usual one.

(e) Finally, for given $f \in L^2(\Omega)$, consider the problem

$$u \in H_0^2(\Omega), \qquad \Delta^2 u = f.$$

Write equivalent variational formulations and minimization principles. Show existence and uniqueness of solutions.

2.10. **More Sobolev spaces** . For $1 \le p \le \infty$, consider the spaces

$$W^{1,p}(\Omega) := \{u \in L^p(\Omega) : \nabla u \in L^p(\Omega)^d\},$$

endowed with the norms

$$\|u\|_{1,p,\Omega} := \|u\|_{L^p(\Omega)} + \|\nabla u\|_{L^p(\Omega)^d}.$$

Note that $W^{1,2}(\Omega) = H^1(\Omega)$, although we have written here a norm that is slightly different. Show that for all p, $W^{1,p}(\Omega)$ is a Banach space.

2.11. **A problem in free space.** Given $f \in L^2(\mathbb{R}^d)$, show that there exists a unique $u \in H^1(\mathbb{R}^d)$ such that $-\Delta u + u = f$.

3

Lipschitz transformations and Lipschitz domains

3.1   Lipschitz transformations of domains 47
3.2   How Lipschitz maps preserve H^1 behavior (*) 49
3.3   Lipschitz domains 52
3.4   Localization and pullback 56
3.5   Normal fields and integration on the boundary 59
Final comments and literature 62
Exercises 62

This is a technical chapter that will ease the way to constructing **the trace operator** in Chapter 4, our main obstacle in the pursuit of the analysis of nonhomogeneous Dirichlet boundary conditions for elliptic problems. We will prepare the way by dealing with four issues:

- Proving that the Sobolev spaces H^1 are mapped naturally to each other under Lipschitz changes of variables.

- Setting up a class of domains on which we will be able to handle nonhomogeneous boundary conditions for elliptic PDE.

- Preparing access to information close to the boundary on the said domains by organizing a set of charts, a cover of the boundary, and an associated partition of unity.

- Describing integration on the boundary of a Lipschitz domain.

3.1   Lipschitz transformations of domains

Throughout this section, we will employ the symbol for the gradient vector to provide a *row vector* with the partial derivatives of a function. On the other hand $\partial_{x_j} F$ will be considered as a *column vector* and DF will be the matrix whose columns are the vectors $\partial_{x_j} F$.

Let $\mathcal{O}, \Omega \subset \mathbb{R}^d$ be open sets. A map $F : \mathcal{O} \to \Omega$ is said to be a **bi-Lipschitz homeomorphism** if it is a bijection and F and F^{-1} are Lipschitz maps, i.e., there are constants $L_1, L_2 > 0$ such that

$$L_1|\mathbf{x} - \mathbf{y}| \le |F(\mathbf{x}) - F(\mathbf{y})| \le L_2|\mathbf{x} - \mathbf{y}| \qquad \forall \mathbf{x}, \mathbf{y} \in \mathcal{O}.$$

Rademacher's theorem (see Appendix A.2) states that Lipschitz maps in $\mathbb{R}^d$ are differentiable almost everywhere, and therefore

$$DF := \left(\frac{\partial F_i}{\partial x_j} \right)_{i,j}$$

is well defined almost everywhere and it is bounded as a function defined from $\mathcal{O}$ to $\mathbb{R}^{d \times d}$ (this follows from the Lipschitz condition). Since DF^{-1} is also almost everywhere differentiable, then there exist two positive quantities C_1 and C_2 such that

$$0 < C_1 \le |\det DF| \le C_2,$$

almost everywhere.

Changes of variables. Suppose $F : \mathcal{O} \to \Omega$ is a bi-Lipschitz homeomorphism; see Figure 3.1. The change of variables formula for F is given by

$$\int_\Omega u(\mathbf{x}) \, d\mathbf{x} = \int_\mathcal{O} (u \circ F)(\tilde{\mathbf{x}})|\det DF(\tilde{\mathbf{x}})| \, d\tilde{\mathbf{x}}. \tag{3.1}$$

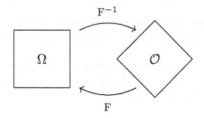

Figure 3.1: A graphical representation of the bi-Lipschitz map F.

In particular, notice that F takes $L^2(\mathcal{O})$ to $L^2(\Omega)$. For a bi-Lipschitz homeomorphism, we state the following distributional chain rule and save the proof for a later section.

Theorem 3.1. *Suppose* $F : \mathcal{O} \to \Omega$ *is a bi-Lipschitz homeomorphism. If* $u \in H^1(\Omega)$, *then*

$$\partial_{x_j}(u \circ F) = (\nabla u \circ F) \cdot \partial_{x_j}F \qquad in \; \mathcal{D}'(\mathcal{O}). \tag{3.2}$$

If we understand the gradient vector as a row vector (which is consistent with the notation we used above for Jacobian matrices) we can have the following compressed statement of Theorem 3.1,

$$\nabla_{\mathcal{O}}\left(u \circ \mathrm{F}\right) = \left(\left(\nabla_{\Omega} u\right) \circ \mathrm{F}\right) D\mathrm{F},$$

where $\nabla_{\mathcal{O}}$ and ∇_{Ω} are the gradients in $\mathcal{O}$ and Ω, respectively. We also arrive at the following useful result.

Corollary 3.1. *If* $\mathrm{F} : \mathcal{O} \to \Omega$ *is a bi-Lipschitz homeomorphism, then* $u \in H^1(\Omega)$ *if and only if* $u \circ \mathrm{F} \in H^1(\mathcal{O})$. *Moreover, the map*

$$H^1(\Omega) \ni u \longmapsto u \circ \mathrm{F} \in H^1(\mathcal{O})$$

is bounded.

Proof. This is a direct consequence of (3.2) and Rademacher's theorem. More specifically, since F is bi-Lipschitz, we can bound $|\det D\mathrm{F}|$ away from zero almost everywhere and the result follows from considering the change of variables formula (3.1) for u and its derivatives. $\qquad\qquad\square$

Proposition 3.1. *If* $\mathrm{F} : \mathcal{O} \to \Omega$ *is a bi-Lipschitz homeomorphism, then* $u \in H^1_0(\Omega)$ *if and only if* $u \circ \mathrm{F} \in H^1_0(\mathcal{O})$.

Proof. Let $u \in H^1_0(\Omega)$ and $\{\varphi_n\}$ be a sequence in $\mathcal{D}(\Omega)$ such that $\varphi_n \to u$ in $H^1(\Omega)$. Since $\mathcal{D}(\Omega) \subset H^1(\Omega)$, Corollary 3.1 tells us that $\varphi_n \circ \mathrm{F} \in H^1(\mathcal{O})$ for each n and $\varphi_n \circ \mathrm{F} \to u \circ \mathrm{F}$ in $H^1(\mathcal{O})$. Furthermore, since each φ_n has compact support in Ω, $\varphi_n \circ \mathrm{F}$ has compact support in $\mathcal{O}$ (note that $\varphi_n \circ \mathrm{F}$ is not smooth, but it is still continuous), showing that $\varphi_n \circ \mathrm{F} \in H^1_0(\mathcal{O})$ for all n by Proposition 2.7. Therefore, since $H^1_0(\mathcal{O})$ is closed, we must have $u \circ \mathrm{F} \in H^1_0(\mathcal{O})$. $\qquad\square$

3.2   How Lipschitz maps preserve H^1 behavior (*)

We devote our energy in this section to the technical proof of Theorem 3.1. The proof is not very 'natural' or inspiring. The reluctant reader is encouraged to move on to the next sections. Throughout this section we will assume that the *bounded open* sets Ω and $\mathcal{O}$, and the Lipschitz map F are the ones in the statement. We will also need to collect some results from Chapter 2 and we encourage the reader to revisit Proposition 2.7 in particular.

Let us introduce the set of functions in $H^1(\Omega)$ that can be approximated by functions in $\mathcal{C}^\infty(\overline{\Omega})$,

$$E(\Omega) := \left\{ u \in H^1(\Omega) : \begin{array}{c} \exists \{u_n\} \text{ in } \mathcal{C}^\infty(\overline{\Omega}) \\ u_n \longrightarrow u \text{ in } H^1(\Omega) \end{array} \right\}.$$

Here we understand
$$\mathcal{C}^\infty(\overline{\Omega}) := \{U|_\Omega \, : \, U \in \mathcal{C}^\infty(\mathbb{R}^d)\}.$$

Let us also recall the space (see Proposition 2.7)
$$H_{\underline{\underline{\,}}}^1(\Omega) := \{u \in H^1(\Omega) \, : \, u \equiv 0 \text{ in a neighborhood of } \partial\Omega\} \subset H_0^1(\Omega).$$

It is true that for many domains $E(\Omega) = H^1(\Omega)$ (we will prove this in Chapter 4), however, this is not a property that is satisfied by every domain. In Exercise 3.4 you are asked to prove that
$$H_{\underline{\underline{\,}}}^1(\Omega) \subset E(\Omega). \tag{3.3}$$

Proposition 3.2. *If $u \in H^1(\Omega)$ and $v \in E(\Omega)$, then*
$$\nabla(u\,v) = u\nabla v + v\nabla u.$$

Proof. Given $v \in E(\Omega)$, we consider a sequence $\{v_n\}$ in $\mathcal{C}^\infty(\overline{\Omega})$ such that
$$v_n \longrightarrow v \text{ in } L^2(\Omega), \qquad \nabla v_n \longrightarrow \nabla v \text{ in } \mathbf{L}^2(\Omega) \equiv L^2(\Omega)^d. \tag{3.4}$$

Note that
$$\nabla(u\,v_n) = u\nabla v_n + v_n\nabla u \tag{3.5}$$

because $v_n \in \mathcal{C}^\infty(\overline{\Omega}) \subset \mathcal{C}^\infty(\Omega)$. Also, (3.4) implies that
$$u\,v_n \longrightarrow u\,v \qquad \text{in } \mathcal{D}'(\Omega).$$

Therefore
$$\partial_{x_j}(u\,v_n) \longrightarrow \partial_{x_j}(u\,v) \qquad \text{in } \mathcal{D}'(\Omega),$$

but at the same time, (3.4) and (3.5) imply that
$$\partial_{x_j}(u\,v_n) = u\,\partial_{x_j}v_n + v_n\,\partial_{x_j}u \longrightarrow u\,\partial_{x_j}v + v\,\partial_{x_j}u \quad \forall j \qquad \text{in } \mathcal{D}'(\Omega).$$

This finishes the proof.   □

Note that Proposition 3.2 says that when $E(\Omega) = H^1(\Omega)$, the product of two functions in $H^1(\Omega)$ is in $L^1(\Omega)$ with derivatives in $L^1(\Omega)$.

Proposition 3.3. *If $u \in H^1(\Omega)$ and*
$$v \in \mathcal{C}(\Omega), \qquad \operatorname{supp} v \subset \Omega, \qquad \nabla v \in L^\infty(\Omega)^d,$$

then $u\,v \in E(\Omega)$.

Proof. First of all, the properties of v imply that $v \in H^1(\Omega)$ and v vanishes in a neighborhood of $\partial\Omega$. By (3.3), it follows that $v \in E(\Omega)$. Also, note that $u\,v \in L^2(\Omega)$ because $v \in L^\infty(\Omega)$. Proposition 3.2 implies that we can apply Leibniz's rule
$$\nabla(u\,v) = u\nabla v + v\nabla u \in \mathbf{L}^2(\Omega),$$

so $u\,v \in H^1(\Omega)$. Finally, we can again apply (3.3) (Proposition 2.7), because $u\,v \equiv 0$ in a neighborhood of $\partial\Omega$ (since v satisfies this property).   □

We next prove a chain rule for functions in $E(\Omega)$ and bi-Lipschitz homeomorphisms.

Proposition 3.4. *If $u \in E(\Omega)$, then*

$$\partial_{x_j}(u \circ F) = (\nabla u \circ F)\partial_{x_j}F \qquad \forall j. \tag{3.6}$$

Therefore $u \circ F \in H^1(\mathcal{O})$.

Proof. The proof of this result is very similar to that of Proposition 3.2. We start by choosing a sequence $\{u_n\}$ in $\mathcal{C}^\infty(\overline{\Omega})$ such that

$$u_n \longrightarrow u \quad \text{in } L^2(\Omega), \qquad \nabla u_n \longrightarrow \nabla u \quad \text{in } \mathbf{L}^2(\Omega).$$

This implies that

$$u_n \circ F \longrightarrow u \circ F \quad \text{in } L^2(\mathcal{O}), \qquad (\nabla u_n) \circ F \longrightarrow (\nabla u) \circ F \quad \text{in } \mathbf{L}^2(\mathcal{O}),$$

and finally

$$\partial_{x_j}(u \circ F) \longleftarrow \partial_{x_j}(u_n \circ F) = (\nabla u_n \circ F)\partial_{x_j}F \longrightarrow (\nabla u \circ F)\partial_{x_j}F,$$

with convergence in the sense of distributions. We have applied that (3.6) holds true for $u \in \mathcal{C}^\infty(\overline{\Omega})$ and that $\partial_{x_j}F \in L^\infty(\mathcal{O})^d$. $\qquad \square$

Lemma 3.1. *If $u \in H^1(\Omega)$ vanishes in a neighborhood of $\partial\Omega$, then*

$$\int_\Omega \partial_{x_j}u(\mathbf{x})\,\mathrm{d}\mathbf{x} = 0 \qquad \forall j.$$

Proof. We first choose $\varphi \in \mathcal{D}(\Omega)$ such that $\varphi \equiv 1$ in the region where u does not vanish (see Theorem 1.2). It follows that

$$\int_\Omega \partial_{x_j}u(\mathbf{x})\,\mathrm{d}\mathbf{x} = \int_\Omega \varphi(\mathbf{x})\,\partial_{x_j}u(\mathbf{x})\,\mathrm{d}\mathbf{x} = \langle \partial_{x_j}u, \varphi \rangle_{\mathcal{D}'(\Omega) \times \mathcal{D}(\Omega)}$$

$$= -\langle u, \partial_{x_j}\varphi \rangle_{\mathcal{D}'(\Omega) \times \mathcal{D}(\Omega)}$$

$$= -\int_\Omega u(\mathbf{x})\,\partial_{x_j}\varphi(\mathbf{x})\,\mathrm{d}\mathbf{x} = 0,$$

because $u\,\partial_{x_j}\varphi \equiv 0$ from our choice of φ. $\qquad \square$

With this last lemma, we are ready to prove our main result.

Proof of Theorem 3.1. As mentioned at the beginning of this section, we only need to prove this result for bounded sets. (Why?) Let $\varphi \in \mathcal{D}(\mathcal{O})$. The function $\varphi \circ F^{-1}$ is continuous in Ω, has compact support in Ω, and has bounded gradient. Therefore, by Propositions 3.3 and 3.4,

$$w := u\,(\varphi \circ F^{-1}) \in E(\Omega), \qquad \text{and} \qquad w \circ F \in H^1(\mathcal{O}).$$

Also, by Proposition 3.2

$$\nabla w = (\varphi \circ F^{-1})\nabla u + u\nabla(\varphi \circ F^{-1}) \qquad (3.7)$$
$$= (\varphi \circ F^{-1})\nabla u + u(\nabla\varphi \circ F^{-1})DF^{-1}.$$

Using Proposition 3.4 we prove that

$$\partial_{x_j}(w \circ F) = (\nabla w \circ F)\partial_{x_j}F$$
$$= \varphi(\nabla u \circ F)\partial_{x_j}F + (u \circ F)\nabla\varphi\underbrace{(DF^{-1} \circ F)\partial_{x_j}F}_{(DF)^{-1}(DF)_j = \mathbf{e}_j} \quad \text{(by (3.7))}$$
$$= \varphi(\nabla u \circ F)\partial_{x_j}F + (u \circ F)\partial_{x_j}\varphi.$$

Since $w \circ F = (u \circ F)\varphi \in H^1(\mathcal{O})$ vanishes in a neighborhood of $\partial\mathcal{O}$, then by Lemma 3.1

$$0 = \int_{\mathcal{O}} \partial_{x_j}(w \circ F)(\mathbf{x})d\mathbf{x}$$
$$= \int_{\mathcal{O}} (\nabla u \circ F)(\mathbf{x})\partial_{x_j}F(\mathbf{x})\,\varphi(\mathbf{x})d\mathbf{x} + \int_{\mathcal{O}} (u \circ F)(\mathbf{x})\,\partial_{x_j}\varphi(\mathbf{x})d\mathbf{x}$$
$$= \langle (\nabla u \circ F)\partial_{x_j}F, \varphi \rangle_{\mathcal{D}'(\mathcal{O}) \times \mathcal{D}(\mathcal{O})} + \langle u \circ F, \partial_{x_j}\varphi \rangle_{\mathcal{D}'(\mathcal{O}) \times \mathcal{D}(\mathcal{O})}.$$

Since this equality holds for all $\varphi \in \mathcal{D}(\mathcal{O})$, this proves that

$$(\nabla u \circ F)\partial_{x_j}F = \partial_{x_j}(u \circ F) \qquad \text{in } \mathcal{D}'(\mathcal{O}).$$

Finally $u \circ F \in L^2(\mathcal{O})$, $\nabla u \circ F \in \mathbf{L}^2(\mathcal{O})$ and $\partial_{x_j}F \in L^\infty(\mathcal{O})^d$, which proves that $u \circ F \in H^1(\mathcal{O})$. $\qquad\square$

3.3   Lipschitz domains

We now shift our attention to characterizing the types of domains for which we will eventually be able to define a trace operator. In what follows we use the notation $B_{d-1}(\mathbf{0}; 1)$ to denote the open unit ball in $\mathbb{R}^{d-1}$. Let Ω be an open, bounded domain, and let $\Gamma = \partial\Omega$. We say that Ω is a **(strong) Lipschitz domain** if for all $\mathbf{x} \in \Gamma$ there exist:

(a) a scaled rigid motion (the composition of a change of scale, an orthogonal transformation, and a translation) $R_{\mathbf{x}} : \mathbb{R}^d \to \mathbb{R}^d$, i.e.,

$$R_{\mathbf{x}}(\mathbf{y}) = c_{\mathbf{x}}Q_{\mathbf{x}}\mathbf{y} + \mathbf{d}_{\mathbf{x}} \qquad c_{\mathbf{x}} > 0, \quad Q_{\mathbf{x}}^\top = Q_{\mathbf{x}}^{-1}, \quad \mathbf{d}_{\mathbf{x}} \in \mathbb{R}^d,$$

(b) a Lipschitz function $h_{\mathbf{x}} : B_{d-1}(\mathbf{0}; 1) \to \mathbb{R}$,

(c) and a positive parameter $\eta_{\mathbf{x}} > 0$,

such that

$$\begin{aligned}
&\mathrm{R}_{\mathbf{x}}(\mathbf{0}, h_{\mathbf{x}}(\mathbf{0})) = \mathbf{x}, \\
&\mathrm{R}_{\mathbf{x}}(\widetilde{\mathbf{y}}, h_{\mathbf{x}}(\widetilde{\mathbf{y}})) \in \Gamma && \forall \widetilde{\mathbf{y}} \in B_{d-1}(\mathbf{0}; 1), \\
&\mathrm{R}_{\mathbf{x}}(\widetilde{\mathbf{y}}, h_{\mathbf{x}}(\widetilde{\mathbf{y}}) + \eta) \in \Omega && \forall \widetilde{\mathbf{y}} \in B_{d-1}(\mathbf{0}; 1), \quad 0 < \eta < \eta_{\mathbf{x}}, \\
&\mathrm{R}_{\mathbf{x}}(\widetilde{\mathbf{y}}, h_{\mathbf{x}}(\widetilde{\mathbf{y}}) - \eta) \in \overline{\Omega}^{c} && \forall \widetilde{\mathbf{y}} \in B_{d-1}(\mathbf{0}; 1), \quad 0 < \eta < \eta_{\mathbf{x}}.
\end{aligned}$$

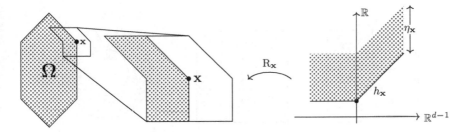

Figure 3.2: Example of the scaled rigid motion $\mathrm{R}_{\mathbf{x}}$ and Lipschitz function $h_{\mathbf{x}}$ for a given point $\mathbf{x}$ on a Lipschitz boundary.

Let us emphasize that the rigid motion $\mathrm{R}_{\mathbf{x}}$ not only depends upon the geometry of Γ but also on the fixed point $\mathbf{x}$ around which the transformation is defined. This family of domains are often also referred to with the expression: Ω is **locally a Lipschitz epigraph**, referring to the fact that we can focus around any point of the boundary of Ω and, up to a change of orientation, see the domain as the region above the graph of a Lipschitz function. (See Figure 3.2 for a cartoonish representation of Lipschitz epigraphs.)

Some examples and nonexamples of Lipschitz domains. Let us build our intuition by discussing some examples of Lipschitz domains, and examples of domains that fail to be Lipschitz. See Figures 3.3 and 3.4 for examples (and nonexamples) in $\mathbb{R}^2$. We can see from Figure 3.3(c) that the inclusion of holes in the domain does not immediately disqualify it from being Lipschitz, but a slit inside the domain as in Figure 3.4(a) or Figure 3.4(b) does because for each $\mathbf{x}$ on the slit, Ω is not locally a Lipschitz epigraph. The point $\mathbf{x}_0$ in Figure 3.4(a) represents a point on the boundary in which there is no Lipschitz function $h_{\mathbf{x}_0}$ which could represent the boundary at this point. If we examine the point $\mathbf{x}_0$ on Figure 3.4(c), we see that it fails both the Lipschitz function representation and the epigraph test. An example of a non-Lipschitz domain in $\mathbb{R}^3$ is a 'double brick' shape. As can be seen in Figure 3.5, at the point $\mathbf{x}_0$ the boundary of the double brick fails to be Lipschitz, analogous to the point in Figure 3.4(c). This example serves to illustrate that while Lipschitz domains at first seem quite general, we may eliminate even some polyhedra from consideration.

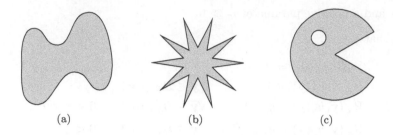

Figure 3.3: Examples of Lipschitz domains in $\mathbb{R}^2$.

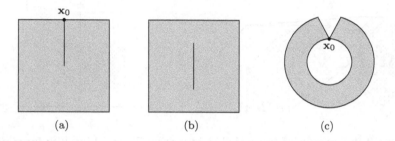

Figure 3.4: Examples of domains which fail to be Lipschitz.

Another perspective. We will refer to the reference configuration (see Figure 3.6) as the following subsets of the unit cylinder in d dimensions:

$$\mho := B_{d-1}(\mathbf{0}; 1) \times (-1, 1),$$
$$\mho^+ := B_{d-1}(\mathbf{0}; 1) \times (0, 1),$$
$$\mho^- := B_{d-1}(\mathbf{0}; 1) \times (-1, 0),$$
$$\Gamma_0 := B_{d-1}(\mathbf{0}; 1) \times \{0\}.$$

Now assume that Ω is a Lipschitz domain. For $\mathbf{x} \in \Gamma = \partial\Omega$, consider the triple $(\mathrm{R}_\mathbf{x}, h_\mathbf{x}, \eta_\mathbf{x})$ (a scaled rigid motion, a local Lipschitz representation of the boundary, and a 'vertical' displacement parameter) as in the definition, and build the function

$$\mathrm{F}_\mathbf{x}(\mathbf{y}) = \mathrm{F}_\mathbf{x}(\widetilde{\mathbf{y}}, y_d) := \mathrm{R}_\mathbf{x}(\widetilde{\mathbf{y}}, h_\mathbf{x}(\widetilde{\mathbf{y}}) + y_d\, \eta_\mathbf{x}).$$

The map

$$B_{d-1}(\mathbf{0}; 1) \times \mathbb{R} \ni \mathbf{y} = (\widetilde{\mathbf{y}}, y_d) \longmapsto (\widetilde{\mathbf{y}}, h_\mathbf{x}(\widetilde{\mathbf{y}}) + y_d\, \eta_\mathbf{x}) \in B_{d-1}(\mathbf{0}; 1) \times \mathbb{R}$$

is Lipschitz with Lipschitz inverse

$$\mathbf{z} = (\widetilde{\mathbf{z}}, z_d) \longmapsto \left(\widetilde{\mathbf{z}}, \frac{1}{\eta_\mathbf{x}}(z_d - h_\mathbf{x}(\widetilde{\mathbf{z}}))\right).$$

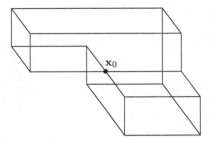

Figure 3.5: The double brick, an example of a non-Lipschitz polyhedron.

This proves that

$$F_{\mathbf{x}} : \mho = B_{d-1}(\mathbf{0}; 1) \times (-1, 1) \longrightarrow \Omega_{\mathbf{x}} := F_{\mathbf{x}}(\mho)$$

is a bi-Lipschitz transformation of the reference domain $\mho$ to an open neighborhood of $\mathbf{x} \in \Gamma$. Therefore, if Ω is a Lipschitz domain, then for all $\mathbf{x} \in \Gamma = \partial\Omega$ there is a bi-Lipschitz homeomorphism $F_{\mathbf{x}} : \mho \to \Omega_{\mathbf{x}}$, where $\Omega_{\mathbf{x}} \subset \mathbb{R}^d$ is an open neighborhood of $\mathbf{x}$ (see Figure 3.7), $F_{\mathbf{x}}(\mathbf{0}) = \mathbf{x}$, and we have

$$F_{\mathbf{x}}(\mho^+) = \Omega \cap \Omega_{\mathbf{x}}, \qquad F_{\mathbf{x}}(\mho^-) = \overline{\Omega}^c \cap \Omega_{\mathbf{x}}, \qquad F_{\mathbf{x}}(\Gamma_0) = \partial\Omega \cap \Omega_{\mathbf{x}}.$$

This second characterization is a slightly more general definition of Lipschitz domain. For instance, the image of a strong Lipschitz domain under a Lipschitz transformation satisfies the requirements of this second characterization, but might fail to be locally a Lipschitz epigraph. This second collection of sets is often referred to as **weakly Lipschitz domains.**

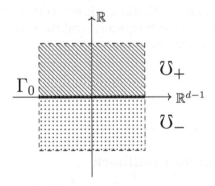

Figure 3.6: The reference configuration $\mho$.

Description with a finite collection of charts. Let Ω be a Lipschitz domain and consider the pairs $(F_{\mathbf{x}}, \Omega_{\mathbf{x}})$ (bi-Lipschitz transformation and open neighborhood) associated to the points $\mathbf{x} \in \Gamma$. Since the boundary Γ is a compact set of $\mathbb{R}^d$ and $\{\Omega_{\mathbf{x}} : \mathbf{x} \in \Gamma\}$ is an open cover, we can choose a finite

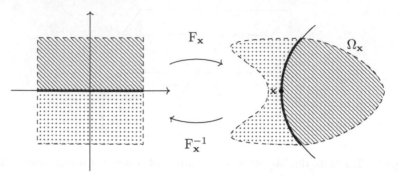

Figure 3.7: A cartoon of the local homeomorphism F_x mapping $\mho$ to a region surrounding a part of the boundary of a Lipschitz domain Ω.

subcover and rename it (losing track of the points $\mathbf{x}$), so that we have an open cover of the boundary $\{\Omega_1, \ldots, \Omega_J\}$ and associated bi-Lipschitz transformations

$$F_j : \mho \to \Omega_j \qquad j = 1, \ldots, J,$$

satisfying

$$F_j(\mho^+) = \Omega_j \cap \Omega, \qquad F_j(\Gamma_0) = \Omega_j \cap \partial\Omega, \qquad F_j(\mho^-) = \Omega_j \cap \overline{\Omega}^c.$$

> **Warning.** In this book we will stick to strong Lipschitz domains. However, in many important parts of the book (for the definition of the trace operator, for instance), using the more general description of domains by local charts will be enough. We will not insist on these details, as technical difficulties easily show up in this theory, and the class of strong Lipschitz domains is already a large one.

3.4   Localization and pullback

Associated partitions of unity. The starting point of the next technical tool for dealing with Lipschitz domains is the open cover of the boundary $\{\Omega_1, \ldots, \Omega_J\}$. We will shortly give the construction of a collection of functions

$\varphi_0, \varphi_1, \ldots, \varphi_J$ with the following properties:

$$\varphi_0 \in \mathcal{D}(\Omega),$$
$$\varphi_j \in \mathcal{D}(\Omega_j) \qquad j = 1, \ldots, J,$$
$$\sum_{j=0}^{J} \varphi_j \equiv 1 \qquad \text{in a neighborhood of } \overline{\Omega}.$$

The (obviously nonunique) construction can be done with different tools. For instance, it can be built using mollification of characteristic functions. We will give here a detailed construction using tilings (as in Section 1.6).

Pullback. Let $u \in H^1(\Omega)$ and consider $\varphi_0, \ldots, \varphi_J$ to be the functions discussed above. Such a smooth partition of unity satisfies

$$\sum_{j=0}^{J} \varphi_j \in \mathcal{D}(\mathbb{R}^d), \qquad \varphi_j|_\Omega \in C^\infty(\overline{\Omega}), \qquad u\varphi_j \in H^1(\Omega) \quad \forall j.$$

The process of transforming $u\varphi_j \in H^1(\Omega_j \cap \Omega)$ (for $j = 1, \ldots, J$) into $(u\varphi_j) \circ F_j \in H_0^1(\mho^+)$ will be referred to as a pullback (to the reference domain). Note that we will leave the interior part $u\varphi_0 \in H_0^1(\Omega)$ untouched in this process, since it is unrelated to anything that happens on the boundary.

Construction of the partitions of unity. The reader who is familiar with partitions of unity (a very common tool in mathematical analysis) is recommended to skip the remainder of this section. We will take some time to construct these functions now. In the details that are to follow it may help the reader to revisit Section 1.6, as we will be using many of the ideas developed in that section.

Building an ε-neighborhood of $\partial\Omega$. Our first step is to construct an open neighborhood around $\partial\Omega$. Given $\mathbf{x} \in \partial\Omega$, we can find an Ω_j from our open cover at the beginning of this section and an $\varepsilon_{\mathbf{x}} > 0$ such that $B(\mathbf{x}; \varepsilon_{\mathbf{x}}) \subset \Omega_j$. Using these open balls, we can construct an open cover of $\partial\Omega$

$$\partial\Omega \subset \bigcup_{\mathbf{x} \in \partial\Omega} B\left(\mathbf{x}; \tfrac{\varepsilon_{\mathbf{x}}}{2}\right).$$

Since $\partial\Omega$ is compact, there exist pairs $\{(\mathbf{x}_\ell, \varepsilon_\ell)\}_{\ell=1}^L$ such that the open balls $B\left(\mathbf{x}_\ell; \tfrac{\varepsilon_\ell}{2}\right)$ are a finite subcover of $\partial\Omega$ and for each ℓ, $B(\mathbf{x}_\ell; \varepsilon_\ell) \subset \Omega_j$ for some j. Setting $\varepsilon := \min_\ell \varepsilon_\ell$ we consider the set

$$N_{\varepsilon/2}(\partial\Omega) := \{\mathbf{x} \in \mathbb{R}^d : \text{dist}(\mathbf{x}, \partial\Omega) < \tfrac{\varepsilon}{2}\} \subset \bigcup_{j=1}^{J} \Omega_j,$$

to be our open neighborhood around $\partial\Omega$.

Building a fine tiling of the space. We next consider the closed d-cubes

$$Q_{\mathbf{n}} := \epsilon \mathbf{n} + \left[-\tfrac{\epsilon}{2}, \tfrac{\epsilon}{2}\right]^d \subset Q_{\mathbf{n}}^{\text{ext}} := \epsilon \mathbf{n} + \left[-\tfrac{3}{4}\epsilon, \tfrac{3}{4}\epsilon\right]^d \qquad \mathbf{n} \in \mathbb{Z}^d,$$

where

$$\tfrac{9}{4}\epsilon^2 d < \tfrac{1}{4}\varepsilon^2, \qquad \text{so that} \qquad \text{diam}\, Q_{\mathbf{n}}^{\text{ext}} < \tfrac{1}{2}\varepsilon.$$

We now recall how to build a partition of unity associated to the overlapped tiling $Q_{\mathbf{n}}^{\text{ext}}$ for $\mathbf{n} \in \mathbb{Z}^d$. Recalling the functions g and h from Chapter 1,

$$g(r) := \begin{cases} \exp\left(\frac{1}{r^2 - 1}\right) & r < 1, \\ 0 & r \geq 1, \end{cases}$$

$$h(r) := \frac{1}{\int_{-\infty}^{\infty} g(2x - 1)\, dx} \int_{-\infty}^{r} g(2x - 1)\, dx,$$

we can define

$$\varphi_{\mathbf{n}}(\mathbf{x}) := \prod_{i=1}^{d} h\left(3 - \tfrac{4}{\epsilon}|x_i - \epsilon n_i|\right),$$

and

$$\Psi_{\mathbf{n}} := \frac{1}{\sum_{\mathbf{m} \in \mathbb{Z}^d} \varphi_{\mathbf{m}}} \varphi_{\mathbf{n}} \qquad \mathbf{n} \in \mathbb{Z}^d.$$

(Note that the function in the denominator belongs to $\mathcal{C}^\infty(\mathbb{R}^d)$, it is strongly positive and ϵ-periodic in all variables.) We observe that for each $\mathbf{n}$, supp $\Psi_{\mathbf{n}} = Q_{\mathbf{n}}^{\text{ext}}$ and $\sum_{\mathbf{n}} \Psi_{\mathbf{n}} \equiv 1$. Thus the functions $\Psi_{\mathbf{n}}$ form a smooth partition of unity of $\mathbb{R}^d$.

Assigning tiles to domains. Assume $Q_{\mathbf{n}}^{\text{ext}} \cap \partial\Omega \neq \emptyset$. We then take a fixed $\mathbf{x} \in Q_{\mathbf{n}}^{\text{ext}} \cap \partial\Omega$ and a general $\mathbf{y} \in Q_{\mathbf{n}}^{\text{ext}}$. By construction, we have that $|\mathbf{x} - \mathbf{y}| < \varepsilon/2$ because we made the diameter of $Q_{\mathbf{n}}^{\text{ext}}$ less than $\varepsilon/2$. Now, we also have that for each $\mathbf{x} \in \partial\Omega$, there exists an index $\ell \in \{1, \ldots, L\}$ such that

$$|\mathbf{x} - \mathbf{x}_\ell| < \tfrac{\varepsilon_\ell}{2},$$

so that $\mathbf{x} \in B(\mathbf{x}_\ell; \varepsilon_\ell) \subset \Omega_j$ for some j. We have seen this inclusion before. Now we have

$$|\mathbf{x}_\ell - \mathbf{y}| < \tfrac{\varepsilon}{2} + \tfrac{\varepsilon_\ell}{2} \leq \varepsilon_\ell,$$

and thus $\mathbf{y} \in B(\mathbf{x}_\ell; \varepsilon_\ell) \subset \Omega_j$ for each $\mathbf{y} \in Q_{\mathbf{n}}^{\text{ext}}$. In other words, if $Q_{\mathbf{n}}^{\text{ext}} \cap \partial\Omega \neq \emptyset$, there exists j such that $Q_{\mathbf{n}}^{\text{ext}} \subset \Omega_j$. From this result we construct the following sets of indices based on collections of cubes

$$I_1 := \{\mathbf{n} \in \mathbb{Z}^d : Q_{\mathbf{n}}^{\text{ext}} \subset \Omega_1\},$$

$$I_2 := \{\mathbf{n} \in \mathbb{Z}^d : Q_{\mathbf{n}}^{\text{ext}} \subset \Omega_2\} \setminus I_1,$$

$$I_3 := \{\mathbf{n} \in \mathbb{Z}^d : Q_{\mathbf{n}}^{\text{ext}} \subset \Omega_3\} \setminus (I_1 \cup I_2),$$

$$\vdots$$

$$I_J := \{\mathbf{n} \in \mathbb{Z}^d : Q_{\mathbf{n}}^{\text{ext}} \subset \Omega_J\} \setminus \left(\bigcup_{j=1}^{J-1} I_j\right).$$

Note that all $\mathbf{n}$ corresponding to cubes $Q_{\mathbf{n}}^{\text{ext}}$ which intersect $\partial\Omega$ are accounted in the index sets $I_1 \cup \ldots \cup I_J$.

Final construction of the partition of unity. We are now ready to construct the functions φ_j. Indeed, letting for $j = 1, \ldots, J$,

$$\varphi_j := \sum_{\mathbf{n} \in I_j} \Psi_{\mathbf{n}} \in \mathcal{D}(\Omega_j), \qquad \varphi_1 + \ldots + \varphi_J \equiv 1 \quad \text{in} \quad N_{\varepsilon/2}(\partial\Omega).$$

The last property can be seen by realizing that all functions $\Psi_{\mathbf{n}}$ corresponding to cubes $Q_{\mathbf{n}}^{\text{ext}}$ which intersect $\partial\Omega$ have been collected in the sum, and every point $\mathbf{y} \in N_{\varepsilon/2}(\partial\Omega)$ is in one of these cubes. All that remains is to define the function $\varphi_0 : \Omega \to \mathbb{R}$, which we do as

$$\varphi_0 := 1 - (\varphi_1 + \ldots + \varphi_J).$$

It is clear that by this construction we have $\varphi_0 \in \mathcal{C}^{\infty}(\Omega)$ and $\varphi_0 \equiv 0$ on $N_{\varepsilon/2}(\partial\Omega) \cap \Omega$. Therefore $\varphi_0 \in \mathcal{D}(\Omega)$ and we have the desired functions $\{\varphi_j\}_{j=0}^{J}$. As mentioned before, this construction is not unique and could also have been completed using mollification.

3.5 Normal fields and integration on the boundary

In this section we give a working definition of what we will understand as integration over Γ. This is quite a technical issue and this section will be the only place in this text where we will openly ask the reader to admit some technical results. Our starting point is the collection of charts $F_j : \mho \to \Omega_j$ from $j = 1, \ldots, J$ that locally desribe a neighborhood of Γ. The functions

$$\mathbf{\Phi}_j : B_{d-1}(0;1) \to \Gamma, \qquad \mathbf{\Phi}_j(\widetilde{\mathbf{y}}) := F_j(\widetilde{\mathbf{y}}, 0),$$

are almost everywhere differentiable (they are restrictions of a Lipschitz function) and can be used to parametrize the patches $\Gamma \cap \Omega_j$ for $j = 1, \ldots, J$. Associated to these parametrizations there is a **hypersurface element**. In the lowest dimensions, this is easy to describe. For $d = 2$ (Γ is a curve in the plane)

$$d\mathbf{\Phi}_j(\widetilde{\mathbf{y}}) := |\partial_{x_1}\mathbf{\Phi}_j(\widetilde{\mathbf{y}})|d\widetilde{\mathbf{y}}, \qquad \widetilde{\mathbf{y}} \in (-1, 1),$$

and for $d = 3$ (surfaces in space)

$$d\mathbf{\Phi}_j(\widetilde{\mathbf{y}}) := |\partial_{x_1}\mathbf{\Phi}_j(\widetilde{\mathbf{y}}) \times \partial_{x_2}\mathbf{\Phi}_j(\widetilde{\mathbf{y}})|d\widetilde{\mathbf{y}}, \qquad \widetilde{\mathbf{y}} \in B_2(0;1).$$

The general case can be expressed like this. If $\widetilde{\mathbf{y}} \in B_{d-1}(0;1)$ is a point where $D\mathbf{\Phi}_j(\widetilde{\mathbf{y}}) \in \mathbb{R}^{d \times (d-1)}$ is well defined, then there exists a (nonnormalized) normal vector $\mathbf{n}_j(\widetilde{\mathbf{y}}) \in \mathbb{R}^d$ such that

$$\det \left[\, D\mathbf{\Phi}_j(\widetilde{\mathbf{y}}) \mid \mathbf{z} \, \right] = \mathbf{n}_j(\widetilde{\mathbf{y}}) \cdot \mathbf{z} \qquad \forall \mathbf{z} \in \mathbb{R}^d.$$

We know that the vector $\mathbf{n}_j(\widetilde{\mathbf{y}})$ exists since the mapping

$$\mathbf{z} \longmapsto \det \left[\ D\mathbf{\Phi}_j(\widetilde{\mathbf{y}}) \mid \mathbf{z}\ \right]$$

is a linear functional, and all linear functionals in $\mathbb{R}^d$ can be represented as dot products. Moreover, the vector $\mathbf{n}_j(\widetilde{\mathbf{y}})$ is orthogonal to the vectors $\partial_{x_i} \mathbf{\Phi}_j(\widetilde{\mathbf{y}})$ for $i = 1, \ldots, d-1$. We thus have

$$d\mathbf{\Phi}_j(\widetilde{\mathbf{y}}) := |\mathbf{n}_j(\widetilde{\mathbf{y}})| d\widetilde{\mathbf{y}}.$$

With the notation of Section 3.3, taking $\mathbf{x} = \mathbf{\Phi}_j(\mathbf{0})$, the local normal vector in terms of coordinates in the parametric domain is

$$\mathbf{n}_j(\widetilde{\mathbf{y}}) = c_{\mathbf{x}}^{d-1} Q_{\mathbf{x}} \begin{bmatrix} -\nabla h_{\mathbf{x}}(\widetilde{\mathbf{y}}) \\ 1 \end{bmatrix},$$

(here we understand the gradient as a column vector) as can be shown with a simple computation. For $\mathbf{y} \in \Gamma$ such that $\mathbf{y} = \mathbf{\Phi}_j(\widetilde{\mathbf{y}})$ for some j and $\widetilde{\mathbf{y}} \in B_{d-1}(\mathbf{0}; 1)$ and such that $\mathbf{\Phi}_j$ is differentiable at $\widetilde{\mathbf{y}}$, we define

$$\mathbf{n}(\mathbf{y}) := \frac{1}{|\mathbf{n}_j(\widetilde{\mathbf{y}})|} \mathbf{n}_j(\widetilde{\mathbf{y}}).$$

In other words, on $\Gamma \cap \Omega_j$ we define $\mathbf{n}$ with the relation

$$\mathbf{n} \circ \mathbf{\Phi}_j = \frac{1}{|\mathbf{n}_j|} \mathbf{n}_j.$$

This defines (Exercise 3.6) a **unit normal** vector field $\mathbf{n} : \Gamma \to \mathbb{R}^d$.

Integration. The above can be used to give a parametric definition of the integral on $\Gamma \cap \Omega_j$:

$$\int_{\Gamma \cap \Omega_j} f(\mathbf{y})\, d\Gamma(\mathbf{y}) = \int_{B_{d-1}(\mathbf{0};1)} (f \circ \mathbf{\Phi}_j)(\widetilde{\mathbf{y}})\, d\mathbf{\Phi}_j(\widetilde{\mathbf{y}}).$$

The full integral is computed by localization (multiplication by the partition of unity) and pullback (parametrization using the patch parametrizations given by $\mathbf{\Phi}_j$ for $j = 1, \ldots, J$). We can then define integrals on Γ by

$$\int_{\Gamma} g(\mathbf{y})\, d\Gamma(\mathbf{y}) = \int_{\Gamma} g(\mathbf{y}) \left(\sum_{j=1}^{J} \varphi_j(\mathbf{y}) \right) d\Gamma(\mathbf{y}) = \sum_{j=1}^{J} \int_{\Gamma \cap \Omega_j} g(\mathbf{y}) \varphi_j(\mathbf{y})\, d\Gamma(\mathbf{y})$$

$$= \sum_{j=1}^{J} \int_{B_{d-1}(\mathbf{0};1)} (g \varphi_j) \circ \mathbf{\Phi}_j(\widetilde{\mathbf{y}})\, d\mathbf{\Phi}_j(\widetilde{\mathbf{y}}).$$

It is easy to prove that

$$\left| \int_{\Gamma} g(\mathbf{y})\, d\Gamma(\mathbf{y}) \right| \leq \int_{\Gamma} |g(\mathbf{y})|\, d\Gamma(\mathbf{y}),$$

and that for $g \in C(\Gamma)$,

$$\int_\Gamma |g(\mathbf{y})| \, \mathrm{d}\Gamma(\mathbf{y}) \le C_\Gamma \max_{\mathbf{x} \in \Gamma} |g(\mathbf{x})| = C_\Gamma \|g\|_{C(\Gamma)}, \quad \text{with} \quad C_\Gamma := \int_\Gamma 1 \, \mathrm{d}\Gamma(\mathbf{y}).$$

The quantity C_Γ is the $(d-1)$-measure of Γ. With this concept of integration, we will identify functions g and h such that $g \circ \mathbf{\Phi}_j = h \circ \mathbf{\Phi}_j$ almost everywhere in $B_{d-1}(\mathbf{0}; 1)$ for all j. The spaces $L^1(\Gamma)$ and $L^2(\Gamma)$ are then defined naturally by mapping back to the reference configuration, after using a cutoff process, and integrating there. The space $L^\infty(\Gamma)$ is similarly defined. We will use the following result often, but the reader is invited to skip its proof, which we will only sketch.

Proposition 3.5. $L^2(\Gamma)$ *is a Hilbert space.*

Proof. Recall the smooth partition of unity on the boundary $\{\varphi_j\}_{j=1}^J$ from Section 3.4, and consider the open sets

$$B_j := \{\widetilde{\mathbf{x}} \in B_{d-1}(\mathbf{0}; 1) \, : \, \varphi_j(\mathbf{\Phi}_j(\widetilde{\mathbf{x}})) > 0\},$$

and the weights $\omega_j := (\varphi_j \circ \mathbf{\Phi}_j)|\mathbf{n}_j|$, which are bounded and positive almost-everywhere in B_j. We then consider the subsets

$$B_{ij} := \{\widetilde{\mathbf{x}} \in B_i \, : \, \mathbf{\Phi}_i(\widetilde{\mathbf{x}}) = \mathbf{\Phi}_j(\widetilde{\mathbf{y}}) \quad \text{for some } \widetilde{\mathbf{y}} \in B_j\},$$

and the bi-Lipschitz transformations $\mathbf{\Psi}_{ji} := \mathbf{\Phi}_j^{-1} \circ \mathbf{\Phi}_i : B_{ij} \to B_{ji}$. The pullback process $g \mapsto (g \circ \mathbf{\Phi}_j)_{j=1}^J$ defines a bounded isomorphism of $L^2(\Gamma)$ and

$$\{(g_1, \dots, g_J) \in \Xi := \prod_{j=1}^J L^2_{\omega_j}(B_j) \, : \, g_i|_{B_{ij}} = g_j \circ \mathbf{\Psi}_{ji} \text{ a.e. } \forall i, j\},$$

where $L^2_{\omega_j}(B_j)$ is the L^2 space corresponding to the weight ω_j, and a.e. stands for almost everywhere. However, Ξ is a closed subspace of a product of Hilbert spaces as follows from the Riesz-Fischer theorem. □

We finish this chapter with the divergence theorem stated on Lispchitz domains for smooth vector fields. This will only be relevant when we compare the two possible definitions of normal traces of vector fields in Chapter 6. For a sketch of the proof, see Exercise 3.7.

Proposition 3.6 (Divergence theorem). *On a Lipschitz domain Ω we have*

$$\int_\Omega (\nabla \cdot \mathbf{p})(\mathbf{x}) \mathrm{d}\mathbf{x} = \int_\Gamma (\mathbf{p} \cdot \mathbf{n})(\mathbf{x}) \mathrm{d}\Gamma(\mathbf{x}) \qquad \forall \mathbf{p} \in C^1(\overline{\Omega}; \mathbb{R}^d).$$

Final comments and literature

We have introduced (strongly) Lipschitz domains because they are the most standard domains in the Sobolev space literature. However, for most of our proofs about H^1 spaces we will use the slightly more general definition using the finite collection of charts. Note that the image of a strong Lipschitz domain by a bi-Lipschitz homeomorphism can fail to be a strong Lipschitz domain. The nitty-gritty details about different types of domains (strong and weak Lipschitz domains, John domains, etc.) is a difficulty we will avoid. We will try to point out when a hypothesis on the domains becomes crucial for a proof to be valid. Pierre Girsvard's celebrated 1985 monograph *Elliptic problems in nonsmooth domains* (see [59] for a recent edition) contains an example of a peculiar ray-shaped polygonal domain that is weakly but not strongly Lipschitz. More on types of domains and their influence in theorems on Sobolev spaces can be found in Jindřich Nečas's very detailed treatment in his book *Les méthodes directes en théorie des équations elliptiques* (see [87] for a recent edition translated to English).

Exercises

3.1. Let $F : \mathcal{O} \to \Omega$ be a bi-Lipschitz homeomorphism. Show that it can be (uniquely) extended to a bi-Lipschitz homeomorphism $F : \overline{\mathcal{O}} \to \overline{\Omega}$. (**Hint.** Use that F transforms Cauchy sequences in $\mathcal{O}$ to Cauchy sequences in Ω.)

3.2. Let $F : \mathcal{O} \to \Omega$ be a bi-Lipschitz homeomorphism, and $u \in L^1_{\mathrm{loc}}(\Omega)$. Show that if $u \equiv 0$ in a neighborhood of $\partial\Omega$, then $u \circ F \equiv 0$ in a neighborhood of $\partial\mathcal{O}$.

3.3. Affine transformations. Let $F(\mathbf{x}) := B\mathbf{x} + \mathbf{c}$, where $B \in \mathbb{R}^{d \times d}$ is invertible and $\mathbf{c} \in \mathbb{R}^d$, let $\mathcal{O} \subset \mathbb{R}^d$ be an open set and let $\Omega = F(\mathcal{O})$. Given $T \in \mathcal{D}'(\Omega)$, we define $T \circ F : \mathcal{D}(\mathcal{O}) \to \mathbb{R}$ by

$$\langle T \circ F, \varphi \rangle := \langle T, |\det B| \, \varphi \circ F^{-1} \rangle_{\mathcal{D}'(\Omega) \times \mathcal{D}(\Omega)}.$$

(a) Show that $T \circ F \in \mathcal{D}'(\mathcal{O})$, and that the definition is compatible with the change of variables for locally integrable functions, i.e., if $T = u \in L^1_{\mathrm{loc}}(\Omega)$, then $T \circ F = u \circ F$.

(b) Prove the associated chain rule

$$\partial_{x_j}(T \circ F) = \sum_{i=1}^{d} b_{ij}(\partial_{x_i} T) \circ F.$$

(c) Show directly, without using Theorem 3.1, that $u \in H^1(\Omega)$ if and only if $u \circ F \in H^1(\mathcal{O})$.

3.4. With the notation of Section 3.2, show that $H^1_{\underline{\equiv}}(\Omega) \subset E(\Omega)$, that is, show that if $u \in H^1(\Omega)$ and $u \equiv 0$ in a set of the form $\{\mathbf{x} \in \Omega : d(\mathbf{x}, \partial\Omega) < \varepsilon\}$, then there exists a sequence $\{u_n\}$ in $C^\infty(\mathbb{R}^d)$ such that $u_n|_\Omega \to u$ in $H^1(\Omega)$.

3.5. **Global C^1 transformations.** Let $F : \mathbb{R}^d \to \mathbb{R}^d$ be a C^1 bijection, with C^1 inverse and such that DF and DF^{-1} are bounded. Show (directly, without using the more general result for bi-Lipschitz maps) that $u \in H^1(\mathbb{R}^d)$ if and only if $u \circ F \in H^1(\mathbb{R}^d)$. (**Hint.** Prove that $C^1(\mathbb{R}^d)$ compactly supported functions are dense in $H^1(\mathbb{R}^d)$ and use this fact.)

3.6. **The normal vector field.** Using the notation of Section 3.5, consider the normalized vector fields

$$\widehat{\mathbf{n}_j} := |\mathbf{n}_j|^{-1}\mathbf{n}_j : B_{d-1}(\mathbf{0}; 1) \to \mathbb{R}^d,$$

the sets

$$B_{ij} := \{\widetilde{\mathbf{x}} \in B_{d-1}(\mathbf{0}; 1) : \boldsymbol{\Phi}_i(\widetilde{\mathbf{x}}) = \boldsymbol{\Phi}_j(\widetilde{\mathbf{y}}) \quad \text{for some } \widetilde{\mathbf{y}}\},$$

and the bi-Lipschitz maps $\boldsymbol{\Psi}_{ji} := \boldsymbol{\Phi}_j^{-1} \circ \boldsymbol{\Phi}_i : B_{ij} \to B_{ji}$. Show that

$$\widehat{\mathbf{n}_i}|_{B_{ij}} = \widehat{\mathbf{n}_j} \circ \boldsymbol{\Psi}_{ji}.$$

3.7. **The divergence theorem on a Lipschitz domain.** Prove Proposition 3.6 by using the following steps:

(a) Prove the result in a domain of the form

$$\Omega = \{\mathbf{x} = (\widetilde{\mathbf{x}}, x_d) \in \mathbb{R}^d : \widetilde{\mathbf{x}} \in B_{d-1}(\mathbf{0}; 1), \quad h(\widetilde{\mathbf{x}}) < x_d < \delta + h(\widetilde{\mathbf{x}})\},$$

where $\delta > 0$ and $h : B_{d-1}(\mathbf{0}; 1) \to \mathbb{R}$ is Lipschitz and where the field $\mathbf{p} \in C^1(\overline{\Omega}; \mathbb{R}^d)$ vanishes in a neighborhood of

$$\partial\Omega \setminus \{(\widetilde{\mathbf{x}}, h(\widetilde{\mathbf{x}})) : \widetilde{\mathbf{x}} \in B_{d-1}(\mathbf{0}; 1)\}.$$

(b) Show that if the divergence theorem (as stated in Proposition 3.6) holds for a domain Ω, then it also holds in the transformed domain $\{c Q\mathbf{x} + \mathbf{d} : \mathbf{x} \in \Omega\}$, where $c > 0$, $Q^\top = Q^{-1}$ and $\mathbf{d} \in \mathbb{R}^d$.

(c) Use localization to prove the divergence theorem based on (a) and (b).

4

The nonhomogeneous Dirichlet problem

4.1 The extension theorem 66
4.2 The trace operator 68
4.3 The range and kernel of the trace operator 70
4.4 The nonhomogeneous Dirichlet problem 73
4.5 General right-hand sides 75
4.6 The Navier-Lamé equations (*) 79
Final comments and literature 83
Exercises 84

In Chapter 2 we explored the Dirichlet problem with homogeneous boundary conditions. It is natural to ask what happens if we want to apply boundary conditions which are not homogeneous. The situation is much more complicated, and we devote this chapter to answering that question. We thus want to explore the solvability theory for a problem of the form

$$-\Delta u = f \quad \text{in } \Omega, \qquad u = g \quad \text{on } \partial\Omega. \tag{4.1}$$

Given the fact that we will be working in the Sobolev space $H^1(\Omega)$, this will require us to give a precise meaning to the concept of restriction to the boundary (we will call it a 'trace' operator). The construction is not entirely obvious and requires some prior technical work, which will help us better understand the Sobolev spaces $H^1(\Omega)$, under some new constraints on what the open set Ω can be. Here is the plan for this chapter:

- We will relate the problems of the extension of $H^1(\Omega)$ functions to $H^1(\mathbb{R}^d)$ functions with the possibility of having a dense subset of $H^1(\Omega)$ comprised of smooth functions.

- Once the density results have been made clear, we will be ready to go to the boundary and take the trace of an $H^1(\Omega)$ function on the boundary of Ω. This will be a good moment to discuss some simple functional analytic tools about image norms.

- We will next identify the kernel of the trace operator with the space $H_0^1(\Omega)$, thus proving that the homogeneous Dirichlet problem is a particular case of the nonhomogeneous problem (this might look like a trivial statement, but it is not) when the trace operator is defined.

65

- Finally, when all the tools are ready, we will be able to describe (4.1) in a rigorous way and prove its unique solvability and well-posedness.

4.1   The extension theorem

Let $\Omega \subset \mathbb{R}^d$ be an open set. We say Ω has the H^1-**extension property** if for every $u \in H^1(\Omega)$ there is a $U \in H^1(\mathbb{R}^d)$ such that $U|_\Omega = u$. The extension property can be described in many equivalent ways. For instance, it is clear that Ω has the extension property if and only if the restriction operator $R_\Omega : H^1(\mathbb{R}^d) \to H^1(\Omega)$ given by

$$R_\Omega U := U|_\Omega$$

is surjective. Before we give another equivalent definition of the H^1-extension property, let us have a look at a very simple lemma from Hilbert space functional analysis.

Lemma 4.1. *If X and Y are Hilbert spaces and $A : X \to Y$ is a bounded, linear, surjective map, then there is a bounded linear right inverse $A^\dagger : Y \to X$.*

Proof. The inverse we seek is known as the Moore-Penrose pseudoinverse. If we say that $x = A^\dagger y$, then x is the unique solution satisfying

$$\|x\|_X = \min! \qquad x \in X, \qquad Ax = y.$$

We decompose the space $X = \ker A \oplus (\ker A)^\perp$. With this decomposition we have that $A^\dagger y \in (\ker A)^\perp$ for each $y \in Y$. Also, the restriction $A|_{(\ker A)^\perp} : (\ker A)^\perp \to Y$ is a bounded bijection, and by the Banach isomorphism theorem has a bounded inverse, i.e., $A^\dagger$. $\qquad\square$

Proposition 4.1. *Let Ω be an open set, not necessarily bounded. The following statements are equivalent:*

(a) Ω has the H^1-extension property.

(b) The restriction $R_\Omega : H^1(\mathbb{R}^d) \to H^1(\Omega)$ given by $u \mapsto u|_\Omega$ is surjective.

(c) There exists a linear and bounded extension operator $E_\Omega : H^1(\Omega) \to H^1(\mathbb{R}^d)$, such that $(E_\Omega u)|_\Omega = u$ for all $u \in H^1(\Omega)$.

Proof. By definition (a) implies (b) and (c) implies (a), so we need only to show that (b) implies (c). The result follows immediately by using the Moore-Penrose pseudoinverse. $\qquad\square$

Proposition 4.2. *If Ω has the H^1-extension property, then $C^\infty(\overline{\Omega}) \cap H^1(\Omega)$ is a dense subset of $H^1(\Omega)$.*

Proof. For $u \in H^1(\Omega)$, let $U \in H^1(\mathbb{R}^d)$ such that $U|_\Omega = u$. Since $\mathcal{D}(\mathbb{R}^d)$ is dense in $H^1(\mathbb{R}^d)$, there is a sequence $\{\varphi_n\}$ in $\mathcal{D}(\mathbb{R}^d)$ such that $\varphi_n \to U$ in $H^1(\mathbb{R}^d)$. Since $\varphi_n|_\Omega \in C^\infty(\overline{\Omega}) \cap H^1(\Omega)$ for each n and $\varphi_n|_\Omega \to U|_\Omega = u$ in $H^1(\Omega)$, we have the result. □

For brevity, we introduce the notation $A \overset{d}{\subset} B$ for the statement that A is a dense subset of B. For bounded Ω with the H^1-extension property, we have the following chain of inclusions:

$$C^\infty(\overline{\Omega}) \cap H^1(\Omega) = C^\infty(\overline{\Omega}) \overset{d}{\subset} C^1(\overline{\Omega}) \overset{d}{\subset} H^1(\Omega) \cap C(\overline{\Omega}) \overset{d}{\subset} H^1(\Omega).$$

Proposition 4.3. *The upper half space $\mathbb{R}^d_+ := \{\mathbf{x} \in \mathbb{R}^d : x_d > 0\}$ has the H^1-extension property.*

Proof. The proof of this property is given in Exercise 4.2. □

Theorem 4.1. *If Ω is Lipschitz, then Ω has the H^1-extension property.*

Proof. Let $F_j : \mho \to \Omega_j$ be the local system of charts associated to a finite open cover of $\partial\Omega$ (see Sections 3.3 and 3.4) and let $\{\varphi_j : 0 \le j \le J\}$ be the associated partition of unity. Given $u \in H^1(\Omega)$, we consider the functions

$$u\varphi_j \circ F_j|_{\mho^+} \in H^1(\mho^+),$$

and note that they vanish near $\partial\mho$. Using the same arguments as in Proposition 2.7, it is easy to see that the function $w_j : \mathbb{R}^d_+ \to \mathbb{R}$ given by

$$w_j := \begin{cases} u\varphi_j \circ F_j & \text{in } \mho^+, \\ 0 & \text{elsewhere}, \end{cases}$$

is in $H^1(\mathbb{R}^d_+)$ and then can be extended by symmetry (Proposition 4.3) to $Ew_j \in H^1(\mathbb{R}^d)$. Moreover,

$$(Ew_j) \circ F_j^{-1} \in H^1_0(\Omega_j),$$

and by Proposition 2.7, we can define the extension by zero from Ω_j to $\mathbb{R}^d$

$$u_j := \widetilde{(Ew_j)} \circ F_j^{-1} \in H^1(\mathbb{R}^d).$$

This construction is done for $j = 1, \ldots, J$. Finally we add the interior part as

$$u_0 := \widetilde{\varphi_0 u}$$

using the fact that $\varphi_0 u \in H^1_0(\Omega)$. The final claim is that the function

$$U = \sum_{j=0}^{J} u_j \in H^1(\mathbb{R}^d)$$

is an extension of u. To see this note that $u_0|_\Omega = \varphi_0\, u$ and

$$u_j|_{\Omega_j \cap \Omega} = u\, \varphi_j,$$

and recall that $\sum_{j=0}^{J} \varphi_j \equiv 1$ in Ω. □

4.2   The trace operator

In this section we will construct the trace operator which acts as a weak restriction to the boundary. We will begin by first constructing this operator on the reference domain and showing that it exists and satisfies the desired properties. Once this has been accomplished, we will show how this relates to the trace operator on arbitrary Lipschitz domains.

Construction of the trace on the reference cylinder. Let us first recall the definitions of the pertinent sets: $\mho^+ = B_{d-1}(0;1) \times (0,1)$ and $\mathbb{R}^d \supset \Gamma_0 = B_{d-1}(0;1) \times \{0\} \cong B_{d-1}(0;1) \subset \mathbb{R}^{d-1}$. In the arguments that follow, we will use the fact $C^1(\overline{\mho^+}) \stackrel{d}{\subset} H^1(\mho^+)$.

Now consider the mapping of $C^1(\overline{\mho^+}) \to L^2(\Gamma_0) \cong L^2(B_{d-1}(0;1))$ given by $u(\mathbf{x}) \mapsto u(\widetilde{\mathbf{x}}, 0)$. We can write the image of this mapping as

$$u(\widetilde{\mathbf{x}}, 0) = u(\widetilde{\mathbf{x}}, x_d) - \int_0^{x_d} \partial_{x_d} u(\widetilde{\mathbf{x}}, t)\, \mathrm{d}t.$$

With the careful use of the triangle and Cauchy-Schwarz inequalities, as well as keeping in mind that for our purposes $0 \le x_d \le 1$, we can bound

$$|u(\widetilde{\mathbf{x}}, 0)|^2 \le 2|u(\widetilde{\mathbf{x}}, x_d)|^2 + 2\left|\int_0^{x_d} \partial_{x_d} u(\widetilde{\mathbf{x}}, t)\, \mathrm{d}t\right|^2$$

$$\le 2|u(\widetilde{\mathbf{x}}, x_d)|^2 + 2x_d \int_0^{x_d} |\partial_{x_d} u(\widetilde{\mathbf{x}}, t)|^2\, \mathrm{d}t$$

$$\le 2|u(\widetilde{\mathbf{x}}, x_d)|^2 + 2\int_0^1 |\partial_{x_d} u(\widetilde{\mathbf{x}}, t)|^2\, \mathrm{d}t.$$

Taking this result and integrating over $B_{d-1}(0;1)$, we obtain

$$\int_{B_{d-1}(0;1)} |u(\widetilde{\mathbf{x}}, 0)|^2\, \mathrm{d}\widetilde{\mathbf{x}} \le 2\int_{B_{d-1}(0;1)} |u(\widetilde{\mathbf{x}}, x_d)|^2\, \mathrm{d}\widetilde{\mathbf{x}} + 2\int_{\mho^+} |\partial_{x_d} u(\mathbf{x})|^2\, \mathrm{d}\mathbf{x},$$

and therefore

$$\int_{B_{d-1}(0;1)} |\widetilde{u}(\widetilde{\mathbf{x}}, 0)|^2\, \mathrm{d}\widetilde{\mathbf{x}} \le 2\int_{\mho^+} |u(\mathbf{x})|^2\, \mathrm{d}\mathbf{x} + 2\int_{\mho^+} |\partial_{x_d} u(\mathbf{x})|^2\, \mathrm{d}\mathbf{x}.$$

We have thus shown that

$$\|u(\cdot,0)\|_{\Gamma_0} \le \sqrt{2}\|u\|_{1,\mho^+}.$$

Proposition 4.4 (Completion process). *Let C be a dense subspace of a Hilbert space H and let $\gamma : C \to V$ be a linear operator satisfying*

$$\|\gamma u\|_V \le \alpha \|u\|_H \qquad \forall u \in C,$$

where V is another Hilbert space. There exists a unique bounded extension $\widetilde{\gamma} : H \to V$ (therefore $\widetilde{\gamma}u = \gamma u$ for all $u \in C$), and $\|\widetilde{\gamma}u\|_V \le \alpha \|u\|_H$ for all $u \in H$.

Proof. If $u \in H$ there exists a sequence $\{u_n\}$ in C such that $u_n \to u$ in H. We also have that γu_n is a Cauchy sequence in V, hence convergent. If we have another sequence $\{v_n\}$ converging to u, mixing it with $\{u_n\}$ we can prove that $\lim_n \gamma u_n = \lim_n \gamma v_n$ in V and we can thus define $\widetilde{\gamma}u = \lim_n \gamma u_n$. Finally, taking a constant sequence shows that $\widetilde{\gamma}$ is an extension of γ. $\qquad\square$

The trace operator on Γ_0. By the completion process, there is a unique linear and bounded operator $\gamma_{\Gamma_0} : H^1(\mho^+) \to L^2(\Gamma_0)$ that satisfies $\gamma_{\Gamma_0}u = u(\widetilde{\mathbf{x}},0)$ for all $u \in \mathcal{C}^1(\overline{\mho^+})$. In addition, we have the bound $\|\gamma_{\Gamma_0}u\|_{L^2(\Gamma_0)} \le \sqrt{2}\|u\|_{H^1(\mho^+)}$ for all $u \in H^1(\mho^+)$. The operator γ_{Γ_0} is the trace operator on Γ_0.

Theorem 4.2 (Construction of the trace operator). *For any bounded Lipschitz open set Ω, there exists a bounded linear operator*

$$\gamma : H^1(\Omega) \to L^2(\Gamma)$$

such that

$$\gamma u = u|_\Gamma \qquad \forall u \in \mathcal{C}(\overline{\Omega}) \cap H^1(\Omega).$$

Proof. Let $u \in H^1(\Omega) \cap \mathcal{C}(\overline{\Omega})$. We can decompose u using the partition of unity by $u = u\,\varphi_0 + \sum_{j=1}^N u\,\varphi_j$. It is then clear that $u\,\varphi_0 \in H_0^1(\Omega) \cap \mathcal{C}(\overline{\Omega})$ and $u\,\varphi_0|_\Gamma = 0$. We can consider $u\,\varphi_j \circ \mathrm{F}_j \in H^1(\mho^+) \cap \mathcal{C}(\overline{\mho^+})$, and restricting to Γ_0, we have

$$(u\,\varphi_j \circ \mathrm{F}_j)\,|_{\Gamma_0} = u\,\varphi_j|_\Gamma \circ \boldsymbol{\Phi}_j,$$

where $\boldsymbol{\Phi}_j = \mathrm{F}_j|_{\Gamma_0}$ are the functions defined in Section 3.5. The above quantities are equal by the continuity of the restriction. By the previous section, we have

$$\int_{\Gamma} |u(\mathbf{x})|^2 d\Gamma(\mathbf{x}) = \sum_j \int_{\Gamma \cap \Omega_j} |u(\mathbf{x})|^2 \varphi_j(\mathbf{x}) d\Gamma(\mathbf{x})$$

$$= \sum_j \int_{B_{d-1}(\mathbf{0};1)} (|u|^2 \varphi_j \circ \mathbf{\Phi}_j)(\mathbf{x}) d\mathbf{\Phi}_j(\mathbf{x})$$

$$\leq C \sum_j \int_{B_{d-1}(\mathbf{0};1)} |u \circ \mathbf{\Phi}_j|^2 (\mathbf{x}) d\mathbf{\Phi}_j(\mathbf{x})$$

$$\leq C \sum_j \|u \circ F_j\|_{1,\mho^+}^2 \leq C \sum_j \|u\|_{1,\Omega \cap \Omega_j}^2$$

$$\leq C \|u\|_{1,\Omega}^2.$$

In all of the above, the constant C is generic and is allowed to change value from one line to the next. We have shown that we have a map from $\mathcal{C}(\overline{\Omega}) \cap H^1(\Omega) \to \mathcal{C}(\Gamma) \subset L^2(\Gamma)$ given by $u \mapsto u|_\Gamma$ which is bounded as

$$\|u|_\Gamma\|_\Gamma \leq C \|u\|_{1,\Omega} \qquad \forall u \in \mathcal{C}(\overline{\Omega}) \cap H^1(\Omega).$$

The result now follows by density using the completion process of Proposition 4.4. $\qquad\square$

4.3   The range and kernel of the trace operator

In the last section, we carefully constructed the trace operator $\gamma : H^1(\Omega) \to L^2(\Gamma)$. On the reference domain $\mho^+$ the construction consisted of extending the map $u(\tilde{\mathbf{x}}, 0) = u|_{\Gamma_0}$ to a bounded linear operator $H^1(\mho^+) \to L^2(\Gamma_0)$. We saw previously that $H_0^1(\Omega) \subset \ker \gamma$. In fact, we will later show that $H_0^1(\Omega) = \ker \gamma$.

The range of γ. Consider the set

$$\{g \in L^2(\Gamma) : g = \gamma u \text{ for some } u \in H^1(\Omega)\} = \text{range } \gamma =: H^{1/2}(\Gamma).$$

When equipped with the norm

$$\|g\|_{1/2,\Gamma} := \inf \{\|u\|_{1,\Omega} : \gamma u = g\},$$

$H^{1/2}(\Gamma)$ is a Hilbert space. We also have the trivial inequality $\|\gamma u\|_{1/2,\Gamma} \leq \|u\|_{1,\Omega}$.

It may seem quite bold to assert that the range of the trace operator equipped with such a norm is a Hilbert Space. We take the time now to justify this claim. Suppose we have a linear operator $\gamma : X \to Y$ where X is a Hilbert space and $\ker \gamma$ is closed. An element $g \in Y$ is in the range of γ if there is a $u \in X$ such that $\gamma u = g$. We can decompose u as $u = u_0 + u_1$ where $u_0 \in (\ker \gamma)^{\perp}$ and $u_1 \in \ker \gamma$. By the orthogonality of the decomposition, we have

$$\|u\|^2 = \|u_0\|^2 + \|u_1\|^2.$$

All solutions of $\gamma u = g$ are of the form $u_0 + \ker \gamma$, and u_0 minimizes the norm of u. The construction $\gamma^{\dagger} : \text{range } \gamma \to X$ is the Moore-Penrose pseudoinverse of γ, and is only defined on the range of the trace. We can then consider the image norm

$$\|g\|_{\text{range } \gamma} = \inf \{\|u\|_X : \gamma u = g\} = \|\gamma^{\dagger} g\|_X.$$

We have now shown that the operators $\gamma : (\ker \gamma)^{\perp} \to \text{range } \gamma$ and $\gamma^{\dagger} : \text{range } \gamma \to (\ker \gamma)^{\perp}$ are isometries and inverses of one another when restricted to the appropriate spaces. The space $(\ker \gamma)^{\perp}$ is a closed subspace of a Hilbert space, and is therefore itself a Hilbert space. Therefore range γ, equipped with this norm, is also a Hilbert space. Putting this argument back in the above context, we have that $H^{1/2}(\Gamma)$ is a Hilbert space.

Proposition 4.5. $\ker \gamma = H_0^1(\Omega)$.

Proof. If $u \in \mathcal{D}(\Omega)$, then clearly $\gamma u = u|_{\Gamma} = 0$. If $u \in H_0^1(\Omega)$, then there is a sequence $\{\varphi_n\}$ in $\mathcal{D}(\Omega)$ such that $\varphi_n \to u$ in $H^1(\Omega)$. Furthermore,

$$0 = \gamma \varphi_n \to \gamma u,$$

and so $\gamma u = 0$ for all $u \in H_0^1(\Omega)$, i.e., $H_0^1(\Omega) \subset \ker \gamma$.

Now let $u \in H^1(\Omega)$ satisfy $\gamma u = 0$. We divide the rest of the argument into four steps: localization, pullback, work on the reference cylinder, and push-forward.

1. First we localize. Let $\{\Omega_j\}_{j=1}^J$ be an open cover of Γ, as in Sections 3.3 and 3.4, and let $\{\varphi_j\}_{j=1}^J$ be the associated partition of unity. By Lemma 4.2 below $\varphi_j u \in \ker \gamma$ for all j. Therefore we do not leave the kernel of γ by localizing.

2. In a second step, we want to show that $\varphi_j u \circ F_j \in \ker \gamma_{\Gamma_0}$, where $F_j : \mho \to \Omega_j$ are the bi-Lipschitz homeomorphisms which were introduced in Section 3.3. This is proved by a density argument. Take a sequence $\{u_n\}$ in $\mathcal{C}^{\infty}(\overline{\Omega})$ such that $u_n \to u$ in $H^1(\Omega)$. We then have the convergence

$$\varphi_j u_n \longrightarrow \varphi_j u \quad \text{in } H^1(\Omega),$$

which by Corollary 3.1 and the continuity of the functions F_j implies

$$\varphi_j u_n \circ F_j \longrightarrow \varphi_j u \circ F_j \quad \text{in } H^1(\mho^+).$$

Turning our attention to the traces of these functions we see that

$$\gamma(\varphi_j u_n) = (\varphi_j u_n)|_{\Gamma} \longrightarrow 0 \quad \text{in } L^2(\Gamma),$$

and

$$\gamma_{\Gamma_0}(\varphi_j u_n \circ F_j) = \varphi_j u_n \circ \Phi_j \longrightarrow \gamma_{\Gamma_0}(\varphi_j u \circ F_j) \quad \text{in } L^2(\Gamma_0), \qquad (4.2)$$

where $\Phi_j = F_j|_{\Gamma_0}$ are the same functions used in Section 3.5. The change of variables derived in Section 3.5 yields

$$0 \longleftarrow \|\gamma(\varphi_j u_n)\|_\Gamma^2 = \|\varphi_j u_n\|_\Gamma^2 = \|\varphi_j u_n\|_{\Gamma \cap \Omega_j}^2$$
$$= \int_{\Gamma_0} |\varphi_j u_n \circ \Phi_j|^2 (\mathbf{x}) d\Phi_j(\mathbf{x}) \longrightarrow \int_{\Gamma_0} |\gamma_{\Gamma_0}(\varphi_j u \circ F_j)(\mathbf{x})|^2 d\Phi_j(\mathbf{x}),$$

which along with (4.2) shows that $\varphi_j u \circ F_j \in \ker \gamma_{\Gamma_0}$ for each j.

3. On the reference cylinder we want to show that if $\breve{u} \in H^1(\mathcal{U}^+)$, $\gamma_{\Gamma_0}\breve{u} = 0$ and $\breve{u} = 0$ in a neighborhood of $\partial \mathcal{U}^+ \setminus \Gamma_0$, then $\breve{u} \in H_0^1(\mathcal{U}^+)$. The details of this nontrivial result are outlined in Exercise 4.3.

4. Finally, taking the function $u\varphi_j \circ F_j \in H_0^1(\mathcal{U}^+)$, by Proposition 3.1 we have

$$(u\varphi_j \circ F_j) \circ F_j^{-1} = u\varphi_j \in H_0^1(\Omega_j \cap \Omega)$$

which shows $u\varphi_j \in H_0^1(\Omega)$ for each j. The result then follows.   □

The previous proof uses the following lemma.

Lemma 4.2. *If $u \in \ker \gamma$ and $\varphi \in C^\infty(\overline{\Omega})$, then $\varphi u \in \ker \gamma$.*

Proof. We first take a sequence $\{u_n\}$ in $C^\infty(\overline{\Omega})$ such that $u_n \to u$ in $H^1(\Omega)$. By the trace theorem (Section 4.2) we have

$$\gamma u_n = u_n|_\Gamma \longrightarrow 0 \quad \text{in } L^2(\Gamma).$$

Using Leibniz's rule and the dominated convergence theorem we obtain the convergence

$$\varphi u_n \longrightarrow \varphi u \quad \text{in } H^1(\Omega),$$

and therefore

$$\gamma(\varphi u_n) = \varphi|_\Gamma u_n|_\Gamma = \varphi|_\Gamma \gamma u_n \longrightarrow 0,$$

since $\varphi|_\Gamma$ is bounded. Appealing to the trace theorem once more, we have $\gamma(\varphi u_n) \to \gamma(\varphi u)$, which using the previous convergence shows that $\varphi u \in \ker \gamma$.   □

4.4   The nonhomogeneous Dirichlet problem

This section begins with the study of the nonhomogeneous Dirichlet problem for the Laplacian:

$$u \in H^1(\Omega),$$
$$-\Delta u = f,$$
$$\gamma u = g.$$

The domain Ω will assumed to be bounded, open, and Lipschitz. For our data, we require that $f \in L^2(\Omega)$ and $g \in H^{1/2}(\Gamma)$. The Laplace operator is implicitly understood in the sense of distributions. We will seek a solution $u \in H^1(\Omega)$ so that the trace is well-defined.

We may be tempted to try to solve this problem by taking a particular $u_g \in H^1(\Omega)$ such that $\gamma u = g$ and working with the homogeneous unknown $u_0 = u - u_g$. We would then have $\gamma u_0 = 0$, and attempt to apply the theory developed for the homogeneous problem. However, if we compute $-\Delta u_0 = -\Delta u + \Delta u_g = f + \Delta u_g$, then the Laplacian of u_g can only be shown to be a distribution, and not an element of $L^2(\Omega)$. While this approach can be shown to work (using weak right-hand sides as explained in Section 4.5), we will deal with the nonhomogeneous boundary condition at the level of the variational formulation.

Variational formulation. As was done for the homogeneous problem, we seek a variational form of the problem to try to better understand what we are working with. We have a so-called 'essential' boundary condition, because the inhomogeneous trace is not part of the solution space $H^1(\Omega)$. Contrast this with the 'natural' boundary conditions found in the homogenous problem, where they were imposed directly in the solution space (i.e., in $H_0^1(\Omega)$). To find the variational form, we begin by computing with the PDE:

$$\langle -\Delta u, \varphi \rangle_{\mathcal{D}' \times \mathcal{D}} = -\sum_j \langle \partial_{x_j}^2 u, \varphi \rangle_{\mathcal{D}' \times \mathcal{D}} = \sum_j \langle \partial_{x_j} u, \partial_{x_j} \varphi \rangle_{\mathcal{D}' \times \mathcal{D}}$$
$$= (\nabla u, \nabla \varphi)_\Omega = \langle f, \varphi \rangle_{\mathcal{D}' \times \mathcal{D}} = (f, \varphi)_\Omega.$$

Filling by density, we find the variational formulation to be

$$u \in H^1(\Omega), \tag{4.3a}$$
$$\gamma u = g, \tag{4.3b}$$
$$(\nabla u, \nabla v)_\Omega = (f, v)_\Omega \quad \forall v \in H_0^1(\Omega). \tag{4.3c}$$

Notice the mismatch between the test space and solution space. Because of this, we cannot apply the Riesz-Fréchet theorem out of the box.

Existence, uniqueness, and stability. To show well-posedness of (4.3), we begin by picking a $u_g \in H^1(\Omega)$ such that $\gamma u_g = g$, and write $u_0 = u - u_g$. We then seek a solution to the problem

$$u_0 \in H_0^1(\Omega),$$

$$(\nabla u_0, \nabla v_0)_\Omega = (f, v)_\Omega - (\nabla u_g, \nabla v)_\Omega \qquad \forall v \in H_0^1(\Omega). \qquad (4.4)$$

Note that our test and trial spaces are the kernel of the trace operator. We may now apply the Riesz-Fréchet theorem with the space $V = H_0^1(\Omega)$. The Dirichlet form defines an inner product that is equivalent to the usual inner product, and we have the Poincaré inequality $\|v\|_{1,\Omega} \leq C_\Omega \|\nabla v\|_\Omega$. The functional $\ell(v) := (f, v)_\Omega - (\nabla u_g, \nabla v)_\Omega$ is linear and bounded in V, and therefore there exists a solution to (4.4). A problem arises when we return to the original unknown $u = u_0 + u_g$. The term u_g depends on the problem, and a different choice of u_g leads to a different u_0. To show uniqueness, we return to (4.3). If we subtract two solutions to this problem, we end up solving

$$w \in H^1(\Omega), \quad \gamma w = 0,$$

$$(\nabla w, \nabla v)_\Omega = 0 \quad \forall v \in H_0^1(\Omega).$$

This problem only has the trivial solution, and therefore the solution to (4.3) is unique. To derive the stability of the solution, we begin again with $u_g \in H^1(\Omega)$ such that $\gamma u_g = g$. Notice that $u_g + \ker \gamma = u_g + H_0^1(\Omega) = \{v \in H^1(\Omega) : \gamma v = g\}$. From the definition of $\ell(v)$, we can bound

$$|\ell(v)| = |(f, v)_\Omega - (\nabla u_g, \nabla v)_\Omega| \leq \|f\|_\Omega \|v\|_\Omega + \|\nabla u_g\|_\Omega \|\nabla v\|_\Omega$$

$$\leq \left(\|f\|_\Omega^2 + \|\nabla u_g\|_\Omega^2 \right)^{1/2} \|v\|_{1,\Omega}.$$

Combining this bound with (4.4), letting $v = u_0$ we arrive at $\|\nabla u_0\|_\Omega \leq C_\Omega \left(\|f\|_\Omega^2 + \|\nabla u_g\|_\Omega^2 \right)^{1/2}$, and therefore $\|u_0\|_{1,\Omega} \leq C_\Omega^2 \left(\|f\|_\Omega^2 + \|\nabla u_g\|^2 \right)^{1/2}$, which shows that the problem for u_0 is well posed. We may then consider the quantity of interest, u, and compute

$$\|u\|_{1,\Omega} = \|u_g + u_0\|_{1,\Omega} \leq C_\Omega^2 \left(\|f\|_\Omega + \|u_g\|_{1,\Omega} \right) + \|u_g\|_{1,\Omega}$$

$$\leq C_\Omega^2 \|f\|_\Omega + (C_\Omega^2 + 1)\|u_g\|_{1,\Omega}.$$

Therefore, since this bound holds for any choice of u_g, we have

$$\|u\|_{1,\Omega} \leq C_\Omega^2 \|f\|_\Omega + (1 + C_\Omega^2) \inf\{\|u_g\|_{1,\Omega} : \gamma u_g = g\}$$

$$= C_\Omega^2 \|f\|_\Omega + (1 + C_\Omega^2)\|g\|_{1/2,\Gamma},$$

which gives the final stability estimate $\|u\|_{1,\Omega} \leq C \left(\|f\|_\Omega + \|g\|_{1/2,\Gamma} \right)$. To see that the solution map is linear, consider the space $W = \{u \in H^1(\Omega) : \Delta u \in L^2(\Omega)\}$ and the map $W \to L^2(\Omega) \times H^{1/2}(\Gamma)$ given by $u \mapsto (-\Delta u, \gamma u)$. This map is linear and invertible, and so the inverse is linear. (This is just another way to refer to the **superposition principle** for the solution operator for uniquely solvable linear problems.)

Associated minimization problem. We saw in Chapter 2 that the homogeneous Dirichlet problem can be cast as a minimization problem. Can we show a similar equivalence here? Indeed, we can prove that

$$u \in H^1(\Omega), \qquad \gamma u = g,$$
$$(\nabla u, \nabla v)_\Omega = (f, v)_\Omega \quad \forall v \in H_0^1(\Omega),$$

is equivalent to

$$\tfrac{1}{2}\|\nabla u\|_\Omega^2 - (f, v) = \min! \qquad u \in H^1(\Omega), \quad \gamma u = g.$$

Because of the 'shifting' imposed by g, this problem is one of minimizing a quadratic functional on an affine manifold. To prove the equivalence between the variational and minimization problems, we just need to apply the following lemma.

Lemma 4.3. *Suppose V and M are vector spaces and $\gamma : V \to M$ is a linear map, $z \in \mathrm{range}\,\gamma$, and $V_0 = \ker \gamma$. In addition, let $a : V \times V \to \mathbb{R}$ be a bilinear and symmetric form that is positive semidefinite on V_0, and let $\ell : V \to \mathbb{R}$. The problem of finding*

$$\tfrac{1}{2}a(u, u) - \ell(u) = \min! \qquad u \in V, \quad \gamma u = z$$

is equivalent to

$$u \in V, \qquad \gamma u = z,$$
$$a(u, v) = \ell(v) \quad \forall v \in V_0.$$

Proof. Let $v \in V_0$ and consider

$$\frac{1}{2}a(u + tv, u + tv) - \ell(u + tv)$$
$$= \left(\frac{1}{2}a(u, u) - \ell(u) \right) + t(a(u, v) - \ell(u)) + \frac{t^2}{2}a(v, v),$$

where the last term is always nonnegative. This quadratic form attains its minimum if and only if $a(u, v) = \ell(v)$ for $u \in V$ such that $\gamma u = z$ and $v \in V_0$. The minimum value is exactly $\tfrac{1}{2}a(u, u) - \ell(u)$. $\square$

4.5 General right-hand sides

In this section we study the problem

$$u \in H^1(\Omega), \qquad -\Delta u = f, \qquad \gamma u = g,$$

for $g \in H^{1/2}(\Gamma)$, allowing f to be as general as possible so that the equality $-\Delta u = f$ holds in $\mathcal{D}'(\Omega)$. We first recall some basic results on adjoint operators in Hilbert spaces and introduce **Gelfand triples**, also known as **Courant triads**.

Adjoints and Hilbert space adjoints. Let V and H be Hilbert spaces and let $A : V \to H$ be a bounded linear operator. The adjoint $A' : H' \to V'$ is defined by

$$H' \ni h' \longmapsto A'h' := h' \circ A \in V'.$$

We can also define the Hilbert space adjoint $A^* : H \to V$ with the identity

$$(A^*h, v)_V = (h, Av)_H \qquad \forall h \in H, v \in V.$$

Note that if $R_H : H \to H'$ and $R_V : V \to V'$ are the Riesz-Fréchet maps

$$R_H h := (h, \cdot)_H, \qquad R_V v := (v, \cdot)_V,$$

(recall that they are isometric isomorphisms), then

$$A' = R_V A^* R_H^{-1}, \qquad A^* = R_V^{-1} A' R_H,$$

and therefore

$$\|A'\| = \|A^*\|.$$

It easy to prove that

$$\|A^*\| \leq \|A\|.$$

Using that $A^{**} = A$ (this is straightforward from the definition), we can then show that

$$\|A'\| = \|A^*\| = \|A\|.$$

Proposition 4.6. *If $A : V \to H$ is injective, then* range A' *is dense in V'. If $A : V \to H$ has dense range, then A' is injective.*

Proof. It is clear that

$$
\begin{aligned}
(\text{range } A)^{\perp} &= \{u \in H \ : \ (u, Av)_H = 0 \quad \forall v \in V\} \\
&= \{u \in H \ : \ (A^*u, v)_V = 0 \quad \forall v \in V\} = \ker A^*
\end{aligned}
$$

and therefore

$$\overline{\text{range } A} = (\ker A^*)^{\perp}.$$

Obviously, this implies that if range A is dense in H, A^* is injective and therefore so is A'. If we use

$$\overline{\text{range } A^*} = (\ker A)^{\perp},$$

we show that if A is injective then the range of A' is dense. $\square$

Gelfand triples. Now let V and H be Hilbert spaces with $V \subset H$ satisfying

$$\|v\|_H \leq C\|v\|_V \qquad \forall v \in V, \qquad \overline{V} = H,$$

i.e., V is dense in H and the embedding of V in H is continuous. The adjoint of the embedding map $i : V \to H$ is the map $i' : H' \to V'$ given by

$$i'h' = h' \circ i = h'|_V \qquad \forall h' \in H'.$$

This means that the adjoint of the embedding map is the restriction map for elements of the dual. Since V is dense in H, it follows that $i' : H' \to V'$ is injective, and we can write

$$H' \subset V',$$

or, more properly speaking, we can identify H' with a subset of V'. Since $i : V \to H$ is injective, the injection $H' \subset V'$ is dense. The Gelfand triple follows from the identification of H with H' via the Riesz-Fréchet map, thus arriving at the structure

$$V \subset H \equiv H' \subset V'.$$

Both embeddings are continuous and dense. An element $h \in H$ defines a functional $(h, \cdot)_H \in H'$ which can be restricted to an element of V' and therefore, we can take H as a subspace of V' via the Riesz-Fréchet map of H. Note the paradoxical situation where we have identified V with a proper dense subspace of V', which seems to contradict the fact that V and V' are isomorphic. This is not really a paradox (just some dangerous bends in the sense of Bourbaki), that can be explained by the fact that the Riesz-Fréchet maps of H and V are different and produce incompatible representations of the spaces.

Two Gelfand triples. We can use the fact that $H^1(\Omega)$ and $H_0^1(\Omega)$ are dense in $L^2(\Omega)$ (both contain $\mathcal{D}(\Omega)$, which is dense in $L^2(\Omega)$ due to the variational lemma), and define the triples

$$H_0^1(\Omega) \subset L^2(\Omega) \subset H^{-1}(\Omega),$$

and

$$H^1(\Omega) \subset L^2(\Omega) \subset \widetilde{H}^{-1}(\Omega).$$

Here we are using the notation

$$H^{-1}(\Omega) := (H_0^1(\Omega))', \qquad \widetilde{H}^{-1}(\Omega) := (H^1(\Omega))',$$

understanding that the dual spaces are represented by what comes from identifying $L^2(\Omega)$ with its dual space. As we will see next, these two weak spaces have very different meanings in the context of PDE in the domain Ω.

Proposition 4.7. *For an open set Ω, the elements of $H^{-1}(\Omega)$ are distributions in Ω.*

Proof. If $T : H_0^1(\Omega) \to \mathbb{R}$ is linear and bounded, then its restriction to $T|_{\mathcal{D}(\Omega)} :$ $\mathcal{D}(\Omega) \to \mathbb{R}$ is linear and bounded with respect to the $H^1(\Omega)$ norm. However, convergence in $\mathcal{D}(\Omega)$ implies convergence in $H^1(\Omega)$ (this is very simple; prove it!) and, therefore $T|_{\mathcal{D}(\Omega)} \in \mathcal{D}'(\Omega)$. Since $\mathcal{D}(\Omega)$ is dense in $H_0^1(\Omega)$, if $T|_{\mathcal{D}(\Omega)} = 0$, then $T = 0$ and we can thus identify $H^{-1}(\Omega)$ with a subspace of $\mathcal{D}'(\Omega)$. □

A remark. The situation is very different for $\widetilde{H}^{-1}(\Omega)$, since typically $\mathcal{D}(\Omega)$ is not dense in $H^1(\Omega)$. If this is the case, we can have several elements of $\widetilde{H}^{-1}(\Omega)$ with the property that their restriction to the test space coincides. For instance, if Ω is a Lipschitz domain the bounded linear map

$$u \longmapsto \int_\Gamma \gamma u(\mathbf{x}) \mathrm{d}\Gamma(\mathbf{x})$$

is an element of $\widetilde{H}^{-1}(\Omega)$, but its restriction to $\mathcal{D}(\Omega)$ is zero.

Proposition 4.8. *If* $u \in L^2(\Omega)$, *then* $\partial_{x_i} u \in H^{-1}(\Omega)$. *Moreover, the map* $\partial_{x_i} : L^2(\Omega) \to H^{-1}(\Omega)$ *is bounded.*

Proof. By definition

$$\langle \partial_{x_i} u, \varphi \rangle = -\int_\Omega u(\mathbf{x}) \, \partial_{x_i} \varphi(\mathbf{x}) \mathrm{d}\mathbf{x}, \qquad (4.5)$$

and therefore

$$|\langle \partial_{x_i} u, \varphi \rangle| \le \|u\|_\Omega \|\varphi\|_{1,\Omega} \qquad \forall \varphi \in \mathcal{D}(\Omega). \qquad (4.6)$$

The integral expression (4.5) gives the element of $H^{-1}(\Omega)$ that we identify with the distribution $\partial_{x_i} u$ and the inequality (4.6) proves that

$$\|\partial_{x_i} u\|_{H^{-1}(\Omega)} \le \|u\|_\Omega \qquad \forall u \in L^2(\Omega).$$

This finishes the proof. □

A general Dirichlet problem. Using Proposition 4.8, it follows that if $u \in H^1(\Omega)$, then $\Delta u \in H^{-1}(\Omega)$. We can then deal with the boundary value problem

$$u \in H^1(\Omega), \quad -\Delta u = f, \quad \gamma u = g, \qquad (4.7)$$

for general $f \in H^{-1}(\Omega)$ and $g \in H^{1/2}(\Gamma)$, with the PDE taken in the sense of distributions in Ω. Problem (4.7) is equivalent to the variational problem

$$u \in H^1(\Omega), \quad \gamma u = g, \qquad (4.8a)$$

$$(\nabla u, \nabla v)_\Omega = \langle f, v \rangle_{H^{-1}(\Omega) \times H_0^1(\Omega)} \quad \forall v \in H_0^1(\Omega), \qquad (4.8b)$$

and (4.8) is well posed by the same argument that we used in Section 4.4. What is slightly less obvious is the associated minimization problem, since $\langle f, u \rangle$ is not a valid duality product and, therefore, the quadratic functional

$$\tfrac{1}{2}\|\nabla u\|_\Omega^2 - \langle f, u \rangle$$

is not well defined. Problem (4.8) is actually equivalent to the following family of very similar minimization problems. Let $\check{f} : H^1(\Omega) \to \mathbb{R}$ be any bounded linear functional that extends $f : H_0^1(\Omega) \to \mathbb{R}$. The quadratic minimization problem

$$\tfrac{1}{2}\|\nabla u\|_\Omega^2 - \langle \check{f}, u \rangle_{\widetilde{H}^{-1}(\Omega) \times H^1(\Omega)} = \min! \qquad u \in H^1(\Omega), \quad \gamma u = g \qquad (4.9)$$

is then equivalent to

$$\begin{aligned} &u \in H^1(\Omega), \quad \gamma u = g, \\ &(\nabla u, \nabla v)_\Omega = \langle \check{f}, v \rangle_{\widetilde{H}^{-1}(\Omega) \times H^1(\Omega)} \quad \forall v \in H_0^1(\Omega), \end{aligned}$$

and, since $\check{f}|_{H_0^1(\Omega)} = f$, to (4.8). This shows that even if (4.9) is meaningful for any $\check{f} \in \widetilde{H}^{-1}(\Omega)$, the solution of this minimization problem depends exclusively on the restriction $f := \check{f}|_{H_0^1(\Omega)}$, which is the only element that can appear in the right-hand side of (4.7).

4.6 The Navier-Lamé equations (*)

In this section we give an example of a system of linear partial differential equations associated to linear elasticity.

The associated functional spaces. Consider the spaces

$$\begin{aligned} \mathbf{H}^1(\Omega) :=& H^1(\Omega; \mathbb{R}^d) \equiv H^1(\Omega)^d \\ =& \{\mathbf{u} = (u_1, \dots, u_d) : \Omega \to \mathbb{R}^d : u_j \in H^1(\Omega) \quad j = 1, \dots, d\}, \\ \mathbf{H}^{1/2}(\Gamma) :=& H^{1/2}(\Gamma; \mathbb{R}^d) \equiv H^{1/2}(\Gamma)^d \\ =& \{\mathbf{g} = (g_1, \dots, g_d) : \Gamma \to \mathbb{R}^d : g_j \in H^{1/2}(\Gamma) \quad j = 1, \dots, d\}, \end{aligned}$$

endowed with the respective product norms

$$\|\mathbf{u}\|_{1,\Omega}^2 = \sum_{j=1}^d \|u_j\|_{1,\Omega}^2, \qquad \|\mathbf{g}\|_{1/2,\Gamma}^2 = \sum_{j=1}^d \|g_j\|_{1/2,\Gamma}^2.$$

Obviously, the associated diagonal trace operator

$$\mathbf{H}^1(\Omega) \ni \mathbf{u} \longmapsto \gamma\mathbf{u} = (\gamma u_1, \dots, \gamma u_d) \in \mathbf{H}^{1/2}(\Gamma)$$

is bounded and surjective.

The material law. Given $\mathbf{u} \in \mathbf{H}^1(\Omega)$, we write

$$\varepsilon(\mathbf{u}) := \tfrac{1}{2}(\nabla\mathbf{u} + (\nabla\mathbf{u})^\top)$$

to denote the symmetric part of the Jacobian of $\mathbf{u}$. In the language of elasticity, $\varepsilon(\mathbf{u})$ is the linearized strain. Note that ε defines a bounded linear operator

$$\varepsilon : \mathbf{H}^1(\Omega) \longrightarrow L^2(\Omega; \mathbb{R}^{d \times d}_{\mathrm{sym}}),$$

where $\mathbb{R}^{d \times d}_{\mathrm{sym}}$ is the space of symmetric matrices. We next consider the stress operator

$$\boldsymbol{\sigma}(\mathbf{u}) := 2\mu\varepsilon(\mathbf{u}) + \lambda \operatorname{tr} \varepsilon(\mathbf{u})\, \mathrm{I} = \mu(\nabla\mathbf{u} + (\nabla\mathbf{u})^\top) + \lambda(\nabla \cdot \mathbf{u})\,\mathrm{I},$$

where μ, λ are constants (they are called the Lamé constants) and I is the $d \times d$ identity matrix. Note that

$$\boldsymbol{\sigma} : \mathbf{H}^1(\Omega) \longrightarrow L^2(\Omega; \mathbb{R}^{d \times d}_{\mathrm{sym}})$$

is a bounded linear operator.

The Dirichlet problem for the Navier-Lamé equations. Given $\mathbf{f} \in \mathbf{L}^2(\Omega) := L^2(\Omega; \mathbb{R}^d) \equiv L^2(\Omega)^d$ and $\mathbf{g} \in \mathbf{H}^{1/2}(\Gamma)$ we look for a solution to

$$\mathbf{u} \in \mathbf{H}^1(\Omega), \quad -\operatorname{div} \boldsymbol{\sigma}(\mathbf{u}) = \mathbf{f}, \quad \gamma\mathbf{u} = \mathbf{g}, \qquad (4.10)$$

where div is the divergence operator (in the sense of distributions) applied to the rows of $\boldsymbol{\sigma}(\mathbf{u})$ and the equation (actually, a system of equations) is satisfied in the sense of distributions. A simple computation shows that

$$\operatorname{div} \boldsymbol{\sigma}(\mathbf{u}) = \mu\Delta\mathbf{u} + (\mu + \lambda)\nabla(\nabla \cdot \mathbf{u}),$$

where the Laplace operator is applied componentwise. To derive a variational formulation, consider $\boldsymbol{\varphi} \in \mathcal{D}(\Omega; \mathbb{R}^d) \equiv \mathcal{D}(\Omega)^d$ and let us observe what the divergence operator does to a generic function $\boldsymbol{\sigma} \in L^2(\Omega; \mathbb{R}^{d \times d}_{\mathrm{sym}})$:

$$-\langle \operatorname{div} \boldsymbol{\sigma}, \boldsymbol{\varphi} \rangle_{\mathcal{D}'(\Omega)^d \times \mathcal{D}(\Omega)^d} = -\sum_{i=1}^{d}\sum_{j=1}^{d} \langle \partial_{x_j}\sigma_{ij}, \varphi_i \rangle$$

$$= \sum_{i=1}^{d}\sum_{j=1}^{d} \langle \sigma_{ij}, \partial_{x_j}\varphi_i \rangle = \sum_{i=1}^{d}\sum_{j=1}^{d} \int_\Omega \sigma_{ij}\, \partial_{x_j}\varphi_i$$

$$= \int_\Omega \boldsymbol{\sigma}(\mathbf{x}) : \nabla\boldsymbol{\varphi}(\mathbf{x})\mathrm{d}\mathbf{x} = \int_\Omega \boldsymbol{\sigma}(\mathbf{x}) : \varepsilon(\boldsymbol{\varphi})(\mathbf{x})\mathrm{d}\mathbf{x}.$$

Here we have used the colon for the Frobenius inner product of matrices

$$\mathrm{A} : \mathrm{B} = \sum_{i,j=1}^{d} a_{ij}b_{ij},$$

and we have applied that if $A \in \mathbb{R}^{d \times d}_{\text{sym}}$, then

$$A : B = A : (\tfrac{1}{2}(B + B^{\top})).$$

This computation shows us what the elastic bilinear form is:

$$a(\mathbf{u}, \mathbf{v}) := \int_{\Omega} \boldsymbol{\sigma}(\mathbf{u}) : \boldsymbol{\varepsilon}(\mathbf{v}) = (\boldsymbol{\sigma}(\mathbf{u}), \boldsymbol{\varepsilon}(\mathbf{v}))_{\Omega} = (\boldsymbol{\sigma}(\mathbf{u}), \nabla \mathbf{v})_{\Omega}$$
$$= 2\mu(\boldsymbol{\varepsilon}(\mathbf{u}), \boldsymbol{\varepsilon}(\mathbf{v}))_{\Omega} + \lambda(\nabla \cdot \mathbf{u}, \nabla \cdot \mathbf{v})_{\Omega}.$$

The last expression (which follows from the fact that $I : \nabla \mathbf{v} = \nabla \cdot \mathbf{v}$) shows that this bilinear form is symmetric for any choice of the parameters μ, λ. The variational formulation of (4.10), which uses the test space $\mathbf{H}_0^1(\Omega) = H_0^1(\Omega)^d$ is easily obtained using a density argument.

Proposition 4.9. *The Dirichlet problem for the Navier-Lamé equations* (4.10) *is equivalent to the variational problem*

$$\mathbf{u} \in \mathbf{H}^1(\Omega), \quad \gamma \mathbf{u} = \mathbf{g},$$
$$a(\mathbf{u}, \mathbf{v}) = (\mathbf{f}, \mathbf{v})_{\Omega} \qquad \forall \mathbf{v} \in \mathbf{H}_0^1(\Omega).$$

The elastic bilinear form $a : \mathbf{H}^1(\Omega) \times \mathbf{H}^1(\Omega) \to \mathbb{R}$ is clearly bounded. Specifically,

$$\begin{aligned} |a(\mathbf{u}, \mathbf{v})| &= |2\mu(\boldsymbol{\varepsilon}(\mathbf{u}), \boldsymbol{\varepsilon}(\mathbf{v}))_{\Omega} + \lambda(\nabla \cdot \mathbf{u}, \nabla \cdot \mathbf{v})_{\Omega}| \\ &\leq 2|\mu||(\boldsymbol{\varepsilon}(\mathbf{u}), \boldsymbol{\varepsilon}(\mathbf{v}))_{\Omega}| + |\lambda||(\nabla \cdot \mathbf{u}, \nabla \cdot \mathbf{v})_{\Omega}| \\ &\leq 2|\mu| \|\boldsymbol{\varepsilon}(\mathbf{u})\|_{\Omega} \|\boldsymbol{\varepsilon}(\mathbf{v})\|_{\Omega} + |\lambda| \|\nabla \cdot \mathbf{u}\|_{\Omega} \|\nabla \cdot \mathbf{v}\|_{\Omega} \\ &\leq (2|\mu| + |\lambda|) \|\mathbf{u}\|_{1,\Omega} \|\mathbf{v}\|_{1,\Omega}. \end{aligned}$$

The next result gives sufficient conditions on the Lamé parameters to prove coercivity. (An improved estimate is given in Exercise 4.10.)

Proposition 4.10 (Korn's first inequality). *If $\mu > 0$ and $\lambda \geq 0$, then*

$$a(\mathbf{u}, \mathbf{u}) \geq 2\mu \|\boldsymbol{\varepsilon}(\mathbf{u})\|_{\Omega}^2 \geq \mu \|\nabla \mathbf{u}\|_{\Omega}^2 \qquad \forall \mathbf{u} \in \mathbf{H}_0^1(\Omega).$$

Proof. By definition of the bilinear form a, we have

$$a(\mathbf{u}, \mathbf{u}) = 2\mu(\boldsymbol{\varepsilon}(\mathbf{u}), \boldsymbol{\varepsilon}(\mathbf{u}))_{\Omega} + \lambda(\nabla \cdot \mathbf{u}, \nabla \cdot \mathbf{u})_{\Omega} \geq 2\mu \|\boldsymbol{\varepsilon}(\mathbf{u})\|_{\Omega}^2,$$

where the last inequality comes from the fact that $\lambda \|\nabla \cdot \mathbf{u}\|_{\Omega}^2$ is always nonnegative. Using the definition of $\boldsymbol{\varepsilon}$, it is easy to see that

$$(\boldsymbol{\varepsilon}(\mathbf{u}), \boldsymbol{\varepsilon}(\mathbf{u}))_{\Omega} = \frac{1}{2} \sum_{i,j=1}^{d} \left(\frac{\partial u_i}{\partial x_j}, \frac{\partial u_i}{\partial x_j} \right)_{\Omega} + \frac{1}{2} \sum_{i,j=1}^{d} \left(\frac{\partial u_i}{\partial x_j}, \frac{\partial u_j}{\partial x_i} \right)_{\Omega}.$$

If $\mathbf{u} \in \mathcal{D}(\Omega)^d$, then

$$(\partial_{x_j} u_i, \partial_{x_i} u_j)_\Omega = \langle \partial_{x_j} u_i, \partial_{x_i} u_j \rangle = -\langle u_i, \partial_{x_j} \partial_{x_i} u_j \rangle$$
$$= \langle \partial_{x_i} u_i, \partial_{x_j} u_j \rangle = (\partial_{x_i} u_i, \partial_{x_j} u_j)_\Omega,$$

and therefore

$$(\varepsilon(\mathbf{u}), \varepsilon(\mathbf{u}))_\Omega = \tfrac{1}{2}(\nabla \mathbf{u}, \nabla \mathbf{u})_\Omega + \tfrac{1}{2}(\nabla \cdot \mathbf{u}, \nabla \cdot \mathbf{u})_\Omega \geq \tfrac{1}{2}\|\nabla \mathbf{u}\|_\Omega^2.$$

With this, the result follows in $\mathcal{D}(\Omega)^d$. To obtain the result in the full space, we take $\mathbf{u} \in \mathbf{H}_0^1(\Omega)$ and a sequence $\{\mathbf{u}_n\}$ in $\mathcal{D}(\Omega)^d$ such that $\mathbf{u}_n \to \mathbf{u}$ in $\mathbf{H}^1(\Omega)$. The result then follows since

$$a(\mathbf{u}, \mathbf{u}) \longleftarrow a(\mathbf{u}_n, \mathbf{u}_n) \geq \mu\|\nabla \mathbf{u}_n\|_\Omega^2 \longrightarrow \mu\|\nabla \mathbf{u}\|_\Omega^2,$$

where we have used the density of $\mathcal{D}(\Omega)^d$ in $\mathbf{H}_0^1(\Omega)$ and the continuity of both the bilinear form a and the norm.   □

The result of the above proposition does not entirely give the coercivity of the bilinear form. We need to apply the Poincaré-Friedrichs inequality to the right-hand side to obtain

$$a(\mathbf{u}, \mathbf{u}) \geq \mu C^2 \|\mathbf{u}\|_{1,\Omega}^2, \tag{4.11}$$

where C is a constant depending only on the domain Ω.

Well-posedness and a minimization problem. With $\mu, \lambda > 0$

$$\|\mathbf{u}\|_{1,\Omega} \leq C(2\mu + \lambda + 1)\mu^{-1}(\|\mathbf{f}\|_\Omega + \|\mathbf{g}\|_{1/2,\Gamma}), \tag{4.12}$$

where once again the constant C depends only on the domain. We arrive at this bound in a similar manner as the stability bound achieved in Section 4.4. First we choose $\mathbf{u_g} \in \mathbf{H}^1(\Omega)$ such that $\gamma \mathbf{u_g} = \mathbf{g}$. Next we solve the problem

$$\mathbf{u}_0 \in \mathbf{H}_0^1(\Omega),$$
$$a(\mathbf{u}_0, \mathbf{v}) = (\mathbf{f}, \mathbf{v})_\Omega - a(\mathbf{u_g}, \mathbf{v}) \quad \forall \mathbf{v} \in \mathbf{H}_0^1(\Omega).$$

Defining $\ell(\mathbf{v}) := (\mathbf{f}, \mathbf{v})_\Omega - a(\mathbf{u_g}, \mathbf{v})$, we can bound

$$|\ell(\mathbf{v})| \leq \|\mathbf{f}\|_\Omega \|\mathbf{v}\|_\Omega + (2\mu + \lambda)\|\mathbf{u_g}\|_{1,\Omega}\|\mathbf{v}\|_{1,\Omega}$$
$$\leq (\|\mathbf{f}\|_\Omega + (2\mu + \lambda)\|\mathbf{u_g}\|_{1,\Omega})\|\mathbf{v}\|_{1,\Omega}.$$

Combining this with the coercivity result (4.11), now in terms of $\mathbf{u}_0$, we obtain

$$\mu C^2\|\mathbf{u}_0\|_{1,\Omega}^2 \leq |a(\mathbf{u}_0, \mathbf{u}_0)| \leq |\ell(\mathbf{u}_0)| \leq (\|\mathbf{f}\|_\Omega + (2\mu + \lambda)\|\mathbf{u_g}\|_{1,\Omega})\|\mathbf{u}_0\|_{1,\Omega},$$

which implies

$$\|\mathbf{u}_0\|_{1,\Omega} \leq C^{-2}\mu^{-1}(\|\mathbf{f}\|_\Omega + (2\mu + \lambda)\|\mathbf{u_g}\|_{1,\Omega}). \tag{4.13}$$

To arrive at the above bound on $\|\mathbf{u}\|_{1,\Omega}$ we first set $\|\mathbf{u}\|_{1,\Omega} = \|\mathbf{u}_0 + \mathbf{u_g}\|_{1,\Omega}$. Now, using (4.13), the definition

$$\|\mathbf{g}\|_{1/2,\Gamma} = \inf\{\|\mathbf{u_g}\|_{1,\Omega} : \gamma\mathbf{u_g} = \mathbf{g}\},$$

and the fact that we are free to use any $\mathbf{u_g}$ which satisfies the boundary condition, we have

$$\|\mathbf{u}\|_{1,\Omega} \leq C\mu^{-1}(\|\mathbf{f}\|_\Omega + (2\mu + \lambda)\|\mathbf{g}\|_{1/2,\Gamma}) + \|\mathbf{g}\|_{1/2,\Gamma},$$

where we are abusing notation for the constant C. From this, the bound in (4.12) follows. The equivalent minimization problem associated to the Navier-Lamé equations is

$$\tfrac{1}{2}(\boldsymbol{\sigma}(\mathbf{u}), \boldsymbol{\varepsilon}(\mathbf{u}))_\Omega - (\mathbf{f}, \mathbf{u})_\Omega = \min! \qquad \mathbf{u} \in \mathbf{H}^1(\Omega), \qquad \gamma\mathbf{u} = \mathbf{g}.$$

To see that this is the correct minimization problem, we need only appeal to Lemma 4.3.

Final comments and literature

The dense inclusion of $\mathcal{C}^1(\overline{\Omega})$ into $H^1(\Omega)$, which holds for any bounded domain with the H^1 extension property, can be used to give an alternative definition of the Sobolev spaces, by completion of $\mathcal{C}^1(\overline{\Omega})$ with respect to the $H^1(\Omega)$ norm. It is actually customary to start with the space

$$\{u \in \mathcal{C}(\overline{\Omega}) : \partial_{x_i} u \in \mathcal{C}(\overline{\Omega})\},$$

which contains $\mathcal{C}^1(\overline{\Omega})$, and take the completion with respect to $\|\cdot\|_{1,\Omega}$. The spaces defined in this form, on bounded domains, are called the Beppo Levi spaces. (Incidentally, Beppo Levi seems to be one of the few mathematicians whose full name, as opposed to only his family name, is attached to his creations.) It is clear from what we have said, that for bounded Lipschitz domains, the Beppo Levi space associated to the $\|\cdot\|_{1,\Omega}$ norm is just the Sobolev space $H^1(\Omega)$. Defining spaces in the Beppo Levi form has several advantages from the point of view of the easiness of the proofs (Nečas's monograph [87] first works out all the theory of the Beppo Levi spaces before approaching Sobolev spaces), but it has the relative disadvantage of defining the space through completion, which means that we do not really know what the elements of the space are, but only that they can be approximated by elements we understand. The reader might think that this is a moot point, but the fact remains that the space defined by the completion of $\mathcal{D}(\mathbb{R}^2)$ with respect to the Dirichlet norm $\|\nabla u\|_{\mathbb{R}^2}$ seems to contain elements that cannot be understood as distributions (see [46, Section 4] and [51, Chapter II] for more details).

We meet a similar problem with our definition of the space $H^{1/2}(\Gamma)$ as the range of the trace operator. Given in this way, it is unclear whether a given function on Γ belongs to the trace space or not. For instance, if Γ is the boundary of a polyhedron and $g \equiv 1$ in one face of Γ, while $g \equiv 0$ everywhere else, it can be shown that $g \notin H^{1/2}(\Gamma)$. However, this is difficult to prove with our definition of the trace space. The way around this involves showing that the Sobolev-Slobodeckij norm (sometimes called the Aronszajn-Slobodeckij norm)

$$\left(\|g\|_\Gamma^2 + \int_\Gamma \int_\Gamma \frac{|g(\mathbf{x}) - g(\mathbf{y})|^2}{|\mathbf{x} - \mathbf{y}|^d} d\Gamma(\mathbf{x}) d\Gamma(\mathbf{y}) \right)^{1/2} \tag{4.14}$$

is an equivalent norm in $H^{1/2}(\Gamma)$ and that the above norm computed on the characteristic function of a face of a polyhedron is unbounded. Alternatively, we would have to show that if we consider the space of functions $g \in L^2(\Gamma)$ such that the norm is the one defined in (4.14), then the trace operator is bounded and surjective on this space. We will briefly revisit this topic in Chapter 13, where we relate Sobolev norms with norms defined with Fourier transforms and Sobolev-Aronszajn-Slobodeckij norms.

Exercises

4.1. If Ω is a Lipschitz domain, show that $\mathbb{R}^d \backslash \overline{\Omega}$ also satisfies the H^1-extension property.

4.2. The extension theorem in half space . Let $h : \mathbb{R} \to \mathbb{R}$ be a smooth version of the Heaviside function

$$h \in \mathcal{C}^\infty(\mathbb{R}), \quad 0 \leq h \leq 1, \quad \operatorname{supp} h = [0, \infty), \quad \operatorname{supp}(1 - h) = (-\infty, 1],$$

and let $h_n(\mathbf{x}) := h(n\, x_d - 1)$. We will write

$$\mathbb{R}^d \ni \mathbf{x} = (\widetilde{\mathbf{x}}, x_d) \longmapsto \check{\mathbf{x}} := (\widetilde{\mathbf{x}}, -x_d),$$

and consider the extension operator for functions $u : \mathbb{R}_+^d \to \mathbb{R}$,

$$(Eu)(\mathbf{x}) := \begin{cases} u(\mathbf{x}), & \text{if } \mathbf{x} \in \mathbb{R}_+^d, \\ u(\check{\mathbf{x}}), & \text{if } x_d < 0. \end{cases}$$

(a) Make a plot of the functions h_n and show that

$$h_n \varphi \in \mathcal{D}(\mathbb{R}_+^d), \qquad h_n \varphi \to \varphi \text{ in } L^2(\mathbb{R}_+^d) \quad \forall \varphi \in \mathcal{D}(\mathbb{R}^d).$$

(b) Show that if $u \in L^2(\mathbb{R}_+^d)$, then

$$\langle Eu, \varphi \rangle = \int_{\mathbb{R}_+^d} u(\mathbf{x})\big(\varphi(\mathbf{x}) + \varphi(\check{\mathbf{x}})\big) d\mathbf{x} \qquad \forall \varphi \in \mathcal{D}(\mathbb{R}^d).$$

(c) By carefully playing with the functions h_n, show that if $u \in H^1(\mathbb{R}_+^d)$, then

$$\partial_{x_j}(Eu) = E(\partial_{x_j}u) \qquad 1 \leq j \leq d-1.$$

(d) Show that

$$(\partial_{x_d}h_n)(\varphi - \varphi(\dot{}\,)) \to 0 \text{ in } L^2(\mathbb{R}_+^d) \quad \forall \varphi \in \mathcal{D}(\mathbb{R}^d).$$

(e) Finally, using (a) and (d), show that if $u \in H^1(\mathbb{R}_+^d)$, then

$$\langle \partial_{x_d}(Eu), \varphi \rangle = \int_{\mathbb{R}_+^d} \partial_{x_d}u(\mathbf{x})\big(\varphi(\mathbf{x}) - \varphi(\check{\mathbf{x}})\big)\,\mathrm{d}\mathbf{x} \qquad \forall \varphi \in \mathcal{D}(\mathbb{R}^d),$$

i.e.,

$$(\partial_{x_d}Eu)(\mathbf{x}) = \begin{cases} (\partial_{x_d}u)(\mathbf{x}) & \mathbf{x} \in \mathbb{R}_+^d, \\ -(\partial_{x_d}u)(\check{\mathbf{x}}) & \mathbf{x} \in \mathbb{R}_-^d. \end{cases}$$

The previous results show that if $u \in H^1(\mathbb{R}_+^d)$, then $Eu \in H^1(\mathbb{R}^d)$, and therefore the upper half space $\mathbb{R}_+^d$ has the H^1-extension property. Why?
Remark: In all of the arguments above you are not allowed to use 'integration by parts,' but you can use that the functions $h_n (\varphi + \varphi(\dot{}\,))$ and $h_n (\varphi - \varphi(\dot{}\,))$ are in $\mathcal{D}(\mathbb{R}_+^d)$.

4.3. A step necessary to show that $\ker \gamma = H_0^1(\Omega)$. The goal of this exercise given $u \in H^1(\mho+)$ such that $\gamma_{\Gamma_0}u = 0$ and $u \equiv 0$ in a neighborhood of $\partial\mho^+ \setminus \Gamma_0$, is to show that $u \in H_0^1(\mho^+)$. This result was used in the proof of Proposition 4.5.

(a) Show that there exists a sequence $\{u_n\}$ in $\mathcal{C}^1(\overline{\mho^+})$ such that $u_n \equiv 0$ in a neighborhood of $\partial\mho^+ \setminus \Gamma_0$, $u_n \to u$ in $H^1(\mho^+)$, and $\|u_n\|_{\Gamma_0} \leq \frac{1}{n}$.

(b) Show that if $\mathbf{x} = (\tilde{\mathbf{x}}, x_d)$ with $x_d \in (0, 2/n)$, then

$$|u_n(\mathbf{x})|^2 \leq 2|u_n(\tilde{\mathbf{x}}, 0)|^2 + 2x_d \int_0^{2/n} |\partial_{x_d}u_n(\tilde{\mathbf{x}}, t)|^2 \, \mathrm{d}t,$$

and therefore

$$\int_{B_{d-1}(0;1)\times\left(\frac{1}{n}, \frac{2}{n}\right)} |u_n(\mathbf{x})|^2 \, \mathrm{d}\mathbf{x} \leq \frac{2}{n^3} + \frac{3}{n^2}\|\nabla u_n\|_{B_{d-1}(0;1)\times(0,2/n)}^2.$$

(c) Now let

$$w_n(\mathbf{x}) := h(nx_d - 1)u_n(\mathbf{x}),$$

where the function h is the smoothened Heaviside function which has been used before. Prove that $w_n \in H_0^1(\mho^+)$ for each n, and $w_n \to u$ in $H^1(\mho^+)$. (**Hint.** Use (b) to estimate the limit in $L^2(\mho^+)$ of $v_n(\mathbf{x}) := nh'(nx_d - 1)u_n(\mathbf{x})$.)

4.4. Assume that $\partial\Omega$ is composed of two disjoint connected parts, Γ_1 and Γ_2, each of them the boundary of a Lipschitz domain (think of an annular domain). Show that

$$H^{1/2}(\Gamma) \equiv H^{1/2}(\Gamma_1) \times H^{1/2}(\Gamma_2).$$

(**Hint.** Use $\varphi_1, \varphi_2 \in \mathcal{D}(\mathbb{R}^d)$ such that $\varphi_1 + \varphi_2 \equiv 1$ in a neighborhood of Ω and such that

$$\operatorname{supp}\varphi_2 \cap \Gamma_1 = \emptyset \quad \text{and} \quad \operatorname{supp}\varphi_1 \cap \Gamma_2 = \emptyset,$$

to separate the boundaries.)

4.5. **The trace operator on part of the boundary.** Let Ω be a Lipschitz domain and $\Gamma_{\mathrm{pc}} \subset \partial\Omega$ a subset of its boundary such that it is possible to integrate on it. Consider the operator $\gamma_{\mathrm{pc}} : H^1(\Omega) \to L^2(\Gamma_{\mathrm{pc}})$ given by

$$\gamma_{\mathrm{pc}} u := (\gamma u)|_{\Gamma_{\mathrm{pc}}}.$$

Show that this operator is the only possible extension of the operator

$$
\begin{array}{ccc}
H^1(\Omega) \cap \mathcal{C}(\overline{\Omega}) & \longrightarrow & L^2(\Gamma_{\mathrm{pc}}) \\
u & \longmapsto & u|_{\Gamma_{\mathrm{pc}}}.
\end{array}
$$

(Note that the restriction operators in the previous formulas are different to each other. Why?)

4.6. **The trace from an exterior domain.** Let Ω_- be a bounded Lipschitz domain and $\Omega_+ := \mathbb{R}^d \setminus \overline{\Omega_-}$. Since both $\Omega_\pm$ satisfy the extension property, we can define different trace operators

$$\gamma^\pm : H^1(\Omega_\pm) \to L^2(\Gamma).$$

(a) Show that if $u \in H^1(\mathbb{R}^d)$, then $\gamma^+ u = \gamma^- u$.

(b) Show that the ranges of both trace operators are the same.

(c) Show that if $u \in H^1(\mathbb{R}^d \setminus \Gamma)$ and $\gamma^+ u = \gamma^- u$, then $u \in H^1(\mathbb{R}^d)$. (**Hint.** Let $u_\pm := u|_{\Omega_\pm}$. Extend u_+ to an element of $H^1(\mathbb{R}^d)$ and show that this extension minus u_- is in $H_0^1(\Omega_-)$.)

4.7. **Reaction-diffusion problems.** On a bounded Lipschitz domain, we consider two coefficients

$$\kappa, c \in L^\infty(\Omega), \qquad \kappa \geq \kappa_0 > 0, \qquad c \geq 0 \qquad \text{(almost everywhere)}$$

and two data functions $(f, g) \in L^2(\Omega) \times H^{1/2}(\Gamma)$. Consider the problem

$$u \in H^1(\Omega), \qquad \gamma u = g, \qquad -\operatorname{div}(\kappa \nabla u) + c u = f.$$

(a) Write its equivalent variational formulation and the associated minimization problem.

(b) Show the well-posedness of this problem.

4.8. The optimal lifting. Consider the operator $\gamma^\dagger : H^{1/2}(\Gamma) \to H^1(\Omega)$, given by $u = \gamma^\dagger g$ where u is the solution of

$$u \in H^1(\Omega), \qquad \gamma u = g, \qquad -\Delta u + u = 0 \quad \text{in } \Omega.$$

Show that it is well defined, linear, and bounded. Write the associated minimization problem and show that $\gamma^\dagger$ is the Moore-Penrose pseudoinverse of the trace $\gamma : H^1(\Omega) \to H^{1/2}(\Gamma)$.

4.9. An isomorphism related to the Dirichlet problem. Show that the map

$$H^1(\Omega) \ni u \longmapsto (\Delta u, \gamma u) \in H^{-1}(\Omega) \times H^{1/2}(\Gamma)$$

is an isomorphism.

4.10. More on the elastic bilinear form. With the notation of Section 4.6, we want to prove that if $\mu > 0$ and $(d+1)\mu + d\lambda > 0$, then

$$(\sigma(\mathbf{u}), \varepsilon(\mathbf{u}))_\Omega \geq C \|\nabla \mathbf{u}\|_\Omega^2 \qquad \forall \mathbf{u} \in \mathbf{H}_0^1(\Omega).$$

Proceed as follows. Consider the operator $T : \mathbb{R}^{d \times d} \to \mathbb{R}^{d \times d}$ given by $T(A) = \mu A + (\lambda + \mu)(A : \mathrm{I})\mathrm{I}$.

(a) Show that T is self-adjoint and that its only eigenvalues are μ and $\mu + d(\lambda + \mu)$.

(b) Show that

$$(\sigma(\mathbf{u}), \varepsilon(\mathbf{u}))_\Omega = (T(\nabla \mathbf{u}), \nabla \mathbf{u})_\Omega \qquad \forall \mathbf{u} \in \mathbf{H}_0^1(\Omega).$$

(c) Use the spectral theorem applied to T to show the ellipticity property with $C = \min\{\mu, (d+1)\mu + d\lambda\}$.

4.11. The space $H^1(a,b)$. In this problem we show that one-dimensional Sobolev spaces (when defined on intervals) are actually very simple. At the end of the problem we will have proved that $H^1(a,b) \subset C[a,b]$ with bounded injection and that $H^1(a,b)$ is a Banach algebra. Note first that a bounded interval (a,b) is a Lipschitz domain and therefore the H^1 extension property holds in (a,b) and therefore $C^\infty[a,b]$ is dense in $H^1(a,b)$.

(a) Using the extension operator from Exercise 4.2 (from $(0,\infty)$ to $\mathbb{R}$), build a bounded extension operator from $H^1(a,b)$ to $H^1(\mathbb{R})$. (**Hint.** Use cutoff functions to separate the two parts of the boundary.)

(b) Prove that for every $u \in C^1[a,b]$

$$|u(x) - u(y)|^2 \leq |x - y| \|u'\|_{(a,b)}^2 \qquad \forall x, y \in (a,b)$$

and

$$|u(x)|^2 \leq 2(b-a)\|u'\|_{(a,b)}^2 + \frac{2}{b-a}\|u\|_{(a,b)}^2 \qquad \forall x \in (a,b).$$

(c) Show that $H^1(a, b) \subset C[a, b]$ with continuous embedding.

(d) Prove that there exists C such that for every $u, v \in C^1[a, b]$

$$\|(u\,v)'\|_{(a,b)} \leq C\|u\|_{1,(a,b)}\|v\|_{1,(a,b)}.$$

(e) Show that if $u, v \in H^1(a, b)$, then $u\,v \in H^1(a, b)$ and that the product of functions is a bounded bilinear operator in $H^1(a, b)$. (This makes $H^1(a, b)$ a Banach algebra.)

Note that (b) can also be used to prove that $H^1(a, b)$ is continuously embedded in the Hölder space

$$C^{0,1/2}[a, b] := \left\{ u \in C[a, b] : \sup_{x \neq y \in [a,b]} \frac{|u(x) - u(y)|}{|x - y|^{1/2}} < \infty \right\}.$$

4.12. An extension operator for $H^2(\mathbb{R}^d_+)$. Given $u \in H^2(\mathbb{R}^d_+)$ (see Exercise 2.3), we define

$$(Eu)(\mathbf{x}) = (Eu)(\tilde{\mathbf{x}}, x_d) := \begin{cases} u(\tilde{\mathbf{x}}, x_d) & \text{if } x_d > 0, \\ 4u(\tilde{\mathbf{x}}, -\tfrac{1}{2}x_d) - 3u(\tilde{\mathbf{x}}, -\tfrac{1}{3}x_d), & \text{if } x_d < 0. \end{cases}$$

(a) Show that $Eu \in H^2(\mathbb{R}^d)$.

(b) Show that $\|Eu\|_{\mathbb{R}^d} \leq C_0\|u\|_{\mathbb{R}^d_+}$ for all $u \in L^2(\mathbb{R}^d_+)$.

(c) Show that $\|Eu\|_{1,\mathbb{R}^d} \leq C_1\|u\|_{1,\mathbb{R}^d_+}$ for all $u \in H^1(\mathbb{R}^d_+)$.

(d) Show that $\|Eu\|_{2,\mathbb{R}^d} \leq C_2\|u\|_{2,\mathbb{R}^d_+}$ for all $u \in H^2(\mathbb{R}^d_+)$.

5

Nonsymmetric and complex problems

5.1   The Lax-Milgram lemma 89
5.2   Convection-diffusion equations 93
5.3   Complex and complexified spaces 95
5.4   The Laplace resolvent equations 98
5.5   The Ritz-Galerkin projection (*) 101
Final comments and literature 103
Exercises 103

In this section we will deal with boundary value problems associated to non-symmetric bilinear forms, like the convection-diffusion equation

$$-\Delta u + \mathbf{b} \cdot \nabla u = f.$$

This will require proving a simple generalization of the Riesz-Fréchet representation theorem, dealing with nonsymmetric bounded and coercive bilinear forms. We will next extend our toolbox to complex vector spaces, working on the complexification of the Sobolev spaces we have defined in previous chapters.

5.1   The Lax-Milgram lemma

The problems we will be interested in studying in this section will have the form of

$$u \in V,$$
$$a(u, v) = \ell(v) \quad \forall v \in V,$$

where, as usual, V is a Hilbert space and $\ell \in V'$ is linear and bounded. The novelty will be in the assumption that $a : V \times V \to \mathbb{R}$ is bilinear bounded, and coercive, but not symmetric. Recall that when we say that a is bounded, we mean that there is a positive constant M, such that

$$|a(u, v)| \leq M \|u\|_V \|v\|_V \quad \forall u, v \in V.$$

Bilinear forms and operators. We can rewrite the bilinear form a, fixing one component, to be a map

$$V \ni u \longmapsto \mathcal{A}u := a(u, \cdot) \in V',$$

which is clearly linear. Since

$$\|\mathcal{A}u\|_{V'} = \sup_{0 \neq v \in V} \frac{|a(u, v)|}{\|v\|_V} \leq M\|u\|_V \qquad \forall u \in V,$$

the operator $\mathcal{A} : V \to V'$ is bounded with

$$\|\mathcal{A}\|_{V \to V'} \leq M.$$

We can then express the equation $a(u, v) = \ell(v)$ as an equation in V':

$$\mathcal{A}u = \ell. \tag{5.1}$$

Our next step will be to rewrite (5.1) as an operator equation in V, not in V'. To do that let us first identify $\mathcal{A}u \in V'$ with an element $Au \in V$ via the Riesz-Fréchet representation theorem

$$\mathcal{A}u = (Au, \cdot)_V = a(u, \cdot),$$

and let us do the same for the right-hand side

$$f_\ell \in V, \qquad \ell = (f_\ell, \cdot)_V.$$

Therefore, (5.1) is equivalent to

$$Au = f_\ell. \tag{5.2}$$

Note that $A : V \to V$ is the composition of $\mathcal{A} : V \to V'$ with the Riesz-Fréchet representation operator $V' \to V$, which is an isometric isomorphism. Therefore

$$\|Au\|_V = \|\mathcal{A}u\|_{V'} \leq M\|u\|_V \qquad \forall u \in V.$$

We have thus moved from a variational problem to an operator equation in V. Coercivity is the missing ingredient that will allow us to prove invertibility of (5.2).

Proposition 5.1 (Lax-Milgram lemma). *If V is a Hilbert space, $a : V \times V \to \mathbb{R}$ is bilinear, bounded, and coercive, i.e., there is an $\alpha > 0$ such that*

$$a(u, u) \geq \alpha\|u\|_V^2 \quad \forall u \in V,$$

then the operator $A : V \to V$ defined by

$$(Au, v)_V = a(u, v) \qquad \forall u, v \in V$$

is invertible and

$$\|A^{-1}\|_{V \to V} \leq 1/\alpha.$$

Therefore, for every $\ell \in V'$, the variational problem

$$u \in V, \tag{5.3a}$$

$$a(u, v) = \ell(v) \quad \forall v \in V, \tag{5.3b}$$

has a unique solution satisfying

$$\|u\|_V \leq (1/\alpha)\|\ell\|_{V'}$$

and the solution operator $\ell \mapsto u$ is linear and bounded.

Proof. We have already shown that A is linear and bounded. The coercivity of a implies that A is injective. Indeed

$$\|Au\|_V \|u\|_V \geq (Au, u)_V = a(u, u) \geq \alpha\|u\|_V^2,$$

which shows that

$$\|Au\|_V \geq \alpha\|u\|_V. \tag{5.4}$$

Hence if $Au = 0$, we must have $u = 0$, and therefore A is injective.

Next we show that the range of A is closed. To see this, we choose a sequence in the range of A, $\{Au_n\}$, and assume that $Au_n \to w$. This implies that $\{Au_n\}$ is a Cauchy sequence, and therefore $\{u_n\}$ is a Cauchy sequence by (5.4). Now we have that there exists $u \in V$ such that $u_n \to u$, and from the boundedness of A it follows that $Au_n \to Au$. Thus $w = Au \in \text{range } A$ and range A is closed.

Since range A is closed, we can decompose

$$V = \text{range } A \oplus (\text{range } A)^\perp.$$

We will finally show that the coercivity hypothesis implies that $(\text{range } A)^\perp = \{0\}$, and therefore $A : V \to V$ is bijective. To show this, let $u \in (\text{range } A)^\perp$, and note that by coercivity

$$0 = (Au, u)_V = a(u, u) \geq \alpha\|u\|_V^2,$$

hence u must be zero. Since $A : V \to V$ is linear, bounded, and invertible, by the Banach isomorphism theorem $A^{-1} : V \to V$ is bounded. Actually, we can obtain a bound for the inverse of A using (5.4), taking $w = A^{-1}u$, so that

$$\|w\|_V = \|AA^{-1}w\|_V \geq \alpha\|A^{-1}w\|_V,$$

which proves the bound for the inverse. The final part of the statement of the proposition, leading with the variational problem (5.3) is a straightforward consequence of the possibility of writing this problem in equivalent form as $Au = f_\ell$, noticing that

$$\alpha\|u\|_V \leq \|f_\ell\|_V = \|\ell\|_{V'},$$

which proves the result. $\qquad\square$

A slight modification to the Lax-Milgram lemma. When dealing with nonhomogeneous Dirichlet conditions, the Lax-Milgram lemma has to be slightly modified as follows.

Proposition 5.2 (Generalized Lax-Milgram lemma). *Let V and M be Hilbert spaces, $a : V \times V \to \mathbb{R}$ a bounded bilinear form, and $\gamma : V \to M$ a linear, bounded, and surjective map with $\ker \gamma = V_0$. Also assume that the bilinear form a is coercive in the kernel of γ, i.e., there is $\alpha > 0$ such that*

$$a(u, u) \geq \alpha \|u\|_V^2 \quad \forall u \in V_0.$$

Given data $(\ell, g) \in V' \times M$, we consider the problem

$$u \in V, \tag{5.5a}$$
$$\gamma u = g, \tag{5.5b}$$
$$a(u, v) = \ell(v) \quad \forall v \in V_0. \tag{5.5c}$$

With the above hypotheses, we have:

  (a) the problem has a unique solution u,

  (b) $\|u\|_V \leq C(\|\ell\|_{V'} + \|g\|_M)$,

  (c) the solution map $(\ell, g) \mapsto u$ is linear and bounded.

Proof. Since $\gamma : V \to M$ is bounded and surjective, there is a bounded right inverse $\gamma^\dagger : M \to V$ such that $\gamma\gamma^\dagger g = g$ for all g in M and

$$\|\gamma^\dagger g\|_V \leq \|\gamma^\dagger\|_{M \to V} \|g\|_M \quad \forall g \in M.$$

We set $u_g = \gamma^\dagger g$ and then consider the unknown $u_0 = u - u_g \in V_0$ that satisfies

$$u_0 \in V_0,$$
$$a(u_0, v) = \ell(v) - a(u_g, v) \quad \forall v \in V_0.$$

By the Lax-Milgram lemma, there is a unique solution u_0 to this problem and

$$\|u_0\|_V \leq \frac{1}{\alpha}\left(\|\ell - a(u_g, \cdot)\|_{V'}\right) \leq \frac{1}{\alpha}\|\ell\|_{V'} + \frac{M}{\alpha}\|u_g\|_V$$
$$\leq \frac{1}{\alpha}\|\ell\|_{V'} + \frac{M}{\alpha}\|\gamma^\dagger\|_{M \to V}\|g\|_M.$$

What remains is relatively simple. The quantity $u = u_0 + u_g$ satisfies the original variational problem but we still need to show that it is unique. To prove uniqueness, consider two solutions u_1 and u_2 to the problem. If we let $w = u_1 - u_2$, then due to the linearity of a, we have that w satisfies

$$w \in V_0, \qquad a(w, v) = 0 \quad \forall v \in V_0.$$

We then need only appeal to the Lax-Milgram lemma once more to see that $\|w\|_V = 0$ and therefore $u_1 = u_2$. Now that we have established the unique solvability of (5.5) we will show that the problem is well posed by showing the bound on the solution (the solution depends continuously on the data). Using the definition of u and the the the bound we achieved on $\|u_0\|_V$, we easily obtain

$$\|u\|_V = \|u_0 + u_g\|_V \leq \frac{1}{\alpha}\|\ell\|_{V'} + \left(\frac{M}{\alpha} + 1\right)\|\gamma^\dagger\|_{M \to V}\|g\|_M,$$

from which (b) follows. Finally, we need to show that the solution operator is linear, which is a simple consequence of the superposition principle. Given two sets of (ℓ_1, g_1) and (ℓ_2, g_2) for (5.5) with solutions u_1 and u_2 respectively, and two constants $c_1, c_2 \in \mathbb{R}$, we want to show that $c_1 u_1 + c_2 u_2$ solves (5.5) with given data $(c_1\ell_1 + c_2\ell_2, c_1 g_1 + c_2 g_2)$. This follows from the linearity of a, ℓ, and γ. The bound on the solution operator comes from part (b). $\qquad \square$

5.2   Convection-diffusion equations

We now return to the context of having a bounded open Lipschitz domain Ω and consider a general PDE with convection, reaction, and diffusion terms given by

$$u \in H^1(\Omega), \qquad \gamma u = g, \tag{5.6a}$$
$$-\operatorname{div}(\kappa \nabla u) + \mathbf{b} \cdot \nabla u + cu = f, \tag{5.6b}$$

where we take as data for the problem $f \in L^2(\Omega)$ and $g \in H^{1/2}(\Gamma)$ and the coefficients satisfy:

$$\kappa, c \in L^\infty(\Omega), \qquad \kappa \geq \kappa_0 > 0 \quad \text{a.e.}, \qquad \mathbf{b} \in L^\infty(\Omega; \mathbb{R}^d).$$

We will add some further hypotheses on the convection and reaction coefficients ($\mathbf{b}$ and c respectively) to ensure well-posedness of (5.6). Equation (5.6b) is actually a convection-diffusion-reaction equation.

Variational formulation Moving towards the variational formulation, just realize that (5.6b) is equivalent to

$$-\sum_{j=1}^{d}\langle \partial_{x_j}(\kappa \partial_{x_j} u), \varphi\rangle + \sum_{j=1}^{d}\langle b_j \partial_{x_j} u, \varphi\rangle + \langle cu, \varphi\rangle = \langle f, \varphi\rangle \quad \forall \varphi \in \mathcal{D}(\Omega),$$

which is equivalent to

$$\sum_{j=1}^{d}\langle \kappa \partial_{x_j} u, \partial_{x_j}\varphi\rangle + \sum_{j=1}^{d}\langle b_j \partial_{x_j} u, \varphi\rangle + \langle c\, u, \varphi\rangle = \langle f, \varphi\rangle \quad \forall \varphi \in \mathcal{D}(\Omega).$$

Using the regularity hypotheses on the coefficients, data, and solution, the duality brackets in the last equality become $L^2(\Omega)$ inner products leading to

$$(\kappa \nabla u, \nabla \varphi)_\Omega + (\mathbf{b} \cdot \nabla u, \varphi)_\Omega + (cu, \varphi)_\Omega = (f, \varphi)_\Omega \qquad \forall \varphi \in \mathcal{D}(\Omega).$$

Setting $a(u, v) := (\kappa \nabla u, \nabla v)_\Omega + (\mathbf{b} \cdot \nabla u, v)_\Omega + (cu, v)_\Omega$ and $\ell(v) := (f, v)_\Omega$, we can begin to verify the hypotheses of the Lax-Milgram lemma. First, the bilinear form a is bounded:

$$|a(u, v)| \leq \|\kappa\|_\infty \|\nabla u\|_\Omega \|\nabla v\|_\Omega + \|\mathbf{b}\|_\infty \|\nabla u\|_\Omega \|v\|_\Omega + \|c\|_\infty \|u\|_\Omega \|v\|_\Omega$$
$$\leq (\|\kappa\|_\infty + \|\mathbf{b}\|_\infty + \|c\|_\infty) \|u\|_{1,\Omega} \|v\|_{1,\Omega}.$$

The right-hand side $\ell(v)$ clearly satisfies $|\ell(v)| \leq \|f\|_\Omega \|v\|_{1,\Omega}$ for all $v \in H^1(\Omega)$. To summarize what we have thus far, if $u \in H^1(\Omega)$ and we have the appropriate hypotheses on coefficients and f, we have shown that, neglecting boundary conditions, the distributional PDE (5.6b) is satisfied if and only if

$$a(u, v) = \ell(v) \quad \forall v \in H_0^1(\Omega),$$

by the density of $\mathcal{D}(\Omega)$ in $H_0^1(\Omega)$. As stated before, we have not yet included the boundary conditions in the variational formulation since the functional $\ell(v)$ does not contain any boundary data. To remedy this, we augment our variational formulation to be

$$u \in H^1(\Omega), \qquad \gamma u = g, \tag{5.7a}$$
$$a(u, v) = \ell(v) \quad \forall v \in H_0^1(\Omega). \tag{5.7b}$$

Coercivity. What remains in order to be able to invoke the Lax-Milgram lemma (or more specifically our slightly more general version) is to show the coercivity of a in the kernel of γ, that is, in $H_0^1(\Omega)$. The Dirichlet form term $(\kappa \nabla u, \nabla v)_\Omega$ can be handled by assuming that $\kappa \geq \kappa_0 > 0$ almost everywhere in Ω. We now need to study the lower order terms

$$(\mathbf{b} \cdot \nabla u, v)_\Omega \quad \text{and} \quad (cu, v)_\Omega$$

to see what conditions are needed to make these terms nonnegative. Considering the convective term first, we see

$$(\mathbf{b} \cdot \nabla \varphi, \varphi)_\Omega = \sum_j \int_\Omega b_j(\mathbf{x}) \partial_{x_j} \varphi(\mathbf{x}) \varphi(\mathbf{x}) \, d\mathbf{x} = \sum_{j=1}^d \int_\Omega b_j(\mathbf{x}) \partial_{x_j} \left(\tfrac{1}{2} \varphi^2(\mathbf{x}) \right) d\mathbf{x}$$
$$= \tfrac{1}{2} \sum_{j=1}^d \langle b_j, \partial_{x_j} \varphi^2 \rangle = -\tfrac{1}{2} \sum_{j=1}^d \langle \partial_{x_j} b_j, \varphi^2 \rangle \qquad \forall \varphi \in \mathcal{D}(\Omega).$$

Combining this computation with the additional term $(c\varphi, \varphi)_\Omega = \langle c, \varphi^2 \rangle$, we see that if we require

$$\langle -\tfrac{1}{2} \nabla \cdot \mathbf{b} + c, \varphi \rangle \geq 0 \qquad \forall \varphi \in \mathcal{D}(\Omega) \text{ such that } \varphi \geq 0,$$

which can also be written (see Exercise 5.1) as

$$c - \tfrac{1}{2}\nabla \cdot \mathbf{b} \geq 0 \quad \text{in } \mathcal{D}'(\Omega),$$

then

$$(\mathbf{b} \cdot \nabla\varphi, \varphi)_\Omega + (c\varphi, \varphi)_\Omega \geq 0 \quad \forall \varphi \in \mathcal{D}(\Omega),$$

or equivalently (by density)

$$(\mathbf{b} \cdot \nabla u, u)_\Omega + (cu, u)_\Omega \geq 0 \quad \forall u \in H_0^1(\Omega),$$

and we can ensure the coercivity of a in the kernel of γ. Before we continue, we should remark that if we assume $\mathbf{b} \in L^\infty(\Omega)^d$, the term $\tfrac{1}{2}\nabla \cdot \mathbf{b}$ can only be thought of as a distribution, and not as a function or even as a regular distribution. If, in addition, we assume that $\nabla \cdot \mathbf{b} \in L^\infty(\Omega)$, we can show that $c - \tfrac{1}{2}\nabla \cdot \mathbf{b} \geq 0$ almost everywhere in Ω (see Exercise 5.1), and so we also require this condition. We now have that a is coercive in $H_0^1(\Omega)$, and so by the Lax-Milgram lemma (Proposition 5.2) there is a unique solution u to the variational problem (5.7) with the stability bound

$$\|u\|_{1,\Omega} \leq C \left(\|f\|_\Omega + \|g\|_{1/2,\Gamma} \right).$$

5.3   Complex and complexified spaces

The complexification of a real space. Suppose V is a real vector space. We define the complexification of V as $V_\mathbb{C} := V + \imath V$. To define a vector space structure on $V_\mathbb{C}$, we need to extend the definitions of addition and scalar multiplication in V. These extensions are defined as follows: for $u, v, \tilde{u}, \tilde{v} \in V$ and $\alpha + \imath\beta \in \mathbb{C}$,

$$(u + \imath v) + (\tilde{u} + \imath\tilde{v}) := (u + \tilde{u}) + \imath(v + \tilde{v}),$$
$$(\alpha + \imath\beta)(u + \imath v) := (\alpha u - \beta v) + \imath(\beta u + \alpha v).$$

There is a natural map

$$V_\mathbb{C} \ni w = u + \imath v \longmapsto u - \imath v =: \overline{w} \in V_\mathbb{C},$$

which is the vector analogue of complex conjugation. This map is a conjugate linear involution from $V_\mathbb{C}$ to $V_\mathbb{C}$, that is,

$$\overline{\overline{w}} = w,$$
$$\overline{w_1 + w_2} = \overline{w_1} + \overline{w_2},$$
$$\overline{\lambda w} = \overline{\lambda}\,\overline{w},$$

for all $w, w_1, w_2 \in V_{\mathbb{C}}$ and $\lambda \in \mathbb{C}$. If V is a real inner product space, we can similarly complexify V to create a complex inner product space $V_{\mathbb{C}}$. The inner product in $V_{\mathbb{C}}$ is defined by

$$
\begin{aligned}
(w_1, w_2)_{V_{\mathbb{C}}} &= (w_1^{re} + \imath w_1^{im}, w_2^{re} + \imath w_2^{im})_{V_{\mathbb{C}}} \\
&:= (w_1^{re}, w_2^{re})_V + (w_1^{im}, w_2^{im})_V + \imath \left((w_1^{im}, w_2^{re})_V - (w_1^{re}, w_2^{im})_V \right),
\end{aligned}
$$

which is a sesquilinear form, linear in the first component and conjugate linear in the second. For a complexified inner product space with inner product $(\cdot, \cdot)_{V_{\mathbb{C}}}$ it is easy to show that

$$
\overline{(w_1, w_2)}_{V_{\mathbb{C}}} = (\overline{w_1}, \overline{w_2})_{V_{\mathbb{C}}} \qquad \forall w_1, w_2 \in V_{\mathbb{C}},
$$

and

$$
\begin{aligned}
\|w\|_{V_{\mathbb{C}}}^2 = \|w^{re} + \imath w^{im}\|_{V_{\mathbb{C}}}^2 &= (w^{re} + \imath w^{im}, w^{re} + \imath w^{im})_{V_{\mathbb{C}}} \\
&= \|w^{re}\|_V^2 + \|w^{im}\|_V^2,
\end{aligned}
$$

which shows that the complexification is topologically equivalent to the space $V \times V$, therefore if V is a Hilbert space, then so is $V_{\mathbb{C}}$.

Example. As an illustration of this concept, consider the complexification of $L^2(\Omega)$, which we will denote $L^2(\Omega; \mathbb{C})$. Using the same process as above, we can consider this space as $L^2(\Omega) + \imath L^2(\Omega)$ with the inner product

$$
(u, v)_\Omega := \int_\Omega u(\mathbf{x}) \, \overline{v(\mathbf{x})} \, d\mathbf{x}.
$$

With this, we can complexify $H^1(\Omega)$ as

$$
H^1(\Omega; \mathbb{C}) := \{ u \in L^2(\Omega; \mathbb{C}) : \nabla u \in L^2(\Omega; \mathbb{C}^d) \},
$$

where the gradient operator acts on the real and imaginary parts of u in the sense of distributions. This space has the complex inner product

$$
(u, v)_{1, \Omega} := \int_\Omega u(\mathbf{x}) \, \overline{v(\mathbf{x})} \, d\mathbf{x} + \int_\Omega \nabla u(\mathbf{x}) \cdot \overline{\nabla v(\mathbf{x})} \, d\mathbf{x}.
$$

We can restate the Riesz-Fréchet representation theorem in this context.

Warning. Whenever we deal with complex inner product spaces, the inner product will be taken to be linear in the first component and antilinear (conjugate linear) in the second component. Given a complex vector space V, we will consider its antidual V^* (not V') to be the space of antilinear bounded functionals from V to $\mathbb{C}$.

Proposition 5.3 (Riesz-Fréchet theorem for complex spaces). *Let V be a complex vector space and let V^* be the space of conjugate linear bounded functionals on V. The map*

$$V \ni u \longmapsto (u, \cdot)_V \in V^*$$

is an isometric isomorphism.

Proof. The proof of this result is very similar to the corresponding one for the real case and is left as Exercise 5.3. $\square$

Quadratic minimization problems. Suppose that V is a complex vector space. If we have a conjugate linear map $\ell : V \to \mathbb{C}$ and a sesquilinear form $a : V \times V \to \mathbb{C}$ that is Hermitian, i.e.,

$$a(u, v) = \overline{a(v, u)} \qquad \forall u, v \in V,$$

and a is positive semidefinite, then the minimization problem

$$\tfrac{1}{2} a(u, u) - \operatorname{Re} \ell(u) = \min! \qquad u \in V$$

is equivalent to (Exercise 5.4) the variational problem

$$u \in V, \qquad a(u, v) = \ell(v) \qquad \forall v \in V.$$

We can also restate the Lax-Milgram lemma for complex Hilbert spaces.

Proposition 5.4 (Lax-Milgram lemma for complex spaces). *Suppose V is a complex Hilbert space and $a : V \times V \to \mathbb{C}$ is a bounded sesquilinear form which is coercive in $\mathbb{C}$, i.e., there is some $\alpha > 0$ such that*

$$|a(u, u)| \geq \alpha \|u\|_V^2 \qquad \forall u \in V.$$

For any $\ell \in V^*$, the variational problem

$$u \in V, \qquad a(u, v) = \ell(v) \qquad \forall v \in V,$$

has a unique solution that satisfies

$$\|u\|_V \leq \frac{1}{\alpha} \|\ell\|_{V^*}.$$

In addition, the solution operator, $\ell \mapsto u$, is linear.

5.4   The Laplace resolvent equations

Suppose that $\Omega \subset \mathbb{R}^d$ is open, bounded, and Lipschitz. In this section, we study the Laplace resolvent equations, which is the family of PDE depending on the complex parameter s given by

$$-\Delta u + su = f, \qquad \gamma u = g, \tag{5.8}$$

for $s \in \mathbb{C} \setminus (-\infty, 0]$ and data $f \in L^2(\Omega; \mathbb{C})$ and $g \in H^{1/2}(\Gamma; \mathbb{C})$. In what follows, we will look for solutions $u \in H^1(\Omega; \mathbb{C})$.

Variational formulation. Moving towards a variational formulation for the Laplace resolvent equations, we work with the PDE distributionally:

$$-\langle \Delta u, \varphi \rangle + s \langle u, \varphi \rangle = \langle f, \varphi \rangle \quad \forall \varphi \in \mathcal{D}(\Omega),$$

which is the same as

$$\sum_{j=1}^{d} \langle \partial_{x_j} u, \partial_{x_j} \varphi \rangle + s \langle u, \varphi \rangle = \langle f, \varphi \rangle \quad \forall \varphi \in \mathcal{D}(\Omega).$$

We rewrite this so that we consider the unknown and data as regular distributions in $L^2(\Omega; \mathbb{C})$,

$$(\nabla u, \nabla \varphi)_\Omega + s(u, \varphi)_\Omega = (f, \varphi)_\Omega \qquad \forall \varphi \in \mathcal{D}(\Omega),$$

or, by density,

$$(\nabla u, \nabla v)_\Omega + s(u, v)_\Omega = (f, v)_\Omega \qquad \forall v \in H_0^1(\Omega).$$

The test functions above are all real-valued because we started off with real-valued test functions in $\mathcal{D}(\Omega)$. However, it follows from an easy linearity argument that the above is equivalent to

$$(\nabla u, \nabla v)_\Omega + s(u, v)_\Omega = (f, v)_\Omega \qquad \forall v \in H_0^1(\Omega; \mathbb{C}),$$

where $H_0^1(\Omega; \mathbb{C}) := H_0^1(\Omega) + \imath H_0^1(\Omega)$, similar to what we have done before. The sesquilinear form

$$a(u, v) := (\nabla u, \nabla v)_\Omega + s(u, v)_\Omega,$$

is s-dependent and the functional

$$\ell(v) := (f, v)_\Omega,$$

is conjugate linear and bounded. The variational form for the original PDE is now

$$u \in H^1(\Omega; \mathbb{C}), \qquad \gamma u = g, \tag{5.9a}$$

$$a(u, v) = \ell(v) \quad \forall v \in H_0^1(\Omega; \mathbb{C}). \tag{5.9b}$$

We now explore the boundedness and coercivity of the sesquilinear form $a :$ $H^1(\Omega; \mathbb{C}) \times H^1(\Omega; \mathbb{C}) \to \mathbb{C}$. To show the boundedness of a, we compute

$$|a(u,v)| = |(\nabla u, \nabla v)_\Omega + s(u,v)_\Omega| \leq \|\nabla u\|_\Omega \|\nabla v\|_\Omega + |s|\|u\|_\Omega \|v\|_\Omega$$
$$\leq \max\{1, |s|\} (\|\nabla u\|_\Omega \|\nabla v\|_\Omega + \|u\|_\Omega \|v\|_\Omega)$$
$$\leq \max\{1, |s|\}\|u\|_{1,\Omega}\|v\|_{1,\Omega}.$$

A careful computation for the coercivity estimate of a actually allows us to prove that $a(u, v)$ is coercive in $H^1(\Omega; \mathbb{C})$. Indeed, multiplying a by $\bar{s}^{1/2}$ and considering only the real part, we observe

$$\mathrm{Re}\left(\bar{s}^{1/2} a(u,u)\right) = \mathrm{Re}\left(\bar{s}^{1/2}\|\nabla u\|_\Omega^2 + \bar{s}^{1/2}(s^{1/2})^2\|u\|_\Omega^2\right)$$
$$= \mathrm{Re}\left(\bar{s}^{1/2}\|\nabla u\|_\Omega^2 + s^{1/2}|s|\|u\|_\Omega^2\right)$$
$$= \mathrm{Re}(s^{1/2})\left(\|\nabla u\|_\Omega^2 + |s|\|u\|^2\right)$$
$$\geq \mathrm{Re}(s^{1/2})\min\{1, |s|\}\|u\|_{1,\Omega}^2 \quad \forall u \in H^1(\Omega; \mathbb{C}).$$

Therefore

$$|\bar{s}^{1/2} a(u,u)| \geq \mathrm{Re}(\bar{s}^{1/2} a(u,u)) \geq \mathrm{Re}(\bar{s}^{1/2})\min\{1, |s|\}\|u\|_{1,\Omega}^2,$$

from which we arrive at the coercivity bound

$$|a(u,u)| \geq \left(\frac{\mathrm{Re}(s^{1/2})}{|s|^{1/2}}\min\{1, |s|\}\right)\|u\|_{1,\Omega}^2.$$

If we instead prove coercivity in $H_0^1(\Omega; \mathbb{C})$, we are able to do away with the $\min\{1, |s|\}$ term, but gain the constant C_Ω which appears from the use of the Poincaré-Friedrich's inequality. An easy argument then shows that (5.8) and (5.9) are well posed problems.

A connection to the Helmholtz equation. Now consider the homogeneous Dirichlet problem for the Laplace resolvent equations:

$$u \in H_0^1(\Omega; \mathbb{C}), \qquad -\Delta u + su = f \quad \text{in } \Omega.$$

We have shown the well-posedness of this problem for $s \in \mathbb{C} \setminus (-\infty, 0]$. For $s = 0$, the equation reduces to Poisson's equation and we can also show existence and uniqueness of solutions. We can also show well-posedness for $s = -k^2$, for $k > 0$ and k 'very small' in a way that will be made precise in what follows. In the case $s = -k^2$, the resolvent equation is the Helmholtz equation

$$\Delta u + k^2 u = -f,$$

with associated weak form

$$(\nabla u, \nabla v)_\Omega - k^2(u,v)_\Omega = (f,v)_\Omega.$$

To prove coercivity, we require

$$a(u, u) = \|\nabla u\|_\Omega^2 - k^2 \|u\|_\Omega^2 \geq \left(1 - k^2 C_{\mathrm{PF}}^2\right) \|\nabla u\|_\Omega^2,$$

where

$$\|u\|_\Omega \leq C_{\mathrm{PF}} \|\nabla u\|_\Omega \quad \forall u \in H_0^1(\Omega; \mathbb{C}).$$

This is true when

$$k \leq \frac{1}{C_{\mathrm{PF}}},$$

i.e., for low enough wave numbers, in what is known as a low frequency regime for the Helmhotz equation. From here we can bound

$$a(u, u) \geq C_\Omega (1 - k^2 C_{\mathrm{PF}}^2) \|u\|_{1,\Omega}^2,$$

which guarantees the coercivity of a.

A more abstract point of view. Consider again the homogeneous resolvent equation

$$u \in H_0^1(\Omega; \mathbb{C}), \qquad -\Delta u + su = f \quad \text{in } \Omega,$$

which has the variational form

$$(\nabla u, \nabla v)_\Omega + s(u, v)_\Omega = (f, v)_\Omega.$$

We can apply the Riesz-Fréchet theorem in the space $H_0^1(\Omega; \mathbb{C})$ to show that there is a linear operator $A(s) : H_0^1(\Omega; \mathbb{C}) \to H_0^1(\Omega; \mathbb{C})$ such that

$$(A(s)u, v)_{1,\Omega} = (\nabla u, \nabla v)_\Omega + s(u, v)_\Omega \quad \forall u, v \in H_0^1(\Omega; \mathbb{C}).$$

We can split

$$A(s) = A(0) + sB,$$

where $B : H_0^1(\Omega; \mathbb{C}) \to H_0^1(\Omega; \mathbb{C})$ is given by

$$(Bu, v)_{1,\Omega} = (u, v)_\Omega \quad \forall u, v \in H_0^1(\Omega; \mathbb{C}).$$

From the definition, we can see that we can bound

$$\|A(s)\|_{H_0^1 \to H_0^1} \leq \max\{1, |s|\}.$$

Furthermore, the map

$$\mathbb{C} \ni s \longmapsto A(s) \in \mathcal{B}(H_0^1(\Omega; \mathbb{C}), H_0^1(\Omega; \mathbb{C}))$$

is affine and therefore analytic. We have shown that $A(s)^{-1}$ exists for all $s \in \mathbb{C} \setminus (-\infty, 0]$, and the resolvent set $\{s : A(s)^{-1} \text{ exists}\}$ is open and contains $\mathbb{C} \setminus (-\infty, 0]$. In particular, since $A(0)$ is invertible, there exists a neighborhood of zero in $\mathbb{C}$ such that $A(s)$ is invertible. This provides an alternative proof of the well-posedness of the Helmholtz equation for small enough wave number.

5.5 The Ritz-Galerkin projection (*)

Approximation of symmetric coercive problems. Consider again a real Hilbert space V, a symmetric bounded and coercive bilinear form $a : V \times V \to \mathbb{R}$

$$\alpha \|u\|_V^2 \leq a(u, u) := \|u\|_a^2 \leq M \|u\|_V^2 \qquad \forall u \in V, \tag{5.10}$$

and $\ell \in V'$. As we know, the variational problem

$$u \in V, \qquad a(u, v) = \ell(v) \qquad \forall v \in V, \tag{5.11}$$

and the minimization problem

$$\tfrac{1}{2} a(u, u) - \ell(u) = \text{min!} \qquad u \in V,$$

are equivalent and uniquely solvable. Now let V_h be a finite-dimensional subspace of V. (Note that tagging 'discrete' spaces in the parameter h is the common usage in the finite element method community, one of the most heavy users of what follows.) We can the consider the restricted minimization problem

$$\tfrac{1}{2} a(u_h, u_h) - \ell(u_h) = \text{min!} \qquad u_h \in V_h,$$

or its equivalent variational formulation

$$u_h \in V_h, \qquad a(u_h, v_h) = \ell(v_h) \qquad \forall v_h \in V_h. \tag{5.12}$$

Problem (5.12) is uniquely solvable, as easily follows from using the Riesz-Fréchet theorem in the space V_h with the inner product defined by a. The element u_h is called the **Ritz projection** of u onto V_h. We can actually relate u and u_h without the right-hand side ℓ being involved in the process, by writing

$$u_h \in V_h, \qquad a(u_h, v_h) = a(u, v_h) \qquad \forall v_h \in V_h,$$

or even better

$$u_h \in V_h, \qquad a(u_h - u, v_h) = 0 \qquad \forall v_h \in V_h,$$

which shows that u_h is the orthogonal projection of u onto V_h when a is used as the inner product in V. Therefore

$$\|u - u_h\|_a = \text{min!} \qquad u_h \in V_h,$$

and thus

$$\begin{aligned}
\|u - u_h\|_V &\leq \alpha^{-1/2} \|u - u_h\|_a \\
&\leq \alpha^{-1/2} \|u - v_h\|_a \\
&\leq (M/\alpha)^{1/2} \|u - v_h\|_V \qquad \forall v_h \in V_h,
\end{aligned}$$

(we have used the constants of (5.10)) or, in other words,

$$\|u - u_h\|_V \leq (M/\alpha)^{1/2} \min_{v_h \in V_h} \|u - v_h\|_V. \tag{5.13}$$

The inequality (5.13) proves that the Ritz projection behaves 'like' the best approximation onto V_h (with respect to the original norm of V), in the sense that, while the Ritz projection does not provide the best approximation, it yields a proportion of it.

The Galerkin approximation. In a few words, the Galerkin projection is the Ritz projection applied to well posed variational problems associated to nonsymmetric (or sometimes noncoercive) bilinear forms. Consider thus a bounded coercive bilinear form $a : V \times V \to \mathbb{R}$, satisfying

$$a(u, u) \geq \alpha\|u\|_V^2, \qquad |a(u, v)| \leq M\|u\|_V \|v\|_V \qquad \forall u, v \in V,$$

and $\ell \in V'$. We again consider the approximation of the unique solution of (5.11) by the unique solution of (5.12). Note that both problems are uniquely solvable by virtue of the Lax-Milgram lemma (applied in different spaces). The map $u \mapsto u_h$ is called the **Galerkin projection** and u_h is called the Galerkin approximation of u in V_h, while the process of approximating a 'continuous' variational problem (5.11) by a 'discrete' one (5.12) is often called a **Galerkin method.** As shown in Exercise 5.7, once a basis of V_h is chosen, equations (5.12) can be equivalently formulated as a linear system of algebraic equations. The property

$$u_h \in V_h, \qquad a(u_h - u, v_h) = 0 \qquad \forall v_h \in V_h, \tag{5.14}$$

is often called **Galerkin orthogonality**. It is actually the orthogonality of $u - u_h$ to V_h with respect to the bilinear form (not an inner product!) a. It is clear that, since (5.14) uniquely defines u_h in terms of u, if we take $u \in V_h$ as 'continuous' data, then $u_h = u$ and therefore the map $u \mapsto u_h$ is a projection with range V_h. Using Galerkin orthogonality (5.14), coercivity and boundedness, we can prove

$$\alpha\|u_h\|_V^2 \leq a(u_h, u_h) = a(u, u_h) \leq M\|u\|_V \|u_h\|_V,$$

and therefore

$$\|u_h\|_V \leq (M/\alpha)\|u\|_V,$$

which shows that the norm of the Galerkin projection

$$V \ni u \mapsto u_h \in V_h \subset V$$

is bounded by the ratio of the boundedness and coercivity constants M/α (see (5.14)). With similar arguments, we can prove that for all $v_h \in V_h$

$$
\begin{aligned}
\alpha\|u - u_h\|_V^2 &\leq a(u - u_h, u - u_h) && \text{(coercivity)} \\
&= a(u - u_h, u - v_h) && \text{(Galerkin orthogonality)} \\
&\leq M\|u - u_h\|_V \|u - v_h\|_V, && \text{(boundedness)}
\end{aligned}
$$

and therefore

$$\|u - u_h\|_V \leq (M/\alpha) \min_{v_h \in V_h} \|u - v_h\|_V. \tag{5.15}$$

The estimate (5.15), comparing the solution to (5.11) with its Galerkin approximation is called **Céa's lemma**. Note that the constant in the right-hand side of (5.15) can be improved in the case of symmetric problems (5.13).

Final comments and literature

The Lax-Milgram lemma is due to Peter Lax and Arthur Milgram [72] and is the tool of choice for numerical analyses of all kinds. While the Lax-Milgram lemma is *per se* a generalization of the Riesz-Fréchet representation theorem for nonsymmetric but coercive bilinear forms, it is common to read that, for symmetric coercive problems (the Dirichlet problem for the Laplacian) the Lax-Milgram lemma is being used even when the use of the equivalent inner product and the Riesz-Fréchet theorem is enough for these purposes.

For a very entertaining historical introduction to the Ritz-Galerkin projections, the article of Martin Gander and Gerhard Wanner [52] is a must read. The finite element method is the prime example of a Galerkin method for elliptic boundary value problems. The estimate (5.15) is usually presented as

$$\|u - u_h\|_V \leq (M/\alpha) \inf_{v_h \in V_h} \|u - v_h\|_V,$$

although the infimum is actually a minimum, provided by the best approximation (orthogonal projection). This nice almost trivial result is due to Jean Céa [31]. It is also referred to as the quasi-optimality of the Galerkin approximation.

While we can use the natural complexifications of the $L^p(\Omega)$ spaces (just take the functions to be complex-valued), the process of complexifying a Banach space is not entirely trivial [82]. (See also Exercise 5.9.)

Exercises

5.1. Nonnegative distributions. Let $T \in \mathcal{D}'(\Omega)$. We say that $T \geq 0$ when

$$\langle T, \varphi \rangle \geq 0 \qquad \forall \varphi \in \mathcal{D}_+(\Omega) := \{\varphi \in \mathcal{D}(\Omega) : \varphi \geq 0\}.$$

Show that this definition is coherent for regular distributions, that is, when $T = f \in L^1_{\text{loc}}(\Omega)$, then $T \geq 0$ is equivalent to $f \geq 0$ almost everywhere. (**Hint.** Read the proof of the variational lemma in Chapter 1.)

5.2. Let Ω be a bounded open set and $\kappa : \Omega \to \mathbb{R}^{d \times d}$ be a matrix valued function satisfying:

$$\kappa_{ij} \in L^\infty(\Omega) \quad \forall i, j,$$

and

$$\sum_{i,j=1}^{d} \kappa_{ij}\xi_i\xi_j \geq \kappa_0 \sum_{j=1}^{d} |\xi_i|^2 \quad \text{a.e.} \quad \forall(\xi_1, \ldots, \xi_d) \in \mathbb{R}^d.$$

(a) Study the well-posedness of the problem

$$u \in H_0^1(\Omega), \qquad (\kappa\nabla u, \nabla v)_\Omega = (f, v)_\Omega \quad \forall v \in H_0^1(\Omega).$$

(b) Write an equivalent boundary value problem.

(c) Show that the components of κ^{-1} are $L^\infty(\Omega)$ functions.

(d) Show that if $\kappa^\top = \kappa$, then there is an associated minimization principle and the expression

$$\|u\|_\kappa^2 := \int_\Omega (\kappa(\mathbf{x})\nabla u(\mathbf{x})) \cdot \nabla u(\mathbf{x}) \, d\mathbf{x},$$

defines an equivalent norm in $H_0^1(\Omega)$.

5.3. Prove the Riesz-Fréchet theorem in the complex case, namely, the map

$$\begin{aligned} V &\longrightarrow V^* \\ u &\longmapsto (u, \cdot)_V, \end{aligned}$$

is an isometric isomorphism between a complex Hilbert space V and its antidual V^*.

5.4. Let V be a complex vector space, $a : V \times V \to \mathbb{C}$ be sesquilinear, Hermitian, and positive semidefinite, and let $\ell : V \to \mathbb{C}$ be conjugate linear. Show that the minimization problem

$$\tfrac{1}{2}a(u, u) - \operatorname{Re}\ell(u) = \min! \qquad u \in V,$$

is equivalent to the variational problem

$$u \in V, \qquad a(u, v) = \ell(v) \qquad \forall v \in V.$$

(**Hint.** Show that the following problem

$$u \in V, \qquad \operatorname{Re}a(u, v) = \operatorname{Re}\ell(v) \qquad \forall v \in V,$$

is equivalent to both problems.)

5.5. Let V be a complex vector space endowed with a conjugate linear involution that we will call conjugation, that is, we have a map $V \to V$, whose action we denote $u \mapsto \bar{u}$ such that

$$\bar{\bar{u}} = u, \qquad \overline{u+v} = \bar{u}+\bar{v}, \qquad \overline{\alpha u} = \bar{\alpha}\,\bar{u}, \qquad \forall u, v \in V, \quad \forall \alpha \in \mathbb{C}.$$

(Note that we are using the overline symbol with two different meanings in the last formula.)

(a) Show that there exists a real vector space W whose complexification is V. (**Hint.** Consider the space $W = \{u \in V : u = \bar{u}\}$ with multiplication by real scalars.)

(b) Assume that V is an inner product space and that

$$\overline{(u,v)_V} = (\bar{u}, \bar{v})_V \qquad \forall u, v \in V.$$

Show that we can endow W with an inner product so that, when we complexify, we recover the inner product of V.

5.6. Consider two functions $f_1, f_2 \in L^2(\Omega)$ and the following system of boundary value problems (here Ω is a bounded set):

$$u_1, u_2 \in H_0^1(\Omega),$$
$$-\Delta u_1 + u_2 = f_1,$$
$$\Delta u_2 + u_1 = f_2.$$

(Note the different signs of the Laplacians.)

(a) Write and show the well-posedness of an equivalent variational formulation working on the space $V = H_0^1(\Omega) \times H_0^1(\Omega)$:

$$(u_1, u_2) \in V,$$
$$a\big((u_1, u_2), (v_1, v_2)\big) = \ell\big((v_1, v_2)\big) \qquad \forall (v_1, v_2) \in V.$$

(b) Now consider the function $u = u_1 + \imath u_2 \in H_0^1(\Omega; \mathbb{C}) =: V_{\mathbb{C}}$. Rewrite the boundary value problem in the variable u, find its equivalent variational formulation and show that it is well posed.

(c) Consider the above problem where the last equation is now

$$-\Delta u_2 + u_1 = f_2.$$

Show the well-posedness of this problem by taking new variables $w_1 := u_1 + u_2$ and $w_2 := u_2$. Write a variational formulation for this problem in $V = H_0^1(\Omega) \times H_0^1(\Omega)$ and show that there exists a linear transformation $R : \mathbb{R}^2 \to \mathbb{R}^2$ such that

$$a((u_1, u_2), R(u_1, u_2)) \ge \alpha \|(u_1, u_2)\|_V^2.$$

5.7. The Galerkin equations. Let $\{\phi_1, \ldots, \phi_N\}$ be a basis for V_h. Show that the Galerkin equations

$$u_h \in V_h, \qquad a(u_h, v_h) = \ell(v_h) \qquad \forall v_h \in V_h,$$

are equivalent to the linear system

$$\sum_{j=1}^{N} a(\phi_j, \phi_i) c_j = \ell(\phi_i) \qquad i = 1, \ldots, N,$$

followed by the reconstruction step

$$u_h = \sum_{j=1}^{N} c_j \phi_j.$$

Show that when a is symmetric and coercive, the associated linear system is symmetric and positive definite.

5.8. The Courant space. Let $\Omega \subset \mathbb{R}^2$ be an open polygon and let $\mathcal{T}_h$ be a partition of Ω into finitely many disjoint open triangles so that

$$\overline{\Omega} = \cup\{\overline{T} : T \in \mathcal{T}_h\},$$

and

> if $\overline{T} \cap \overline{T'} \neq \emptyset$, then $\overline{T} \cap \overline{T'}$ is either a common vertex or a common edge of both triangles.

Let $u : \Omega \to \mathbb{R}$ be such that

$$u|_T \in \mathcal{P}_1 := \text{span}\,\{1, x_1, x_2\} \qquad \forall T \in \mathcal{T}_h.$$

Show that $u \in H^1(\Omega)$ if and only if whenever $e = \overline{T} \cap \overline{T'}$ is a common edge of two elements of the partition, then

$$\gamma_T u|_e = \gamma_{T'} u|_e,$$

where $\gamma_T : H^1(T) \to H^{1/2}(\partial T)$ is the associated local trace operator.

5.9. Complexification of Banach spaces. Let X be a real Banach space. Show that

$$\|(x^{re}, x^{im})\| := \max_{t \in \mathbb{R}} \| \cos t\, x^{re} - \sin t\, x^{im} \|_X$$

defines a norm in $X \times X$ which is equivalent to $\|x^{re}\| + \|x^{im}\|$. Show that this norm can be used to define a norm in the complexification of X such that conjugation is an isometry. (**Hint.** The only difficulty is related to scalar multiplication.)

6

Neumann boundary conditions

6.1   Duality on the boundary 107
6.2   Normal components of vector fields 108
6.3   Neumann boundary conditions 111
6.4   Impedance boundary conditions 114
6.5   Transmission problems (*) 116
6.6   Nonlocal boundary conditions (*) 118
6.7   Mixed boundary conditions (*) 120
Final comments and literature 122
Exercises 123

In this chapter we give a weak interpretation of the normal derivative $\nabla u \cdot \mathbf{n}$ for $u \in H^1(\Omega)$ with $\Delta u \in L^2(\Omega)$. This will be done using duality on the trace space $H^{1/2}(\Gamma)$ (as usual $\Gamma := \partial\Omega$) and Green's first identity as the *definition* of the normal derivative. The process will be done gradually by first working on what we understand by the normal component on the boundary of a vector field $\mathbf{p} \in \mathbf{L}^2(\Omega) := L^2(\Omega; \mathbb{R}^d)$ such that $\nabla \cdot \mathbf{p} \in L^2(\Omega)$. We will use the definition of the weak normal derivative to explore Neumann boundary conditions on several coercive problems. The Neumann problem for the Laplacian

$$-\Delta u = f \quad \text{in } \Omega, \qquad \nabla u \cdot \mathbf{n} = h \quad \text{on } \Gamma,$$

will have to wait until Chapter 7 and so will some problems for which it is less clear that the associated bilinear form is coercive in the entire space $H^1(\Omega)$. The reason for this postponement is the need to prove a family of Poincaré type inequalities, which are derived from some compact embeddings.

6.1   Duality on the boundary

We start this section by proving that if Γ is the boundary of a Lipschitz domain, the injection $H^{1/2}(\Gamma) \subset L^2(\Gamma)$, is dense. This is a slightly technical proof, which we will break into a series of statements. First note that

$$\|u|_\Gamma\|_\Gamma \leq C\|u\|_{1,\Omega} \qquad \forall u \in \mathcal{C}(\overline{\Omega}) \cap H^1(\Omega)$$

and therefore, by density,

$$\|\gamma u\|_\Gamma \leq C\|u\|_{1,\Omega} \qquad \forall u \in H^1(\Omega).$$

This implies the inequality

$$\|g\|_\Gamma \leq C\|g\|_{1/2,\Gamma} \qquad \forall g \in H^{1/2}(\Gamma),$$

which proves that the injection of $H^{1/2}(\Gamma)$ into $L^2(\Gamma)$ is bounded.

Proposition 6.1. *The trace space $H^{1/2}(\Gamma)$ is dense in $L^2(\Gamma)$.*

Proof. We will admit that the space $\mathcal{C}(\Gamma)$ is dense in $L^2(\Gamma)$. The proof of this statement is proposed as Exercise 6.1. Now consider the set

$$\mathcal{C}^1(\Gamma) := \{U|_\Gamma : U \in \mathcal{C}^1(\mathbb{R}^d)\}.$$

It is clear that $\mathcal{C}^1(\Gamma) \subset H^{1/2}(\Gamma)$ and $\mathcal{C}^1(\Gamma) \subset \mathcal{C}(\Gamma)$. It is simple to see that $\mathcal{C}^1(\Gamma)$ is an algebra (it is a subspace that is closed by multiplication) containing constant functions and separating points of Γ (given $\mathbf{x} \neq \mathbf{y}$ in Γ, take $\varphi \in \mathcal{D}(\mathbb{R}^d)$ such that $\varphi(\mathbf{x}) = 1$ and $\varphi(\mathbf{y}) = 0$). By the Stone-Weierstrass theorem, $\mathcal{C}^1(\Gamma)$ is dense in $\mathcal{C}(\Gamma)$.

Since $\mathcal{C}(\Gamma) \subset L^2(\Gamma)$ with dense and continuous injection, and $\mathcal{C}^1(\Gamma)$ is a dense subspace of $\mathcal{C}(\Gamma)$, it follows that the space $\mathcal{C}^1(\Gamma)$ is dense in $L^2(\Gamma)$. Finally, since $\mathcal{C}^1(\Gamma) \subset H^{1/2}(\Gamma)$, the result follows. $\square$

The dense and continuous embedding of $H^{1/2}(\Gamma)$ into $H^0(\Gamma) := L^2(\Gamma)$ allows us to define a corresponding Gelfand triple

$$H^{1/2}(\Gamma) \subset H^0(\Gamma) \subset H^{-1/2}(\Gamma).$$

This representation of the dual space of $H^{1/2}(\Gamma)$ will be the key space where we will impose Neumann boundary conditions. The dual norm in $H^{-1/2}(\Gamma)$ will be denoted

$$\|h\|_{-1/2,\Gamma} := \sup_{0 \neq g \in H^{1/2}(\Gamma)} \frac{|\langle h, g\rangle_\Gamma|}{\|g\|_{1/2,\Gamma}},$$

where the angled bracket $\langle h, g\rangle_\Gamma$ is used to denote the action of $h \in H^{-1/2}(\Gamma)$ on $g \in H^{1/2}(\Gamma)$. When $h \in L^2(\Gamma) \subset H^{-1/2}(\Gamma)$, we have

$$\langle h, g\rangle_\Gamma = \int_\Gamma h(\mathbf{x})g(\mathbf{x})d\Gamma(\mathbf{x}).$$

6.2   Normal components of vector fields

In the space

$$\mathbf{H}(\mathrm{div}, \Omega) := \left\{\mathbf{p} : \Omega \to \mathbb{R}^d : \mathbf{p} \in \mathbf{L}^2(\Omega)^d, \ \nabla \cdot \mathbf{p} \in L^2(\Omega)\right\},$$

we define the inner product

$$(\mathbf{p}, \mathbf{q})_{\mathrm{div},\Omega} = (\mathbf{p}, \mathbf{q})_\Omega + (\nabla \cdot \mathbf{p}, \nabla \cdot \mathbf{q})_\Omega.$$

Proposition 6.2. *For any open set* Ω, $\mathbf{H}(\mathrm{div}, \Omega)$ *is a Hilbert space.*

Proof. (The proof of this result is very similar to the proof that $H^1(\Omega)$ is a Hilbert space.) We obviously only need to prove that this space is complete. Let $\{\mathbf{p}_n\}$ be a Cauchy sequence in $\mathbf{H}(\mathrm{div}, \Omega)$. Thus $\{\mathbf{p}_n\}$ is $\mathbf{L}^2(\Omega)$ Cauchy, so the sequence converges to some $\mathbf{p}$ in $\mathbf{L}^2(\Omega)$, and $\mathbf{p}_n \to \mathbf{p}$ in $\mathcal{D}'(\Omega)^d$. Therefore the sequence of divergences converges $\nabla \cdot \mathbf{p}_n \to \nabla \cdot \mathbf{p}$ as distributions. We also have that $\nabla \cdot \mathbf{p}_n$ is $L^2(\Omega)$ Cauchy, so it converges to an element $v \in L^2(\Omega)$. It is clear that $v = \nabla \cdot \mathbf{p}$. This completes the proof. $\qquad\square$

A step-by-step construction of the normal component. We now give a construction of the normal component of a vector field $\mathbf{p} \in \mathbf{H}(\mathrm{div}, \Omega)$ on the boundary of Ω. We will state a theorem that summarizes the following results at the end of this section. At this moment, we are going to slowly build up to a weak variational definition of what we mean by $\mathbf{p} \cdot \mathbf{n}$.

(1) First, consider the bilinear form $t : \mathbf{H}(\mathrm{div}, \Omega) \times H^1(\Omega) \to \mathbb{R}$ given by

$$t(\mathbf{p}, u) := (\mathbf{p}, \nabla u)_\Omega + (\nabla \cdot \mathbf{p}, u)_\Omega.$$

The following easy computation

$$\begin{aligned}
|t(\mathbf{p}, u)| &\le \|\mathbf{p}\|_\Omega \|\nabla u\|_\Omega + \|\nabla \cdot \mathbf{p}\|_\Omega \|u\|_\Omega \\
&\le \left(\|\mathbf{p}\|_\Omega^2 + \|\nabla \cdot \mathbf{p}\|_\Omega^2\right)^{1/2} \left(\|\nabla u\|_\Omega^2 + \|u\|_\Omega^2\right)^{1/2} \\
&= \|\mathbf{p}\|_{\mathrm{div},\Omega} \|u\|_{1,\Omega},
\end{aligned}$$

shows that t is bounded.

(2) If we take $\varphi \in \mathcal{D}(\Omega)$, then

$$\begin{aligned}
t(\mathbf{p}, \varphi) &= (\mathbf{p}, \nabla \varphi)_\Omega + (\nabla \cdot \mathbf{p}, \varphi)_\Omega \\
&= \sum_{j=1}^d \langle p_j, \partial_{x_j} \varphi \rangle + \sum_{j=1}^d \langle \partial_{x_j} p_j, \varphi \rangle = 0.
\end{aligned}$$

Therefore, by density,

$$t(\mathbf{p}, u) = 0 \qquad \forall \mathbf{p} \in \mathbf{H}(\mathrm{div}, \Omega) \quad u \in H_0^1(\Omega).$$

We can understand this statement as the proof that the bilinear form t does not really depend on u but on its trace: if $u_1, u_2 \in H^1(\Omega)$ satisfy $\gamma u_1 = \gamma u_2$, then $\gamma(u_1 - u_2) \in H_0^1(\Omega)$, and therefore $t(\mathbf{p}, u_1) = t(\mathbf{p}, u_2)$.

(3) Now consider the bilinear form $t_\Gamma : \mathbf{H}(\mathrm{div}, \Omega) \times H^{1/2}(\Gamma) \to \mathbb{R}$ given by

$$t_\Gamma(\mathbf{p}, g) := t(\mathbf{p}, u) = (\mathbf{p}, \nabla u)_\Omega + (\nabla \cdot \mathbf{p}, u)_\Omega,$$

where u is any element of $H^1(\Omega)$ such that $\gamma u = g$. Let us now show that the bilinear form t_Γ is bounded. If $\mathbf{p} \in \mathbf{H}(\mathrm{div}, \Omega)$ and $g \in H^{1/2}(\Gamma)$, then

$$|t_\Gamma(\mathbf{p}, g)| = |t(\mathbf{p}, u)| \leq \|\mathbf{p}\|_{\mathrm{div},\Omega} \|u\|_{1,\Omega} \qquad \forall u \in H^1(\Omega), \qquad \gamma u = g.$$

Taking the infimum on the above inequality

$$|t_\Gamma(\mathbf{p}, g)| \leq \|\mathbf{p}\|_{\mathrm{div},\Omega} \inf\{\|u\|_{1,\Omega} \,:\, u \in H^1(\Omega), \quad \gamma u = g\}$$
$$= \|\mathbf{p}\|_{\mathrm{div},\Omega} \|g\|_{1/2,\Gamma}. \tag{6.1}$$

We then define

$$\mathbf{p} \cdot \mathbf{n} := t_\Gamma(\mathbf{p}, \cdot) : H^{1/2}(\Gamma) \to \mathbb{R},$$

which is a linear and bounded functional on $H^{1/2}(\Gamma)$, that is, $\mathbf{p} \cdot \mathbf{n} \in H^{-1/2}(\Gamma)$. Using the notation for the $H^{-1/2}(\Gamma) \times H^{1/2}(\Gamma)$ duality product, we can write

$$\langle \mathbf{p} \cdot \mathbf{n}, g \rangle_\Gamma = (\mathbf{p}, \nabla u)_\Omega + (\nabla \cdot \mathbf{p}, u)_\Omega, \qquad \gamma u = g,$$

or equivalently

$$\langle \mathbf{p} \cdot \mathbf{n}, \gamma u \rangle_\Gamma = (\mathbf{p}, \nabla u)_\Omega + (\nabla \cdot \mathbf{p}, u)_\Omega.$$

Note that (6.1) implies that

$$\|\mathbf{p} \cdot \mathbf{n}\|_{-1/2,\Gamma} \leq \|\mathbf{p}\|_{\mathrm{div},\Omega}. \tag{6.2}$$

(4) In the final step, we observe that the normal trace operator

$$\mathbf{H}(\mathrm{div}, \Omega) \ni \mathbf{p} \longmapsto \mathbf{p} \cdot \mathbf{n} \in H^{-1/2}(\Gamma),$$

is linear (this is almost trivial to prove) and bounded because of (6.2).

We have thus proved the following theorem.

Theorem 6.1. *For $\mathbf{p} \in \mathbf{H}(\mathrm{div}, \Omega)$, the expression*

$$\langle \mathbf{p} \cdot \mathbf{n}, \gamma u \rangle_\Gamma := (\mathbf{p}, \nabla u)_\Omega + (\nabla \cdot \mathbf{p}, u)_\Omega \qquad u \in H^1(\Omega),$$

defines a bounded functional $\mathbf{p} \cdot \mathbf{n} \in H^{-1/2}(\Gamma)$ and $\|\mathbf{p} \cdot \mathbf{n}\|_{-1/2,\Gamma} \leq \|\mathbf{p}\|_{\mathrm{div},\Omega}$.

We next show that for smooth enough vector fields, the notation $\mathbf{p} \cdot \mathbf{n}$ is justified by a classical use of the divergence theorem. Proposition 6.3 will not be used until much further in this book, when we start comparing strong and weak normal traces of vector fields. The reader is invited to first revisit the introduction of the unit normal vector field $\mathbf{n} \in L^\infty(\Gamma)^d \equiv L^\infty(\Gamma; \mathbb{R}^d)$ in Section 3.5.

Proposition 6.3. *On a Lipschitz domain Ω we have the equality*

$$\mathbf{p} \cdot \mathbf{n} = \gamma \mathbf{p} \cdot \mathbf{n} \in L^2(\Gamma) \qquad \forall \mathbf{p} \in \mathbf{H}^1(\Omega) := H^1(\Omega; \mathbb{R}^d) \equiv H^1(\Omega)^d.$$

Proof. Recall first that (see Proposition 3.6)

$$\int_\Omega (\nabla \cdot \mathbf{q})(\mathbf{x}) \mathrm{d}\mathbf{x} = \int_\Gamma \mathbf{q}(\mathbf{x}) \cdot \mathbf{n}(\mathbf{x}) \, \mathrm{d}\Gamma(\mathbf{x}) \qquad \forall \mathbf{q} \in \mathcal{C}^1(\overline{\Omega}; \mathbb{R}^d).$$

For given

$$\mathbf{p} \in \mathcal{C}^1(\overline{\Omega})^d \subset \mathbf{H}(\mathrm{div}, \Omega), \qquad u \in \mathcal{C}^1(\overline{\Omega}) \subset H^1(\Omega),$$

we can compute

$$
\begin{aligned}
\langle \mathbf{p} \cdot \mathbf{n}, \gamma u \rangle_\Gamma &= (\mathbf{p}, \nabla u)_\Omega + (\nabla \cdot \mathbf{p}, u)_\Omega \\
&= \int_\Omega \mathbf{p}(\mathbf{x}) \cdot (\nabla u)(\mathbf{x}) \mathrm{d}\mathbf{x} + \int_\Omega (\nabla \cdot \mathbf{p})(\mathbf{x}) u(\mathbf{x}) \mathrm{d}\mathbf{x} \\
&= \int_\Omega \nabla \cdot (u\mathbf{p})(\mathbf{x}) \mathrm{d}\mathbf{x} = \int_\Gamma (u\mathbf{p})(\mathbf{x}) \cdot \mathbf{n}(\mathbf{x}) \mathrm{d}\Gamma(\mathbf{x}) \\
&= \int_\Gamma (\gamma u)(\mathbf{x})(\gamma \mathbf{p} \cdot \mathbf{n})(\mathbf{x}) \mathrm{d}\Gamma(\mathbf{x}) = \langle \gamma \mathbf{p} \cdot \mathbf{n}, \gamma u \rangle_\Gamma,
\end{aligned}
$$

since $\gamma \mathbf{p} \cdot \mathbf{n} \in L^2(\Gamma)$. By a simple density argument, the result follows. $\qquad \square$

6.3 Neumann boundary conditions

Having developed the tools to understand the normal derivative in a distributional sense, we now turn our attention to some simple Neumann problems for equations of the second order. When $u \in H^1(\Omega)$ and $\Delta u \in L^2(\Omega)$, it is clear that $\nabla u \in \mathbf{H}(\mathrm{div}, \Omega)$, and we can define

$$\partial_n u := \nabla u \cdot \mathbf{n} \in H^{-1/2}(\Gamma).$$

Note that, by definition

$$\langle \partial_n u, \gamma v \rangle_\Gamma = (\nabla u, \nabla v)_\Omega + (\Delta u, v)_\Omega \qquad \forall v \in H^1(\Omega),$$

and

$$\|\partial_n u\|_{-1/2,\Gamma} \le \|\nabla u\|_{\mathrm{div},\Omega} = (\|\nabla u\|_\Omega^2 + \|\Delta u\|_\Omega^2)^{1/2}.$$

We then consider the problem

$$u \in H^1(\Omega), \tag{6.3a}$$
$$-\Delta u + cu = f, \tag{6.3b}$$
$$\partial_n u = h. \tag{6.3c}$$

The reaction coefficient is a function $c \in L^\infty(\Omega)$ satisfying $c \geq c_0 > 0$ almost everywhere, while the data are $f \in L^2(\Omega)$ and $h \in H^{-1/2}(\Gamma)$. We will need to wait until the next chapter and the introduction of Poincaré inequalities to be able to study problems where the reaction coefficient c can vanish or is completely absent. Since $\nabla \cdot \nabla u = \Delta u = cu - f \in L^2(\Omega)$, we conclude that $\nabla u \in \mathbf{H}(\mathrm{div}, \Omega)$. In this way, the boundary condition only makes sense because the PDE is satisfied.

Variational formulation. To derive the variational formulation of the Neumann problem (6.3), we begin with the boundary condition:

$$\langle h, \gamma v \rangle_\Gamma = \langle \nabla u \cdot \mathbf{n}, \gamma v \rangle_\Gamma = (\nabla u, \nabla v)_\Omega + (\nabla \cdot \nabla u, v)_\Omega$$
$$= (\nabla u, \nabla v)_\Omega + (\Delta u, v)_\Omega$$
$$= (\nabla u, \nabla v)_\Omega + (cu - f, v)_\Omega.$$

From this, we find the variational formulation to be

$$u \in H^1(\Omega), \tag{6.4a}$$
$$(\nabla u, \nabla v)_\Omega + (cu, v)_\Omega = (f, v)_\Omega + \langle h, \gamma v \rangle_\Gamma \qquad \forall v \in H^1(\Omega). \tag{6.4b}$$

The variational form includes both the PDE and the boundary conditions, and the duality product $\langle \cdot, \cdot \rangle_\Gamma$ is only an integral when $h \in L^2(\Gamma)$. To show that the variational form (6.4) implies the PDE, we test with $\varphi \in \mathcal{D}(\Omega)$ to prove that

$$(\nabla u, \nabla \varphi)_\Omega + (cu, \varphi)_\Omega = (f, \varphi)_\Omega,$$

which is equivalent to

$$\sum_{j=1}^d \langle \partial_{x_j} u, \partial_{x_j} \varphi \rangle + \langle cu, \varphi \rangle = \langle f, \varphi \rangle,$$

and therefore to

$$-\sum_{j=1}^d \langle \partial_{x_j}^2 u, \varphi \rangle + \langle cu, \varphi \rangle = \langle f, \varphi \rangle.$$

All the above brackets are distributional. This shows that $-\Delta u + cu = f$ and therefore $\nabla u \in \mathbf{H}(\mathrm{div}, \Omega)$, which allows us to define $\partial_n u = \nabla u \cdot \mathbf{n} \in H^{-1/2}(\Gamma)$. Substituting $\Delta u = cu - f$ in the definition of $\partial_n u$ and using the variational formulation (6.4) again, we have

$$\langle \partial_n u, \gamma v \rangle_\Gamma = (\nabla u, \nabla v)_\Omega + (\Delta u, v)_\Omega$$
$$= (\nabla u, \nabla v)_\Omega - (f, v)_\Omega + (cu, v)_\Omega$$
$$= \langle h, \gamma v \rangle_\Gamma \quad \forall v \in H^1(\Omega).$$

Since the set of all traces of functions in $H^1(\Omega)$ is $H^{1/2}(\Gamma)$, this shows that $\langle \partial_n u, g \rangle_\Gamma = \langle h, g \rangle_\Gamma$ for all $g \in H^{1/2}(\Gamma)$, i.e., $\partial_n u = h$ as elements of $H^{-1/2}(\Gamma)$. Note that an intermediate equivalent formulation for the problem is

$$u \in H^1(\Omega),$$
$$-\Delta u + cu = f,$$
$$(\nabla u, \nabla v)_\Omega + (\Delta u, v)_\Omega = \langle h, \gamma v \rangle_\Gamma \quad \forall v \in H^1(\Omega).$$

This formulation ignores the definition of $\partial_n u$ and writes the effect of substituting the PDE in the variational equation (6.4).

Well-posedness. At this point, showing the well-posedness of the problem is simple. The bilinear form $a(u, v) = (\nabla u, \nabla v)_\Omega + (cu, v)_\Omega$ is bounded and coercive in $H^1(\Omega)$. The linear functional $\ell(v) = (f, v)_\Omega + \langle h, \gamma v \rangle_\Gamma$ is bounded:

$$|\ell(v)| \le \|f\|_\Omega \|v\|_\Omega + \|h\|_{-1/2,\Gamma} \|\gamma v\|_{1/2,\Gamma}$$
$$\le \left(\|f\|_\Omega + \|h\|_{-1/2,\Gamma} \right) \|v\|_{1,\Omega}.$$

Well-posedness then follows by the Lax-Milgram lemma.

Equivalent minimization principle. The bilinear form a is symmetric, and coercivity implies it is positive semidefinite, which means the minimization problem associated to the PDE is then the search for $u \in H^1(\Omega)$ such that

$$\tfrac{1}{2} \left(\|\nabla u\|_\Omega^2 + (cu, u)_\Omega \right) - \ell(u) = \min!$$

If we take $h = 0$ and $c = 1$, then the minimization problem is

$$\tfrac{1}{2} \left(\|\nabla u\|_\Omega^2 + \|u\|_\Omega^2 \right) - (f, u)_\Omega = \min! \qquad u \in H^1(\Omega).$$

Once again, this problem is equivalent to

$$u \in H^1(\Omega), \qquad -\Delta u + u = f, \qquad \partial_n u = 0.$$

We say that the homogeneous Neumann condition in this case is a natural boundary condition, since it arises from the variational form and is not imposed in the solution space. This is contrasted with the Dirichlet problem:

$$\tfrac{1}{2} \left(\|\nabla u\|_\Omega^2 + \|u\|_\Omega^2 \right) - (f, u)_\Omega = \min! \qquad u \in H_0^1(\Omega),$$

where the boundary conditions are imposed in the search space. This is an example of an essential boundary condition.

Another simple observation. The map from $\mathbf{H}(\mathrm{div}, \Omega) \to H^{-1/2}(\Gamma)$ given by $\mathbf{p} \mapsto \mathbf{p} \cdot \mathbf{n}$ is surjective. To show this, we consider the boundary value problem

$$u \in H^1(\Omega), \qquad -\Delta u + u = 0, \qquad \partial_n u = h,$$

and define $\mathbf{p} := \nabla u$. The surjectivity follows from the well-posedness of this problem for any $h \in H^{-1/2}(\Gamma)$.

6.4   Impedance boundary conditions

We will begin looking at impedance boundary conditions through the associated minimization problem. Consider the minimization problem

$$\tfrac{1}{2}\left(\|\nabla u\|_\Omega^2 + (c\,u, u)_\Omega + \langle \alpha\,\gamma u, \gamma u\rangle_\Gamma\right) - (f, u)_\Omega - \langle h, \gamma u\rangle_\Gamma = \min! \quad u \in H^1(\Omega).$$

We take the data $c \in L^\infty(\Omega)$ such that $c \geq c_0 > 0$ almost everywhere, $\alpha \in L^\infty(\Gamma)$, $\alpha \geq 0$ almost everywhere, $f \in L^2(\Omega)$ and $h \in H^{-1/2}(\Gamma)$. We have already encountered the space $L^\infty(\Gamma)$ in Section 3.5. For some of what comes later we will need the norm

$$\|g\|_{\infty,\Gamma} := \max_{j=1,\dots,J} \|g \circ \mathbf{\Phi}_j\|_{L^\infty(B_{d-1}(0;1))},$$

where $\mathbf{\Phi}_j$ are the parametrizations of the patches of Γ. With this definition, it follows that

$$\|f\,g\|_\Gamma \leq \|f\|_{\infty,\Gamma}\|g\|_\Gamma \qquad f \in L^\infty(\Gamma), \quad g \in L^2(\Gamma).$$

The variational form of this minimization problem is

$$u \in H^1(\Omega), \tag{6.5}$$
$$(\nabla u, \nabla v)_\Omega + (cu, v)_\Omega + \langle \alpha\gamma u, \gamma v\rangle_\Gamma = (f, v)_\Omega + \langle h, \gamma v\rangle_\Gamma \qquad \forall v \in H^1(\Omega).$$

Testing the above with elements of $\mathcal{D}(\Omega)$, it follows that

$$-\Delta u + cu = f, \tag{6.6}$$

in the sense of distributions and, therefore, with equality as functions in $L^2(\Omega)$. If we now substitute (6.6) in (6.5) and simplify, we obtain

$$(\Delta u, v)_\Omega + (\nabla u, \nabla v)_\Omega + \langle \alpha\gamma u, \gamma v\rangle_\Gamma = \langle h, \gamma v\rangle_\Gamma \qquad \forall v \in H^1(\Omega). \tag{6.7}$$

However, (6.7) is equivalent to

$$\langle \partial_n u, \gamma v\rangle_\Gamma + \langle \alpha\gamma u, \gamma v\rangle_\Gamma = \langle h, \gamma v\rangle_\Gamma \qquad \forall v \in H^1(\Omega).$$

We have thus proved that (6.5) is equivalent to the boundary value problem

$$u \in H^1(\Omega), \qquad -\Delta u + cu = f, \qquad \partial_n u + \alpha\gamma u = h.$$

To go in the other direction and show the boundary value problem implies the variational form, we begin with the definition of the normal derivative as in the Neumann problem, apply the boundary conditions and PDE, and verify that $\nabla u \in \mathbf{H}(\mathrm{div}, \Omega)$. This is left as an exercise for the reader.

Boundedness of the bilinear form. We continue with the routine computations of showing the bilinear form and linear functional are bounded in the appropriate spaces. It is clear that the part of the bilinear form involving volume terms,

$$(\nabla u, \nabla v)_\Omega + (cu, v)_\Omega,$$

is bounded. We turn our attention to the boundary term

$$|\langle \alpha\gamma u, \gamma v\rangle_\Gamma| \leq \|\alpha\gamma u\|_{-1/2,\Gamma}\|\gamma v\|_{1/2,\Gamma} \leq C\|\alpha\gamma u\|_\Gamma\|v\|_{1,\Omega}$$
$$\leq C^2\|\alpha\|_{\infty,\Gamma}\|\gamma u\|_\Gamma\|v\|_{1,\Omega} \leq C^2\|\alpha\|_{\infty,\Gamma}\|u\|_{1,\Omega}\|v\|_{1,\Omega}.$$

The constant C in the above inequalities is the one for the trace theorem, namely, it is the constant $C > 0$ such that

$$\|\gamma u\|_\Gamma \leq C\|u\|_{1,\Omega} \qquad \forall u \in H^1(\Omega).$$

We have seen in Section 6.1 that this is the same constant such that

$$\|g\|_\Gamma \leq C\|g\|_{1/2,\Gamma} \qquad \forall g \in H^{1/2}(\Gamma),$$

i.e., the norm of $\gamma : H^1(\Omega) \to L^2(\Gamma)$ is the norm of the bounded injection $H^{1/2}(\Gamma) \to L^2(\Gamma)$. Moreover,

$$\|h\|_{-1/2,\Gamma} = \sup_{0 \neq g \in H^{1/2}(\Gamma)} \frac{|\langle h, g\rangle_\Gamma|}{\|g\|_{1/2,\Gamma}}$$
$$\leq \|h\|_\Gamma \sup_{0 \neq g \in H^{1/2}(\Gamma)} \frac{\|g\|_\Gamma}{\|g\|_{1/2,\Gamma}} \leq C\|h\|_\Gamma \qquad \forall h \in L^2(\Gamma),$$

which shows that this is the same constant for the 'adjoint inclusion' of $L^2(\Gamma)$ into $H^{-1/2}(\Gamma)$.

Coercivity. The final piece of the puzzle is to show the coercivity of the bilinear form from the impedance problem. To this end, we compute

$$(\nabla u, \nabla u)_\Omega + (cu, u)_\Omega + \langle \alpha\gamma u, \gamma u\rangle_\Gamma$$
$$\geq \|\nabla u\|_\Omega^2 + c_0\|u\|_\Omega^2 + \int_\Gamma \alpha(\mathbf{x})|\gamma u(\mathbf{x})|^2 d\Gamma(\mathbf{x}).$$

If $\alpha \geq 0$, then we have coercivity and therefore well-posedness. If α is allowed to be negative, we can bound more carefully

$$\|\nabla u\|_\Omega^2 + c_0\|u\|_\Omega^2 + \int_\Gamma \alpha(\mathbf{x})|\gamma u(\mathbf{x})|^2 d\Gamma(\mathbf{x}) \geq \|\nabla u\|_\Omega^2 + c_0\|u\|_\Omega^2 - C_\alpha\|\gamma u\|_\Gamma^2$$

$$\geq \min\{1, c_0\}\|u\|_{1,\Omega}^2 - C_\alpha C^2\|u\|_{1,\Omega}^2,$$

where $\alpha \geq -C_\alpha$ and C is the constant of the trace theorem. This shows that we have some room for negative values of α while still maintaining coercivity.

6.5   Transmission problems (*)

In this section we consider some problems set in free space with a bounded interface. These kinds of problems will become relevant once more when we discuss potential theory tools in Chapter 14. We now consider a Lipschitz domain Ω_- (we do not need Ω_- to be connected) with boundary Γ. The unbounded exterior domain

$$\Omega_+ := \mathbb{R}^d \setminus \overline{\Omega_-},$$

shares the same boundary. We will keep the normal vector pointing from Ω_- into Ω_+. In practice this will mean that in the definition of the normal component a sign change will be needed.

Two-sided traces. First of all, we can define a single double-sided trace

$$\gamma : H^1(\mathbb{R}^d) \to H^{1/2}(\Gamma).$$

Actually, we can prove (see Exercise 4.6) that if we have two trace operators, one from each side of Γ

$$\gamma^\pm : H^1(\Omega_\pm) \to H^{1/2}(\Gamma),$$

then

$$H^1(\mathbb{R}^d) \equiv \{(u_-, u_+) \in H^1(\Omega_-) \times H^1(\Omega_+) : \gamma^- u_- = \gamma^+ u_+\}.$$

We can also prove that all the above trace operators are surjective onto $H^{1/2}(\Gamma)$.

Two-sided normal components. Now consider $u \in H^1(\mathbb{R}^d \setminus \Gamma)$ such that $\Delta u \in L^2(\mathbb{R}^d \setminus \Gamma)$, where the Laplacian is applied in the sense of distributions in the open set $\mathbb{R}^d \setminus \Gamma$. (There is a serious difference with applying the Laplacian in $\mathbb{R}^d$ as we will see in Exercise 6.5.) Note that Γ has zero measure, so in principle $L^2(\mathbb{R}^d) \equiv L^2(\mathbb{R}^d \setminus \Gamma)$, but we will still write $\Delta u \in L^2(\mathbb{R}^d \setminus \Gamma)$ to emphasize the fact that we applied the Laplacian in $\mathbb{R}^d \setminus \Gamma$. We then have two normal derivatives

$$\langle \partial_n^- u, \gamma^- v \rangle_\Gamma = (\nabla u, \nabla v)_{\Omega_-} + (\Delta u, v)_{\Omega_-} \qquad \forall v \in H^1(\Omega_-), \qquad \text{(6.8a)}$$

$$\langle \partial_n^+ u, \gamma^+ v \rangle_\Gamma = -(\nabla u, \nabla v)_{\Omega_+} - (\Delta u, v)_{\Omega_+} \qquad \forall v \in H^1(\Omega_+). \qquad \text{(6.8b)}$$

Once again, note the minus sign for the exterior normal derivative, due to the fact that we have fixed one direction for the normal vector field on Γ, which makes this field point inwards when we are thinking of the exterior domain Ω_+. In particular, if $u \in H^1(\mathbb{R}^d)$ (so no jump in the trace) satisfies $\Delta u \in L^2(\mathbb{R}^d \setminus \Gamma)$, we have

$$\langle \partial_n^- u - \partial_n^+ u, \gamma v \rangle_\Gamma = (\nabla u, \nabla v)_{\mathbb{R}^d} + (\Delta u, v)_{\mathbb{R}^d \setminus \Gamma} \qquad \forall v \in H^1(\mathbb{R}^d). \qquad \text{(6.9)}$$

Note how we emphasized that the gradient is taken in $\mathcal{D}'(\mathbb{R}^d)$ and the Laplacian is taken in $\mathcal{D}'(\mathbb{R}^d \setminus \Gamma)$.

The transmission problem. Now let $f \in L^2(\mathbb{R}^d)$ and $h \in H^{-1/2}(\Gamma)$, and consider the variational problem

$$u \in H^1(\mathbb{R}^d), \qquad \text{(6.10a)}$$

$$(\nabla u, \nabla v)_{\mathbb{R}^d} + (u, v)_{\mathbb{R}^d} = (f, v)_{\mathbb{R}^d} + \langle h, \gamma v \rangle_\Gamma \qquad \forall v \in H^1(\mathbb{R}^d). \qquad \text{(6.10b)}$$

This problem is clearly uniquely solvable, since we are just looking for the Riesz-Fréchet representative in $H^1(\mathbb{R}^d)$ of the functional $v \mapsto (f, v)_{\mathbb{R}^d} + \langle h, \gamma v \rangle_\Gamma$. This problem is equivalent to the minimization problem

$$\tfrac{1}{2}\|u\|_{1,\mathbb{R}^d}^2 - (f, u)_{\mathbb{R}^d} - \langle h, \gamma u \rangle_\Gamma = \text{min!} \qquad u \in H^1(\mathbb{R}^d).$$

Using test functions in $\mathcal{D}(\mathbb{R}^d \setminus \Gamma) \subset H^1(\mathbb{R}^d)$ in (6.10b) we can easily see that the solution of (6.10) satisfies

$$-\Delta u + u = f \qquad \text{in } \mathcal{D}'(\mathbb{R}^d \setminus \Gamma),$$

and therefore (substituting the above in (6.10b))

$$(\nabla u, \nabla v)_{\mathbb{R}^d} + (\Delta u, v)_{\mathbb{R}^d \setminus \Gamma} = \langle h, \gamma v \rangle_\Gamma \qquad \forall v \in H^1(\mathbb{R}^d). \qquad \text{(6.11)}$$

This equation and (6.9) (the weak expression for the jump of the normal derivative of u across Γ) show that

$$\langle \partial_n^- u - \partial_n^+ u, \gamma v \rangle_\Gamma = \langle h, \gamma v \rangle_\Gamma \qquad \forall v \in H^1(\mathbb{R}^d),$$

but, as we know that the trace operator from $H^1(\mathbb{R}^d)$ to $H^{1/2}(\Gamma)$ is surjective, this is equivalent to stating that

$$\partial_n^- u - \partial_n^+ u = h,$$

with equality as elements of $H^{-1/2}(\Gamma)$. We have thus proved that the unique solution of (6.10) also solves the transmission problem

$$u \in H^1(\mathbb{R}^d), \tag{6.12a}$$

$$-\Delta u + u = f \quad \text{in } \mathbb{R}^d \setminus \Gamma, \tag{6.12b}$$

$$\partial_n^- u - \partial_n^+ u = h. \tag{6.12c}$$

For practice, let us show why a solution of (6.12) is a solution of (6.10). While in informal textbooks this is often presented as a multiplication of (6.12b) by test functions followed by integration by parts and substitution of the 'natural' boundary condition (6.12c), the rigorous argument starts directly with the transmission condition. The definition of the jump of the normal derivative (6.9) shows that (6.12c) can be written as (6.11). Substituting the PDE (6.12b) in (6.11) we reach (6.10b), which finishes the proof. The transmission problem (6.12) can be equivalently written as

$$u \in H^1(\mathbb{R}^d \setminus \Gamma), \qquad -\Delta u + u = f \quad \text{in } \mathbb{R}^d \setminus \Gamma,$$
$$\gamma^- u - \gamma^+ u = 0, \qquad \partial_n^- u - \partial_n^+ u = h,$$

which makes the continuity of the trace operator across Γ more evident. (Before it was hidden in the fact that we demand $u \in H^1(\mathbb{R}^d)$, even if the equation is taking place on $\mathbb{R}^d \setminus \Gamma$.)

6.6   Nonlocal boundary conditions (*)

In this section we explore a problem very similar to the one in Section 6.5, but reduce the entire problem to one defined exclusively on the interior domain by using a **nonlocal** boundary condition that deals with the exterior domain. We place ourselves in the same geometric setting as in Section 6.5 although with slightly different notation: we have a bounded Lipschitz domain Ω, whose boundary is denoted Γ, and we consider the unbounded surrounding domain $\Omega_+ := \mathbb{R}^d \setminus \overline{\Omega}$. Exterior traces and normal derivatives will be tagged with the $+$ superscript, while interior traces and normal derivatives will be unscripted.

An exterior Dirichlet-to-Neumann operator. Consider the operator $S : H^{1/2}(\Gamma) \to H^{-1/2}(\Gamma)$ given by $Sg := \partial_n^+ w$, where

$$w \in H^1(\Omega_+), \qquad \gamma^+ w = g, \qquad -\Delta w + w = 0. \tag{6.13}$$

We collect three important properties of this operator in the following proposition (note that Exercise 6.6 deals with a similar operator in the interior domain).

Proposition 6.4. *The operator* $S : H^{1/2}(\Gamma) \to H^{-1/2}(\Gamma)$ *(defined by $Sg :=$ $\partial_n^+ w$, where (6.13) holds) is linear, bounded, and satisfies:*

$$\langle Sg, g' \rangle_\Gamma = \langle Sg', g \rangle_\Gamma \qquad \forall g, g' \in H^{1/2}(\Gamma),$$
$$-\langle Sg, g \rangle_\Gamma \geq C_\Gamma \|g\|_{1/2,\Gamma}^2 \qquad \forall g \in H^{1/2}(\Gamma),$$

for some $C_\Gamma > 0$.

Proof. If w is the solution to (6.13) and w' is the solution to the same problem with g' as data, then $Sg = \partial_n^+ w$ and $g' = \gamma^+ w'$ and therefore (use (6.8))

$$-\langle Sg, g' \rangle_\Gamma = -\langle \partial_n^+ w, \gamma^+ w' \rangle_\Gamma = (\nabla w, \nabla w')_{\Omega_+} + (\Delta w, w')_{\Omega_+}$$
$$= (\nabla w, \nabla w')_{\Omega_+} + (w, w')_{\Omega_+}.$$

This proves symmetry, while the bound

$$-\langle Sg, g \rangle_\Gamma = \|w\|_{1,\Omega_+}^2 \geq C_\Gamma \|\gamma^+ w\|_{1/2,\Gamma}^2 = C_\Gamma \|g\|_{1/2,\Gamma}^2,$$

proves coercivity. □

A problem with nonlocal boundary conditions. Given $f \in L^2(\Omega)$ we consider the boundary value problem

$$u \in H^1(\Omega), \qquad -\Delta u + u = f, \qquad \partial_n u - S\gamma u = 0, \qquad (6.14)$$

its variational formulation

$$u \in H^1(\Omega), \qquad\qquad\qquad\qquad\qquad\qquad\qquad\qquad (6.15a)$$
$$(\nabla u, \nabla v)_\Omega + (u, v)_\Omega - \langle S\gamma u, \gamma v \rangle_\Gamma = (f, v)_\Omega \qquad \forall v \in H^1(\Omega), \qquad (6.15b)$$

and the associated minimization problem

$$\tfrac{1}{2}\|u\|_{1,\Omega}^2 - \tfrac{1}{2}\langle S\gamma u, \gamma u \rangle_\Gamma - (f, u)_\Omega = \min! \qquad u \in H^1(\Omega). \qquad (6.16)$$

The equivalence between (6.14) and (6.15) follows from the definition of the normal derivative. Problem (6.15) is uniquely solvable, since the bilinear form is bounded and coercive (coercivity is due to Proposition 6.4). Finally, the equivalence between the variational formulation (6.15) and the minimization principle (6.16) is due to the symmetry of the boundary bilinear form given in Proposition 6.4.

The equivalent problem in free space. Now let u be the solution to (6.14) and let v be the unique solution to

$$v \in H^1(\Omega_+), \qquad -\Delta v + v = 0 \quad (\text{in } \Omega_+), \qquad \gamma^+ v = \gamma u.$$

We then glue these two functions to create a single function in free space:

$$U := \begin{cases} u & \text{in } \Omega, \\ v & \text{in } \Omega_+. \end{cases}$$

By construction $\gamma U = \gamma^+ U$ and therefore $U \in H^1(\mathbb{R}^d)$. Moreover,

$$\partial_n^+ U = \partial_n^+ v = S\gamma u = \partial_n u = \partial_n U,$$

which proves that U is the unique solution of

$$U \in H^1(\mathbb{R}^d) \qquad -\Delta U + U = \tilde{f} \quad \text{in } \mathbb{R}^d,$$

where

$$\tilde{f} := \begin{cases} f & \text{in } \Omega, \\ 0 & \text{in } \Omega_+. \end{cases}$$

6.7   Mixed boundary conditions (*)

In this section we sketch the treatment of boundary value problems with Dirichlet and Neumann boundary conditions on complementary parts of the boundary. The reader is asked in Exercise 6.8 to fill in the gaps by proving all the statements made in this section.

Let Ω be a bounded Lipschitz domain with boundary Γ, let $\Gamma_D \subset \Gamma$ be a relatively open subset of the boundary with positive $(d-1)$-dimensional measure, i.e.,

$$\int_{\Gamma_D} d\Gamma(\mathbf{x}) > 0,$$

and let $\Gamma_N := \Gamma \setminus \overline{\Gamma_D}$. Consider the space

$$V_D := \{ u \in H^1(\Omega) : \gamma u = 0 \quad \text{on } \Gamma_D \}.$$

It is clear that V_D is a closed subspace of $H^1(\Omega)$.

Two trace spaces on part of the boundary. We now consider the space

$$H^{1/2}(\Gamma_N) := \{ \rho|_{\Gamma_N} : \rho \in H^{1/2}(\Gamma) \},$$

endowed with the image norm

$$\|\xi\|_{1/2,\Gamma_N} := \inf \{ \|\rho\|_{1/2,\Gamma} : \rho|_{\Gamma_N} = \xi \}.$$

This norm can be shown to be equal to the norm

$$\|\xi\| := \inf \{ \|u\|_{1,\Omega} : \gamma u|_{\Gamma_N} = \xi \},$$

which is the image norm of the trace-and-restriction operator $H^1(\Omega) \to L^2(\Gamma_N)$. Moreover, there exists a bounded extension operator $H^{1/2}(\Gamma_N) \to H^{1/2}(\Gamma)$.

Additionally, we consider the space (the three spaces on the right-hand side of the formula below coincide)

$$\tilde{H}^{1/2}(\Gamma_N) := \{\xi \in L^2(\Gamma_N) : \xi = \gamma u|_{\Gamma_N}, \quad u \in V_D\}$$
$$= \{\rho|_{\Gamma_N} : \rho \in H^{1/2}(\Gamma), \quad \rho|_{\Gamma_D} = 0\}$$
$$= \{\xi \in H^{1/2}(\Gamma_N) : \tilde{\xi} \in H^{1/2}(\Gamma)\},$$

where we have used the extension-by-zero operator

$$L^2(\Gamma_N) \ni \xi \longmapsto \tilde{\xi} \in L^2(\Gamma), \qquad \tilde{\xi} := \begin{cases} \xi & \text{in } \Gamma_N, \\ 0 & \text{in } \Gamma_D. \end{cases}$$

In this space we choose the norm

$$\|\xi\|_{\tilde{H}^{1/2}(\Gamma_N)} := \|\tilde{\xi}\|_{1/2,\Gamma} = \inf\{\|u\|_{1,\Omega} : \gamma u = \tilde{\xi}\}.$$

It can be proved that

$$\|\xi\|_{1/2,\Gamma_N} \le \|\xi\|_{\tilde{H}^{1/2}(\Gamma_N)} \qquad \forall \xi \in \tilde{H}^{1/2}(\Gamma_N),$$

and

$$\tilde{H}^{1/2}(\Gamma_N) \subset H^{1/2}(\Gamma_N) \subset L^2(\Gamma_N),$$

with dense and bounded injections.

More duality on the boundary. The dual spaces for the two possible trace spaces on Γ_N are defined so that the following

$$H^{1/2}(\Gamma_N) \subset L^2(\Gamma_N) \subset \tilde{H}^{-1/2}(\Gamma_N),$$
$$\tilde{H}^{1/2}(\Gamma_N) \subset L^2(\Gamma_N) \subset H^{-1/2}(\Gamma_N)$$

are Gelfand triples. We will formally write

$$\tilde{H}^{-1/2}(\Gamma_N) := H^{1/2}(\Gamma_N)', \qquad H^{-1/2}(\Gamma_N) := \tilde{H}^{1/2}(\Gamma_N)'.$$

The expression

$$\langle (\mathbf{p} \cdot \mathbf{n})|_{\Gamma_N}, \xi \rangle_{H^{-1/2}(\Gamma_N) \times \tilde{H}^{1/2}(\Gamma_N)} := (\mathbf{p}, \nabla v)_\Omega + (\nabla \cdot \mathbf{p}, v)_\Omega,$$

where $v \in V_D$ is such that $\gamma v|_{\Gamma_N} = \xi$, defines a bounded linear map $\mathbf{H}(\mathrm{div}, \Omega) \to H^{-1/2}(\Gamma_N)$. Also

$$\langle (\mathbf{p} \cdot \mathbf{n})|_{\Gamma_N}, \xi \rangle_{H^{-1/2}(\Gamma_N) \times \tilde{H}^{1/2}(\Gamma_N)} = \langle \mathbf{p} \cdot \mathbf{n}, \tilde{\xi} \rangle_\Gamma,$$

where $\widetilde{\xi}$ is the extension by zero of ξ. In much of the literature, the space $\widetilde{H}^{1/2}(\Gamma_N)$ is denoted $H_{00}^{1/2}(\Gamma_N)$ (this space is known as the Lions-Magenes space). Confusion reigns when denoting duals.

Problems with mixed boundary conditions. Let

$$f \in L^2(\Omega), \qquad g \in H^{1/2}(\Gamma_D), \qquad h \in H^{-1/2}(\Gamma_N).$$

For simplicity, given $v \in V_D$, we will write

$$\langle h, \gamma v \rangle_{\Gamma_N} := \langle h, (\gamma v)|_{\Gamma_N} \rangle_{H^{-1/2}(\Gamma_N) \times \widetilde{H}^{1/2}(\Gamma_N)}.$$

The variational problem

$$u \in H^1(\Omega), \quad \gamma u|_{\Gamma_D} = g,$$
$$(\nabla u, \nabla v)_\Omega + (u, v)_\Omega = (f, u)_\Omega + \langle h, \gamma v \rangle_{\Gamma_N} \qquad \forall v \in V_D$$

is uniquely solvable. This variational problem is equivalent to the boundary value problem

$$u \in H^1(\Omega), \quad -\Delta u + u = f, \quad \gamma u = g \ \text{ on } \Gamma_D, \quad \partial_n u = h \ \text{ on } \Gamma_N,$$

where the later equality has to be understood as the equality $(\nabla u \cdot \mathbf{n})|_{\Gamma_N} = h$. When $g = 0$, these problems are equivalent to

$$\tfrac{1}{2}\|u\|_{1,\Omega}^2 - (f, u)_\Omega - \langle h, \gamma u \rangle_{\Gamma_D} = \min! \qquad u \in V_D.$$

We will be able to handle the associated mixed boundary value problem

$$u \in H^1(\Omega), \quad -\Delta u = f, \quad \gamma u = g \ \text{ on } \Gamma_D, \quad \partial_n u = h \ \text{ on } \Gamma_N,$$

once we have proved the generalized Poincaré conditions in the next chapter.

Final comments and literature

The space $\mathbf{H}(\mathrm{div}, \Omega)$ is one of the many Sobolev spaces for vector fields adapted to particular differential operators. These spaces are more popular in the literature of numerical methods for partial differential equations (see for instance the classic on mixed methods by Boffi, Brezzi, and Fortin [14] or the bible on finite element methods for Stokes flow by Girault and Raviart [57]) than those found in PDE-oriented texts.

The normal component of a vector field of the form $\kappa \nabla u$, where $\kappa : \Omega \to \mathbb{R}^{d \times d}$ is a general diffusion coefficient (possibly nonsymmetric) is often called the conormal derivative of u. Impedance boundary conditions are often called Robin boundary conditions or boundary conditions of the third kind (Dirichlet and Neumann being first and second kind respectively).

Nonlocal boundary conditions based on Dirichlet-to-Neumann or Neumann-to-Dirichlet operators are the bread and butter of practitioners of numerical PDEs willing to reduce unbounded domains to the boundary of a bounded domain. It is common to call the DtN and NtD maps Poincaré-Steklov operators. The monograph of Gatica and Hsiao [53] contains a collection of equivalent formulations where the exterior nonlocal operator is realized via integral operators on the boundary. We will come back to this topic once we study the layer potentials for some elliptic operators in Chapter 14.

Exercises

6.1. Prove that the space $\mathcal{C}(\Gamma)$ is dense in $L^2(\Gamma)$. (**Hint.** Use pullbacks and the fact that $\mathcal{D}(B_{d-1}(\mathbf{0}; 1))$ is dense in $L^2(B_{d-1}(\mathbf{0}; 1))$.)

6.2. Let Ω be a bounded Lipschitz domain. Show that the divergence operator

$$\mathrm{div} : \mathbf{H}(\mathrm{div}, \Omega) \longrightarrow L^2(\Omega),$$

is surjective. (**Hint.** Solve a Laplacian and take a gradient.)

6.3. **The normal derivative operator.** Let Ω be a bounded Lipschitz domain with boundary Γ. Consider the space

$$H^1_\Delta(\Omega) := \{u \in H^1(\Omega) : \Delta u \in L^2(\Omega)\},$$

endowed with the norm

$$\|u\|^2_{H^1_\Delta(\Omega)} := \|u\|^2_\Omega + \|\nabla u\|^2_\Omega + \|\Delta u\|^2_\Omega.$$

(a) Show that it is a Hilbert space. (This includes finding the inner product.)

(b) Show that $\nabla : H^1_\Delta(\Omega) \to \mathbf{H}(\mathrm{div}, \Omega)$ is bounded.

(c) Show that the normal derivative map $\partial_n : H^1_\Delta(\Omega) \to H^{-1/2}(\Gamma)$, given by $\partial_n u := (\nabla u) \cdot \mathbf{n}$ is bounded and surjective.

(d) Show that if $u \in H^2(\Omega)$, then $\partial_n u = \gamma \nabla u \cdot \mathbf{n}$.

6.4. Consider the boundary value problem

$$-\mathrm{div}(\kappa \nabla u) + \mathbf{b} \cdot \nabla u + cu = f, \qquad (\kappa \nabla u) \cdot \mathbf{n} + \alpha \gamma u = h.$$

We require the coefficients to satisfy: $\mathbf{b} \in L^\infty(\Omega)^d$, $\kappa \in L^\infty(\Omega)$ such that $\kappa \geq \kappa_0 > 0$ almost everywhere, $c \in L^\infty(\Omega)$, and $\alpha \in L^\infty(\Gamma)$. The data are $f \in L^2(\Omega)$ and $h \in H^{-1/2}(\Gamma)$. Note that the boundary condition requires that $\kappa \nabla u \in \mathbf{H}(\mathrm{div}, \Omega)$. Find general hypotheses on the coefficients $\alpha, \mathbf{b}$, and c to ensure that the associated bilinear form is coercive. (**Remark.** With the results of Chapter 7, we will be able to generalize these hypotheses.)

6.5. Fully nonhomogeneous transmission problems. Consider the geometric configuration of Section 6.5 and the notation of two-sided traces and normal derivatives therein. Let $f \in L^2(\mathbb{R}^d \setminus \Gamma)$, $h \in H^{-1/2}(\Gamma)$, and $g \in H^{1/2}(\Gamma)$. Show that the transmission problem

$$u \in H^1(\mathbb{R}^d \setminus \Gamma), \qquad -\Delta u + u = f \quad \text{in } \mathbb{R}^d \setminus \Gamma,$$
$$\gamma^- u - \gamma^+ u = g, \qquad \partial_n^- u - \partial_n^+ u = h,$$

is well posed. Write an equivalent minimization problem. Finally show that if $g = 0$ and $h = 0$, then the solution of the transmission problem satisfies

$$-\Delta u + u = f \qquad \text{in } \mathbb{R}^d,$$

that is, the distribution $-\Delta u + u \in \mathcal{D}'(\mathbb{R}^d)$ is regular and equal to f.

6.6. A Neumann-to-Dirichlet operator. Let Ω be a bounded Lipschitz domain and consider the operator $T : H^{-1/2}(\Gamma) \to H^{1/2}(\Gamma)$ given by $Th := \gamma u$, where

$$u \in H^1(\Omega), \qquad -\Delta u + u = 0, \qquad \partial_n u = h.$$

Prove the following:

(a) T is linear and bounded.

(b) T is self-adjoint:

$$\langle h, Th' \rangle_\Gamma = \langle h', Th \rangle_\Gamma \qquad \forall h, h' \in H^{-1/2}(\Gamma).$$

(c) T is strongly positive definite

$$\langle h, Th \rangle_\Gamma \geq c_\Gamma \|h\|_{1/2,\Gamma}^2 \qquad \forall h \in H^{1/2}(\Gamma),$$

i.e., the bilinear form associated to T is coercive.

(d) T is invertible.

(**Hint.** Write a variational formulation for the problem $h \mapsto u$. When dealing with $h, h' \in H^{-1/2}(\Gamma)$ consider the associated $u, u' \in H^1(\Omega)$ that solve the corresponding Neumann problems. There are many ways to approach the invertibility of T.)

6.7. Let Ω be a bounded Lipschitz domain with boundary Γ. and consider a bounded linear operator $\Phi : H^{1/2}(\Gamma) \to H^{-1/2}(\Gamma)$ satisfying

$$\langle \Phi g, g \rangle_\Gamma \geq 0 \qquad \forall g \in H^{1/2}(\Gamma),$$

and the problem

$$u \in H^1(\Omega), \qquad -\mathrm{div}(\kappa \nabla u) + c u = f, \qquad (\kappa \nabla u) \cdot \mathbf{n} + \Phi \gamma u = h,$$

where $\kappa, c \in L^\infty(\Omega)$ satisfy $\kappa \geq \kappa_0 > 0$, $c \geq c_0 > 0$ almost everywhere (for some constants κ_0 and c_0), $f \in L^2(\Omega)$ and $h \in H^{-1/2}(\Gamma)$. Show that this problem is well posed. Give an additional hypothesis on Φ so that the problem is equivalent to a minimization problem in $H^1(\Omega)$.

6.8. Prove all the assertions of Section 6.7.

7

Poincaré inequalities and Neumann problems

7.1   Compactness 126
7.2   The Rellich-Kondrachov theorem 128
7.3   The Deny-Lions theorem 129
7.4   The Neumann problem for the Laplacian 132
7.5   Compact embedding in the unit cube 133
7.6   Korn's inequalities (*) 137
7.7   Traction problems in elasticity (*) 142
Final comments and literature 144
Exercises 145

The goal of this chapter is simple: we want to build a theory that allows us to prove the well-posedness of the Neumann problem for the Laplace operator

$$-\Delta u = f, \qquad \partial_n u = h.$$

The solution to this problem is clearly nonunique (all the operators in the left-hand side vanish when applied to constant functions) and it is clear that for the problem to have a unique solution we need

$$\langle h, 1 \rangle_\Gamma = \langle \partial_n u, \gamma 1 \rangle_\Gamma = (\Delta u, 1)_\Omega = -(f, 1)_\Omega,$$

that is, the data have to be compatible. Proving some sort of coercivity condition will require us to show the Poincaré inequality

$$\inf_{c \in \mathbb{R}} \|u - c\|_\Omega \leq C \|\nabla u\|_\Omega \qquad \forall u \in H^1(\Omega).$$

We will be able to transform this inequality into other similar (and equivalent) ones that prove coercivity of the Dirichlet form $(\nabla u, \nabla v)_\Omega$ in different subspaces of $H^1(\Omega)$. However, the proof of the Poincaré inequality that we will give requires the concept of compactness and, specifically, the compact embedding of $H^1(\Omega)$ into $L^2(\Omega)$.

7.1   Compactness

Compact operators. Let X and Y be two Hilbert spaces. A linear operator $A : X \to Y$ is said to be compact when the image of the unit ball

$$\{Ax \: : \: x \in X, \; \|x\|_X = 1\},$$

is relatively compact in Y, that is, when the closure of this set is compact in Y. Since relatively compact sets are bounded, every linear compact operator is automatically bounded and therefore continuous. Moreover, by a simple linearity argument, it is easy to see that the image of any bounded set in X under the action of a compact operator $A : X \to Y$ is a relatively compact set of Y.

Taking advantage of the fact that in complete metric spaces (such as Hilbert spaces) compact and sequentially compact sets are the same, we can characterize a compact operator in the following sequential form: for every bounded sequence $\{x_n\}$ in X there exists a subsequence $\{x_{n_k}\}$ such that $\{Ax_{n_k}\}$ is convergent in Y.

Weak convergence. One of the nicest features of Hilbert spaces lies in the fact that compactness can be characterized by how weakly convergent sequences are transformed into strongly convergent sequences. Let us first review some basic facts about weak convergence. We say that a sequence $\{x_n\}$ in a Hilbert space X is weakly convergent to x (and we write $x_n \rightharpoonup x$), when

$$(x_n, z)_X \longrightarrow (x, z)_X \qquad \forall z \in X,$$

or, equivalently (thanks to the Riesz-Fréchet theorem), when

$$\ell(x_n) \longrightarrow \ell(x) \qquad \forall \ell \in X'.$$

With the first definition (based on inner products), it is easy to prove that the same sequence cannot have two weak limits. Now let us collect some key properties of weakly convergent sequences in the following proposition. Note that its proof requires the use of some important theorems of operators in Banach spaces.

Proposition 7.1. *Let X and Y be Hilbert spaces.*

(a) *If $x_n \to x$, then $x_n \rightharpoonup x$.*

(b) *If $\{x_n\}$ is weakly convergent, then it is bounded.*

(c) *If $\{x_n\}$ is bounded, it has a weakly convergent subsequence.*

(d) *If $A : X \to Y$ is linear and bounded, then*

$$x_n \rightharpoonup x \qquad \Longrightarrow \qquad Ax_n \rightharpoonup Ax.$$

Proof. Strongly convergent sequences are weakly convergent by the Cauchy-Schwarz inequality (if we look at the inner product definition of weak convergence) or by the fact that bounded linear functionals are (sequentially) continuous.

The statement in (b) is a consequence of the Banach-Steinhaus theorem (or uniform boundedness principle). Consider the sequence of functionals

$$\ell_n := (x_n, \,\cdot\,)_X \in X'. \tag{7.1}$$

For every x, we have that the set $\{\ell_n(x) : n \geq 1\}$ is bounded, and therefore, the set of functionals $\{\ell_n : n \geq 1\}$ is bounded in X'. Since, by the Cauchy-Schwarz inequality $\|x_n\|_X = \|\ell_n\|_{X'}$, the result follows.

The statement in (c) is a rephrasing of the Banach-Alaouglu theorem: considering the bounded set of functionals ℓ_n defined in (7.1), we can use that the bounded subsets of X' contain weak-$*$ convergent subsequences, that is, there exists $\{\ell_{n_k}\}$ such that $\ell_{n_k}(z) \to \ell(z)$ for all z. If $x \in X$ is given by $\ell = (x, \,\cdot\,)_X$, then $x_{n_k} \rightharpoonup x$.

Finally, (d) is a simple consequence of the definition of Hilbert space adjoint of an operator, since

$$(Ax_n, y)_Y = (A^*y, x_n)_X \to (A^*y, x)_X = (Ax, y)_Y \quad \forall y \in Y.$$

This finishes the proof. $\qquad\qquad\qquad\qquad\qquad\qquad\qquad\qquad\qquad\qquad$ □

Proposition 7.2. *Let $A : X \to Y$ be a bounded linear operator between two Hilbert spaces. The operator A is compact if and only if $x_n \rightharpoonup x$ implies $Ax_n \longrightarrow Ax$.*

Proof. The proof of this result is a simple consequence of Proposition 7.1. Suppose the operator transforms weakly convergent sequences to strongly convergent sequences. Given a bounded sequence $\{x_n\}$ we can find a weakly convergent subsequence $x_{n_k} \rightharpoonup x$ and therefore $Ax_{n_k} \to Ax$. This is sequential compactness of A, which is equivalent to compactness of A.

Reciprocally, let $x_n \rightharpoonup x$ and note that by continuity of A we have $Ax_n \rightharpoonup Ax$. Since $\{x_n\}$ is bounded, any subsequence $\{x_{n_k}\}$ is also bounded and therefore $\{Ax_{n_k}\}$ contains a strongly convergent subsequence. This convergent subsequence must converge to Ax, because strong convergence implies weak convergence and we already know the weak limit of Ax_n to be Ax. This finishes the proof. $\qquad\qquad\qquad\qquad\qquad\qquad\qquad\qquad$ □

Proposition 7.2 gives a striaghtforward proof of the following result, which can also be easily proved by using direct arguments on sequential compactness or on the relative compactness of the image of the unit ball.

Corollary 7.1. *If $A : X \to Y$ is compact and $B : Y \to Z$ is bounded, then $BA : X \to Z$ is compact. Similarly, if $A : X \to Y$ is bounded and $B : Y \to Z$ is compact, then BA is compact.*

7.2   The Rellich-Kondrachov theorem

In this section we study under what conditions on Ω the embeddings

$$H_0^1(\Omega) \subset L^2(\Omega) \quad \text{and} \quad H^1(\Omega) \subset L^2(\Omega)$$

are compact. The compactness of an embedding operator is equivalent (Proposition 7.2) to the fact that weakly convergent sequences in the stronger norm are strongly convergent sequences in the weaker norm.

We start with a result whose proof we postpone to Section 7.5. We will be able to derive all the compact embedding results from this prototypical one.

Proposition 7.3. *If $Q := (-M, M)^d$ is a d-dimensional cube, then $H^1(Q)$ is compactly embedded into $L^2(Q)$.*

Proposition 7.4 (Rellich-Kondrachov). *If Ω is a bounded domain, then $H_0^1(\Omega)$ is compactly embedded into $L^2(\Omega)$.*

Proof. Let Q be a d-cube containing Ω. Consider the following sequence of operators

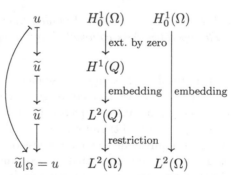

which can be considered as a factorization of the embedding operator from $H_0^1(\Omega)$ to $L^2(\Omega)$. The extension-by-zero operator on the left is bounded because of Proposition 2.7, while the central embedding operator is compact by Proposition 7.3. This and Corollary 7.1 imply that the composed operator above is compact, which finishes the proof. $\square$

Proposition 7.5. *If Ω is a bounded domain with the H^1 extension property, then $H^1(\Omega)$ is compactly embedded into $L^2(\Omega)$.*

Proof. Consider first the operator

$$\begin{array}{ccccc} H^1(\Omega) & \longrightarrow & H^1(\mathbb{R}^d) & \longrightarrow & H^1(Q) \\ u & \longmapsto & Eu & \longmapsto & Eu|_Q \end{array},$$

where E is the extension operator and Q is a d-cube containing Ω. This shows that there is a bounded extension operator $H^1(\Omega) \to H^1(Q)$. The rest of the proof closely follows the argument proving the Rellich-Kondrachov theorem (Proposition 7.4): we compose a bounded extension operator, with the compact embedding, and the restriction operator

$$H^1(\Omega) \xrightarrow{\text{extension}} H^1(Q) \xrightarrow{\text{embedding}} L^2(Q) \xrightarrow{\text{restriction}} L^2(\Omega),$$

to have a factorization of the embedding with one of the factors being compact. $\square$

The proof of the following generalization of Proposition 7.5 is left as an exercise (Exercise 7.1).

Proposition 7.6. *Let* $\Omega_1, \ldots, \Omega_M$ *be domains such that* $H^1(\Omega_j)$ *is compactly embedded into* $L^2(\Omega_j)$ *for all* j, *and let* $\Omega := \Omega_1 \cup \ldots \cup \Omega_M$. *The embedding of* $H^1(\Omega)$ *into* $L^2(\Omega)$ *is compact.*

7.3 The Deny-Lions theorem

Theorem 7.1 (Deny-Lions). *Suppose* Ω *is a bounded, open, connected set such that* $H^1(\Omega)$ *is compactly embedded into* $L^2(\Omega)$. *Let* $j : H^1(\Omega) \to \mathbb{R}$ *be a linear and bounded functional such that* $j(1) \neq 0$. *There exists* $C > 0$ *such that*

$$\|u\|_\Omega^2 \leq C \left(\|\nabla u\|_\Omega^2 + |j(u)|^2 \right) \qquad \forall u \in H^1(\Omega). \tag{7.2}$$

Proof. Assume by way of contradiction that the inequality (7.2) does not hold. Taking $C = n \in \mathbb{N}$, we can find $u_n \in H^1(\Omega)$ such that

$$\|u_n\|_\Omega = 1 \qquad \|\nabla u_n\|_\Omega^2 + |j(u_n)|^2 < \frac{1}{n}. \tag{7.3}$$

Therefore

$$\|u_n\|_{1,\Omega}^2 = \|\nabla u_n\|_\Omega^2 + 1,$$

which means that $\{u_n\}$ is bounded in $H^1(\Omega)$. We can then find a subsequence $\{u_{n_k}\}$ such that

$$u_{n_k} \rightharpoonup u \quad \text{in } H^1(\Omega),$$

and therefore

$$u_{n_k} \longrightarrow u \quad \text{in } L^2(\Omega),$$

by the compact embedding. We now have that $u_{n_k} \to u$ in $\mathcal{D}'(\Omega)$ and therefore $\nabla u_{n_k} \to \nabla u \in \mathcal{D}'(\Omega)^d$. On the other hand, we know that $\nabla u_n \to \mathbf{0}$ in $\mathbf{L}^2(\Omega)$ by (7.3). Uniqueness of the distributional limit implies that $\nabla u = \mathbf{0}$ and, since Ω is connected, this implies that $u_{n_k} \rightharpoonup c$ weakly in $H^1(\Omega)$, where c is a constant function. Since the functional j is bounded, it follows that $j(u_{n_k}) \to j(c)$. Given that $j(u_n) \to 0$ (see (7.3)), it is clear that $c = 0$. In summary, we have found that $u_{n_k} \to 0$ in $L^2(\Omega)$, but this contradicts the first condition in (7.3). $\qquad \square$

This theorem provides us with some useful corollaries.

Corollary 7.2 (Poincaré-Friedrich's inequality). *In the hypotheses of Theorem 7.1, there exists a constant $C > 0$ such that*

$$\|u\|_\Omega \leq C \|\nabla u\|_\Omega \qquad \forall u \in H_0^1(\Omega).$$

Proof. Take $j(u) := \int_\Gamma \gamma u$ in the Deny-Lions theorem. Note that the Poincaré-Friedrichs inequality holds on much more general domains, as proved in Chapter 2. $\qquad \square$

Corollary 7.3 (Poincaré's inequality). *In the hypotheses of Theorem 7.1, there exists a constant $C > 0$ such that*

$$\left\| u - \frac{1}{|\Omega|} \int_\Omega u(\mathbf{x}) \, d\mathbf{x} \right\|_\Omega \leq C \|\nabla u\|_\Omega \qquad \forall u \in H^1(\Omega).$$

Proof. Use

$$j(u) := \frac{1}{|\Omega|} \int_\Omega u(\mathbf{x}) d\mathbf{x},$$

and notice that

$$\nabla \left(u - \frac{1}{|\Omega|} \int_\Omega u(\mathbf{x}) d\mathbf{x} \right) = \nabla u \quad \text{and} \quad j\left(u - \frac{1}{|\Omega|} \int_\Omega u(\mathbf{x}) d\mathbf{x} \right) = 0.$$

$\qquad \square$

Corollary 7.4 (Generalized Poincaré's inequality). *In the hypotheses of Theorem 7.1, if V is a closed subspace of $H^1(\Omega)$ such that*

$$\mathcal{P}_0(\Omega) \cap V = \{0\},$$

(i.e., V does not contain constants or equivalently $u \in V$ satisfying $\nabla u = 0$ has to vanish), the Dirichlet form defines a norm equivalent to the usual $H^1(\Omega)$ norm in V.

Proof. Let $P : H^1(\Omega) \to V$ be the orthogonal projection onto V. Note that since constant functions are not in V we have $w := 1 - P1 \neq 0$. Consider the bounded linear operator

$$j(u) := (w, u - Pu)_{1,\Omega}.$$

It is clear that $j(1) = \|1 - P1\|_{1,\Omega}^2 \neq 0$ and $j(u) = 0$ for all $u \in V$ (since $Pu = u$ for all $u \in V$). The Deny-Lions theorem proves then the result. Note that the functional j can equivalently be written as

$$j(u) = (1 - P1, u - Pu)_{1,\Omega} = (1, u - Pu)_{1,\Omega} = (1, u - Pu)_\Omega,$$

since $I - P$ is also an orthogonal projection. $\qquad\square$

Examples. Poincaré inequalities are available in the following subspaces:

(a) $V = H_0^1(\Omega)$. (This is, once again, the Poincaré-Friedrichs inequality.)

(b) $V = \{u \in H^1(\Omega) : \int_\Omega u = 0\} = \mathcal{P}_0(\Omega)^\perp$.

(c) $V = \{u \in H^1(\Omega) : \gamma u = 0 \text{ on } \Gamma_D\}$, where Γ_D is a part of Γ such that

$$\int_{\Gamma_D} \mathrm{d}\Gamma(\mathbf{x}) \neq 0.$$

Impedance problems revisited. The bilinear form

$$a(u, v) = (\kappa \nabla u, \nabla v)_\Omega + c(u, v)_\Omega + \langle \alpha \gamma u, \gamma v \rangle_\Gamma,$$

plays a key role for a problem of the form

$$-\mathrm{div}(\kappa \nabla u) + c\, u = f, \qquad (\kappa \nabla u) \cdot \mathbf{n} + \alpha \gamma u = h,$$

(see Section 6.4). We now prove that the bilinear form a is coercive in $H^1(\Omega)$ when $\kappa, c \in L^\infty(\Omega)$, $\alpha \in L^\infty(\Gamma)$, $\kappa \geq \kappa_0 > 0$, $c \geq 0$ and $\alpha \geq 0$, with the additional requirement that c or α not be identically zero, so that one of

$$j_c(u) = \int_\Omega c(\mathbf{x}) u(\mathbf{x}) \mathrm{d}\mathbf{x},$$

$$j_\alpha(u) = \int_\Gamma \alpha(\mathbf{x}) \gamma u(\mathbf{x}) \mathrm{d}\Gamma(\mathbf{x}),$$

does not vanish on constant functions. These are weaker conditions than the ones given in Section 6.4, where strict positivity of the reaction coefficient c was used to ensure coercivity. We will thus now prove that in the above hypotheses, there exists $C > 0$ such that

$$a(u, u) \geq C\|u\|_{1,\Omega}^2 \qquad \forall u \in H^1(\Omega). \tag{7.4}$$

We will prove the estimate (7.4) when $j_c(1) > 0$. The other case follows similarly. By positivity of κ and nonnegativity of α, we can bound

$$a(u, u) \geq \kappa_0 \|\nabla u\|_\Omega^2 + (cu, u)_\Omega.$$

Additionally, we can bound

$$\left| \int_\Omega c(\mathbf{x}) u(\mathbf{x}) d\mathbf{x} \right|^2 \leq \left(\int_\Omega c(\mathbf{x}) d\mathbf{x} \right) \left(\int_\Omega c(\mathbf{x}) u^2(\mathbf{x}) d\mathbf{x} \right) = j_c(1)(cu, u)_\Omega,$$

and therefore

$$a(u, u) \geq \kappa_0 \|\nabla u\|_\Omega^2 + \frac{1}{j_c(1)} |j_c(u)|^2$$

$$\geq \frac{\kappa_0}{2} \|\nabla u\|_\Omega^2 + \frac{\kappa_0}{2} \left(\|\nabla u\|_\Omega^2 + |j(u)|^2 \right),$$

where

$$j(u) := \sqrt{\frac{2}{\kappa_0 j_c(1)}} \, j_c(u).$$

A direct application of the Deny-Lions theorem finishes the proof.

7.4   The Neumann problem for the Laplacian

In the first examples of Neumann problems we presented for the Laplacian, there was always some kind of reaction term (an L^2 inner product in the variational formulation) that guaranteed coercivity of the bilinear form. We will now tackle problems that lack this characteristic, and therefore may introduce a kernel into the problem. Consider the simplest problem of this type, the Poisson equation with Neumann boundary conditions:

$$-\Delta u = f, \qquad \partial_n u = h. \tag{7.5}$$

We require that Ω be an open, bounded, and Lipschitz domain, $f \in L^2(\Omega)$, and $h \in H^{-1/2}(\Gamma)$. Note that $-\Delta u = f \in L^2(\Omega)$ and therefore $\nabla u \in \mathbf{H}(\mathrm{div}, \Omega)$. A routine calculation shows that the variational form is

$$u \in H^1(\Omega), \qquad (\nabla u, \nabla v)_\Omega = (f, v)_\Omega + \langle h, \gamma v \rangle_\Gamma \qquad \forall v \in H^1(\Omega).$$

We can similarly show the variational form can be used to deduce the PDE. The associated minimization problem is

$$\tfrac{1}{2} \|\nabla u\|_\Omega^2 - (f, u)_\Omega - \langle h, \gamma u \rangle_\Gamma = \min! \qquad u \in H^1(\Omega).$$

Notice that if $u = 1$, then $\Delta u = 0$, and so there may not be a unique solution, and we will need to impose some conditions to ensure that a solution exists.

Proposition 7.7. *If we consider the boundary value problem (7.5), then the following hold:*

(a) *The condition $(f,1)_\Omega + \langle h,1 \rangle_\Gamma = 0$ is a necessary and sufficient condition for the existence of a solution.*

(b) *The solution is unique up to an additive constant.*

(c) *We have the stability estimate $\|\nabla u\|_\Omega \leq C \left(\|f\|_\Omega + \|h\|_{-1/2,\Gamma} \right)$.*

Proof. We will prove everything at the level of the variational formulation. Taking $v = 1$ shows that $(f,1)_\Omega + \langle h,1 \rangle_\Gamma = 0$ is a necessary condition for a solution. To show sufficiency, assume $(f,1)_\Omega + \langle h,1 \rangle_\Gamma = 0$, then the variational form can be shown to be equivalent to

$$u \in H^1(\Omega), \qquad (\nabla u, \nabla v)_\Omega = (f,v)_\Omega + \langle h, \gamma v \rangle_\Gamma \qquad \forall v \in V,$$

with $V = \{v \in H^1(\Omega) : \int_\Omega v = 0\}$. For an element $v \in H^1(\Omega)$, we can decompose

$$v = \underbrace{v - \frac{1}{|\Omega|} \int_\Omega v(\mathbf{x}) d\mathbf{x}}_{=:w \in V} + \underbrace{\frac{1}{|\Omega|} \int_\Omega v(\mathbf{x}) d\mathbf{x}}_{=:c \in \mathcal{P}_0(\Omega)} .$$

Testing the variational form with v gives

$$\begin{aligned}
(\nabla u, \nabla v)_\Omega = (\nabla u, \nabla w)_\Omega &= (f,w)_\Omega + \langle h, \gamma w \rangle_\Gamma \\
&= (f,w)_\Omega + \langle h, \gamma w \rangle_\Gamma + c\,((f,1)_\Omega + \langle h, \gamma 1 \rangle_\Gamma) \\
&= (f, w+c)_\Omega + \langle h, \gamma(w+c) \rangle_\Gamma \\
&= (f,v)_\Omega + \langle h, \gamma v \rangle_\Gamma.
\end{aligned}$$

If we now consider the modified variational problem

$$\tilde{u} \in V, \qquad (\nabla \tilde{u}, \nabla v)_\Omega = (f,v)_\Omega + \langle h, \gamma v \rangle_\Gamma \qquad \forall v \in V,$$

we can use Deny-Lions to verify that this is a coercive problem, and therefore there is a unique solution that depends continuously on the data. This also shows that $\tilde{u} + c$ is a solution to (7.5) for any constant $c \in \mathbb{R}$, proving (b). The statement (c) is the result of the standard calculations which have been done in previous chapters, which we leave to the reader to verify. $\qquad \square$

7.5 Compact embedding in the unit cube

In this section we will prove that if $Q := (0,1)^d$, then $H^1(Q)$ is compactly embedded into $L^2(Q)$. With a simple affine transformation, this result will prove Proposition 7.3, which is the same assertion written for a more general d-cube. The proof will be carried out using some arguments of Fourier analysis that, in a way, we will be borrowing from the future, since we will be dealing with the spectral system for the Neumann problem for the Laplacian in Q. While the proof of this result can be considered to be somewhat technical, the reader is recommended to peruse it, since it introduces techniques on orthogonal sets that will be useful later in the book.

We will use the following result for compact operators.

Proposition 7.8. *Every $A : X \to Y$ linear bounded operator with finite-dimensional range is compact. Furthermore, the set*

$$\mathcal{K}(X,Y) := \{A : X \to Y : A \text{ is linear and compact}\},$$

is closed in $\mathcal{B}(X,Y)$, i.e., every convergent sequence of compact operators converges to a compact operator.

Proof. The proof of the first assertion is simple. Using the characterization of compact operators with convergent sequences, we notice that since A is bounded, $x_n \rightharpoonup x$ implies that $Ax_n \rightharpoonup Ax$, but due to the finite dimensionality of the range of A, weak and strong convergence are equivalent concepts. Seen in another way, the image of the unit ball of X is a bounded set in range A, which is a finite-dimensional space. However, bounded sets in finite dimensions are relatively compact. (This is another way of writing the Heine-Borel theorem.)

Now let $\{A_m\}$ in $\mathcal{K}(X,Y)$ satisfy $A_m \to A$. Let $\{u_n\}$ be a sequence in X such that $u_n \rightharpoonup u$ in X. In particular, there exists $C > 0$ such that $\|u_n\|_X \leq C$ for all n. For a fixed $\varepsilon > 0$, we can find M such that $\|A_M - A\| \leq \varepsilon$ in the operator norm, and therefore

$$\|Au_n - Au\|_Y \leq \|(A - A_M)u_n\|_Y + \|A_M u_n - A_M u\|_Y + \|A_M u - Au\|_Y$$
$$\leq \varepsilon(C + \|u\|_X) + \|A_M u_n - A_M u\|_Y.$$

Since A_M is compact, we have $A_M u_n \to A_M u$ and therefore, there exists n_0 such that

$$\|A_M u_n - A_M u\|_Y < \varepsilon \qquad \forall n \geq n_0.$$

This concludes the proof. $\qquad\qquad\qquad\qquad\qquad\qquad\qquad\qquad\qquad\qquad\qquad\square$

Countable complete orthogonal sets. We will also need the concept of complete orthogonal sequences. Let $\{u_n\}$ be a sequence of nonzero vectors in a Hilbert space H such that

$$(u_n, u_m)_H = \delta_{nm} \|u_n\|_H^2 \qquad \|u_n\|_H \neq 0 \qquad \forall n, m.$$

We say that this sequence is **complete orthogonal** when

$$(u_n, v)_H = 0 \quad \forall n \qquad \Longrightarrow \qquad v = 0,$$

that is, $\{u_n\}$ cannot be expanded to a larger orthogonal set. In Appendix A.4 we have a proof that, given an orthogonal sequence, the following assertions are equivalent:

(a) $\{u_n\}$ is complete orthogonal;

(b) span$\{u_n : n \geq 1\}$ is dense in H;

(c) for all $v \in H$

$$\|v\|_H^2 = \sum_{n=1}^{\infty} \frac{|(v, u_n)_H|^2}{\|u_n\|_H^4} = \sum_{n=1}^{\infty} \frac{|(v, u_n)_H|^2}{|(u_n, u_n)_H|^2};$$

(d) for all $v \in H$, we have the equality

$$v = \sum_{n=1}^{\infty} \frac{1}{\|u_n\|_H^2} (v, u_n)_H u_n,$$

with convergence in H.

The norm equality in (c) is called **Parseval's identity**. Complete orthogonal sequences (also called countable Hilbert bases) exist in a Hilbert space H if and only if H is separable.

The proof. Consider the family of functions indexed by a multi-index α

$$\phi_\alpha(\mathbf{x}) = \cos(\pi\alpha_1 x_1)\cos(\pi\alpha_2 x_2)\ldots\cos(\pi\alpha_d x_d).$$

1. It is simple to show that the functions $\{\phi_\alpha\}$ are orthogonal in $L^2(Q)$. We next show that $\{\phi_\alpha\}$ is $L^2(Q)$ complete, or equivalently, the set

$$\mathbb{T} := \text{span}\{\phi_\alpha : \alpha \in \mathbb{N}^d\},$$

is dense in $L^2(Q)$. Note that $\mathbb{T}$ is an algebra, as easily follows from the trigonometric identity

$$\cos(\pi nt)\cos(\pi mt) = \tfrac{1}{2}\cos(\pi(n+m)t) + \tfrac{1}{2}\cos(\pi|n-m|t).$$

Constant functions are elements of $\mathbb{T}$ and if $\mathbf{x} \neq \mathbf{y}$ are elements of $\overline{Q}$, then there exists $i \in \{1,\ldots,d\}$ such that $x_i \neq y_i$ which implies that

$$\phi_{e_i}(\mathbf{x}) = \cos(\pi x_i) \neq \cos(\pi y_i) = \phi_{e_i}(\mathbf{y}),$$

where e_i is the multi-index with 1 in the i-th position and zeros in all other. By the Stone-Weierstrass theorem $\mathbb{T}$ is dense in $\mathcal{C}(\overline{Q})$ (with the uniform norm), but $\mathcal{D}(Q) \subset \mathcal{C}(\overline{Q}) \subset L^2(Q)$ with dense inclusions (this is due to the variational lemma), which implies that $\mathbb{T}$ is dense in $L^2(Q)$.

2. It is also simple to see that

$$-\Delta\phi_\alpha + \phi_\alpha = (1 + \pi^2\|\alpha\|^2)\phi_\alpha, \qquad \|\alpha\|^2 := |\alpha_1|^2 + \cdots + |\alpha_d|^2,$$

and

$$\nabla\phi_\alpha \cdot \mathbf{n} = 0 \qquad \text{on } \partial Q.$$

Therefore, for all $v \in H^1(Q)$

$$
\begin{aligned}
(\phi_\alpha, v)_{1,Q} &= (\nabla\phi_\alpha, \nabla v)_Q + (\phi_\alpha, v)_Q \\
&= (\nabla\phi_\alpha, \nabla v)_Q + (\Delta\phi_\alpha, v)_Q + (1 + \pi^2\|\alpha\|^2)(\phi_\alpha, v)_Q \\
&= \langle \nabla\phi_\alpha \cdot \mathbf{n}, \gamma v \rangle_{\partial Q} + (1 + \pi^2\|\alpha\|^2)(\phi_\alpha, v)_Q \\
&= (1 + \pi^2\|\alpha\|^2)(\phi_\alpha, v)_Q.
\end{aligned}
\tag{7.6}
$$

(This identity can also be proved using traditional integration by parts for $v \in \mathcal{C}^1(\overline{Q})$ and then extended to $v \in H^1(Q)$ using density.) In particular, this implies that the sequence $\{\phi_\alpha\}$ is also complete orthogonal in $H^1(Q)$ and

$$\|\phi_\alpha\|_{1,Q}^2 = (1 + \pi^2\|\alpha\|^2)\|\phi_\alpha\|_Q^2. \tag{7.7}$$

3. Finally, we define the sequence of bounded operators $I_N : H^1(Q) \to L^2(Q)$

$$I_N u := \sum_{\|\alpha\| \leq N} \frac{1}{\|\phi_\alpha\|_Q^2} (u, \phi_\alpha)_Q \phi_\alpha.$$

Note that the range of I_N is finite and, therefore, $I_N : H^1(Q) \to L^2(Q)$ is compact. Since $\{\phi_\alpha\}$ is complete orthogonal in $L^2(Q)$, we can write the remainder (of Parseval's identity)

$$
\begin{aligned}
\|u - I_N u\|_Q^2 &= \sum_{\|\alpha\| > N} \frac{1}{\|\phi_\alpha\|_Q^2} |(u, \phi_\alpha)_Q|^2 \\
&= \sum_{\|\alpha\| > N} \frac{1}{\|\phi_\alpha\|_Q^2} \frac{1}{(1 + \pi^2\|\alpha\|^2)^2} |(u, \phi_\alpha)_{1,Q}|^2 \quad \text{(by (7.6))} \\
&= \sum_{\|\alpha\| > N} \frac{|(u, \phi_\alpha)_{1,Q}|^2}{\|\phi_\alpha\|_{1,Q}^2} \frac{1}{1 + \pi^2\|\alpha\|^2} \quad\quad \text{(by (7.7))} \\
&\leq \frac{1}{1 + \pi^2 N^2} \|u\|_{1,Q}^2,
\end{aligned}
$$

where in the last inequality we have applied that $\{\phi_\alpha\}$ is orthogonal in $H^1(Q)$ (completeness is actually not needed for this argument). This proves that in operator norm $H^1(Q) \to L^2(Q)$, the sequence I_N converges to the inclusion operator and, therefore, the inclusion operator is compact.

7.6 Korn's inequalities (*)

This section is the linear elasticity counterpart of the section where we proved Poincaré's inequality. Unfortunately the corresponding inequality (named Korn's second inequality) is much more complicated to prove and we will leave an important and highly nontrivial step unproved in the process. Let us recall some notation. We consider the space $\mathbf{H}^1(\Omega) := H^1(\Omega; \mathbb{R}^d) \equiv H^1(\Omega)^d$ and the linear strain operator (symmetric gradient)

$$\varepsilon(\mathbf{u}) := \tfrac{1}{2}(\nabla\mathbf{u} + (\nabla\mathbf{u})^\top).$$

We have already seen that

$$\|\mathbf{u}\|_\Omega \le C\|\varepsilon(\mathbf{u})\|_\Omega \qquad \forall \mathbf{u} \in \mathbf{H}_0^1(\Omega) \equiv H_0^1(\Omega)^d.$$

This inequality holds in every domain where the Poincaré-Friedrichs inequality holds, that is, on every open domain which is bounded in at least one direction. However, when we look for the Neumann problem for linear elasticity (the Navier equations with imposed normal traction on all the boundary; see Section 7.7), we need a different type of inequality, more in the vein of Poincaré's inequality or its generalizations, to deal with coercivity issues. In particular, we want to know what we need to add to the seminorm $\|\varepsilon(\mathbf{u})\|_\Omega$ to make it an equivalent norm in $\mathbf{H}^1(\Omega)$.

Some preliminaries. Recall that $H^{-1}(\Omega)$ is the representation of the dual space of $H_0^1(\Omega)$ in the Gelfand triple

$$H_0^1(\Omega) \subset L^2(\Omega) \subset H^{-1}(\Omega),$$

and that the distributional derivative can be understood as a bounded linear operator $\partial_{x_i} : L^2(\Omega) \to H^{-1}(\Omega)$ using the formula

$$\langle \partial_{x_i} u, v \rangle_{H^{-1}(\Omega) \times H_0^1(\Omega)} = -(u, \partial_{x_i} v)_\Omega \qquad v \in H_0^1(\Omega).$$

We will use the notation $\|\cdot\|_{-1,\Omega}$ for the $H^{-1}(\Omega)$ or $H^{-1}(\Omega)^d$ norms. We start our path towards Korn's second inequality with a simple technical lemma.

Proposition 7.9. *For all* $\mathbf{u} \in \mathbf{H}^1(\Omega)$,

$$\|\partial_{x_i}\partial_{x_j} u_\ell\|_{-1,\Omega} \le \|\varepsilon_{i\ell}(\mathbf{u})\|_\Omega + \|\varepsilon_{ij}(\mathbf{u})\|_\Omega + \|\varepsilon_{j\ell}(\mathbf{u})\|_\Omega \qquad i,j,\ell = 1,\dots,d,$$

where $\varepsilon_{ij}(\mathbf{u}) = \tfrac{1}{2}(\partial_{x_i} u_j + \partial_{x_j} u_i)$ *are the entries of* $\varepsilon(\mathbf{u})$.

Proof. The result is almost trivial to prove once we realize that

$$\partial_{x_i}\partial_{x_j} u_\ell = \partial_{x_j}\varepsilon_{i\ell}(\mathbf{u}) + \partial_{x_i}\varepsilon_{j\ell}(\mathbf{u}) - \partial_{x_\ell}\varepsilon_{ij}(\mathbf{u}),$$

and therefore

$$\langle \partial_{x_i}\partial_{x_j}u_\ell, v \rangle_{H^{-1}(\Omega) \times H_0^1(\Omega)} = -(\varepsilon_{i\ell}(\mathbf{u}), \partial_{x_j}v)_\Omega - (\varepsilon_{j\ell}(\mathbf{u}), \partial_{x_i}v)_\Omega$$
$$+ (\varepsilon_{ij}(\mathbf{u}), \partial_{x_\ell}v)_\Omega \qquad \forall v \in H_0^1(\Omega).$$

$\square$

The next ingredient is a deep result related to the gradient operator $\nabla :$ $L^2(\Omega) \to H^{-1}(\Omega)^d$. We will come back to this important and nontrivial result in Section 10.7, but for the moment being, we will just accept the result. The averaging functional we consider is

$$\mu(u) := \frac{1}{|\Omega|}\int_\Omega u(\mathbf{x})\mathrm{d}\mathbf{x}.$$

[Proof not provided]

Proposition 7.10. *Let Ω be a bounded connected Lipschitz domain. There exists $C > 0$ such that*

$$\|u - \mu(u)\|_\Omega \le C\|\nabla u\|_{-1,\Omega} \qquad \forall u \in L^2(\Omega).$$

Propositions 7.9 and 7.10 and Poincaré's inequality (Corollary 7.3) are the building blocks for our first version of Korn's second inequality.

Proposition 7.11. *There exists a constant $C > 0$ such that*

$$\|\mathbf{u}\|_{1,\Omega} \le C\left(\sum_{i=1}^d |\mu(u_i)| + \sum_{i,j=1}^d |\mu(\partial_{x_i}u_j - \partial_{x_j}u_i)| + \|\boldsymbol{\varepsilon}(\mathbf{u})\|_\Omega\right) \qquad \forall \mathbf{u} \in \mathbf{H}^1(\Omega).$$

Proof. For all $i, j \in \{1, \ldots, d\}$ we have

$$\begin{aligned}
\|\partial_{x_i}u_j\|_\Omega &\le C_1(|\mu(\partial_{x_i}u_j)| + \|\nabla\partial_{x_i}u_j\|_{-1,\Omega}) && \text{(by Proposition 7.10)}\\
&\le C_2(|\mu(\partial_{x_i}u_j)| + \|\boldsymbol{\varepsilon}(\mathbf{u})\|_\Omega) && \text{(by Proposition 7.9)}\\
&\le C_3(|\mu(\partial_{x_i}u_j - \partial_{x_j}u_i)| + \|\boldsymbol{\varepsilon}(\mathbf{u})\|_\Omega),
\end{aligned}$$

where in the last inequality we have used that

$$\begin{aligned}
|\mu(\partial_{x_i}u_j)| &\le |\mu(\varepsilon_{ij}(\mathbf{u}))| + \tfrac{1}{2}|\mu(\partial_{x_i}u_j - \partial_{x_j}u_i)|\\
&\le |\Omega|^{-1/2}\|\boldsymbol{\varepsilon}(\mathbf{u})\|_\Omega + \tfrac{1}{2}|\mu(\partial_{x_i}u_j - \partial_{x_j}u_i)|.
\end{aligned}$$

This proves that

$$|\mathbf{u}|_{1,\Omega} \le C\left(\sum_{i,j=1}^d |\mu(\partial_{x_i}u_j - \partial_{x_j}u_i)| + \|\boldsymbol{\varepsilon}(\mathbf{u})\|_\Omega\right) \qquad \forall \mathbf{u} \in \mathbf{H}^1(\Omega).$$

Finally, Poincaré's inequality proves that

$$\|\mathbf{u}\|_\Omega \le C \left(\sum_{i=1}^d |\mu(u_i)| + |\mathbf{u}|_{1,\Omega} \right) \qquad \mathbf{u} \in \mathbf{H}^1(\Omega),$$

which completes the proof. $\qquad\qquad\square$

An equivalent norm. Before we move on, let us first notice that the right-hand side in the inequality of Proposition 7.11 might look misleading in containing $d + d^2 + 1$ terms, although all diagonal terms in the double sum vanish and off-diagonal terms can be paired. To avoid duplications, we consider the index set

$$\mathcal{I} := \{1, \ldots, d\} \cup \{(i,j) : 1 \le i < j \le d\}, \qquad \#\mathcal{I} = m := \tfrac{1}{2}d(d+1),$$

and number the functionals in the right-hand side using a single index in the set $\mathcal{I}$

$$r \in \mathcal{I} \longmapsto \mu_r(u) := \begin{cases} \mu(u_i), & r = i, \\ \mu(\partial_{x_i} u_j - \partial_{x_j} u_i), & r = (i,j). \end{cases}$$

We finally collect all functionals in a single bounded linear operator

$$\mathbf{H}^1(\Omega) \ni \mathbf{u} \longmapsto \boldsymbol{\mu}(\mathbf{u}) := (\mu_r(\mathbf{u}))_{r \in \mathcal{I}} \in \mathbb{R}^m.$$

As a simple consequence of Proposition 7.11 we have

$$C_1 \|\mathbf{u}\|_{1,\Omega}^2 \le |\boldsymbol{\mu}(\mathbf{u})|^2 + \|\boldsymbol{\varepsilon}(\mathbf{u})\|_\Omega^2 \le C_2 \|\mathbf{u}\|_{1,\Omega}^2. \qquad (7.8)$$

This is already a perfectly good working form of Korn's second inequality. We will, however, derive three more statements that are more ready to use.

Infinitesimal rigid motions. We consider the finite-dimensional space

$$\mathcal{M} := \{\mathbf{u}(\mathbf{x}) := \mathbf{a} + A\mathbf{x} : \mathbf{a} \in \mathbb{R}^d, \quad A \in \mathbb{R}^{d \times d}, \quad A^\top = -A\}.$$

Note that

$$\dim \mathcal{M} = \#\mathcal{I} = m = \tfrac{1}{2}d(d+1).$$

The elements of $\mathcal{M}$ can be shown to be linearized forms of rigid motions on $\mathbb{R}^d$, although this is not relevant for this study. The following simple lemma is though.

Lemma 7.1. *The 'interpolation' operator* $P : \mathbf{H}^1(\Omega) \to \mathcal{M}$ *given by*

$$P\mathbf{u} \in \mathcal{M} \qquad \boldsymbol{\mu}(P\mathbf{u}) = \boldsymbol{\mu}(\mathbf{u}),$$

is well-defined and bounded

$$\|P\mathbf{u}\|_{1,\Omega} \le C\|\mathbf{u}\|_{1,\Omega} \qquad \forall \mathbf{u} \in \mathbf{H}^1(\Omega).$$

Proof. Consider the linear map

$$\mathcal{M} \ni \mathbf{m} \longmapsto \boldsymbol{\mu}(\mathbf{m}) \in \mathbb{R}^m, \tag{7.9}$$

between vector spaces of the same dimension. We first prove this is injective, and therefore invertible. To show this, we write

$$\mathbf{m}(\mathbf{x}) = \sum_{i=1}^{d} c_i \mathbf{e}_i + \sum_{i<j} c_{ij} \left((\mathbf{e}_j \cdot \mathbf{x})\mathbf{e}_i - (\mathbf{e}_i \cdot \mathbf{x})\mathbf{e}_j) \right)$$

$$= \sum_{i=1}^{d} c_i \mathbf{e}_i + \sum_{i<j} c_{ij} (\mathbf{E}_{ij} - \mathbf{E}_{ji})\mathbf{x},$$

where

$$\mathbf{E}_{ij} = \mathbf{e}_i \mathbf{e}_j^\top, \qquad (\mathbf{E}_{ij})_{lk} = \delta_{il}\delta_{jk},$$

i.e., $\mathbf{E}_{ij}$ is the $d \times d$ matrix with zero entries except in the (i, j) position where we place the unit. We can also write this componentwise as

$$m_i(\mathbf{x}) = c_i + \sum_{j \neq i} c_{ij}(x_j - x_i) \qquad i = 1, \ldots, d.$$

If we assume that $\mu_r(\mathbf{m}) = 0$ for all $r \in \mathcal{I}$, then

$$0 = \mu_{(i,j)}(\mathbf{m}) = \mu(\partial_{x_i} m_j - \partial_{x_j} m_i) = -2c_{ij} \qquad 1 \leq i < j \leq d.$$

Now if $\boldsymbol{\mu}(\mathbf{m}) = \mathbf{0}$ it follows that $\mathbf{m} = \mathbf{0}$. Therefore the map (7.9) is invertible and we can find $\{\mathbf{m}_r : r \in \mathcal{I}\}$ such that

$$\mu_r(\mathbf{m}_s) = \delta_{rs} \qquad r, s \in \mathcal{I}.$$

(This is the Lagrange basis for the abstract interpolation operator defined by the functionals μ_r in the space $\mathcal{M}$.) We then just need to define

$$P\mathbf{u} = \sum_{r \in \mathcal{I}} \mu_r(\mathbf{u})\mathbf{m}_r.$$

Since $\mu_r : \mathbf{H}^1(\Omega) \to \mathbb{R}$ are bounded linear functionals with

$$|\mu_r(\mathbf{u})| \leq \|\mathbf{u}\|_{1,\Omega} \qquad r \in \mathcal{I},$$

the result follows readily, with the bound given by

$$\|P\mathbf{u}\|_{1,\Omega} \leq (\max_{r \in \mathcal{I}} \|\mathbf{m}_r\|_{1,\Omega}) \|\mathbf{u}\|_{1,\Omega}.$$

This finishes the proof. □

Proposition 7.12 (Korn's second inequality, v1). *There exists $C > 0$ such that*
$$\|\mathbf{u} - P\mathbf{u}\|_{1,\Omega} \leq C\|\boldsymbol{\varepsilon}(\mathbf{u})\|_\Omega \qquad \forall \mathbf{u} \in \mathbf{H}^1(\Omega),$$
where $P : \mathbf{H}^1(\Omega) \to \mathcal{M}$ is the interpolation operator of Lemma 7.1. Therefore $\boldsymbol{\varepsilon}(\mathbf{u}) = 0$ if and only if $\mathbf{u} \in \mathcal{M}$.

Proof. By Proposition 7.11 (see also (7.8)), we can bound
$$\|\mathbf{u} - P\mathbf{u}\|_{1,\Omega}^2 \leq C\left(|\boldsymbol{\mu}(\mathbf{u} - P\mathbf{u})|^2 + \|\boldsymbol{\varepsilon}(\mathbf{u} - P\mathbf{u})\|_\Omega^2\right),$$
but $\boldsymbol{\mu}(P\mathbf{u}) = \boldsymbol{\mu}(\mathbf{u})$ by construction of P and $\boldsymbol{\varepsilon}(P\mathbf{u}) = 0$, since $\boldsymbol{\varepsilon}(\mathbf{m}) = 0$ for all $\mathbf{m} \in \mathcal{M}$. $\qquad\square$

Proposition 7.13 (Korn's second inequality, v2). *If $\boldsymbol{\chi} : \mathbf{H}^1(\Omega) \to \mathbb{R}^m$ ($m = \dim \mathcal{M}$) is a bounded linear operator such that $\ker \boldsymbol{\chi} \cap \mathcal{M} = \{\mathbf{0}\}$, i.e.,*
$$\boldsymbol{\chi}(\mathbf{m}) = \mathbf{0}, \quad \mathbf{m} \in \mathcal{M} \qquad \Longrightarrow \qquad \mathbf{m} = \mathbf{0}, \tag{7.10}$$
then there exists $C > 0$ such that
$$\|\mathbf{u}\|_{1,\Omega} \leq C(|\boldsymbol{\chi}(\mathbf{u})| + \|\boldsymbol{\varepsilon}(\mathbf{u})\|_\Omega) \qquad \forall \mathbf{u} \in \mathbf{H}^1(\Omega). \tag{7.11}$$

Proof. If (7.11) does not hold, we can find a sequence with the following properties
$$\|\mathbf{u}_n\|_{1,\Omega} = 1 \quad \forall n, \qquad \|\boldsymbol{\varepsilon}(\mathbf{u}_n)\|_\Omega \to 0, \qquad |\boldsymbol{\chi}(\mathbf{u}_n)| \to 0.$$
We can then take a subsequence $\{\mathbf{u}_{n_k}\}$ such that $\mathbf{u}_{n_k} \rightharpoonup \mathbf{u} \in \mathbf{H}^1(\Omega)$. This implies that $\boldsymbol{\varepsilon}(\mathbf{u}_{n_k}) \rightharpoonup \boldsymbol{\varepsilon}(\mathbf{u})$ in $L^2(\Omega; \mathbb{R}^{d\times d})$, while $\boldsymbol{\varepsilon}(\mathbf{u}_{n_k}) \to 0$ in the same space. Therefore $\mathbf{u} \in \mathcal{M}$. At the same time, the weak convergence implies that $\boldsymbol{\chi}(\mathbf{u}_{n_k}) \to \boldsymbol{\chi}(\mathbf{u})$ and $\boldsymbol{\chi}(\mathbf{u}_{n_k}) \to \mathbf{0}$, which implies that $\mathbf{u} = \mathbf{0}$, by hypothesis (7.10).

Finally, we have that $\mathbf{u}_{n_k} \rightharpoonup \mathbf{0}$ in $\mathbf{H}^1(\Omega)$, which implies that $\boldsymbol{\mu}(\mathbf{u}_{n_k}) \to \mathbf{0}$ and therefore (Proposition 7.11) $\|\mathbf{u}_{n_k}\|_{1,\Omega} \to 0$, which contradicts the fact that $\|\mathbf{u}_n\|_{1,\Omega} = 1$ for all n. $\qquad\square$

Proposition 7.14 (Korn's second inequality, v3). *If V is a closed subspace of $\mathbf{H}^1(\Omega)$ such that $V \cap \mathcal{M} = \{\mathbf{0}\}$, then there exists $C > 0$ such that*
$$\|\mathbf{u}\|_{1,\Omega} \leq C\|\boldsymbol{\varepsilon}(\mathbf{u})\|_\Omega \qquad \forall \mathbf{u} \in V. \tag{7.12}$$

Proof. If (7.12) does not hold, we can build a sequence
$$\mathbf{u}_n \in V \qquad \|\mathbf{u}_n\|_{1,\Omega} = 1, \qquad \|\boldsymbol{\varepsilon}(\mathbf{u}_n)\|_\Omega \to 0.$$
Taking a weakly convergent subsequence $\mathbf{u}_{n_k} \rightharpoonup \mathbf{u}$, and proceeding as in the proof of Proposition 7.13, we can show that $\mathbf{u} \in \mathcal{M}$. However,
$$0 = (\mathbf{u}_{n_k}, \mathbf{v}) \longrightarrow (\mathbf{u}, \mathbf{v}) \qquad \forall \mathbf{v} \in V^\perp,$$
which implies that $\mathbf{u} \in V^{\perp\perp} = V$. Since $\mathcal{M} \cap V = \{\mathbf{0}\}$ it follows that $\mathbf{u} = \mathbf{0}$ and therefore $\boldsymbol{\mu}(\mathbf{u}_{n_k}) \to \mathbf{0}$. Using Proposition 7.11 it follows that $\|\mathbf{u}_{n_k}\|_{1,\Omega} \to 0$, which is a contradiction with the fact that these functions have unit norm. $\qquad\square$

7.7 Traction problems in elasticity (*)

In Section 4.6 we explored the Dirichlet problem for the linear elasticity equations (the Navier-Lamé system). In this section we will study the associated Neumann problem. We will first generalize the strain-to-stress relation to deal with more general linear material laws.

General material laws. We consider $d^4 = d^2 \times d^2$ material coefficients

$$C_{ijkl} \in L^\infty(\Omega) \qquad i,j,k,l = 1,\ldots,d,$$

with the following symmetry conditions

$$C_{ijkl} = C_{jikl} = C_{klij} \qquad \text{a.e.} \qquad \forall i,j,k,l, \tag{7.13}$$

and the positivity condition

$$\sum_{i,j,k,l=1}^{d} C_{ijkl}\epsilon_{ij}\epsilon_{kl} \geq c_0 \sum_{i,j=1}^{d} \epsilon_{ij}^2 \qquad \text{a.e.} \qquad \forall \epsilon_{ij} = \epsilon_{ji}. \tag{7.14}$$

The above coefficients constitute the 4-index stiffness tensor, which provides the strain-to-stress relation by the formula

$$\sigma_{ij} = \sum_{k,l=1}^{d} C_{ijkl}\varepsilon_{kl}.$$

To simplify notation, we will consider the operator

$$C : \mathbb{R}^{d \times d} \to L^\infty(\Omega; \mathbb{R}^{d \times d}),$$

given by

$$(CM)_{ij} = \sum_{k,l=1}^{d} C_{ijkl} M_{kl}.$$

The symmetry conditions (7.13) are equivalent to the following properties:

$$CM \in \mathbb{R}^{d \times d}_{\text{sym}} \qquad \text{a.e.} \qquad \forall M \in \mathbb{R}^{d \times d},$$
$$(CM) : N = M : (CN) \qquad \text{a.e.} \qquad \forall M, N \in \mathbb{R}^{d \times d},$$

where $\mathbb{R}^{d \times d}_{\text{sym}}$ is the space of symmetric matrices. These properties imply that $CM = 0$ if M is skew-symmetric and therefore

$$CM = C(\tfrac{1}{2}(M + M^\top)) \qquad (CM) : N = (CM) : (\tfrac{1}{2}(N + N^\top)).$$

The positivity property (7.14) is equivalent to

$$(\mathrm{CM}) : \mathrm{M} \geq c_0 \|\mathrm{M}\|^2 \qquad \text{a.e.} \qquad \forall \mathrm{M} \in \mathbb{R}^{d\times d}_{\mathrm{sym}}, \qquad (7.15)$$

where $\|\mathrm{M}\|^2 = \mathrm{M} : \mathrm{M}$.

Simpler cases. When there exist two functions $\lambda, \mu \in L^\infty(\Omega)$ such that we can write the material coefficients using the following expression in terms of Kronecker symbols,

$$C_{ijkl} = \lambda \delta_{ij}\delta_{kl} + \mu(\delta_{ik}\delta_{jl} + \delta_{il}\delta_{jk}),$$

we say that the material is isotropic. In this case

$$\boldsymbol{\sigma} = 2\mu\boldsymbol{\varepsilon}(\mathbf{u}) + \lambda(\nabla \cdot \mathbf{u})\mathrm{I}.$$

When the coefficients C_{ijkl} are constant, we say that the material is homogeneous. The simplest case, homogeneous isotropic materials, corresponds to material laws (stiffness tensors) that can be expressed in terms of the two constant Lamé parameters.

The normal traction. The elastic bilinear form associated to the material law $\boldsymbol{\sigma} = \mathrm{C}\boldsymbol{\varepsilon}$ is given by

$$a(\mathbf{u}, \mathbf{v}) := (\mathrm{C}\boldsymbol{\varepsilon}(\mathbf{u}), \boldsymbol{\varepsilon}(\mathbf{v}))_\Omega = (\mathrm{C}\boldsymbol{\varepsilon}(\mathbf{u}), \nabla\mathbf{v})_\Omega = (\boldsymbol{\varepsilon}(\mathbf{u}), \mathrm{C}\boldsymbol{\varepsilon}(\mathbf{v}))_\Omega.$$

This is a bounded bilinear form in $\mathbf{H}^1(\Omega)$ thanks to the boundedness of the coefficients of the stiffness tensor C. The 'integration by parts' formula (Betti's formula in the jargon of elasticity theory) provides a definition of $\boldsymbol{\sigma}(\mathbf{u})\mathbf{n} \in H^{-1/2}(\Gamma)^d$

$$\langle \boldsymbol{\sigma}(\mathbf{u})\mathbf{n}, \gamma\mathbf{v}\rangle_\Gamma := (\boldsymbol{\sigma}(\mathbf{u}), \boldsymbol{\varepsilon}(\mathbf{v}))_\Omega + (\mathrm{div}\,\boldsymbol{\sigma}(\mathbf{u}), \mathbf{v})_\Omega$$
$$= (\boldsymbol{\sigma}(\mathbf{u}), \nabla\mathbf{v})_\Omega + (\mathrm{div}\,\boldsymbol{\sigma}(\mathbf{u}), \mathbf{v})_\Omega \qquad \mathbf{v} \in \mathbf{H}^1(\Omega),$$

for all $\mathbf{u} \in \mathbf{H}^1(\Omega)$ such that $\mathrm{C}\,\nabla\mathbf{u} = \mathrm{C}\boldsymbol{\varepsilon}(\mathbf{u}) \in \mathbf{H}(\mathrm{div}, \Omega)^d$. Note that we are just taking the normal component of the rows of the stress $\boldsymbol{\sigma}(\mathbf{u})$.

Coercivity. The positivity condition (7.15) implies that

$$a(\mathbf{u}, \mathbf{u}) = (\mathrm{C}\boldsymbol{\varepsilon}(\mathbf{u}), \boldsymbol{\varepsilon}(\mathbf{u}))_\Omega \geq c_0 \|\boldsymbol{\varepsilon}(\mathbf{u})\|^2_\Omega \qquad \forall\mathbf{u} \in \mathbf{H}^1(\Omega).$$

Coercivity of a is then just a consequence of Korn's second inequality.

The pure traction problem. The data functions are

$$\mathbf{f} \in \mathbf{L}^2(\Omega), \qquad \mathbf{h} \in H^{-1/2}(\Gamma)^d,$$

the minimization problem is

$$\tfrac{1}{2}a(\mathbf{u}, \mathbf{u}) - (\mathbf{f}, \mathbf{u})_\Omega - \langle \mathbf{h}, \gamma\mathbf{u}\rangle_\Gamma = \min! \qquad \mathbf{u} \in \mathbf{H}^1(\Omega), \qquad (7.16)$$

the variational formulation is

$$\mathbf{u} \in \mathbf{H}^1(\Omega), \qquad a(\mathbf{u}, \mathbf{v}) = (\mathbf{f}, \mathbf{v})_\Omega + \langle \mathbf{h}, \gamma \mathbf{v} \rangle_\Gamma \qquad \forall \mathbf{v} \in \mathbf{H}^1(\Omega), \qquad (7.17)$$

and the boundary value problem is

$$\mathbf{u} \in \mathbf{H}^1(\Omega), \qquad -\mathrm{div}\, \boldsymbol{\sigma}(\mathbf{u}) = \mathbf{f}, \qquad \boldsymbol{\sigma}(\mathbf{u})\mathbf{n} = \mathbf{h}. \qquad (7.18)$$

Proposition 7.15. *Problems* (7.16), (7.17), *and* (7.18) *are equivalent. They are solvable if and only if*

$$(\mathbf{f}, \mathbf{m})_\Omega + \langle \mathbf{h}, \gamma \mathbf{m} \rangle_\Gamma = 0 \qquad \forall \mathbf{m} \in \mathcal{M}.$$

The solution is unique up to an element of $\mathcal{M}$. *The only solution of* (7.16)-(7.18) *which satisfies*

$$(\mathbf{u}, \mathbf{m})_\Omega = 0 \qquad \forall \mathbf{m} \in \mathcal{M},$$

can be bounded by

$$\|\mathbf{u}\|_{1,\Omega} \le C(\|\mathbf{f}\|_\Omega + \|\mathbf{h}\|_{-1/2,\Gamma}).$$

Final comments and literature

Our approach to proving any of the many versions of Poincaré's inequality is using a compactness argument, based on the Rellich-Kondrachov theorem. This theorem is typically stated on a general bounded open set Ω in $\mathbb{R}^d$ as the compact injection of $H_0^1(\Omega)$ into $L^2(\Omega)$. The proof for the compact injection of $H^1(\Omega)$ into $L^2(\Omega)$ follows, with some restrictions on the type of domain, from an extension and restriction argument. Our approach has been to prove the compact injection of $H^1(\Omega)$ into $L^2(\Omega)$ for a hypercube, using tools from Fourier analysis (essentially, the spectral set for the Neumann problem for the Laplacian) and then proceed to prove everything else by extension. There are other ways to prove the Poincaré inequality with a more quantitative knowledge of the constants involved, for instance [56, Section 7.8] contains a proof on convex domains.

We prove the traditional Poincaré inequality (Corollary 7.3) as a consequence of the Deny-Lions theorem (Theorem 7.1) although, knowing the Rellich-Kondrachov theorem, we could first prove the corollary and then follow with the theorem. There is not much agreement on how to refer to the many versions of Poincaré's and Friedrichs' inequalities. In the finite element community, it is common to refer to the Bramble-Hilbert lemma [20], which is a rephrasing of the Deny-Lions theorem (see Exercise 7.6).

Korn's inequalities are another can of worms. The first inequality, holding in $\mathbf{H}_0^1(\Omega)$ (Proposition 4.10), is very easy to prove, but the many equivalent ones in $\mathbf{H}^1(\Omega)$ are a completely different story. We derive it from Proposition 7.10 due to Nečas [86] (see [57, Chapter 1, Theorem 2.2] and the comments therein). An entire chapter on Korn's inequalties can be found in [45], while the relation of Korn's inequalities with the divergence operator is very well explained in the monograph by Acosta and Durán [1], which we will discuss again in Chapter 10.

Exercises

7.1. Prove Proposition 7.6, namely, if $\Omega_1, \ldots, \Omega_M$ are domains such that $H^1(\Omega_j)$ is compactly embedded into $L^2(\Omega_j)$ for all j, and $\Omega := \Omega_1 \cup \ldots \cup \Omega_M$, then the embedding of $H^1(\Omega)$ into $L^2(\Omega)$ is compact.

7.2. Let Ω be a bounded Lipschitz domain with boundary Γ. Let $\alpha \in L^\infty(\Gamma)$ be nonnegative. Use the Deny-Lions theorem to show that the bilinear form

$$(\nabla u, \nabla v)_\Omega + \langle \alpha \gamma u, \gamma v \rangle_\Gamma$$

is coercive in $H^1(\Omega)$ if and only if $\alpha \neq 0$. (Note that we have assumed $\alpha \geq 0$.)

7.3. **A variant of the Deny-Lions theorem.** Let $a : H^1(\Omega) \times H^1(\Omega) \to \mathbb{R}$ be bilinear, bounded, and such that

$$a(u, u) \geq C \|\nabla u\|_\Omega^2 \qquad \forall u \in H^1(\Omega), \qquad a(1, 1) \neq 0.$$

Show that a is coercive in $H^1(\Omega)$. (**Hint.** Assume that it is not coercive, and follow the proof of the Deny-Lions theorem.)

7.4. **Convection-diffusion.** We want to find conditions on β and c ensuring that the bilinear form

$$(\nabla u, \nabla v)_\Omega + (\boldsymbol{\beta} \cdot \nabla u, v)_\Omega + (c u, v)_\Omega$$

is bounded and coercive in $H^1(\Omega)$.

(a) Show that $c \in L^\infty(\Omega), \boldsymbol{\beta} \in \mathcal{C}^1(\overline{\Omega})^d$ satisfying

$$\int_\Omega (c(\mathbf{x}) - \frac{1}{2} \nabla \cdot \boldsymbol{\beta}(\mathbf{x}))(u(\mathbf{x}))^2 d\mathbf{x} + \frac{1}{2} \int_\Gamma (\boldsymbol{\beta}(\mathbf{x}) \cdot \mathbf{n}(\mathbf{x})) (\gamma u(\mathbf{x}))^2 d\mathbf{x} \geq 0,$$

and

$$\int_\Omega c(\mathbf{x}) \, d\mathbf{x} > 0,$$

are sufficient conditions for coercivity.

(b) Show that $c \in L^\infty(\Omega), \beta \in L^\infty(\Omega)^d$ satisfying $\nabla \cdot \beta \in L^\infty(\Omega)$, $\beta \cdot \mathbf{n} = 0$ (this normal trace is taken in the sense of $\mathbf{H}(\text{div}, \Omega)$) and

$$c - \tfrac{1}{2}\nabla \cdot \beta \geq 0, \qquad \text{and} \qquad \int_\Omega c(\mathbf{x}) \, d\mathbf{x} > 0,$$

are also sufficient conditions for coercivity.

(**Hint.** At a crucial moment, the result of the previous exercise is quite useful.)

7.5. Let Ω be a set such that $H^1(\Omega)$ is compactly embedded into $L^2(\Omega)$. Show that $H^2(\Omega)$ is compactly embedded into $H^1(\Omega)$. (**Hint.** Show that if $u_n \rightharpoonup u$ in $H^2(\Omega)$, then $u_n \rightharpoonup u$ in $H^1(\Omega)$ and $\partial_{x_i} u_n \rightharpoonup \partial_{x_i} u$ in $H^1(\Omega)$ for $i = 1, \ldots, d$.)

7.6. **More Poincaré inequalities.** For a nonnegative integer k and an open set Ω, consider the space

$$H^k(\Omega) := \{u \in L^2(\Omega) : \partial^\alpha u \in L^2(\Omega) \quad |\alpha| \leq k\},$$

endowed with the norm

$$\|u\|_{k,\Omega}^2 := \sum_{|\alpha| \leq k} \|\partial^\alpha u\|_\Omega^2,$$

where the sum runs over multi-indices $\alpha \in \mathbb{N}^d$.

(a) Show that $H^k(\Omega)$ is a Hilbert space, that the injection $H^k(\Omega) \subset H^{k-1}(\Omega)$ is continuous, and that the derivative operators $\partial_{x_i} : H^k(\Omega) \to H^{k-1}(\Omega)$ are bounded.

(b) Prove that if Ω is such that $H^1(\Omega)$ is compactly embedded into $L^2(\Omega)$ (for instance, Ω is the union of bounded sets with the H^1 extension property), then $H^k(\Omega)$ is compactly embedded into $H^{k-1}(\Omega)$.

(c) Consider the Sobolev seminorm

$$|u|_{k,\Omega}^2 := \sum_{|\alpha|=k} \|\partial^\alpha u\|_\Omega^2.$$

Assume that Ω is bounded. Show that $|u|_{k,\Omega} = 0$ if and only if

$$u \in \mathcal{P}_{k-1}(\Omega) := \text{span}\,\{x_1^{\alpha_1} \ldots x_d^{\alpha_d} : |\alpha| \leq k-1\}.$$

For the next questions, we assume that the hypothesis of (b) holds and Ω is connected.

(d) Let $\pi : L^2(\Omega) \to \mathcal{P}_{k-1}(\Omega)$ be the orthogonal projection onto the space of polynomials. Show that there exists $C > 0$ such that

$$\|u - \pi u\|_\Omega \leq C|u|_{k,\Omega} \quad \forall u \in H^k(\Omega).$$

(e) **The Bramble-Hilbert lemma.** Let $Q : H^k(\Omega) \to H^m(\Omega)$ for $k \geq m$ be a bounded linear operator such that $Qu = u$ for all $u \in \mathcal{P}_{k-1}(\Omega)$. Show that there exists $C > 0$ such that

$$\|u - Qu\|_{m,\Omega} \leq C|u|_{k,\Omega} \qquad \forall u \in H^k(\Omega).$$

7.7. Yet another version of Korn's inequality. Show that there exists $C > 0$ such that

$$\|\mathbf{u}\|_{1,\Omega} \leq C(\|\mathbf{u}\|_\Omega + \|\varepsilon(\mathbf{u})\|_\Omega) \qquad \forall \mathbf{u} \in \mathbf{H}^1(\Omega).$$

7.8. The pure traction problem in elasticity. Prove Proposition 7.15. (**Hint.** Use the same process as in the study of the Neumann problem for the Laplacian, considering the space

$$V := \{\mathbf{u} \in \mathbf{H}^1(\Omega) : (\mathbf{u}, \mathbf{m})_\Omega = 0 \quad \forall \mathbf{m} \in \mathcal{M}\},$$

and one of the many versions of Korn's second inequality.)

7.9. Periodic boundary conditions. Let Ω be an open rectangular domain with boundary

$$\Gamma = \overline{\Gamma_N} \cup \overline{\Gamma_S} \cup \overline{\Gamma_E} \cup \overline{\Gamma_W},$$

where this partition represents the top, bottom, right, and left boundaries of the rectangle respectively. We define the space

$$H^1_\#(\Omega) := \{u \in H^1(\Omega) : \gamma_N u \equiv \gamma_S u, \gamma_W u \equiv \gamma_E u\},$$

to be the space of H^1 functions on Ω with periodic boundary conditions, where the various γ operators are trace operators on the respective parts of Γ, and the symbol '$\equiv$' denotes that the traces are the same up to a translation.

(a) Show that $H^1_\#(\Omega)$ is closed in $H^1(\Omega)$.

(b) Show that $u \in H^1_\#(\Omega)$ if and only if $P_\# u \in H^1_{\text{loc}}(\mathbb{R}^d)$, where $P_\#$ is the periodic extension of u (just keep making copies of u in Ω and tile all of $\mathbb{R}^d$).

(c) Show that for $f \in L^2(\Omega)$, the problem

$$u \in H^1_\#(\Omega),$$
$$(\nabla u, \nabla v)_\Omega = (f, v)_\Omega \qquad \forall v \in H^1_\#(\Omega),$$

where we require

$$\int_\Omega u(\mathbf{x}) \, d\mathbf{x} = \int_\Omega f(\mathbf{x}) \, d\mathbf{x} = 0,$$

is well posed.

(d) Derive conditions on $\partial_n u$ for the solution of (c). For simplicity, assume initially that $\partial_n u \in L^2(\Gamma)$.

7.10. Consider the unit ball $B(\mathbf{0}; 1) \subset \mathbb{R}^2$ and the sets $\Omega_j := B(\mathbf{0}; 1) \setminus \Xi_j$, where

$$\Xi_1 := (-\tfrac{1}{2}, \tfrac{1}{2}) \times \{0\}, \qquad \Xi_2 := (0, 1) \times \{0\}.$$

Show that the Rellich-Kondrachov theorem holds in these sets, that is, that the space $H^1(\Omega_j)$ is compactly embedded into $L^2(\Omega_j)$. (**Hint.** Separate the unit ball into a positive and negative part and use the continuity of the restriction operators.)

8

Compact perturbations of coercive problems

8.1   Self-adjoint Fredholm theorems 150
8.2   The Helmholtz equation 152
8.3   Compactness on the boundary 156
8.4   Neumann and impedance problems revisited 157
8.5   Kirchhoff plate problems (*) 159
8.6   Fredholm theory: the general case 162
8.7   Convection-diffusion revisited 165
8.8   Impedance conditions for Helmholtz (*) 167
8.9   Galerkin projections and compactness (*) 169
Final comments and literature 173
Exercises 173

Simplifying, we could say that the goal of this chapter is the study of the Helmholtz equation

$$\Delta u + k^2 u = f,$$

with different sets of boundary conditions. In reality, we are going to develop the use of the Fredholm theory for problems that can be understood as a compact perturbation of a coercive problem. This will allow us to consider low order terms as perturbations and derive a theory that deals with existence and uniqueness by looking at: (a) well-posedness of a problem without the perturbations; (b) uniqueness of solution. We will derive the Fredholm alternative and its applications to variational problems in two steps. In the first step, we will deal with self-adjoint problems, using a very simple proof of the Fredholm alternative that holds for self-adjoint problems. Examples will include the Helmholtz equation with Dirichlet or Neumann boundary conditions, and we will revisit Neumann and Robin boundary value problems for the Laplace equation, allowing for lack of uniqueness of solutions and deriving compatibility conditions on the data in a systematic way. In a second step, we will deal with more general operators and variational formulations, which will include the convection-diffusion equation and the Helmholtz equation with impedance boundary conditions.

8.1   Self-adjoint Fredholm theorems

Proposition 8.1. *If $K : H \to H$ is a compact linear operator in a Hilbert space H, then $\ker(I - K)$ is finite-dimensional.*

Proof. Assume that $\ker(I - K)$ is infinite-dimensional. Using the Gram-Schmidt orthogonalization method, we can find an orthonormal sequence $\{e_n\}$ in $\ker(I - K)$. Since orthonormal sequences satisfy the Bessel inequality (see Appendix A.4)

$$\sum_{n=1}^{\infty} |(e_n, u)|^2 \leq \|u\|_H^2 \qquad \forall u \in H,$$

then $e_n \rightharpoonup 0$. By compactness of K and since $e_n - Ke_n = 0$, it follows that

$$e_n = Ke_n \longrightarrow 0,$$

which contradicts the fact that $\|e_n\|_H = 1$. □

Proposition 8.2. *If $K : H \to H$ is a compact linear operator in a Hilbert space H, then $\mathrm{range}\,(I - K)$ is closed.*

Proof. We are going to show that there exists $C > 0$ such that

$$\|u - Ku\|_H \geq C\|u\|_H \qquad \forall u \in \ker(I - K)^{\perp}, \tag{8.1}$$

which implies that $\mathrm{range}\,(I - K)$ is closed. To see why, take a Cauchy sequence $\{z_n\}$ in $\mathrm{range}\,(I - K)$. We can then write $z_n = u_n - Ku_n$, where $\{u_n\}$ is a sequence in $\ker(I - K)^{\perp}$. The inequality (8.1) along with the fact that $\{z_n\}$ is Cauchy, implies that $\{u_n\}$ is itself a Cauchy sequence and therefore there exists a $u \in \ker(I - K)^{\perp}$ such that $u_n \to u$, implying that $z_n = u_n - Ku_n \to u - Ku \in \mathrm{range}\,(I - K)$, which proves the result.

Assume now that (8.1) does not hold. We can then find a sequence $\{u_n\}$ in $\ker(I - K)^{\perp}$ such that

$$\|u_n\|_H = 1, \qquad \|u_n - Ku_n\|_H \leq n^{-1}.$$

Since the sequence is bounded, we can find a weakly convergent subsequence $\{u_{n_k}\}$. Hence, there is a $u \in \ker(I - K)^{\perp}$ such that

$$u_{n_k} \rightharpoonup u \quad \text{and therefore} \quad Ku_{n_k} \longrightarrow Ku.$$

This implies

$$0 \longleftarrow u_{n_k} - Ku_{n_k} \rightharpoonup u - Ku,$$

and therefore $u \in \ker(I - K)$. Since $u_{n_k} \in \ker(I - K)^{\perp}$ (by construction), we have

$$0 = (u_{n_k}, u)_H \longrightarrow \|u\|_H^2,$$

and therefore $u = 0$. However,

$$u_{n_k} = u_{n_k} - K u_{n_k} + K u_{n_k} \longrightarrow 0 + Ku = 0,$$

which is a contradiction with the fact that $\|u_{n_k}\|_H = 1$ for all k. $\qquad\square$

Corollary 8.1. *If $K : H \to H$ is a compact self-adjoint linear operator in a Hilbert space H, i.e., K is compact and*

$$(Ku, v)_H = (u, Kv)_H \qquad \forall u, v \in H,$$

then

$$\ker(I - K)^{\perp} = \mathrm{range}\,(I - K).$$

Proof. For a general bounded linear operator, we have

$$\overline{\mathrm{range}\, A^*} = \ker A^{\perp}.$$

If $A = I - K$ and K is self-adjoint, then range $A^* = $ range A is closed (Proposition 8.2) and the proof is finished. $\qquad\square$

Theorem 8.1 (Self-adjoint Fredholm alternative). *If $A : H \to H$ is a bounded self-adjoint operator in a Hilbert space H satisfying*

$$(Au, u)_H \geq \alpha \|u\|_H^2 \qquad \forall u \in H,$$

for some $\alpha > 0$, and $K : H \to H$ is compact and self-adjoint, then

(a) *$A + K$ is injective if and only if it is surjective.*

(b) *$Range(A + K) = \ker(A + K)^{\perp}$, i.e.,*

$$Au + Ku = f,$$

is solvable if and only if

$$(f, \phi) = 0 \qquad \forall \phi \in \ker(A + K),$$

and $\ker(A + K)$ is finite-dimensional.

Proof. Since A is bounded, self-adjoint, and the associated bilinear form is coercive, the expression

$$(u, v)_A := (Au, v)_H \qquad u, v \in H,$$

defines an equivalent inner product in H. Note that the hypotheses imply the invertibility of A, as follows from the Riesz-Fréchet representation theorem. Now consider the decomposition

$$A + K = A(I + A^{-1}K) = A(I - \widehat{K}) \qquad \text{with} \quad \widehat{K} := -A^{-1}K.$$

The operator $\widehat{K}$ is compact and, since

$$(\widehat{K}u, v)_A = -(Ku, v)_H = -(u, Kv)_H$$
$$= -(Au, A^{-1}Kv)_H = (u, \widehat{K}v)_A,$$

we have that $\widehat{K}$ is A-self-adjoint. We can then apply Corollary 8.1 to $\widehat{K}$. In particular

$$A + K \text{ is injective} \iff I - \widehat{K} \text{ is injective}$$
$$\iff I - \widehat{K} \text{ is surjective}$$
$$\iff A + K \text{ is surjective},$$

which proves (a). Since $\ker(A + K) = \ker(I - \widehat{K})$, it follows that $\ker(A + K)$ is finite-dimensional, by Proposition 8.1. Finally, the following equivalences follow from Corollary 8.1 and elementary arguments:

$$f \in \text{range}\,(A + K) \iff A^{-1}f \in \text{range}\,(I - \widehat{K})$$
$$\iff A^{-1}f \perp_A \ker(I - \widehat{K})$$
$$\iff f \perp \ker(I - \widehat{K}) = \ker(A + K).$$

This finishes the proof. $\qquad\qquad\qquad\qquad\qquad\qquad\qquad\qquad\qquad\quad\square$

8.2   The Helmholtz equation

In this section we study the following boundary value problem (note the 'wrong' sign on the Laplacian)

$$u \in H_0^1(\Omega), \qquad \Delta u + k^2 u = f. \tag{8.2}$$

Here Ω is a bounded domain and $f \in L^2(\Omega)$. The boundedness of the domain guarantees the validity of the Poincaré-Friedrichs inequality (and therefore the Dirichlet form defines an equivalent inner product) and the compactness of the embedding of $H_0^1(\Omega)$ into $L^2(\Omega)$. The Helmholtz equation is typically written as a complex-valued problem: the equation itself is related to a real operator, but the data are often complex-valued. In this case, with homogeneous Dirichlet boundary conditions, there is no need to resort to complex-valued functions and we will keep everything real. When we explore 'first order absorbing boundary conditions' for the Helmholtz equation at the end of this chapter, we will need to take into account the complexification of the Sobolev spaces. Recall that in Chapter 5 we have already explored unique solvability of (8.2) for small enough k^2, and of

$$u \in H_0^1(\Omega), \qquad -\Delta u + \lambda u = 0,$$

for $\lambda \in \mathbb{C} \setminus (-\infty, 0]$. This means that we only need to worry about $k \in \mathbb{R}$ in (8.2). Problem (8.2) is equivalent to its variational formulation

$$u \in H_0^1(\Omega), \qquad (\nabla u, \nabla v)_\Omega - k^2 (u, v)_\Omega = -(f, v)_\Omega \qquad \forall v \in H_0^1(\Omega). \quad (8.3)$$

Rewriting the problem. We are now going to rewrite (8.3) in the form

$$Au + Ku = w_f, \qquad (8.4)$$

for appropriate operators A and K satisfying the hypotheses of Theorem 8.1. Note that there will be more than one form of doing this and that the operators themselves are not relevant, since at the end of the day we will need to translate all results and conclusions to our original problem (8.2)-(8.3). We first define the operator $A : H_0^1(\Omega) \to H_0^1(\Omega)$ by

$$(Au, v)_{1,\Omega} = (\nabla u, \nabla v)_\Omega \qquad \forall u, v \in H_0^1(\Omega). \quad (8.5)$$

The operator A is well-defined (Au is the Riesz-Fréchet representative of the functional $v \mapsto (\nabla u, \nabla v)_\Omega$), it is self-adjoint (the right-hand side of (8.5) is clearly symmetric), and it satisfies

$$(Au, u)_{1,\Omega} \geq C\|u\|_{1,\Omega}^2 \qquad \forall u \in H_0^1(\Omega),$$

by the Poincaré-Friedrichs inequality. We similarly define $K : H_0^1(\Omega) \to H_0^1(\Omega)$ by

$$(Ku, v)_{1,\Omega} = -k^2 (u, v)_\Omega \qquad \forall u, v \in H_0^1(\Omega),$$

and note that K is bounded and self-adjoint. Moreover,

$$\|Ku\|_{1,\Omega} = \sup_{0 \neq v \in H_0^1(\Omega)} \frac{|(Ku, v)_{1,\Omega}|}{\|v\|_{1,\Omega}}$$

$$= k^2 \sup_{0 \neq v \in H_0^1(\Omega)} \frac{|(u, v)_\Omega|}{\|v\|_{1,\Omega}} \leq k^2 \|u\|_\Omega \qquad \forall u \in H_0^1(\Omega).$$

Therefore, if $u_n \rightharpoonup u$ in $H_0^1(\Omega)$, by the Rellich-Kondrachov theorem (Proposition 7.4) it follows that $u_n \to u$ in $L^2(\Omega)$ and the above estimate for the continuity of K implies that $Ku_n \to Ku$ in $H^1(\Omega)$. This implies that K is compact. Finally, we consider $w_f \in H_0^1(\Omega)$ satisfying

$$(w_f, v)_{1,\Omega} = -(f, v)_\Omega \qquad \forall v \in H_0^1(\Omega)$$

and note that

$$\|w_f\|_{1,\Omega} \leq \|f\|_\Omega,$$

which can be easily proved by taking $v = w_f$ or using the fact that w_f is just a Riesz-Fréchet representative of the functional $v \mapsto -(f, v)_\Omega$.

The well posed case. Note now that $u \in \ker(A + K)$ is equivalent to

$$u \in H_0^1(\Omega), \qquad (\nabla u, \nabla v)_\Omega - k^2 (u, v)_\Omega = 0 \qquad \forall v \in H_0^1(\Omega),$$

or, in PDE form

$$u \in H_0^1(\Omega), \qquad \Delta u + k^2 u = 0. \tag{8.6}$$

The latter problem is often written in 'eigenvalue form' (see Chapter 9):

$$u \in H_0^1(\Omega), \qquad -\Delta u = k^2 u.$$

If problem (8.6) has only the trivial solution, then (8.2), (8.3), and (8.4) are uniquely solvable. Since the inverse operator of $A + K$ is bounded, there exists a constant (which we cannot easily estimate in terms of constants from known inequalities or from the coefficient k^2) such that

$$\|u\|_{1,\Omega} \le C\|w_f\|_{1,\Omega} \le C'\|f\|_\Omega.$$

The resonant case. If $\ker(A + K)$ is nontrivial, we can find a basis $\{u_1, \ldots, u_N\}$ of the space of solutions to (8.6). The problem $Au + Ku = w_f$ is solvable (and therefore so is (8.2)) if and only if

$$0 = (w_f, u_j)_{1,\Omega} = -(f, u_j)_\Omega \qquad j = 1, \ldots, N.$$

The solution of (8.2)-(8.3) is unique up to a linear combination of the functions $\{u_1, \ldots, u_N\}$. The data function f has to be $L^2(\Omega)$-orthogonal to all these functions for the solution to be unique. In this case, since the operator

$$\ker(A + K)^\perp \ni u \longmapsto (A + K)u \in \ker(A + K)^\perp,$$

is a bounded isomorphism and we can thus obtain an estimate

$$\|u\|_{1,\Omega} \le C\|f\|_\Omega,$$

for the unique solution of (8.2)-(8.3) that is orthogonal to $\{u_1, \ldots, u_N\}$.

A different representation. There is another way of rewriting (8.3) in the language of the Fredholm alternative. To do that, consider the inner product $(\nabla u, \nabla v)_\Omega$ in $H_0^1(\Omega)$. We then define the operator $K_\circ : H_0^1(\Omega) \to H_0^1(\Omega)$ given by

$$(\nabla K_\circ u, \nabla v)_\Omega = -k^2 (u, v)_\Omega \qquad u, v \in H_0^1(\Omega).$$

This operator is once again self-adjoint and compact. (The proof is very similar to the one for K and is left to the reader.) We then define $\omega_f \in H_0^1(\Omega)$ with the problem

$$(\nabla \omega_f, \nabla v)_\Omega = -(f, v)_\Omega \qquad \forall v \in H_0^1(\Omega),$$

which leads to writing (8.3) in the form

$$(\nabla(u + K_o u), \nabla v)_\Omega = (\nabla \omega_f, \nabla v)_\Omega \qquad \forall v \in H_0^1(\Omega),$$

or, equivalently,

$$u + K_o u = \omega_f.$$

As we can easily see, the operators are different, but the conclusions extracted from this representation and the Fredholm alternative will be identical.

Working with bilinear forms. We can do perfectly fine without the introduction of the operators A and K (and the modified right-hand side) by working directly with bilinear forms. The context of Section 8.1, specifically Theorem 8.1, can be expressed with bilinear forms in the following way. Let us consider a Hilbert space V and the variational problem:

$$u \in V, \qquad a(u, v) + b(u, v) = \ell(v) \qquad \forall v \in V, \tag{8.7}$$

where

(a) The sesquilinear form $a : V \times V \to \mathbb{C}$ is bounded, Hermitian, and coercive.

(b) The sesquilinear form $b : V \times V \to \mathbb{C}$ is Hermitian and admits a bound of the form

$$|b(u, v)| \le C \|u\|_H \|v\|_V \qquad \forall u, v \in V,$$

where H is a Hilbert space such that the injection $V \subset H$ is compact.

(c) $\ell \in V^*$ is a bounded conjugate linear functional.

The conditions given on b ensure that the associated operator $K : V \to V$ defined by

$$(Ku, v)_V = b(u, v) \qquad \forall u, v \in V,$$

satisfies

$$\|Ku\|_V \le C \|u\|_H \qquad \forall u \in V,$$

and is therefore compact. (Take a weakly convergent sequence $\{u_n\}$ in V, note that the same sequence is strongly convergent in H and therefore $\{Ku_n\}$ is strongly convergent.) The Fredholm alternative says then that, either

$$a(u, v) + b(u, v) = 0 \qquad \forall v \in V, \tag{8.8}$$

admits only the trivial solution and then (8.7) is a well posed problem, or (8.8) admits a finite number of linearly independent solutions $\{u_1, \ldots, u_N\}$ and then (8.7) is solvable (uniquely modulo linear combinations of the homogeneous solutions) if and only if

$$\ell(u_j) = 0 \qquad j = 1, \ldots, N.$$

8.3   Compactness on the boundary

This short section shows that the trace operator is compact when the target space is taken to be $L^2(\Gamma)$ and that, as a consequence, $H^{1/2}(\Gamma)$ is compactly embedded into $L^2(\Gamma)$.

Proposition 8.3. *On a bounded Lipschitz domain, the trace operator*

$$\gamma : H^1(\Omega) \to L^2(\Gamma),$$

is compact.

Proof. The process of taking the trace can be factored into the following pieces: localization and pullback, trace in the reference element, and push-forward (including adding the functions defined on pieces of the boundary and naturally extended by zero to the rest of the boundary):

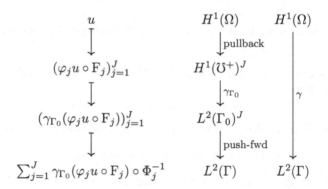

Here $\{\Omega_1, \ldots, \Omega_J\}$ is the cover of the boundary, $\{\varphi_1, \ldots, \varphi_J\}$ is the associated smooth partition of unity, $F_j : \mho^+ := B_{d-1}(\mathbf{0}; 1) \times (0, 1) \to \Omega \cap \Omega_j$ are the local charts, $\Gamma_0 = B_{d-1}(\mathbf{0}; 1) \times \{0\}$ is the lower boundary of $\mho^+$, and $\Phi_j := F_j|_{\Gamma_0} : \Gamma_0 \to \Gamma \cap \Gamma_j$ are the local parametrizations of the pieces of the boundary.

The result holds once we prove that

$$\gamma_{\Gamma_0} : H^1(\mho^+) \to L^2(\Gamma_0),$$

is compact. This is done in Exercise 8.6.   □

Corollary 8.2. *If Γ is the boundary of a Lipschitz domain, then $H^{1/2}(\Gamma)$ is compactly embedded into $L^2(\Gamma)$.*

Proof. Consider a bounded right inverse of the trace operator $\gamma^\dagger : H^{1/2}(\Gamma) \to H^1(\Omega)$ and the composition

$$H^{1/2}(\Gamma) \xrightarrow{\gamma^\dagger} H^1(\Omega) \xrightarrow{\gamma} L^2(\Gamma),$$

which factors the inclusion operator. Since the trace is compact from $H^1(\Omega)$ to $L^2(\Gamma)$, the result follows. $\qquad\qquad\qquad\qquad\qquad\qquad\qquad\qquad\qquad\qquad\qquad$ $\square$

A simple application. Let $\alpha \in L^\infty(\Gamma)$ and consider the bilinear form $b : H^1(\Omega) \times H^1(\Omega) \to \mathbb{R}$ and its associated operator $K : H^1(\Omega) \to H^1(\Omega)$ given by

$$b(u, v) := \langle \alpha \gamma u, \gamma v \rangle_\Gamma = (Ku, v)_{1,\Omega}.$$

We can easily prove K is compact: we first bound

$$\|Ku\|_{1,\Omega} = \sup_{0 \neq v \in H^1(\Omega)} \frac{|\langle \alpha \gamma u, \gamma v \rangle_\Gamma|}{\|v\|_{1,\Omega}} \leq \sup_{0 \neq v \in H^1(\Omega)} \frac{\|\alpha\|_{L^\infty(\Gamma)} \|\gamma u\|_\Gamma \|\gamma v\|_\Gamma}{\|v\|_{1,\Omega}}$$

$$\leq C \|\alpha\|_{L^\infty(\Gamma)} \|\gamma u\|_\Gamma,$$

and then note that if $u_n \rightharpoonup u$ in $H^1(\Omega)$, then $\gamma u_n \to \gamma u$ in $L^2(\Gamma)$ (see Proposition 8.3) and therefore $Ku_n \to Ku$ in $H^1(\Omega)$.

8.4 Neumann and impedance problems revisited

The Neumann problem for the Laplacian. We once again consider the boundary value problem from Section 7.4

$$-\Delta u = f, \qquad \partial_n u = h, \tag{8.9}$$

with data $f \in L^2(\Omega)$ and $h \in H^{-1/2}(\Gamma)$, which has the variational form

$$u \in H^1(\Omega), \qquad (\nabla u, \nabla v)_\Omega = (f, v)_\Omega + \langle h, \gamma v \rangle_\Gamma \qquad \forall v \in H^1(\Omega).$$

This can be cast in the framework of compact perturbations of coercive problems using the bilinear forms

$$a(u, v) := (\nabla u, \nabla v)_\Omega + (u, v)_\Omega, \qquad b(u, v) := -(u, v)_\Omega.$$

The operator associated to a is the identity, while, since

$$|b(u, v)| \leq \|u\|_\Omega \|v\|_\Omega,$$

the operator K associated to b is compact. Both operators are self-adjoint, since the bilinear forms are symmetric. The associated operator has a one-dimensional kernel

$$\ker(I + K) = \{u \in H^1(\Omega) : (\nabla u, \nabla v)_\Omega = 0 \quad \forall v \in H^1(\Omega)\} = \mathcal{P}_0(\Omega), S$$

and therefore, there is a solution of (8.9), unique up to additive constants, if and only if

$$(f, v)_\Omega + \langle h, \gamma v \rangle_\Gamma = 0 \qquad \forall v \in \mathcal{P}_0(\Omega),$$

which is the compatibility condition that we met in Chapter 7.

Impedance boundary conditions. Consider the elliptic problem with impedance boundary conditions:

$$-\Delta u + cu = f, \qquad \partial_n u + \alpha \gamma u = h, \tag{8.10}$$

with coefficients $c \in L^\infty(\Omega)$, $\alpha \in L^\infty(\Gamma)$ and data $f \in L^2(\Omega)$, $h \in H^{-1/2}(\Gamma)$. We will first only assume that

$$c \geq 0, \qquad \alpha \geq 0. \tag{8.11}$$

We can decompose the bilinear form associated to the variational formulation of (8.10) as the sum of these two bounded symmetric bilinear forms.

$$a(u, v) := (\nabla u, \nabla v)_\Omega + (u, v)_\Omega + \langle \alpha \gamma u, \gamma v \rangle_\Gamma + (c\, u, v)_\Omega,$$
$$b(u, v) := -(u, v)_\Omega.$$

The bilinear form a is clearly coercive and b is associated to a compact operator. Uniqueness is ensured by the following condition:

$$\int_\Omega c(\mathbf{x}) d\mathbf{x} + \int_\Gamma \alpha(\mathbf{x}) d\Gamma(\mathbf{x}) > 0. \tag{8.12}$$

To prove this note that if

$$a(u, v) + b(u, v) = 0 \qquad \forall v \in H^1(\Omega),$$

then

$$0 = a(u, u) + b(u, u) = \|\nabla u\|_\Omega^2 + (c\, u, u)_\Omega + \langle \alpha \gamma u, \gamma u \rangle_\Gamma \geq \|\nabla u\|_\Omega^2,$$

and therefore $u \in \mathcal{P}_0(\Omega)$. If $u \equiv m$, then

$$a(u, u) + b(u, u) = m^2 \left(\int_\Omega c(\mathbf{x}) d\mathbf{x} + \int_\Gamma \alpha(\mathbf{x}) d\Gamma(\mathbf{x}) \right),$$

which shows that injectivity is a direct consequence of (8.12). Note also that coercivity of the full bilinear form can be derived using the Deny-Lions theorem (see Section 7.3). The sign conditions (8.11) can be completely eliminated at the risk of losing uniqueness, while keeping the Fredholm character. We could have actually started with a different decomposition of the bilinear form

$$a(u, v) := (\nabla u, \nabla v)_\Omega + (u, v)_\Omega,$$
$$b_1(u, v) := ((c - 1)\, u, v)_\Omega = (K_1 u, v)_{1,\Omega},$$
$$b_2(u, v) := \langle \alpha \gamma u, \gamma v \rangle_\Gamma = (K_2 u, v)_{1,\Omega}.$$

The compactness of K_2 was discussed in Section 8.3, as a simple consequence of the compactness of $\gamma : H^1(\Omega) \to L^2(\Gamma)$. More details about this problem are requested from the reader in Exercise 8.4.

8.5 Kirchhoff plate problems (*)

In this section we examine several problems associated to a Kirchhoff plate or Kirchhoff-Love plate model, which is related to the biharmonic equation

$$\Delta^2 u = f.$$

Although the Kirchhoff plate model is a two-dimensional problem, we will deal with this equation in d dimensions. The associated one-dimensional equation is the so called Euler-Bernoulli beam equation. We will assume that Ω is a Lipschitz connected open set in all that follows.

The bilinear form. Consider the bilinear form $c : H^2(\Omega) \times H^2(\Omega) \to \mathbb{R}$ given by

$$c(u, v) := \nu(\Delta u, \Delta v)_\Omega + (1 - \nu)(D^2 u, D^2 v)_\Omega,$$

where $(D^2 u)_{ij} = \partial_{x_i}\partial_{x_j} u$ is the Hessian matrix, and $\nu \in (0, 1)$ is a given parameter. Physically, for the two-dimensional case, $\nu \in (0, 1/2]$ is referred to as Poisson's ratio. The bilinear form c is clearly bounded in $H^2(\Omega)$. Note also that we can decompose

$$c(u, v) = a(u, v) + b(u, v),$$

where

$$a(u, v) := \nu(\Delta u, \Delta v)_\Omega + (1 - \nu)(D^2 u, D^2 v)_\Omega + (u, v)_{1,\Omega},$$
$$b(u, v) := -(u, v)_{1,\Omega}.$$

It is clear that a is coercive in $H^2(\Omega)$, since

$$a(u, v) \geq (1 - \nu)\|D^2 u\|_\Omega^2 + \|u\|_{1,\Omega}^2 \geq (1 - \nu)\|u\|_{2,\Omega}^2.$$

Also,

$$|b(u, v)| \leq \|u\|_{1,\Omega}\|v\|_{1,\Omega} \qquad \forall u, v \in H^2(\Omega),$$

and, since $H^2(\Omega)$ is compactly embedded into $H^1(\Omega)$ (see Exercise 7.5), this bilinear form will produce a compact operator. Moreover, given $f \in L^2(\Omega)$ and $u \in H^2(\Omega)$

$$c(u, \varphi) = (f, \varphi)_\Omega \qquad \forall \varphi \in \mathcal{D}(\Omega),$$

is equivalent to $\Delta^2 u = f$, since

$$c(u, \varphi) = \nu \sum_{i,j=1}^d \langle \partial_{x_i}^2 u, \partial_{x_j}^2 \varphi \rangle + (1 - \nu) \sum_{i,j=1}^d \langle \partial_{x_i}\partial_{x_j} u, \partial_{x_i}\partial_{x_j}\varphi \rangle$$

$$= \nu \sum_{i,j=1}^d \langle \partial_{x_i}^2 \partial_{x_j}^2 u, \varphi \rangle + (1 - \nu) \sum_{i,j=1}^d \langle \partial_{x_i}^2 \partial_{x_j}^2 u, \varphi \rangle$$

$$= \langle \Delta^2 u, \varphi \rangle.$$

This shows that the parameter ν does not influence the associated PDE, so it can only influence the boundary conditions.

Four spaces. The trace operator γ is well-defined on $H^2(\Omega)$. For the moment being, we will consider this as an operator $\gamma : H^2(\Omega) \to L^2(\Gamma)$. The normal derivative operator ∂_n is well-defined on $H^2(\Omega) \subset \{u \in H^1(\Omega) : \nabla u \in \mathbf{H}(\mathrm{div}, \Omega)\}$, but due to Proposition 6.3, since $\nabla u \in \mathbf{H}^1(\Omega) = H^1(\Omega; \mathbb{R}^d)$, then we have $\partial_n u = (\gamma \nabla u) \cdot \mathbf{n} \in L^2(\Gamma)$. We collect these two operators in a single trace operator

$$(\gamma, \partial_n) : H^2(\Omega) \to L^2(\Gamma) \times L^2(\Gamma).$$

The range of this operator is a complicated issue that we will deal with later, although only partially. We are going to consider four spaces, associated to imposing homogeneous boundary conditions

$$\begin{aligned}
\mathcal{V}_{\mathrm{A}} &:= \{u \in H^2(\Omega) : \gamma u = \partial_n u = 0\}, \\
\mathcal{V}_{\mathrm{B}} &:= \{u \in H^2(\Omega) : \gamma u = 0\} = H^2(\Omega) \cap H^1_0(\Omega), \\
\mathcal{V}_{\mathrm{C}} &:= \{u \in H^2(\Omega) : \partial_n u = 0\}, \\
\mathcal{V}_{\mathrm{D}} &:= H^2(\Omega).
\end{aligned}$$

All four spaces defined above are closed subspaces of $H^2(\Omega)$. The bounded bilinear form c defines bounded self-adjoint operators $C_\circ : \mathcal{V}_\circ \to \mathcal{V}_\circ$ for $\circ \in \{\mathrm{A}, \mathrm{B}, \mathrm{C}, \mathrm{D}\}$. Moreover, note that

$$c(u, u) = 0 \quad \Longleftrightarrow \quad D^2 u = 0 \quad \Longleftrightarrow \quad u \in \mathcal{P}_1(\Omega),$$

where $\mathcal{P}_1 = \mathrm{span}\{1, x_1, \dots, x_d\}$. Therefore,

$$\begin{aligned}
\ker C_\circ &= \{u \in \mathcal{V}_\circ : c(u, v) = 0 \quad \forall v \in \mathcal{V}_\circ\} = \{u \in \mathcal{V}_\circ : c(u, u) = 0\} \\
&= \mathcal{V}_\circ \cap \mathcal{P}_1(\Omega),
\end{aligned}$$

where we have used that c is a symmetric positive semidefinite bilinear form and, therefore, the Cauchy-Schwarz inequality holds for c. This shows that

$$\begin{aligned}
\ker C_{\mathrm{A}} = \ker C_{\mathrm{B}} &= \{0\}, \\
\ker C_{\mathrm{C}} &= \mathcal{P}_0(\Omega), \\
\ker C_{\mathrm{D}} &= \mathcal{P}_1(\Omega).
\end{aligned}$$

Four problems. Consider the minimization problem

$$\tfrac{1}{2} c(u, u) - (f, u)_\Omega = \min! \qquad u \in \mathcal{V}_\circ,$$

that is,

$$\frac{\nu}{2} \|\Delta u\|_\Omega^2 + \frac{1-\nu}{2} \|D^2 u\|_\Omega^2 - (f, u)_\Omega = \min! \qquad u \in \mathcal{V}_\circ,$$

or their equivalent variational formulations

$$u \in \mathcal{V}_\circ, \qquad c(u, v) = (f, v)_\Omega \qquad \forall v \in \mathcal{V}_\circ, \qquad (8.13)$$

where $\circ \in \{A, B, C, D\}$. In all cases, we can decompose $C_\circ = A_\circ + K_\circ$, where $A_\circ$ is coercive and $K_\circ$ is compact. We can thus easily state results for solvability of (8.13). The problems in $\mathcal{V}_A$ and $\mathcal{V}_B$ are uniquely solvable for all $f \in L^2(\Omega)$ and

$$\|u\|_{2,\Omega} \le C\|f\|_\Omega. \qquad (8.14)$$

The problem in $\mathcal{V}_C$ is uniquely solvable up to additive constants if and only if

$$\int_\Omega f(\mathbf{x}) \mathrm{d}\mathbf{x} = 0,$$

while the problem in $\mathcal{V}_D = H^2(\Omega)$ is uniquely solvable modulo $\mathcal{P}_1(\Omega)$ if and only if

$$\int_\Omega f(\mathbf{x}) p(\mathbf{x}) \mathrm{d}\mathbf{x} = 0 \qquad \forall p \in \mathcal{P}_1(\Omega),$$

which is a set of $d + 1$ linearly independent constraints. In the latter cases, uniqueness and a continuity bound (8.14) can be attained (Prove it!) in the spaces

$$\{u \in \mathcal{V}_C : (u, 1)_\Omega = 0\},$$
$$\{u \in H^2(\Omega) : (u, p)_\Omega = 0 \quad \forall p \in \mathcal{P}_1(\Omega)\}.$$

The clamped plate. Understanding problems (8.13) as boundary value problems is not an easy task. For the class of Lipschitz domains, it can be proved (we will not do it) that

$$\{u \in H^2(\Omega) : \gamma u = \partial_n u = 0\} = H_0^2(\Omega),$$

where $H_0^2(\Omega)$ is the closure of $\mathcal{D}(\Omega)$ in the $H^2(\Omega)$ norm. In this case, (8.13) (for $\circ = A$) is equivalent to

$$u \in H^2(\Omega), \qquad \Delta^2 u = f, \qquad \gamma u = 0, \qquad \partial_n u = 0.$$

A nonhomogeneous version of this problem

$$u \in H^2(\Omega), \qquad \Delta^2 u = f, \qquad \gamma u = g, \qquad \partial_n u = h,$$

can be easily studied assuming that $(g, h) \in \text{range}\,(\gamma, \partial_n)$, i.e., assuming that there exists $v \in H^2(\Omega)$ such that $\gamma v = g$ and $\partial_n v = h$. In principle, there is no guarantee that $\text{range}\,(\gamma, \partial_n) = \text{range}\,\gamma \times \text{range}\,\partial_n$, which means that: (a) we might not be allowed to choose these two data separately; (b) the norm for the data is a joint norm, and not the sum of the norms of the separate data. We will deal with this and more problems, assuming some additional conditions in Exercise 8.8.

8.6   Fredholm theory: the general case

The general case of Fredholm's theory studies the solvability of equations of the form

$$Au + Ku = f,$$

where $A : H \to H$ is invertible and $K : H \to H$ is compact and at least one of these operators is not self-adjoint. This class of operators will be used often in the sequel: when an operator can be written as the sum of an invertible and a compact operator, we will say that it is **Fredholm of index zero.** We start with two very simple results about weak convergence and compactness.

Proposition 8.4. *In a Hilbert space H, if $u_n \rightharpoonup u$ and $v_n \longrightarrow v$, then $(u_n, v_n)_H \to (u, v)_H$.*

Proof. Note that

$$|(u_n, v_n)_H - (u, v)_H| \leq |(u_n - u, v)_H| + \|u_n\|_H \|v_n - v\|_H,$$

and recall that weakly convergent sequences are bounded.   □

Proposition 8.5. *Let $K : X \to Y$ be a bounded linear operator between Hilbert spaces and let $K^* : Y \to X$ be its adjoint. If K is compact, then K^* is compact.*

Proof. Let $y_n \rightharpoonup y$ in Y, which implies that $K^* y_n \rightharpoonup K^* y$, since K^* is bounded, and

$$KK^* y_n \longrightarrow KK^* y,$$

since K is compact. Therefore

$$0 \longleftarrow (KK^*(y_n - y), y_n - y)_Y = \|K^*(y_n - y)\|_X^2,$$

and the result is proved.   □

Using the results of Section 8.1, we have that

$$\dim \ker(I - K) < \infty, \qquad \dim \ker(I - K^*) < \infty,$$

and range $(I - K)$ is closed. Therefore

$$\text{range}\,(I - K) = \ker(I - K^*)^{\perp}.$$

From this, we prove two more (nontrivial) propositions which will lead to the Fredholm alternative in the general case.

Proposition 8.6. *Let H be a Hilbert space and $K : H \to H$ be a compact operator. If $I - K$ is injective, then it is surjective.*

Proof. Assume that range $(I - K) \neq H$ and define the spaces

$$H_0 := H, \qquad H_n = \text{range}\,(I - K)^n.$$

Note that $(I - K)^n = I - C_n$, where C_n is compact and, therefore, H_n is closed. Note also that

$$H_{n+1} \subset H_n, \qquad (I - K)H_n = \{(I - K)u : u \in H_n\} \subset H_{n+1}.$$

Assume for a moment that $H_n = H_{n+1}$ and take $z \in H$ such that $z \notin$ range $(I - K)$. We then define

$$u = (I - K)^n z \in \text{range}\,(I - K)^n = \text{range}\,(I - K)^{n+1},$$

and find $w \in H$ such that

$$u = (I - K)^{n+1}w = (I - K)^n z,$$

and therefore

$$(I - K)^n (z - (I - K)w) = u - u = 0.$$

Since $(I - K)^n$ is injective, this proves that $z = (I - K)w \in$ range $(I - K)$, which contradicts our choice of z. This means that H_{n+1} is a proper subspace of H_n for all n.

We can thus select

$$u_n \in H_n, \qquad \|u_n\|_H = 1, \qquad u_n \in H_{n+1}^\perp.$$

Note then that

$$Ku_n - Ku_{n+m} = \quad u_n - \underbrace{(I - K)u_n}_{\in H_{n+1}} - \underbrace{u_{n+m}}_{\in H_{n+m}} + \underbrace{(I - K)u_{n+m}}_{\in H_{n+m+1}}$$

$$= \quad u_n - v_n, \qquad \text{with} \quad v_n \in H_{n+1}.$$

Since $u_n \in H_{n+1}^\perp$, we can compute

$$\|Ku_n - Ku_{n+m}\|_H^2 = \|u_n - v_n\|_H^2 = \|u_n\|_H^2 + \|v_n\|_H^2 \geq 1,$$

which proves that the sequence $\{Ku_n\}$ does not contain Cauchy (and therefore convergent) subsequences. However, K is compact and $\{u_n\}$ is bounded and we arrive at a contradiction. □

The missing difficult part of what we will need to prove is the content of the following proposition.

Proposition 8.7. *If $K : H \to H$ is a compact operator in a Hilbert space, then*

$$\dim \ker(I - K) = \dim \ker(I - K^*).$$

Proof. We will prove that

$$\dim \ker(I - K) \geq \dim \ker(I - K^*). \tag{8.15}$$

Since $K^{**} = K$, this proposition follows from (8.15).

To prove (8.15), assume to the contrary that

$$\dim \ker(I - K) < \dim \operatorname{range}(I - K)^{\perp} = \dim \ker(I - K^*).$$

Since these are finite-dimensional spaces, we can build a linear operator

$$A : \ker(I - K) \longrightarrow \operatorname{range}(I - K)^{\perp},$$

which is injective but not surjective. Let then $P : H \to \ker(I - K)$ be the orthogonal projection and finally consider the operator $AP : H \to H$, which is bounded and has the same range as A. That makes AP compact.

Now let $u \in \ker(I - AP - K)$. It follows that

$$(I - K)u = APu \in \operatorname{range}(I - K) \cap \operatorname{range} A \subset \operatorname{range}(I - K) \cap \operatorname{range}(I - K)^{\perp},$$

which shows that $(I - K)u = 0$. Therefore $Pu = u$ and

$$0 = (I - K)u = APu = Au,$$

but, given that A is injective, it follows that $u = 0$. This shows that $I - AP - K$ is injective and, by Proposition 8.6, $I - AP - K$ is surjective.

Finally take

$$0 \neq v \in \operatorname{range}(I - K)^{\perp} \cap \operatorname{range} A^{\perp} = (\operatorname{range}(I - K) \oplus \operatorname{range} A)^{\perp}.$$

The invertibility of $I - AP - K$ shows that there exists $u \in H$ satisfying

$$v = (I - AP - K)u = (I - K)u - APu \in \operatorname{range}(I - K) \oplus \operatorname{range} A,$$

but this implies that $v = 0$, which is a contradiction. $\square$

Theorem 8.2 (The Fredholm alternative). *Suppose H is a Hilbert space, $A : H \to H$ is invertible, and $K : H \to H$ is compact. Either*

- *$\ker(A + K) = 0$, and then $(A + K)u = f$ has a unique solution for all f, or*

- *$\ker(A + K) = \operatorname{span}\{\phi_1, \ldots, \phi_n\}$ and $\ker(A^* + K^*) = \operatorname{span}\{\varphi_1, \ldots, \varphi_n\}$, and so $(A + K)u = f$ has a solution, unique up to a linear combination of elements of $\ker(A + K)$, if and only if $f \perp \varphi_j$ for $j = 1, \ldots, n$.*

Proof. We can write $A + K = A(I + A^{-1}K)$. The 'either' implication is trivial.

To show the 'or' implication, we know that $\ker(A+K) = \ker(I+A^{-1}K)$, and so

$$
\begin{aligned}
\dim\ker(A+K) &= \dim\ker(I+A^{-1}K) \\
&= \dim\ker(I+K^*(A^{-1})^*) \\
&= \dim\ker((A^*+K^*)(A^{-1})^*) \\
&= \dim\ker(A^*+K^*).
\end{aligned}
$$

The rest follows because $\mathrm{range}(A+K)$ is closed. $\qquad\square$

8.7 Convection-diffusion revisited

Consider the convection-diffusion problem

$$
u \in H_0^1(\Omega), \qquad -\nabla\cdot(\kappa\nabla u) + \boldsymbol{\beta}\cdot\nabla u = f, \tag{8.16}
$$

where for our coefficients we assume

$$
\kappa \in L^\infty(\Omega), \qquad \kappa \geq \kappa_0 > 0, \qquad \boldsymbol{\beta} \in L^\infty(\Omega)^d,
$$

and for the data we take $f \in L^2(\Omega)$. The variational formulation for this PDE is

$$
u \in H_0^1(\Omega), \qquad (\kappa\nabla u, \nabla v)_\Omega + (\boldsymbol{\beta}\cdot\nabla u, v)_\Omega = (f, v)_\Omega \qquad \forall v \in H_0^1(\Omega).
$$

From this bilinear form we can define invertible and compact operators by

$$
(Au, v)_{1,\Omega} = (\kappa\nabla u, \nabla v)_\Omega, \qquad (Ku, v)_{1,\Omega} = (\boldsymbol{\beta}\cdot\nabla u, v)_\Omega.
$$

We show that A is invertible by using the coercivity of the Dirichlet form in $H_0^1(\Omega)$. We will now demonstrate the compactness of K. First, we will attempt to prove the compactness of K with our usual approach, which will fail. From this we will use a new approach that will allow us to succeed in our endeavor.

Compactness. Our first attempt follows our usual strategy estimating the norm of the operator

$$
\|Ku\|_{1,\Omega} = \sup_{0 \neq v \in H_0^1(\Omega)} \frac{|(Ku, v)_{1,\Omega}|}{\|v\|_{1,\Omega}} = \sup_{0 \neq v \in H_0^1(\Omega)} \frac{|(\boldsymbol{\beta}\cdot\nabla u, v)_\Omega|}{\|v\|_{1,\Omega}} \leq \|\boldsymbol{\beta}\|_\infty\|\nabla u\|_\Omega,
$$

We fail to gain the desired bound because this estimate is wasteful. We now try another approach, making use of the adjoint K^*. If we show compactness of K^*, then we will have compactness of K. To prove this, we estimate

$$\|K^*v\|_{1,\Omega} = \sup_{0 \neq u \in H_0^1(\Omega)} \frac{|(K^*v, u)_{1,\Omega}|}{\|u\|_{1,\Omega}} = \sup_{0 \neq u \in H_0^1(\Omega)} \frac{|(Ku, v)_{1,\Omega}|}{\|u\|_{1,\Omega}} \leq \|\beta\|_\infty \|v\|_\Omega,$$

which shows that K^* is compact, and so K is as well.

The Fredholm alternative for the convection-diffusion equation. We have demonstrated that we can write our original PDE (8.16) as $A + K$ for invertible A and compact K. We now need to study the adjoint problem $A^* + K^*$. It is clear that A is Hermitian. To get a handle on K^*, we consider the adjoint problem

$$((A^* + K^*)v, u)_{1,\Omega} = ((A + K)u, v)_{1,\Omega}$$
$$= (\kappa \nabla u, \nabla v)_\Omega + (\beta \cdot \nabla u, v)_\Omega,$$

where now the quantity v is our unknown and u is a test function. Therefore

$$\ker(A + K) = \{u \in H_0^1(\Omega) : (\kappa \nabla u, \nabla v)_\Omega + (\beta \cdot \nabla u, v)_\Omega = 0 \quad \forall v \in H_0^1(\Omega)\},$$

and

$$\ker(A^* + K^*) = \{v \in H_0^1(\Omega) : (\kappa \nabla u, \nabla v)_\Omega + (\beta \cdot \nabla u, v)_\Omega = 0 \quad \forall u \in H_0^1(\Omega)\}$$
$$= \{v \in H_0^1(\Omega) : \langle \kappa \nabla v, \nabla \varphi \rangle + \langle \beta v, \nabla \varphi \rangle = 0 \quad \forall \varphi \in \mathcal{D}(\Omega)\}$$
$$= \{v \in H_0^1(\Omega) : -\nabla \cdot (\kappa \nabla v + \beta v) = 0\}.$$

Therefore, either $u \in H_0^1(\Omega)$ satisifying the PDE

$$-\nabla \cdot (\kappa \nabla u) + \beta \cdot \nabla u = 0,$$

is identically zero, and the problem (8.16) is well posed for all right-hand sides $f \in L^2(\Omega)$, or if $v \in H_0^1(\Omega)$ satisfies the homogeneous PDE

$$\nabla \cdot (\kappa \nabla v) + \nabla \cdot (v\beta) = 0,$$

then it is in the null space spanned by $\{\psi_1, \ldots, \psi_n\}$, and if $u \in H_0^1(\Omega)$ solves

$$-\nabla \cdot (\kappa \nabla u) + \beta \cdot \nabla u = 0,$$

then it is in the null space spanned by $\{\phi_1, \ldots, \phi_n\}$. We can then conclude that the original problem (8.16) is uniquely solvable (modulo elements of the kernel) if and only if $f \perp \psi_j$ for $j = 1, \ldots, n$, with orthogonality in $L^2(\Omega)$.

The one-dimensional case. In one dimension, the homogeneous adjoint convection-diffusion problem reduces to the search for $v \in H_0^1(a, b)$ such that

$$(\kappa v' + \beta v)' = 0.$$

Therefore, $\kappa v' + \beta v = c$ for some constant $c \in \mathbb{R}$, or equivalently $v' + \frac{\beta}{\kappa} v = \frac{c}{\kappa}$. Using integrating factors, we define

$$P(x) := \int_a^x \frac{\beta(y)}{\kappa(y)} \, dy,$$

and rewrite the equation as

$$e^{P(x)} v(x) = \int_a^x e^{P(t)} \frac{c}{\kappa(t)} \, dt + d.$$

Applying the boundary conditions we get that $v(a) = 0$ implies that $d = 0$, while $v(b) = 0$ implies that $c = 0$. Therefore $v \equiv 0$ and in all cases the convection-diffusion equation in one dimension is uniquely solvable.

8.8   Impedance conditions for Helmholtz (*)

In this section, we study the following problem

$$u \in H^1(\Omega; \mathbb{C}), \qquad \Delta u + k^2 u = f, \qquad \partial_n u + \imath k \gamma u = h,$$

where $f \in L^2(\Omega; \mathbb{C})$, $h \in H^{-1/2}(\Gamma; \mathbb{C})$, and $0 \neq k \in \mathbb{R}$. The k-dependent imaginary-valued impedance coefficient in the boundary condition is often used in first order approximations of absorbing boundary conditions. We will show that this problem is always well posed. The most difficult part will be proving uniqueness of the solution.

Let us start with the equivalent variational formulation

$$u \in H^1(\Omega; \mathbb{C}),$$
$$(\nabla u, \nabla v)_\Omega - k^2 (u, v)_\Omega + \imath k \langle \gamma u, \gamma v \rangle_\Gamma = -(f, v)_\Omega + \langle h, \gamma v \rangle_\Gamma$$
$$\forall v \subset H^1(\Omega; \mathbb{C}).$$

We can decompose the sesquilinear form as the sum of two bounded sesquilinear forms

$$a(u, v) := (\nabla u, \nabla v)_\Omega + (u, v)_\Omega + \imath k \langle \gamma u, \gamma v \rangle_\Gamma,$$
$$b(u, v) := -(k^2 + 1)(u, v)_\Omega,$$

where the first one is coercive as

$$\operatorname{Re} a(u, u) = \|u\|_{1,\Omega}^2 \qquad \forall u \in H^1(\Omega).$$

The sesquilinear form b corresponds to a compact perturbation. Note that the sesquilinear form $\langle \gamma u, \gamma v \rangle_\Gamma$ is associated to a compact operator since

$$|\langle \gamma u, \gamma v \rangle_\Gamma| \le C \|\gamma u\|_\Gamma \|v\|_{1,\Omega},$$

and the operator $\gamma : H^1(\Omega) \to L^2(\Gamma)$ is compact by Proposition 8.3. This means that we could do a different decomposition of the bilinear form, with the boundary term as part of the compact perturbation.

Uniqueness. Let

$$u \in H^1(\Omega; \mathbb{C}), \tag{8.17a}$$

$$(\nabla u, \nabla v)_\Omega - k^2(u, v)_\Omega + \imath k \langle \gamma u, \gamma v \rangle_\Gamma = 0 \qquad \forall v \in H^1(\Omega; \mathbb{C}). \tag{8.17b}$$

Testing with $v = u$, we have

$$\|\nabla u\|_\Omega^2 - k^2 \|u\|_\Omega^2 + \imath k \|\gamma u\|_\Gamma^2 = 0.$$

Thinking of the above equation in terms of real and imaginary parts, we see that $\gamma u = 0$ and $\partial_n u = -\imath k \gamma u = 0$. We thus have that (8.17) is equivalent to

$$u \in H^1(\Omega), \qquad \Delta u + k^2 u = 0, \qquad \gamma u = 0, \qquad \partial_n u = 0. \tag{8.18}$$

We just need to prove that $u = 0$ is the unique solution of (8.18). This can be done with techniques of potential theory, however, we will prove it in another way. Define the extension by zero

$$U := \begin{cases} u, & \text{in } \Omega, \\ 0, & \text{in } \mathbb{R}^d \setminus \overline{\Omega}, \end{cases}$$

and note that since $\gamma u = 0$ (i.e., $u \in H_0^1(\Omega)$), we have $U \in H^1(\mathbb{R}^d)$. Moreover, for all $\varphi \in \mathcal{D}(\mathbb{R}^d)$,

$$-\langle \Delta U + k^2 U, \varphi \rangle = (\nabla U, \nabla \overline{\varphi})_{\mathbb{R}^d} - k^2(U, \overline{\varphi})_{\mathbb{R}^d} = (\nabla u, \nabla \overline{\varphi})_\Omega - k^2(u, \overline{\varphi})_\Omega = 0,$$

by (8.17) and the fact that $\gamma u = 0$. Therefore $U \in H^1(\mathbb{R}^d)$ vanishes identically in $\mathbb{R}^d \setminus \Omega$ and satisfies

$$\Delta U + k^2 U = 0 \qquad \text{in } \mathbb{R}^d.$$

Once again, we can use more complicated results about analytic continuation of the solutions of the Helmholtz equation to show that this implies that $U = 0$. Instead, we will again borrow future results from spectral theory to make an argument showing that $U = 0$ in a cube $Q = (-M, M)^d$ containing Ω. We define the functions

$$\psi_\alpha(\mathbf{x}) := \prod_{i=1}^d \cos\left(\frac{\pi}{2M} \alpha_i(x_i - M)\right)$$

$$= \phi_\alpha\left(\tfrac{1}{2M}(\mathbf{x} - M\mathbf{e})\right) \qquad \mathbf{e} = (1, \dots, 1), \qquad \alpha \in \mathbb{N}^d,$$

where ϕ_α are the functions defined in Section 7.5. By a simple scaling argument (or rephrasing a proof given in Section 7.5) we can easily prove that $\{\psi_\alpha\}$ is a complete orthogonal system in $L^2(Q)$. It is also simple to show that

$$\Delta\psi_\alpha = -\left(\frac{\|\alpha\|\pi}{2M}\right)^2 \psi_\alpha, \qquad \|\alpha\|^2 = |\alpha_1|^2 + \ldots + |\alpha_d|^2,$$

and therefore

$$(\nabla U, \nabla\psi_\alpha)_Q = \left(\frac{\|\alpha\|\pi}{2M}\right)^2 (U, \psi_\alpha)_Q, \qquad (8.19)$$

since $U \in H^1(Q)$ (recall that U vanishes outside Ω). At the same time

$$(\nabla U, \nabla\psi_\alpha)_Q = k^2(U, \psi_\alpha)_Q, \qquad (8.20)$$

since $\Delta U + k^2 U = 0$ and $\partial_n U = 0$ on ∂Q. Equating (8.19) and (8.20), we have

$$\left(\left(\frac{\|\alpha\|\pi}{2M}\right)^2 - k^2\right)(U, \psi_\alpha)_Q = 0 \qquad \forall\alpha \in \mathbb{N}^d. \qquad (8.21)$$

We just now need to choose M so that it is large enough to guarantee that $\Omega \subset Q$ and satisfies

$$M \neq \frac{\|\alpha\|\pi}{2k} \qquad \forall\alpha \in \mathbb{N}^d.$$

With this, (8.21) and the completeness of the orthogonal system $\{\psi_\alpha\}$ implies that $U = 0$, which finishes the proof.

A remark. For those readers acquainted with spectral properties of the Laplacian, our goal was to show that a function cannot be simultaneously a Dirichlet and Neumann eigenfunction of the negative Laplacian for the same eigenvalue. By moving to free space, we have easily shown the same result on a cube by making it impossible to have k^2 as an eigenvalue.

8.9 Galerkin projections and compactness (*)

Review. In Chapter 5 (in Section 5.5, to be more precise) we introduced Galerkin projections for coercive problems, namely, we considered a well posed coercive variational problem

$$u \in V, \qquad a(u, v) = \ell(v) \qquad \forall v \in V, \qquad (8.22)$$

and an approximation based on the choice of a finite-dimensional subspace $V_h \subset V$

$$u_h \in V_h, \qquad a(u_h, v_h) = \ell(v_h) \qquad \forall v_h \in V_h. \qquad (8.23)$$

Coercivity shows that the 'discrete' problem (8.23) is uniquely solvable and we can bound

$$\alpha\|u_h\|_V^2 \leq a(u_h, u_h) = \ell(u_h) \leq \|\ell\|_{V'}\|u_h\|_V,$$

which proves Galerkin stability

$$\|u_h\|_V \leq (1/\alpha)\|\ell\|_{V'}. \tag{8.24}$$

We can also bound the discrete solution in terms of the exact solution

$$\alpha\|u_h\|_V^2 \leq a(u_h, u_h) = a(u, u_h) \leq M\|u\|_V\|u_h\|_V,$$

which proves the estimate

$$\|u_h\|_V \leq (M/\alpha)\|u\|_V.$$

We also have the bound

$$\alpha\|u - u_h\|_V^2 \leq a(u - u_h, u - u_h) = a(u - u_h, u - \pi_h u) \leq M\|u - u_h\|_V\|u - \pi_h u\|_V,$$

i.e., Galerkin quasi-optimality

$$\|u - u_h\|_V \leq (M/\alpha)\|u - \pi_h u\|_V,$$

where $\pi_h : V \to V_h$ is the orthogonal projection (best approximation operator).

Galerkin stability in different words. Our main goal will be to explore Galerkin stability and quasi-optimality for variational problems associated to a coercive bilinear form with a compact perturbation. Before we do that, let us write (8.23) and (8.24) in a different way. The exact problem (8.22) can be rewritten in operator form as

$$Au = f_\ell,$$

where $A : V \to V$ satisfies $(Au, v)_V = a(u, v)$ for all $u, v \in V$, and $\ell = (f_\ell, \cdot)_V$. The discrete equations (8.23) can be rewritten as

$$u_h \in V_h, \qquad \pi_h A u_h = \pi_h f_\ell = \pi_h A u,$$

and Galerkin stability (8.24) is equivalent to

$$\|u_h\|_V \leq C\|\pi_h A u_h\|_V \qquad \forall u_h \in V_h, \tag{8.25}$$

with the same constant $C = 1/\alpha$. (Prove this!) Since $\pi_h A u_h \in V_h$, by the Cauchy-Schwarz inequality, we have

$$\|\pi_h A u_h\|_V = \sup_{0 \neq v \in V_h} \frac{(\pi_h A u_h, v)_V}{\|v\|_V} = \sup_{0 \neq v \in V_h} \frac{(A u_h, v)_V}{\|v\|_V} = \sup_{0 \neq v \in V_h} \frac{a(u_h, v)}{\|v\|_V},$$

and we can write (8.25) in the form

$$\|u_h\|_V \le C \sup_{0 \ne v \in V_h} \frac{a(u_h, v_h)}{\|v_h\|_V} \qquad \forall u_h \in V_h.$$

Compact perturbations. Now consider a variational problem of the form

$$u \in V, \qquad a(u, v) + b(u, v) = \ell(v) \qquad \forall v \in V, \qquad (8.26)$$

where:

(a) a is bounded and coercive

(b) the bilinear form b is related to a compact operator $K : V \to V$ in the form

$$b(u, v) = (Ku, v)_V \qquad \forall u, v \in V,$$

(c) problem (8.26) is well posed, or equivalently, the operator $A+K : V \to V$ is injective (and therefore surjective), i.e.,

$$a(u, v) + b(u, v) = 0 \qquad \forall v \in V \qquad \Longrightarrow \qquad u = 0.$$

We now consider a sequence of finite-dimensional subspaces $V_h \subset V$ directed in a real parameter $h \to 0$ and such that

$$\pi_h u \longrightarrow u \qquad \forall u \in V. \qquad (8.27)$$

Using these spaces, we define Galerkin approximations of problem (8.26)

$$u_h \in V_h, \qquad a(u_h, v_h) + b(u_h, v_h) = \ell(v_h) \qquad \forall v_h \in V_h. \qquad (8.28)$$

The next result shows that once the approximation property (8.27) kicks in, problems (8.28) are uniquely solvable and we have Galerkin stability. For readers not comfortable with 'sequences' directed in a parameter, think of $h = 1/n$ with $n \in \mathbb{N}$.

Proposition 8.8. *In the hypotheses* (a)-(c) *for the bilinear forms, there exists* $C > 0$ *and* $h_0 > 0$ *such that*

$$\|u_h\|_V \le C\|\pi_h(A+K)u_h\|_V \qquad \forall u_h \in V_h, \qquad \forall h \le h_0. \qquad (8.29)$$

Proof. If (8.29) does not hold, then we can find a sequence $\{u_h\}$ such that

$$\|u_h\|_V = 1, \qquad \pi_h(A+K)u_h \longrightarrow 0, \qquad u_h \rightharpoonup u.$$

This implies

$$(\pi_h(A+K)u_h, v)_V = ((A+K)u_h, \pi_h v)_V \longrightarrow ((A+K)u, v)_V \qquad \forall v \in V,$$

given the fact that $(A + K)u_h \rightharpoonup (A + K)u$ and $\pi_h v \to v$ for all v. Therefore $\pi_h(A + K)u \rightharpoonup (A + K)u$, which implies that $(A + K)u = 0$. Invertibility of $A + K$ then implies that $u = 0$ and $u_h \rightharpoonup 0$. Compactness of K implies that $Ku_h \to 0$ and therefore $\pi_h Ku_h \to 0$. We can then go back to the fact that $\pi_h(A + K)u_h \to 0$ and prove that

$$1 = \|u_h\|_V \leq C_A \|\pi_h Au_h\|_V \longrightarrow 0,$$

which is a contradiction. $\qquad\square$

Another way to state the above result is that we have the bound

$$\|u_h\|_V \leq C \sup_{0 \neq v_h \in V_h} \frac{a(u_h, v_h) + b(u_h, v_h)}{\|v_h\|_V} \qquad \forall v_h \in V_h.$$

The Galerkin projector. Assume that we are in the situation of Proposition 8.8 and $h \leq h_0$. In this case, (8.28) is uniquely solvable for all $\ell \in V'$. To prove that, note that due to the finite dimensionality of (8.28) we only need to deal with the homogeneous problem

$$u_h \in V_h, \qquad a(u_h, v_h) + b(u_h, v_h) = 0 \qquad \forall v_h \in V_h$$

or, in other words,

$$u_h \in V_h, \qquad \pi_h(A + K)u_h = 0.$$

The stability inequality (8.29) implies that $u_h = 0$ and therefore (8.28) is uniquely solvable for any right-hand side. Now consider the operator $G_h : V \to V_h$ given by $G_h u = u_h$, where

$$u_h \in V_h, \qquad \pi_h(A + K)u_h = \pi_h(A + K)u.$$

The operator G_h is a projection onto V_h, since if $u \in V_h$, then by unique solvability $G_h u = u$. Therefore

$$
\begin{aligned}
\|G_h u\|_V &\leq C \|\pi_h(A + K)G_h u\|_V \qquad \text{(Proposition 8.8)} \\
&= C \|\pi_h(A + K)u\|_V \qquad \text{(def of } G_h) \\
&\leq C \|A + K\| \, \|u\|_V,
\end{aligned}
$$

which proves that the operator G_h is bounded independently of the parameter h directing the approximating sequence. Finally

$$
\begin{aligned}
\|u - G_h u\|_V &\leq \|u - \pi_h u\|_V + \|\pi_h u - G_h u\|_V \\
&\leq \|u - \pi_h u\|_V + \|G_h(u - \pi_h u)\|_V \qquad (G_h \pi_h = \pi_h) \\
&\leq (1 + C\|A + K\|)\|u - \pi_h u\|_V,
\end{aligned}
$$

or, in terms of the solutions of (8.26) and (8.28), we have a Céa estimate

$$\|u - u_h\|_V \leq (1 + C\|A + K\|) \inf_{v_h \in V_h} \|u - v_h\|_V.$$

Final comments and literature

The Fredholm alternative is a classic result of functional analysis for researchers interested in time-harmonic wave propagation (or scattering) problems, of which the Helmholtz equation is the simplest case. We offer a very simple proof of the alternative in the self-adjoint case, a result that is enough to handle some of the most basic problems. Rainer Kress's book [69] on *Linear Integral Equations* includes a very clean presentation of Fredholm theory on dual pairs of normed spaces, which includes the general case on Hilbert spaces. We will only deal with Fredholm operators of index zero (those of the form invertible operator plus compact operator), but there is a general concept of Fredholm operators of any integer index which is of interest in applications. The interested reader is invited to learn about this in [69].

We note that until the proof of the compact injection of $H^{1/2}(\Gamma)$ into $L^2(\Gamma)$, we did not have any mathematical evidence of the fact that $H^{1/2}(\Gamma)$ is a proper (dense) subspace of $L^2(\Gamma)$. With other equivalent definitions of $H^{1/2}(\Gamma)$, this is obvious, but then you need to prove that $H^{1/2}(\Gamma)$ is the range of the trace operator. We will see more about this in Chapter 13.

The result (Proposition 8.8) on Galerkin approximation of invertible operator equations of the form $A + K$, where A is self-adjoint and strongly positive (the associated bilinear form is coercive), is part of the folklore on numerical analysis of integral equations and has been attributed to different authors. It seems that the paper of Stefan Hildebrandt and Ernst Wienholtz [63] might be one of the oldest sources, although the result is presented in very different language.

Exercises

8.1. Let Ω be a Lipschitz domain. Study the well-posedness and conditions of solvability for

$$u \in H^1(\Omega), \qquad \Delta u + k^2 u = f, \qquad \partial_n u = h,$$

for $f \in L^2(\Omega)$ and $h \in H^{-1/2}(\Gamma)$.

8.2. Let Ω be a Lipschitz domain. Show that the problem

$$u \in H^1(\Omega), \qquad \Delta u + k^2 u = f, \qquad \gamma u = g,$$

with $f \in L^2(\Omega)$ and $g \in H^{1/2}(\Gamma)$ is well posed if and only if the associated homogeneous problem is uniquely solvable. Write sovability conditions in terms of f and the harmonic extension of g, i.e., the solution of

$$u \in H^1(\Omega), \qquad \Delta u = 0, \qquad \gamma u = g,$$

in case there are homogeneous solutions.

8.3. **Finite dimensionality of eigenfunction spaces.** Let Ω be a Lipschitz domain. A Dirichlet eigenvalue of the Laplacian in Ω is $\lambda \in \mathbb{C}$ such that the problem

$$u \in H^1(\Omega), \qquad -\Delta u = \lambda u, \qquad \gamma u = 0,$$

has nontrivial solutions. Show that the set of solutions of this problem is finite-dimensional. Repeat the argument for Neumann eigenfunctions, that is, solutions of

$$u \in H^1(\Omega), \ \lambda \in \mathbb{R}, \qquad -\Delta u = \lambda u, \qquad \partial_n u = 0.$$

8.4. Let Ω be a Lipschitz domain, and $\kappa \in L^\infty(\Omega)$ satisfy $\kappa \geq \kappa_0 > 0$ almost everywhere. Also let $c \in L^\infty(\Omega)$ and $\alpha \in L^\infty(\Gamma)$. (No sign conditions are assumed on these two coefficients.) Show that the boundary value problem

$$u \in H^1(\Omega), \qquad -\nabla \cdot (\kappa \nabla u) + cu = f, \qquad \kappa \nabla u \cdot \mathbf{n} + \alpha \gamma u = h,$$

is well posed (with arbitrary data $f \in L^2(\Omega)$, $h \in H^{-1/2}(\Gamma)$) if and only if

$$\left. \begin{array}{c} u \in H^1(\Omega) \\ -\nabla \cdot (\kappa \nabla u) + cu = 0 \\ \kappa \nabla u \cdot \mathbf{n} + \alpha \gamma u = 0 \end{array} \right] \Longrightarrow u = 0.$$

8.5. Let Ω be a Lipschitz domain and $\kappa \in L^\infty(\Omega)$ satisfy $\kappa \geq \kappa_0 > 0$ almost everywhere. Let $\boldsymbol{\beta} \in L^\infty(\Omega)^d$ and $c \in L^\infty(\Omega)$. Show that the problem

$$u \in H^1(\Omega), \qquad -\nabla \cdot (\kappa \nabla u) + \boldsymbol{\beta} \cdot \nabla u + cu = f, \qquad \gamma u = g,$$

is well posed (data are arbitrary functions $f \in L^2(\Omega)$, $g \in H^{1/2}(\Gamma)$) if and only if

$$\left. \begin{array}{c} u \in H_0^1(\Omega) \\ -\nabla \cdot (\kappa \nabla u) + \boldsymbol{\beta} \cdot \nabla u + cu = 0 \end{array} \right] \Longrightarrow u = 0.$$

8.6. **A compactness result for the trace operator.** Let $Q := (0,1)^d$, $\square := (0,1)^{d-1} \equiv (0,1)^{d-1} \times \{0\}$, and $\gamma : H^1(Q) \to L^2(\square)$ be the associated trace operator. The goal of this exercise is to prove that γ is compact. Consider the functions

$$\phi_\alpha(\mathbf{x}) := \prod_{j=1}^{d} \cos(\alpha_j \pi x_j) \quad \alpha \in \mathbb{N}^d,$$

$$\psi_\beta(\mathbf{x}) := \prod_{j=1}^{d-1} \cos(\beta_j \pi x_j) \quad \beta \in \mathbb{N}^{d-1}.$$

Note that
$$\gamma\phi_{(\beta,m)} = \psi_\beta \qquad \beta \in \mathbb{N}^{d-1}, \ m \in \mathbb{N} \cup \{0\}.$$

Finally, consider the projection
$$P_\beta u := \sum_{m=0}^{\infty} \frac{(u, \phi_{(\beta,m)})_{1,Q}}{\|\phi_{(\beta,m)}\|_{1,Q}^2} \phi_{(\beta,m)}.$$

(a) Show that
$$u = \sum_{\beta \in \mathbb{N}^{d-1}} P_\beta u \qquad \text{in } H^1(Q), \qquad \forall u \in H^1(Q).$$

(Note that the series is an orthogonal series.)

(b) Show that the operator $\gamma P_\beta : H^1(Q) \to L^2(\square)$ is compact.

(c) Show that
$$\left\| \gamma u - \sum_{|\beta| \le N} \gamma P_\beta u \right\|_\square^2 \le C_N \|u\|_{1,Q}^2 \qquad \forall u \in H^1(Q),$$

where $C_N \to 0$ as $N \to \infty$. Prove that $\gamma : H^1(Q) \to L^2(\square)$ is compact. (**Hint.** Show that
$$\|\gamma P_\beta u\|_\square \le C_N \|P_\beta u\|_{1,Q},$$

where $C_N \to 0$ ans $N \to \infty$.)

8.7. Let Ω be a bounded Lipschitz domain with boundary Γ, and let $X(\Gamma) := \{\xi \in L^2(\Gamma) : \xi = \gamma u, \ u \in H^2(\Omega)\}$, endowed with the image norm. Show that $X(\Gamma)$ is compactly embedded into $H^{1/2}(\Gamma)$. (**Remark.** The space $X(\Gamma)$ is sometimes denoted $H^{3/2}(\Gamma)$, although this notation is misleading when Γ is not smooth enough, due to conflicts with some associated Hilbert scales of Sobolev spaces on Γ.)

8.8. **Boundary conditions for the Kirchhoff plate.**

Consider the spaces
$$\begin{aligned} X(\Gamma) &:= \{\gamma u : u \in H^2(\Omega)\} \subset H^{1/2}(\Gamma), \\ Y(\Gamma) &:= \{\partial_n u : u \in H^2(\Omega)\} \subset L^2(\Gamma). \end{aligned}$$

We will assume that the domain Ω is such that
$$\text{range}\,(\gamma, \partial_n) = X(\Gamma) \times Y(\Gamma), \qquad \ker(\gamma, \partial_n) = H_0^2(\Omega).$$

(a) Consider the space $W := \{u \in H^2(\Omega) : \Delta^2 u \in L^2(\Omega)\}$. Prove that there are bounded linear operators
$$m_3 : W \to X(\Gamma)', \qquad m_2 : W \to Y(\Gamma)',$$

such that

$$\langle m_2 u, \partial_n v \rangle_{Y' \times Y} - \langle m_3 u, \gamma v \rangle_{X' \times X} = c(u, v) - (\Delta^2 u, v)_\Omega$$

for all $u \in W$ and $v \in H^2(\Omega)$, where

$$c(u, v) := \nu(\Delta u, \Delta v)_\Omega + (1 - \nu)(D^2 u, D^2 v)_\Omega.$$

(**Hint.** The construction is very similar to the one for the normal component. Note that $(X(\Gamma) \times Y(\Gamma))' \equiv X(\Gamma)' \times Y(\Gamma)'$.)

(b) Write down equivalent boundary value problems for the four problems in (8.13).

(c) Show that the operators

$$(\gamma, m_2) : W \longrightarrow X(\Gamma) \times Y(\Gamma)',$$
$$(m_3, \partial_n) : W \longrightarrow X(\Gamma)' \times Y(\Gamma),$$
$$(m_3, m_2) : W \longrightarrow X(\Gamma)' \times Y(\Gamma)',$$

are surjective. (**Hint.** Use a boundary value problem associated to the equation

$$\Delta^2 u - \Delta u + u = 0,$$

to avoid problems with lack of injectivity.)

Remark. For plane domains with a smooth boundary, the hypotheses above hold and we can identify the spaces $X(\Gamma)$ and $Y(\Gamma)$. Both of them are dense in $L^2(\Gamma)$, which produces two Gelfand triples. It happens that $Y(\Gamma) = H^{1/2}(\Gamma)$. In this case, m_2 and m_3 can be given an explicit formula involving normal and tangential derivatives.

9

Eigenvalues of elliptic operators

9.1   Dirichlet and Neumann eigenvalues 177
9.2   Eigenvalues of compact self-adjoint operators 180
9.3   The Hilbert-Schmidt theorem 182
9.4   Proof of the Hilbert-Schmidt theorem (*) 185
9.5   Spectral characterization of Sobolev spaces 188
9.6   Classical Fourier series 192
9.7   Steklov eigenvalues (*) 195
9.8   A glimpse of interpolation (*) 198
Final comments and literature 200
Exercises 201

The prototypical problems for the study of eigenvalues and eigenfunctions of elliptic operators are the search for Dirichlet eigenpairs for the negative Laplacian

$$u \in H_0^1(\Omega), \ \lambda \in \mathbb{R}, \qquad -\Delta u = \lambda u,$$

and their Neumann counterparts,

$$u \in H^1(\Omega), \ \lambda \in \mathbb{R}, \qquad -\Delta u = \lambda u, \qquad \partial_n u = 0.$$

We will next develop formulations for these two problems and place them in the context of the spectral orthogonal decomposition for compact self-adjoint operators, also known as the Hilbert-Schmidt theorem. We will also study how the associated orthogonal (Fourier) series associated to the eigenfunctions can be used to characterize some Sobolev spaces.

9.1   Dirichlet and Neumann eigenvalues

We are interested in studying the eigenvalues and eigenfunctions for operators $G : X \to X$ where X is a Hilbert space, i.e., in finding scalars μ and nontrivial $\phi \in X$ such that

$$G\phi = \mu\phi.$$

For now we will assume that G is compact, self-adjoint, and positive definite. If X is a real Hilbert space, we will have to initially admit $\mu \in \mathbb{C}$ and $\phi \in X_{\mathbb{C}}$, the complexification of X.

Green's operator for the Dirichlet problem. Consider the real Hilbert space $X = L^2(\Omega)$ and the operator $G : L^2(\Omega) \to L^2(\Omega)$, $f \mapsto u$ by solving the PDE

$$u \in H_0^1(\Omega), \qquad -\Delta u = f,$$

which has the variational form

$$u \in H_0^1(\Omega), \qquad (\nabla u, \nabla v)_\Omega = (f, v)_\Omega \qquad \forall v \in H_0^1(\Omega), \tag{9.1}$$

and the stability estimate

$$\|u\|_{1,\Omega} \le C\|f\|_\Omega.$$

This is independent of the regularity of $\partial\Omega$, since we are imposing homogeneous Dirichlet boundary conditions. The stability bound of the solution shows that

$$\|Gf\|_{1,\Omega} \le C\|f\|_\Omega \qquad \forall f \in L^2(\Omega). \tag{9.2}$$

This proves compactness of $G : L^2(\Omega) \to L^2(\Omega)$. To elaborate on this, let us explain the compactness of G in two different ways. First, suppose that $\{f_n\}$ is a weakly convergent sequence in $L^2(\Omega)$, $f_n \rightharpoonup f$ in $L^2(\Omega)$, then $Gf_n \rightharpoonup Gf$ in $H_0^1(\Omega)$ by the boundedness of G as an operator from $L^2(\Omega)$ to $H_0^1(\Omega)$, that is, (9.2). Therefore, by the Rellich-Kondrachov theorem (Proposition 7.4), $Gf_n \to Gf$ in $L^2(\Omega)$. The other explanation is actually much simpler. We simply notice that G can be factored as the solution operator for the well posed problem (9.1), followed by the Rellich-Kondrachov embedding theorem:

$$L^2(\Omega) \longrightarrow H_0^1(\Omega) \longrightarrow L^2(\Omega).$$

Self-adjointness of G follows from the following simple argument: if $f_1, f_2 \in L^2(\Omega)$ and $u_1 = Gf_1$, $u_2 = Gf_2$, then

$$(f_1, Gf_2)_\Omega = (f_1, u_2)_\Omega = (\nabla u_1, \nabla u_2)_\Omega = (u_1, f_2)_\Omega = (Gf_1, f_2)_\Omega,$$

where it is visible that what matters is the symmetry of the bilinear form that defines the Dirichlet problem. We can also write this argument by simply showing that

$$(f_1, Gf_2)_\Omega = (\nabla Gf_1, \nabla Gf_2)_\Omega \qquad \forall f_1, f_2 \in L^2(\Omega).$$

Note also that G is injective (if $Gf = u = 0$, then $f = -\Delta u = 0$) and positive definite, since

$$(f, Gf)_\Omega = (\nabla Gf, \nabla Gf)_\Omega = \|\nabla Gf\|_\Omega^2 \ge C_{\mathrm{PF}}\|f\|_\Omega^2 \qquad \forall f \in L^2(\Omega).$$

Dirichlet eigenvalues. If a pair (μ, ϕ) satisfies $G\phi = \mu\phi$, then, since $u = G\phi$, we have $-\Delta(\mu\phi) = \phi$, or,

$$\phi \in H_0^1(\Omega), \qquad -\Delta\phi = \mu^{-1}\phi.$$

If we set $\lambda = \mu^{-1}$, then we say that λ is a Dirichlet eigenvalue of the Laplacian and ϕ is the corresponding eigenfunction.

A formal setup for Neumann eigenvalues. We can consider the problem of looking for nontrivial solutions of

$$\phi \in H^1(\Omega), \ \lambda \in \mathbb{R}, \qquad -\Delta\phi = \lambda\phi, \qquad \partial_n\phi = 0, \qquad (9.3)$$

or, in variational form,

$$\phi \in H^1(\Omega), \ \lambda \in \mathbb{R}, \qquad (9.4\text{a})$$
$$(\nabla\phi, \nabla v)_\Omega = \lambda\,(\phi, v)_\Omega \qquad \forall v \in H^1(\Omega). \qquad (9.4\text{b})$$

Note that problem (9.4) can be considered in more general domains than problem (9.3), which needs the definition of the normal derivative, and hence the trace operator. We will be able to handle problem (9.4) in any domain where $H^1(\Omega)$ is compactly embedded into $L^2(\Omega)$. Given the fact that the Neumann problem for the Laplace equation has solvability issues, we need to make some adjustments to create a Green's operator whose eigenvalues are related to the Neumann eigenvalues. As a first attempt, we work on the Hilbert space

$$L_\circ^2(\Omega) := \{f \in L^2(\Omega) : \int_\Omega f(\mathbf{x})\mathrm{d}\mathbf{x} = 0\},$$

and consider the well-defined operator $G : L_\circ^2(\Omega) \to L_\circ^2(\Omega)$ given by $u = Gf$ being the solution of the coercive problem (recall Poincaré's inequality)

$$u \in H^1(\Omega) \cap L_\circ^2(\Omega),$$
$$(\nabla u, \nabla v)_\Omega = (f, v)_\Omega \qquad \forall v \in H^1(\Omega) \cap L_\circ^2(\Omega).$$

When the normal derivative operator is well-defined, this is equivalent to

$$u \in H^1(\Omega), \qquad -\Delta u = f, \qquad \partial_n u = 0, \qquad (u, 1)_\Omega = 0,$$

a problem which has a unique solution only when $(f, 1)_\Omega = 0$. Since

$$\|Gf\|_{1,\Omega} \leq \|f\|_\Omega \qquad \forall f \in L_\circ^2(\Omega),$$

it follows that G is compact (by the compact embedding of $H^1(\Omega)$ into $L^2(\Omega)$), while the formula

$$(f_1, Gf_2)_\Omega = (\nabla Gf_1, \nabla Gf_2)_\Omega \qquad \forall f_1, f_2 \in L_\circ^2(\Omega),$$

easily shows that G is self-adjoint and positive definite. As happened in the Dirichlet case, the eigenvalues of G are the inverses of the Neumann eigenvalues. Note that with this formalization of the problem the zero Neumann eigenvalue (corresponding to constant eigenfunctions) is missing.

A different formalization for Neumann eigenvalues. Instead of working in $L_\circ^2(\Omega)$, we can displace the eigenvalues and think of the following eigenvalue problem

$$\phi \in H^1(\Omega), \ \xi \in \mathbb{R}, \qquad -\Delta\phi + \phi = \xi\phi, \qquad \partial_n\phi = 0,$$

or in variational form

$$u \in H^1(\Omega), \ \xi \in \mathbb{R},$$
$$(\nabla u, \nabla v)_\Omega + (u, v)_\Omega = \xi(u, v)_\Omega \qquad \forall v \in H^1(\Omega).$$

The associated Green's operator $G : L^2(\Omega) \to L^2(\Omega)$, given by $Gf = u$, where

$$u \in H^1(\Omega), \qquad (\nabla u, \nabla v)_\Omega + (u, v)_\Omega = (f, v)_\Omega \qquad \forall v \in H^1(\Omega),$$

is compact, self-adjoint and positive definite. Note that if $G\phi = \mu\phi$, then

$$(\nabla u, \nabla v)_\Omega = (\mu^{-1} - 1)(u, v)_\Omega \qquad \forall v \in H^1(\Omega),$$

which transfers the eigenvalues and eigenfunctions of G to Neumann eigenfunctions.

A remark. In principle, we should be considering the possibility of having complex eigenvalues and the corresponding eigenfunctions, by working on the spaces $L^2(\Omega; \mathbb{C})$ and $H^1(\Omega; \mathbb{C})$. As we will shortly see, this will not be needed.

9.2   Eigenvalues of compact self-adjoint operators

For the following, suppose that X is a Hilbert space and $G : X \to X$ is a compact self-adjoint operator. If X is a real space, we consider its complexification $X_\mathbb{C}$ and the operator $G : X_\mathbb{C} \to X_\mathbb{C}$ that acts separately on the real and imaginary parts. This operator is again compact and self-adjoint. This means that, without loss of generality, we can work on complex Hilbert spaces. The first results are based on very elementary linear algebra. They are recalled here for practice.

Proposition 9.1. *The eigenvalues of a self-adjoint operator are real.*

Proof. If G is self-adjoint and $G\phi = \mu\phi$, then $(G\phi, \phi)_X = \mu(\phi, \phi)_X$. Since $(G\phi, \phi)_X$ is real (G is self-adjoint) and $(\phi, \phi)_X = \|\phi\|_X^2$, then $\mu \in \mathbb{R}$ as well. $\qquad\square$

Proposition 9.2. *Eigenfunctions for different eigenvalues of a self-adjoint operator are orthogonal.*

Proof. Let $\lambda \neq \mu$ be (real) eigenvalues for G. If $G\phi = \lambda\phi$ and $G\psi = \mu\psi$, then

$$\lambda(\phi, \psi)_X = (G\phi, \psi)_X = (G\psi, \phi)_X = \mu(\phi, \psi)_X.$$

Therefore $(\phi, \psi)_X = 0$. $\qquad\qquad\qquad\qquad\qquad\qquad\qquad\qquad\qquad\qquad\quad\square$

Proposition 9.3. *If $\lambda \neq 0$ is an eigenvalue of a compact operator G, then* $\dim \ker(\lambda I - G)$ *is finite.*

Proof. We write $\lambda I - G = \lambda(I - \lambda^{-1}G)$, which is the identity plus a compact operator. Therefore standard Fredholm theory applies. $\qquad\qquad\qquad\square$

Proposition 9.4. *If $\{\mu_n\}$ is a sequence of pairwise different eigenvalues of a compact operator G, and $\mu_n \to \mu$, then $\mu = 0$.*

Proof. Consider normalized eigenfunctions corresponding to the eigenvalues:

$$G\phi_n = \mu_n\phi_n, \qquad \|\phi_n\|_X = 1.$$

Since $\{\phi_n\}$ is an orthonormal sequence in the Hilbert space X, we have $\phi_n \rightharpoonup 0$. By the compactness of G, we have $G\phi_n \to 0$ and therefore $\mu_n\phi_n \to 0$. By taking the norm of this convergent sequence, it follows that $|\mu_n| \to 0$. $\qquad\square$

Some conclusions about the spectrum. Before we continue, let us rephrase some of what we have already proved. Given an infinite-dimensional complex Hilbert space X and a compact self-adjoint operator $G : X \to X$, we extract two conclusions from Fredholm's alternative:

(a) $G : X \to X$ cannot be invertible. If it were, then $I = G^{-1}G$ would be compact, but we are working in infinite dimensions and the identity operator is not compact.

(b) If $\lambda \neq 0$, either $\lambda I - G$ has a nontrivial kernel (and λ is an eigenvalue of G), or $\lambda I - G : X \to X$ is invertible.

The resolvent set for G is the set

$$\rho(G) := \{\lambda \in \mathbb{C} : (\lambda I - G)^{-1} \quad \text{exists}\},$$

and its complement is called the spectrum and denoted $\sigma(G)$. We know that $0 \in \sigma(G)$ and that if $0 \neq \lambda \in \sigma(G)$, then λ is an eigenvalue. Note that if $G\phi = \mu\phi$, then $|\mu| \leq \|G\|$, that is, the spectrum is a bounded set. We have additionally seen that:

(c) The spectrum of G is contained in the interval $[-\|G\|, \|G\|]$.

(d) The spectrum of G is finite or countable. This follows from the following argument and Proposition 9.4. In the set $[-1, -1/n] \cup [1/n, 1]$, there can only be a finite number of eigenvalues, since if there were an infinite number of them, they would have an accumulation point, but Proposition 9.4 would prove that this point is 0.

Let us note that some of the properties above are not exclusive of compact self-adjoint operators (only (c) needs G to be self-adjoint) and can be extended to general compact operators. What is very characteristic of compact self-adjoint operators is the actual existence of eigenvalues.

Proposition 9.5. *If G is a compact self-adjoint operator, then either $\|G\|$ or $-\|G\|$ is an eigenvalue.*

Proof. Recall that

$$\|G\| = \sup_{0 \neq x \in X} \frac{\|Gx\|_X}{\|x\|_X} = \sup_{\|x\|_X = 1} \|Gx\|_X.$$

Thus there is a sequence $\{x_n\}$ with unit norm such that $\|Gx_n\|_X \to \|G\|$. The sequence $\{x_n\}$ is bounded, and therefore has a weakly convergent subsequence $x_{n_k} \rightharpoonup x$, for some $x \in X$. Since G is compact, it follows that $Gx_{n_k} \to Gx$. This implies that $\|Gx\|_X = \|G\|$ and ensures that $x \neq 0$. Let us now look at the following computation

$$\|G^2 x - \|G\|^2 x\|_X^2 = (G^2 x, G^2 x)_X - 2\|G\|^2 \mathrm{Re}(G^2 x, x)_X + \|G\|^4 \|x\|_X^2$$
$$= \|G^2 x\|_X^2 - 2\|G\|^2 \|Gx\|_X^2 + \|G\|^4 \|x\|_X^2$$
$$\leq \|G\|^2 \|Gx\|_X^2 - 2\|G\|^2 \|Gx\|_X^2 + \|G\|^4 = 0.$$

In particular $G^2 x = \|G\|^2 x$ and therefore

$$G^2 x - \|G\|^2 x = (G + \|G\|I)(G - \|G\|I)x = 0,$$

which implies that either $Gx = \|G\|x$ and $\|G\|$ is an eigenvalue, or $y = Gx - \|G\|x \neq 0$ and $Gy = -\|G\|y$ and $-\|G\|$ is an eigenvalue. $\qquad\square$

9.3   The Hilbert-Schmidt theorem

Theorem 9.1 (Hilbert-Schmidt). *If X is a real or complex Hilbert space, $G : X \to X$ is a compact, self-adjoint, and positive semidefinite linear operator such that $\mathrm{range}\, G$ is infinite-dimensional, then there is a nonincreasing* *sequence of numbers $\mu_n > 0$ with $\mu_n \to 0$ and an orthonormal sequence $\{\phi_n\}$ in X such that*

$$G = \sum_{n=1}^{\infty} \mu_n(\cdot, \phi_n)_X \phi_n, \tag{9.5}$$

with convergence in the operator norm. In addition, for all $f \in X$

$$f = \sum_{n=1}^{\infty} (f, \phi_n)_X \phi_n + Qf, \tag{9.6}$$

where $Q : X \to \ker G$ is the orthogonal projection onto $\ker G$.

Remarks and consequences. The Hilbert-Schmidt theorem admits another formulation where the operator is not required to be positive semidefinite. In that case $|\mu_n|$ is nonincreasing. When range G is finite-dimensional, the theorem still holds but in this case the sums in (9.5) and (9.6) contain a finite number of terms. There are some conclusions that can be extracted from this formulation of the Hilbert-Schmidt theorem (although some of these conclusions are used as intermediate steps in the proof).

(a) From (9.5) it follows that $G\phi_n = \mu_n \phi_n$ for all n.

(b) The set $\{\phi_n\}$ is complete orthonormal in $(\ker G)^\perp = $ range G. This follows from (9.6), since

$$f - Qf = \sum_{n=1}^{\infty} (f, \phi_n)_X \phi_n,$$

is the orthogonal projection onto $(\ker G)^\perp$.

(c) If $\ker G = \{0\}$, then $\{\phi_n\}$ is a complete orthonormal set in X and, therefore, X is a separable Hilbert space. In particular, if G is a self-adjoint compact positive definite operator, then $\ker G = \{0\}$, range G has to be infinite-dimensional, and X has to be separable.

(d) If $G\phi = \mu\phi$ with $\phi \neq 0$ and $\mu \neq 0$, then $\mu = \mu_k$ for at least one k and ϕ is a finite linear combination of the functions $\{\phi_n : \mu_n = \mu_k\}$. This can be proved by expanding

$$\phi = \sum_{n=1}^{\infty} (\phi, \phi_n)_X \phi_n,$$

(note that ϕ is orthogonal to the elements of the kernel, which are eigenfunctions for a different eigenvalue) and comparing the series $G\phi$ and $\mu\phi$:

$$\sum_{n=1}^{\infty} (\mu_n - \mu)(\phi, \phi_n)_X \phi_n = 0.$$

In particular, this shows that $\mu_1 = \|G\|$.

(e) The operator G is the uniform limit of the sequence

$$G_N := \sum_{n=1}^{N} \mu_n(\,\cdot\,,\phi_n)_X \phi_n,$$

of self-adjoint operators with finite-dimensional range.

Note that the expansion (9.5) does not include the eigenfunctions corresponding to the zero eigenvalue, since there might be an uncountable number of those. The spectral decomposition (9.5) needs not be unique: eigenfunctions can be multiplied by numbers with unit absolute value, and eigenfunctions corresponding to the same eigenvalue can be mixed to provide new orthonormal sequences.

Dirichlet eigenvalues. Suppose that Ω is an open, bounded set and therefore $H_0^1(\Omega)$ is compactly embedded in $L^2(\Omega)$. We can consider the map (see Section 9.1) $G : L^2(\Omega) \to L^2(\Omega)$ given by $Gf = u$ where

$$u \in H_0^1(\Omega), \qquad -\Delta u = f, \tag{9.7}$$

or, equivalently,

$$u \in H_0^1(\Omega), \qquad (\nabla u, \nabla v)_\Omega = (f,v)_\Omega \qquad \forall v \in H_0^1(\Omega).$$

The operator G is compact, self-adjoint, and positive definite, hence injective. Therefore there exists a complete orthonormal set $\{\phi_n\}$ in $L^2(\Omega)$ and a nonincreasing sequence of positive numbers $\{\mu_n\}$ with $\mu_n \to 0$ such that

$$G = \sum_{n=1}^{\infty} \mu_n(\,\cdot\,,\phi_n)_\Omega \phi_n.$$

Given $f \in L^2(\Omega)$,

$$u = Gf = \sum_{n=1}^{\infty} \mu_n(f,\phi_n)_\Omega \phi_n,$$

is the unique solution of (9.7). The quantities

$$0 < \lambda_n := \mu_n^{-1} \longrightarrow \infty,$$

are the only Dirichlet eigenvalues. Note that we have found a complete orthonormal sequence $\{\phi_n\}$ satisfying

$$\phi_n \in H_0^1(\Omega), \qquad -\Delta \phi_n = \lambda_n \phi_n,$$

and apart from possible linear combinations of eigenfunctions for multiple eigenvalues, we have localized all possible eigenfunctions for the Dirichlet problem for the Laplacian. Note also that

$$(\nabla \phi_n, \nabla \phi_m)_\Omega = \lambda_n\,(\phi_n,\phi_m)_\Omega = \lambda_n\,\delta_{nm} \qquad \forall n,m.$$

This proves that $\{\phi_n\}$ are also orthogonal in $H_0^1(\Omega)$. We will deal with more properties of this sequence when we explore (Section 9.5) the different forms of convergence of the spectral series

$$f = \sum_{n=1}^{\infty} (f, \phi_n)_\Omega \phi_n,$$

which, in principle, converges in $L^2(\Omega)$.

Neumann eigenvalues. Now consider the operator $G : L^2(\Omega) \to L^2(\Omega)$ given by $u = Gf$ being the solution of

$$u \in H^1(\Omega), \qquad (\nabla u, \nabla v)_\Omega + (u, v)_\Omega = (f, v)_\Omega \qquad \forall v \in H^1(\Omega). \qquad (9.8)$$

When Ω is a bounded set such that $H^1(\Omega)$ is compactly embedded into $L^2(\Omega)$ (see Section 7.2), the operator G is compact, self-adjoint, and positive definite. We then have an $L^2(\Omega)$ complete orthonormal sequence $\{\psi_n\}$ such that

$$G = \sum_{n=1}^{\infty} \mu_n (\cdot, \psi_n)_\Omega \psi_n.$$

Note that if $Gf = \mu f$, from (9.8) it follows that

$$(1 - \mu)\|f\|_\Omega^2 = \|\nabla Gf\|_\Omega^2 \geq 0.$$

Therefore $\mu = 1$ is the largest possible eigenvalue, corresponding to constant f and with a one-dimensional eigenspace. The quantities

$$0 \leq \lambda_n := \mu_n^{-1} - 1 \longrightarrow \infty$$

are the Neumann eigenvalues, which, for Lipschitz domains, are the solutions of

$$\phi_n \in H^1(\Omega), \qquad -\Delta\phi_n = \lambda_n \phi_n, \qquad \partial_n \phi_n = 0.$$

As already mentioned, $\lambda_1 = 0$ and $\lambda_n > 0$ for $n \geq 2$.

9.4 Proof of the Hilbert-Schmidt theorem (*)

We will prove the Hilbert-Schmidt theorem using a deflation argument, collecting all possible eigenfunctions for the largest eigenvalues.

Collecting eigenfunctions. We will develop the tools for deflation of a self-adjoint operator so that we can prove the Hilbert-Schmidt theorem. Suppose that G is a self-adjoint operator for which we have nonzero eigenvalues and an orthonormal system with their associated eigenfunctions

$$G\phi_n = \mu_n\phi_n \qquad n = 1,\ldots,N \qquad (\phi_n,\phi_m)_X = \delta_{nm},$$

Consider the degenerate (finite rank) operator

$$G_N := \sum_{n=1}^{N} \mu_n(\,\cdot\,,\phi_n)_X\phi_n.$$

If $P_N : X \to \text{span}\{\phi_1,\ldots,\phi_N\}$ is the orthogonal projection onto the space of eigenfunctions

$$P_N u = \sum_{n=1}^{N} (u,\phi_n)_X\phi_n,$$

then

$$P_N G u = \sum_{n=1}^{N} (Gu,\phi_n)_X\phi_n \;=\; \sum_{n=1}^{N} (u,G\phi_n)_X\phi_n$$

$$= \sum_{n=1}^{N} \mu_n(u,\phi_n)_X\phi_n = G_N u,$$

from which it is easy to see that

$$G_N = P_N G = G P_N = P_N G P_N.$$

Now consider the remainder

$$R_N := G - \sum_{n=1}^{N} \mu_n(\,\cdot\,,\phi_n)_X\phi_n = (I - P_N)G(I - P_N).$$

We then have the following result on the spectrum of R_N.

Lemma 9.1. *If $R_N\phi = \mu\phi$, then either $\mu = 0$ or $G\phi = \mu\phi$ and $\phi \perp \phi_n$ for all n.*

Proof. If $R_N u = \mu u$, then

$$Gu - \mu u = G_N u \in \text{span}\{\phi_1,\ldots,\phi_N\},$$

and we write

$$Gu - \mu u = \sum_{n=1}^{N} (Gu - \mu u,\phi_n)_X\phi_n = \sum_{n=1}^{N} (u, G\phi_n - \mu\phi_n)_X\phi_n$$

$$= \sum_{n=1}^{N} (\mu_n - \mu)(u,\phi_n)_X\phi_n.$$

We compare the latter expression with $G_N u$ and note that this implies that

$$\mu(u, \phi_n)_X = 0 \qquad n = 1, \ldots, N.$$

If $\mu = 0$ the proof is finished. Otherwise $(u, \phi_n)_X = 0$ for all n and then $G_N u = 0$, which means that $R_N u = Gu = \mu u$. $\qquad\qquad\square$

Deflation to the limit for self-adjoint operators. If G is a nontrivial operator on a Hilbert space X, and G is compact, self-adjoint, and positive semidefinite, we can make a countable list of eigenpairs

$$G\phi_n = \mu_n \phi_n, \qquad (\phi_m, \phi_n)_X = \delta_{mn},$$

where $\{\mu_n\}$ is nonincreasing. (This is done by taking an orthonormal basis for the finite-dimensional spaces $\ker(\mu I - G)$ when μ is an eigenvalue. There are two options for the sequence μ_n. Either $\{\mu_n\}$ is a finite list of eigenvalues, or it is decreasing to zero. We take

$$G_N \phi = \sum_{n=1}^{N} \mu_n (\phi, \phi_n)_X \phi_n, \qquad P_N \phi = \sum_{n=1}^{N} (\phi, \phi_n)_X \phi_n,$$

and write $R_N = G - G_N = (I - P_N)G(I - P_N)$. We have already seen that R_N is self-adjoint and compact. We claim that it is also positive semidefinite. To see this, test R_N by an element ϕ and compute

$$\begin{aligned}
(R_N \phi, \phi)_X &= ((I - P_N)G(I - P_N)\phi, \phi)_X \\
&= (G(I - P_N)\phi, (I - P_N)\phi)_X \geq 0,
\end{aligned}$$

where the last inequality follows since G itself is positive semidefinite. We can classify the eigenvalues of R_N as the nonzero eigenvalues of G and zero, regardless of whether zero is an eigenvalue of G or not. We also have $\|R_N\| = \mu_{N+1}$. In the case when G has only finitely many eigenvalues, we can take N large enough so that $\mu_{N+1} = 0$ and therefore

$$G\phi = \sum_{n=1}^{N} \mu_n (\phi, \phi_n)_X \phi_n,$$

which shows that G is a degenerate operator (it has finite rank). If G has a countable sequence of eigenpairs with $\mu_n \to 0$, then $\|R_N\| = \mu_{N+1} \to 0$ as $n \to \infty$. Since $R_N = G - G_N$, we have $\|G - G_N\| \to 0$. This shows that $G = \sum_n \mu_n(\cdot, \phi_n)_X \phi_n$ with convergence in norm. This proves the first part of the Hilbert-Schmidt theorem. We now prove the second part of the theorem.

The end of the proof. We need to show that the operator Q defined by

$$Q\phi := \phi - \sum_{n=1}^{\infty} (\phi, \phi_n)_X \phi_n,$$

is the orthogonal projection onto $\ker G$. In the degenerate case, the sum will only be from $n = 1$ to $n = N$. First, note that

$$GQ\phi = G\phi - G\left(\sum_{n=1}^{\infty}(\phi, \phi_n)_X \phi_n\right) = G\phi - \sum_{n=1}^{\infty}(\phi, \phi_n)_X G\phi_n$$

$$= G\phi - \sum_{n=1}^{\infty}\mu_n(\phi, \phi_n)_X \phi_n = 0,$$

and therefore $Q\phi \in \ker G$. From this, we see that

$$Q\phi - \phi = -\sum_{n=1}^{\infty}(\phi, \phi_n)_X \phi_n,$$

is a convergent series in the closed space $(\ker G)^\perp$ (ϕ_n is an eigenfunction for a nonzero eigenvalue), then $Q\phi - \phi \in (\ker G)^\perp$, which proves that Q is the orthogonal projection onto $\ker G$.

9.5   Spectral characterization of Sobolev spaces

In this section we are going to look at convergence properties for the spectral series associated to an elliptic eigenvalue problem. We will carry out all the details for a particular example, the Neumann eigenvalues for a nonconstant diffusion parameter.

A Neumann eigenvalue problem. Let Ω be a Lipschitz domain and let $\kappa \in L^\infty(\Omega)$ be strongly positive. The variational equations

$$u \in H^1(\Omega), \qquad (\kappa\nabla u, \nabla v)_\Omega + (u, v)_\Omega = (f, v)_\Omega \qquad \forall v \in H^1(\Omega) \qquad (9.9)$$

are equivalent to the boundary value problem

$$u \in H^1(\Omega), \qquad -\nabla \cdot (\kappa\nabla u) + u = f, \qquad (\kappa\nabla u) \cdot \mathbf{n} = 0. \qquad (9.10)$$

For the moment being, let us write

$$a(u, v) := (\kappa\nabla u, \nabla v)_\Omega + (u, v)_\Omega$$

to denote the bilinear form associated to (9.9). This bilinear form defines an inner product in $H^1(\Omega)$ that is equivalent to the usual one. The operator $G : L^2(\Omega) \to L^2(\Omega)$ defined by $Gf = u$, where u is the solution of (9.9) and (9.10) is linear and admits the bound

$$\|Gf\|_{1,\Omega} \leq C\|f\|_\Omega \qquad \forall f \in L^2(\Omega).$$

From this, it follows that G is compact (see Section 9.1). Also, by the variational formulation (9.9), we have

$$a(Gf_1, Gf_2) = (f_1, Gf_2)_\Omega \qquad \forall f_1, f_2 \in L^2(\Omega),$$

which shows that G is self-adjoint and positive definite. Therefore there exists a nonincreasing sequence of positive numbers $\{\mu_n\}$ and a complete orthonormal sequence $\{\phi_n\}$ in $L^2(\Omega)$ such that

$$G = \sum_{n=0}^{\infty} \mu_n \, (\,\cdot\,, \phi_n)_\Omega \phi_n. \tag{9.11}$$

Note that in (9.11), we have decided to start counting from $n = 0$. The reason to do this is to isolate the largest eigenvalue $\mu_0 = 1$ corresponding to constant eigenfunctions and we can take $\phi_0 \equiv |\Omega|^{-1/2}$ as the first element of the orthonormal sequence, showing that all other elements of the sequence have zero average over Ω. The eigenvalue $\mu_0 = 1$ is simple and is the largest possible eigenvalue. (Prove it.) The eigenvalue properties

$$G\phi_n = \mu_n \phi_n \qquad n \geq 0,$$

can be rewritten in terms of the diffusion problem as follows (with $\lambda_n :=$ $\mu_n^{-1} - 1$)

$$\phi_n \in H^1(\Omega), \qquad -\nabla \cdot (\kappa \nabla \phi_n) = \lambda_n \phi_n, \qquad (\kappa \nabla \phi_n) \cdot \mathbf{n} = 0. \tag{9.12}$$

Note that $\lambda_0 = 0$ and $\lambda_n > 0$ for all n defines a nondecreasing sequence that diverges to infinity.

Fourier characterization of the energy space. The space $H^1(\Omega)$ is the space where we look for the solution of (9.9). However, an inspection of (9.10) shows that u automatically satisfies additional properties that we will explore when we study range G. Recall that $\{\phi_n\}$ is a complete orthonormal set in $L^2(\Omega)$. The property $G\phi_n = \mu_n \phi_n$ is equivalent to

$$a(\phi_n, v) = \mu_n^{-1} (\phi_n, v)_\Omega \qquad \forall v \in H^1(\Omega), \tag{9.13}$$

and therefore

$$a(\phi_n, \phi_m) = \mu_n^{-1} (\phi_n, \phi_m)_\Omega = \mu_n^{-1} \delta_{nm} \qquad \forall n, m \geq 0.$$

This implies that the functions $\psi_n := \mu_n^{1/2} \phi_n$ are orthonormal in $H^1(\Omega)$, when using the bilinear form a as the inner product, that is,

$$(\kappa \nabla \psi_n, \nabla \psi_m)_\Omega + (\psi_n, \psi_m)_\Omega = \delta_{nm} \qquad \forall n, m \geq 0.$$

This orthonormal system (a rescaling of $\{\phi_n\}$ so that they have unit energy norm) is complete because

$$a(\phi_n, v) = 0 \quad \forall n \qquad \Longrightarrow \qquad (\phi_n, v)_\Omega = 0 \quad \forall n,$$

by (9.13), and this implies that $v = 0$. Since $\{\psi_n\}$ is a complete orthonormal sequence in $H^1(\Omega)$, we can use Parseval's identity and write

$$a(u, u) = \sum_{n=0}^{\infty} |a(u, \psi_n)|^2.$$

However,

$$a(u, \psi_n) = \mu_n^{-1}(u, \psi_n)_\Omega = \mu_n^{-1/2}(u, \phi_n)_\Omega \qquad \forall n,$$

and therefore

$$(\kappa \nabla u, \nabla u)_\Omega + (u, u)_\Omega = a(u, u) = \sum_{n=0}^{\infty} \mu_n^{-1} |(u, \phi_n)_\Omega|^2.$$

Using the fact that $\{\phi_n\}$ is orthonormal complete in $L^2(\Omega)$, we can also write

$$(\kappa \nabla u, \nabla u)_\Omega = \sum_{n=0}^{\infty} \mu_n^{-1} |(u, \phi_n)_\Omega|^2 - \sum_{n=0}^{\infty} |(u, \phi_n)_\Omega|^2,$$

and therefore, recalling the Neumann eigenvalues, we have

$$(\kappa \nabla u, \nabla u)_\Omega = \sum_{n=0}^{\infty} \lambda_n |(u, \phi_n)_\Omega|^2. \tag{9.14}$$

Note that the sum in (9.14) starts actually in $n = 1$, since $\lambda_0 = 0$. This computation can be wrapped up nicely in a proposition that also looks at some converse properties.

Proposition 9.6. *If $\{(\lambda_n, \phi_n)\}$ is the set of all solutions of (9.12), where $\{\phi_n\}$ is taken to be $L^2(\Omega)$ orthonormal, then, the following statements are equivalent:*

(a) $u \in H^1(\Omega)$.

(b) $\sum_{n=0}^{\infty} \lambda_n |(u, \phi_n)_\Omega|^2 < \infty$.

(c) *The series $\sum_{n=0}^{\infty} (u, \phi_n)_\Omega \phi_n$ converges to u in $H^1(\Omega)$.*

Proof. We have already proved that (a) implies (b). Assume that (b) holds. Without loss of generality we can suppose that $(u, \phi_0)_\Omega = 0$, since that does not affect convergence and we can always substract a constant from u without modifying its smoothness properties. Take the sequence

$$u_N := \sum_{n=1}^{N} (u, \phi_n) \phi_n.$$

We know that $u_N \to u$ in $L^2(\Omega)$ and that $u_N \in H^1(\Omega)$, since it is a linear combination of eigenfunctions. Moreover

$$(\kappa \nabla u_N, \nabla u_N)_\Omega = \sum_{n=1}^{N} \lambda_n |(u, \phi_n)_\Omega|^2,$$

which can easily be used to prove that $\{u_N\}$ is Cauchy in $H^1(\Omega)$ and therefore convergent in $H^1(\Omega)$. Since convergence in $H^1(\Omega)$ implies convergence in $L^2(\Omega)$, (c) follows. The fact that (c) implies (a) is straightforward. □

Fourier characterization of the range. Finally, we discuss the space range $G \subset H^1(\Omega)$. It is easy to verify that

$$\text{range}\, G = \{u \in H^1(\Omega) \,:\, \kappa \nabla u \in \mathbf{H}(\text{div}, \Omega), \quad (\kappa \nabla u) \cdot \mathbf{n} = 0\},$$

as follows by observing that $u = Gf$ is the solution of (9.10). However, the spectral representation of G and Picard's criterion (Theorem 9.2 below) show that

$$\text{range}\, G = \{u \in L^2(\Omega) \,:\, \sum_{n=0}^{\infty} \lambda_n^2 |(u, \phi_n)_\Omega|^2 < \infty\}.$$

We thus have that the Fourier series associated to the Neumann problem

$$u = \sum_{n=0}^{\infty} (u, \phi_n)_\Omega \phi_n,$$

can be used to characterize the two spaces:

$$H^1(\Omega) = \{u \in L^2(\Omega) \,:\, \sum_{n=0}^{\infty} \lambda_n |(u, \phi_n)_\Omega|^2 < \infty\},$$

$$\text{range}\, G = \{u \in L^2(\Omega) \,:\, \sum_{n=0}^{\infty} \lambda_n^2 |(u, \phi_n)_\Omega|^2 < \infty\}.$$

Theorem 9.2 (Picard's criterion). *If $G : X \to X$ is a compact, self-adjoint, positive definite operator with spectral decomposition*

$$G = \sum_{n=1}^{\infty} \mu_n (\,\cdot\,, \phi_n)_X \phi_n,$$

then

$$\text{range}\, G = \left\{ u \in X \,:\, \sum_{n=1}^{\infty} \mu_n^{-2} |(u, \phi_n)_X|^2 < \infty \right\}.$$

Proof. Let $u = Gv \in \text{range}\, G$. We can compare the two associated series

$$\sum_{n=1}^{\infty} (u, \phi_n)_X \phi_n = u = Gv = \sum_{n=1}^{\infty} \mu_n (v, \phi_n)_X \phi_n,$$

to show that $(u, \phi_n)_X = \mu_n (v, \phi_n)_X$. Therefore

$$\|v\|_X^2 = \sum_{n=1}^{\infty} |(v, \phi_n)_X|^2 = \sum_{n=1}^{\infty} \mu_n^{-2} |(u, \phi_n)_X|^2.$$

Now let $u \in X$ satisfy

$$\sum_{n=1}^{\infty} \mu_n^{-2} |(u, \phi_n)_X|^2 < \infty$$

and define

$$v := \sum_{n=1}^{\infty} \mu_n^{-1} (u, \phi_n)_X \phi_n.$$

We then have $Gv = u \in \text{range } G$, which finishes the proof. $\qquad\qquad \square$

Dirichlet eigenvalues. The Dirichlet eigensystem

$$\phi_n \in H_0^1(\Omega), \qquad -\nabla \cdot (\kappa \nabla \phi_n) = \lambda_n \phi_n,$$

with $\{\lambda_n\}$ positive nondecreasing and diverging to infinity, and $\{\phi_n\}$ complete orthonormal in $L^2(\Omega)$. The norm associated to the bilinear form can be written in terms of the Dirichlet spectral decomposition

$$(\kappa \nabla u, \nabla u)_\Omega = \sum_{n=1}^{\infty} \lambda_n |(u, \phi_n)_\Omega|^2.$$

We can characterize the energy space $H_0^1(\Omega)$ (the space where we look for the solution of the Dirichlet problem) as

$$H_0^1(\Omega) = \{u \in L^2(\Omega) : \sum_{n=1}^{\infty} \lambda_n |(u, \phi_n)_\Omega|^2 < \infty\},$$

and the range of the associated Green's operator is

$$\text{range } G = \{u \in H_0^1(\Omega) : \nabla \cdot (\kappa \nabla u) \in L^2(\Omega)\}$$

$$= \{u \in L^2(\Omega) : \sum_{n=1}^{\infty} \lambda_n^2 |(u, \phi_n)_\Omega|^2 < \infty\}.$$

9.6   Classical Fourier series

In this section we will examine the classical Fourier series that are related to one-dimensional eigenvalue problems of the form

$$-u'' = \lambda u \qquad \text{in } (0, L), \tag{9.15}$$

with different combinations of boundary conditions. All of the details of this section are left to the reader. We start with a general argument. Let V be a closed subspace of $H^1(0, L)$ and recall that $H^1(0, L)$ is continuously embedded into $\mathcal{C}[0, L]$. If

$$u \in V, \quad \lambda \in \mathbb{R}, \qquad (u', v')_{(0,L)} = \lambda \, (u, v)_{(0,L)} \qquad \forall v \in V, \qquad (9.16)$$

then (9.15) holds in the sense of distributions. If $\lambda = 0$, then $u \in \mathcal{P}_1$, that is, $u(x) = a_0 + a_1 x$ for some $a_0, a_1 \in \mathbb{R}$. If $\lambda \neq 0$, then it is easy to show (note that u is continuous and it is the right-hand side of the equation) that $u \in \mathcal{C}^2[0, L]$ and (9.15) holds in a classical sense. Therefore, the spectral theory for the one-dimensional Laplacian includes the classical Sturm-Liouville problem. We will use this to derive simple proofs involving the convergence of Fourier series.

The zero eigenvalue. If $V \cap \mathcal{P}_0 = \{0\}$, problem (9.16) is the 'inverse eigenvalue problem' to the problem of finding eigenvalues for the Green's operator $G : L^2(0, L) \to L^2(0, L)$ defined by $Gf = u$ being the solution of

$$u \in V, \qquad (u'v')_{(0,L)} = (f, v)_{(0,L)} \qquad \forall v \in V. \qquad (9.17)$$

In this case, all eigenvalues associated to (9.16) are strictly positive and diverge to infinity. If $\mathcal{P}_0 \subset V$, we need to slightly modify (9.17) adding an L^2-product to the bilinear form:

$$u \in V, \qquad (u'v')_{(0,L)} + (u, v)_{(0,L)} = (f, v)_{(0,L)} \qquad \forall v \in V.$$

This changes the relation between the eigenvalues of G and the solutions to (9.16), which are not just related by inversion but by inversion and shifting. (See the Neumann eigenvalues in the past sections.) The eigenvalue $\lambda = 0$ (with associated constant eigenfunctions) appears only when $\mathcal{P}_0 \subset V$.

One-dimensional trace and normal derivative Note that the trace operator can be understood as a bounded linear operator

$$H^1(0, L) \ni u \longmapsto (u(0), u(L)) \in \mathbb{R}^2,$$

where we use the fact that $H^1(0, L) \subset \mathcal{C}[0, L]$ with continuous embedding. Because the boundary of $\Omega = (0, L)$ is just the set $\partial\Omega = \Gamma = \{0, L\}$, it is clear that we can identify $L^2(\Gamma) = \mathbb{R}^2 = H^{1/2}(\Gamma)$ and the norm on this space is not relevant (all of them are equivalent). The space $\mathbf{H}(\mathrm{div}, \Omega)$ when $\Omega = (0, L)$ is just $H^1(0, L)$ and the normal component operator is a signed trace

$$H^1(0, L) \ni p \longmapsto (-p(0), p(L)) \in \mathbb{R}^2,$$

using the formula

$$(p', u)_{(0,L)} + (p, u')_{(0,L)} = -p(0)u(0) + p(L)u(L),$$

valid for all $p \in H^1(0, L) = \mathbf{H}(\mathrm{div}, (0, L))$ and $u \in H^1(0, L)$. The normal derivative is therefore defined in

$$H^2(0, L) = \{u \in H^1(0, L) : u' \in H^1(0, L)\},$$

as $\partial_n u = (-u'(0), u'(L))$.

Four problems. Different choices of V yield different pairs of boundary conditions and different eigenvalue problems. First, we just list the problems and the range of the associated Green's operator. All of the associated boundary conditions can be seen in the range of the Green's operator

$$W := \text{range}\, G = \{Gf \ : \ f \in L^2(0, L)\}.$$

Next, we will come to the eigenvalues, eigenfunctions, and Fourier series.

(a) The case $V = H_0^1(0, L) = \{u \in H^1(0, L) \ : \ u(0) = u(1) = 0\}$ includes the boundary conditions in the definition of the space. Note that constant functions are not in V. The range of the Green's operator is

$$W = H^2(0, L) \cap H_0^1(0, L) = \{u \in H^2(0, L) \ : \ u(0) = u(1) = 0\}.$$

(b) The space $V = H^1(0, L)$ includes constant functions and, in this case

$$W = \{u \in H^2(0, L) \ : \ u'(0) = u'(L) = 0\}.$$

(c) When we choose $V = \{u \in H^1(0, L) \ : \ u(0) = 0\}$, we exclude constants, and yield the range

$$W = \{u \in H^2(0, L) \ : \ u(0) = u'(L) = 0\}.$$

The case $V = \{u \in H^1(0, L) \ : \ u(L) = 0\}$ is very similar and we will not consider it separately.

(d) Finally, when $V = \{u \in H^1(0, L) \ : \ u(0) = u(L)\}$, we allow constants in V again. The associated range of G is

$$W = \{u \in H^2(0, L) \ : \ u(0) = u(L), \quad u'(0) = u'(L)\}.$$

Dirichlet eigenvalues and sine series. In case (a) above, the eigenvalues and eigenfunctions can be computed by hand:

$$\lambda_n := \left(\frac{n\pi}{L}\right)^2 \qquad \phi_n(x) := \sqrt{\frac{2}{L}} \sin\left(\frac{n\pi x}{L}\right) \qquad n \geq 1.$$

This means that the sine functions are a Hilbert basis of $L^2(0, L)$ and are orthogonal complete in $H_0^1(0, L)$. The associated Fourier series can be written as

$$u = \sum_{n=1}^{\infty} (\,\cdot\,, \phi_n)_{(0,L)} \phi_n = \sum_{n=1}^{\infty} \widehat{u}_n \sin(n\pi \cdot /L), \tag{9.18}$$

where

$$\widehat{u}_n = \frac{2}{L} \int_0^L u(x) \sin(n\pi x/L)\mathrm{d}x.$$

The second expression is the more commonly seen in elementary textbooks on differential equations, with the normalization factor moved to the integral coefficient. Note that we can now characterize the energy space V and $W = $ range G in terms of convergence of the sine Fourier series:

$$H_0^1(0, L) = \{u \in L^2(0, L) : \sum_{n=1}^{\infty} n^{-2} |\widehat{u}_n|^2 < \infty\}, \qquad (9.19a)$$

$$H_0^1(0, L) \cap H^2(0, L) = \{u \in L^2(0, L) : \sum_{n=1}^{\infty} n^{-4} |\widehat{u}_n|^2 < \infty\}. \qquad (9.19b)$$

In these expressions we are not using the actual coefficients $(u, \phi_n)_{(0,L)}$ and eigenvalues λ_n, but sequences that behave exactly like them. We also have that if $u \in H_0^1(0, L)$, then the sine Fourier series (9.18) converges in $H^1(0, L)$ and therefore uniformly.

Neumann eigenvalues and cosine series. In case (b), we start counting the eigenvalues from $n = 0$

$$\lambda_n = \left(\frac{n\pi}{L}\right)^2 \qquad n \geq 0,$$

and write the $L^2(0, L)$ normalized eigenfunctions as follows

$$\phi_0(x) = \sqrt{\frac{1}{L}}, \qquad \phi_n(x) = \sqrt{\frac{2}{L}} \cos\left(\frac{n\pi x}{L}\right) \qquad n \geq 1.$$

The associated Fourier series is traditionally written in the form

$$u = \frac{1}{2}\widehat{u}_0 + \sum_{n=1}^{\infty} \widehat{u}_n \cos(n\pi \cdot /L), \qquad (9.20)$$

where now

$$\widehat{u}_n = \frac{2}{L} \int_0^L u(x) \cos(n\pi x/L) dx \qquad n \geq 0. \qquad (9.21)$$

The factor $1/2$ in front of $\widehat{u}_0$ in (9.20) is written so as to unify the expression of the Fourier cosine coefficients (9.21). Formulas like (9.19) now characterize the space $H^1(0, L)$ and $\{u \in H^2(0, L) : u' \in H_0^1(0, L)\}$. See Exercise 9.3 for the two remaining cases, with mixed and periodic boundary conditions.

9.7 Steklov eigenvalues (*)

The Steklov eigenvalue problem is a peculiar differential eigenvalue problem, where the eigenvalue appears in the boundary condition:

$$u \in H^1(\Omega), \quad \lambda \in \mathbb{R}, \qquad \Delta u = 0, \qquad \partial_n u = \lambda \gamma u. \qquad (9.22)$$

Note that all solutions to this problem are harmonic functions. An equivalent variational formulation is

$$u \in H^1(\Omega), \quad \lambda \in \mathbb{R},$$
$$(\nabla u, \nabla v)_\Omega = \lambda \langle \gamma u, \gamma v \rangle_\Gamma \qquad \forall v \in H^1(\Omega).$$

Constant functions $u \in \mathcal{P}_0(\Omega)$ are eigenfunctions for $\lambda = 0$. We will eliminate them by working on the space

$$V := \{u \in H^1(\Omega) \, : \, \gamma u \in L^2_\circ(\Gamma)\},$$

where

$$L^2_\circ(\Gamma) := \{g \in L^2(\Gamma) \, : \, \langle 1, g \rangle_\Gamma = 0\}.$$

Operator reformulation. One way to see that this is quite a different eigenvalue problem is that it is better reformulated as the spectral decomposition of an operator defined exclusively on the boundary. We thus define $N : L^2_\circ(\Gamma) \to L^2_\circ(\Gamma)$, $Nh := \gamma u$, where

$$u \in V, \qquad (\nabla u, \nabla v)_\Omega = \langle h, \gamma v \rangle_\Gamma \qquad \forall v \in V. \tag{9.23}$$

Due to the generalized Poincaré inequality Corollary 7.4 (V does not contain constants), (9.23) is a coercive problem and N is well-defined. Since

$$\|\gamma u\|_\Gamma \le C_1 \|u\|_{1,\Omega} \le C_2 \|h\|_{-1/2,\Gamma} \le C_3 \|h\|_\Gamma \qquad \forall h \in L^2_\circ(\Gamma),$$

it follows that N is bounded, and since $\gamma : H^1(\Omega) \to L^2(\Gamma)$ is compact (see Section 8.3), then N is compact. Also, given $h, g \in L^2_\circ(\Gamma)$ and the corresponding solutions $u, v \in V$ to (9.23) (so that $\gamma u = Nh$, $\gamma v = Ng$), we have

$$\langle h, Ng \rangle_\Gamma = \langle h, \gamma v \rangle_\Gamma = (\nabla u, \nabla v)_\Omega,$$

which proves symmetry and positive semidefiniteness. We also have $\langle h, Nh \rangle_\Gamma = \|\nabla u\|_\Omega^2$. If $Nh = 0$, then $u = 0$ (recall that the gradient defines an equivalent norm in V) and therefore

$$\langle h, \gamma v \rangle_\Gamma = 0 \qquad \forall v \in H^1(\Omega),$$

($\langle h, 1 \rangle_\Gamma = 0$ by hypothesis). This implies that $h = 0$ as $H^{1/2}(\Gamma)$ is dense in $L^2(\Gamma)$. We thus have an orthonormal basis of $L^2_\circ(\Gamma)$, $\{g_n\}_{n \ge 1}$, and associated eigenvalues $\mu_n > 0$, decreasing to zero. We add back the zero eigenvalue and $g_0 \equiv |\Gamma|^{-1/2}$, so that we have a full Hilbert basis for $L^2(\Gamma) = \mathcal{P}_0(\Gamma) \oplus L^2_\circ(\Gamma)$ and the orthogonal decompositions

$$g = \tfrac{1}{|\Gamma|} \langle g, 1 \rangle_\Gamma + \sum_{n=1}^\infty \langle g, g_n \rangle_\Gamma g_n \qquad \forall g \in L^2(\Gamma). \tag{9.24}$$

The operator N can be thus expressed in series form as

$$Ng = \sum_{n=1}^{\infty} \mu_n \langle g, g_n \rangle_\Gamma g_n,$$

with the understanding that it has been extended to vanish on constant functions.

The Steklov eigenfunctions. So far we have worked on an operator from the boundary to the boundary. We now need to associate solutions to the Steklov eigenvalue problem (9.22) to the eigenfunctions $\{g_n\}$ of the operator N. To this end, let $\phi_n \in V$ be the harmonic function such that $N g_n = \gamma \phi_n$. Since $N g_n = \mu_n g_n$, and we have $g_n = \mu_n^{-1} \gamma \phi_n$ and therefore

$$\phi_n \in H^1(\Omega), \qquad \Delta \phi_n = 0, \qquad \partial_n \phi_n = \mu_n^{-1} \gamma \phi_n.$$

We add the eigenfunction $\phi_0 := |\Gamma|^{-1/2}$, corresponding to the zero eigenvalue. If we write $\lambda_0 = 0$ and $\lambda_n = \mu_n^{-1}$ we find that (9.22) has a countable sequence of real eigenvalues, diverging to infinity. We will now study the properties of the associated eigenfunctions $\{\phi_n\}$. Note that for $n \geq 1$

$$\phi_n \in H^1(\Omega), \qquad \gamma \phi_n \in L_\circ^2(\Gamma), \qquad (\nabla \phi_n, \nabla v)_\Omega = \langle g_n, \gamma v \rangle_\Gamma \qquad \forall v \in H^1(\Omega),$$

and therefore

$$(\nabla \phi_n, \nabla \phi_m)_\Omega = \langle g_n, N g_m \rangle_\Gamma = \mu_n \delta_{nm}, \qquad n, m \geq 1.$$

This means that $\{\phi_n\}_{n \geq 0}$ is orthogonal with respect to the inner product

$$a(u, v) := (\nabla u, \nabla v)_\Omega + \langle 1, \gamma u \rangle_\Gamma \langle 1, \gamma v \rangle_\Gamma,$$

equivalent to the usual inner product in $H^1(\Omega)$. If $v \in H^1(\Omega)$ is a-orthogonal to all ϕ_n, then $\gamma v \in L_\circ^2(\Gamma)$ (orthogonality with ϕ_0) and

$$(\nabla v, \nabla \phi_n)_\Omega = \langle g_n, \gamma v \rangle_\Gamma = 0 \qquad \forall n \geq 0.$$

Therefore $\gamma v = 0$ and $v \in H_0^1(\Omega)$. Reciprocally, if $v \in H_0^1(\Omega)$, then (recall that $\Delta \phi_n = 0$)

$$a(\phi_n, v) = (\nabla \phi_n, \nabla v)_\Omega = \langle \partial_n \phi_n, \gamma v \rangle_\Gamma = 0 \qquad \forall n \geq 0.$$

The closed space

$$\mathcal{H} := \{u \in H^1(\Omega) : a(u, v) = 0 \quad \forall v \in H_0^1(\Omega)\}$$
$$= \{u \in H^1(\Omega) : (\nabla u, \nabla v)_\Omega = 0 \quad \forall v \in H_0^1(\Omega)\} = \{u \in H^1(\Omega) : \Delta u = 0\}$$

is thus the closure of the span of the Steklov eigenfunctions and the functions $\psi_0 := \phi_0$ and $\psi_n := \mu_n^{-1/2} \phi_n$ form a Hilbert basis of $\mathcal{H}$.

Revisiting the trace space. We now come back to the boundary and study more properties of the convergence of the orthogonal series (9.24). We take $u \in \mathcal{H}$ and decompose it, with convergence in $H^1(\Omega)$, as

$$u = \sum_{n=0}^{\infty} a(u, \psi_n)\psi_n = |\Gamma|^{-1}\langle 1, \gamma u\rangle_\Gamma + \sum_{n=1}^{\infty} \mu_n^{-1}(\nabla u, \nabla \phi_n)_\Omega \phi_n$$

$$= |\Gamma|^{-1}\langle 1, \gamma u\rangle_\Gamma + \sum_{n=1}^{\infty} \mu_n^{-1}\langle g_n, \gamma u\rangle_\Gamma \phi_n,$$

since $\partial_n \phi_n = \mu_n^{-1}\gamma\phi_n = \mu_n^{-1}Ng_n = g_n$. We now take the trace, again use that $\mu_n^{-1}\gamma\phi_n = g_n$, and have convergence in $H^{1/2}(\Gamma)$ for the following series

$$\gamma u = |\Gamma|^{-1}\langle 1, \gamma u\rangle_\Gamma + \sum_{n=1}^{\infty}\langle g_n, \gamma u\rangle_\Gamma g_n.$$

This proves that $g \in H^{1/2}(\Gamma)$ if and only if the above series, which is an orthogonal series in $L^2(\Gamma)$, converges in $H^{1/2}(\Gamma)$.

9.8 A glimpse of interpolation (*)

Since the previous sections have introduced a collection of orthogonal series that converge simultaneously in more than one space, we are going to take advantage of the situation to give a flavor of interpolation theorems in some Hilbert spaces. We start with a Hilbert space X_0, with inner product denoted $(u, v)_0$, and for which $\{\phi_n\}$ is a Hilbert basis. (Therefore X_0 is separable.) We let $\{\lambda_n\}$ be a positive, nondecreasing, divergent sequence of real numbers. For $s \in (0, 1]$ we define the spaces

$$X_s := \{u \in X_0 : \sum_{n=1}^{\infty} \lambda_n^{2s}|(u, \phi_n)_0|^2 < \infty\},$$

with norms

$$\|u\|_s^2 := \sum_{n=1}^{\infty} \lambda_n^{2s}|(u, \phi_n)_0|^2.$$

It is simple to prove that the spaces X_s are Hilbert spaces, continuously and compactly embedded into X_0. The sequence $\{\phi_n\}$ is orthogonal (not orthonormal) and complete in all the spaces X_s. Moreover, given $u \in X_0$, we have that $u \in X_s$ if and only if the series

$$u = \sum_{n=1}^{\infty}(u, \phi_n)_0 \phi_n$$

converges in the norm $\|\cdot\|_s$. The goal of this section is to prove the following result.

Proposition 9.7. *If $A : X_0 \to X_0$ is bounded, $Au \in X_1$ for all $u \in X_1$, $A : X_1 \to X_1$ is also bounded, then $A : X_{1/2} \to X_{1/2}$ is bounded and*

$$\|A\|_{X_{1/2} \to X_{1/2}} \le \|A\|_{X_0 \to X_0}^{1/2} \|A\|_{X_1 \to X_1}^{1/2}. \tag{9.25}$$

A similar result can be obtained for the other 'intermediate spaces.' (See Exercise 9.10.) Note that we have examples where $X_0 = L^2(\Omega)$, $X_1 = \text{range } G$ (G is a Green's operator) and $X_{1/2}$ is a Sobolev space associated to the variational formulation that was used to define G.

Let $\mathbb{T}_n := \text{span}\{\phi_1, \ldots, \phi_n\}$ and $\Lambda : X_0 \to X_{1/2}$ be given by

$$\Lambda u := \sum_{n=1}^{\infty} \lambda_n^{-1/2}(u, \phi_n)_0 \phi_n,$$

so that $\|\Lambda u\|_{1/2} = \|u\|_0$. It follows that Λ is an isometric isomporphism, $\Lambda \phi_n = \lambda_n^{-1/2} \phi_n$ and therefore the spaces $\mathbb{T}_n$ are invariant under the action of Λ. We also have $\|\Lambda^2 u\|_1 = \|u\|_0$. Consider the truncation operators (orthogonal projections) $P_n : X_0 \to \mathbb{T}_n$

$$P_n u := \sum_{j=1}^{n}(u, \phi_j)_0 \phi_j$$

and note that $\|P_n u\|_1 \le \|u\|_1$ for all $u \in X_1$. Consider the operators

$$A_n := P_n A|_{\mathbb{T}_n} : \mathbb{T}_n \to \mathbb{T}_n,$$

which are linear and bounded, in any norm, since we are in finite dimensions.

Lemma 9.2. *For all $n \ge 1$, we have*

$$\|A_n u\|_{1/2} \le \|A\|_{X_0 \to X_0}^{1/2} \|A\|_{X_1 \to X_1}^{1/2} \|u\|_{1/2} \qquad \forall u \in \mathbb{T}_n.$$

Proof. We identify elements of $\mathbb{T}_n$ with vectors of coefficients

$$\mathbb{T}_n \ni u = \sum_{j=1}^{n}(u, \phi_j)_0 \phi_j \longleftrightarrow \mathbf{u} = \{(u, \phi_j)_0\}_{j=1}^{n} \in \mathbb{C}^n,$$

noting that $\|u\|_0 = |\mathbf{u}|$. We now take A to be the matrix representation of A_n and D to be the matrix representation (diagonal and positive real) of $\Lambda^{-1}|_{\mathbb{T}_n}$ with respect to the basis $\{\phi_1, \ldots, \phi_n\}$. For the following argument, we use the spectral norm of a matrix $\|B\|^2 = \rho(B^*B)$, where B^* is the conjugate transpose and ρ is the spectral radius. We can thus write

$$\sup_{0\neq u\in\mathbb{T}_n}\frac{\|A_nu\|_{1/2}}{\|u\|_{1/2}}=\sup_{0\neq u\in\mathbb{T}_n}\frac{\|\Lambda^{-1}A_nu\|_0}{\|u\|_{1/2}}=\sup_{0\neq v\in\mathbb{T}_n}\frac{\|\Lambda^{-1}A_n\Lambda^{-1}v\|_0}{\|v\|_0}$$

$$=\sup_{0\neq v\in\mathbb{C}^n}\frac{|DAD\mathbf{v}|}{|\mathbf{v}|}=\|DAD\|=\rho(DA^*D^2AD)^{1/2}$$

$$=\rho(A^*D^2AD^2)^{1/2}\leq\|A^*D^2AD^2\|^{1/2}$$

$$\leq\|A^*\|^{1/2}\|D^2AD^2\|^{1/2}=\|A\|^{1/2}\|D^2AD^2\|^{1/2},$$

where we have used that $\rho(\mathrm{B})\leq\|\mathrm{B}\|$. However

$$\|A\|=\sup_{0\neq\mathbf{u}\in\mathbb{C}^n}\frac{|\mathbf{Au}|}{|\mathbf{u}|}=\sup_{0\neq u\in\mathbb{T}_n}\frac{\|A_nu\|_0}{\|u\|_0}\leq\sup_{0\neq u\in\mathbb{T}_n}\frac{\|Au\|_0}{\|u\|_0}\leq\|A\|_{X_0\to X_0},$$

and

$$\|D^2AD^2\|=\sup_{0\neq\mathbf{u}\in\mathbb{C}^n}\frac{|D^2AD^2\mathbf{u}|}{|\mathbf{u}|}=\sup_{0\neq u\in\mathbb{T}_n}\frac{\|\Lambda^{-2}A_n\Lambda^{-2}u\|_0}{\|u\|_0}$$

$$=\sup_{0\neq v\in\mathbb{T}_n}\frac{\|A_nv\|_1}{\|v\|_1}\leq\sup_{0\neq v\in\mathbb{T}_n}\frac{\|Av\|_1}{\|v\|_1}\leq\|A\|_{X_1\to X_1},$$

which finishes the proof. ◻

Proof of Proposition 9.7. If $u\in\mathbb{T}_n$, then $A_nu=P_nAu\to Au$ in $X_{1/2}$ and therefore, by the above lemma,

$$\|Au\|_{1/2}\leq\|A\|_{X_0\to X_0}^{1/2}\|A\|_{X_1\to X_1}^{1/2}\|u\|_{1/2}\qquad\forall u\in\mathbb{T}_n.$$

Finally, by density we have (9.25). ◻

Final comments and literature

The Dirichlet and Neumann eigenvalue problems for the Laplacian on a bounded domain (strongly Lipschitz in the case of Neumann eigenvalues) are classic examples of the theory of compact self-adjoint positive operators on Hilbert spaces, represented in all its glory in the famous Hilbert-Schmidt theorem. The spectral characterization of Sobolev spaces (Section 9.5) is very well-known for one-dimensional problems and many more Sobolev spaces can be characterized with these series. With the spectral series for any compact self-adjoint positive definite operator we can define an associated Hilbert scale (sometimes called a Sobolev tower), which is a collection of Hilbert spaces parametrized by a real number. This is possibly the simplest case of Hilbert space interpolation, which itself is the 'simplest case' of the equivalent J-theory and K-theory of real interpolation of Banach spaces [98]. We take advantage of having introduced these ideas in an elementary form to give a basic result on how to prove boundedness of some operators by interpolation (Section 9.8). The proof of Proposition 9.7 (generalized in Exercise 9.10) is due to Michel Crouzeix, and was given to the first author when he was still a graduate student struggling to understand these concepts.

Finally, Stekloff eigenvalues (see, for instance, [70]) offer a nice non-standard example of an eigenvalue problem for the boundary condition attached to an elliptic operator. This seems to be a problem of increasing interest in the community of inverse problems for scattering.

Exercises

9.1. Dirichet eigenvalues and the Poincaré-Friedrichs inequality. Let λ_1 be the minimum Dirichlet eigenvalue for the Laplacian in Ω, show that

$$\|u\|_\Omega \le \sqrt{\lambda_1}\|\nabla u\|_\Omega \qquad \forall u \in H_0^1(\Omega).$$

9.2. Generalized Dirichlet and Neumann eigenvalues. Study the following eigenvalue problems:

$$u \in H_0^1(\Omega), \ \lambda \in \mathbb{R}, \qquad -\nabla \cdot (\kappa \nabla u) = \lambda \rho u,$$

and

$$u \in H^1(\Omega), \ \lambda \in \mathbb{R}, \qquad -\nabla \cdot (\kappa \nabla u) = \lambda \rho u, \qquad (\kappa \nabla u) \cdot \mathbf{n} + \alpha \gamma u = 0,$$

where $\kappa, \rho \in L^\infty(\Omega)$ are strongly positive, and $\alpha \in L^\infty(\Gamma)$ is nonnegative. (**Hint.** Work on the space $L^2(\Omega)$ with the equivalent weighted norm $(\rho u, u)_\Omega^{1/2}$.)

9.3. One-dimensional eigenvalue problems. Extract all the information that the Hilbert-Schmidt theory provides on the following eigenvalue problems. In particular, characterize Sobolev spaces in terms of the corresponding Fourier series. (Note that all eigenvalues and eigenfunctions can be computed in these cases and the theory gives additional insight on the convergence of the different Fourier series.)

(a) One-dimensional Dirichlet eigenvalues and sine series:

$$-u'' = \lambda u \quad \text{in } (0,1), \qquad u(0) = u(1) = 0.$$

(b) One-dimensional Neumann eigenvalues and cosine series:

$$-u'' = \lambda u \quad \text{in } (0,1), \qquad u'(0) = u'(1) = 0.$$

(c) One-dimensional mixed eigenvalues and half-sine series:

$$-u'' = \lambda u \quad \text{in } (0,1), \qquad u(0) = u'(1) = 0.$$

(d) Periodic problem and sine-and-cosine series:

$$-u'' = \lambda u \qquad u(0) = u(1), \qquad u'(0) = u'(1).$$

9.4. Series solution for the Helmholtz equation. Let $\{\lambda_n; \phi_n\}$ be a complete orthonormal eigensystem for the Laplacian on a bounded domain with Dirichlet boundary conditions

$$\phi_n \in H_0^1(\Omega), \qquad -\Delta\phi_n = \lambda_n\phi_n, \qquad (\phi_n, \phi_m)_\Omega = \delta_{nm}.$$

Let $k^2 \neq \lambda_n$ for all n. Show that for $f \in L^2(\Omega)$, the series

$$u = \sum_{n=1}^{\infty} \frac{1}{k^2 - \lambda_n}(f, \phi_n)_\Omega\phi_n$$

converges in $H_0^1(\Omega)$ to the solution of

$$u \in H_0^1(\Omega), \qquad \Delta u + k^2 u = f.$$

9.5. A reciprocal of the Hilbert-Schmidt theorem. Let μ_n be a nonincreasing sequence of positive real numbers converging to zero. Let $\{\phi_n\}$ be an X-orthonormal sequence. Show that the series

$$\sum_{n=1}^{\infty} \mu_n(\cdot, \phi_n)_X\phi_n$$

converges in the space of bounded linear operators $X \to X$ to a compact, self-adjoint and positive definite operator. Show that $\ker G$ is the orthogonal of $\text{span}\{\phi_n : n \geq 1\}$.

9.6. The singular value decomposition. Let X and Y be Hilbert spaces and $G : X \to Y$ be a compact operator such that $(\ker G)^\perp$ is infinite-dimensional.

(a) Show that $\ker G^*G = \ker G$ and $\ker GG^* = \ker G^*$.

(b) Show that we can find an X-orthonormal sequence, $\{\phi_n\}$, and a sequence of positive nonincreasing numbers, converging to zero, such that

$$G^*G = \sum_{n=1}^{\infty} \mu_n(\cdot, \phi_n)_X\phi_n,$$

with convergence in the sense of bounded operators $X \to X$.

(c) Now let $\sigma_n := \sqrt{\mu_n}$ and $\psi_n := \sigma_n^{-1} G \phi_n$. Show that

$$G^* \psi_n = \sigma_n \phi_n, \qquad G \phi_n = \sigma_n \psi_n, \qquad GG^* \psi_n = \mu_n \psi_n.$$

Prove that $\{\psi_n\}$ is Y−orthonormal.

(d) Now let

$$R := G - \sum_{n=1}^{\infty} \sigma_n (\cdot, \phi_n)_X \psi_n.$$

Show that R is well-defined and it is a compact operator $X \to Y$. Show that $G^* R = 0$ and that $R \phi \perp \ker G^*$ for all ϕ. From this, prove that $R = 0$, that is,

$$G = \sum_{n=1}^{\infty} \sigma_n (\cdot, \phi_n)_X \psi_n.$$

This decomposition is called the **singular value decomposition** (SVD) of G.

(e) Show that

$$G^* = \sum_{n=1}^{\infty} \sigma_n (\cdot, \psi_n)_Y \phi_n.$$

9.7. Dirichlet eigenvalues on a d-box. In elementary PDE courses you get to compute Dirichlet eigenvalues and eigenfunctions for the domain

$$Q := (a_1, b_1) \times \ldots \times (a_d, b_d)$$

by using separation of variables. Show that you do not miss any eigenvalue by doing this.

9.8. The spectral series for the Dirichlet problem. Consider the Dirichlet eigensystem for the Laplacian:

$$\phi_n \in H_0^1(\Omega), \qquad -\Delta \phi_n = \lambda_n \phi_n,$$

with $L^2(\Omega)$ orthonormal eigenfunctions $\{\phi_n\}$.

(a) Show that

$$(\nabla u, \nabla v)_\Omega = \sum_{n=1}^{\infty} \lambda_n (u, \phi_n)_\Omega (\phi_n, v)_\Omega \qquad \forall u, v \in H_0^1(\Omega).$$

(b) Let $\{c_n\}$ be a sequence of real numbers such that

$$\sum_{n=1}^{\infty} \lambda_n^{-1} |c_n|^2 < \infty.$$

Show that $\ell : H^1(\Omega) \to \mathbb{R}$ given by

$$\ell(v) := \sum_{n=1}^{\infty} c_n (u, \phi_n)_\Omega$$

is well-defined, continuous, and $c_n = \ell(\phi_n)$.

(c) If $\ell \in H^{-1}(\Omega)$, show that

$$\sup_{0 \neq v \in H_0^1(\Omega)} \frac{\ell(v)}{\|\nabla v\|_\Omega} = \left(\sum_{n=1}^{\infty} \lambda_n^{-1} |\ell(\phi_n)|^2 \right)^{1/2},$$

defines an equivalent norm in $H^{-1}(\Omega)$.

(d) Show that the space

$$D_\Delta^{\mathrm{dir}} := \{ u \in H_0^1(\Omega) \, : \, \Delta u \in L^2(\Omega) \}$$

is a Hilbert space with any of the three equivalent norms

$$\|\Delta u\|_\Omega \leq \left(\|u\|_\Omega^2 + \|\Delta u\|_\Omega^2 \right)^{1/2} \leq \left(\|u\|_{1,\Omega}^2 + \|\Delta u\|_\Omega^2 \right)^{1/2}.$$

(e) Show that the operator formally defined by

$$Au := \sum_{n=1}^{\infty} \lambda_n (u, \phi_n)_\Omega \phi_n$$

is a bounded isomorphism from D_Δ^{dir} to $L^2(\Omega)$ and from $H_0^1(\Omega)$ to $H^{-1}(\Omega)$. Show that $A = -\Delta$.

(f) Show that D_Δ^{dir} is compactly embedded into $H_0^1(\Omega)$. (**Hint.** Consider the sequence of operators

$$P_N u := \sum_{n=1}^{N} (u, \phi_n)_\Omega \, \phi_n,$$

as bounded operators from D_Δ^{dir} to $H_0^1(\Omega)$ and look for their limit.)

9.9. **The harmonic extension and the Steklov eigenvalues.** Consider the space $\mathcal{H} = \{ u \in H^1(\Omega) \, : \, \Delta u = 0 \}$ and note that $\gamma : \mathcal{H} \to H^{1/2}(\Gamma)$ is an isomorphism. Its inverse $H : H^{1/2}(\Gamma) \to \mathcal{H}$ is called the harmonic extension. Show that if a is the bilinear form of Section 9.7, $a(Hg, Hg)^{1/2}$ defines an equivalent norm in $H^{1/2}(\Gamma)$ and the functions $\{g_n\}_{n \geq 0}$ are orthogonal and complete in $H^{1/2}(\Gamma)$.

9.10. **More interpolation results.** Consider two separable Hilbert spaces X_0 and Y_0 with respective given Hilbert bases $\{\phi_n\}$ and $\{\psi_n\}$ and two positive nondecreasing divergent sequences $\{\lambda_n\}$ and $\{\mu_n\}$. Also, consider the spaces for $\theta \in (0,1)$:

$$X_\theta := \{u \in X_0 : \|u\|_{X_\theta} < \infty\}, \quad \|u\|_{X_\theta}^2 := \sum_{n=1}^\infty \lambda_n^{2\theta} |(u, \phi_n)|^2,$$

$$Y_\theta := \{v \in Y_0 : \|v\|_{Y_\theta} < \infty\}, \quad \|v\|_{Y_\theta}^2 := \sum_{n=1}^\infty \lambda_n^{2\theta} |(v, \psi_n)|^2.$$

(a) Prove that if the restriction of a linear bounded operator $A : X_0 \to Y_0$ to X_1 satisfies that $A : X_1 \to Y_1$ is bounded, then

$$\|A\|_{X_{1/2} \to Y_{1/2}} \le \|A\|_{X_0 \to X_0}^{1/2} \|A\|_{X_1 \to Y_1}^{1/2}.$$

(b) **A lemma.** Let A be a square matrix and let D_1, D_2 be square diagonal matrices with positive diagonal. Show that

$$\|D_1^\theta A D_2^\theta\| \le \|D_1 A D_2\|^\theta \|A\|^{1-\theta}.$$

(**Hint.** Prove it first for $\theta = k/2^m$ with integer k.)

(c) Under the same hypothesis as (a), show that

$$\|A\|_{X_\theta \to Y_\theta} \le \|A\|_{X_0 \to X_0}^{1-\theta} \|A\|_{X_1 \to Y_1}^\theta.$$

Part II

Extensions and Applications

10

Mixed problems

10.1   Surjectivity 210
10.2   Systems with mixed structure 213
10.3   Weakly imposed Dirichlet conditions 217
10.4   Saddle point problems 221
10.5   The mixed Laplacian 223
10.6   Darcy flow 226
10.7   The divergence operator 228
10.8   Stokes flow 234
10.9   Stokes-Darcy flow 236
10.10   Brinkman flow 242
10.11   Reissner-Mindlin plates 245
Final comments and literature 248
Exercises 249

This chapter deals with variational problems with the following mixed structure

$$
\begin{aligned}
&(u,p) \in V \times M,\\
&a(u,v) + b(v,p) = \ell(v) && \forall v \in V,\\
&b(u,q) = \chi(q) && \forall q \in M,
\end{aligned}
$$

where V and M are Hilbert spaces, a and b are bounded bilinear forms, and $\ell \in V'$, $\chi \in M'$ are generic data. We will start by studying the (not unique) solvability of the equation $b(u,q) = \chi(q)$ and by discussing different forms of surjectivity of the associated operator $B : V \to M$ in terms of the bilinear form. This will just be a rewritten form of a corollary of the Banach closed range theorem. We will then state the necessary and sufficient Babuška-Brezzi conditions for the well-posedness of problems with mixed structure and relate some of them to constrained minimization problems. The rest of the chapter is devoted to classical examples of problems with mixed structure: Stokes, Darcy, and Brinkman flow, or a two-field formulation of the Reissner-Mindlin plate equations. The Stokes problem brings along the interesting and important nontrivial issue of discovering the range of the divergence operator when restricted to $\mathbf{H}_0^1(\Omega) := H_0^1(\Omega)^d$. This is actually related to the result which we left unproved in Section 7.6 on the closedness of the range of the gradient operator restricted to $L^2(\Omega)$. We will give more details about this problem and show how the result can be derived from a (still nontrivial) result on right inverses for the divergence operator. Finally, we will use the Brinkman and Reissner-Mindlin models to show how the solutions of some parameter-dependent model equations converge weakly to the solution of the reduced limit model, of which they are a singular perturbation.

10.1   Surjectivity

In this section we will slowly prove a characterization of surjectivity for bounded linear operators between Hilbert spaces. This will be a key ingredient for our analysis of variational problems with mixed structure. Let V and M be Hilbert spaces, $B : V \to M$ be a bounded linear operator, $B^* : M \to V$ be its Hilbert space adjoint, and

$$b(u, p) := (Bu, p)_M = (u, B^*p)_V$$

be the associated bounded bilinear form. We want to show that B is surjective if and only if there exists $\beta > 0$ such that

$$\|B^*p\|_V \geq \beta \|p\|_M \qquad \forall p \in M.$$

We will do this in three simple steps.

Proposition 10.1. *Let $B : V \to M$ be a bounded linear operator between Hilbert spaces. The following statements are equivalent:*

(a) *Range B is closed.*

(b) *Range $B = (\ker B^*)^\perp$.*

(c) *There exists $\beta > 0$ such that*

$$\|Bu\|_M \geq \beta \|u\|_V \qquad \forall u \in (\ker B)^\perp.$$

(d) *There exists $\beta > 0$ such that*

$$\|Bu\|_M \geq \beta \|u - Pu\|_V \qquad \forall u \in V,$$

where $P : V \to \ker B$ is the orthogonal projection on the kernel of B.

The constant β in (c) *and* (d) *can be taken to be the same.*

Proof. We can already see that

$$(\text{range } B)^\perp = \ker B^*,$$

and therefore (a) and (b) are equivalent. It is also straightforward to prove that (c) and (d) are equivalent.

Now consider the invertible operator

$$(\ker B)^{\perp} \ni u \longmapsto Cu := Bu \in \text{range } B.$$

If range B is closed, then C is a bounded isomorphism between Hilbert spaces and has a bounded inverse by the Banach isomorphism theorem. Therefore, there exists a constant such that

$$\|C^{-1}p\|_V \leq (1/\beta)\|p\|_M \qquad \forall p \in \text{range } B.$$

Given $u \in (\ker B)^{\perp}$ we have $p = Bu = Cu \in \text{range } B$ and therefore (c) follows.

Finally, assume that (c) holds. If $\{p_n\}$ is a convergent sequence in range B, then we can write $p_n = Bu_n$ with $u_n \in (\ker B)^{\perp}$. We now have that $\{Bu_n\}$ is Cauchy in M, which implies that $\{u_n\}$ is Cauchy in V. Since V is a Hilbert space, $u_n \to u$ for some u and $Bu_n \to Bu$. This proves that range B is closed. □

Proposition 10.2. *Given $B : V \to M$ linear bounded between Hilbert spaces, range B is closed if and only if range B^* is closed.*

Proof. Since $B^{**} = B$, we clearly only need to prove that if range B is closed, then so is range B^*. Let $p \in (\ker B^*)^{\perp} = \text{range } B$ and note that

$$\|B^*p\|_V = \sup_{0 \neq u \in V} \frac{(B^*p, u)_V}{\|u\|_V} \qquad \text{(Cauchy-Schwarz)}$$

$$\geq \sup_{0 \neq u \in (\ker B)^{\perp}} \frac{(p, Bu)_M}{\|u\|_V}$$

$$\geq \beta \sup_{0 \neq u \in (\ker B)^{\perp}} \frac{(p, Bu)_M}{\|Bu\|_M} \qquad \text{(Proposition 10.1(c))}$$

$$= \beta \sup_{0 \neq q \in \text{range } B} \frac{(p, q)_M}{\|q\|_M}$$

$$= \beta\|p\|_M. \qquad (p \in \text{range } B)$$

Using the characterization (c) of Proposition 10.1, it follows that range B^* is closed. Note that the constant β for the lower bound of B^* is the same as the one for B. □

Proposition 10.3. *Let $B : V \to M$ be a bounded linear operator between Hilbert spaces. The following statements are equivalent:*

(a) *B is surjective.*

(b) *B^* is injective and has closed range.*

(c) *There exists $\beta > 0$ such that*

$$\|B^*p\|_V \geq \beta \|p\|_M \qquad \forall p \in M.$$

Moreover, there exists a right inverse of B, $B^\dagger : M \to V$ with $\|B^\dagger\| \leq 1/\beta$.

Proof. If B is surjective, then it has closed range which implies (Proposition 10.2) that range B^* is closed and $\ker B^* = (\text{range } B)^\perp = \{0\}$. This shows that (a) implies (b). If range B^* is closed, then so is range B (Proposition 10.2 again) and range $B = (\ker B^*)^\perp$, which shows that injectivity implies surjectivity. (In other words, if range B and range B^* are closed, B is surjective if and only if B^* is injective.)

Finally, by Proposition 10.1, we have that (b) implies (c), while (c) clearly shows that B^* is injective and range B^* is closed. $\qquad\square$

The inf-sup condition. The characterization (c) in Proposition 10.3 can be written in terms of the bilinear form $b(u,p) = (Bu, p)_M = (u, B^*p)_V$ as

$$\sup_{0 \neq u \in V} \frac{b(u,p)}{\|u\|_V} \geq \beta \|p\|_M \qquad \forall p \in M, \tag{10.1}$$

since

$$\|B^*p\|_V = \sup_{0 \neq u \in V} \frac{(u, B^*p)_V}{\|u\|_V} = \sup_{0 \neq u \in V} \frac{b(u,p)}{\|u\|_V}.$$

We can also write (10.1) as

$$\inf_{0 \neq p \in M} \sup_{0 \neq u \in V} \frac{b(u,p)}{\|p\|_M \|u\|_V} > 0. \tag{10.2}$$

In fact, if β is the value of the left-hand side of (10.2), then (10.1) holds (Exercise 10.3).

Invertibility. As a very simple consequence of the above, if $A : V \to V$ is a bounded linear operator associated to a bilinear form $a : V \times V \to \mathbb{R}$ in the form

$$a(u,v) = (Au, v)_V \qquad \forall u, v \in V,$$

then A is invertible if and only if the following two conditions hold:

(a) There exists $\alpha > 0$ such that

$$\sup_{0 \neq u \in V} \frac{|a(u,v)|}{\|u\|_V} \geq \alpha \|v\|_V \qquad \forall v \in V.$$

(b) For all nonzero $u \in V$

$$\sup_{v \in V} |a(u,v)| > 0.$$

Note that condition (a) is surjectivity and condition (b) is equivalent to $\|Au\|_V \neq 0$ for all nonzero u. Also by Proposition 10.3, it follows that $\|A^{-1}\| \leq 1/\alpha$, since in this case any right inverse of A is the inverse of A.

10.2 Systems with mixed structure

The goal of this section is the study of a general problem of the form

$$(u, p) \in V \times M, \tag{10.3a}$$
$$a(u, v) + b(v, p) = \ell(v) \qquad \forall v \in V, \tag{10.3b}$$
$$b(u, q) = \chi(q) \qquad \forall q \in M, \tag{10.3c}$$

where V and M are Hilbert spaces,

$$a : V \times V \to \mathbb{R}, \qquad b : V \times M \to \mathbb{R}$$

are bounded bilinear forms, and $(\ell, \chi) \in V' \times M' \equiv (V \times M)'$ are arbitrary data. We can introduce operators $A : V \to V$, $B : V \to M$ (and its Hilbert space adjoint $B^* : M \to V$) associated to the bilinear forms,

$$a(u, v) = (Au, v)_V, \quad b(u, p) = (Bu, p)_M = (u, B^*p)_V \quad \forall u, v \in V, \quad p \in M,$$

and write (10.3) in the equivalent form

$$(u, p) \in V \times M, \qquad \begin{bmatrix} A & B^* \\ B & 0 \end{bmatrix} \begin{bmatrix} u \\ p \end{bmatrix} = \begin{bmatrix} w_\ell \\ r_\chi \end{bmatrix}, \tag{10.4}$$

where $(w_\ell, r_\chi) \in V \times M$ is the pair of Riesz-Fréchet representatives of the original data. We thus have to deal with the invertibility of the matrix of operators in (10.4). Note that a *necessary condition* is the surjectivity of B, since $r_\chi \in M$ is arbitrary. If B is invertible, then the problem is solved in a trivial form

$$u = B^{-1}r_\chi, \qquad p = (B^*)^{-1}(w_\ell - Au).$$

We want to explore the general situation where B is just surjective. The theory for this problem is not complicated (see Theorem 10.1 below), but we will take some time to learn about matrices of operators and related topics, and also to pay attention to how different inf-sup inequalities affect the norm of the global operator.

Matrices of operators and their norms. Consider Hilbert spaces $V_1, \ldots, V_n$ and $M_1, \ldots, M_m$ and bounded linear operators

$$B_{ij} : V_j \to M_i, \qquad i = 1, \ldots, m, \qquad j = 1, \ldots, n.$$

In the product spaces

$$\mathbb{V} := V_1 \times \ldots \times V_n, \qquad \mathbb{M} := M_1 \times \ldots \times M_m,$$

we are going to consider the simple product norms

$$\|(u_1,\ldots,u_n)\|_{\mathbb{V}}^2 := \sum_{j=1}^{n} \|u_j\|_{V_j}^2,$$

$$\|(p_1,\ldots,p_m)\|_{\mathbb{M}}^2 := \sum_{i=1}^{m} \|p_i\|_{M_i}^2.$$

We can then define the operator

$$\mathbb{B} := [B_{ij}] : \mathbb{V} \to \mathbb{M},$$

and the associated matrix of norms

$$N_{ij} := \|B_{ij}\|_{V_j \to M_i} \qquad i = 1,\ldots,m, \qquad j = 1,\ldots,n.$$

Furthermore, (see Exercise 10.4)

$$\|\mathbb{B}\|_{\mathbb{V}\to\mathbb{M}} \leq |N|_2 := \sup_{0\neq \mathbf{z}\in\mathbb{R}^n} \frac{|N\mathbf{z}|_2}{|\mathbf{z}|_2},$$

where $|\cdot|_2$ denotes the Euclidean norm in $\mathbb{R}^k$ for any k. The matrix norm $|N|_2$ can be computed using eigenvalues of $N^\top N$, namely,

$$|N|_2^2 = \rho(N^\top N) = \max\{\lambda : \lambda \text{ is an eigenvalue of } N^\top N\}.$$

However, there are two matrix norms that are easier to handle

$$|N|_1 := \sup_{0\neq \mathbf{z}\in\mathbb{R}^n} \frac{|N\mathbf{z}|_1}{|\mathbf{z}|_1} = \max_j \sum_{i=1}^{n} |N_{ij}|, \tag{10.5a}$$

$$|N|_\infty := \sup_{0\neq \mathbf{z}\in\mathbb{R}^n} \frac{|N\mathbf{z}|_\infty}{|\mathbf{z}|_\infty} = \max_i \sum_{j=1}^{m} |N_{ij}| = |N^\top|_1, \tag{10.5b}$$

where

$$|\mathbf{z}|_1 := \sum_{i=1}^{k} |z_i| \qquad |\mathbf{z}|_\infty := \max_i |z_i|$$

are norms in $\mathbb{R}^k$. (The proof of the equalities in (10.5), which is a very simple exercise, can be found in any introductory textbook concerning numerical linear algebra.) We can then relate these matrix norms with the inequalities (see Exercise 10.4):

$$|N|_2 \leq \sqrt{n}\,|N|_1, \qquad |N|_2 \leq \sqrt{m}\,|N|_\infty. \tag{10.6}$$

Instead of using (10.6), in the proof of Theorem 10.1, we will use the following very tight estimate

$$|N|_2 \leq |N|_1^{1/2}|N|_\infty^{1/2}. \tag{10.7}$$

This upper bound is one of the simplest forms of the Riesz-Thorin theorem. While we will not attempt to prove this result, the use of (10.7) will give us a nicely rounded estimate in Theorem 10.1 and, for once, we will indulge in the use of a much deeper result.

Theorem 10.1 (Babuška-Brezzi conditions). *Let $a : V \times V \to \mathbb{R}$ and $b : V \times M \to \mathbb{R}$ be bounded bilinear forms in the Hilbert spaces V and M,*

$$|a(u,v)| \leq C_a \|u\|_V \|v\|_V, \quad |b(u,p)| \leq C_b \|u\|_V \|p\|_M \quad \forall u, v \in V, \quad p \in M,$$

and

$$V_0 := \{u \in V : b(u,p) = 0 \quad \forall p \in M\}.$$

Assume that

(a) *There exists $\beta > 0$ such that*

$$\sup_{0 \neq u \in V} \frac{|b(u,p)|}{\|u\|_V} \geq \beta \|p\|_M \qquad \forall p \in M.$$

(b) *There exists $\alpha > 0$ such that*

$$\sup_{0 \neq u \in V_0} \frac{|a(u,v)|}{\|u\|_V} \geq \alpha \|v\|_V \qquad \forall v \in V_0.$$

(c) *For all $u \in V_0$*

$$a(u,v) = 0 \qquad \forall v \in V_0 \qquad \implies \qquad u = 0.$$

With these conditions, the variational problem

$$
\begin{align}
(u,p) &\in V \times M, & &\text{(10.8a)} \\
a(u,v) + b(v,p) &= \ell(v) & \forall v \in V, &\text{(10.8b)} \\
b(u,q) &= \chi(q) & \forall q \in M &\text{(10.8c)}
\end{align}
$$

is uniquely solvable for arbitrary $\ell \in V'$ and $\chi \in M'$. Moreover,

$$\|(u,p)\|_{V \times M} \leq \left(1 + \frac{C_a}{\beta}\right)^2 \max\left\{\frac{1}{\alpha}, \frac{1}{\beta}\right\} \|(\ell, \chi)\|_{V' \times M'}.$$

Proof. Instead of rewriting (10.8) as an operator equation in $V \times M$ as in (10.4), we will dig deeper and find a 3×3 matrix of operators that represents (10.8) and at the same time shows its invertibility. If we consider the operator $B : V \to M$ associated to the bilinear form b, we can easily notice that $\ker B = V_0$. We define $V_1 := V_0^\perp = (\ker B)^\perp$, and the operators

$$A_{ij} : V_j \to V_i \qquad i, j \in \{0, 1\}$$

by the relations

$$(A_{ij}u_j, v_i)_V = a(u_j, v_i) \qquad u_j \in V_j, \quad v_i \in V_i,$$

and the operator $B_1 := B|_{V_1} : V_1 \to M$. Before we come back to (10.8), let us make some quick observations about these operators. First of all $B_1 : V_1 \to M$ is invertible by the hypothesis (a) that implies surjectivity (Proposition 10.3), since we have restricted the action of B to the orthogonal of its kernel. Moreover

$$\|B_1^{-1}\|_{M \to V_0} = \|(B_1^*)^{-1}\|_{V_0 \to M} \le 1/\beta.$$

Second, hypotheses (b) and (c) are equivalent to $A_{00} : V_0 \to V_0$ being invertible with

$$\|A_{00}^{-1}\|_{V_0 \to V_0} \le 1/\alpha.$$

We can then write (10.8) in the equivalent form

$$\begin{bmatrix} A_{00} & A_{01} & 0 \\ A_{10} & A_{11} & B_1^* \\ 0 & B_1 & 0 \end{bmatrix} \begin{bmatrix} u_0 \\ u_1 \\ p \end{bmatrix} = \begin{bmatrix} w_0 \\ w_1 \\ r \end{bmatrix}, \tag{10.9}$$

where $(u_0 + u_1, p)$ is the solution of (10.8) with the first unknown decomposed in the orthogonal sum $V = V_0 \oplus V_1$, and

$$\begin{aligned} (w_0, v_0)_V &= \ell(v_0) & \forall v_0 \in V_0, \\ (w_1, v_1)_V &= \ell(v_1) & \forall v_1 \in V_1, \\ (r, q)_M &= \chi(q) & \forall q \in M \end{aligned}$$

are Riesz-Fréchet representatives of $\ell|_{V_0}$, $\ell|_{V_1}$ and χ. Note that a simple argument shows that

$$\|w_0\|_V^2 + \|w_1\|_V^2 = \|\ell\|_{V'}^2.$$

We thus just need to show that the matrix of operators in (10.9) is invertible. This is actually quite simple once we notice that we can solve (10.9) in the following order:

$$\begin{aligned} B_1 u_1 &= r, \\ A_{00} u_0 &= w_0 - A_{01}u_1, \\ B_1^* p &= w_1 - A_{10}u_0 - A_{11}u_1. \end{aligned}$$

A compact way to write this inversion is the following: we factor the operator in the form

$$\begin{bmatrix} I & A_{01}B_1^{-1} & 0 \\ 0 & 0 & I \\ 0 & I & 0 \end{bmatrix} \begin{bmatrix} A_{00} & 0 & 0 \\ 0 & B_1 & 0 \\ 0 & 0 & C_1 \end{bmatrix} \begin{bmatrix} I & 0 & 0 \\ 0 & I & 0 \\ C_1^{-1}A_{10} & C_1^{-1}A_{11} & I \end{bmatrix},$$

(we have written $C_1 = B_1^*$ to shorten some expressions) and then invert each matrix separately

$$
\begin{bmatrix}
I & 0 & 0 \\
0 & I & 0 \\
-C_1^{-1}A_{10} & -C_1^{-1}A_{11} & I
\end{bmatrix}
\begin{bmatrix}
A_{00}^{-1} & 0 & 0 \\
0 & B_1^{-1} & 0 \\
0 & 0 & C_1^{-1}
\end{bmatrix}
\begin{bmatrix}
I & 0 & -A_{01}B_1^{-1} \\
0 & 0 & I \\
0 & I & 0
\end{bmatrix}.
$$

The factorization is just a way of writing the effect of doing Gauss-Jordan elimination on the matrix of operators, that is, applying row and column elimination to find an equivalent diagonal form. The norms of these three matrices can be bounded above by

$$
1 + \frac{C_a}{\beta}, \qquad \max\{1/\alpha, 1/\beta\}, \qquad 1 + \frac{C_a}{\beta},
$$

respectively. (The only one that might not be obvious is the first one. The precise bound is left as an exercise.) This finishes the proof. $\qquad\square$

Several remarks. Assume that we are dealing with a variational problem of the form (10.8), where a, b are bounded bilinear forms.

(1) The hypotheses of Theorem 10.1 are necessary for well-posedness. This can be seen from the proof.

(2) The coercivity condition

$$
a(u, u) \geq \alpha \|u\|_V^2 \qquad \forall u \in V_0
$$

implies (b) and (c) and is thus sufficient, in addition to (a), to ensure well-posedness. If the bilinear form a is symmetric and positive semidefinite in V_0, coercivity is actually a necessary condition (see Exercise 10.2).

(3) If

$$
a(u, u) \geq \alpha \|u\|_V^2 \qquad \forall u \in V,
$$

then the problem is well posed if and only if condition (a) (surjectivity of the operator B) holds.

10.3 Weakly imposed Dirichlet conditions

As a first example of a mixed formulation, let us revisit the nonhomogeneous Dirichlet problem

$$
u \in H^1(\Omega), \qquad -\nabla \cdot (\kappa \nabla u) = f, \qquad \gamma u = g, \qquad (10.10)
$$

where $\kappa \in L^\infty(\Omega)$ is strongly positive $f \in L^2(\Omega)$ and $g \in H^{1/2}(\Gamma)$. Recall that this problem is equivalent to the constrained minimization problem

$$\tfrac{1}{2}(\kappa \nabla u, \nabla u)_\Omega - (f, u)_\Omega = \min! \quad u \in H^1(\Omega), \qquad \gamma u = g.$$

Let us introduce a new unknown to the system

$$h := -(\kappa \nabla u) \cdot \mathbf{n} \in H^{-1/2}(\Gamma), \tag{10.11}$$

and recall that, by definition of the normal component, we have

$$(\kappa \nabla u, \nabla v)_\Omega + (\nabla \cdot (\kappa \nabla u), v)_\Omega = -\langle h, \gamma v \rangle_\Gamma \quad \forall v \in H^1(\Omega).$$

It is then clear that if u is the solution of (10.10) and h is defined by (10.11), then

$$(u, h) \in H^1(\Omega) \times H^{-1/2}(\Gamma), \tag{10.12a}$$

$$(\kappa \nabla u, \nabla v)_\Omega + \langle h, \gamma v \rangle_\Gamma = (f, v)_\Omega \quad \forall v \in H^1(\Omega), \tag{10.12b}$$

$$\langle \mu, \gamma u \rangle_\Gamma \qquad\qquad = \langle \mu, g \rangle_\Gamma \quad \forall \mu \in H^{-1/2}(\Gamma). \tag{10.12c}$$

Note that (10.12c) is equivalent to the Dirichlet boundary condition $\gamma u = g$. This is due to the fact that in a Hilbert space H

$$p = q \quad \Longleftrightarrow \quad \ell(p) = \ell(q) \quad \forall \ell \in H'. \tag{10.13}$$

(This holds by the Riesz-Fréchet theorem, although the result also holds in Banach spaces due to the Hahn-Banach theorem.) Second, testing (10.12b) with $v \in \mathcal{D}(\Omega)$ we prove that $-\nabla \cdot (\kappa \nabla u) = f$ in the sense of distributions. Finally, substituting the equation back in (10.12b) we have

$$\begin{aligned}
\langle (\kappa \nabla u) \cdot \mathbf{n}, \gamma v \rangle_\Gamma &= (\kappa \nabla u, \nabla v)_\Omega + (\nabla \cdot (\kappa \nabla u), v)_\Omega \\
&= (\kappa \nabla u, \nabla v)_\Omega - (f, v)_\Omega \\
&= -\langle h, \gamma v \rangle_\Gamma \quad \forall v \in H^1(\Omega),
\end{aligned}$$

which shows that $h = -(\kappa \nabla u) \cdot \mathbf{n}$, since $\gamma : H^1(\Omega) \to H^{1/2}(\Gamma)$ is surjective and $H^{-1/2}(\Gamma)$ is the dual of $H^{1/2}(\Gamma)$. This shows that (10.12) is equivalent to (10.10) and (10.11).

Theoretical study. Problem (10.12) fits in the framework of Section 10.2 with the bilinear forms

$$a(u, v) := (\kappa \nabla u, \nabla v)_\Omega, \qquad b(u, h) := \langle h, \gamma v \rangle_\Gamma.$$

Before dealing with the surjectivity problems associated to the bilinear form b (which we will do in equivalent but slightly different ways), let us deal with the easy aspects of this problem. First of all, we can recognize

$$\begin{aligned}
V_0 &= \{ u \in H^1(\Omega) : \langle \mu, \gamma u \rangle_\Gamma = 0 \quad \forall \mu \in H^{-1/2}(\Gamma) \} \\
&= \{ u \in H^1(\Omega) : \gamma u = 0 \} \qquad\qquad \text{(recall (10.13))} \\
&= H_0^1(\Omega).
\end{aligned}$$

We notice that a is coercive in $H_0^1(\Omega) = V_0$ by the Poincaré-Friedrichs inequality, which proves conditions (b) and (c) of Theorem 10.1. Therefore, we just need to show that the bilinear form b satisfies condition (a) of Theorem 10.1 or, equivalently, that the associated operator $B : H^1(\Omega) \to H^{-1/2}(\Gamma)$ is surjective.

Verification of the inf-sup condition. There are many equivalent forms of dealing with the condition

$$\sup_{0 \neq u \in H^1(\Omega)} \frac{\langle h, \gamma u \rangle_\Gamma}{\|u\|_{1,\Omega}} \geq \beta \|h\|_{-1/2,\Gamma} \qquad \forall h \in H^{-1/2}(\Gamma). \tag{10.14}$$

We will explore several of them. The first option is using the bilinear form and solving a little problem. Take $h \in H^{-1/2}(\Gamma)$ and solve the problem

$$u \in H^1(\Omega), \qquad -\Delta u + u = 0, \qquad \partial_n u = h.$$

This problem is well posed and we have

$$\langle h, \gamma u \rangle_\Gamma = (\nabla u, \nabla u)_\Omega + (u, u)_\Omega = \|u\|_{1,\Omega}^2. \tag{10.15}$$

We also know that

$$\begin{aligned} |\langle \partial_n u, \gamma v \rangle_\Gamma| &= |\langle \nabla u \cdot \mathbf{n}, \gamma v \rangle_\Gamma| \\ &= |(\nabla u, \nabla v)_\Omega + (u, v)_\Omega| \qquad (\Delta u = \nabla \cdot \nabla u = u) \\ &\leq \|u\|_{1,\Omega} \|v\|_{1,\Omega}, \end{aligned}$$

and therefore, taking the infimum over v with $\gamma v = g$

$$|\langle \partial_n u, g \rangle_\Gamma| \leq \|u\|_{1,\Omega} \|g\|_{1/2,\Gamma} \qquad \forall g \in H^{1/2}(\Gamma),$$

which translates into the inequality

$$\|h\|_{-1/2,\Gamma} = \|\partial_n u\|_{-1/2,\Gamma} \leq \|u\|_{1,\Omega}, \tag{10.16}$$

and therefore (combine (10.15) and (10.16)) to a proof of (10.14) with $\beta = 1$. For the second proof of (10.14), we will use the fact that there exists a bounded right inverse of the trace

$$\gamma^\dagger : H^{1/2}(\Gamma) \to H^1(\Omega), \qquad \gamma \gamma^\dagger g = g \qquad \forall g \in H^{1/2}(\Gamma).$$

By construction of the $H^{1/2}(\Gamma)$ norm (see Chapter 4) we can have

$$\|\gamma^\dagger g\|_{1,\Omega} = \|g\|_{1/2,\Gamma} \qquad \forall g \in H^{1/2}(\Gamma). \tag{10.17}$$

Therefore

$$\begin{aligned} \|h\|_{-1/2,\Gamma} &= \sup_{0 \neq g \in H^{1/2}(\Gamma)} \frac{\langle h, g \rangle_\Gamma}{\|g\|_{1/2,\Gamma}} \\ &= \sup_{0 \neq g \in H^{1/2}(\Gamma)} \frac{\langle h, \gamma \gamma^\dagger g \rangle_\Gamma}{\|\gamma^\dagger g\|_{1,\Omega}} \qquad \text{(by (10.17))} \\ &\leq \sup_{0 \neq u \in H^1(\Omega)} \frac{\langle h, \gamma u \rangle_\Gamma}{\|u\|_{1,\Omega}}, \end{aligned}$$

which gives an alternative proof of (10.14).

Identifying the hidden operator. The bilinear form

$$b : H^1(\Omega) \times H^{-1/2}(\Gamma) \to \mathbb{R},$$

hides two operators: one of them $B : H^1(\Omega) \to H^{-1/2}(\Gamma)$ is related to the inner product in $H^{-1/2}(\Gamma)$, while the other one

$$H^1(\Omega) \ni u \longmapsto \mathcal{B}u := \langle \, \cdot \, , \gamma u \rangle_\Gamma \in H^{-1/2}(\Gamma)',$$

uses the dual of $H^{-1/2}(\Gamma)$. However, Hilbert spaces are reflexive (see below) and we can identify $H^{-1/2}(\Gamma)' \equiv H^{1/2}(\Gamma)$ with the norm estimate

$$\|g\|_{1/2,\Gamma} = \sup_{0 \neq h \in H^{-1/2}(\Gamma)} \frac{\langle h, g \rangle_\Gamma}{\|h\|_{-1/2,\Gamma}},$$

which is a way of saying that $H^{1/2}(\Gamma)$ and $H^{-1/2}(\Gamma)$ are dual to each other. This happens in all Hilbert spaces and is shown in Proposition 10.4 below. It also shows that, if we identify $H^{-1/2}(\Gamma)'$ with $H^{1/2}(\Gamma)$, then $\mathcal{B} = \gamma$, which is surjective. While we have developed the theory for inf-sup conditions (Proposition 10.3) with the Riesz-Fréchet representative $B : V \to M$, it can be equally developed with an equivalent operator $\mathcal{B} : V \to M'$. In other words, we can recognize $\mathcal{B}$ in the following diagram, which shows that B is surjective if and only if γ is surjective:

$$H^1(\Omega) \xrightarrow{\quad B \quad} H^{-1/2}(\Gamma)$$

with γ going to $H^{1/2}(\Gamma) \equiv H^{-1/2}(\Gamma)'$ and a Riesz-Fréchet arrow.

Proposition 10.4. *If H is a Hilbert space and H' is its dual, then*

$$\|u\|_H = \sup_{0 \neq \ell \in H'} \frac{\ell(u)}{\|\ell\|_{H'}} \qquad \forall u \in H,$$

and the map

$$H \ni u \longmapsto \langle \, \cdot \, , u \rangle_{H' \times H} \in H'' := (H')' \qquad (10.18)$$

is an isometric isomorphism. In other words, every Hilbert space is reflexive.

Proof. The fact that the map (10.18) is an isometry holds for every Banach space, in virtue of the Hahn-Banach theorem. In the Hilbert space case, this can be proved using the Riesz-Fréchet theorem. First of all

$$\|\langle \, \cdot \, , u \rangle\|_{H' \times H} = \sup_{0 \neq \ell \in H'} \frac{|\langle \ell, u \rangle_{H' \times H}|}{\|\ell\|_{H'}} \leq \|u\|_H \qquad \forall u \in H.$$

Moreover, if $\ell := (\,\cdot\,, u)_H$, then $\|\ell\|_{H'} = \|u\|_H$ and $\langle \ell, u \rangle_{H' \times H} = \|u\|_H^2$, which proves that

$$\|\langle \,\cdot\,, u \rangle\|_{H' \times H} = \|u\|_H \qquad \forall u \in H.$$

To prove surjectivity we need to apply the Riesz-Fréchet theorem in H and in H'. For $\rho \in H''$, and using the Riesz-Fréchet theorem in H' we can find $\ell \in H'$ such that

$$(\ell, \tau)_{H'} = \rho(\tau) \qquad \forall \tau \in H'.$$

Now fixing $\tau \in H'$ as well, we can find the Riesz-Fréchet representatives of ℓ and τ in H and therefore

$$\begin{aligned} \rho(\tau) &= (\ell, \tau)_{H'} \\ &= (u_\ell, u_\tau)_H && \text{(RF is a linear isometry } H \to H') \\ &= \langle \tau, u_\ell \rangle_{H' \times H} && (\tau = (u_\tau, \,\cdot\,)_H), \end{aligned}$$

which finishes the proof, since $\rho = \langle \,\cdot\,, u_\ell \rangle_{H' \times H}$. $\qquad\square$

10.4 Saddle point problems

Linearly constrained quadratic optimization. The problem of Section 10.3 could be understood as a minimization problem subject to a linear constraint. The additional variable h created the mixed structure and, as we will now see, acts as a Lagrange multiplier of the problem. To fix ideas, consider two vector spaces V and M, a symmetric positive semidefinite bilinear form $a : V \times V \to \mathbb{R}$ and a bilinear form $b : V \times M \to \mathbb{R}$. Given data $\ell \in V'$ and $\chi \in M'$, we look for a solution of the constrained minimization problem

$$\tfrac{1}{2}a(u, u) - \ell(u) = \min! \qquad u \in V, \qquad b(u, \,\cdot\,) = \chi. \tag{10.19}$$

The constraint can be written using the operator $\mathcal{B} : V \to M'$ given by $\mathcal{B}u := b(u, \,\cdot\,)$ in the form $\mathcal{B}u = \chi$, or with an operator $B : V \to M$ and a Riesz-Fréchet representation of the functional χ. Obviously, for problem (10.19) to make sense, we need the admissible set to be nonempty, that is, we need $\chi \in \text{range } \mathcal{B}$. We then associate the Lagrangian $\mathcal{L} : V \times M \to \mathbb{R}$ defined as

$$\mathcal{L}(u, p) := \tfrac{1}{2}a(u, u) - \ell(u) + b(u, p) - \chi(p). \tag{10.20}$$

The following elementary result ties some saddle points of $\mathcal{L}$ (minimum in the V direction and maximum in the M direction) with possible solutions of the minimization problem (10.19). As opposed to what happens in finite dimensions, the existence of a Lagrange multiplier p is not guaranteed unless some additional hypotheses on the spaces and bilinear forms are imposed.

Proposition 10.5. *If $(u, p) \in V \times M$ is a saddle point for the Lagrangian defined in (10.20) in the sense that*

$$\mathcal{L}(u, q) \leq \mathcal{L}(u, p) \leq \mathcal{L}(v, p) \qquad \forall (v, q) \in V \times M, \qquad (10.21)$$

then u is a solution of the minimization problem (10.19).

Proof. Let (u, p) be a saddle point of $\mathcal{L}$ and consider the linear functional $\xi := b(u, \cdot) - \chi$. The leftmost inequality in (10.21) is equivalent to

$$\xi(q) \leq \xi(p) \qquad \forall q \in M,$$

which is equivalent to $\xi = 0$ and, therefore, u satisfies the constraint in (10.19). Note that if $v \in V$ satisfies $b(v, \cdot) - \chi = 0$, then

$$\mathcal{L}(v, p) = \tfrac{1}{2} a(v, v) - \ell(v),$$

and therefore, the rightmost inequality in (10.21) implies that u is a minimum subject to the constraint. $\qquad \square$

Note that Proposition 10.5 does not need any symmetry or sign condition on the bilinear forms a and b. The next result identifies saddle points with solutions of a variational problem with mixed structure. Symmetry and nonnegativity of a are now added to the set of hypotheses.

Proposition 10.6. *If V and M are vector spaces, $a : V \times V \to \mathbb{R}$ is bilinear, symmetric and positive semidefinite, $b : V \times M \to \mathbb{R}$ is bilinear, and $\ell : V \to \mathbb{R}$ and $\chi : M \to \mathbb{R}$ are linear maps, then (u, p) is a saddle point of the Lagrangian $\mathcal{L}$ given in (10.20) if and only if*

$$a(u, v) + b(v, p) = \ell(v) \qquad \forall v \in V, \qquad (10.22a)$$
$$b(u, q) \qquad\quad = \chi(q) \qquad \forall q \in M. \qquad (10.22b)$$

Proof. We have already seen in the proof of Proposition 10.5 that

$$\mathcal{L}(u, q) \leq \mathcal{L}(u, p) \qquad \forall q \in M$$

is equivalent to (10.22b). Note then that if (10.22b) holds, then

$$\mathcal{L}(u, p) \leq \mathcal{L}(v, p) \qquad \forall v \in V$$

is equivalent to

$$\mathcal{L}(u, p) \leq \mathcal{L}(u + t v, p) \qquad \forall t \in \mathbb{R}, \qquad \forall v \in V.$$

After simplification, using the symmetry of a, this is equivalent to

$$t\big(a(u, v) + b(v, p) - \ell(v)\big) + t^2 a(v, v) \geq 0 \qquad \forall t \in \mathbb{R}, \qquad \forall v \in V.$$

Finally, due to the nonnegativity of a, this is equivalent to (10.22a) and the proof is finished. $\qquad \square$

To finish this section, let us point out that if we add more hypotheses to the variational equations (10.22), we can show that the minimization problem is equivalent to the saddle point problem.

Proposition 10.7. *Let V and M be Hilbert spaces, $a : V \times V \to \mathbb{R}$ and $b : V \times M \to \mathbb{R}$ be bounded bilinear forms. Assume that*

(a) *The bilinear form b defines a surjective operator $B : V \to M$.*

(b) *The bilinear form a is symmetric, positive semidefinite in V and coercive in $V_0 := \{u \in V : b(u, \cdot) = 0\}$.*

For arbitrary $\ell \in V'$, $\chi \in M'$, the minimization problem (10.19) is equivalent to the saddle point problem (10.21) for the associated Lagrangian, in the sense that if u is the unique solution to (10.19), then there exists a unique $p \in M$, such that (u, p) is the unique saddle point (10.21).

Proof. This is a simple exercise using the fact that (10.22) is now uniquely solvable and so is (10.19). $\qquad\square$

Going back to the problem of Section 10.3, note that the conormal derivative $h = -(\kappa \nabla u) \cdot \mathbf{n}$ acts as the Lagrange multiplier for this augmented Dirichlet problem.

10.5 The mixed Laplacian

Consider the following first order system set up in a Lipschitz domain Ω:

$$\mathbf{q} + \kappa \nabla u = 0, \qquad \nabla \cdot \mathbf{q} = f, \qquad \gamma u = g, \qquad (10.23)$$

for a diffusion coefficient $\kappa \in L^\infty(\Omega)$ with $\kappa \geq \kappa_0 > 0$ almost everywhere and with data $f \in L^2(\Omega)$ and $g \in H^{1/2}(\Gamma)$. Note that this is a first order formulation (with the additional field $\mathbf{q} = -\kappa \nabla u$) of the second order elliptic problem

$$-\nabla \cdot (\kappa \nabla u) = f, \qquad \gamma u = g. \qquad (10.24)$$

There is going to be a key difference in our treatment of these two problems: in (10.24) we look for $u \in H^1(\Omega)$ and are thus allowed to use $f \in H^{-1}(\Omega)$ (see Section 4.5), while in (10.23) we will look for $\mathbf{q} \in \mathbf{H}(\mathrm{div}, \Omega)$, which will force f to be in $L^2(\Omega)$.

Mixed variational formulation. Let $\mathbf{r} \in \mathbf{H}(\mathrm{div}, \Omega)$ be a test function. If $(\mathbf{q}, u)$ is a solution of (10.23), then we have

$$
\begin{aligned}
\langle \mathbf{r} \cdot \mathbf{n}, g \rangle_\Gamma &= \langle \mathbf{r} \cdot \mathbf{n}, \gamma u \rangle_\Gamma && \text{(Dirichlet B.C.)} \\
&= (\nabla \cdot \mathbf{r}, u)_\Omega + (\mathbf{r}, \nabla u)_\Omega && \text{(Definition of } \mathbf{r} \cdot \mathbf{n}) \\
&= (\nabla \cdot \mathbf{r}, u)_\Omega - (\kappa^{-1} \mathbf{q}, \mathbf{r})_\Omega && (\mathbf{q} = -\kappa \nabla u).
\end{aligned}
$$

Formally, we have taken the differential equation

$$\kappa^{-1}\mathbf{q} + \nabla u = 0,$$

multiplied by a test function $\mathbf{r}$ and integrated by parts, although in full rigor, we have just used the definition of the normal component of the test function $\mathbf{r}$. It is then simple to see that a solution of (10.24) also solves the problem

$$(\mathbf{q}, u) \in \mathbf{H}(\mathrm{div}, \Omega) \times L^2(\Omega), \tag{10.25a}$$

$$(\kappa^{-1}\mathbf{q}, \mathbf{r})_\Omega - (\nabla \cdot \mathbf{r}, u)_\Omega = -\langle \mathbf{r} \cdot \mathbf{n}, g \rangle_\Gamma \qquad \forall \mathbf{r} \in \mathbf{H}(\mathrm{div}, \Omega), \tag{10.25b}$$

$$(\nabla \cdot \mathbf{q}, v)_\Omega \qquad\qquad = (f, v)_\Omega \qquad \forall v \in L^2(\Omega). \tag{10.25c}$$

There is an interesting effect in (10.25) about the regularity of the unknowns $(\mathbf{q}, u)$. In (10.23) we need to have $u \in H^1(\Omega)$, because we are taking a trace, but just by assuming that $\mathbf{q} \in \mathbf{L}^2(\Omega) := L^2(\Omega; \mathbb{R}^d)$ and $u \in L^2(\Omega)$, it automatically happens that $u \in H^1(\Omega)$. This is similar to what happens in the Neumann problem, where we look for $u \in H^1(\Omega)$ but, only after the equation is imposed, that is we have $\Delta u \in L^2(\Omega)$, are we are allowed to take the normal derivative of u. Similarly, in (10.25), the regularity of u has been reduced to the space $L^2(\Omega)$, but the Dirichlet condition has moved to the right-hand side of the equation and is no longer imposed separately.

Equivalence of variational formulation and first order system. Now take $\boldsymbol{\varphi} \in \mathcal{D}(\Omega)^d$ as test function in (10.25b), and note that the resulting identity is equivalent to the equation $\kappa^{-1}\mathbf{q} + \nabla u = \mathbf{0}$ in the sense of distributions. In particular $u \in H^1(\Omega)$. Substituting $\kappa^{-1}\mathbf{q} = -\nabla u$ in the left-hand side of (10.25b), we have

$$(\nabla u, \mathbf{r})_\Omega + (u, \nabla \cdot \mathbf{r})_\Omega = \langle \mathbf{r} \cdot \mathbf{n}, g \rangle_\Gamma \qquad \forall \mathbf{r} \in \mathbf{H}(\mathrm{div}, \Omega),$$

or, equivalently,

$$\langle \mathbf{r} \cdot \mathbf{n}, \gamma u \rangle_\Gamma = \langle \mathbf{r} \cdot \mathbf{n}, g \rangle_\Gamma \qquad \forall \mathbf{r} \in \mathbf{H}(\mathrm{div}, \Omega),$$

or, equivalently (recall that the normal trace operator is surjective)

$$\langle h, \gamma u \rangle_\Gamma = \langle h, g \rangle_\Gamma \qquad \forall h \in H^{-1/2}(\Gamma).$$

Since $H^{1/2}(\Gamma) = H^{-1/2}(\Gamma)'$ (this is reflexivity, once again), this is equivalent to $\gamma u = g$. This proves that a solution of (10.25) is a solution of (10.23). (The equation $\nabla \cdot \mathbf{q} = f$ is obviously equivalent to (10.25c).)

Well-posedness. Problem (10.25) has a mixed structure with spaces $V = \mathbf{H}(\mathrm{div}, \Omega)$ and $M = L^2(\Omega)$, bilinear forms

$$a(\mathbf{q}, \mathbf{r}) := (\kappa^{-1}\mathbf{q}, \mathbf{r})_\Omega, \qquad b(\mathbf{q}, v) := -(\nabla \cdot \mathbf{q}, v)_\Omega,$$

and right-hand sides

$$\ell(\mathbf{r}) := -\langle \mathbf{r} \cdot \mathbf{n}, g \rangle_\Gamma, \qquad \chi(v) := -(f, v)_\Omega.$$

We can easily identify the operator $B : \mathbf{H}(\mathrm{div}, \Omega) \to L^2(\Omega)$ such that

$$(B\mathbf{q}, v)_\Omega = -(\nabla \cdot \mathbf{q}, v)_\Omega \qquad \forall \mathbf{q} \in \mathbf{H}(\mathrm{div}, \Omega), \qquad \forall v \in L^2(\Omega),$$

as $B\mathbf{q} = -\nabla \cdot \mathbf{q}$. However, we already know that the divergence operator from $\mathbf{H}(\mathrm{div}, \Omega)$ to $L^2(\Omega)$ is surjective, which proves the corresponding inf-sup condition for the bilinear form b (Proposition 10.3). The kernel of B is the space

$$V_0 = \{\mathbf{q} \in \mathbf{H}(\mathrm{div}, \Omega) : \nabla \cdot \mathbf{q} = 0\},$$

and in this space a is coercive, since

$$\begin{aligned}
a(\mathbf{q}, \mathbf{q}) &= (\kappa^{-1}\mathbf{q}, \mathbf{q})_\Omega \\
&\geq \alpha \|\mathbf{q}\|_\Omega^2 && (\kappa^{-1} \geq \alpha > 0) \\
&= \alpha(\|\mathbf{q}\|_\Omega^2 + \|\nabla \cdot \mathbf{q}\|_\Omega^2) && (\nabla \cdot \mathbf{q} = 0) \\
&= \alpha \|\mathbf{q}\|_{\mathrm{div},\Omega}^2 && \forall \mathbf{q} \in V_0.
\end{aligned}$$

This proves that (10.25) is a well posed problem and there exists $C > 0$ such that

$$\|\mathbf{q}\|_{\mathrm{div},\Omega} + \|u\|_\Omega \leq C(\|f\|_\Omega + \|g\|_{1/2,\Gamma}),$$

as follows from easy estimates for the norms of the functionals ℓ and χ.

Minimization problem and Lagrangian. Following the results of Section 10.4, the solution of (10.25) solves the constrained minimization problem

$$\tfrac{1}{2}(\kappa^{-1}\mathbf{q}, \mathbf{q})_\Omega + \langle \mathbf{q} \cdot \mathbf{n}, g \rangle_\Gamma = \min! \qquad \mathbf{q} \in \mathbf{H}(\mathrm{div}, \Omega), \qquad \nabla \cdot \mathbf{q} = f. \quad (10.26)$$

It is interesting to compare this minimization problem with the one satisfied by the solution of (10.24)

$$\tfrac{1}{2}(\kappa \nabla u, \nabla u)_\Omega - (f, u)_\Omega = \min! \qquad u \in H^1(\Omega), \qquad \gamma u = g, \quad (10.27)$$

with the roles of the data reversed. If we think of problems (10.24) and (10.27) as our primal problems, we can think of (10.26) as a dual minimization problem. It also has a variational formulation:

$$\mathbf{q} \in \mathbf{H}(\mathrm{div}, \Omega), \qquad \nabla \cdot \mathbf{q} = f, \quad (10.28a)$$

$$(\kappa^{-1}\mathbf{q}, \mathbf{r})_\Omega = -\langle \mathbf{r} \cdot \mathbf{n}, g \rangle_\Gamma \qquad \forall \mathbf{r} \in \mathbf{H}_0, \quad (10.28b)$$

where $\mathbf{H}_0 = \{\mathbf{r} \in \mathbf{H}(\mathrm{div}, \Omega) : \nabla \cdot \mathbf{r} = 0\}$ is the kernel of the side condition that is imposed in (10.26). (See Exercise 10.6 for more on (10.28).) Also, using the results of Section 10.4, it follows that the solution of (10.25) is a saddle point of the Lagrangian

$$\mathcal{L}(\mathbf{q}, u) := \tfrac{1}{2}(\kappa^{-1}\mathbf{q}, \mathbf{q})_\Omega + \langle \mathbf{q} \cdot \mathbf{n}, g \rangle_\Gamma - (\nabla \cdot \mathbf{q}, u)_\Omega + (f, u)_\Omega.$$

10.6   Darcy flow

A simple change of boundary conditions in the problem of Section 10.5 leads to some new interesting questions about the space $\mathbf{H}(\mathrm{div}, \Omega)$. The system of equations

$$(\mathbf{q}, u) \in \mathbf{H}(\mathrm{div}, \Omega) \times L^2(\Omega), \qquad \mathbf{q} + \kappa \nabla u = 0, \tag{10.29a}$$
$$\nabla \cdot \mathbf{q} = f, \tag{10.29b}$$
$$\mathbf{q} \cdot \mathbf{n} = h, \tag{10.29c}$$

is often associated to a linearized model for flow in saturated porous media. It is clearly a first order formulation of the Neumann problem

$$u \in H^1(\Omega), \qquad -\nabla \cdot (\kappa \nabla u) = f, \qquad (\kappa \nabla u) \cdot \mathbf{n} = -h. \tag{10.30}$$

As we saw in Chapters 7 and 8, the data $f \in L^2(\Omega)$ and $h \in H^{-1/2}(\Gamma)$ must satisfy the compatibility condition

$$(f, 1)_\Omega - \langle h, 1 \rangle_\Gamma = 0 \tag{10.31}$$

for (10.30) to be solvable and the solution of (10.30) is unique up to an additive constant.

The kernel of the normal component operator. In the Dirichlet problem for the Laplacian, the fact that $H_0^1(\Omega)$ (the closure of $\mathcal{D}(\Omega)$ in $H^1(\Omega)$) is the kernel of the trace operator plays a central role in showing that the boundary value problem

$$u \in H^1(\Omega), \qquad \gamma u = g, \qquad -\Delta u = f$$

is equivalent to its variational formulation

$$u \in H^1(\Omega), \qquad \gamma u = g, \qquad (\nabla u, \nabla v)_\Omega = (f, v)_\Omega \qquad \forall v \in H_0^1(\Omega).$$

In a similar vein, but with Dirichlet and Neumann boundary conditions having switched places in the mixed formulation, the next result will be central to providing a variational formulation for (10.29).

Proposition 10.8 (The space $\mathbf{H}_0(\mathrm{div}, \Omega)$). *The space $\mathcal{D}(\Omega; \mathbb{R}^d)$ is dense in*

$$\{\mathbf{p} \in \mathbf{H}(\mathrm{div}, \Omega) : \mathbf{p} \cdot \mathbf{n} = 0\}.$$

This space is denoted $\mathbf{H}_0(\mathrm{div}, \Omega)$.

Proof. If $\varphi \in \mathcal{D}(\Omega; \mathbb{R}^d)$, then

$$\langle \varphi \cdot \mathbf{n}, \gamma v \rangle_\Gamma = (\nabla v, \varphi)_\Omega + (v, \nabla \cdot \varphi)_\Omega$$
$$= \langle \nabla v, \varphi \rangle_{\mathcal{D}' \times \mathcal{D}} + \langle v, \nabla \cdot \varphi \rangle_{\mathcal{D}' \times \mathcal{D}} = 0 \qquad \forall v \in H^1(\Omega),$$

and therefore $\varphi \cdot \mathbf{n} = 0$. To show that the closure of $\mathcal{D}(\Omega; \mathbb{R}^d)$ in $\mathbf{H}(\mathrm{div}, \Omega)$ is the kernel of the normal component operator, we will show that

$$\mathcal{D}(\Omega; \mathbb{R}^d)^{\perp} \subset \{\mathbf{p} \in \mathbf{H}(\mathrm{div}, \Omega) : \mathbf{p} \cdot \mathbf{n} = 0\}^{\perp}, \tag{10.32}$$

where the orthogonal complement is taken with respect to the $\mathbf{H}(\mathrm{div}, \Omega)$ inner product. This will identify the two orthogonal complements (note that the reverse inclusion follows from the inclusion of $\mathcal{D}(\Omega; \mathbb{R}^d)$ in the kernel of the normal component operator), and therefore, by taking the orthogonal complement again

$$\overline{\mathcal{D}(\Omega; \mathbb{R}^d)} = \mathcal{D}(\Omega; \mathbb{R}^d)^{\perp\perp} = \{\mathbf{p} \in \mathbf{H}(\mathrm{div}, \Omega) : \mathbf{p} \cdot \mathbf{n} = 0\},$$

since the latter space is closed. The proof of (10.32) is quite simple though. We first note that $\mathbf{p}$ is orthogonal to $\mathcal{D}(\Omega; \mathbb{R}^d)$ in $\mathbf{H}(\mathrm{div}, \Omega)$ if and only if

$$(\mathbf{p}, \varphi)_\Omega + (\nabla \cdot \mathbf{p}, \nabla \cdot \varphi)_\Omega = 0 \qquad \forall \varphi \in \mathcal{D}(\Omega; \mathbb{R}^d),$$

which is equivalent to $\mathbf{p} = \nabla(\nabla \cdot \mathbf{p})$. Let $u = \nabla \cdot \mathbf{p}$ and note that $\nabla u = \mathbf{p}$ and therefore $u \in H^1(\Omega)$. If $\mathbf{q} \cdot \mathbf{n} = 0$, then

$$(\mathbf{p}, \mathbf{q})_{\mathrm{div},\Omega} = (\nabla u, \mathbf{q})_\Omega + (u, \nabla \cdot \mathbf{q})_\Omega = \langle \mathbf{q} \cdot \mathbf{n}, \gamma u \rangle_\Gamma = 0.$$

This shows (10.32) and finishes the proof of the result. $\qquad\square$

Variational formulation. Thanks to Proposition 10.8 we can easily show that the first order system (10.29) is equivalent to

$$(\mathbf{q}, u) \in \mathbf{H}(\mathrm{div}, \Omega) \times L^2(\Omega), \tag{10.33a}$$

$$\mathbf{q} \cdot \mathbf{n} = h, \tag{10.33b}$$

$$(\kappa^{-1}\mathbf{q}, \mathbf{p})_\Omega - (\nabla \cdot \mathbf{p}, u)_\Omega = 0 \qquad \forall \mathbf{p} \in \mathbf{H}_0(\mathrm{div}, \Omega), \tag{10.33c}$$

$$(\nabla \cdot \mathbf{q}, v)_\Omega \qquad\qquad = (f, v)_\Omega \qquad \forall v \in L^2(\Omega). \tag{10.33d}$$

Let us emphasize how the boundary condition (a 'natural' Neumann boundary condition in the second order formulation (10.30)) has now become an 'essential' boundary condition that is imposed separately from the variational formulation and is compensated by testing on the kernel of the boundary condition operator. Problem (10.33) cannot be uniquely solvable, because it is clear that $(\mathbf{q}, u) = (0, 1)$ is a solution of the homogeneous problem. To select a solution, we will use the space

$$L_\circ^2(\Omega) := \{u \in L^2(\Omega) : (u, 1)_\Omega = 0\}. \tag{10.34}$$

Note that $L^2(\Omega) = \mathcal{P}_0(\Omega) \oplus L_\circ^2(\Omega)$ is an orthogonal decomposition and

$$\langle h, 1 \rangle_\Gamma = \langle \mathbf{q} \cdot \mathbf{n}, 1 \rangle_\Gamma = (\nabla \cdot \mathbf{q}, 1)_\Omega = (f, 1)_\Omega,$$

which shows that equation (10.33d) is equivalent to the compatibility condition (10.31) and the equation

$$(\nabla \cdot \mathbf{q}, v)_\Omega = (f, v)_\Omega \qquad \forall v \in L^2_\circ(\Omega).$$

This leads to the equivalent reduced formulation

$$(\mathbf{q}, u) \in \mathbf{H}(\mathrm{div}, \Omega) \times L^2_\circ(\Omega), \tag{10.35a}$$

$$\mathbf{q} \cdot \mathbf{n} = h, \tag{10.35b}$$

$$(\kappa^{-1}\mathbf{q}, \mathbf{p})_\Omega - (\nabla \cdot \mathbf{p}, u)_\Omega = 0 \qquad \forall \mathbf{p} \in \mathbf{H}_0(\mathrm{div}, \Omega), \tag{10.35c}$$

$$(\nabla \cdot \mathbf{q}, v)_\Omega \qquad\qquad = (f, v)_\Omega \qquad \forall v \in L^2_\circ(\Omega). \tag{10.35d}$$

Well-posedness of (10.35) is proposed as Exercise 10.7. Problems (10.33) and (10.35) are not pure mixed problems, since they contain a side condition with the boundary condition. Therefore, they are not trivially equivalent to a saddle point problem.

10.7   The divergence operator

This section is the first of several sections devoted to the study of problems related to viscous incompressible flow. Since the main unknown will be a velocity field, we will use the shortened notation

$$\mathbf{H}^1(\Omega) := H^1(\Omega; \mathbb{R}^d) \equiv H^1(\Omega)^d, \qquad \mathbf{H}^1_0(\Omega) := H^1_0(\Omega)^d.$$

In many cases the pressure variable will be determined up to a constant, which will make the space $L^2_\circ(\Omega)$, defined in (10.34), relevant. The Gelfand triple

$$H^1_0(\Omega) \subset L^2(\Omega) \subset H^{-1}(\Omega)$$

will have a vector-valued counterpart

$$\mathbf{H}^1_0(\Omega) \subset \mathbf{L}^2(\Omega) \subset \mathbf{H}^{-1}(\Omega).$$

The dual norm of $H^{-1}(\Omega)$ and $\mathbf{H}^{-1}(\Omega) \equiv H^{-1}(\Omega)^d$ will be equally denoted as $\|\cdot\|_{-1,\Omega}$. In Section 4.5 and more particularly Proposition 4.8 we have seen how the gradient defines a bounded operator $\nabla : L^2(\Omega) \to \mathbf{H}^{-1}(\Omega)$.

An inf-sup condition for the divergence operator. Before we embark on the study of the Stokes problem, let us investigate the inf-sup condition that will end up being equivalent to the well-posedness of the problem. We first recall a result we saw without proof in Section 7.6, as we looked for a proof of Korn's second inequality. As already mentioned in Chapter 7, the proof of the following result is unfortunately beyond the level of difficulty of this textbook: If Ω is a bounded connected Lipschitz domain, then there exists $C > 0$ such that

$$\left\| u - \frac{1}{|\Omega|} \int_\Omega u(\mathbf{x}) d\mathbf{x} \right\|_\Omega \leq C \|\nabla u\|_{-1,\Omega} \qquad \forall u \in L^2(\Omega). \qquad (10.36)$$

We will first characterize the inequality (10.36) (in a given bounded domain Ω) in different ways. We will then give a sketch of the proof of an equivalent form, leaving some of the most intricate details out.

Proposition 10.9. *On a bounded connected open set* Ω, *the following statements are equivalent:*

(a) *There exists* $c > 0$ *such that*

$$\sup_{0 \neq \mathbf{v} \in \mathbf{H}_0^1(\Omega)} \frac{(\nabla \cdot \mathbf{v}, q)_\Omega}{\|\mathbf{v}\|_{1,\Omega}} \geq c \|q\|_\Omega \qquad \forall q \in L_\circ^2(\Omega).$$

(b) *There exists* $c > 0$ *such that*

$$\|\nabla q\|_{-1,\Omega} \geq c \|q\|_\Omega \qquad \forall q \in L_\circ^2(\Omega).$$

(c) *The range of* $\nabla : L^2(\Omega) \to \mathbf{H}^{-1}(\Omega)$ *is closed.*

(d) $\mathrm{div} : \mathbf{H}_0^1(\Omega) \to L_\circ^2(\Omega)$ *is surjective.*

Proof. Before we start with the proof itself, let us prepare the way with some comments. First of all, if $\mathbf{v} \in \mathbf{H}_0^1(\Omega)$, then $\nabla \cdot \mathbf{v} \in L_\circ^2(\Omega)$. Also, the kernel of the gradient is the set of constant functions (here is where we use that Ω is connected). Let us finally introduce the bounded bilinear form $b : \mathbf{H}_0^1(\Omega) \times L_\circ^2(\Omega) \to \mathbb{R}$ given by

$$b(\mathbf{v}, q) := (\nabla \cdot \mathbf{v}, q)_\Omega = -\langle \nabla q, \mathbf{v} \rangle_{\mathbf{H}^{-1}(\Omega) \times \mathbf{H}_0^1(\Omega)}.$$

Since

$$\|\nabla q\|_{-1,\Omega} = \sup_{0 \neq \mathbf{v} \in \mathbf{H}_0^1(\Omega)} \frac{b(\mathbf{v}, q)}{\|\mathbf{v}\|_{1,\Omega}},$$

it is clear that (a) and (b) are equivalent. We have also seen (see Proposition 10.3) that (b) is equivalent to $\nabla : L_\circ^2(\Omega) \to \mathbf{H}^{-1}(\Omega)$ being injective with closed range. However, $L^2(\Omega) = L_\circ^2(\Omega) \oplus \ker \nabla$, and this proves the equivalence of (b) and (c). Finally, the bilinear form b defines a bounded operator $B : \mathbf{H}_0^1(\Omega) \to L_\circ^2(\Omega)$ via the Riesz-Fréchet theorem, but this operator is just the divergence. Therefore the equivalence of (a) and (d) follows from Proposition 10.3 (see (10.1) too). $\qquad \square$

Proposition 10.10. *Let* Ω_1 *and* Ω_2 *be non-disjoint bounded open domains and* $\Omega := \Omega_1 \cup \Omega_2$. *If* $\mathrm{div} : \mathbf{H}_0^1(\Omega_j) \to L_\circ^2(\Omega_j)$ *is surjective for* $j = 1, 2$, *then* $\mathrm{div} : \mathbf{H}_0^1(\Omega) \to L_\circ^2(\Omega)$ *is surjective.*

Proof. We take $f \in L^2_\circ(\Omega)$, an open ball $B \subset \Omega_1 \cap \Omega_2$, and $c := |B|^{-1}(f,1)_{\Omega_1}$. We then define

$$f_1 := f\chi_{\Omega_1} - c\chi_B \in L^2_\circ(\Omega_1), \qquad f_2 := (1 - \chi_{\Omega_1})\chi_{\Omega_2}f + c\chi_B \in L^2_\circ(\Omega_2),$$

where we have used the fact that

$$(f_2, 1)_{\Omega_2} = (f,1)_{\Omega_2} - (f,1)_{\Omega_1 \cap \Omega_2} + (f,1)_{\Omega_1} = (f,1)_\Omega = 0.$$

Note that by definition $f = \widetilde{f}_1 + \widetilde{f}_2$. We now take $\mathbf{u}_j \in \mathbf{H}^1_0(\Omega_j)$ such that $\nabla \cdot \mathbf{u}_j = f_j$ for $j = 1, 2$, and note that $\mathbf{u} := \widetilde{\mathbf{u}_1} + \widetilde{\mathbf{u}_2} \in \mathbf{H}^1_0(\Omega)$ satisfies

$$\nabla \cdot \mathbf{u} = \widetilde{\nabla \cdot \mathbf{u}_1} + \widetilde{\nabla \cdot \mathbf{u}_2} = \widetilde{f}_1 + \widetilde{f}_2 = f,$$

which finishes the proof. $\qquad\square$

The Bogovskiĭ operator. For the moment being, we will consider Ω to be star-shaped with respect to a ball $B \subset \Omega$, that is, for any $\mathbf{x} \in \Omega$ and $\mathbf{y} \in B$, the straight segment joining $\mathbf{x}$ and $\mathbf{y}$ is contained in Ω. We choose a fixed $\omega \in \mathcal{D}(\mathbb{R}^d)$ such that $\operatorname{supp}\omega \subset B$ and

$$\int_{\mathbb{R}^d} \omega(\mathbf{x})\mathrm{d}\mathbf{x} = 1.$$

We now consider the operator

$$Bf(\mathbf{x}) := \int_\Omega \alpha(\mathbf{y}, \mathbf{x} - \mathbf{y})f(\mathbf{y})\,(\mathbf{x} - \mathbf{y})\mathrm{d}\mathbf{y},$$

where

$$\alpha(\mathbf{y}, \mathbf{v}) := \int_1^\infty \omega(\mathbf{y} + t\mathbf{v})t^{d-1}\mathrm{d}t = \int_0^1 \omega(\mathbf{y} + s^{-1}\mathbf{v})s^{-d-1}\mathrm{d}s$$

$$= \frac{1}{|\mathbf{v}|^d}\int_{|\mathbf{v}|}^\infty \omega(\mathbf{y} + \tau|\mathbf{v}|^{-1}\mathbf{v})\tau^{d-1}\mathrm{d}\tau.$$

The proof of the following result is long and technical, requiring some non-trivial tools of harmonic analysis. We will give a precise reference for where to find its proof in the final comments of the chapter. For readers willing to try the proof of this result, we mention that the difficult part of the proof consists of showing that $\mathbf{u} \in \mathbf{H}^1(\mathbb{R}^d)$ and bounding its $\mathbf{H}^1(\mathbb{R}^d)$ norm in terms of f.

[Proof not provided]

Theorem 10.2 (Bogovskiĭ). *If $f \in L^\infty(\Omega)$ and $(f,1)_\Omega = 0$, then $\mathbf{u} := Bf$ satisfies: $\mathbf{u} \in \mathcal{C}(\mathbb{R}^d; \mathbb{R}^d)$, $\mathbf{u} \equiv \mathbf{0}$ outside Ω, $\mathbf{u} \in \mathbf{H}^1(\mathbb{R}^d)$, and $\nabla \cdot \mathbf{u} = f$ in Ω. Moreover, there exists a constant c such that*

$$\|Bf\|_{1,\Omega} = \|Bf\|_{1,\mathbb{R}^d} \le c\|f\|_\Omega.$$

Corollary 10.1. *If Ω is star-shaped with respect to a ball B, then* div $:$ $\mathbf{H}_0^1(\Omega) \to L_\circ^2(\Omega)$ *is surjective.*

Proof. The Bogovskiĭ operator defines a linear map $B : L^\infty(\Omega) \cap L_\circ^2(\Omega) \to \mathbf{H}_0^1(\Omega)$ that is a right inverse of the divergence operator and is continuous $L^2(\Omega) \to \mathbf{H}_0^1(\Omega)$. Since $L^\infty(\Omega) \cap L_\circ^2(\Omega)$ is dense in $L_\circ^2(\Omega)$ (this is easy to prove), we can extend B in a unique way to $L_\circ^2(\Omega)$, and thus obtain a bounded right inverse of the divergence operator. $\qquad\square$

We now show an interesting result on Lipschitz domains. To avoid having to go back and forth to Chapter 3 (and specifically to Section 3.3), here we give a quick review of the definition of Lipschitz domain. For every boundary point of a Lipschitz domain $\mathbf{x} \in \Gamma$, there exists an invertible affine transformation $R_\mathbf{x}$ (the composition of a dilation, a translation and a proper orthogonal transformation), a Lipschitz function $h_\mathbf{x} : B_{d-1}(\mathbf{0}; 1) \to \mathbb{R}$

$$|h_\mathbf{x}(\widetilde{\mathbf{y}}) - h_\mathbf{x}(\widetilde{\mathbf{z}})| \leq L_\mathbf{x} |\widetilde{\mathbf{y}} - \widetilde{\mathbf{z}}|,$$

and a thickness parameter $\delta_\mathbf{x}$ such that the map $R_\mathbf{x}$ transforms:

(a) the point $(\mathbf{0}, h_\mathbf{x}(\mathbf{0}))$ to $\mathbf{x}$,

(b) the graph $\mathcal{G}_\mathbf{x} := \{(\widetilde{\mathbf{y}}, h_\mathbf{x}(\widetilde{\mathbf{y}})) : |\widetilde{\mathbf{y}}| < 1\}$ to a relatively open part of Γ,

(c) the epigraph $\mathcal{G}_\mathbf{x} + (0, 4\delta_\mathbf{x})\mathbf{e}_d$ to a subset of Ω,

(d) the hypograph $\mathcal{G}_\mathbf{x} + (-4\delta_\mathbf{x}, 0)\mathbf{e}_d$ to a subset of the exterior of Ω.

(Note that in comparison with Section 3.3, we are writing $\delta_\mathbf{x} := \eta_\mathbf{x}/4$, as this will simplify several expressions to come.) In Figure 10.1 we give a cartoon of the situation being described in Lemma 10.1, where we reduce the tubular domain of the definition to a subdomain which is star-shaped with respect to a cylinder. This figure does not make use of the transformation $R_\mathbf{x}$ and so Lemma 10.1 (b) is not illustrated.

Lemma 10.1. *In the above notation, if we take*

$$\rho_\mathbf{x} \leq \min\left\{1, \frac{\delta_\mathbf{x}}{3L_\mathbf{x}}\right\}$$

such that

$$|h_\mathbf{x}(\widetilde{\mathbf{x}}) - h_\mathbf{x}(\mathbf{0})| \leq \delta_\mathbf{x} \qquad \forall \widetilde{\mathbf{x}} \in B_{d-1}(\mathbf{0}; \rho_\mathbf{x}),$$

and define

$$\mathcal{U}_\mathbf{x} := \{(\widetilde{\mathbf{x}}, x_d) : |\widetilde{\mathbf{x}}| < \rho_\mathbf{x}, \ h_\mathbf{x}(\widetilde{\mathbf{x}}) - h_\mathbf{x}(\mathbf{0}) - \delta_\mathbf{x} < x_d < \delta_\mathbf{x}\},$$

then

(a) $B_{d-1}(\mathbf{0}; \rho_\mathbf{x}) \times (0, \delta_\mathbf{x}) \subset \mathcal{U}_\mathbf{x}$,

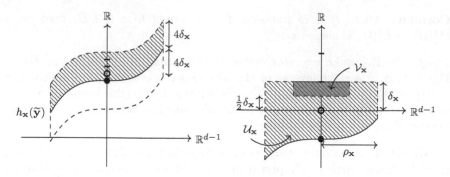

Figure 10.1: Given a strong Lipschitz domain and a point $\mathbf{x}$ on the boundary, we show the graph of $h_{\mathbf{x}}$ and the tubular domain in the x_d direction with width $8\delta_{\mathbf{x}}$. The epigraph (which up to dilation, translation, and rotation, is in Ω) is show shaded. On the right we show the sets $\mathcal{U}_{\mathbf{x}}$ and $\mathcal{V}_{\mathbf{x}}$ for an appropriate choice of $\rho_{\mathbf{x}}$ as in Lemma 10.1.

(b) $R_{\mathbf{x}}(\mathbf{y} + (h_{\mathbf{x}}(\mathbf{0}) + \delta_{\mathbf{x}})\mathbf{e}_d) \in \Omega$ $\forall \mathbf{y} \in \mathcal{U}_{\mathbf{x}}$,

(c) $\mathcal{U}_{\mathbf{x}}$ *is star-shaped with respect to* $\mathcal{V}_{\mathbf{x}} := B_{d-1}(\mathbf{0}; \tfrac{1}{2}\rho_{\mathbf{x}}) \times (\tfrac{1}{2}\delta_{\mathbf{x}}, \delta_{\mathbf{x}})$.

Proof. First, we note that (a) is a simple consequence of the definition of $\mathcal{U}_{\mathbf{x}}$ since $h_{\mathbf{x}}(\tilde{\mathbf{x}}) - h_{\mathbf{x}}(\mathbf{0}) - \delta_{\mathbf{x}} \le 0$ for every $\tilde{\mathbf{x}} \in B_{d-1}(\mathbf{0}; \rho_{\mathbf{x}})$. To show (b), let $\mathbf{y} \in \mathcal{U}_{\mathbf{x}}$, then

$$\mathbf{y} + (h_{\mathbf{x}}(\mathbf{0}) + \delta_{\mathbf{x}})\mathbf{e}_d = (\tilde{\mathbf{x}}, x_d + h_{\mathbf{x}}(\mathbf{0}) + \delta_{\mathbf{x}}) = (\tilde{\mathbf{x}}, h_{\mathbf{x}}(\tilde{\mathbf{x}}) + \eta),$$

where $|\tilde{\mathbf{x}}| < \rho_{\mathbf{x}}$ and

$$0 < \eta = x_d + h_{\mathbf{x}}(\mathbf{0}) + \delta_{\mathbf{x}} - h_{\mathbf{x}}(\tilde{\mathbf{x}}) < 3\delta_{\mathbf{x}}.$$

Therefore $\eta < 4\delta_{\mathbf{x}}$ and (b) holds by the definition of $R_{\mathbf{x}}$. Now we turn our attention to the main result (c). We decompose $\mathcal{U}_{\mathbf{x}} = B_{d-1}(\mathbf{0}; \rho_{\mathbf{x}}) \times [0, \delta_{\mathbf{x}}) \cup \mathcal{U}_{\mathbf{x}} \setminus (B_{d-1}(\mathbf{0}; \rho_{\mathbf{x}}) \times [0, \delta_{\mathbf{x}}))$, noting that $\mathcal{V}_{\mathbf{x}} \subset B_{d-1}(\mathbf{0}; \rho_{\mathbf{x}}) \times [0, \delta_{\mathbf{x}})$ and this set is convex and hence star-shaped with respect to any subset, in particular $\mathcal{V}_{\mathbf{x}}$. Now, if the remainder of $\mathcal{U}_{\mathbf{x}}$ is not star-shaped with respect to $\mathcal{V}_{\mathbf{x}}$, then there exists $(\tilde{\mathbf{x}}, x_d) \in \mathcal{V}_{\mathbf{x}}$ such that a ray originating at $(\tilde{\mathbf{x}}, x_d)$ intersects the set

$$\{(\tilde{\mathbf{x}}, h_{\mathbf{x}}(\tilde{\mathbf{x}}) - h_{\mathbf{x}}(\mathbf{0}) - \delta_{\mathbf{x}}) : |\tilde{\mathbf{x}}| < \rho_{\mathbf{x}}\}$$

twice. If $(\tilde{\mathbf{y}}_1, h_{\mathbf{x}}(\tilde{\mathbf{y}}_1) - h_{\mathbf{x}}(\mathbf{0}) - \delta_{\mathbf{x}}) \ne (\tilde{\mathbf{y}}_2, h_{\mathbf{x}}(\tilde{\mathbf{y}}_2) - h_{\mathbf{x}}(\mathbf{0}) - \delta_{\mathbf{x}})$ are two points from this set on the ray, then the vectors $(\tilde{\mathbf{y}}_1 - \tilde{\mathbf{y}}_2, h_{\mathbf{x}}(\tilde{\mathbf{y}}_1) - h_{\mathbf{x}}(\tilde{\mathbf{y}}_2))$ and $(\tilde{\mathbf{y}}_1 - \tilde{\mathbf{x}}, h_{\mathbf{x}}(\tilde{\mathbf{y}}_1) - h_{\mathbf{x}}(\mathbf{0}) - \delta_{\mathbf{x}} - x_d)$ are parallel. We also have, however, that

$$L_{\mathbf{x}} \ge \frac{|h_{\mathbf{x}}(\tilde{\mathbf{y}}_1) - h_{\mathbf{x}}(\tilde{\mathbf{y}}_2)|}{|\tilde{\mathbf{y}}_1 - \tilde{\mathbf{y}}_2|} = \frac{|h_{\mathbf{x}}(\tilde{\mathbf{y}}_1) - h_{\mathbf{x}}(\mathbf{0}) - \delta_{\mathbf{x}} - x_d|}{|\tilde{\mathbf{y}}_1 - \tilde{\mathbf{x}}|} > \frac{\tfrac{1}{2}\delta_{\mathbf{x}}}{\tfrac{3}{2}\rho_{\mathbf{x}}} \ge L_{\mathbf{x}}.$$

From this contradiction, we see that (c) must hold, and this finishes the proof. $\qquad\square$

Proposition 10.11. *Every Lipschitz domain is the union of finitely many subdomains that are star-shaped with respect to a ball. Moreover, the closure of a Lipschitz domain can be covered with finitely many open sets whose intersection with the domain is start-shaped with respect to a ball.*

Proof. Let Ω be a strong Lipschitz domain. For each $\mathbf{x} \in \Gamma$, we define $R_{\mathbf{x}}, c_{\mathbf{x}}, \delta_{\mathbf{x}}$, and $h_{\mathbf{x}}$ as above, $\rho_{\mathbf{x}}$ and $\mathcal{U}_{\mathbf{x}}$ as in Lemma 10.1, and

$$\mathcal{W}_{\mathbf{x}} = \{(\widetilde{\mathbf{x}}, x_d) : |\widetilde{x}| < \rho_{\mathbf{x}}, \quad h_{\mathbf{x}}(\widetilde{\mathbf{x}}) - h_{\mathbf{x}}(\mathbf{0}) - 2\delta_{\mathbf{x}} < x_d < \delta_{\mathbf{x}}\}.$$

We note that $\mathcal{U}_{\mathbf{x}} \subset \mathcal{W}_{\mathbf{x}}$. Now we define the open sets $\Omega_{\mathbf{x}} := R_{\mathbf{x}}((h_{\mathbf{x}}(\mathbf{0}) + \delta_{\mathbf{x}})\mathbf{e}_d + \mathcal{W}_{\mathbf{x}})$. This last vertical translation relocates the domain $\mathcal{U}_{\mathbf{x}}$ (again, this is part of $\mathcal{W}_{\mathbf{x}}$) to the interior of the tubular domain. We can choose a finite number of points $\mathbf{x}_j$ such that $\Gamma \subset \bigcup_{j=1}^{N} \Omega_{\mathbf{x}_j}$, and consider the set

$$\Omega_{\text{int}} = \{\mathbf{x} \in \Omega : \mathbf{x} \notin \Omega_{\mathbf{x}_j} \quad \forall j\}.$$

Noticing that $\Omega_{\text{int}} = \left(\overline{\Omega}^c \cup \left(\bigcup_{j=1}^{N} \Omega_{\mathbf{x}_j}\right)\right)^c$, we have that Ω_{int} is closed and bounded, hence compact, so we can cover it with a finite number of balls of any given radius. Furthermore, since this set is separated from Γ, we can choose these balls to be strictly contained in Ω. This finishes the proof since each $\Omega_{\mathbf{x}_j} \cap \Omega$ is star-shaped with respect to a ball (see Lemma 10.1), and a ball is obviously also star-shaped with respect to a ball. $\square$

Corollary 10.2. *If Ω is a connected bounded Lipschitz domain, then* div : $\mathbf{H}_0^1(\Omega) \to L_\circ^2(\Omega)$ *is surjective.*

Proof. By Corollary 10.1 the result holds for strongly star-shaped domains. By Proposition 10.10, the result holds for connected open domains that can be written as the union of star-shaped domains, since we can make progressive unions of domains, given the fact that Ω is connected and we will always have an overlap. Finally, Proposition 10.11 shows that connected bounded Lipschitz domains are in the class covered by the previous argument. $\square$

As a corollary, it is also simple to see that div : $\mathbf{H}^1(\Omega) \to L^2(\Omega)$ is surjective using an extension argument (Exercise 10.8) . The final result of this section deals with the case of partially homogeneous Dirichlet conditions.

Proposition 10.12. *Let Ω be Lipschitz and $\Sigma \subset \Gamma$ be such that there exist $\mathbf{x} \in \Gamma \setminus \Sigma$ and $\varepsilon > 0$ with $B(\mathbf{x}; \varepsilon) \cap \Sigma = \emptyset$. The divergence operator*

$$\text{div} : \mathbf{H}_\Sigma^1(\Omega) := \{\mathbf{u} \in \mathbf{H}^1(\Omega) : \gamma\mathbf{u} = 0 \text{ on } \Sigma\} \to L^2(\Omega)$$

is surjective.

Proof. Assume that we have found $\mathbf{u}_0 \in \mathbf{H}_\Sigma^1(\Omega)$ such that $\nabla \cdot \mathbf{u}_0 \notin L_\circ^2(\Omega)$, i.e., $(\nabla \cdot \mathbf{u}_0, 1)_\Omega \neq 0$. Given $f \in L^2(\Omega)$, there exists $\mathbf{v} \in \mathbf{H}_0^1(\Omega) \subset \mathbf{H}_\Sigma^1(\Omega)$ such that

$$\nabla \cdot \mathbf{v} = f - \frac{(f, 1)_\Omega}{(\nabla \cdot \mathbf{u}_0, 1)_\Omega} \nabla \cdot \mathbf{u}_0 \in L_\circ^2(\Omega),$$

and therefore

$$\mathbf{u} := \mathbf{v} + \frac{(f, 1)_\Omega}{(\nabla \cdot \mathbf{u}_0, 1)_\Omega} \mathbf{u}_0 \in \mathbf{H}^1_\Sigma(\Omega)$$

satisfies $\nabla \cdot \mathbf{u} = f$. Now we need only show that such a $\mathbf{u}_0 \in \mathbf{H}^1_\Sigma(\Omega)$ exists. To do this take the point $\mathbf{x} \in \Gamma \setminus \Sigma$ and a ball $B := B(\mathbf{x}; \varepsilon)$ from the statement of the proposition. Let us now consider two balls of the same radius

$$B_+ \subset B \cap \Omega, \qquad B_- \subset B \cap (\mathbb{R}^d \setminus \Omega),$$

and the function $f = \chi_{B_+} - \chi_{B_-} \in L^2_\circ(B)$. (Figure 10.2 shows a graphic representation of this construction.) We can then find $\mathbf{v} \in \mathbf{H}^1_0(B)$ such that $\nabla \cdot \mathbf{v} = f$. The function

$$\mathbf{u}_0 = \begin{cases} \mathbf{v}, & \text{in } B \cap \Omega, \\ \mathbf{0}, & \text{elsewhere}, \end{cases}$$

satisfies the requirements.   □

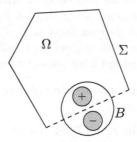

Figure 10.2: A sketch of the geometric construction for the proof of Proposition 10.12.

10.8   Stokes flow

The Stokes problem. For $\mathbf{f} \in \mathbf{L}^2(\Omega)$, we consider the problem

$$(\mathbf{u}, p) \in \mathbf{H}^1_0(\Omega) \times L^2_\circ(\Omega), \tag{10.37a}$$

$$-\nu \Delta \mathbf{u} + \nabla p = \mathbf{f}, \tag{10.37b}$$

$$\nabla \cdot \mathbf{u} = 0. \tag{10.37c}$$

The parameter $\nu > 0$ (the kinematic viscosity) does not play any central role in this problem, due to its linearity (we can easily scale $\mathbf{u}$ to eliminate it), but we will keep it in the formulation due to its relevance for extensions like the Brinkman flow problem (Section 10.10) and the Navier-Stokes equations (Chapter 12). It is clear that the pressure p is underdetermined in equation (10.37b) (the only one where it appears), which justifies the choice of requesting the side condition $(p, 1)_\Omega = 0$ in the formulation. Note also that

$$(\nabla \cdot \mathbf{u}, 1)_\Omega = 0 \qquad \forall \mathbf{u} \in \mathbf{H}_0^1(\Omega),$$

(this can easily be proved by density) and therefore (10.37c) is equivalent to

$$(\nabla \cdot \mathbf{u}, q)_\Omega = 0 \qquad \forall q \in L_\circ^2(\Omega),$$

since $L^2(\Omega) = L_\circ^2(\Omega) \oplus \mathcal{P}_0(\Omega)$. Therefore, the Stokes problem (10.37) is equivalent to the variational formulation

$$(\mathbf{u}, p) \in \mathbf{H}_0^1(\Omega) \times L_\circ^2(\Omega), \tag{10.38a}$$

$$\nu(\nabla \mathbf{u}, \nabla \mathbf{v})_\Omega - (\nabla \cdot \mathbf{v}, p)_\Omega = (\mathbf{f}, \mathbf{v})_\Omega \qquad \forall \mathbf{v} \in \mathbf{H}_0^1(\Omega), \tag{10.38b}$$

$$(\nabla \cdot \mathbf{u}, q)_\Omega \qquad\qquad = 0 \qquad \forall q \in L_\circ^2(\Omega). \tag{10.38c}$$

Problem (10.38) fits readily in the framework of mixed problems, with a bilinear form $a : \mathbf{H}_0^1(\Omega) \times \mathbf{H}_0^1(\Omega) \to \mathbb{R}$

$$a(\mathbf{u}, \mathbf{v}) := \nu(\nabla \mathbf{u}, \nabla \mathbf{v})_\Omega,$$

which is coercive in the entire space $\mathbf{H}_0^1(\Omega)$, due to the Poincaré-Friedrichs inequality. The other bilinear form $b : \mathbf{H}_0^1(\Omega) \times L_\circ^2(\Omega) \to \mathbb{R}$ given by

$$b(\mathbf{u}, p) := -(\nabla \cdot \mathbf{u}, p)_\Omega,$$

satisfies the necessary inf-sup condition due to Proposition 10.9 and Corollary 10.2. Well-posedness of (10.38) is then guaranteed by the abstract theory of mixed problems. The associated Lagrangian

$$\frac{\nu}{2}\|\nabla \mathbf{u}\|_\Omega^2 - (\mathbf{f}, \mathbf{u})_\Omega - (\nabla \cdot \mathbf{u}, p)_\Omega$$

shows the role of the pressure p as a Lagrange multiplier in the problem. The associated constrained minimization problem can be set on the space of **solenoidal** vector fields

$$\mathbf{V}_0 := \{\mathbf{u} \in \mathbf{H}_0^1(\Omega) : \nabla \cdot \mathbf{u} = 0\}$$

as

$$\frac{\nu}{2}\|\nabla \mathbf{u}\|_\Omega^2 - (\mathbf{f}, \mathbf{u})_\Omega = \min! \qquad \mathbf{u} \in \mathbf{V}_0.$$

A variant of the above problem, consisting of adding a convection term to the Stokes problem, leads to the Oseen equations (see Exercise 10.10).

Hydrodynamic stress formulation. An equivalent (but different) formulation of the Stokes problem (10.37) can also be proposed for the Dirichlet problem for the Stokes equations. The difference of bilinear forms will be relevant when other types of boundary conditions are handled. If, as we did with the elasticity equations in past chapters, we consider the divergence operator to act *on the rows* on a matrix-valued function, a simple computation shows that for every vector-valued distribution

$$\mathrm{div}(\nabla \mathbf{u})^\top = \nabla(\nabla \cdot \mathbf{u}),$$

and we can write the Stokes equations using the symmetric gradient

$$\varepsilon(\mathbf{u}) = \tfrac{1}{2}(\nabla \mathbf{u} + (\nabla \mathbf{u})^\top),$$

as

$$\mathbf{u} \in H_0^1(\Omega), \qquad -2\nu \, \mathrm{div} \, \varepsilon(\mathbf{u}) + \nabla p = \mathbf{f}, \qquad \nabla \cdot \mathbf{u} = 0,$$

or

$$\mathbf{u} \in H_0^1(\Omega), \qquad -\mathrm{div} \, \boldsymbol{\sigma} = \mathbf{f}, \qquad \boldsymbol{\sigma} = 2\nu \varepsilon(\mathbf{u}) - p\mathrm{I}, \qquad \nabla \cdot \mathbf{u} = 0,$$

where we have introduced a hydrodynamic stress tensor, which takes values in a space of symmetric matrices. Using the associated integration by parts formula (the reader should review the introduction of the elasticity equations in Section 4.6), we can find an equivalent variational formulation

$$\mathbf{u} \in \mathbf{H}_0^1(\Omega), \; p \in L_\circ^2(\Omega),$$
$$2\nu \, (\varepsilon(\mathbf{u}), \varepsilon(\mathbf{v}))_\Omega - (\nabla \cdot \mathbf{v}, p)_\Omega = (\mathbf{f}, \mathbf{v})_\Omega \qquad \forall \mathbf{v} \in \mathbf{H}_0^1(\Omega),$$
$$(\nabla \cdot \mathbf{u}, q)_\Omega = 0 \qquad\qquad\qquad\qquad \forall q \in L_\circ^2(\Omega).$$

The inf-sup condition of the Stokes problem remains unchanged, while the diagonal bilinear form is still coercive in the full space, due to Korn's first inequality (Proposition 4.10)

$$2\|\varepsilon(\mathbf{u})\|_\Omega^2 \geq \|\nabla \mathbf{u}\|_\Omega^2 \qquad \forall \mathbf{u} \in \mathbf{H}_0^1(\Omega),$$

and the Poincaré-Friedrichs inequality.

10.9   Stokes-Darcy flow

We devote this section to the study of a model problem that combines Stokes and Darcy flow on different parts of a domain. We will formulate the problem using joint velocity and pressure fields on the entire domain, and we will come back to it in Section 11.5 using separate spaces in the separate subdomains; see Figure 10.3.

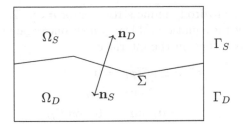

Figure 10.3: A sketch of the Stokes-Darcy geometrical setting. All three domains involved (Ω_S, Ω_D, and the global domain) are assumed to be Lipschitz. Two opposing normal fields appear in the interface Σ.

Geometric setting. A bounded Lipschitz domain $\Omega \subset \mathbb{R}^d$ is separated by an interface Σ into two nonoverlapping Lipschitz open sets Ω_S and Ω_D so that $\overline{\Omega} = \overline{\Omega_S} \cup \overline{\Omega_D}$ and $\Sigma := \partial\Omega_D \cap \partial\Omega_S$. The rest of the boundary is separated into two relatively open sets Γ_S and Γ_D, so that

$$\overline{\Gamma_S} \cup \overline{\Gamma_D} = \partial\Omega, \qquad \overline{\Gamma_S} \cup \overline{\Sigma} = \partial\Omega_S, \qquad \overline{\Gamma_D} \cup \overline{\Sigma} = \partial\Omega_D,$$

The normal vector field on $\partial\Omega_D$ points out and is called $\mathbf{n}_D$. On $\partial\Omega_S$ we call it $\mathbf{n}_S$. Clearly, on Σ we have two opposing normal vector fields and therefore $\mathbf{n}_S + \mathbf{n}_D = \mathbf{0}$. We will identify functions with pairs of their restrictions to each subdomain

$$\mathbf{u} \equiv (\mathbf{u}_S, \mathbf{u}_D) : \Omega \to \mathbb{R}^d, \qquad p \equiv (p_S, p_D) : \Omega \to \mathbb{R}.$$

Model equations. We first present the equations in a not entirely rigorous fashion. The data are $\mathbf{f}_S \in \mathbf{L}^2(\Omega_S)$ and $f_D \in L^2(\Omega_D)$, although we will see how an additional condition on f_D is needed. In Ω_S, we have the Stokes equations with homogeneous Dirichlet conditions everywhere but on the interface

$$-2\nu \operatorname{div} \varepsilon(\mathbf{u}_S) + \nabla p_s = \mathbf{f}_S, \qquad (10.39\text{a})$$

$$\nabla \cdot \mathbf{u}_S = 0, \qquad (10.39\text{b})$$

$$\gamma \mathbf{u}_S = \mathbf{0} \quad \text{on } \Gamma_S. \qquad (10.39\text{c})$$

Here we have used the symmetric gradient

$$\varepsilon(\mathbf{u}_S) = \tfrac{1}{2}(\nabla\mathbf{u}_S + (\nabla\mathbf{u}_S)^\top).$$

We will write the first equation in the equivalent form

$$-\operatorname{div}(2\nu\varepsilon(\mathbf{u}_S) - p_S\mathbf{I}) = \mathbf{f}_S,$$

where we can see the hydrodynamic stress tensor $\boldsymbol{\sigma}_S := 2\nu\boldsymbol{\varepsilon}(\mathbf{u}_S) - p_S\mathbf{I}$ being used. In Ω_D we have the equations of Darcy flow, once again with homogeneous essential conditions except on the interface

$$\kappa^{-1}\mathbf{u}_D + \nabla p_D = \mathbf{0}, \tag{10.39d}$$

$$\nabla \cdot \mathbf{u}_D = f_D, \tag{10.39e}$$

$$\mathbf{u}_D \cdot \mathbf{n}_D = 0 \quad \text{on } \Gamma_D. \tag{10.39f}$$

We will clarify what we mean with the last condition later on. On the interface we will have two transmission conditions

$$\gamma\mathbf{u}_S \cdot \mathbf{n}_S + \mathbf{u}_D \cdot \mathbf{n}_D = 0 \qquad \text{on } \Sigma, \tag{10.39g}$$

$$(2\nu\boldsymbol{\varepsilon}(\mathbf{u}_S) - p_S\mathbf{I})\mathbf{n}_S + \mathbf{c}\pi_T\mathbf{u}_S = (\gamma p_D)\mathbf{n}_D \qquad \text{on } \Sigma. \tag{10.39h}$$

We have used the tangential trace operator

$$\pi_T\mathbf{u}_S := \gamma\mathbf{u}_S - (\gamma\mathbf{u}_S \cdot \mathbf{n}_S)\mathbf{n}_S.$$

Normalization for the pressure will be done with the condition

$$(p_D, 1)_{\Omega_D} = 0. \tag{10.39i}$$

We could also normalize in the entire space $p \in L^2_\circ(\Omega)$, but we will learn something from this process anyway. The coefficients are: a constant kinematic viscosity $\nu > 0$; a permeability tensor $\kappa : \Omega \to \mathbb{R}^{d\times d}_{\text{sym}}$, uniformly positive definite and with components in $L^\infty(\Omega)$; an interface positive semidefinite tensor $\mathbf{c} : \Sigma \to \mathbb{R}^{d\times d}_{\text{sym}}$ satisfying $\mathbf{c}\mathbf{n} = \mathbf{0}$ and with components in $L^\infty(\Gamma)$. (Typically this tensor is written in relation to κ and ν, but this is not needed for the theoretical treatment of the equations.) The first transmission condition (10.39g) is conservation of mass. The second one (10.39h) can be separated as compensation of normal stress

$$\mathbf{n}_S \cdot (\boldsymbol{\sigma}_S\mathbf{n}_S) = \gamma p_D,$$

and the unilateral condition

$$\mathbf{t} \cdot (\boldsymbol{\sigma}_S\mathbf{n}_S + \mathbf{c}\gamma\mathbf{u}_S) = 0,$$

where $\mathbf{t} : \Sigma \to \mathbb{R}^d$ is **any tangential field** on Σ. This is often called the Beavers-Joseph-Saffman condition.

Formalization. A solution to the problem will have

$$\mathbf{u} = (\mathbf{u}_S, \mathbf{u}_D) \in \mathbf{H}^1_{\Gamma_S}(\Omega_S) \times \mathbf{H}(\text{div}, \Omega_D),$$

where

$$\mathbf{H}^1_{\Gamma_S}(\Omega_S) := \{\mathbf{u} \in \mathbf{H}^1(\Omega_S) : \gamma\mathbf{u} = \mathbf{0} \text{ on } \Gamma_S\},$$

and

$$p = (p_S, p_D) \in Q := L^2(\Omega_S) \times L^2_\circ(\Omega_D).$$

Note that for a solution of the equations, we will have $p_D \in H^1(\Omega_D)$, and this makes γp_D in the interface condition makes sense. We also have that

$$\boldsymbol{\sigma}_S \in \mathbf{H}(\mathrm{div}, \Omega_S; \mathbb{R}_{\mathrm{sym}}^{d \times d}) := \{\boldsymbol{\tau} \in L^2(\Omega; \mathbb{R}_{\mathrm{sym}}^{d \times d}) : \mathrm{div}\,\boldsymbol{\tau} \in \mathbf{L}^2(\Omega)\},$$

which justifies taking the normal component. The Darcy boundary condition and the first transmission condition will be considered as a single boundary condition on $\partial\Omega_D$

$$\mathbf{u}_D \cdot \mathbf{n}_D = -\widetilde{\gamma\mathbf{u}_S|_\Sigma \cdot \mathbf{n}_S},$$

with equality as elements of $H^{-1/2}(\partial\Omega_D)$, where the tilde operator is the extension by zero of $\gamma\mathbf{u}_S|_\Sigma \cdot \mathbf{n}_S \in L^2(\Sigma)$ to an element of $L^2(\partial\Omega_D) \subset H^{-1/2}(\partial\Omega_D)$. This is equivalent to

$$(\mathbf{u}_D, \nabla v)_{\Omega_D} + (\nabla \cdot \mathbf{u}_D, v)_{\Omega_D} = -\langle \gamma\mathbf{u}_S \cdot \mathbf{n}_S, \gamma v\rangle_\Sigma \qquad \forall v \in H^1(\Omega_D). \quad (10.40)$$

The second transmission condition is read as follows (we write it in terms of the stress $\boldsymbol{\sigma}_S = 2\nu\boldsymbol{\varepsilon}(\mathbf{u}_S) - p_S\mathbf{I}$)

$$\begin{aligned}
(\boldsymbol{\sigma}_S, \boldsymbol{\varepsilon}(\mathbf{v}))_{\Omega_S} + (\mathrm{div}\,\boldsymbol{\sigma}_S, \mathbf{v})_{\Omega_S} &= \langle \boldsymbol{\sigma}_S\mathbf{n}_S, \gamma\mathbf{v}\rangle_{\partial\Omega_S} \\
&= -\langle c\pi_T\mathbf{u}_S, \gamma\mathbf{v}\rangle_\Sigma + \langle(\gamma p_D)\mathbf{n}_D, \gamma\mathbf{v}\rangle_\Sigma \\
&= -\langle c\pi_T\mathbf{u}_S, \pi_T\mathbf{v}\rangle_\Sigma + \langle \gamma p_D, \gamma\mathbf{v} \cdot \mathbf{n}_D\rangle_\Sigma \\
&\qquad\qquad \forall \mathbf{v} \in \mathbf{H}_{\Gamma_S}^1(\Omega_S).
\end{aligned}$$

We will collect the two velocity fields in the space

$$\mathbf{V} := \{\mathbf{u} \in \mathbf{H}_{\Gamma_S}^1(\Omega_S) \times \mathbf{H}(\mathrm{div}, \Omega_D) : \mathbf{u}_D \cdot \mathbf{n}_D = -\widetilde{\gamma\mathbf{u}_S|_\Sigma \cdot \mathbf{n}_S}\},$$

which is a Hilbert space when endowed with the norm

$$\|\mathbf{v}\|_{\mathbf{V}}^2 := \|\mathbf{v}_S\|_{1,\Omega_S}^2 + \|\mathbf{v}_D\|_{\mathrm{div},\Omega_D}^2.$$

Proposition 10.13. *We have*

$$\mathbf{H}_0^1(\Omega_S) \times \mathbf{H}_0(\mathrm{div}, \Omega_D) \subset \mathbf{V} \subset \mathbf{H}(\mathrm{div}, \Omega)$$

and $\mathbf{H}_0^1(\Omega) \subset \mathbf{V}$. *Moreover, for every* $\mathbf{v}_S \in \mathbf{H}_{\Gamma_S}^1(\Omega_S)$ *there exists* $\mathbf{v}_D$ *such that* $(\mathbf{v}_S, \mathbf{v}_D) \in \mathbf{V}$.

Proof. The first inclusion is simple. To see the second one, take $\varphi \in \mathcal{D}(\Omega)$ and compute the divergence in the sense of distributions

$$\begin{aligned}
\langle \nabla \cdot \mathbf{u}, \varphi\rangle &= -\langle \mathbf{u}, \nabla\varphi\rangle = -(\mathbf{u}_S, \nabla\varphi)_{\Omega_S} - (\mathbf{u}_D, \nabla\varphi)_{\Omega_D} \\
&= (\nabla \cdot \mathbf{u}_S, \varphi)_{\Omega_S} - \langle\gamma\mathbf{u}_S \cdot \mathbf{n}_S, \gamma\varphi\rangle_\Sigma - (\mathbf{u}_D, \nabla\varphi)_{\Omega_D} \\
&= (\nabla \cdot \mathbf{u}_S, \varphi)_{\Omega_S} + (\nabla \cdot \mathbf{u}_D, \varphi)_{\Omega_D},
\end{aligned}$$

which implies that $\nabla \cdot \mathbf{u} \in L^2(\Omega)$. If $\mathbf{u} \in \mathbf{H}_0^1(\Omega)$, then

$$(\mathbf{u}, \nabla v)_\Omega + (\nabla \cdot \mathbf{u}, v)_\Omega = 0 \qquad \forall v \in H^1(\Omega),$$

and therefore

$$(\mathbf{u}_D, \nabla v)_{\Omega_D} + (\nabla \cdot \mathbf{u}_D, v)_{\Omega_D} = -\langle \gamma \mathbf{u}_S \cdot \mathbf{n}_S, \gamma v \rangle_{\partial \Omega_S}$$
$$= -\langle \gamma \mathbf{u}_S \cdot \mathbf{n}_S, \gamma v \rangle_\Sigma \qquad \forall v \in H^1(\Omega).$$

However, the map $H^1(\Omega) \ni v \mapsto v|_{\Omega_D} \in H^1(\Omega_D)$ is onto and this proves that $\mathbf{u} \in \mathbf{V}$. Finally, we need to use that the map $\mathbf{H}_0^1(\Omega) \ni \mathbf{v} \mapsto \mathbf{v}|_{\Omega_S} \in \mathbf{H}_{\Gamma_S}^1(\Omega_S)$ is onto. (See Exercise 10.12.) □

Towards a variational formulation. We start with the second transmission condition (10.39h), and test with $\mathbf{v} = (\mathbf{v}_S, \mathbf{v}_D) \in \mathbf{V}$ to obtain

$$2\nu(\boldsymbol{\varepsilon}(\mathbf{u}_S), \boldsymbol{\varepsilon}(\mathbf{v}))_{\Omega_S} - (p_S, \nabla \cdot \mathbf{v}_S)_{\Omega_S} + \langle c\pi_T \mathbf{u}_S, \pi_T \mathbf{v} \rangle_\Sigma$$
$$= (\mathbf{f}_S, \mathbf{v}_S)_{\Omega_S} + \langle \gamma p_D, \gamma \mathbf{v}_S \cdot \mathbf{n}_D \rangle_\Sigma.$$

Recalling the transmission condition that makes $\mathbf{v} \in \mathbf{V}$, using p_D as test function in (10.40), and substituting $\nabla p_D = -\kappa^{-1} \mathbf{u}_D$, we have

$$\langle \gamma p_D, \gamma \mathbf{v}_S \cdot \mathbf{n}_D \rangle_\Sigma = (\nabla p_D, \mathbf{v}_D)_{\Omega_D} + (p_D, \nabla \cdot \mathbf{v}_D)_{\Omega_D}$$
$$= -(\kappa^{-1} \mathbf{u}_D, \mathbf{v}_D)_{\Omega_D} + (p_D, \nabla \cdot \mathbf{v}_D)_{\Omega_D}.$$

We define the bounded bilinear forms $a : \mathbf{V} \times \mathbf{V} \to \mathbb{R}$ and $b : \mathbf{V} \times L^2(\Omega) \to \mathbb{R}$

$$a(\mathbf{u}, \mathbf{v}) := 2\nu(\boldsymbol{\varepsilon}(\mathbf{u}_S), \boldsymbol{\varepsilon}(\mathbf{v}))_{\Omega_S} + \langle c\pi_T \mathbf{u}_S, \pi_T \mathbf{v} \rangle_\Sigma + (\kappa^{-1} \mathbf{u}_D, \mathbf{v}_D)_{\Omega_D},$$
$$b(\mathbf{u}, q) := (\nabla \cdot \mathbf{u}, q)_\Omega = (\nabla \cdot \mathbf{u}_S, q_S)_{\Omega_S} + (\nabla \cdot \mathbf{u}_D, q_D)_{\Omega_D}.$$

In a first approximation we ignore the normalization for the pressure and write the formulation

$$(\mathbf{u}, p) \in \mathbf{V} \times L^2(\Omega), \tag{10.41a}$$
$$a(\mathbf{u}, \mathbf{v}) - b(\mathbf{v}, p) = (\mathbf{f}_S, \mathbf{v}_S)_{\Omega_S} \qquad \forall \mathbf{v} \in \mathbf{V}, \tag{10.41b}$$
$$b(\mathbf{u}, q) = (f_D, q_D)_{\Omega_D} \qquad \forall q \in L^2(\Omega). \tag{10.41c}$$

Proposition 10.14. *The condition*

$$(f_D, 1)_{\Omega_D} = 0 \tag{10.42}$$

is necessary for existence of solutions to (10.41). *If* (10.42) *holds, the problem* (10.41) *is equivalent to*

$$(\mathbf{u}, p) \in \mathbf{V} \times Q, \tag{10.43a}$$
$$a(\mathbf{u}, \mathbf{v}) - b(\mathbf{v}, p) = (\mathbf{f}_S, \mathbf{v}_S)_{\Omega_S} \qquad \forall \mathbf{v} \in \mathbf{V}, \tag{10.43b}$$
$$b(\mathbf{u}, q) = (f_D, q_D)_{\Omega_D} \qquad \forall q \in Q. \tag{10.43c}$$

up to any constant that can be added to p_D *in* (10.41).

Proof. If $\mathbf{u} \in \mathbf{V}$, then

$$b(\mathbf{u},1) = (\nabla \cdot \mathbf{u}_S, 1)_{\Omega_S} + (\nabla \cdot \mathbf{u}_D, 1)_{\Omega_D}$$
$$= \langle \gamma \mathbf{u}_S \cdot \mathbf{n}_S, 1 \rangle_\Sigma + \langle \mathbf{u}_D \cdot \mathbf{n}_D, 1 \rangle_{\partial\Omega_D} = 0,$$

and therefore (10.42) is a necessary condition for solutions. We can write

$$L^2(\Omega) = Q \oplus \mathcal{P}_0(\Omega) = Q \oplus \mathrm{span}\{\chi_{\Omega_D}\},$$

(the second decomposition is orthogonal, the first one is not) and note that if the condition (10.42) is satisfied, then both problems are equivalent. $\square$

Recovery of the transmission problem. Testing (10.41b) with elements of $\mathcal{D}(\Omega_S)^d \times \{\mathbf{0}\}$ and $\{\mathbf{0}\} \times \mathcal{D}(\Omega_D)^d$, we obtain the equations

$$-\mathrm{div}(2\nu\,\boldsymbol{\varepsilon}(\mathbf{u}_S) - p_S I) = \mathbf{f}_S, \qquad \kappa^{-1}\mathbf{u}_D + \nabla p_D = \mathbf{0} \qquad (10.44)$$

and as a consequence $p_D \in H^1(\Omega_D)$. Now testing (10.41c) with a general element of $\mathcal{D}(\Omega_S) \times \{0\}$ and $\{0\} \times \mathcal{D}(\Omega_D)$ we get

$$\nabla \cdot \mathbf{u}_S = 0, \qquad \nabla \cdot \mathbf{u}_D = f_D.$$

The condition $\mathbf{u} \in \mathbf{V}$ incorporates the boundary condition for the Stokes velocity on Γ_S, the boundary condition for Darcy on Γ_D, and the first transmission condition. Substituting (10.44) in the first equation of (10.41), we now have

$$(\boldsymbol{\sigma}_S, \boldsymbol{\varepsilon}(\mathbf{v}_S))_{\Omega_S} + (\mathrm{div}\,\boldsymbol{\sigma}_S, \mathbf{v}_S) + \langle c\pi_T \mathbf{u}_S, \gamma \mathbf{v}_S \rangle_\Sigma$$
$$- (\nabla p_D, \mathbf{v}_D)_{\Omega_D} - (p_D, \nabla \cdot \mathbf{v}_D)_{\Omega_D} = 0 \qquad \forall \mathbf{v} \in \mathbf{V}$$

or equivalently

$$\langle \boldsymbol{\sigma}_S \mathbf{n}_S, \gamma \mathbf{v}_S \rangle_{\partial\Omega_S} + \langle c\pi_T \mathbf{u}_S, \gamma \mathbf{v}_S \rangle_\Sigma = -\langle (\gamma p_D)\mathbf{n}_S, \gamma \mathbf{v}_S \rangle_\Sigma \qquad \forall \mathbf{v} \in \mathbf{V}.$$

However, by Proposition 10.13, this implies that

$$\langle \boldsymbol{\sigma}_S \mathbf{n}_S, \gamma \mathbf{v}_S \rangle_{\partial\Omega_S} + \langle c\pi_T \mathbf{u}_S, \gamma \mathbf{v} \rangle_\Sigma = \langle (\gamma p_D)\mathbf{n}_D, \gamma \mathbf{v} \rangle_\Sigma \qquad \forall \mathbf{v} \in \mathbf{H}^1_{\Gamma_S}(\Omega_S),$$

which is the second transmission condition.

Well-posedness. We have already seen that $b(\mathbf{v},1) = 0$ for all $\mathbf{v} \in \mathbf{V} \subset \mathbf{H}(\mathrm{div},\Omega)$ and thus $\mathrm{div} : \mathbf{V} \to L^2_\circ(\Omega)$ is bounded. However, $\mathbf{H}^1_0(\Omega) \subset \mathbf{V}$ and $\mathrm{div} : \mathbf{H}^1_0(\Omega) \to L^2_\circ(\Omega)$ is surjective, which implies that

$$\sup_{0 \neq \mathbf{v} \in \mathbf{V}} \frac{b(\mathbf{v},q)}{\|\mathbf{v}\|_\mathbf{V}} \geq c\|q\|_\Omega \qquad \forall q \in L^2_\circ(\Omega).$$

The map

$$Q \ni q \longmapsto Cq := q - \tfrac{1}{|\Omega|}(q,1)_\Omega \in L^2_\circ(\Omega)$$

is bounded and invertible, its inverse being

$$C^{-1}q := q - \tfrac{1}{|\Omega_D|}(q,1)_{\Omega_D}.$$

Since

$$b(\mathbf{v},q) = b(\mathbf{v},q+c) \qquad \forall c \in \mathbb{R},$$

we have

$$\sup_{0 \neq \mathbf{v} \in \mathbf{V}} \frac{b(\mathbf{v},Cq)}{\|\mathbf{v}\|_{\mathbf{V}}} \geq c\|Cq\|_{\Omega} \geq c'\|q\|_{\Omega} \qquad \forall q \in Q,$$

which is the inf-sup condition. The kernel associated to the bilinear form b is

$$\mathbf{V}_0 = \{\mathbf{u} \in \mathbf{V} : \nabla \cdot \mathbf{u} = 0\} \subset \mathbf{H}^1_{\Gamma_S}(\Omega_S) \times \{\mathbf{u}_D \in \mathbf{H}(\mathrm{div},\Omega_D) : \nabla \cdot \mathbf{u}_D = 0\},$$

and in the latter set

$$\|\varepsilon(\mathbf{v}_S)\|^2_{\Omega_S} + \|\mathbf{v}_D\|^2_{\Omega_D}$$

is equivalent to the $\mathbf{V}$ norm (here we use Korn's second inequality, Proposition 7.13), so the bilinear form a is coercive in $\mathbf{V}_0$. This finishes the verification of the Babuška-Brezzi conditions (Theorem 10.1) and therefore problem (10.43) is well posed.

10.10   Brinkman flow

In this section we study a simple variant of the Stokes problem, which poses no difficulty in itself, but we now take care of the dependence of the solution with respect to the viscosity parameter ν and, in particular, we examine what happens when $\nu \to 0$. This will require using ν-weighted norms, tailored to the problem. We are given $\mathbf{f} \in \mathbf{L}^2(\Omega)$ and we look for a pair, tagged in the kinematic viscosity $\nu > 0$,

$$(\mathbf{u}_\nu, p_\nu) \in \mathbf{H}^1_0(\Omega) \times L^2_\circ(\Omega), \quad -\nu\Delta\mathbf{u}_\nu + \mathbf{u}_\nu + \nabla p_\nu = \mathbf{f}, \quad \nabla \cdot \mathbf{u}_\nu = 0. \quad (10.45)$$

A priori this is even simpler than the Stokes problem, since the diagonal bilinear form in its variational formulation will just be the $\mathbf{H}^1(\Omega)$ inner product, weighted with ν. Our interest is seeing how the solution of this problem is related to a Darcy-like problem (take $\nu = 0$ above) and note that the spaces will have to be changed

$$(\mathbf{u}_0, p_0) \in \mathbf{H}_0(\mathrm{div},\Omega) \times L^2_\circ(\Omega), \quad \mathbf{u}_0 + \nabla p_0 = \mathbf{f}, \quad \nabla \cdot \mathbf{u}_0 = 0. \quad (10.46)$$

This problem is also well posed by the theory of mixed problems. Since we only care about the limit, we will restrict our attention to $\nu \leq 1$ to make some inequalities simpler.

Weighted norms. We consider the norm

$$\|\mathbf{u}\|_\nu^2 := \nu\|\nabla\mathbf{u}\|_\Omega^2 + \|\mathbf{u}\|_\Omega^2,$$

for which we have

$$\nu^{1/2}\|\mathbf{u}\|_{1,\Omega} \le \|\mathbf{u}\|_\nu \le \|\mathbf{u}\|_{1,\Omega} \qquad \forall \mathbf{u} \in \mathbf{H}^1(\Omega).$$

We also consider the following 'peculiar' operational norm in $L_\circ^2(\Omega)$

$$[p]_\nu := \sup_{0\ne\mathbf{u}\in\mathbf{H}_0^1(\Omega)} \frac{(p, \nabla\cdot\mathbf{u})_\Omega}{\|\mathbf{u}\|_\nu}.$$

This norm is equivalent to the L^2 norm in the following way

$$C\|p\|_\Omega \le [p]_\nu \le \nu^{-1/2}\|p\|_\Omega \qquad \forall p \in L_\circ^2(\Omega).$$

On one hand

$$[p]_\nu \ge \sup_{0\ne\mathbf{u}\in\mathbf{H}_0^1(\Omega)} \frac{(p, \nabla\cdot\mathbf{u})_\Omega}{\|\mathbf{u}\|_{1,\Omega}} \ge C\|p\|_\Omega$$

by the surjectivity of the divergence operator, and on the other hand,

$$[p]_\nu \le \sup_{0\ne\mathbf{u}\in\mathbf{H}_0^1(\Omega)} \frac{(p, \nabla\cdot\mathbf{u})_\Omega}{\nu^{1/2}\|\mathbf{u}\|_{1,\Omega}} \le \nu^{-1/2}\|p\|_\Omega.$$

The bounded bilinear forms

$$a_\nu(\mathbf{u}, \mathbf{v}) := \nu(\nabla\mathbf{u}, \nabla\mathbf{v})_\Omega + (\mathbf{u}, \mathbf{v})_\Omega,$$
$$b(\mathbf{u}, q) := (p, \nabla\cdot\mathbf{u})_\Omega,$$

satisfy

$$|a_\nu(\mathbf{u}, \mathbf{v})| \le \|\mathbf{u}\|_\nu\|\mathbf{v}\|_\nu, \qquad a_\nu(\mathbf{u}, \mathbf{u}) = \|\mathbf{u}\|_\nu^2 \qquad \forall \mathbf{u}, \mathbf{v} \in \mathbf{H}^1(\Omega),$$

and

$$|b(\mathbf{u}, q)| \le \|\mathbf{u}\|_\nu[q]_\nu, \qquad \sup_{0\ne\mathbf{u}\in\mathbf{H}_0^1(\Omega)} \frac{b(\mathbf{u}, q)}{\|\mathbf{u}\|_\nu} = [q]_\nu \qquad \forall q \in L_\circ^2(\Omega),$$

that is, all four constants in the Babuška-Brezzi conditions (Theorem 10.1) are one. Note also that

$$|(\mathbf{f}, \mathbf{v})_\Omega| \le \|\mathbf{f}\|_\Omega\|\mathbf{v}\|_\Omega \le \|\mathbf{f}\|_\Omega\|\mathbf{v}\|_\nu \qquad \forall \mathbf{v} \in \mathbf{H}_0^1(\Omega).$$

By Theorem 10.1 the solution of

$$\begin{aligned}
(\mathbf{u}_\nu, p_\nu) &\in \mathbf{H}_0^1(\Omega)\times L_\circ^2(\Omega),\\
a_\nu(\mathbf{u}_\nu, \mathbf{v}) - b(\mathbf{v}, p_\nu) &= (\mathbf{f}, \mathbf{v})_\Omega \qquad \forall \mathbf{v} \in \mathbf{H}_0^1(\Omega),\\
b(\mathbf{u}_\nu, q) &= 0 \qquad\qquad \forall q \in L_\circ^2(\Omega),
\end{aligned}$$

a problem that is clearly equivalent to (10.45), satisfies

$$\|\mathbf{u}_\nu\|_\nu + [p_\nu]_\nu \leq C\|\mathbf{f}\|_\Omega,$$

with a constant C independent of $\nu \in (0,1]$.

The weak limit. Note now that

$$\|\mathbf{u}_\nu\|_{\mathrm{div},\Omega} + \|p_\nu\|_\Omega = \|\mathbf{u}_\nu\|_\Omega + \|p_\nu\|_\Omega \leq C(\|\mathbf{u}_\nu\|_\nu + [p_\nu]_\nu) \leq C'\|\mathbf{f}\|_\Omega.$$

Therefore we can find a sequence $\nu_n \to 0$ such that

$$\mathbf{u}_n := \mathbf{u}_{\nu_n} \rightharpoonup \mathbf{u}_0 \quad \text{in } \mathbf{H}(\mathrm{div}, \Omega), \qquad p_n := p_{\nu_n} \rightharpoonup p_0 \quad \text{in } L^2(\Omega),$$

for some $\mathbf{u}_0 \in \mathbf{H}_0(\mathrm{div}, \Omega)$ and $p_0 \in L_\circ^2(\Omega)$. (The condition $\mathbf{u}_0 \cdot \mathbf{n} = 0$ is due to the fact that $\gamma \mathbf{u}_n = \mathbf{0}$ for all n. Similarly, we prove that $(p_0, 1)_\Omega = 0$.) We also have, for all $\mathbf{v} \in \mathbf{H}_0^1(\Omega)$

$$|\nu_n(\nabla \mathbf{u}_n, \nabla \mathbf{v})_\Omega| \leq \nu_n^{1/2}\|\nabla \mathbf{v}\|_\Omega\|\mathbf{u}_n\|_{\nu_n} \leq C\nu_n^{1/2}\|\nabla \mathbf{v}\|_\Omega\|\mathbf{f}\|_\Omega \longrightarrow 0,$$

and

$$(\mathbf{u}_n, \mathbf{v})_\Omega \longrightarrow (\mathbf{u}_0, \mathbf{v})_\Omega, \qquad (\nabla \cdot \mathbf{v}, p_n)_\Omega \longrightarrow (\nabla \cdot \mathbf{v}, p_0)_\Omega,$$

and therefore

$$(\mathbf{u}_0, \mathbf{v})_\Omega - (p_0, \nabla \cdot \mathbf{v})_\Omega = (\mathbf{f}, \mathbf{v})_\Omega \qquad \forall \mathbf{v} \in \mathbf{H}_0^1(\Omega), \qquad (10.47\mathrm{a})$$

$$(\nabla \cdot \mathbf{u}_0, q)_\Omega \qquad\qquad = 0 \qquad \forall q \in L_\circ^2(\Omega). \qquad (10.47\mathrm{b})$$

Since $\mathcal{D}(\Omega)^d \subset \mathbf{H}_0^1(\Omega) \subset \mathbf{H}_0(\mathrm{div}, \Omega)$, we can extend the testing in (10.47a) to $\mathbf{v} \in \mathbf{H}_0(\mathrm{div}, \Omega)$. This implies that $(\mathbf{u}_0, p_0)$ is the unique solution to (10.46). To prove that we have $(\mathbf{u}_\nu, p_\nu) \rightharpoonup (\mathbf{u}_0, p_0)$ in $\mathbf{H}(\mathrm{div}, \Omega) \times L^2(\Omega)$, we use the general argument given in the next lemma.

Lemma 10.2. *Let H be a Hilbert space and $\|u_\nu\|_H \leq C$ for $\nu > 0$ satisfy: there exists $u_0 \in H$ such that if $\nu_n \to 0$ and $\{u_{\nu_n}\}$ converges weakly in H, then it converges weakly to u_0. In these conditions, $u_\nu \rightharpoonup u_0$ in H as $\nu \to 0$.*

Proof. If the conclusion does not hold, there exists a sequence $\nu_n \to 0$ such that $\{u_{\nu_n}\}$ does not converge weakly to u_0, that is, there exists $v \in H$ such that $(u_{\nu_n}, v)_H$ does not converge to $(u_0, v)_H$. Therefore, there exists $\varepsilon > 0$ and a subsequence (we do not change its name) such that

$$\nu_n \to 0, \qquad |(u_{\nu_n}, v)_H - (u_0, v)_H| \geq \varepsilon \qquad \forall n.$$

Since $\|u_{\nu_n}\|_H \leq C$ we can extract a weakly convergent subsequence and this subsequence has to converge to u_0 by the hypothesis. This contradiction finishes the proof. $\qquad\qquad\square$

10.11 Reissner-Mindlin plates

This section is about a simplified plate model, also depending on a small parameter, which in this case is related to the width of the plate. The physical model works on a two-dimensional domain $\Omega \subset \mathbb{R}^2$, but since the equations are perfectly meaningful in any space dimension, we will present the results for a bounded Lipschitz domain $\Omega \subset \mathbb{R}^d$. We will play the same game as in Brinkman, taking the parameter to zero. The main novelty here will be related to dealing with a fourth order model and with having two divergence operators, one acting on vector fields and the other on symmetric tensors, at the same time. We will keep the same notation as in the rest of the book and write $\nabla \cdot \mathbf{u}$ for the divergence of a vector field and $\operatorname{div} \boldsymbol{\sigma}$ for the divergence of a matrix-valued function (distribution). The matrix-divergence will be applied to the rows, although in this particular problem matrix-valued functions will always be symmetric. The following space

$$\mathbb{H}(\Omega) := \{ \boldsymbol{\sigma} \in L^2(\Omega; \mathbb{R}^{d \times d}_{\mathrm{sym}}) \ : \ \operatorname{div} \boldsymbol{\sigma} \in \mathbf{L}^2(\Omega) \},$$

will play a significant role in what follows. It is clearly a Hilbert space endowed with its natural norm.

A simple form of the plate equations. The unknowns will be a scalar field $u_t : \Omega \to \mathbb{R}$ (vertical displacement of the plate) and a matrix-valued field $M_t : \Omega \to \mathbb{R}^{d \times d}_{\mathrm{sym}}$, satisfying the equations, for given $f \in L^2(\Omega)$,

$$\nabla \cdot \operatorname{div} M_t = f, \qquad M_t - \boldsymbol{\varepsilon}(\nabla u_t + t^2 \operatorname{div} M_t) = 0 \qquad (10.48\mathrm{a})$$

$$\gamma u_t = 0, \qquad \boldsymbol{\gamma}(\nabla u_t + t^2 \operatorname{div} M_t) = \mathbf{0}. \qquad (10.48\mathrm{b})$$

To be precise, in principle we only require the regularity conditions (here we are also including the boundary conditions)

$$u_t \in H^1_0(\Omega), \qquad \nabla u_t + t^2 \operatorname{div} M_t \in \mathbf{H}^1_0(\Omega).$$

However, the equations imply that $M_t \in L^2(\Omega; \mathbb{R}^{d \times d}_{\mathrm{sym}})$ and $\operatorname{div} M_t \in \mathbf{L}^2(\Omega)$, and therefore $M_t \in \mathbb{H}(\Omega)$. Equations (10.48) are equivalent to the following variational formulation

$$M_t \in \mathbb{H}(\Omega), \quad u_t \in H^1_0(\Omega), \qquad (10.49\mathrm{a})$$

$$(M_t, \Theta)_\Omega + t^2 (\operatorname{div} M_t, \operatorname{div} \Theta)_\Omega + (\nabla u_t, \operatorname{div} \Theta)_\Omega = 0 \quad \forall \Theta \in \mathbb{H}(\Omega), \quad (10.49\mathrm{b})$$

$$- (\operatorname{div} M_t, \nabla z)_\Omega = (f, z)_\Omega \qquad \forall z \in H^1_0(\Omega). \quad (10.49\mathrm{c})$$

The work to show the equivalence is related to the boundary condition. We know (see Section 7.7, dealing with the traction conditions for elasticity) that the normal component operator $\mathbb{H}(\Omega) \ni \boldsymbol{\sigma} \longmapsto \boldsymbol{\sigma}\mathbf{n} \in \mathbf{H}^{-1/2}(\Gamma)$, given by

$$\langle \boldsymbol{\sigma}\mathbf{n}, \gamma\mathbf{v} \rangle_\Gamma = (\boldsymbol{\sigma}, \boldsymbol{\varepsilon}(\mathbf{v}))_\Omega + (\operatorname{div} \boldsymbol{\sigma}, \mathbf{v})_\Omega \qquad \forall \mathbf{v} \in \mathbf{H}^1(\Omega)$$

is surjective. Equation (10.49b) implies the differential equation

$$M_t - \nabla(\nabla u_t + t^2 \operatorname{div} M_t) = 0,$$

but, since M_t is symmetric, we also get

$$M_t - \varepsilon(\nabla u_t + t^2 \operatorname{div} M_t) = 0.$$

This proves that $\mathbf{v} := \nabla u_t + t^2 \operatorname{div} M_t \in \mathbf{H}^1(\Omega)$ and, going back to (10.49b), we have

$$\langle \Theta \mathbf{n}, \gamma \mathbf{v} \rangle_\Gamma = (\operatorname{div} \Theta, \mathbf{v})_\Omega + (\varepsilon(\mathbf{v}), \Theta)_\Omega = 0 \qquad \forall \Theta \in \mathbb{H}(\Omega),$$

which implies (this is where we use that the normal component operator is surjective) that $\gamma \mathbf{v} = \mathbf{0}$.

Well-posedness. The analysis of (10.49) can clearly be reduced to the verification of an inf-sup condition, since the diagonal bilinear form is coercive in the entire space $\mathbb{H}(\Omega)$. Nevertheless, we will give all the details for 'correctly weighted' norms, so that we can study the weak limit as $t \to 0$. We consider the following norm in $\mathbb{H}(\Omega)$

$$\|M\|_t^2 := \|M\|_\Omega^2 + t^2 \|\operatorname{div} M\|_\Omega^2,$$

the gradient norm $\|\nabla \cdot \|_\Omega$ in $H_0^1(\Omega)$, and the bounded bilinear forms:

$$a(M, \Theta) := (M, \Theta)_\Omega + t^2 (\operatorname{div} M, \operatorname{div} \Theta)_\Omega,$$
$$b(M, z) := (\operatorname{div} M, \nabla z)_\Omega.$$

The bilinear form a is uniformly bounded and uniformly elliptic (it is the inner product associated to the norm we have defined in $\mathbb{H}(\Omega)$. Note that

$$\operatorname{div} : \mathbb{H}(\Omega) \to \mathbf{L}^2(\Omega), \qquad \operatorname{div} : \mathbf{L}^2(\Omega) \to H^{-1}(\Omega)$$

are surjective operators and therefore so is their composition. We also have that

$$-\langle \nabla \cdot \operatorname{div} M, z \rangle_{H^{-1}(\Omega) \times H_0^1(\Omega)} = b(M, z),$$

and therefore, the surjectivity of $\nabla \cdot \operatorname{div}$ is equivalent to an inf-sup condition

$$\sup_{0 \neq M \in \mathbb{H}(\Omega)} \frac{b(M, z)}{\left(\|M\|_\Omega^2 + \|\operatorname{div} M\|_\Omega^2\right)^{1/2}} \geq c \|\nabla z\|_\Omega \qquad \forall z \in H_0^1(\Omega),$$

which implies that

$$\sup_{0 \neq M \in \mathbb{H}(\Omega)} \frac{b(M, z)}{\|M\|_t} \geq c \|\nabla z\|_\Omega \qquad \forall z \in H_0^1(\Omega), \qquad t \in (0, 1].$$

At the time of applying the bounds of Theorem 10.1 it is important to realize that the norm of the bilinear form b (which behaves like t^{-1} with our scaled norm in $\mathbb{H}(\Omega)$) does not play any role in the upper bound for the norm of the inverse operator, while all other relevant quantities are independent of t. We thus have that (10.49) is uniquely solvable and

$$\|M_t\|_\Omega + t\,\|\mathrm{div}\,M_t\|_\Omega + \|\nabla u_t\|_\Omega \le C\|f\|_\Omega,$$

with C independent of t.

Weak limit. We now sketch the proof of the weak convergence as $t \to 0$

$$M_t \rightharpoonup M_0 \quad \text{in } L^2(\Omega; \mathbb{R}^{d \times d}_{\mathrm{sym}}), \qquad u_t \rightharpoonup u_0 \quad \text{in } H^1_0(\Omega),$$

where u_0 is the only solution of the clamped Kirchhoff plate equations and M_0 is its Hessian matrix:

$$u_0 \in H^2_0(\Omega), \qquad \Delta^2 u_0 = f, \qquad M_0 = D^2 u_0.$$

Since the details in this case are exactly like those of the weak limit of the Brinkman flow solution, we will just verify the key hypotheses. Assume therefore that $t_n \to 0$ and

$$M_n := M_{t_n} \rightharpoonup M_0 \quad \text{in } L^2(\Omega; \mathbb{R}^{d \times d}_{\mathrm{sym}}), \qquad u_n := u_{t_n} \rightharpoonup u_0 \quad \text{in } H^1_0(\Omega).$$

Taking the limits in (10.49) (note that $t_n^2 \mathrm{div}\,M_n \to 0$), we have that the limit satisfies

$$(M_0, \Theta)_\Omega + (\nabla u_0, \mathrm{div}\,\Theta)_\Omega = 0 \qquad \forall \Theta \in \mathbb{H}(\Omega), \qquad (10.50a)$$
$$(\mathrm{div}\,M_0, \nabla z)_\Omega \qquad\qquad = -(f, z)_\Omega \qquad \forall z \in H^1_0(\Omega). \qquad (10.50b)$$

Therefore,

$$M_0 - \varepsilon(\nabla u_0) = 0, \qquad \nabla \cdot \mathrm{div}\,M_0 = f,$$

but since $\varepsilon(\nabla u_0) = D^2 u_0$ and $\nabla \cdot \mathrm{div}\,D^2 u_0 = \Delta^2 u_0$, it follows that

$$\Delta^2 u_0 = f, \qquad M_0 = D^2 u_0,$$

and in particular $u_0 \in H^2(\Omega) \cap H^1_0(\Omega)$. Substituting in (10.50a) we have

$$(D^2 u_0, \Theta)_\Omega + (\nabla u_0, \mathrm{div}\,\Theta)_\Omega = 0 \qquad \forall \Theta \in \mathbb{H}(\Omega),$$

which is equivalent to

$$\langle \Theta \mathbf{n}, \gamma \nabla u_0 \rangle_\Gamma = 0 \qquad \forall \Theta \in \mathbb{H}(\Omega).$$

This implies $\gamma \nabla u_0 = \mathbf{0}$ and therefore $u_0 \in H^2_0(\Omega)$.

Final comments and literature

The section on surjectivity can be understood as a rephrasing of Banach's closed range theorem in terms of bilinear forms. This has gone through many different formats. The pairings of an inf-sup condition with an injectivity condition to express invertibility (the conditions for $a : V_0 \times V_0 \to \mathbb{R}$ in Theorem 10.1, see Exercise 10.1 below) is frequently attributed to Jindrich Nečas (they appear in [85]) and also to Ivo Babuška, as it can be found in the much cited classnotes of Babuška and Abdul Kadir Aziz [8]. The full set of hypotheses applied to mixed problems can be found in Franco Brezzi's work [24]. The story is quite complicated and there is much confusion about what we are referring to on top of the different opinions about how to call the conditions. Alexandre Ern and Jean-Luc Guermond's textbook on finite elements [49] opted for renaming the invertibility conditions as Banach-Nečas-Babuška, but it is doubtful that the name will catch on. To make it more difficult, the inf-sup condition for the divergence operator in $\mathbf{H}_0^1(\Omega)$, needed for the analysis of the Stokes problem (the contents of Section 10.7), is often called the Ladyzhenskaya-Babuška-Brezzi (or LBB for short) condition, adding Olga Ladyzhenskaya's name to the pool, in a complicated honorific attribution, which is often misreferenced to [71]. A short history of the inf-sup condition as it appears in the literature is given in [91].

In Section 10.7 we have sketched a proof of the surjectivity of the divergence operator. Full details (namely the missing proof of Theorem 10.2) can be found in Chapter 2 of the monograph of Gabriel Acosta and Ricardo Durán [1] devoted to the divergence operator and to associated inequalities, in particular, Korn's inequality. The proof is based on the use of Mikhail Bogovskiĭ's operator [15] on a strongly star-shaped or starlike domain, and needs some results on Calderón-Zygmund theory of singular integrals and the Hardy-Littlewood maximal operator. Acosta and Durán provide a different proof working on a more general class of domains (see also [47]). An alternative proof by James Bramble can be found in [21]. The interest in finding the correct constants for these kinds of inequalities has been revived in recent years [13, 36, 40].

The approach we have taken for the Stokes-Darcy problem presented as a simple mixed problem is taken from [77, 78]. We will come back to this problem in Chapter 11, taking a different appproach. The study of Brinkman flow is based on a formulation given in [66].

Exercises

10.1. General theory of well posed variational problems. Let $a : V \times V \to \mathbb{R}$ be a bounded bilinear form.

(a) Show that the problem

$$u \in V,$$
$$a(u, v) = \ell(v) \qquad \forall v \in V,$$

is well posed (it has a unique solution for arbitrary $\ell \in V'$ and the solution is a continuous function of ℓ) if and only if there exists $\alpha > 0$ such that

$$\sup_{0 \neq u \in V} \frac{|a(u, v)|}{\|u\|_V} \geq \alpha \|v\|_V \qquad \forall v \in V,$$

and

$$\sup_{0 \neq v \in V} \frac{|a(u, v)|}{\|v\|_V} \geq \alpha \|u\|_V \qquad \forall u \in V.$$

Show that in this case $\|u\|_V \leq (1/\alpha)\|\ell\|_{V'}$.

(b) Show that if conditions (b) and (c) of the statement of Theorem 10.1 hold (substitute V_0 by V), then the above conditions hold with the same $\alpha > 0$.

10.2. Let $a : V \times V \to \mathbb{R}$ be a bounded symmetric and positive semidefinite bilinear form. Show that the operator $V \ni u \mapsto Au := a(u, \cdot) \in V'$ is invertible if and only if a is coercive. (**Hint.** Use the Cauchy-Schwarz inequality to prove that $\|Au\|_{V'} \leq \|A\|^{1/2} a(u, u)^{1/2}$.)

10.3. Let $b : V \times M \to \mathbb{R}$ be a bounded bilinear form such that

$$\beta := \inf_{0 \neq p \in M} \sup_{0 \neq u \in V} \frac{b(u, p)}{\|p\|_M \|u\|_V} > 0.$$

Show that

$$\sup_{0 \neq u \in V} \frac{b(u, p)}{\|u\|_V} \geq \beta \|p\|_M \qquad \forall p \in M.$$

10.4. Matrix norms. Consider the norms in $\mathbb{R}^k$

$$|\mathbf{z}|_1 := \sum_{i=1}^{k} |z_i|, \qquad |\mathbf{z}|_2 := \left(\sum_{i=1}^{k} |z_i|^2 \right)^{1/2}, \qquad |\mathbf{z}|_\infty := \max_{i=1,\ldots,k} |z_i|,$$

and the associated operator norms for a matrix $B \in \mathbb{R}^{m \times n}$:

$$|B|_p := \sup_{0 \neq \mathbf{z} \in \mathbb{R}^n} \frac{|B\mathbf{z}|_p}{|\mathbf{z}|_p}.$$

(a) Show that

$$|\mathbf{z}|_\infty \leq |\mathbf{z}|_2 \leq |\mathbf{z}|_1 \leq \sqrt{k}|\mathbf{z}|_2 \leq k|\mathbf{z}|_\infty \qquad \forall \mathbf{z} \in \mathbb{R}^k.$$

(b) Show that

$$|B|_2 \leq \sqrt{n}|B|_1, \qquad |B|_2 \leq \sqrt{m}|B|_\infty \qquad \forall B \in \mathbb{R}^{m\times n}.$$

(c) Show that

$$|B|_1 = \max_j \sum_{i=1}^m |B_{ij}|.$$

(**Hint.** One inequality is straightforward. For the second one, take $\mathbf{z}$ to be any of the vectors of the canonical basis.)

(d) Show that $|B|_\infty = |B^\top|_1$. (**Hint.** One inequality requires taking vectors with ± 1 entries matching the signs of the elements B_{ij}.)

(e) Use a spectral decomposition of $B^\top B$ to prove that $|B|_2$ is the square root of the spectral radius of $B^\top B$.

10.5. Consider an operator $\mathbb{B} : \mathbb{V} \to \mathbb{M}$, where

$$\mathbb{V} := V_1 \times \ldots \times V_n, \qquad \mathbb{M} := M_1 \times \ldots \times M_m$$

are product spaces of Hilbert spaces endowed with the product norm. Show that if $B_{ij} : V_j \to M_i$ are the components of $\mathbb{B}$ and N is the $\mathbb{R}^{m\times n}$ matrix with entries $N_{ij} := \|B_{ij}\|_{V_j \to M_i}$, then

$$\|\mathbb{B}\|_{\mathbb{V}\to\mathbb{M}} \leq |N|_2.$$

(**Hint.** Use the vector with entries $x_i := \|u_i\|_{V_i}$, noticing that $\|(u_1, \ldots, u_n)\|_{\mathbb{V}} = |\mathbf{x}|_2$, and the operator definition of the matrix norm of N.)

10.6. **A dual formulation for the Dirichlet problem.** Let $f \in L^2(\Omega)$ and $g \in H^{1/2}(\Gamma)$ be data for a problem with coefficient $\kappa \in L^\infty(\Omega)$, $\kappa \geq \kappa_0 > 0$. Consider the problem looking for $\mathbf{q} \in \mathbf{H}(\mathrm{div}, \Omega)$ such that

$$\nabla \cdot \mathbf{q} = f, \qquad (\kappa^{-1}\mathbf{q}, \mathbf{r})_\Omega = -\langle \mathbf{r} \cdot \mathbf{n}, g \rangle_\Gamma \qquad \forall \mathbf{r} \in \mathbf{H}_0, \qquad (10.51)$$

where $\mathbf{H}_0 = \{\mathbf{r} \in \mathbf{H}(\mathrm{div}, \Omega) : \nabla \cdot \mathbf{r} = 0\}$. Show that this problem is well posed. Show that if u solves

$$u \in H^1(\Omega), \qquad -\nabla \cdot (\kappa \nabla u) = f, \qquad \gamma u = g,$$

then $\mathbf{q} = -\kappa \nabla u$ solves (10.51).

10.7. Darcy flow. Show that the problem

$$(\mathbf{q}, u) \in \mathbf{H}(\mathrm{div}, \Omega) \times L^2_\circ(\Omega),$$
$$\mathbf{q} \cdot \mathbf{n} = h,$$
$$(\kappa^{-1}\mathbf{q}, \mathbf{p})_\Omega - (\nabla \cdot \mathbf{p}, u)_\Omega = 0 \qquad \forall \mathbf{p} \in \mathbf{H}_0(\mathrm{div}, \Omega),$$
$$(\nabla \cdot \mathbf{q}, v)_\Omega = (f, v)_\Omega \qquad \forall v \in L^2_\circ(\Omega),$$

where $L^2_\circ(\Omega) = \{u \in L^2(\Omega) : (u, 1)_\Omega = 0\}$ is well posed for arbitrary $f \in L^2(\Omega)$ and $h \in H^{-1/2}(\Gamma)$ if $\kappa \in L^\infty(\Omega)$ is strongly positive. (Note that the compatibility condition is not needed for well-posedness, but to show equivalence between this problem and the original variational formulation.) In the case $h = 0$, write the equivalent saddle point problem and the associated constrained minimization problem in the variable $\mathbf{q}$.

10.8. Surjectivity of the divergence operator. Show that

$$\mathrm{div} : \mathbf{H}^1(\Omega) \longrightarrow L^2(\Omega)$$

is surjective for any bounded Lipschitz domain. (**Hint.** To find a function whose divergence is constant, work on a larger domain.)

10.9. Hydrodynamic stress and Stokes flow. Consider the problem

$$(\mathbf{u}, p) \in \mathbf{H}^1_{\Gamma_0}(\Omega) \times L^2(\Omega),$$
$$2\nu(\varepsilon(\mathbf{u}), \varepsilon(\mathbf{v}))_\Omega - (p, \nabla \cdot \mathbf{v})_\Omega = (\mathbf{f}, \mathbf{v})_\Omega \qquad \forall \mathbf{v} \in \mathbf{H}^1_{\Gamma_0}(\Omega),$$
$$(\nabla \cdot \mathbf{u}, q)_\Omega = 0 \qquad \forall q \in L^2(\Omega),$$

where $\Gamma_0 \subset \Gamma$ is such that

$$\mathbf{H}^1_{\Gamma_0}(\Omega) := \{\mathbf{u} \in \mathbf{H}^1(\Omega) : \gamma\mathbf{u} = 0 \quad \text{on } \Gamma_0\}$$

differs from $\mathbf{H}^1(\Omega)$ and $\mathbf{H}^1_0(\Omega)$. Show that the variational problem is well posed and write an equivalent boundary value problem in terms of the hydrodynamic stress $\boldsymbol{\sigma} := 2\nu\varepsilon(\mathbf{u}) - p\mathrm{I}$.

10.10. Oseen flow. Give a variational formulation and prove its well-posedness for the problem pf looking for $\mathbf{u} \in \mathbf{H}^1_0(\Omega)$ and $p \in L^2_\circ(\Omega)$ such that
$$-\nu\Delta\mathbf{u} + (\nabla\mathbf{u})\mathbf{b} + \nabla p = \mathbf{f}, \qquad \nabla \cdot \mathbf{u} = 0.$$
Here $\mathbf{f} \in \mathbf{L}^2(\Omega)$ and $\mathbf{b} \in \mathbf{L}^\infty(\Omega)$ satisfies $\nabla \cdot \mathbf{b} = 0$. (Note that it is customary to write $\mathbf{b} \cdot \nabla\mathbf{u}$ when presenting the Oseen equations. This represents the term $\mathbf{b} \cdot \nabla u_i$ in the i-th equation of the system.)

10.11. A singular perturbation of the Stokes problem. Consider the problem looking for $\mathbf{u}_\varepsilon \in \mathbf{H}^1_0(\Omega)$ and $p_\varepsilon \in H^1_\star(\Omega) = H^1(\Omega) \cap L^2_\circ(\Omega)$ such that
$$-\Delta\mathbf{u}_\varepsilon + \nabla p_\varepsilon = \mathbf{f}, \qquad \nabla \cdot \mathbf{u}_\varepsilon - \varepsilon\Delta p_\varepsilon = 0, \qquad \partial_n p_\varepsilon = 0.$$

Show that these equations are uniquely solvable and that

$$\|\nabla u_\varepsilon\|_\Omega + \varepsilon^{1/2}\|\nabla p_\varepsilon\|_\Omega + \|p_\varepsilon\|_\Omega \leq C\|\mathbf{f}\|_\Omega.$$

Use this to prove weak convergence of $(\mathbf{u}_\varepsilon, p_\varepsilon)$ to the solution of the Stokes equations. (**Hint.** Use coercivity and an inf-sup condition to control $\|p_\varepsilon\|_\Omega$.)

10.12. Using the geometric configuration of Section 10.9, prove that the restriction map $\mathbf{H}_0^1(\Omega) \to \mathbf{H}_{\Gamma_S}^1(\Omega_S)$ is surjective.(**Hint.** There is no need to deal with vector fields. Show first that the space of $H^1(\Omega_S)$ functions vanishing in a neighborhood of Γ_S is dense in $H_{\Gamma_S}^1(\Omega_S)$ and then build an extension operator for that space.)

10.13. **The clamped Kirchhoff plate problem.** In this exercise we will use the bracket $[\cdot, \cdot]$ to denote the $H^{-1}(\Omega) \times H_0^1(\Omega)$ and the $\mathbf{H}^{-1}(\Omega) \times \mathbf{H}_0^1(\Omega)$ duality products.

(a) Show that $\mathbf{H}^{-1}(\mathrm{div}, \Omega) := \{\mathbf{p} \in \mathbf{H}^{-1}(\Omega) : \nabla \cdot \mathbf{p} \in H^{-1}(\Omega)\}$, endowed with the norm

$$\|\mathbf{p}\|_{-1,\mathrm{div},\Omega}^2 := \|\mathbf{p}\|_{-1,\Omega}^2 + \|\nabla \cdot \mathbf{p}\|_{-1,\Omega}^2,$$

is a Hilbert space containing $\mathcal{D}(\Omega)^d$.

(b) Given $f \in H^{-1}(\Omega)$, we look for

$$w \in H_0^1(\Omega), \quad \boldsymbol{\theta} \in \mathbf{H}_0^1(\Omega), \quad \boldsymbol{\gamma} \in \mathbf{H}^{-1}(\mathrm{div}, \Omega) \qquad (10.52\mathrm{a})$$

such that

$$[\nabla \cdot \boldsymbol{\gamma}, v] = [f, v] \qquad \forall v \in H_0^1(\Omega), \qquad (10.52\mathrm{b})$$
$$(\boldsymbol{\varepsilon}(\boldsymbol{\theta}), \boldsymbol{\varepsilon}(\boldsymbol{\psi}))_\Omega + [\boldsymbol{\gamma}, \boldsymbol{\psi}] = 0 \qquad \forall \boldsymbol{\psi} \in \mathbf{H}_0^1(\Omega), \qquad (10.52\mathrm{c})$$
$$[\nabla \cdot \boldsymbol{\eta}, w] + [\boldsymbol{\eta}, \boldsymbol{\theta}] \qquad = 0 \qquad \forall \boldsymbol{\eta} \in \mathbf{H}^{-1}(\mathrm{div}, \Omega). \qquad (10.52\mathrm{d})$$

Prove an inf-sup condition for the bilinear form

$$b((w, \boldsymbol{\theta}), \boldsymbol{\eta}) := [\nabla \cdot \boldsymbol{\eta}, w] + [\boldsymbol{\eta}, \boldsymbol{\theta}].$$

(**Hint.** Decompose it as the sum of two.) Prove that the associated kernel is a subspace of

$$V := \{(w, \boldsymbol{\theta}) \in H_0^1(\Omega) \times \mathbf{H}_0^1(\Omega) : \boldsymbol{\theta} = \nabla w\}.$$

(c) Show well-posedness by checking that the diagonal bilinear form is coercive in V (**Hint.** Use the first Korn inequality.)

(d) Show that if $(w, \boldsymbol{\theta}, \boldsymbol{\gamma})$ solves (10.52), then

$$w \in H_0^2(\Omega), \qquad \Delta^2 w = f, \qquad \boldsymbol{\theta} = \nabla w, \qquad \boldsymbol{\gamma} = \mathrm{div}\,\boldsymbol{\varepsilon}(\boldsymbol{\theta}).$$

(e) Assuming that you have proved that $\mathcal{D}(\Omega)^d$ is dense in $\mathbf{H}^{-1}(\mathrm{div}, \Omega)$, prove that the solution of (d) solves the variational problem (10.52).

11

Advanced mixed problems

11.1   Mixed form of reaction-diffusion problems 253
11.2   More indefinite problems 255
11.3   Mixed form of convection-diffusion problems 259
11.4   Double restrictions 264
11.5   A partially uncoupled Stokes-Darcy formulation 266
11.6   Galerkin methods for mixed problems 273
Final comments and literature 275
Exercises 275

This chapter deals with generalized forms of mixed problems, which can be expressed in terms of operators in the following forms:

$$\begin{bmatrix} A & B^* \\ B & -D \end{bmatrix}, \qquad \begin{bmatrix} A & C^* \\ B & 0 \end{bmatrix}, \qquad \begin{bmatrix} A & B_1^* & B_2^* \\ B^1 & 0 & 0 \\ B^2 & 0 & 0 \end{bmatrix}.$$

The first two appear naturally when considering mixed formulations of reaction-diffusion and convection-diffusion problems. The third structure can be considered as a simple mixed problem where the side restriction takes values in a product space. As the cherry on top of this chapter we will include a partially uncoupled formulation of the Stokes-Darcy flow problem, which we will approach using two reorderings of the associated matrix of operators.

11.1   Mixed form of reaction-diffusion problems

First order reaction-diffusion problem. In this section we will explore a first order formulation for the reaction-diffusion equation

$$u \in H^1(\Omega), \qquad -\nabla \cdot (\kappa \nabla u) + c u = f, \qquad \gamma u = g,$$

with the usual requirements on the coefficients ($\kappa, c \in L^\infty(\Omega)$, $\kappa \geq \kappa_0 > 0$ and $c \geq 0$) and data ($f \in L^2(\Omega), g \in H^{1/2}(\Gamma)$). If we introduce the variable

$\mathbf{q} := -\kappa \nabla u$, we can write the equivalent problem

$$(\mathbf{q}, u) \in \mathbf{H}(\text{div}, \Omega) \times L^2(\Omega), \tag{11.1a}$$

$$\kappa^{-1}\mathbf{q} + \nabla u = 0, \tag{11.1b}$$

$$\nabla \cdot \mathbf{q} + cu = f, \tag{11.1c}$$

$$\gamma u = g. \tag{11.1d}$$

Note that, while in the first line we have only demanded $u \in L^2(\Omega)$, the 'state equation' (11.1b) implies that $u \in H^1(\Omega)$ and we can therefore impose a trace condition on u. As we have already mentioned in the last chapter, this moves the Dirichlet condition to a natural condition (it will be included in the variational formulation) that needs an equation to be satisfied before it can be imposed, similarly to what happens to the Neumann condition for second order problems.

A new mixed structure. It is a simple exercise (do it) to prove that problem (11.1) is equivalent to the following variational formulation

$$(\mathbf{q}, u) \in \mathbf{H}(\text{div}, \Omega) \times L^2(\Omega), \tag{11.2a}$$

$$(\kappa^{-1}\mathbf{q}, \mathbf{r})_\Omega - (\nabla \cdot \mathbf{r}, u)_\Omega = -\langle \mathbf{r} \cdot \mathbf{n}, g \rangle_\Gamma \quad \forall \mathbf{r} \in \mathbf{H}(\text{div}, \Omega), \tag{11.2b}$$

$$(\nabla \cdot \mathbf{q}, v)_\Omega + (cu, v)_\Omega = (f, v)_\Omega \quad \forall v \in L^2(\Omega). \tag{11.2c}$$

A simple change of sign makes problem (11.2) symmetric:

$$(\mathbf{q}, u) \in \mathbf{H}(\text{div}, \Omega) \times L^2(\Omega), \tag{11.3a}$$

$$(\kappa^{-1}\mathbf{q}, \mathbf{r})_\Omega - (\nabla \cdot \mathbf{r}, u)_\Omega = -\langle \mathbf{r} \cdot \mathbf{n}, g \rangle_\Gamma \quad \forall \mathbf{r} \in \mathbf{H}(\text{div}, \Omega), \tag{11.3b}$$

$$-(\nabla \cdot \mathbf{q}, v)_\Omega - (cu, v)_\Omega = -(f, v)_\Omega \quad \forall v \in L^2(\Omega). \tag{11.3c}$$

This problems fits in the following general framework: we have two Hilbert spaces V and M, three bounded bilinear forms

$$a : V \times V \to \mathbb{R}, \quad b : V \times M \to \mathbb{R}, \quad d : M \times M \to \mathbb{R},$$

right-hand sides $\ell \in V'$, $\chi \in M'$, and a general structure

$$(u, p) \in V \times M, \tag{11.4a}$$

$$a(u, v) + b(v, p) = \ell(v) \quad \forall v \in V, \tag{11.4b}$$

$$b(u, q) - d(p, q) = \chi(q) \quad \forall q \in M. \tag{11.4c}$$

Note the change of notation to mimic the one for the abstract problems in Chapter 10, so that in this abstract problem (11.4) the variables are $(u, p) \in V \times M$, while in the particular example (11.3) they are $(\mathbf{q}, u) \in \mathbf{H}(\text{div}, \Omega) \times L^2(\Omega)$. We will spend Section 11.2 proving the well-posedness of (11.4) subject to the following hypotheses on the bilinear forms:

(a) a and d are symmetric and positive semidefinite;

(b) b satisfies an inf-sup condition

$$\sup_{0 \neq u \in V} \frac{b(u,p)}{\|u\|_V} \geq \beta \|p\|_M \qquad \forall p \in M,$$

that is, the associated operator $B : V \to M$ is surjective;

(c) a is coercive in the kernel of B, that is,

$$a(u,u) \geq \alpha \|u\|_V^2 \qquad \forall u \in V, \qquad b(u,p) = 0 \qquad \forall p \in M.$$

The proof that the bilinear forms of problem (11.3) satisfy conditions (a)-(c) is straightforward. Just recall that the inf-sup condition (b) is equivalent to the surjectivity of div : $\mathbf{H}(\mathrm{div}, \Omega) \to L^2(\Omega)$ and in $V_0 = \ker B = \{\mathbf{p} \in \mathbf{H}(\mathrm{div}, \Omega) : \nabla \cdot \mathbf{p} = 0\}$, the $\mathbf{L}^2(\Omega)$ norm is equivalent to the $\mathbf{H}(\mathrm{div}, \Omega)$ norm, which proves the coercivity estimate (c). Well-posedness includes a bound of the form

$$\|\mathbf{p}\|_{\mathrm{div},\Omega} + \|u\|_\Omega \leq C(\|g\|_{1/2,\Gamma} + \|f\|_\Omega).$$

11.2 More indefinite problems

For the analysis of (11.4) we will rewrite the problem as an operator equation in $V \times M$. The process of the proof is, however, far from intuitive and requires some fine estimates on the inverses of matrices of operators. In passing, we will offer an estimate on the inverse of the norm associated to the operator in (11.4). While this will not be relevant for our study of well-posedness, such estimates are important in the study of Galerkin schemes. We will write four lemmas, slowly building to the main result of this section (Proposition 11.1 below).

Lemma 11.1. *Let X be a Hilbert space and $G, F : X \to X$ be self-adjoint positive semidefinite operators. If F is uniformly positive definite*

$$(Fx, x)_X \geq \beta \|x\|_X^2 \qquad \forall x \in X,$$

then $I + GF$ is invertible and

$$\|(I + GF)^{-1}\| \leq (\|F\|/\beta)^{1/2}. \tag{11.5}$$

Proof. Note that F is invertible by the Riesz-Fréchet representation theorem, since

$$(x, y)_F := (Fx, y)_X$$

defines an equivalent inner product in X for whose associated norm we have the bounds

$$\beta^{1/2}\|u\|_X \le \|u\|_F \le \|F\|^{1/2}\|u\|_X \qquad \forall x \in X.$$

We then decompose

$$I + GF = F^{-1}(F + FGF).$$

This shows invertibility, since FGF is a self-adjoint positive semidefinite operator and therefore $F + FGF$ is uniformly positive definite. However, we will need the precise estimate (11.5) in Lemma 11.2. Writing

$$((I + GF)u, u)_F = (Fu, u)_X + (GFu, Fu)_X \ge \|u\|_F^2$$

(recall that G is positive semidefinite), we obtain the estimate

$$\|u\|_F \le \|(I + GF)u\|_F,$$

and therefore,

$$\beta^{1/2}\|u\|_X \le \|u\|_F \le \|(I + GF)u\|_F \le \|F\|^{1/2}\|(I + GF)u\|_X.$$

This finishes the proof.   □

Lemma 11.2. *Let X be a Hilbert space and $E, G : X \to X$ be bounded self-adjoint and positive semidefinite operators. The operator $I + GE$ is invertible and*

$$\|(I + GE)^{-1}\| \le 2 + 4\|E\|\,\|G\|.$$

Proof. Given any $\alpha > 0$, we can write $I + GE + \alpha G = I + G(\alpha I + E)$, where now $F := \alpha I + E$ is uniformly positive definite with

$$(Fx, x)_X \ge \alpha\|x\|_X^2, \qquad \|F\| \le \alpha + \|E\|.$$

By Lemma 11.1, the operator $I + GE + \alpha G$ is invertible for any $\alpha > 0$ and

$$\|(I + GE + \alpha G)^{-1}\| \le \left(\frac{\alpha + \|E\|}{\alpha}\right)^{1/2}. \tag{11.6}$$

We then factor

$$\begin{aligned}
I + GE &= (I + GE + \alpha G) - \alpha G \\
&= (I + GE + \alpha G)\left(I - \alpha(I + GE + \alpha G)^{-1}G\right), \tag{11.7}
\end{aligned}$$

which shows that we just need to prove that a smart choice of α makes the operator $\alpha(I + GE + \alpha G)^{-1}G$ have a norm less than one. To do that, note that

$$\|\alpha(I + GE + \alpha G)^{-1}G\| \le \alpha\left(\frac{\alpha + \|E\|}{\alpha}\right)^{1/2}\|G\| = \left(\alpha\|G\|^2(\alpha + \|E\|)\right)^{1/2}.$$

The choice

$$\alpha_0 := \frac{1}{2}\left(\left(\|E\|^2 + \|G\|^{-2}\right)^{1/2} - \|E\|\right)$$

makes

$$\|\alpha_0(I + GE + \alpha G)^{-1}G\|^2 \leq \alpha_0\|G\|^2(\alpha_0 + \|E\|) = \frac{1}{4},$$

and therefore

$$\|(I - \alpha_0(I + GE + \alpha G)^{-1}G)^{-1}\| \leq 2.$$

The decomposition (11.7) and the bound (11.6) yield then

$$\|(I + GE)^{-1}\| \leq 2\sqrt{\frac{\alpha_0 + \|E\|}{\alpha_0}} = \frac{1}{\alpha_0\|G\|} = \frac{2}{\sqrt{1 + \|G\|^2\|E\|^2} - \|G\|\,\|E\|}$$

$$= 2\left(\sqrt{1 + \|G\|^2\|E\|^2} + \|G\|\,\|E\|\right) \leq 2 + 4\|G\|\,\|E\|.$$

This finishes the proof. □

Lemma 11.3. *In the hypotheses of Lemma 11.2, the operator*

$$\begin{bmatrix} G & I \\ I & -E \end{bmatrix} : X \times X \to X \times X$$

is invertible and the norm of its inverse is bounded by

$$2\sqrt{2}(1 + \|G\|)(1 + \|E\|)(1 + 2\|G\|\,\|E\|).$$

Proof. If we factor

$$\begin{bmatrix} G & I \\ I & -E \end{bmatrix} = \begin{bmatrix} G & I \\ I & 0 \end{bmatrix}\begin{bmatrix} I & -E \\ 0 & I + GE \end{bmatrix},$$

a simple application of Lemma 11.2 proves that the inverse can be computed with the formula

$$\begin{bmatrix} G & I \\ I & -E \end{bmatrix}^{-1} = \begin{bmatrix} I & E(I+GE)^{-1} \\ 0 & (I+GE)^{-1} \end{bmatrix}\begin{bmatrix} 0 & I \\ I & -G \end{bmatrix}.$$

The bound for the inverse is based on this factorization (and on Lemma 11.2) and left as an exercise. □

Lemma 11.4. *Let H and M be Hilbert spaces,*

$$G : H \to H, \qquad B : H \to M, \qquad D : M \to M$$

be bounded operators with B invertible and G, D self-adjoint and positive semidefinite. The operator

$$\begin{bmatrix} G & B^* \\ B & -D \end{bmatrix} : H \times M \to H \times M$$

is invertible.

Proof. The result is a simple consequence of the factorization

$$\begin{bmatrix} G & B^* \\ B & -D \end{bmatrix} = \begin{bmatrix} I & 0 \\ 0 & B \end{bmatrix} \begin{bmatrix} G & I \\ I & -B^{-1}D(B^{-1})^* \end{bmatrix} \begin{bmatrix} I & 0 \\ 0 & B^* \end{bmatrix}$$

and Lemma 11.3. □

Proposition 11.1. *Let V and M be Hilbert spaces, and*

$$a : V \times V \to \mathbb{R}, \qquad b : V \times M \to \mathbb{R}, \qquad d : M \times M \to \mathbb{R}$$

be bounded bilinear forms satisfying:

(a) *a and d are symmetric and positive semidefinite;*

(b) *there exists $\beta > 0$ such that*

$$\sup_{0 \neq u \in V} \frac{b(u,p)}{\|u\|_V} \geq \beta \|p\|_M \qquad \forall p \in M;$$

(c) *there exists $\alpha > 0$ such that*

$$a(u,u) \geq \alpha \|u\|_V^2 \qquad \forall u \in V_0 := \{v \in V : b(v, \cdot) = 0\}.$$

For arbitrary $\ell \in V'$ and $\chi \in M'$, the variational problem

$$(u,p) \in V \times M, \tag{11.8a}$$
$$a(u,v) + b(v,p) = \ell(v) \qquad \forall v \in V, \tag{11.8b}$$
$$b(u,q) - d(p,q) = \chi(q) \qquad \forall q \in M, \tag{11.8c}$$

has a unique solution and there exists $C > 0$ such that

$$\|u\|_V + \|p\|_M \leq C(\|\ell\|_{V'} + \|\chi\|_{M'}).$$

Proof. The proof starts in very much the same spirit as the proof of Theorem 10.1. We consider $V_1 := V_0^\perp$ and identify the orthogonal decomposition with the product space $V = V_0 \oplus V_1 \equiv V_0 \times V_1$. We then assign operators

$$A_{ij} : V_j \to V_i, \qquad B_1 : V_1 \to M, \qquad D : M \to M$$

to the bilinear forms, noting that: B_1 is invertible, D is self-adjoint and positive semidefinite, A_{00} is self-adjoint and uniformly positive definite, A_{11} is self-adjoint and positive semidefinite, and $A_{01}^* = A_{10}$. Moreover,

$$\begin{bmatrix} A_{00} & A_{01} \\ A_{10} & A_{11} \end{bmatrix}$$

is self-adjoint and positive semidefinite, and therefore

$$G := A_{11} - A_{10}A_{00}^{-1}A_{01}$$

is self-adjoint and positive semidefinite. The well-posedness of (11.8) is equivalent to the invertibility of

$$
\begin{bmatrix}
A_{00} & A_{01} & 0 \\
A_{10} & A_{11} & B_1^* \\
0 & B_1 & -D
\end{bmatrix},
$$

and this matrix of operators can be factored as

$$
\begin{bmatrix}
I & 0 & 0 \\
A_{10}A_{00}^{-1} & I & 0 \\
0 & 0 & I
\end{bmatrix}
\begin{bmatrix}
A_{00} & 0 & 0 \\
0 & G & B_1^* \\
0 & B_1 & -D
\end{bmatrix}
\begin{bmatrix}
I & A_{00}^{-1}A_{01} & 0 \\
0 & I & 0 \\
0 & 0 & I
\end{bmatrix}.
$$

The central operator of this factorization is invertible thanks to Lemma 11.4, which finishes the proof. □

11.3 Mixed form of convection-diffusion problems

A first order formulation of convection-diffusion problems. In this section we will study a first order formulation of the problem

$$
u \in H^1(\Omega), \qquad -\nabla \cdot (\nabla u + u\boldsymbol{\beta}) = f, \qquad \gamma u = g \tag{11.9}
$$

where $\boldsymbol{\beta} \in L^\infty(\Omega; \mathbb{R}^d)$ is such that

$$
(\boldsymbol{\beta} \cdot \nabla u, u)_\Omega \geq 0 \qquad \forall u \in H_0^1(\Omega), \tag{11.10}
$$

$f \in L^2(\Omega)$ and $g \in H^{1/2}(\Gamma)$. Recall that (11.10) is implied by the hypothesis $\nabla \cdot \boldsymbol{\beta} \leq 0$ in the sense of distributions (see Chapter 5). Problem (11.9) is a convection-diffusion problem written in divergence form. Note that the product rule

$$
\nabla \cdot (u\boldsymbol{\beta}) = \boldsymbol{\beta} \cdot \nabla u + u\,(\nabla \cdot \boldsymbol{\beta})
$$

is not applicable unless we require some more regularity for $\boldsymbol{\beta}$, which means that, even if $\nabla \cdot \boldsymbol{\beta} = 0$, this problem is not necessarily equivalent to the one studied in Chapter 5. The case of variable diffusion is left as Exercise 11.2. We now introduce a new variable

$$
\mathbf{q} := -(\nabla u + u\boldsymbol{\beta}),
$$

and rewrite (11.9) in the equivalent first order form

$$
(\mathbf{q}, u) \in \mathbf{H}(\mathrm{div}, \Omega) \times L^2(\Omega), \qquad \mathbf{q} + \nabla u + u\boldsymbol{\beta} = \mathbf{0}, \tag{11.11a}
$$
$$
\nabla \cdot \mathbf{q} = f, \tag{11.11b}
$$
$$
\gamma u = g. \tag{11.11c}
$$

Variational formulation. Problem (11.11) is equivalent to the following nonsymmetric mixed variational formulation

$$(\mathbf{q}, u) \in \mathbf{H}(\mathrm{div}, \Omega) \times L^2(\Omega), \tag{11.12a}$$

$$(\mathbf{q}, \mathbf{r})_\Omega - (\nabla \cdot \mathbf{r} - \boldsymbol{\beta} \cdot \mathbf{r}, u)_\Omega = -\langle \mathbf{r} \cdot \mathbf{n}, g \rangle_\Gamma \quad \forall \mathbf{r} \in \mathbf{H}(\mathrm{div}, \Omega), \tag{11.12b}$$

$$(\nabla \cdot \mathbf{q}, v)_\Omega \qquad\qquad = (f, v)_\Omega \quad \forall v \in L^2(\Omega). \tag{11.12c}$$

We will devote the rest of this section to proving that this problem fits in the framework of the following generalization of Theorem 10.1.

Proposition 11.2. *Let V and M be Hilbert spaces,*

$$a : V \times V \to \mathbb{R}, \qquad b : V \times M \to \mathbb{R}, \qquad c : V \times M \to \mathbb{R}$$

be bilinear forms, and

$$V_b := \{u \in V : b(u, \cdot) = 0\}, \qquad V_c := \{u \in V : c(u, \cdot) = 0\},$$

be the respective kernels. The variational problem

$$(u, p) \in V \times M, \tag{11.13a}$$

$$a(u, v) + c(v, p) = \ell(v) \quad \forall v \in V, \tag{11.13b}$$

$$b(u, q) \qquad\quad = \chi(q) \quad \forall q \in M, \tag{11.13c}$$

is well posed (it has a unique solution for arbitrary right-hand sides $\ell \in V'$ and $\chi \in M'$, with the solution bounded in terms of the data) if and only if the following conditions hold:

(a) *b and c satisfy inf-sup conditions, that is, there exist positive quantities β and γ such that*

$$\sup_{0 \neq u \in V} \frac{b(u, p)}{\|u\|_V} \geq \beta \|p\|_M \quad \forall p \in M, \tag{11.14a}$$

$$\sup_{0 \neq u \in V} \frac{c(u, p)}{\|u\|_V} \geq \gamma \|p\|_M \quad \forall p \in M, \tag{11.14b}$$

(b) *the bilinear form $a : V_b \times V_c \to \mathbb{R}$ defines an invertible operator.*

Proof. We will only sketch the proof. The reader is requested to fill in the details in Exercise 11.1. Note that the surjectivity of the operator $B : V \to M$ associated to the bilinear form b (equivalently, condition (11.14a)) is necessary for well-posedness. Similarly, transposing the problem, it is clear that condition (11.14b) is also necessary. The proof is then based on rewriting problem (11.13) as an operator equation associated to the operator

$$\begin{bmatrix} A_{00} & A_{01} & 0 \\ A_{10} & A_{11} & C_1^* \\ 0 & B_1 & 0 \end{bmatrix} : V_b \times V_b^\perp \times M \to V_c \times V_c^\perp \times M,$$

where $B_1 : V_b^\perp \to M$ and $C_1 : V_c^\perp \to M$ are invertible. $\qquad\square$

A remark. Using the results of Section 10.1, we can equivalently write condition (b) with the two conditions

(c) There exists $\alpha > 0$ such that

$$\sup_{0 \neq u \in V_b} \frac{|a(u,v)|}{\|u\|_V} \geq \alpha \|v\|_V \qquad \forall v \in V_c.$$

(d) For all $u \in V_b$,

$$a(u,v) = 0 \qquad \forall v \in V_c \qquad \Longrightarrow \qquad u = 0.$$

The roles of the spaces V_b and V_c can be reversed.

Preparing the way. As we will see, the verification of conditions (a) (inf-sup conditions for the bilinear forms b and c) is going to be relatively simple. The invertibility condition associated to the bilinear form $a : V_b \times V_c \to \mathbb{R}$ will be somewhat more complicated, due to the fact that we will be dealing with two slightly similar spaces under an $\mathbf{L}^2(\Omega)$ inner product. We start with some preparatory work: we consider a general vector field $\boldsymbol{\alpha} \in L^\infty(\Omega; \mathbb{R}^d)$ with the property $\nabla \cdot \boldsymbol{\alpha} \leq 0$. We will use the results that we will derive next for $\boldsymbol{\alpha} = \boldsymbol{\beta}$ and for $\boldsymbol{\alpha} = \mathbf{0}$.

Proposition 11.3. *Let* $\boldsymbol{\alpha} \in L^\infty(\Omega; \mathbb{R}^d)$ *satisfy*

$$(\boldsymbol{\alpha} \cdot \nabla u, u)_\Omega \geq 0 \qquad \forall u \in H_0^1(\Omega),$$

and let

$$V_\alpha := \{\mathbf{p} \in \mathbf{H}(\mathrm{div}, \Omega) : \nabla \cdot \mathbf{p} = \boldsymbol{\alpha} \cdot \mathbf{p}\}.$$

The following properties hold:

(1) *There exists* $c > 0$ *such that*

$$\sup_{0 \neq \mathbf{p} \in \mathbf{H}(\mathrm{div}, \Omega)} \frac{(\nabla \cdot \mathbf{p} - \boldsymbol{\alpha} \cdot \mathbf{p}, u)_\Omega}{\|\mathbf{p}\|_{\mathrm{div}, \Omega}} \geq c \|u\|_\Omega \qquad \forall u \in L^2(\Omega).$$

(2) *In* V_α, *the* $\mathbf{H}(\mathrm{div}, \Omega)$ *norm is equivalent to the* $\mathbf{L}^2(\Omega)$ *norm.*

(3) *If we consider the differential operator* $\nabla_\alpha u := \nabla u + u\boldsymbol{\alpha}$, *then*

$$\{\mathbf{q} \in \mathbf{H}(\mathrm{div}, \Omega) : (\mathbf{q}, \mathbf{p})_\Omega = 0 \quad \forall \mathbf{p} \in V_\alpha\} = \nabla_\alpha H_0^1(\Omega).$$

Proof. Due to Proposition 10.1, Property (1) is equivalent to the surjectivity of the operator

$$\mathbf{H}(\mathrm{div}, \Omega) \ni \mathbf{p} \longmapsto \mathrm{div}_\alpha \mathbf{p} := \nabla \cdot \mathbf{p} - \boldsymbol{\alpha} \cdot \mathbf{p} \in L^2(\Omega).$$

To prove this surjectivity, given $f \in L^2(\Omega)$, solve

$$u \in H_0^1(\Omega), \qquad \text{div}_\alpha \nabla u = f,$$

(note that $\text{div}_\alpha \nabla u = \Delta u - \boldsymbol{\alpha} \cdot \nabla u$) or equivalently, the coercive problem

$$u \in H_0^1(\Omega), \qquad (\nabla u, \nabla v)_\Omega + (\boldsymbol{\alpha} \cdot \nabla u, v)_\Omega = -(f, v)_\Omega \qquad \forall v \in H_0^1(\Omega).$$

The vector field $\mathbf{p} := \nabla u \in \mathbf{L}^2(\Omega)$ then satisfies $\nabla \cdot \mathbf{p} = f + \boldsymbol{\alpha} \cdot \nabla u \in L^2(\Omega)$ and $\text{div}_\alpha \mathbf{p} = f$. This proves (1).

In V_α we have

$$\|\mathbf{p}\|_{\text{div},\Omega}^2 = \|\mathbf{p}\|_\Omega^2 + \|\nabla \cdot \mathbf{p}\|_\Omega^2 = \|\mathbf{p}\|_\Omega^2 + \|\boldsymbol{\alpha} \cdot \mathbf{p}\|_\Omega^2 \leq c_\alpha \|\mathbf{p}\|_\Omega^2,$$

which proves the equivalence of norms in (2).

The proof of (3) needs some intermediate steps. First, we prove that

$$\mathbf{L}^2(\Omega) \ni \mathbf{p} \longmapsto \text{div}_\alpha \mathbf{p} := \nabla \cdot \mathbf{p} - \boldsymbol{\alpha} \cdot \mathbf{p} \in H^{-1}(\Omega)$$

is surjective. To do that, given $f \in H^{-1}(\Omega)$ we find the solution of

$$u \in H_0^1(\Omega),$$
$$(\nabla u, \nabla v)_\Omega + (\boldsymbol{\alpha} \cdot \nabla u, v)_\Omega = -\langle f, v \rangle_{H^{-1}(\Omega) \times H_0^1(\Omega)} \qquad \forall v \in H_0^1(\Omega),$$

and then define $\mathbf{p} := \nabla u \in \mathbf{L}^2(\Omega)$ and check that $\text{div}_\alpha \mathbf{p} = f$. Note that the bilinear form associated to the surjective operator $\text{div}_\alpha : \mathbf{L}^2(\Omega) \to H^{-1}(\Omega)$ is

$$\langle \nabla \cdot \mathbf{p} - \boldsymbol{\alpha} \cdot \mathbf{p}, u \rangle_{H^{-1}(\Omega) \times H_0^1(\Omega)} = -(\mathbf{p}, \nabla u + u\boldsymbol{\alpha})_\Omega = -(\mathbf{p}, \nabla_\alpha u)_\Omega,$$

and therefore $\nabla_\alpha : H_0^1(\Omega) \to \mathbf{L}^2(\Omega)$ is injective and has closed range (Proposition 10.3). In other words, we have a generalized Poincaré-Friedrichs inequality

$$\|\nabla_\alpha u\|_\Omega = \|\nabla u + u\boldsymbol{\alpha}\|_\Omega \geq c\|u\|_{1,\Omega} \qquad \forall u \in H_0^1(\Omega).$$

Finally, given $\mathbf{q} \in \mathbf{L}^2(\Omega)$ we have

$$
\begin{aligned}
\mathbf{q} \in V_\alpha \quad &\Longleftrightarrow \quad \text{div}_\alpha \mathbf{q} = 0 \\
&\Longleftrightarrow \quad (\mathbf{q}, \nabla\varphi)_\Omega + (\boldsymbol{\alpha} \cdot \mathbf{q}, \varphi)_\Omega = 0 \qquad \forall \varphi \in \mathcal{D}(\Omega), \\
&\Longleftrightarrow \quad (\mathbf{q}, \nabla_\alpha u)_\Omega = 0 \qquad\qquad\quad \forall u \in H_0^1(\Omega).
\end{aligned}
$$

This means that V_α is the $\mathbf{L}^2(\Omega)$ orthogonal complement of the range of ∇_α (as an operator acting on $H_0^1(\Omega)$). Given that this range is closed, the equality in (3) follows. $\qquad\square$

The missing link. Proposition 11.3 proves the two inf-sup conditions needed to show the well-posedness of (11.12) as requested by Proposition 11.2. The missing step is the study of the reduced bilinear form

$$V_0 \times V_\beta \ni (\mathbf{p}, \mathbf{q}) \longmapsto (\mathbf{p}, \mathbf{q})_\Omega \in \mathbb{R}. \tag{11.15}$$

We will do it by identifying the associated operator $P : V_0 \to V_\beta$ given by the Riesz-Fréchet theorem

$$P\mathbf{p} \in V_\beta, \qquad (P\mathbf{p}, \mathbf{q})_\Omega = (\mathbf{p}, \mathbf{q})_\Omega \qquad \forall \mathbf{q} \in V_\beta,$$

where in V_β we are using the equivalent norm $\| \cdot \|_\Omega$ (see Proposition 11.3(2)). The operator P is just the restriction to V_0 of the $\mathbf{L}^2(\Omega)$-orthogonal projection onto V_β.

Note that if $P\mathbf{p} = \mathbf{0}$, then, by Proposition 11.3(3), we have $\mathbf{p} = \nabla_\beta u$ with $u \in H_0^1(\Omega)$ and $\nabla \cdot \mathbf{p} = \mathrm{div}_0 \mathbf{p} = 0$. Therefore $\nabla \cdot \nabla_\beta u = 0$, which implies that $u = 0$ and thus $\mathbf{p} = \mathbf{0}$. This proves that P is injective.

Given $\mathbf{p} \in V_0$, we can decompose orthogonally and use Proposition 11.3(3) again,

$$\mathbf{p} = P\mathbf{p} + (\mathbf{p} - P\mathbf{p}) = P\mathbf{p} + \nabla_\beta u \in V_\beta + \nabla_\beta H_0^1(\Omega).$$

Taking the divergence on both sides, we have that

$$\nabla \cdot \nabla_\beta u = -\nabla \cdot P\mathbf{p},$$

and therefore u is the only solution of the coercive problem

$$u \in H_0^1(\Omega), \quad (\nabla u, \nabla v)_\Omega + (u, \boldsymbol{\beta} \cdot \nabla v)_\Omega = (P\mathbf{p}, \nabla v)_\Omega \quad \forall v \in H_0^1(\Omega). \quad (11.16)$$

(Note that $-\langle \nabla \cdot P\mathbf{p}, v \rangle_{H^{-1}(\Omega) \times H_0^1(\Omega)} = (P\mathbf{p}, \nabla v)_\Omega$.) Therefore

$$
\begin{aligned}
\|\mathbf{p}\|_\Omega &\le \|P\mathbf{p}\|_\Omega + \|\nabla_\beta u\|_\Omega \\
&\le \|P\mathbf{p}\|_\Omega + C\|u\|_{1,\Omega} \\
&\le C'\|P\mathbf{p}\|_\Omega, \qquad \text{(well-posedness of (11.16))}
\end{aligned}
$$

which shows that P has closed range. In particular, we have the inf-sup condition

$$\sup_{\mathbf{0} \ne \mathbf{q} \in V_\beta} \frac{(\mathbf{p}, \mathbf{q})_\Omega}{\|\mathbf{q}\|_{\mathrm{div},\Omega}} \ge c\|\mathbf{p}\|_{\mathrm{div},\Omega} \qquad \forall \mathbf{p} \in V_0,$$

where we have restituted the $\| \cdot \|_{\mathrm{div},\Omega}$ norms (see Proposition 11.3(2) again). Finally, to be completely done with this bilinear form, we show that if $\mathbf{q} \in V_\beta$ and

$$(\mathbf{p}, \mathbf{q})_\Omega = 0 \quad \forall \mathbf{p} \in V_0,$$

then $\mathbf{q} = \nabla u$ with $u \in H_0^1(\Omega)$ and $\mathrm{div}_\beta \nabla u = 0$, which proves that $u = 0$ and hence $\mathbf{q} = \mathbf{0}$. This is the second of the conditions needed to show that the bilinear form in (11.15) defines an invertible operator or, equivalently, to show that the operator P defined above is invertible.

Conclusions. By virtue of Proposition 11.2, the mixed formulation (11.12), associated to the first order system (11.11), is well posed and we can bound the solution as

$$\|\mathbf{q}\|_\Omega + \|u\|_\Omega \le C(\|f\|_\Omega + \|g\|_{1/2,\Gamma}).$$

11.4   Double restrictions

The verification of inf-sup conditions in product spaces (that is, when there is more than one condition imposed on the same field) can be simplified with this simple result.

Proposition 11.4. *If $B_j : V \to M_j$ are bounded operators between Hilbert spaces and*

$$B = (B_1, B_2) : V \to M_1 \times M_2,$$

then B is surjective if and only if

(a) B_1 and B_2 are surjective,

(b) $V = \ker B_1 + \ker B_2.$

Proof. Clearly the surjectivity of B implies those of B_1 and B_2. If $u \in V$, there exists $v \in V$ such that

$$(B_1 v, B_2 v) = (B_1 u, 0).$$

Therefore $v \in \ker B_2$ and $u - v \in \ker B_1$, which proves that $u \in \ker B_1 + \ker B_2$.

Reciprocally, let $B_j^\dagger : M_j \to V$ be right inverses of B_1 and B_2. If $(p_1, p_2) \in M_1 \times M_2$, we decompose

$$B_2^\dagger p_2 - B_1^\dagger p_1 = u_1 + u_2 \qquad u_j \in \ker B_j,$$

and define $u := u_1 + B_1^\dagger p_1 = -u_2 + B_2^\dagger p_2$. It is then clear that

$$B_1 u = p_1, \qquad B_2 u = p_2,$$

which shows the surjectivity of B. $\qquad\square$

A problem with two restrictions. As an example, let us consider a non-homogeneous Neumann problem written in mixed form and with weak imposition of the boundary condition. The variational problem

$$(\mathbf{q}, u, \eta) \in \mathbf{H}(\mathrm{div}, \Omega) \times L_\circ^2(\Omega) \times H^{1/2}(\Gamma), \tag{11.17a}$$

$$(\kappa^{-1}\mathbf{q}, \mathbf{r})_\Omega - (\nabla \cdot \mathbf{r}, u)_\Omega + \langle \mathbf{r} \cdot \mathbf{n}, \eta \rangle_\Gamma = 0 \qquad \forall \mathbf{r} \in \mathbf{H}(\mathrm{div}, \Omega), \tag{11.17b}$$

$$(\nabla \cdot \mathbf{q}, v)_\Omega \qquad\qquad = (f, v)_\Omega \qquad \forall v \in L_\circ^2(\Omega), \tag{11.17c}$$

$$\langle \mathbf{q} \cdot \mathbf{n}, \mu \rangle_\Gamma \qquad\qquad = \langle h, \mu \rangle_\Gamma \qquad \forall \mu \in H^{1/2}(\Gamma), \tag{11.17d}$$

is equivalent to the equations

$$\kappa^{-1}\mathbf{q} + \nabla u = \mathbf{0}, \qquad \nabla \cdot \mathbf{q} = f, \qquad \mathbf{q} \cdot \mathbf{n} = h,$$

and yields $\eta := \gamma u$, if we assume the compatibility condition (which is needed for existence of solutions anyway)

$$(f, 1)_\Omega + \langle h, 1 \rangle_\Gamma = 0. \tag{11.18}$$

In order to prove that

$$(\nabla \cdot \mathbf{q}, v)_\Omega = (f, v)_\Omega \qquad \forall v \in L^2(\Omega),$$

we use the decomposition $L^2(\Omega) = L_\circ^2(\Omega) \oplus \mathcal{P}_0(\Omega)$ and the equality

$$(\nabla \cdot \mathbf{p}, 1)_\Omega = \langle \mathbf{p} \cdot \mathbf{n}, 1 \rangle_\Gamma = -\langle h, 1 \rangle_\Gamma = (f, 1)_\Omega$$

that follows from the compatibility condition (11.18).

Well-posedness. Let us start with the inf-sup condition, which we will decompose in two pieces as suggested by Proposition 11.4. We associate operators to the off-diagonal bilinear forms

$$(B_1 \mathbf{p}, v)_\Omega = (\nabla \cdot \mathbf{p}, v)_\Omega \qquad \mathbf{p} \in \mathbf{H}(\text{div}, \Omega), \qquad v \in L_\circ^2(\Omega),$$

$$(B_2 \mathbf{p}, \mu)_{1/2,\Gamma} = \langle \mathbf{p} \cdot \mathbf{n}, \mu \rangle_\Gamma \qquad \mathbf{p} \in \mathbf{H}(\text{div}, \Omega), \qquad \mu \in H^{1/2}(\Gamma).$$

Note that $B_1 \mathbf{p} \neq \nabla \cdot \mathbf{p}$, since $\nabla \cdot \mathbf{p}$ is not in $L_\circ^2(\Omega)$ in general. However, we have

$$\sup_{0 \neq \mathbf{q} \in \mathbf{H}(\text{div}, \Omega)} \frac{(\nabla \cdot \mathbf{q}, v)_\Omega}{\|\mathbf{q}\|_{\text{div}, \Omega}} \geq c \|v\|_\Omega \qquad \forall v \in L^2(\Omega),$$

because div $: \mathbf{H}(\text{div}, \Omega) \to L^2(\Omega)$ is surjective, and therefore this inequality holds for $v \in L_\circ^2(\Omega)$, proving that B_1 is surjective. For B_2, we use that $\mathbf{H}(\text{div}, \Omega) \ni \mathbf{q} \mapsto \mathbf{q} \cdot \mathbf{n} \in H^{-1/2}(\Gamma)$ is surjective, since B_2 is this normal component operator composed with a Riesz-Fréchet representation. Next note that

$$\ker B_1 = \{\mathbf{q} \in \mathbf{H}(\text{div}, \Omega) : \nabla \cdot \mathbf{q} \in \mathcal{P}_0(\Omega)\}, \qquad \ker B_2 = \mathbf{H}_0(\text{div}, \Omega). \tag{11.19}$$

If we can prove that

$$\mathbf{H}(\text{div}, \Omega) = \ker B_1 + \ker B_2, \tag{11.20}$$

Proposition 11.4 will give us the inf-sup condition for the joint bilinear form

$$b(\mathbf{p}, (v, \mu)) := (\nabla \cdot \mathbf{p}, v)_\Omega + \langle \mathbf{p} \cdot \mathbf{n}, \mu \rangle_\Gamma.$$

The proof of (11.20) is easy though. We take $\mathbf{q} \in \mathbf{H}(\text{div}, \Omega)$ and solve

$$-\Delta u = -\nabla \cdot \mathbf{q} + |\Omega|^{-1} (\nabla \cdot \mathbf{q}, 1)_\Omega, \qquad \partial_n u = 0, \qquad (u, 1)_\Omega = 0.$$

We have subtracted the average to have a compatibility condition for the data of the Neumann problem. We now define $\mathbf{p} := \nabla u \in \mathbf{H}(\text{div}, \Omega)$ and note that $\mathbf{p} \cdot \mathbf{n} = \partial_n u = 0$ (that is, $\mathbf{p} \in \mathbf{H}_0(\text{div}, \Omega)$) and $\nabla \cdot (\mathbf{q} - \mathbf{p})$ is constant, and therefore $\mathbf{p} - \mathbf{q} \in \ker B_1$. This proves (11.20). We finally turn our attention to the diagonal bilinear form. The common kernel of the operators B_1 and B_2 is

$$\ker B_1 \cap \ker B_2 = \{\mathbf{q} \in \mathbf{H}_0(\mathrm{div}, \Omega) \; : \; \nabla \cdot \mathbf{q} = 0\},$$

as follows from (11.19) and from the fact that $(\nabla \cdot \mathbf{q}, 1)_\Omega = \langle \mathbf{q} \cdot \mathbf{n}, 1 \rangle_\Gamma$. The diagonal bilinear form in the variational problem (11.17) is clearly coercive in $\ker B_1 \cap \ker B_2$, and this finishes the verification of the Babuška-Brezzi conditions for problem (11.17).

As a final note, we could have dealt with (11.17) using the space $\mathbf{H}(\mathrm{div}, \Omega) \times L^2(\Omega) \times H^{1/2}(\Gamma)$, without the conditon $(u, 1)_\Omega = 0$. In this case, we can add a compact term and use the Fredholm alternative. The associated operator has a one-dimensional kernel $\mathrm{span}\{(0, 1, 1)\}$ and, since the problem (11.17) can be written with a symmetric bilinear form, the compatibility condition is the cancellation of the right-hand side on the kernel, from which we recover (11.18).

11.5   A partially uncoupled Stokes-Darcy formulation

In this section we come back to the Stokes-Darcy problem, but we formulate it by keeping the Stokes and Darcy fields in separate spaces. The geometric setup will be the same as in Section 10.9. Let us quickly review the equations. We have velocity and pressure fields $(\mathbf{u}_S, p_S) : \Omega_S \to \mathbb{R}^{d+1}$ satisfying the Stokes equations in Ω_S,

$$-2\nu \, \mathrm{div} \, \varepsilon(\mathbf{u}_S) + \nabla p_S = \mathbf{f}_S, \qquad \nabla \cdot \mathbf{u}_S = 0, \tag{11.21a}$$

and velocity and pressure fields $(\mathbf{u}_D, p_D) : \Omega_D \to \mathbb{R}^{d+1}$ satisfying the Darcy equations in Ω_D

$$\kappa^{-1} \mathbf{u}_D + \nabla p_D = \mathbf{0}, \qquad \nabla \cdot \mathbf{u}_D = f_D. \tag{11.21b}$$

We have homogeneous boundary conditions on the boundary of the total domain

$$\gamma \mathbf{u}_S = \mathbf{0} \quad \text{on } \Gamma_S, \qquad \mathbf{u}_D \cdot \mathbf{n}_D = 0 \quad \text{on } \Gamma_D, \tag{11.21c}$$

two transmission conditions on the common interface

$$\gamma \mathbf{u}_S \cdot \mathbf{n}_S + \mathbf{u}_D \cdot \mathbf{n} = 0 \qquad\qquad \text{on } \Sigma, \tag{11.21d}$$

$$(2\nu\varepsilon(\mathbf{u}_S) - p_S \mathrm{I})\mathbf{n}_S + c\pi_T \mathbf{u}_S = (\gamma p_D)\mathbf{n}_D \qquad \text{on } \Sigma, \tag{11.21e}$$

and a normalization condition for the pressure

$$(p_D, 1)_{\Omega_D} = 0. \tag{11.21f}$$

Instead of forcing the first interface condition in the definition of a joint space for the velocity fields, we add a new unknown

$$\phi := \gamma p_D|_\Sigma \in H^{1/2}(\Sigma),$$

which will act as a Lagrange multiplier for the first transmission condition. As we did in Section 10.9, we will assume the necessary compatibility condition $f_D \in L^2_\circ(\Omega_D)$.

A wealth of spaces. Here we need to recover the language of trace spaces on parts of the boundary, which we introduced in Section 6.7, dealing with mixed boundary conditions. If Γ_0 is a nontrivial part of $\Gamma = \partial\Omega$ we have two trace spaces

$$H^{1/2}(\Gamma_0) := \{\phi|_{\Gamma_0} : \phi \in H^{1/2}(\Gamma)\} = \{\gamma u|_{\Gamma_0} : u \in H^1(\Omega)\},$$

$$\widetilde{H}^{1/2}(\Gamma_0) := \{\gamma u_{\Gamma_0} : u \in H^1(\Omega), \quad \gamma u = 0 \text{ in } \Gamma \setminus \Gamma_0\}.$$

We endow these spaces with image norms and set up two Gelfand triples

$$H^{1/2}(\Gamma_0) \subset L^2(\Gamma_0) \subset \widetilde{H}^{-1/2}(\Gamma_0), \qquad \widetilde{H}^{1/2}(\Gamma_0) \subset L^2(\Gamma_0) \subset H^{-1/2}(\Gamma_0).$$

The Darcy boundary condition ($\mathbf{u}_D \cdot \mathbf{n}_D = 0$ on Γ_D) can be understood as an equality in $H^{-1/2}(\Gamma_D) = \widetilde{H}^{1/2}(\Gamma_D)'$, that is,

$$(\mathbf{u}_D, \nabla v)_{\Omega_D} + (\nabla \cdot \mathbf{u}_D, v)_{\Omega_D} = 0 \qquad \forall v \in H^1(\Omega_D), \qquad \gamma v|_\Sigma = 0. \quad (11.22)$$

We consider the spaces

$$\mathbf{V}_S = \mathbf{H}^1_{\Gamma_S}(\Omega_S) := \{\mathbf{u}_S \in \mathbf{H}^1(\Omega) : \gamma \mathbf{u}_S = \mathbf{0} \text{ on } \Gamma_S\},$$

$$\mathbf{V}_D = \mathbf{H}_{\Gamma_D}(\mathrm{div}, \Omega_D) := \{\mathbf{u}_D \in \mathbf{H}(\mathrm{div}, \Omega_D) : \mathbf{u}_D \cdot \mathbf{n}_D = 0 \text{ on } \Gamma_D\}$$

$$= \{\mathbf{u}_D \in \mathbf{H}(\mathrm{div}, \Omega_D) : (11.22) \text{ holds}\},$$

and $Q_S := L^2(\Omega_S)$, $Q_D := L^2_\circ(\Omega_D)$ (which incorporates the normalization condition). For $\mathbf{u}_D \in \mathbf{H}_{\Gamma_D}(\mathrm{div}, \Omega_D)$, we can consider $\mathbf{u}_D \cdot \mathbf{n}_D \in \widetilde{H}^{-1/2}(\Sigma)$ as the functional

$$\langle \mathbf{u}_D \cdot \mathbf{n}_D, \gamma v \rangle_{\widetilde{H}^{-1/2}(\Sigma) \times H^{1/2}(\Sigma)} = (\mathbf{u}_D, \nabla v)_{\Omega_D} + (\nabla \cdot \mathbf{u}_D, v)_{\Omega_D} \qquad \forall v \in H^1(\Omega_D),$$

since the right-hand side does not change if we add functions whose trace vanishes on Σ, as (11.22) holds.

Bilinear forms. We are now ready to introduce the six bilinear forms that will be used in the formulation

$$a_S(\mathbf{u}_S, \mathbf{v}_S) := 2\nu(\boldsymbol{\varepsilon}(\mathbf{u}_S), \boldsymbol{\varepsilon}(\mathbf{v}_S))_{\Omega_S} + \langle c\,\pi_T \mathbf{u}_S, \pi_T \mathbf{v}_S \rangle_\Sigma,$$

$$b_S(\mathbf{u}_S, q_S) := (\nabla \cdot \mathbf{u}_S, q_S)_{\Omega_S},$$

$$a_D(\mathbf{u}_D, \mathbf{v}_D) := (\kappa^{-1}\mathbf{u}_D, \mathbf{v}_D)_{\Omega_D},$$

$$b_D(\mathbf{u}_D, q_D) := (\nabla \cdot \mathbf{u}_D, q_D)_{\Omega_D},$$

$$c_S(\mathbf{u}_S, \psi) := \langle \psi, \gamma \mathbf{u}_S \cdot \mathbf{n}_S \rangle_\Sigma,$$

$$c_D(\mathbf{u}_D, \psi) := \langle \mathbf{u}_D \cdot \mathbf{n}_D, \psi \rangle_{\widetilde{H}^{-1/2}(\Sigma) \times H^{1/2}(\Sigma)}.$$

The brackets above tagged with Σ (they appear in a_S and c_S) are $L^2(\Sigma)$ inner products and no duality extension is needed there. In the bilinear form c_D we have the $H^{1/2}(\Sigma) \subset L^2(\Sigma) \times \tilde{H}^{-1/2}(\Sigma)$ duality, and we will shorten notation by using the subscript Σ as well for this extension of the $L^2(\Sigma)$ inner product. (Note that we have defined two Gelfand triples around $L^2(\Sigma)$, and were the second one to appear in this formulation, it would be advisable to use a different notation for both. We will encounter this problem in Chapter 16.)

An uncoupled variational formulation. The basic idea is rather simple. We will keep ϕ as a coupling variable that feeds into separate variational formulations, one in Ω_S and one in Ω_D. The solution of those problems is then fed into a coupling condition. We thus look for $\phi \in H^{1/2}(\Sigma)$ such that the solution of

$$
\begin{align}
&(\mathbf{u}_S, p_S) \in \mathbf{V}_S \times Q_S, &&&& \text{(11.23a)} \\
&a_S(\mathbf{u}_S, \mathbf{v}_S) - b_S(\mathbf{v}_S, p_S) + c_S(\mathbf{v}_S, \phi) = (\mathbf{f}_S, \mathbf{v}_S)_{\Omega_s} && \forall \mathbf{v}_S \in \mathbf{V}_S, && \text{(11.23b)} \\
&b_S(\mathbf{u}_S, q_S) \qquad\qquad\qquad\qquad\quad = 0 && \forall q_S \in Q_S, && \text{(11.23c)}
\end{align}
$$

and the solution of

$$
\begin{align}
&(\mathbf{u}_D, p_D) \in \mathbf{V}_D \times Q_D, &&&& \text{(11.24a)} \\
&a_D(\mathbf{u}_D, \mathbf{v}_D) - b_D(\mathbf{v}_D, p_D) + c_D(\mathbf{v}_D, \phi) = 0 && \forall \mathbf{v}_D \in \mathbf{V}_D, && \text{(11.24b)} \\
&b_D(\mathbf{u}_D, q_D) \qquad\qquad\qquad\qquad = (f_D, q_D)_{\Omega_D} && \forall q_D \in Q_D, && \text{(11.24c)}
\end{align}
$$

satisfy the additional equation

$$
c_D(\mathbf{u}_D, \psi) + c_S(\mathbf{u}_S, \psi) = 0 \qquad \forall \psi \in H^{1/2}(\Sigma). \tag{11.25}
$$

Let us just sketch how equations (11.23)-(11.25) imply the coupled equations (11.21) with the additional equation $\phi = \gamma p_D|_\Sigma$. We will just prove the more delicate steps, since at this stage we count on the reader to be able to easily find a way through equations in the sense of distributions. The coupling condition (11.25) is equivalent to

$$
\mathbf{u}_D \cdot \mathbf{n}_D + \gamma \mathbf{u}_S|_\Sigma \cdot \mathbf{n}_S = 0 \qquad \text{in } \tilde{H}^{-1/2}(\Sigma).
$$

The missing testing constant in the divergence condition of the Darcy flow needs to be recovered using the coupling condition and the divergence-free condition of the Stokes flow:

$$
\begin{align}
(\nabla \cdot \mathbf{u}_D, 1)_{\Omega_D} &= \langle \mathbf{u}_D \cdot \mathbf{n}_D, 1 \rangle_{\partial\Omega_D} = \langle \mathbf{u}_D \cdot \mathbf{n}_D, 1 \rangle_\Sigma \\
&= - \langle \gamma \mathbf{u}_S \cdot \mathbf{n}_S, 1 \rangle_\Sigma = -\langle \gamma \mathbf{u}_S \cdot \mathbf{n}_S, 1 \rangle_{\partial\Omega_S} \\
&= - (\nabla \cdot \mathbf{u}_S, 1)_{\Omega_s} = 0.
\end{align}
$$

Once we have settled these preliminary issues, the proof would proceed as follows: (a) we prove that the equations (11.21a) and (11.21b) hold by testing with smooth compactly supported functions and using that $(\nabla \cdot \mathbf{u}_D, 1)_{\Omega_D} = 0$; (b) the boundary conditions (11.21c) and the normalization condition (11.21f) are part of the requirements in the definitions of the spaces where we look for solutions; (c) we have already established the transmission condition (11.21d); (d) we now substitute (11.21a) in (11.23b), getting

$$\langle(2\nu\varepsilon(\mathbf{u}_S) - p_S I)\mathbf{n}_S + c\pi_T \mathbf{u}_S, \gamma\mathbf{v}_S\rangle_{\partial\Omega_s} + \langle\phi\,\mathbf{n}_S, \gamma\mathbf{v}_S\rangle_\Sigma = 0 \qquad \forall \mathbf{v}_S \in \mathbf{V}_S,$$

which is the condition

$$(2\nu\varepsilon(\mathbf{u}_S) - p_S I)\mathbf{n}_S + c\pi_T \mathbf{u}_S = \phi\,\mathbf{n}_D \qquad \text{in } \mathbf{H}^{-1/2}(\Sigma);$$

(e) we finally substitute (11.21b) in (11.24b) proving that

$$\langle\mathbf{v}_D \cdot \mathbf{n}_D, \gamma p_D\rangle_{\partial\Omega_D} = \langle\mathbf{v}_D \cdot \mathbf{n}_D, \phi\rangle_\Sigma \qquad \forall \mathbf{v}_D \in \mathbf{V}_D,$$

which is equivalent to

$$\gamma p_D|_\Sigma = \phi \qquad \text{in } H^{1/2}(\Sigma),$$

since $H^{1/2}(\Sigma) = \widetilde{H}^{-1/2}(\Sigma)'$ and the latter is the space where $\mathbf{v}_D \cdot \mathbf{n}_D$ lives when $\mathbf{v}_D \in \mathbf{V}_D$. This finishes the proof.

Operator form. The analysis of the formulation (11.23)-(11.25) can be done in many different ways. All of them benefit from some quick reformulations of the bilinear forms as operators. We assign operators to the bilinear forms using the inner products of the associated spaces. The associated operators are:

$$(A_S \mathbf{u}_S, \mathbf{v}_S)_{1,\Omega_s} = a_S(\mathbf{u}_S, \mathbf{v}_S),$$
$$(A_D \mathbf{u}_D, \mathbf{v}_D)_{\text{div},\Omega_D} = a_D(\mathbf{u}_D, \mathbf{v}_D),$$
$$(B_S \mathbf{u}_S, q_S)_{\Omega_S} = (\mathbf{u}_S, B^* q_S)_{1,\Omega_s} = b_S(\mathbf{u}_S, q_S),$$
$$(B_D \mathbf{u}_D, q_D)_{\Omega_D} = (\mathbf{u}_D, B^* q_D)_{\text{div},\Omega_D} = b_D(\mathbf{u}_D, q_D),$$
$$(C_S \mathbf{u}_S, \psi)_{1/2,\Sigma} = (\mathbf{u}_S, C_S^* \psi)_{1,\Omega_s} = c_S(\mathbf{u}_S, \psi),$$
$$(C_D \mathbf{u}_D, \psi)_{1/2,\Sigma} = (\mathbf{u}_D, C_D^* \psi)_{\text{div},\Omega_D} = c_D(\mathbf{u}_D, \psi).$$

Note that the inner product in $H^{1/2}(\Sigma)$ makes an appearance, but that it is just needed from a theoretical point of view, which means that the reader can happily ignore the fact that we defined the associated norm as an image norm.

Proposition 11.5. *The operators above satisfy the following properties:*

(a) A_S *and* A_D *are self-adjoint, and* A_S *is strongly positive definite in* $\mathbf{V}_S$,

(b) C_S *is self-adjoint and compact,*

(c) B_D *and* B_S *are surjective,*

(d) C_D *is surjective.*

Proof. Properties (a) and (b) are simple consequences of the definitions. The compactness of C_S follows from the compactness of the trace as an operator from $H^1(\Omega_S)$ to $L^2(\partial\Omega_S)$ (see Section 8.3). The strong positivity of A_S is the coercivity of a_S which follows from a generalized Poincaré inequality.

We have that $B_S \mathbf{u}_S = \nabla \cdot \mathbf{u}_S$ and therefore the surjectivity of $B_S : \mathbf{V}_S \to L^2(\Omega_S) = Q_S$ is Proposition 10.12. It is easy to prove that

$$B_D \mathbf{u}_D = \nabla \cdot \mathbf{u}_D - |\Omega_D|^{-1}(\nabla \cdot \mathbf{u}_D) \in L^2_\circ(\Omega_D) = Q_D.$$

For all $q_D \in L^2_\circ(\Omega_D)$, we have

$$\sup_{\mathbf{0} \neq \mathbf{u}_D \in \mathbf{V}_D} \frac{(\nabla \cdot \mathbf{u}_D, q_D)_{\Omega_D}}{\|\mathbf{u}_D\|_{\mathrm{div},\Omega_D}} \geq \sup_{\mathbf{0} \neq \mathbf{u}_D \in \mathbf{H}_0(\mathrm{div},\Omega_D)} \frac{(\nabla \cdot \mathbf{u}_D, q_D)_{\Omega_D}}{\|\mathbf{u}_D\|_{\mathrm{div},\Omega_D}} \geq c \|q_D\|_{\Omega_D},$$

since the latter inequality is equivalent to the surjectivity of div : $\mathbf{H}_0(\mathrm{div},\Omega_D) \to L^2_\circ(\Omega_D)$. This finishes the proof of (c).

Given $\xi \in \tilde{H}^{-1/2}(\Sigma)$, we solve the equation

$$u \in H^1(\Omega_D), \qquad (\nabla u, \nabla v)_{\Omega_D} + (u, v)_{\Omega_D} = \langle \xi, \gamma v \rangle_\Sigma \qquad \forall v \in H^1(\Omega_D),$$

and define $\mathbf{v} := \nabla u$. It is easy to prove that $\mathbf{v} \in \mathbf{V}_D$ and

$$\langle \mathbf{v} \cdot \mathbf{n}_D, \gamma v \rangle_\Sigma = \langle \xi, \gamma v \rangle_\Sigma \qquad \forall v \in H^1(\Omega_D),$$

and hence

$$\mathbf{v} \cdot \mathbf{n}_D = \xi, \qquad \|\mathbf{v}\|_{\mathrm{div},\Omega_D} \leq C \|\xi\|_{\tilde{H}^{-1/2}(\Sigma)}.$$

Therefore

$$\|\psi\|_{H^{1/2}(\Sigma)} = \sup_{0 \neq \xi \in \tilde{H}^{-1/2}(\Sigma)} \frac{\langle \xi, \psi \rangle_\Sigma}{\|\xi\|_{\tilde{H}^{-1/2}(\Sigma)}} \leq C^{-1} \sup_{\mathbf{0} \neq \mathbf{u}_D \in \mathbf{V}_D} \frac{\langle \mathbf{u}_D \cdot \mathbf{n}_D, \psi \rangle_\Sigma}{\|\mathbf{u}_D\|_{\mathrm{div},\Omega_D}},$$

which shows surjectivity of C_D.   □

We also have

$$\ker B_S = \{\mathbf{u}_S \in \mathbf{V}_S : \nabla \cdot \mathbf{u}_S = 0\},$$
$$\ker B_D = \{\mathbf{u}_D \in \mathbf{V}_D : \nabla \cdot \mathbf{u}_D \in \mathcal{P}_0(\Omega_D)\},$$
$$\ker C = \mathbf{V},$$

where $C(\mathbf{u}_S, \mathbf{u}_D) = C_S \mathbf{u}_S + C_D \mathbf{u}_D$. We now give several approaches to proving the well-posedness of equations (11.23)-(11.25).

First approach. We want to prove the invertibility of the operator

$$\begin{bmatrix} A_S & -B_S^* & & & C_S^* \\ B_S & & & & \\ & & A_D & -B_D^* & C_D^* \\ & & B_D & & \\ C_S & & C_D & & \end{bmatrix}, \tag{11.26}$$

from $\mathbf{V}_S \times Q_S \times \mathbf{V}_D \times Q_D \times H^{1/2}(\Sigma)$ to itself. This ordering of unknowns is the one that we have given to describe (11.23)-(11.25): first Stokes, then Darcy, finally coupling. The operator in (11.26) can be seen as the operator for a mixed problem with a single restriction, that is, we take $V = \mathbf{V}_S \times Q_S \times \mathbf{V}_D \times Q_D$ and $M = H^{1/2}(\Sigma)$. The operator

$$\begin{bmatrix} C_S & 0 & C_D & 0 \end{bmatrix} : \mathbf{V}_S \times Q_S \times \mathbf{V}_D \times Q_D \to H^{1/2}(\Sigma)$$

is surjective, since C_D is surjective. Its kernel is

$$\{(\mathbf{u}_S, p_S, \mathbf{u}_D, p_D) : (\mathbf{u}_S, \mathbf{u}_D) \in \mathbf{V}\},$$

and the block operator restricted to this kernel is the same as the operator we studied in Section 10.9. (Be careful to understand that the kernel is not the product of four spaces, but the $\mathbf{u}$ components are tied to each other. Therefore, there is no 4×4 matrix representation of the operator restricted to this kernel.) This finishes the verification of the Babuška-Brezzi conditions. Note also that the operator (11.26) is a compact perturbation of the operator

$$\begin{bmatrix} A_S & -B_S^* & & & \\ B_S & & & & \\ & & A_D & -B_D^* & C_D^* \\ & & B_D & & \\ & & C_D & & \end{bmatrix},$$

where a Stokes problem and a Darcy problem are decoupled. The Stokes problem has Dirichlet condition on Γ_S and a free condition on Σ (corresponding to $p_D = 0$ in the transmission conditions). The Darcy problem has homogeneous essential boundary conditions on Γ_D and nonhomogeneous natural condition on Σ (first transmission condition ignoring the Stokes field). We can then approach the analysis with Fredholm theory too: since a compact perturbation of the operator (11.26) (eliminate C_D) is invertible, the operator (11.26) is invertible if and only if it is injective. With this approach we just need to prove the injectivity of the operator (11.26).

A second approach. In this second view of the problem, we first consider velocity fields, then pressure fields, and finally coupling:

$$\begin{bmatrix} A_S & & -B_S^* & & C_S^* \\ & A_D & & -B_D & C_D^* \\ B_S & & & & \\ & B_D & & & \\ C_S & C_D & & & \end{bmatrix}. \tag{11.27}$$

If we look at this as a mixed problem related to the spaces

$$V := \mathbf{V}_S \times \mathbf{V}_D \times Q_S \times Q_D, \qquad M := H^{1/2}(\Sigma),$$

we land on exactly the same analysis we did in our first approach. This ordering of variables and equations emphasizes that there is a mixed problem whose diagonal operator is again mixed. If we look at the operator in (11.27) as the operator associated to a mixed problem with spaces

$$V := \mathbf{V}_S \times \mathbf{V}_D, \qquad M := Q_S \times Q_D \times H^{1/2}(\Sigma),$$

we have a mixed problem with multiple restrictions. The key operator is

$$\begin{bmatrix} B_S \\ & B_D \\ C_S & C_D \end{bmatrix} = \begin{bmatrix} B \\ C \end{bmatrix}. \tag{11.28}$$

Its kernel is

$$\{(\mathbf{u}_S, \mathbf{u}_D) \in \mathbf{V} : \nabla \cdot \mathbf{u}_S = 0, \ \nabla \cdot \mathbf{u}_D = 0\}.$$

(In principle $B_D \mathbf{u}_D = 0$ implies that $\nabla \cdot \mathbf{u}_D$ is constant. We saw the argument of why $\nabla \cdot \mathbf{u}_D = 0$ when we related the variational formulation to the transmission problem. It uses the fact that $C_S \mathbf{u}_S + C_D \mathbf{u}_D = 0$, which is the first transmission condition.) The diagonal operator is clearly coercive in $\{(\mathbf{u}_S, \mathbf{u}_D) \in \mathbf{V}_S \times \mathbf{V}_D : \nabla \cdot \mathbf{u}_D = 0\}$, which contains the kernel. Therefore, we are left with the verification of the surjectivity of the operator (11.28). The operators B and C are surjective by Proposition 11.5. We finally need to decompose

$$\mathbf{V}_S \times \mathbf{V}_D = \ker B + \ker C = (\ker B_S \times \ker B_D) + \mathbf{V}, \tag{11.29}$$

and then Proposition 11.4 will imply that the operator (11.28) is surjective. We start by fixing $\mathbf{u}_S^\star \in \mathbf{V}_S$ such that $\nabla \cdot \mathbf{u}_S^\star = 1$ and then extending it to $\mathbf{u}^\star = (\mathbf{u}_S^\star, \mathbf{u}_D^\star) \in \mathbf{V}$, as allowed by Proposition 10.13. Given $\mathbf{u} = (\mathbf{u}_S, \mathbf{u}_D) \in \mathbf{V}_S \times \mathbf{V}_D$, we define $c_S := |\Omega_S|^{-1}(\nabla \cdot \mathbf{u}_S, 1)_{\Omega_S}$ and then find

$$\mathbf{u}_S^0 \in \mathbf{H}_0^1(\Omega_S) \qquad \nabla \cdot \mathbf{u}_S^0 = \nabla \cdot \mathbf{u}_S - c_S \in L_\circ^2(\Omega_S),$$

define $c_D := |\Omega_D|^{-1}(\nabla \cdot (\mathbf{u}_D - c_S \mathbf{u}_D^\star), 1)_{\Omega_D}$ and then find

$$\mathbf{u}_D^0 \in \mathbf{H}_0(\text{div}, \Omega_D) \qquad \nabla \cdot \mathbf{u}_D^0 = \nabla \cdot (\mathbf{u}_D - c_S \mathbf{u}_D^\star) - c_D \in L_\circ^2(\Omega_D).$$

Since

$$(\mathbf{u}_S^0, \mathbf{u}_D^0) + c_S \mathbf{u}^\star \in \mathbf{H}_0^1(\Omega_S) \times \mathbf{H}_0(\text{div}, \Omega_D) + \mathbf{V} \subset \mathbf{V},$$

and

$$\mathbf{u} - (\mathbf{u}_S^0, \mathbf{u}_D^0) - c_S \mathbf{u}^\star \in \ker B,$$

this proves (11.29). Therefore the operator (11.28) is surjective and the operator (11.27) is invertible.

11.6 Galerkin methods for mixed problems

We finish this set of two chapters on mixed problems (we will re-encounter mixed formulations in Chapter 16 for instance) with some comments on Galerkin methods for problems with mixed structure. As opposed to Galerkin discretization of coercive problems and of compact perturbations thereof, in the case of mixed problems, arbitrary choices of the finite-dimensional subspaces do not guarantee well-posedness of the discrete problem or a good quasi-optimality estimate. Consider a well posed mixed problem on a couple of Hilbert spaces $V \times M$:

$$(u, p) \in V \times M, \tag{11.30a}$$
$$a(u, v) + b(v, p) = \ell(v) \qquad \forall v \in V, \tag{11.30b}$$
$$b(u, q) \qquad\quad = \chi(q) \qquad \forall q \in M. \tag{11.30c}$$

If we choose finite-dimensional subspaces

$$V_h \subset V, \qquad M_h \subset M,$$

the Galerkin approximation of (11.30) is the problem

$$(u_h, p_h) \in V_h \times M_h, \tag{11.31a}$$
$$a(u_h, v_h) + b(v_h, p_h) = \ell(v_h) \qquad \forall v_h \in V_h, \tag{11.31b}$$
$$b(u_h, q_h) \qquad\quad = \chi(q_h) \qquad \forall q_h \in M_h, \tag{11.31c}$$

which is equivalent to a square system of linear equations. Unique solvability of (11.31) follows from Theorem 10.1 as long as the associated Babuška-Brezzi conditions are verified. It has to be understood that the conditions of Theorem 10.1 for the bilincar forms in the continuous spaces $V \times M$ do not imply the conditions in the discrete spaces $V_h \times M_h$. For starters, the kernel associated to the restriction in (11.31)

$$V_{h,0} := \{u_h \in V_h : b(u_h, q_h) = 0 \quad \forall q_h \in M_h\}$$

might not be a subspace of $V_0 := \{u \in V : b(u, \cdot) = 0\}$. Necessary and sufficient conditions (Theorem 10.1 and comments thereafter) are: there exist constants $\alpha_h > 0$ and $\beta_h > 0$ such that

$$\sup_{0 \neq u_h \in V_{h,0}} \frac{a(u_h, v_h)}{\|u_h\|_V} \geq \alpha_h \|v_h\|_V \qquad \forall v_h \in V_{h,0}, \tag{11.32a}$$

$$\sup_{0 \neq u_h \in V_h} \frac{b(u_h, q_h)}{\|u_h\|_V} \geq \beta_h \|q_h\|_M \qquad \forall q_h \in M_h. \tag{11.32b}$$

We do not need another condition for the bilinear form since the first condition implies

$$\sup_{0 \neq v_h \in V_{h,0}} \frac{a(u_h, v_h)}{\|v_h\|_V} \geq \alpha_h \|u_h\|_V \qquad \forall u_h \in V_{h,0},$$

with exactly the same constant. Using the theory of mixed methods

$$\|(u_h, p_h)\|_{V \times M} \leq \left(1 + \frac{C_a}{\beta_h}\right)^2 \max\{\alpha_h^{-1}, \beta_h^{-1}\} \|(\ell, \chi)\|_{V'_h \times M'_h}, \qquad (11.33)$$

where

$$\|(\ell, \chi)\|^2_{V'_h \times M'_h} = \left(\sup_{0 \neq v_h \in V_h} \frac{\ell(v_h)}{\|v_h\|_V}\right)^2 + \left(\sup_{0 \neq q_h \in M_h} \frac{\chi(q_h)}{\|q_h\|_M}\right)^2 \leq \|(\ell, \chi)\|_{V' \times M'}.$$

We want to see how relevant the constants α_h and β_h are in a sharp stability estimate like (11.33). Let us choose the best possible (largest possible) constants in (11.32), that is,

$$\alpha_h := \inf_{0 \neq v_h \in V_{h,0}} \left(\sup_{0 \neq u_h \in V_{h,0}} \frac{a(u_h, v_h)}{\|u_h\|_V \|v_h\|_V}\right),$$

$$\beta_h := \inf_{0 \neq q_h \in M_h} \left(\sup_{0 \neq u_h \in V_h} \frac{b(u_h, q_h)}{\|u_h\|_V \|q_h\|_M}\right),$$

and consider the best possible constant $C_h > 0$ to deliver

$$\|(u_h, p_h)\|_{V \times M} \leq C_h \|(\ell, \chi)\|_{V'_h \times M'_h},$$

that is, C_h is the norm of the operator $(\ell|_{V_h}, \chi|_{M_h}) \mapsto (u_h, p_h)$. Note that since we are in finite dimensions, and all infimums and supremums are taken over the unit ball, all of them can be susbtituted by minimums and maximums. The following result shows that the best (largest) estimate of α_h and β_h must remain bounded as we follow an h-tagged sequence of subspaces $V_h \times M_h$ to define Galerkin approximations.

Proposition 11.6. *In the above notation, we have*

$$C_h \geq \max\{\alpha_h^{-1}, \beta_h^{-1}\}.$$

Proof. Let us choose (recall that the infimum is a minimum)

$$p_h \in M_h, \qquad \|p_h\|_M = 1, \qquad \sup_{0 \neq v_h \in V_h} \frac{b(v_h, p_h)}{\|v_h\|_V} = \beta_h,$$

and then define $\ell := b(\cdot, p_h)$, $\chi = 0$, so that

$$\|(\ell, \chi)\|_{V'_h \times M'_h} = \beta_h.$$

The solution to (11.31) with these data is $(0, p_h)$ and therefore $1 \leq C_h \beta_h$. Now take

$$u_h \in V_{h,0}, \qquad \|u_h\|_V = 1, \qquad \sup_{0 \neq v_h \in V_{h,0}} \frac{a(u_h, v_h)}{\|v_h\|_V} = \alpha_h,$$

and then solve

$$p_h \in M_h, \qquad b(v_h, p_h) = -a(u_h, v_h) \qquad \forall v_h \in V_{h,0}^\perp,$$

which exists and is unique. We then define $\ell := a(u_h, \cdot) + b(\cdot, p_h)$ and $\chi := 0$ and note that

$$\sup_{0 \neq v_h \in V_h} \frac{\ell(v_h)}{\|v_h\|_V} = \sup_{0 \neq v_h \in V_{h,0}} \frac{a(u_h, v_h)}{\|v_h\|_V} = \alpha_h,$$

because $\ell(v_h) = 0$ for all $v_h \in V_{h,0}^\perp$. Therefore,

$$1 \leq \|(u_h, p_h)\|_{V \times M} \leq C_h \alpha_h,$$

and the proof is finished. $\qquad\qquad\qquad\qquad\qquad\qquad\qquad\qquad\qquad\qquad$ $\square$

Final comments and literature

Problems with the penalized mixed structure studied by Proposition 11.1 are ubiquitous in mechanical applications, especially in problems with singular perturbation parameters. The finite element textbook of Dietrich Braess [19] contains several important examples of this. The proof of Proposition 11.1 is distilled from [19, Chapter III.4], where it is shown in the context of singular perturbations. The abstract problem handled by Proposition 11.2 appears, probably for the first time, in a paper by Christine Bernardi, Claudio Canuto, and Yvon Maday [12]. Double side conditions were handled in the weak imposition of essential boundary conditions in mixed formulations by Ivo Babuška and Gabriel Gatica [9]. The systematic approach of Proposition 11.4 was given in [55]. The formulation of the Stokes-Darcy problem given in Section 11.5 appears in the work of William Layton, Friedhelm Schieweck, and Ivan Yotov [73]. Obtaining sharp estimates for Galerkin discretizations of mixed problems is of great relevance in the finite element community and has led to careful presentations for general and restricted cases, taking care of all discretization constants [10, 103]. For more results on mixed structures (including the popular augmented Lagrangian formulations), we again recommend the monograph [25].

Exercises

11.1. Prove Proposition 11.2.

11.2. More convection-diffusion problems. Give a variational formulation and study its well-posedness, for a problem that looks for $(\mathbf{q}, u) \in \mathbf{H}(\mathrm{div}, \Omega) \times L^2(\Omega)$ satisfying

$$\mathbf{q} + \kappa \nabla u + u\beta = \mathbf{0}, \qquad \nabla \cdot \mathbf{q} = f, \qquad \gamma u = g.$$

Here $\kappa \in L^\infty(\Omega)$ is strongly positive and $\beta \in L^\infty(\Omega; \mathbb{R}^d)$ satisfies $\nabla \cdot \beta \leq 0$. (**Hint.** Modify the statement and proof of Proposition 11.3 to include the weight κ^{-1} in the $\mathbf{L}^2(\Omega)$ inner product. This weight should also appear in the inner product leading to the definition of the operator P.)

11.3. A realization of the Neumann-to-Dirichlet operator. Consider the space $H_0^{-1/2}(\Gamma) := \{h \in H^{-1/2}(\Gamma) : \langle h, 1 \rangle_\Gamma = 0\}$ and the operator $H_0^{-1/2}(\Gamma) \ni h \longmapsto Th := \eta \in H^{1/2}(\Gamma)$, where

$$(\mathbf{q}, u, \eta) \in \mathbf{H}(\mathrm{div}, \Omega) \times L_\circ^2(\Omega) \times H^{1/2}(\Gamma),$$

$$(\kappa^{-1}\mathbf{q}, \mathbf{r})_\Omega - (\nabla \cdot \mathbf{r}, u)_\Omega + \langle \mathbf{r} \cdot \mathbf{n}, \eta \rangle_\Gamma = 0 \qquad \forall \mathbf{r} \in \mathbf{H}(\mathrm{div}, \Omega),$$

$$(\nabla \cdot \mathbf{q}, v)_\Omega \qquad\qquad = 0 \qquad \forall v \in L_\circ^2(\Omega),$$

$$\langle \mathbf{q} \cdot \mathbf{n}, \mu \rangle_\Gamma \qquad\qquad = \langle h, \mu \rangle_\Gamma \qquad \forall \mu \in H^{1/2}(\Gamma).$$

Show that

$$\langle h, Tg \rangle_\Gamma = \langle g, Th \rangle_\Gamma \qquad \forall h, g \in H_0^{-1/2}(\Gamma)$$

and

$$\langle h, Th \rangle_\Gamma \geq c\|h\|_{-1/2,\Gamma}^2 \qquad \forall h \in H_0^{-1/2}(\Gamma).$$

What is the range of T?

11.4. A regular perturbation of the Stokes problem. Consider the problem looking for $\mathbf{u}_\varepsilon \in \mathbf{H}_0^1(\Omega)$ and $p_\varepsilon \in L_\circ^2(\Omega)$ such that

$$-\nu \Delta \mathbf{u}_\varepsilon + \nabla p_\varepsilon = \mathbf{f}, \qquad \nabla \cdot \mathbf{u}_\varepsilon + \varepsilon p_\varepsilon = 0,$$

for given $\mathbf{f} \in \mathbf{L}^2(\Omega)$. Show that it is well posed and that its solution converges strongly to the solution of the Stokes problem as $\varepsilon \to 0$. (**Hint.** Prove first that there exists C independent of ε such that $\|p_\varepsilon\|_\Omega \leq C\|\mathbf{f}\|_\Omega$ by checking carefully how the constant C in Proposition 11.1 depends on ε. To study the limit, look at the Stokes-like problem satisfied by the pair $(\mathbf{u}_\varepsilon - \mathbf{u}_0, p_\varepsilon - p)$.)

12

Nonlinear problems

12.1   Lipschitz strongly monotone operators   277
12.2   An embedding theorem   279
12.3   Laminar Navier-Stokes flow   282
12.4   A nonlinear diffusion problem   286
12.5   The Browder-Minty theorem   289
12.6   A nonlinear reaction-diffusion problem   292
Final comments and literature   293
Exercises   293

In this chapter we choose three model problems to explore extensions of the theory of elliptic PDE on Lipschitz domains to some nonlinear operators. We will handle: the Navier-Stokes problem with small data (or large viscosity) using the Banach fixed point theorem; a nonlinear diffusion problem which can be rewritten in terms of a Lipschitz strongly monotone operator; and a reaction-diffusion problem with cubic reaction as an application of the Browder-Minty theorem. Two of these problems will motivate us to take a stroll through the dense forest of theorems involving the embedding of Sobolev spaces in L^p spaces.

12.1   Lipschitz strongly monotone operators

We begin this section with the very well-known Banach fixed point theorem, applicable for contraction maps in complete metric spaces, which we will here present in the case of contractions in Banach spaces.

Theorem 12.1 (Banach fixed point theorem). *Let X be a Banach space and let $\Phi : X \to X$ be a function such that there exists $C < 1$ satisfying*

$$\|\Phi(u) - \Phi(v)\| \leq C\|u - v\| \qquad \forall u, v \in X. \qquad (12.1)$$

In these conditions, there exists a unique $u \in X$ such that $u = \Phi(u)$.

Proof. First of all, let us remark that contractivity (12.1) implies continuity of Φ. Uniqueness of a fixed point (in case it exists) is also a straightforward consequence of the strict contractivity of Φ. Finally, existence is proved by building a sequence through repeated iteration of the map Φ: we start with an arbitrary $u_0 \in X$ and then define

$$u_{n+1} := \Phi(u_n) \qquad n \geq 0. \tag{12.2}$$

Contractivity easily proves that for all $n \geq 0$,

$$\|u_{n+1} - u_n\| = \|\Phi(u_n) - \Phi(u_{n-1})\| \leq C\|u_n - u_{n-1}\| \leq C^n\|u_1 - u_0\|,$$

and therefore

$$\|u_{n+m} - u_n\| \leq \|u_{n+m} - u_{n+m-1}\| + \ldots + \|u_{n+2} - u_{n+1}\| + \|u_{n+1} - u_n\|$$
$$\leq (C^{n+m-1} + \ldots C^{n+1} + C^n)\|u_1 - u_0\| \leq \frac{C^n}{1-C}\|u_1 - u_0\|.$$

This implies that the sequence $\{u_n\}$ is Cauchy and therefore convergent to some $u \in X$. Finally, taking the limit on both sides of (12.2), it follows that u is a fixed point of Φ. $\qquad\square$

We will use Banach's fixed point theorem directly to prove the existence of solutions to the Navier-Stokes equations with some restrictions (on the size of the data and the viscosity parameter), but, before that, let us show how this theorem can be used to prove well-posedness of a class of operator equations.

Theorem 12.2. *Let V be a Hilbert space and $F : V \to V$ be uniformly Lipschitz*

$$\|F(u) - F(v)\| \leq L\|u - v\| \qquad \forall u, v \in V, \tag{12.3}$$

and strongly monotone

$$(F(u) - F(v), u - v) \geq \alpha\|u - v\|^2 \qquad \forall u, v \in V, \tag{12.4}$$

for some positive L and α. The map F is invertible and the inverse map $F^{-1} : V \to V$ is uniformly Lipschitz.

Proof. Note that it follows from the conditions (12.3) and (12.4) that necessarily $\alpha \leq L$. Strong monotonocity (12.4) implies injectivity of F and we only need to prove surjectivity. Let $w \in V$ and let us consider the continuous map

$$\Phi(u) := u + \frac{\alpha}{L^2}(w - F(u)).$$

We note that a fixed point of Φ is a solution of the equation $F(u) = w$. Since

$$\Phi(u) - \Phi(v) = u - v - \frac{\alpha}{L^2}(F(u) - F(v)),$$

a simple computation, using hypotheses (12.3) and (12.4), shows that

$$\|\Phi(u) - \Phi(v)\|^2 = \|u - v\|^2 - \frac{2\alpha}{L^2}(F(u) - F(v), u - v) + \frac{\alpha^2}{L^4}\|F(u) - F(v)\|^2$$
$$\leq \left(1 - \frac{2\alpha}{L} + \frac{\alpha^2}{L^2}\right)\|u - v\|^2 = \left(1 - \frac{\alpha}{L}\right)^2\|u - v\|^2.$$

Since $0 < \alpha \le L$, we have that $1 - \alpha/L < 1$, which proves that Φ is a contraction and therefore it has a unique fixed point. This proves the surjectivity of F. Using (12.4) again, we prove that

$$\alpha \|F^{-1}(u) - F^{-1}(v)\|^2 \le (u - v, F^{-1}(u) - F^{-1}(v)) \le \|u - v\| \, \|F^{-1}(u) - F^{-1}(v)\|,$$

and therefore F^{-1} is Lipschitz with parameter $1/\alpha$. $\qquad\square$

Note that the last part of the proof can be detached from the rest, and it asserts that if $F : V \to V$ is surjective and strongly monotone, then F^{-1} exists and is uniformly Lipschitz.

12.2   An embedding theorem

In this section we are going to prove the following theorem, which is a very particular case of a collection of so-called Sobolev embedding theorems.

Theorem 12.3. *If $\Omega \subset \mathbb{R}^d$ is a bounded domain with the H^1-extension property and $d \le 3$, then $H^1(\Omega) \subset L^4(\Omega)$ with continuous inclusion.*

Before we go for the proof, which is relatively involved in the three-dimensional case, let us make some quick comments:

(a) The result in the one-dimensional case ($d = 1$) holds for any open interval (the H^1-extension property holds for any interval) as we have already proved (Exercise 4.11) that

$$H^1(a, b) \subset \mathcal{C}[a, b],$$

with continuous inclusion. This proves that $H^1(a, b) \subset L^p(a, b)$ for any $1 \le p \le \infty$, with continuous inclusion.

(b) In the three-dimensional case, we will prove that $H^1(\mathbb{R}^3) \subset L^6(\mathbb{R}^3)$ with continuous inclusion and then we will use the extension property and the boundedness of Ω to prove the result. Here we will obtain essentially an 'optimal' result.

(c) We will take a shortcut in the two-dimensional case proving the continuous inclusion of $H_0^1(Q)$ into $L^4(Q)$ for any $Q = (-M, M)^2$. This result can be improved (we can embed $H^1(\Omega)$ into $L^p(\Omega)$ for a larger value of p), but since we will only need L^4 spaces for the analysis of the Navier-Stokes equations, we will stick to this particular case.

Lemma 12.1. *The following inequality holds*

$$\left(\int_{\mathbb{R}^3} |u(\mathbf{x})|^{3/2} d\mathbf{x} \right)^{2/3} \le \int_{\mathbb{R}^3} |\nabla u(\mathbf{x})| d\mathbf{x} \qquad \forall u \in \mathcal{D}(\mathbb{R}^3).$$

Proof. It is clear that for all $\mathbf{x} = (x_1, x_2, x_3) \in \mathbb{R}^3$,

$$|u(\mathbf{x})| \leq \int_{-\infty}^{\infty} |\partial_{x_1} u(t_1, x_2, x_3)| dt_1,$$

and using similar expressions for the other two variables, we have

$$|u(\mathbf{x})|^{\frac{3}{2}} \leq \left(\int_{-\infty}^{\infty} |\partial_{x_1} u(t_1, x_2, x_3)| dt_1 \right)^{\frac{1}{2}}$$
$$\left(\int_{-\infty}^{\infty} |\partial_{x_2} u(x_1, t_2, x_3)| dt_2 \right)^{\frac{1}{2}} \left(\int_{-\infty}^{\infty} |\partial_{x_3} u(x_1, x_2, t_3)| dt_3 \right)^{\frac{1}{2}}.$$

Integrating in the variable x_1 and using the Cauchy-Schwarz inequality we can the prove that

$$\int_{-\infty}^{\infty} |u(x_1, x_2, x_3)|^{\frac{3}{2}} dx_1 \leq \left(\int_{-\infty}^{\infty} |\partial_{x_1} u(t_1, x_2, x_3)| dt_1 \right)^{\frac{1}{2}}$$
$$\left(\int_{-\infty}^{\infty} \int_{-\infty}^{\infty} |\partial_{x_2} u(t_1, t_2, x_3)| dt_1 dt_2 \right)^{\frac{1}{2}}$$
$$\left(\int_{-\infty}^{\infty} \int_{-\infty}^{\infty} |\partial_{x_3} u(t_1, x_2, t_3)| dt_1 dt_3 \right)^{\frac{1}{2}}.$$

Integrating now in the variable x_2 and again using the Cauchy-Schwarz inequality it follows that

$$\int_{-\infty}^{\infty} \int_{-\infty}^{\infty} |u(x_1, x_2, x_3)|^{\frac{3}{2}} dx_1 dx_2 \leq \left(\int_{-\infty}^{\infty} \int_{-\infty}^{\infty} |\partial_{x_1} u(t_1, t_2, x_3)| dt_1 dt_2 \right)^{\frac{1}{2}}$$
$$\left(\int_{-\infty}^{\infty} \int_{-\infty}^{\infty} |\partial_{x_2} u(t_1, t_2, x_3)| dt_1 dt_2 \right)^{\frac{1}{2}}$$
$$\left(\int_{\mathbb{R}^3} |\partial_{x_3} u(\mathbf{t})| d\mathbf{t} \right)^{\frac{1}{2}}.$$

Finally, the same argument in the variable x_3 shows that

$$\int_{\mathbb{R}^3} |u(\mathbf{x})|^{\frac{3}{2}} d\mathbf{x} \leq \left(\int_{\mathbb{R}^3} |\partial_{x_1} u(\mathbf{x})| d\mathbf{x} \right)^{\frac{1}{2}} \left(\int_{\mathbb{R}^3} |\partial_{x_2} u(\mathbf{x})| d\mathbf{x} \right)^{\frac{1}{2}} \left(\int_{\mathbb{R}^3} |\partial_{x_3} u(\mathbf{x})| d\mathbf{x} \right)^{\frac{1}{2}}$$
$$\leq \left(\int_{\mathbb{R}^3} |\nabla u(\mathbf{x})| d\mathbf{x} \right)^{\frac{3}{2}},$$

and the result follows. $\qquad\square$

Proposition 12.1. *The following inequality holds*

$$\|u\|_{L^6(\mathbb{R}^3)} \leq 4\|\nabla u\|_{\mathbb{R}^3} \qquad \forall u \in H^1(\mathbb{R}^3).$$

In particular $H^1(\mathbb{R}^3)$ is continuously embedded into $L^6(\mathbb{R}^3)$.

Proof. The inequality only needs to be proved for $u \in \mathcal{D}(\mathbb{R}^3)$, as the result then follows by density. We just apply Lemma 12.1 to u^4, noticing that

$$\|u^4\|_{L^{3/2}(\mathbb{R}^3)} = \|u\|_{L^6(\mathbb{R}^3)}^4,$$

$$\|\nabla u^4\|_{L^1(\mathbb{R}^3)} = 4\|u^3 \nabla u\|_{L^1(\mathbb{R}^3)} \le 4\|u^3\|_{\mathbb{R}^3}\|\nabla u\|_{\mathbb{R}^3},$$

$$\|u^3\|_{\mathbb{R}^3} = \|u\|_{L^6(\mathbb{R}^3)}^3,$$

as can be proved with elementary computations. $\qquad\square$

Proposition 12.2. *If $\Omega \subset \mathbb{R}^3$ is a bounded domain with the H^1-extension property, then there exists $C = C(\Omega, p) > 0$ such that*

$$\|u\|_{L^p(\Omega)} \le C\|u\|_{1,\Omega} \qquad \forall u \in H^1(\Omega), \quad p \in [1,6].$$

Proof. Using an extension operator and Proposition 12.1, we have that the diagram

$$H^1(\Omega) \to H^1(\mathbb{R}^3) \to L^6(\mathbb{R}^3) \to L^6(\Omega)$$

writes the embedding of $H^1(\Omega)$ into $L^6(\Omega)$ as the composition of a bounded extension, the embedding of Proposition 12.1, and the restriction operator from $\mathbb{R}^3$ to Ω, which proves the result for $p = 6$. Since in a bounded domain $L^6(\Omega) \subset L^p(\Omega)$ for $1 \le p < 6$ (this follows from Hölder's inequality, due to the fact that constant functions are in $L^p(\Omega)$ for all p), the result follows. *Note that this proves Theorem 12.3 in the case $d = 3$.* $\qquad\square$

Proposition 12.3. *If $\Omega \subset \mathbb{R}^2$ is a bounded domain with the H^1-extension property, then there exists $C > 0$ such that*

$$\|u\|_{L^4(\Omega)} \le C\|u\|_{1,\Omega} \qquad \forall u \in H^1(\Omega).$$

Proof. An argument very similar (while simpler) to the one in Lemma 12.1 shows that

$$\|u\|_{\mathbb{R}^2} \le \|\nabla u\|_{L^1(\mathbb{R}^2)} \qquad \forall u \in \mathcal{D}(\mathbb{R}^2).$$

Applying this to u^2, where $u \in \mathcal{D}(Q)$ and $Q = (-M, M)^2$, we have

$$\|u\|_{L^4(Q)}^2 = \|u^2\|_{\mathbb{R}^2} \le 2\|u \nabla u\|_{L^1(Q)}$$

$$\le 2\|u\|_Q \|\nabla u\|_Q \le C_Q \|\nabla u\|_Q^2 \qquad \forall u \in \mathcal{D}(Q),$$

where in the last inequality we have applied the Poincaré-Friedrichs inequality in Q. This and a density argument prove that $H_0^1(Q)$ is continuously embedded into $L^4(Q)$. Now using an extension operator to a sufficiently large box Q (and multiplying by a cutoff function after extending to $H^1(\mathbb{R}^2)$), we have the following sequence of embeddings

$$H^1(\Omega) \to H^1(\mathbb{R}^2) \to H_0^1(Q) \to L^4(Q) \to L^4(\Omega),$$

from which we can easily prove the embedding property. $\qquad\square$

12.3   Laminar Navier-Stokes flow

In this section we are going to explore the existence and uniqueness of solutions to the steady-state Navier-Stokes equations with some restrictions on the size of a parameter or the data. We will develop the theory for a general bounded Lipschitz domain $\Omega \subset \mathbb{R}^d$ with $d \leq 3$, with homogeneous Dirichlet boundary conditions. The equations involve a positive parameter $\nu > 0$ (the kinematic viscosity) and the nonlinear differential operator: given a vector field $\mathbf{w} \in \mathbf{L}^2(\Omega) = L^2(\Omega; \mathbb{R}^d)$, we consider the differential operator

$$\mathbf{w} \cdot \nabla : \mathbf{H}^1(\Omega) = H^1(\Omega; \mathbb{R}^d) \to \mathbf{L}^1(\Omega)$$

given by

$$(\mathbf{w} \cdot \nabla \mathbf{u})_i = \sum_{j=1}^d w_j \partial_{x_j} u_i = \mathbf{w} \cdot \nabla u_i.$$

Given $\mathbf{f} \in \mathbf{L}^2(\Omega)$ we look for

$$(\mathbf{u}, p) \in \mathbf{H}_0^1(\Omega) \times L_0^2(\Omega), \tag{12.5a}$$
$$-\nu \Delta \mathbf{u} + \mathbf{u} \cdot \nabla \mathbf{u} + \nabla p = \mathbf{f}, \tag{12.5b}$$
$$\nabla \cdot \mathbf{u} = 0. \tag{12.5c}$$

Note first that these equations are a generalization of the Stokes equations (Section 10.8) and that the Oseen equations (Exercise 10.10) can formally be understood as a linearization of (12.5) as well. Note also that if $\mathbf{u} \in \mathbf{H}^1(\Omega)$ and $p \in L^2(\Omega)$ the left-hand side of (12.5b) is the sum of a distribution with values in $\mathbf{H}^{-1}(\Omega)$ and another one (the nonlinear-linear term) with values in $\mathbf{L}^1(\Omega)$. As a matter of fact, because of Theorem 12.3, the function $\mathbf{u} \cdot \nabla \mathbf{u}$ takes values in the stronger space $\mathbf{L}^{4/3}(\Omega)$ (this is where we need $d \leq 3$ in our arguments), as can be proved using Hölder's inequality.

A trilinear form. Let us now introduce the trilinear form $a : \mathbf{H}^1(\Omega) \times \mathbf{H}^1(\Omega) \times \mathbf{H}^1(\Omega) \to \mathbb{R}$ given by

$$a(\mathbf{u}, \mathbf{v}; \mathbf{w}) := (\mathbf{w} \cdot \nabla \mathbf{u}, \mathbf{v})_\Omega = \sum_{i,j=1}^d \int_\Omega w_j(\mathbf{x}) \partial_{x_j} u_i(\mathbf{x}) v_i(\mathbf{x}) d\mathbf{x}$$

$$= \sum_{i,j=1}^d (\partial_{x_j} u_i, v_i w_j)_\Omega = (\nabla \mathbf{u}, \mathbf{v} \otimes \mathbf{w})_\Omega,$$

where in the last expression we have used the algebraic notation for a tensor product $\mathbf{v} \otimes \mathbf{w} = \mathbf{v} \mathbf{w}^\top$ and $(\nabla \mathbf{u})_{i,j} = \partial_{x_j} u_i$. A simple application of the Cauchy-Schwarz inequality shows that the product of two functions in $L^4(\Omega)$ is in $L^2(\Omega)$ and therefore the above integrals are meaningful.

Lemma 12.2. *There exists $C > 0$ such that*

$$|a(\mathbf{u}, \mathbf{w}; \mathbf{w})| \leq C\|\mathbf{u}\|_{1,\Omega}\|\mathbf{v}\|_{1,\Omega}\|\mathbf{w}\|_{1,\Omega} \qquad \forall \mathbf{u}, \mathbf{v}, \mathbf{w} \in \mathbf{H}^1(\Omega).$$

Proof. Using the Cauchy-Schwarz inequality twice it is simple to see that

$$\begin{aligned}
|(u, v\,w)_\Omega| &\leq \|u\|_\Omega \|v\,w\|_\Omega \leq \|u\|_\Omega \|v^2\|_\Omega^{1/2} \|w^2\|_\Omega^{1/2} \\
&= \|u\|_\Omega \|v\|_{L^4(\Omega)} \|w\|_{L^4(\Omega)} \qquad \forall u \in L^2(\Omega), \quad v, w \in L^4(\Omega).
\end{aligned}$$

Applying this inequality to each term of the sum that defines the trilinear form a, and recalling the bounded inclusion $H^1(\Omega) \subset L^4(\Omega)$ (Theorem 12.3), the result follows. Note that the constant C depends on the constant of the bounded inclusion. □

Before we go for a variational formulation of (12.5), let us show the following identity for the trilinear form.

Lemma 12.3. *If $\mathbf{w} \in \mathbf{H}^1(\Omega)$ satisfies $\nabla \cdot \mathbf{w} = 0$, then*

$$a(\mathbf{u}, \mathbf{v}; \mathbf{w}) + a(\mathbf{v}, \mathbf{u}; \mathbf{w}) = 0 \qquad \forall \mathbf{u}, \mathbf{v} \in \mathbf{H}_0^1(\Omega).$$

Proof. If $\mathbf{u}, \mathbf{v} \in \mathcal{D}(\Omega)^d$, then it is easy to show that

$$\begin{aligned}
a(\mathbf{u}, \mathbf{v}; \mathbf{w}) + a(\mathbf{v}, \mathbf{u}; \mathbf{w}) &= \sum_{i,j=1}^d \int_\Omega w_i(\mathbf{x}) \left(v_j(\mathbf{x})\partial_{x_i} u_j(\mathbf{x}) + u_j(\mathbf{x})\partial_{x_i} v_j(\mathbf{x})\right) d\mathbf{x} \\
&= \sum_{i,j=1}^d \int_\Omega w_i(\mathbf{x})\partial_{x_i}(u_j v_j)(\mathbf{x}) d\mathbf{x} \\
&= \sum_{j=1}^d \langle \mathbf{w}, \nabla(u_j v_j)\rangle_{\mathcal{D}'(\Omega)^d \times \mathcal{D}(\Omega)^d} \\
&= -\sum_{j=1}^d \langle \nabla \cdot \mathbf{w}, u_j v_j\rangle_{\mathcal{D}'(\Omega)\times\mathcal{D}(\Omega)} = 0.
\end{aligned}$$

The result for $\mathbf{u}, \mathbf{v} \in \mathbf{H}_0^1(\Omega)$ then follows by density, using Lemma 12.2 to show that the bilinear form $a(\cdot, \cdot; \mathbf{w})$ is bounded. □

Variational formulation. Testing the Navier-Stokes equations with $\mathbf{v} \in \mathcal{D}(\Omega)^d$ and applying Lemma 12.2 it is clear that any solution of (12.5) satisfies

$$\begin{aligned}
(\mathbf{u}, p) &\in \mathbf{H}_0^1(\Omega) \times L_\circ^2(\Omega), & &\text{(12.6a)} \\
\nu(\nabla\mathbf{u}, \nabla\mathbf{v})_\Omega + a(\mathbf{u}, \mathbf{v}; \mathbf{u}) - (p, \nabla \cdot \mathbf{v})_\Omega &= (\mathbf{f}, \mathbf{v})_\Omega & \forall \mathbf{v} \in \mathbf{H}_0^1(\Omega), & &\text{(12.6b)} \\
(\nabla \cdot \mathbf{u}, q)_\Omega &= 0 & \forall q \in L_\circ^2(\Omega). & &\text{(12.6c)}
\end{aligned}$$

We recall from Section 10.8 that (12.6c) implies that $\nabla \cdot \mathbf{u} = 0$ because of the homogeneous boundary condition for $\mathbf{u}$. It is then simple to show that (12.6) implies (12.5), and therefore (12.5) and (12.6) are equivalent.

A reduced problem. Instead of working on the pair $(\mathbf{u}, p)$, we will now derive a formulation where only $\mathbf{u}$ appears as an unknown. The recovery of the pressure field will be done using the same inf-sup condition from the Stokes problem (see Section 10.8). We will detail this at the end of the section. For the moment being, consider the space of solenoidal vector fields

$$V := \{\mathbf{u} \in \mathbf{H}_0^1(\Omega) \ : \ \nabla \cdot \mathbf{u} = 0\},$$

which is a closed subspace of $\mathbf{H}_0^1(\Omega)$. It is clear that if $(\mathbf{u}, p)$ solves (12.6), then $\mathbf{u}$ solves

$$\mathbf{u} \in V, \qquad \nu(\nabla \mathbf{u}, \nabla \mathbf{v})_\Omega + a(\mathbf{u}, \mathbf{v}; \mathbf{u}) = (\mathbf{f}, \mathbf{v})_\Omega \qquad \forall \mathbf{v} \in V. \qquad (12.7)$$

Our next step is finding an operational form for (12.7). Given $\mathbf{w} \in \mathbf{H}^1(\Omega)$, we can define the bounded linear operator $A(\mathbf{w}) : V \to V$ such that

$$(A(\mathbf{w})\mathbf{u}, \mathbf{v})_{1,\Omega} = \nu(\nabla \mathbf{u}, \nabla \mathbf{v})_\Omega + a(\mathbf{u}, \mathbf{v}; \mathbf{w}) \qquad \forall \mathbf{u}, \mathbf{v} \in V.$$

This can be easily done using the Riesz-Fréchet representation theorem in V and Lemma 12.2. We can also find $\mathbf{b} \in V$ such that

$$(\mathbf{b}, \mathbf{v})_{1,\Omega} = (\mathbf{f}, \mathbf{v})_\Omega \qquad \forall \mathbf{v} \in V,$$

and note that

$$\|\mathbf{b}\|_{1,\Omega} \leq \|\mathbf{f}\|_\Omega. \qquad (12.8)$$

The reduced nonlinear variational problem (12.7) is then equivalent to

$$\mathbf{u} \in V, \qquad A(\mathbf{u})\mathbf{u} = \mathbf{b}. \qquad (12.9)$$

Lemma 12.4. *Let $C_{\mathrm{PF}} > 0$ be the constant from the Poincaré-Friedrichs inequality in Ω, written in the following form*

$$C_{\mathrm{PF}}\|\mathbf{u}\|_{1,\Omega} \leq \|\nabla \mathbf{u}\|_\Omega \qquad \forall \mathbf{u} \in \mathbf{H}_0^1(\Omega). \qquad (12.10)$$

For all $\mathbf{w} \in V$, the operator $A(\mathbf{w}) : V \to V$ is invertible and

$$\|A(\mathbf{w})^{-1}\|_{V \to V} \leq \frac{1}{\nu C_{\mathrm{PF}}^2}.$$

Proof. By Lemma 12.3 and (12.10), we have

$$(A(\mathbf{w})\mathbf{u}, \mathbf{u})_{1,\Omega} = \nu\|\nabla \mathbf{u}\|_\Omega^2 \geq \nu C_{\mathrm{PF}}^2 \|\mathbf{u}\|_{1,\Omega} \qquad \forall \mathbf{u} \in V.$$

This and Lemma 12.2 prove that the bilinear form

$$V \times V \ni (\mathbf{u}, \mathbf{v}) \longmapsto (A(\mathbf{w})\mathbf{u}, \mathbf{v})_{1,\Omega} \in \mathbb{R}$$

is bounded and coercive. By applying the Lax-Milgram lemma, the result follows. □

Proposition 12.4. *Let $C > 0$ be the constant from Lemma 12.2 and $C_{PF} > 0$ be the constant from the Poincaré-Friedrichs inequality (12.10). If $f \in L^2(\Omega)$ satisfies*

$$\frac{C}{\nu^2 C_{PF}^4}\|f\|_\Omega < 1, \tag{12.11}$$

then equations (12.7) (or equivalently (12.9)) are uniquely solvable.

Proof. Consider the function $\Phi : V \to V$ given by $\Phi(u) := A(u)^{-1}b$. This nonlinear function is well-defined by Lemma 12.4. Note now that by Lemma 12.2

$$|((A(w_1) - A(w_2))u, v)_{1,\Omega}| = |a(u, v; w_1 - w_2)|$$
$$\leq C\|u\|_{1,\Omega}\|v\|_{1,\Omega}\|w_1 - w_2\|_{1,\Omega},$$

and therefore

$$\|A(w_1) - A(w_2)\|_{V \to V} \leq C\|w_1 - w_2\|_{1,\Omega} \qquad \forall w_1, w_2 \in H^1(\Omega),$$

which proves that $A : H^1(\Omega) \to \mathcal{B}(V)$ is Lipschitz. Now using Lemma 12.4 and the identity

$$A(w_1)^{-1} - A(w_2)^{-1} = A(w_1)^{-1}(A(w_2) - A(w_1))A(w_2)^{-1} \qquad \forall w_1, w_2 \in V,$$

we obtain the bound

$$\|A(w_1)^{-1} - A(w_2)^{-1}\|_{V \to V} \leq \frac{C}{\nu^2 C_{PF}^2}\|w_1 - w_2\|_{1,\Omega} \qquad \forall w_1, w_2 \in V, \tag{12.12}$$

which proves that $A(\cdot)^{-1} : V \to \mathcal{B}(V)$ is also Lipschitz. Finally, by (12.12) and (12.8)

$$\|\Phi(w_1) - \Phi(w_2)\|_{1,\Omega} = \|(A(w_1)^{-1} - A(w_2)^{-1})b\|_{1,\Omega}$$
$$\leq \frac{C}{\nu^2 C_{PF}^2}\|w_1 - w_2\|_{1,\Omega}\|b\|_{1,\Omega}$$
$$\leq \frac{C}{\nu^2 C_{PF}^2}\|f\|_\Omega\|w_1 - w_2\|_{1,\Omega} \qquad \forall w_1, w_2 \in V.$$

This proves that if the inequality (12.11) holds, $\Phi : V \to V$ is a contraction and therefore has (as per Banach's fixed point theorem) a unique fixed point, i.e., there exists a unique

$$u \in V, \qquad \Phi(u) = u. \tag{12.13}$$

However, (12.13) is equivalent to (12.9), which finishes the proof. $\square$

Proposition 12.5. *If the conditions of Proposition 12.4 hold, equations (12.5) have a unique solution.*

Proof. We first consider $\mathbf{u}$ that solves (12.7) and define the bounded linear form in $\mathbf{H}_0^1(\Omega)$:

$$\ell(\mathbf{v}) := (\mathbf{v}, \mathbf{v})_\Omega - \nu(\nabla \mathbf{u}, \nabla \mathbf{v})_\Omega - a(\mathbf{u}, \mathbf{v}; \mathbf{u})$$

and associate the unique element

$$\mathbf{w} \in \mathbf{H}_0^1(\Omega), \qquad (\mathbf{w}, \mathbf{v})_{1,\Omega} = \ell(\mathbf{v}) \qquad \forall \mathbf{v} \in \mathbf{H}_0^1(\Omega).$$

The fact that $\mathbf{u}$ solves (12.7) is equivalent to ℓ vanishing on V and also to $\mathbf{w} \in V^\perp$. Now let $B : \mathbf{H}_0^1(\Omega) \to L_\circ^2(\Omega)$ be the operator given by the equality

$$(B\mathbf{v}, p)_\Omega = (p, \nabla \cdot \mathbf{v})_\Omega \qquad \forall \mathbf{v} \in \mathbf{H}_0^1(\Omega), \qquad p \in L_\circ^2(\Omega).$$

The inf-sup condition (see Proposition 10.9) is equivalent to the surjectivity of B and also (recall Propositions 10.1 and 10.3, dealing with inf-sup conditions and surjectivity) to $B^* : L_\circ^2(\Omega) \to (\ker B)^\perp$ being an isomorphism. However, $\ker B = V$ (we have already seen this) and therefore, there exists a unique $p \in L_\circ^2(\Omega)$ such that $B^* p = -\mathbf{w}$, or equivalently such that (12.6b) holds. This proves that (12.6) (and therefore (12.5)) has a solution.

To prove uniqueness we observe that if $(\mathbf{u}_1, p_1)$ and $(\mathbf{u}_2, p_2)$ are two solutions of (12.5), then $\mathbf{u}_1$ and $\mathbf{u}_2$ are two solutions of (12.7). However, (12.7) is uniquely solvable and therefore $\mathbf{u}_1 = \mathbf{u}_2$. Using the operator B defined above, we now have $B^* p_1 = B^* p_2$, but since B^* is injective, this proves that $p_1 = p_2$. □

Inequality (12.11) can be seen as an upper bound on the size of data or as a lower bound on the viscosity. Taking into account the two geometric constants (one related to the continuous injection of $H^1(\Omega)$ into $L^4(\Omega)$ and the other coming from the Poincaré-Friedrichs inequality), (12.11) says that on a given domain, the Navier-Stokes equations have a unique solution if $\|\mathbf{f}\|_\Omega$ is small enough for a given fixed viscosity ν, or if ν is large enough for given data $\mathbf{f}$. The larger values of ν physically correspond to laminar flow, while very small values are related to turbulence. (We do not make any claim that Proposition 12.4 implies that the given restriction (11.1) separates these two regimes.) In practical applications, the Navier-Stokes equations are always given in terms of the Reynolds number, which is proportional to the inverse of the viscosity.

12.4   A nonlinear diffusion problem

In this section we apply the results of Section 12.1 to the nonlinear problem

$$u \in H_0^1(\Omega), \qquad -\nabla \cdot (\kappa(|\nabla u|) \nabla u) = f, \tag{12.14}$$

for arbitrary $f \in L^2(\Omega)$ under some conditions on the nonlinear diffusivity parameter $\kappa : [0, \infty) \to \mathbb{R}$ that we will set up next. We assume that:

(a) $\kappa \in \mathcal{C}^1([0,\infty))$,

(b) there exist $C_1, C_2 > 0$ such that

$$C_1 \leq \kappa(r) \leq C_2 \qquad \forall r \in [0,\infty),$$

(c) there exists $C_3 \in (0, C_1/\sqrt{d})$ such that

$$|r\kappa'(r)| \leq C_3 \qquad \forall r \in [0,\infty).$$

An example of such a function is

$$\kappa(r) := \alpha_0 + \frac{\alpha_1}{\beta + r},$$

where

$$\alpha_0, \beta > 0, \qquad |\alpha_1| < \tfrac{4}{5}\alpha_0\beta.$$

Before we start studying the solvability of (12.14), let us have a look at the equations and deal with some technical issues. First of all, note that $\kappa(|\nabla u|) \in L^\infty(\Omega)$, since ∇u is measurable and κ is continuous and bounded. Therefore, if $u \in H^1(\Omega)$, we have that $\kappa(|\nabla u|)\nabla u \in \mathbf{L}^2(\Omega)$ and the equations (12.14) make sense.

For the next set of results, consider the function

$$\mathbf{k}(\mathbf{p}) := \kappa(|\mathbf{p}|)\mathbf{p}, \tag{12.15}$$

so that we can write the equation in (12.14) as

$$-\nabla \cdot \mathbf{k}(\nabla u) = f.$$

Lemma 12.5. *The function* $\mathbf{k} : \mathbb{R}^d \to \mathbb{R}^d$ *defined in* (12.15) *is* $\mathcal{C}^1$, *uniformly Lipschitz, and satisfies*

$$(\mathbf{k}(\mathbf{p}) - \mathbf{k}(\mathbf{q})) \cdot (\mathbf{p} - \mathbf{q}) \geq (C_1 - C_3\sqrt{d})|\mathbf{p} - \mathbf{q}|^2 \qquad \forall \mathbf{p}, \mathbf{q} \in \mathbb{R}^d.$$

Proof. It is clear that $\mathbf{k}$ is continuous. Note also that

$$\partial_{p_j} k_i(\mathbf{p}) = \kappa(|\mathbf{p}|)\delta_{ij} + \frac{\kappa'(|\mathbf{p}|)}{|\mathbf{p}|}p_j p_i.$$

In particular,

$$|\partial_{p_j} k_i(\mathbf{p})| \leq C_2 + C_3 \qquad \forall \mathbf{p} \in \mathbb{R}^d,$$

and by the mean value theorem in $\mathbb{R}^d$

$$|k_i(\mathbf{p}) - k_i(\mathbf{q})| \leq (C_2 + C_3)|\mathbf{p} - \mathbf{q}| \qquad \forall \mathbf{p}, \mathbf{q} \in \mathbb{R}^d, \qquad i = 1, \ldots, d.$$

This proves that $\mathbf{k}$ is uniformly Lipschitz.

Now take points $\mathbf{c}_i \in \mathbb{R}^d$ $(i = 1, \ldots, d)$ and an arbitrary vector $\mathbf{v} \in \mathbb{R}^d$. Since (we write c_{ii} to denote the i-th component of $\mathbf{c}_i$)

$$\nabla k_i(\mathbf{c}_i) \cdot \mathbf{v} = \kappa(|\mathbf{c}_i|)v_i + \frac{\kappa'(|\mathbf{c}_i|)}{|\mathbf{c}_i|}(\mathbf{c}_i \cdot \mathbf{v})c_{ii},$$

we have

$$\sum_{i=1}^{d} v_i(\nabla k_i(\mathbf{c}_i) \cdot \mathbf{v}) = \sum_{i=1}^{d} \kappa(|\mathbf{c}_i|)v_i^2 + \sum_{i=1}^{d} \frac{\kappa'(|\mathbf{c}_i|)}{|\mathbf{c}_i|}(\mathbf{c}_i \cdot \mathbf{v})c_{ii}v_i$$
$$\geq C_1|\mathbf{v}|^2 - C_3\sqrt{d}|\mathbf{v}|^2,$$

by the hypotheses on κ. Using the mean value theorem, we then have

$$(\mathbf{k}(\mathbf{p}) - \mathbf{k}(\mathbf{q})) \cdot (\mathbf{p} - \mathbf{q}) = \sum_{i=1}^{d}(k_i(\mathbf{p}) - k_i(\mathbf{q}))(p_i - q_i)$$
$$= \sum_{i=1}^{d} \left(\nabla k_i(\mathbf{c}_i) \cdot (\mathbf{p} - \mathbf{q})\right)(p_i - q_i),$$

which finishes the proof. □

A nonlinear operator. For each $u \in H_0^1(\Omega)$, there exists (by the Riesz-Fréchet theorem) a unique $F(u) \in H_0^1(\Omega)$ such that

$$(F(u), v)_{1,\Omega} = (\kappa(|\nabla u|)\nabla u, \nabla v)_\Omega = (\mathbf{k}(\nabla u), \nabla v)_\Omega \qquad \forall v \in H_0^1(\Omega).$$

We thus have defined a nonlinear operator $F : H_0^1(\Omega) \to H_0^1(\Omega)$ associated to the boundary value problem (12.14). We can also find $w \in H_0^1(\Omega)$ such that

$$(w, v)_{1,\Omega} = (f, v)_\Omega \qquad \forall v \in H_0^1(\Omega).$$

It is clear that (12.14) is equivalent to

$$u \in H_0^1(\Omega), \qquad (\kappa(|\nabla u|)\nabla u, \nabla v)_\Omega = (f, v)_\Omega \qquad \forall v \in H_0^1(\Omega),$$

and therefore to the operator equation

$$u \in H_0^1(\Omega), \qquad F(u) = w.$$

Proposition 12.6. *The operator F is uniformly Lipschitz and strongly monotone. Therefore (12.14) is uniquely solvable and the solution operator $f \mapsto u$ is uniformly Lipschitz.*

Proof. By the Riesz-Fréchet theorem we have

$$\|F(u) - F(v)\|_{1,\Omega} \leq C\|\mathbf{k}(\nabla u) - \mathbf{k}(\nabla v)\|_\Omega,$$

where the constant C is that of the Poincaré-Friedrichs inequality. Since by Lemma 12.5 there exists $L > 0$ such that

$$|\mathbf{k}(\nabla u) - \mathbf{k}(\nabla v)| \leq L|\nabla u - \nabla v| \qquad \text{a.e.,}$$

it follows that F is uniformly Lipschitz. Also by Lemma 12.5 we have

$$(\mathbf{k}(\nabla u) - \mathbf{k}(\nabla v)) \cdot (\nabla u - \nabla v) \geq \alpha|\nabla u - \nabla v|^2 \qquad \text{a.e.,}$$

the operator F is strongly monotone. Theorem 12.2 then proves the unique solvability of (12.14) and the Lipschitz character of the solution operator. $\square$

12.5 The Browder-Minty theorem

We will admit the following fixed point theorem of modern analysis, whose proof is nontrivial.

> **[Proof not provided]**
>
> **Theorem 12.4** (Brouwer's fixed point theorem)**.** *If $B \subset \mathbb{R}^N$ is a closed Euclidean ball and $F : B \to B$ is continuous, then F has a fixed point.*

We will use this result to prove a fixed point theorem for a class of operator equations in separable Hilbert spaces. The first result is a simple consequence of Brouwer's fixed point theorem applied in the context of continuous operators in finite-dimensional spaces.

Lemma 12.6. *If V is a finite-dimensional real inner product space and $F : V \to V$ is continuous and coercive, i.e.,*

$$\lim_{\|u\| \to \infty} \frac{(F(u), u)}{\|u\|} = \infty,$$

then F is surjective.

Proof. Let $w \in V$. Our goal is to find $u \in V$ such that $F(u) = w$. We define

$$G(u) := F(u) - w$$

(so that now we look for a root of G) and note that

$$\frac{(G(u), u)}{\|u\|} = \frac{(F(u), u)}{\|u\|} - \frac{(w, u)}{\|u\|} \geq \frac{(F(u), u)}{\|u\|} - \|w\|,$$

which implies that G is coercive. We can thus choose $R > 0$ such that

$$(G(u), u) \geq 0 \qquad \forall u \text{ s.t. } \|u\| = R. \tag{12.16}$$

Now consider the set $B_V(R) := \{u \in V : \|u\| < R\}$. If there exists $u \in \overline{B_V(R)}$ such that $G(u) = 0$, we are done. Let us assume that $G(u) \neq 0$ for all $u \in \overline{B_V(R)}$ and define

$$H(u) := -\frac{R}{\|G(u)\|}G(u).$$

Clearly $H : \overline{B_V(R)} \to \partial B_V(R)$ is continuous. By Brouwer's theorem (Theorem 12.4) there exists $u \in \overline{B_V(R)}$ such that $H(u) = u$. (Note that the ball $B_V(R)$ can be identified with a Euclidean ball by using an orthonormal basis of V.) However, this implies that $u \in \partial B_V(R)$ and we have

$$G(u) = -\frac{\|G(u)\|}{R}u,$$

and thus

$$(G(u), u) = -\frac{\|G(u)\|}{R}\|u\|^2 = -\|G(u)\|R < 0,$$

which contradicts our choice of R in (12.16). □

Warning. The concept of coercive operator used in Lemma 12.6 and Theorem 12.5 is more general than the one we use for bilinear forms. Namely, if F is linear and the bilinear form $(F(u), v)$ is coercive, then so is F, but the reciprocal statement does not hold.

Theorem 12.5 (Browder-Minty). *If V is a real separable Hilbert space and $F : V \to V$ is continuous, bounded, coercive and strictly monotone, i.e.,*

$$(F(u) - F(v), u - v) > 0 \qquad \forall u \neq v \in V,$$

then F is invertible.

Proof. If F is strictly monotone, it is clear that F is injective, so we only need to prove surjectivity. We will do it using Galerkin's method. Let $\{\phi_n\}$ be a Hilbert basis of the separable space V. For $n \geq 1$ we define

$$V_n := \text{span}\{\phi_1, \ldots, \phi_n\},$$

and consider the orthogonal projection $P_n : V \to V_n$. We then define

$$F_n : V_n \to V_n, \qquad F_n := P_n F|_{V_n},$$

which is clearly continuous, coercive since

$$\frac{(F_n(u), u)}{\|u\|} = \frac{(F(u), u)}{\|u\|} \qquad \forall u \in V_n,$$

and strictly monotone since

$$(F_n(u) - F_n(v), u - v) = (F(u) - F(v), u - v) \qquad \forall u, v \in V_n.$$

Therefore (Lemma 12.6) F_n is invertible. We now take $w \in V$ and build the sequence $\{u_n\}$ where

$$u_n \in V_n, \qquad F_n(u_n) = P_n w \qquad \forall n \geq 1.$$

Note that u_n is the only solution of the nonlinear variational equations

$$u_n \in V_n, \qquad (F(u_n), v) = (w, v) \qquad \forall v \in V_n,$$

and therefore

$$\frac{(F(u_n), u_n)}{\|u_n\|} \leq \|w\|.$$

Since F is coercive the sequence $\{u_n\}$ is bounded, and since F is bounded, so is $\{F(u_n)\}$. Therefore, we can find a subsequence and two elements $u, \tilde{u} \in V$ such that

$$u_{n_k} \rightharpoonup u, \qquad F(u_{n_k}) \rightharpoonup \tilde{u}.$$

Since $\{\phi_n\}$ is a Hilbert basis of V, we have that $P_n v \to v$ for all $v \in V$ and thus

$$(\tilde{u}, v) \longleftarrow (F(u_{n_k}), P_{n_k} v) = (w, P_{n_k} v) \longrightarrow (w, v) \qquad \forall v \in V,$$

which implies that $\tilde{u} = w$, so that $u_{n_k} \rightharpoonup u$ and $F(u_{n_k}) \rightharpoonup w$. We thus have

$$(F(u_{n_k}) - F(v), u_{n_k} - v) \geq 0 \qquad\qquad \forall v \in V,$$
$$(F(u_{n_k}), u_{n_k}) = (w, u_{n_k}) \longrightarrow (w, u),$$
$$(F(v), u_{n_k}) \longrightarrow (F(v), u) \qquad \forall v \in V,$$
$$(F(u_{n_k}), v) \longrightarrow (w, u),$$

and therefore

$$(w - F(v), u - v) \geq 0 \qquad \forall v \in V. \tag{12.17}$$

We then take $u - t v$ in place of v in (12.17) and we obtain

$$t\, (w - F(u - t v), v) \geq 0 \qquad \forall v \in V, \qquad \forall t \in \mathbb{R}.$$

Taking the limit as $t \to 0^+$ and using that F is continuous, it follows that

$$(w - F(u), v) \geq 0 \qquad \forall v \in V.$$

This clearly implies that $F(u) = w$. Since this construction can be done for any $w \in V$, we have proved that F is surjective. $\qquad\square$

We note that Browder-Minty's theorem holds for general Hilbert spaces, but the well-known proof we have given uses the separability in a strong way.

12.6   A nonlinear reaction-diffusion problem

As a simple model problem to apply the Browder-Minty theorem, consider a bounded open set $\Omega \subset \mathbb{R}^3$ with the H^1-extension property. The restriction of working in three dimensions is due to the fact that we want $H^1(\Omega) \subset L^6(\Omega)$ with bounded injection (see Proposition 12.2). We have not proved this result in domains in the plane, although it also holds. We then consider $f \in L^2(\Omega)$ as data and look for a solution to

$$u \in H_0^1(\Omega), \qquad -\Delta u + u^3 = f. \tag{12.18}$$

First of all, we have that

$$\|u^3\|_\Omega \le \|u\|_{L^6(\Omega)}^3 \le C\|u\|_{1,\Omega} \qquad \forall u \in H^1(\Omega), \tag{12.19}$$

as follows from Proposition 12.2. We will also use the following result (Exercise 12.8).

Lemma 12.7. *The map $H^1(\Omega) \ni u \mapsto u^3 \in L^2(\Omega)$ is continuous.*

Now consider the function $F : H_0^1(\Omega) \to H_0^1(\Omega)$ given by

$$F(u) \in H_0^1(\Omega), \qquad (F(u), v)_{1,\Omega} = (\nabla u, \nabla v)_\Omega + (u^3, v)_\Omega \qquad \forall v \in H_0^1(\Omega),$$

that is, with the usual inner product in $H^1(\Omega)$ (note that we are free to use the Dirichlet form as well), we are solving

$$F(u) \in H_0^1(\Omega), \qquad -\Delta F(u) + F(u) = -\Delta u + u^3.$$

Next, we are going to verify the hypotheses of the Browder-Minty theorem (Theorem 12.5) for the function F in the space $H_0^1(\Omega)$. First of all, by (12.19), we have that for all $u, v \in H_0^1(\Omega)$ we can bound

$$|(F(u), v)_{1,\Omega}| \le (\|\nabla u\|_\Omega^2 + \|u^3\|_\Omega^2)^{1/2}\|v\|_{1,\Omega} \le \|u\|_{1,\Omega}(1 + C^2\|u\|_{1,\Omega}^4)^{1/2}\|v\|_{1,\Omega}$$

and therefore

$$\|F(u)\|_{1,\Omega} \le \|u\|_{1,\Omega}(1 + C^2\|u\|_{1,\Omega}^4)^{1/2},$$

which shows that F is bounded. Also

$$(F(u), u)_{1,\Omega} = \|\nabla u\|_\Omega^2 + (u^3, u)_\Omega \ge \|\nabla u\|_\Omega^2,$$

which proves that F is coercive. Since

$$(u^3 - v^3)(u - v) = (u - v)^2(u^2 + uv + v^2) \ge (u - v)^2(|u|^2 - |v|^2)^2 \ge 0,$$

it follows that

$$(F(u) - F(v), u - v)_{1,\Omega} = (\nabla u - \nabla v, \nabla u - \nabla v)_\Omega + (u^3 - v^3, u - v)_\Omega \ge \|\nabla u - \nabla v\|^2,$$

and F is strictly monotone. Finally, by the Cauchy-Schwarz inequality, we can estimate

$$\|F(u) - F(v)\|_{1,\Omega} \leq (\|\nabla u - \nabla v\|_{\Omega}^2 + \|u^3 - v^3\|_{\Omega}^2)^{1/2},$$

and the continuity of F is now an easy consequence of Lemma 12.7. Equation (12.18) can be written in the form

$$F(u) = \omega_f,$$

where

$$\omega_f \in H_0^1(\Omega), \qquad -\Delta\omega_f + \omega_f = f,$$

and therefore it has a unique solution, since F is invertible.

Final comments and literature

L.E.J. Brower's fixed point theorem is one of the most popular theorems of point set topology and one that has been proved in surprisingly many different and independent ways. A quick visit to the corresponding Wikipedia article can convince the reader that we are in the presence of a rich field of interactions of many different branches of pure and applied mathematics. The proof we offer of the Browder-Minty theorem (named after Felix Browder and George Minty) for separable Hilbert spaces is adapted from the monograph of John Tinsley Oden [88] (see also [53]).

For more on the amazingly complex world of the Navier-Stokes equation, see the monographs of Giovanni Galdi [51, 50], Roger Temam [100], and Girault and Raviart [57], for instance. The embedding theorem(s) of Section 12.2 are the tip of yet another iceberg of embedding theorems of Sobolev spaces in L^p spaces. The proofs that we give here for the very particular cases we are interested in, are distilled from the third volume of Michael Taylor's treatise on partial differential equations [99].

Exercises

12.1. Céa's estimate for Lipschitz strongly monotone operators. Let $F : V \to V$ be Lipschitz and strongly monotone in a Hilbert space V and let $V_h \subset V$ be a finite-dimensional subspace. Show that for any $w \in V$, the problem

$$u_h \in V_h, \qquad (F(u_h), v) = (w, v) \qquad \forall v \in V_h,$$

has a unique solution and

$$\|u - u_h\| \leq \left(1 + \frac{L}{\alpha}\right) \inf_{v_h \in V_h} \|u - v_h\|.$$

(**Hint.** Denoting $P : V \to V_h$ for the orthogonal projection onto V_h, prove that the problem is equivalent to $F_h(u_h) = P_h w$, where $F_h := P_h F|_{V_h}$ is Lipschitz and strongly monotone in V_h. Prove then that

$$\|u_h - v_h\| \leq \frac{L}{\alpha} \|u - v_h\| \qquad \forall v_h \in V_h,$$

using the hypotheses on F.)

12.2. Finish the details of the proof of Proposition 12.3.

12.3. Show that we can put $\mathbf{f} \in \mathbf{H}^{-1}(\Omega)$ as data of the Navier-Stokes equation (12.5) and derive a tight condition on the size of $\mathbf{f}$ guaranteeing the existence and uniqueness of solutions.

12.4. Prove that if $F : V \to V$ is Lipschitz, then F is bounded.

12.5. Prove that if $F : V \to V$ is strongly monotone in the Hilbert space V, then F is coercive. (This exercise and the previous one prove that the hypotheses of Theorem 12.2 imply the hypotheses of the Browder-Minty theorem.)

12.6. Prove unique solvability of the problem

$$u \in H^1(\Omega), \qquad -\nabla \cdot (\kappa(|\nabla u|)\nabla u) = f, \qquad \gamma u = g,$$

where $f \in L^2(\Omega)$, $g \in H^{1/2}(\Gamma)$ and κ is as in Section 12.4.

12.7. Let $\kappa : \Omega \times \mathbb{R} \to \mathbb{R}$ be given by

$$\kappa(\mathbf{x}, r) := a_0(\mathbf{x}) + \frac{a_1(\mathbf{x})}{1 + r},$$

where $a_0, a_1 \in L^\infty(\Omega)$, with a_0 strongly positive. Give hypotheses on the coefficient a_1 guaranteeing the well-posedness of the problem

$$u \in H_0^1(\Omega), \qquad -\nabla \cdot (\kappa(\cdot, |\nabla u|)\nabla u) = f,$$

for arbitrary $f \in L^2(\Omega)$.

12.8. Prove Lemma 12.7. (**Hint.** If you write

$$u^3 - v^3 = (u - v)^3 + 3uv(u - v),$$

it is easy to show that the map $L^6(\Omega) \ni u \mapsto u^3 \in L^2(\Omega)$ is continuous.)

13

Fourier representation of Sobolev spaces

13.1   The Fourier transform in the Schwartz class 296
13.2   A first mix of Fourier and Sobolev 300
13.3   An introduction to H^2 regularity 302
13.4   Topology of the Schwartz class 307
13.5   Tempered distributions 311
13.6   Sobolev spaces by Fourier transforms 314
13.7   The trace space revisited 318
13.8   Interior regularity 321
Final comments and literature 323
Exercises 323

The Schwartz class is a vector space of rapidly decaying smooth functions defined in all $\mathbb{R}^d$. As we will see in the coming sections, the Schwartz class is a complete metric space (it fits in the general construction of Fréchet spaces), which contains $\mathcal{D}(\mathbb{R}^d)$ as a dense subset. Because of this density, the dual space of the Schwartz class will be identifiable to a subset of the space of distributions: they will be called tempered distributions, because, in a way, they are allowed to 'grow' at infinity in a moderate way that can be tackled by the quick decay of the test functions. Perhaps the main use of the Schwartz class (and, by extension, of its dual space) is due to the fact that in it the Fourier transform is an isomorphism. We will proceed with the introduction of this material (which brings together harmonic analysis and PDE theory) very slowly, following the program:

(a) Study the Schwartz class and the Fourier transform in it, absent of any topological (functional analysis) structure.

(b) Extend the Fourier transform to an isometric isomorphism in $L^2(\mathbb{R}^d)$ and recognize functions in $H^1(\mathbb{R}^d)$ and $H^2(\mathbb{R}^d)$ through their Fourier transforms. At this point we will spend a little time in examining the concept of H^2 regularity of the Laplacian.

(c) Study the metrizable topology of the Schwartz class (as an example of Fréchet spaces, which we will introduce as well) and of its dual space, whose elements are called tempered distributions.

(d) Define a class of Sobolev spaces tagged in a real parameter, and show that for positive integer values of the parameter we recover the spaces $H^m(\mathbb{R}^d)$.

(e) Digress on two interesting topics which help explain why we called the trace space $H^{1/2}(\Gamma)$ and how this space exists 'independently' of the trace operator.

(f) Finally, look into additional interior (away from the boundary) regularity of solutions to the Laplace equation.

Warning. In this chapter all functions will be **complex-valued** and we will not warn about it anymore. In particular, the L^p and H^m spaces that we consider now are those whose elements are complex-valued functions. We will give a different name to the newly defined Sobolev spaces, but then prove that they are the same.

13.1   The Fourier transform in the Schwartz class

Notation. The following polynomials will be highly relevant in the arguments that follow:

$$m_\beta(\mathbf{x}) := (-2\pi i)^{|\beta|} \prod_{i=1}^d x_i^{\beta_i} = \prod_{i=1}^d (-2\pi i x_i)^{\beta_i} \qquad \beta = (\beta_1, \ldots, \beta_d) \in \mathbb{N}^d,$$

$$m_i(\mathbf{x}) := m_{\mathbf{e}_i}(\mathbf{x}) = -2\pi i x_i \qquad\qquad i = 1, \ldots, d.$$

Note that

$$m_\beta = \prod_{i=1}^d m_i^{\beta_i}.$$

The Schwartz class. We say that $\varphi : \mathbb{R}^d \to \mathbb{C}$ is an element of the Schwartz class, and we write $\varphi \in \mathcal{S}(\mathbb{R}^d)$, when:

(a) $\varphi \in \mathcal{C}^\infty(\mathbb{R}^d)$,

(b) $p\,\partial^\alpha \varphi \in L^\infty(\mathbb{R}^d)$ for all $\alpha \in \mathbb{N}^d$ and $p \in \mathcal{P}(\mathbb{R}^d)$.

Here we have used $\mathcal{P}(\mathbb{R}^d)$ to denote the space of polynomials of d variables. At this stage we do not need a topology or a meaning of convergence for $\mathcal{S}(\mathbb{R}^d)$, although it is clear that $\mathcal{S}(\mathbb{R}^d)$ is a subspace of $\mathcal{C}^\infty(\mathbb{R}^d)$ and that it contains

$\mathcal{D}(\mathbb{R}^d)$. (The topology of the Schwartz class will be carefully explained in Section 13.4.) The following properties of the Schwartz class are simple to prove and left to the reader as Exercise 13.1.

Proposition 13.1. (a) *The function $\phi(\mathbf{x}) := \exp(-|\mathbf{x}|^2)$ is in the Schwartz class. Therefore $\mathcal{D}(\mathbb{R}^d)$ is a proper subspace of $\mathcal{S}(\mathbb{R}^d)$.*

(b) *If $\varphi \in \mathcal{S}(\mathbb{R}^d)$, then $p\partial^\alpha \varphi \in \mathcal{S}(\mathbb{R}^d)$ for all $p \in \mathcal{P}(\mathbb{R}^d)$ and $\alpha \in \mathbb{N}^d$.*

(c) *If $\varphi \in \mathcal{C}^\infty(\mathbb{R}^d)$, then*

$$\varphi \in \mathcal{S}(\mathbb{R}^d) \qquad \Longleftrightarrow \qquad m_\beta \partial^\alpha \varphi \in L^\infty(\mathbb{R}^d) \qquad \forall \alpha, \beta \in \mathbb{N}^d.$$

We also have the following property concerning integrability and the Schwartz class:

Proposition 13.2. *If $\varphi \in \mathcal{S}(\mathbb{R}^d)$, then*

$$p\partial^\alpha \varphi \in L^1(\mathbb{R}^d) \cap L^2(\mathbb{R}^d) \qquad \forall p \in \mathcal{P}(\mathbb{R}^d), \qquad \forall \alpha \in \mathbb{N}^d.$$

Proof. Considering the function

$$f(\mathbf{x}) := |\mathbf{x}|^{2d} p(\mathbf{x})(\partial^\alpha \varphi)(\mathbf{x}),$$

which is bounded. It is clear that

$$|p(\mathbf{x})(\partial^\alpha \varphi)(\mathbf{x})| \leq \frac{1}{R^{2d}}\|f\|_{L^\infty(\mathbb{R}^d)} \qquad |\mathbf{x}| \geq R.$$

This can easily be used to prove the statement. $\qquad\qquad\square$

The Fourier transform and the inverse Fourier transform. Given $\varphi \in \mathcal{S}(\mathbb{R}^d)$, we define the functions $\mathcal{F}\{\varphi\}, \mathcal{F}^*\{\varphi\} : \mathbb{R}^d \to \mathbb{C}$ by the expressions

$$\mathcal{F}\{\varphi\}(\boldsymbol{\xi}) := \int_{\mathbb{R}^d} \exp(-2\pi\imath \mathbf{x} \cdot \boldsymbol{\xi})\varphi(\mathbf{x})\mathrm{d}\mathbf{x},$$

$$\mathcal{F}^*\{\varphi\}(\boldsymbol{\xi}) := \int_{\mathbb{R}^d} \exp(2\pi\imath \mathbf{x} \cdot \boldsymbol{\xi})\varphi(\mathbf{x})\mathrm{d}\mathbf{x} = \mathcal{F}\{\varphi\}(-\boldsymbol{\xi}) = \overline{\mathcal{F}\{\overline{\varphi}\}(\boldsymbol{\xi})}.$$

Proposition 13.3. *If $\varphi \in \mathcal{S}(\mathbb{R}^d)$, we have:*

(a) $|\mathcal{F}\{\varphi\}(\boldsymbol{\xi})| \leq \|\varphi\|_{L^1(\mathbb{R}^d)}$ *for all $\boldsymbol{\xi} \in \mathbb{R}^d$, and therefore $\mathcal{F}\{\varphi\}$ is bounded,*

(b) $\mathcal{F}\{\varphi\} \in \mathcal{C}(\mathbb{R}^d)$,

(c) $\partial_{\xi_i}\mathcal{F}\{\varphi\} = \mathcal{F}\{m_i\varphi\}$ *for all i,*

(d) $\mathcal{F}\{\partial_{x_i}\varphi\} = -m_i\mathcal{F}\{\varphi\}$ *for all i.*

Proof. We will just give hints for the proofs and will let the reader to prove them as Exercise 13.2. Property (a) follows from Proposition 13.2. To prove (b), use the dominated convergence theorem and to prove (c), the theorem about differentiation under integral sign. Note that $m_i\varphi \in \mathcal{S}(\mathbb{R}^d)$ by Proposition 13.1(b) and therefore the right-hand side of the equality makes sense. To prove (d) first show that for fixed $\boldsymbol{\xi} \in \mathbb{R}^d$ and $\psi \in \mathcal{S}(\mathbb{R}^d)$ we have

$$\lim_{R\to\infty} \left| \int_{\partial B(\mathbf{0};R)} \exp(-2\pi\imath\mathbf{x} \cdot \boldsymbol{\xi})\psi(\mathbf{x})n_i(\mathbf{x})\mathrm{d}\Gamma(\mathbf{x}) \right| = 0,$$

and

$$\mathcal{F}\{\psi\}(\boldsymbol{\xi}) = \lim_{R\to\infty} \int_{B(\mathbf{0};R)} \exp(-2\pi\imath\mathbf{x} \cdot \boldsymbol{\xi})\psi(\mathbf{x})\mathrm{d}\mathbf{x}.$$

Use these to prove the result. $\qquad\square$

Proposition 13.4. *We have $\mathcal{F}\{\varphi\} \in \mathcal{S}(\mathbb{R}^d)$ for all $\varphi \in \mathcal{S}(\mathbb{R}^d)$.*

Proof. Using induction on Proposition 13.3(c) (and Proposition 13.1 that allows you to keep on multiplying elements of the Schwartz class by polynomials), it is easy to show that

$$\partial^\alpha \mathcal{F}\{\varphi\} = \mathcal{F}\{m_\alpha\varphi\} \qquad \forall \alpha \in \mathbb{N}^d. \tag{13.1}$$

By Proposition 13.3(b), it follows that $\mathcal{F}\{\varphi\} \in \mathcal{C}^\infty(\mathbb{R}^d)$. Now using (13.1) and induction based on Proposition 13.3(d) it follows that

$$m_\beta \partial^\alpha \mathcal{F}\{\varphi\} = (-1)^{|\beta|} \mathcal{F}\{\partial^\beta(m_\alpha\varphi)\}.$$

Noting that $\partial^\beta(m_\alpha\varphi) \in \mathcal{S}(\mathbb{R}^d)$, we have from Proposition 13.3(a) that $m_\beta \partial^\alpha \mathcal{F}\{\varphi\}$ is bounded for all α and β, which implies (Proposition 13.1(c)) that $\mathcal{F}\{\varphi\} \in \mathcal{S}(\mathbb{R}^d)$, as we wanted to prove. $\qquad\square$

Lemma 13.1. *The function $\phi(\mathbf{x}) := \exp(-\pi|\mathbf{x}|^2)$ is a fixed point of the Fourier transform, i.e., $\mathcal{F}\{\phi\} = \phi$.*

Proof. This is a well-known clever computation. It all starts with the one-dimensional version $f(t) := \exp(-\pi t^2)$ and with its Fourier transform $g := \mathcal{F}\{f\} \in \mathcal{S}(\mathbb{R})$. First, note that f is the unique solution to the initial value problem

$$f'(t) = -2\pi t f(t) \qquad t \geq 0, \qquad f(0) = 1.$$

We can write the above as

$$f' = -\imath m f \qquad (m(t) := -2\pi\imath t),$$

and then take the Fourier transform on both sides (using the differentiation and multiplication rules of Proposition 13.3(c) and (d)) to obtain

$$-mg = -\imath g',$$

and therefore

$$g'(\xi) = -2\pi\xi g(\xi) \qquad \xi \in \mathbb{R},$$

which proves that $g = g(0)f$. However,

$$g(0) = \int_{-\infty}^{\infty} e^{-\pi t^2} dt = 1, \tag{13.2}$$

which proves that $\mathcal{F}\{f\} = f$, i.e.,

$$\int_{-\infty}^{\infty} e^{-2\pi i \xi t} e^{-\pi t^2} dt = e^{-\pi\xi^2} \qquad \forall \xi \in \mathbb{R}. \tag{13.3}$$

Applying (13.3) d times, we have

$$\mathcal{F}\{\phi\}(\boldsymbol{\xi}) = \prod_{i=1}^{d} \left(\int_{-\infty}^{\infty} e^{-2\pi i \xi_i x_i} e^{-\pi x_i^2} dx_i \right) = \prod_{i=1}^{d} e^{-\pi\xi_i^2} = \phi(\boldsymbol{\xi}) \qquad \forall \boldsymbol{\xi} \in \mathbb{R}^d,$$

which proves the result. The reader who has never seen the computation (13.2) might require an explanation of this tricky, but well-known integral: we can write

$$g(0)^2 = \left(\int_{-\infty}^{\infty} e^{-\pi x_1^2} dx_1 \right) \left(\int_{-\infty}^{\infty} e^{-\pi x_2^2} dx_2 \right) = \int_{\mathbb{R}^2} e^{-\pi(x_1^2 + x_2^2)} dx_1 dx_2,$$

and then use a change to polar coordinates to compute the right-hand side of the above. $\qquad \square$

Proposition 13.5. *We have*

$$\mathcal{F}^*\{\mathcal{F}\{\varphi\}\} = \varphi \qquad \forall \varphi \in \mathcal{S}(\mathbb{R}^d), \tag{13.4}$$

and therefore $\mathcal{F} : \mathcal{S}(\mathbb{R}^d) \to \mathcal{S}(\mathbb{R}^d)$ *is invertible with* $\mathcal{F}^{-1} = \mathcal{F}^*$.

Proof. We first prove that for all $\varphi \in \mathcal{S}(\mathbb{R}^d)$

$$\int_{\mathbb{R}^d} \mathcal{F}\{\varphi\}(\boldsymbol{\xi}) e^{2\pi i \boldsymbol{\xi} \cdot \mathbf{x}} d\boldsymbol{\xi} = \varphi(\mathbf{x}) \qquad \forall \mathbf{x} \in \mathbb{R}^d. \tag{13.5}$$

Let us fix $\mathbf{x} \in \mathbb{R}^d$ and write

$$\int_{\mathbb{R}^d} \mathcal{F}\{\varphi\}(\boldsymbol{\xi}) e^{2\pi i \boldsymbol{\xi} \cdot \mathbf{x}} d\boldsymbol{\xi} = \lim_{\varepsilon \to 0} \int_{\mathbb{R}^d} \mathcal{F}\{\varphi\}(\boldsymbol{\xi}) e^{2\pi i \boldsymbol{\xi} \cdot \mathbf{x}} e^{-\pi|\varepsilon\boldsymbol{\xi}|^2} d\boldsymbol{\xi},$$

which can easily be justified using the dominated convergence theorem. With a simple couple of changes of variables and Lemma 13.1, we prove that

$$\int_{\mathbb{R}^d} \mathcal{F}\{\varphi\}(\boldsymbol{\xi}) e^{2\pi i \boldsymbol{\xi} \cdot \mathbf{x}} e^{-\pi|\varepsilon\boldsymbol{\xi}|^2} d\boldsymbol{\xi}$$

$$= \int_{\mathbb{R}^d} \int_{\mathbb{R}^d} \varphi(\mathbf{y}) e^{-2\pi i \boldsymbol{\xi} \cdot (\mathbf{y} - \mathbf{x})} e^{-\pi|\varepsilon\boldsymbol{\xi}|^2} d\boldsymbol{\xi} d\mathbf{y}$$

$$= \varepsilon^{-d} \int_{\mathbb{R}^d} \varphi(\mathbf{y}) \left(\int_{\mathbb{R}^d} e^{-2\pi i \boldsymbol{\xi} \cdot \frac{1}{\varepsilon}(\mathbf{y} - \mathbf{x})} e^{-\pi|\boldsymbol{\xi}|^2} d\boldsymbol{\xi} \right) d\mathbf{y}$$

$$= \varepsilon^{-d} \int_{\mathbb{R}^d} \varphi(\mathbf{y}) e^{-\pi|\frac{1}{\varepsilon}(\mathbf{y} - \mathbf{x})|^2} d\mathbf{y} = \int_{\mathbb{R}^d} \varphi(\mathbf{x} + \varepsilon\mathbf{z}) e^{-\pi|\mathbf{z}|^2} d\mathbf{z}.$$

Therefore,

$$\int_{\mathbb{R}^d} \mathcal{F}\{\varphi\}(\boldsymbol{\xi})e^{2\pi \imath \boldsymbol{\xi} \cdot \mathbf{x}}\mathrm{d}\boldsymbol{\xi} = \lim_{\varepsilon \to 0} \int_{\mathbb{R}^d} \varphi(\mathbf{x} + \varepsilon \mathbf{z})e^{-\pi|\mathbf{z}|^2}\mathrm{d}\mathbf{z} = \varphi(\mathbf{x}),$$

after once again using the dominated convergence theorem. This proves (13.5), and therefore (13.4), which implies that $\mathcal{F}^*$ is a left inverse of $\mathcal{F}$. Changing $\boldsymbol{\xi} \leftrightarrow -\boldsymbol{\xi}$ in (13.5) (recall that $\mathcal{F}^*\{\varphi\}(\boldsymbol{\xi}) = \mathcal{F}\{\varphi\}(-\boldsymbol{\xi})$) shows that $\mathcal{F}^*$ is also a right-inverse and therefore $\mathcal{F}$ is invertible and $\mathcal{F}^*$ is its inverse. $\qquad\square$

13.2   A first mix of Fourier and Sobolev

Note that

$$\mathcal{D}(\mathbb{R}^d) \subset \mathcal{S}(\mathbb{R}^d) \subset L^2(\mathbb{R}^d)$$

and therefore $\mathcal{S}(\mathbb{R}^d)$ is a dense subspace of $L^2(\mathbb{R}^d)$. This will allow a first extension of the Fourier transform to the space $L^2(\mathbb{R}^d)$ using a conjugation property (referred to as Plancherel's identity) that will actually justify why we used the symbol for adjoint in $\mathcal{F}^*$. Recall that all our functions are complex-valued and that in inner products we take the second component to be conjugate linear.

Proposition 13.6. *For arbitrary $\varphi, \psi \in \mathcal{S}(\mathbb{R}^d)$, we have the identities :*

$$(\mathcal{F}\{\varphi\}, \psi)_{\mathbb{R}^d} = (\varphi, \mathcal{F}^*\{\psi\})_{\mathbb{R}^d}, \tag{13.6a}$$

$$(\mathcal{F}\{\varphi\}, \mathcal{F}\{\psi\})_{\mathbb{R}^d} = (\varphi, \psi)_{\mathbb{R}^d}, \tag{13.6b}$$

$$(\mathcal{F}^*\{\varphi\}, \mathcal{F}^*\{\psi\})_{\mathbb{R}^d} = (\varphi, \psi)_{\mathbb{R}^d}. \tag{13.6c}$$

Proof. Both sides of (13.6a) are equal to

$$\int_{\mathbb{R}^d} \int_{\mathbb{R}^d} \varphi(\mathbf{x})e^{-2\pi \imath \mathbf{x} \cdot \boldsymbol{\xi}}\overline{\psi(\boldsymbol{\xi})}\mathrm{d}\mathbf{x}\mathrm{d}\boldsymbol{\xi},$$

since the use of Fubini's theorem is fully justified by the integrability of φ and ψ. Note that (13.6a) means that $\mathcal{F}^*$ is the formal $L^2(\mathbb{R}^d)$ adjoint of $\mathcal{F}$. (More about this later.) Finally, by Proposition 13.5

$$(\mathcal{F}\{\varphi\}, \mathcal{F}\{\psi\})_{\mathbb{R}^d} = (\varphi, \mathcal{F}^*\{\mathcal{F}\{\psi\}\})_{\mathbb{R}^d} = (\varphi, \psi)_{\mathbb{R}^d},$$

and thus (13.6b) is proved. The proof of (13.6c) is straighforward using that $\mathcal{F}^* = \mathcal{F}^{-1}$. $\qquad\square$

Proposition 13.7. *The Fourier transform $\mathcal{F} : \mathcal{S}(\mathbb{R}^d) \to \mathcal{S}(\mathbb{R}^d)$ and its inverse $\mathcal{F}^* : \mathcal{S}(\mathbb{R}^d) \to \mathcal{S}(\mathbb{R}^d)$ admit unique extensions to (equally denoted) maps*

$$\mathcal{F}, \mathcal{F}^* : L^2(\mathbb{R}^d) \to L^2(\mathbb{R}^d)$$

that are isometric, invertible, reciprocally adjoint, and reciprocally inverse.

Proof. Proposition 13.6 proves that $\mathcal{F}$ and $\mathcal{F}^*$ are isometries with respect to the norm of $L^2(\mathbb{R}^d)$, defined in a dense subspace. Therefore, they admit a unique extension which is equally isometric. All the other statements follow by density and the fact that the extension is unique. $\qquad\square$

An important weight. We now consider the quadratic polynomial

$$\omega := 1 + |m_1|^2 + \ldots + |m_d|^2 = 1 - m_1^2 - \ldots - m_d^2 \in \mathcal{P}(\mathbb{R}^d),$$

that is, $\omega(\mathbf{x}) = 1 + |2\pi\mathbf{x}|^2$. We also consider the following set

$$L_\omega^2(\mathbb{R}^d) := \{v \in L^2(\mathbb{R}^d) \ : \ \omega^{1/2}v \in L^2(\mathbb{R}^d)\},$$

which is a Hilbert space when endowed with the norm

$$\|v\|_\omega := \|\omega^{1/2}v\|_{\mathbb{R}^d}.$$

The fact that $L_\omega^2(\mathbb{R}^d)$ is complete follows from the Riesz-Fischer theorem (completeness of L^p spaces). Note also that

$$\|v\|_{\mathbb{R}^d} \leq \|v\|_\omega \qquad \forall v \in L_\omega^2(\mathbb{R}^d)$$

(this is due to the fact that $\omega \geq 1$ everywhere), so that in the chain of inclusions

$$\mathcal{D}(\mathbb{R}^d) \subset L_\omega^2(\mathbb{R}^d) \subset L^2(\mathbb{R}^d)$$

the right-most one corresponds to a continuous embedding. Since $\mathcal{D}(\mathbb{R}^d)$ is dense in $L^2(\mathbb{R}^d)$, so is $L_\omega^2(\mathbb{R}^d)$. The following result shows how $H^1(\mathbb{R}^d)$ and $L_\omega^2(\mathbb{R}^d)$ are isometrically isomorphic via the Fourier transform.

Proposition 13.8. *If $u \in L^2(\mathbb{R}^d)$, then*

$$u \in H^1(\mathbb{R}^d) \qquad \Longleftrightarrow \qquad \mathcal{F}\{u\} \in L_\omega^2(\mathbb{R}^d),$$

and

$$\|u\|_{1,\mathbb{R}^d} = \|\mathcal{F}\{u\}\|_\omega \qquad \forall u \in H^1(\mathbb{R}^d).$$

Furthermore,

$$\mathcal{F}\{\partial_{x_i} u\} = -m_i \mathcal{F}\{u\} \qquad \forall u \in H^1(\mathbb{R}^d) \qquad i = 1, \ldots, d. \qquad (13.7)$$

Proof. Using Propositions 13.3(d) and 13.6, it is clear that

$$\|\varphi\|_{1,\mathbb{R}^d}^2 = \|\mathcal{F}\{\varphi\}\|_{\mathbb{R}^d}^2 + \sum_{i=1}^d \|\mathcal{F}\{\partial_{x_i}\varphi\}\|_{\mathbb{R}^d}^2$$

$$= \|\mathcal{F}\{\varphi\}\|_{\mathbb{R}^d}^2 + \sum_{i=1}^d \|m_i \mathcal{F}\{\varphi\}\|_{\mathbb{R}^d}^2 = \|\mathcal{F}\{\varphi\}\|_\omega^2 \qquad \forall \varphi \in \mathcal{S}(\mathbb{R}^d).$$

This means that the Fourier transform as a map $\mathcal{S}(\mathbb{R}^d) \to L^2_\omega(\mathbb{R}^d)$ is an isometry when we take the $H^1(\mathbb{R}^d)$ norm in $\mathcal{S}(\mathbb{R}^d)$. However, because of the inclusion

$$\mathcal{D}(\mathbb{R}^d) \subset \mathcal{S}(\mathbb{R}^d) \subset H^1(\mathbb{R}^d),$$

and the density of $\mathcal{D}(\mathbb{R}^d)$ in $H^1(\mathbb{R}^d)$ (seen back in Chapter 4), the Fourier transform can be extended to an isometry between $H^1(\mathbb{R}^d)$ and $L^2_\omega(\mathbb{R}^d)$. This unique extension has to be the same as the Fourier transform defined in $L^2(\mathbb{R}^d)$, as the two following maps

$$
\begin{array}{ccc}
& L^2_\omega(\mathbb{R}^d) & \xrightarrow{\text{inc}} L^2(\mathbb{R}^d) \\
\text{ext of } \mathcal{F} \nearrow & & \\
H^1(\mathbb{R}^d) & & \\
\text{inc} \searrow & & \\
& L^2(\mathbb{R}^d) & \xrightarrow{\mathcal{F}} L^2(\mathbb{R}^d)
\end{array}
$$

coincide in the dense set $\mathcal{S}(\mathbb{R}^d)$.

Now let $u \in L^2(\mathbb{R}^d)$ satisfy $\mathcal{F}\{u\} \in L^2_\omega(\mathbb{R}^d)$. We then take a sequence $\{\varphi_n\}$ in $\mathcal{D}(\mathbb{R}^d)$ such that $\varphi_n \to \mathcal{F}\{u\}$ in $L^2_\omega(\mathbb{R}^d)$ (see Exercise 13.3) and note that this is equivalent to

$$\varphi_n \to \mathcal{F}\{u\}, \qquad m_i\varphi_n \to m_i\mathcal{F}\{u\} \qquad i = 1,\dots,d,$$

with all convergences in $L^2(\mathbb{R}^d)$. Now let $\psi_n := \mathcal{F}^{-1}\{\varphi_n\} \in \mathcal{S}(\mathbb{R}^d)$ and note that by Propositions 13.3(d) and 13.7 ($\mathcal{F}^{-1}$ is an isometry in $L^2(\mathbb{R}^d)$) we have

$$\psi_n \to u, \qquad -\partial_{x_i}\psi_n = \mathcal{F}^{-1}\{m_i\varphi_n\} \to \mathcal{F}^{-1}\{m_i\mathcal{F}\{u\}\} \in L^2(\mathbb{R}^d) \qquad \forall i,$$

with all convergences taking place in $L^2(\mathbb{R}^d)$. However, $\partial_{x_i}\psi_n \to \partial_{x_i}u$ in $\mathcal{D}'(\mathbb{R}^d)$ and therefore

$$\partial_{x_i}u = -\mathcal{F}^{-1}\{m_i\mathcal{F}\{u\}\} \in L^2(\mathbb{R}^d) \qquad i = 1,\dots,d.$$

This implies that $u \in H^1(\mathbb{R}^d)$ as well as (13.7). $\qquad\square$

13.3   An introduction to H^2 regularity

Recall the space (see Exercise 2.3)

$$H^2(\mathbb{R}^d) := \{u \in \mathcal{D}'(\mathbb{R}^d) : \partial^\alpha u \in L^2(\mathbb{R}^d) \qquad |\alpha| \le 2\}$$
$$= \{u \in H^1(\mathbb{R}^d) : \partial_{x_i}u \in H^1(\mathbb{R}^d) \quad i = 1,\dots,d\},$$

and its norm

$$\|u\|_{2,\mathbb{R}^d}^2 = \sum_{|\alpha|\le 2} \|\partial^\alpha u\|_{\mathbb{R}^d}^2,$$

which can be substituted by the equivalent norm

$$\|u\|_{\mathbb{R}^d}^2 + \sum_{i=1}^{d} \|\partial_{x_i} u\|_{1,\mathbb{R}^d}^2.$$

We now show how to use Proposition 13.8 to give an alternative characterization of $H^2(\mathbb{R}^d)$ by looking at the Fourier transform of its elements.

Proposition 13.9. *If $u \in L^2(\mathbb{R}^d)$, then*

$$u \in H^2(\mathbb{R}^d) \qquad \Longleftrightarrow \qquad \omega \mathcal{F}\{u\} \in L^2(\mathbb{R}^d),$$

and

(a) $m_i m_j \mathcal{F}\{u\} = \mathcal{F}\{\partial_{x_i} \partial_{x_j} u\}$ *for all i, j,*

(b) *there exist constants such that*

$$c_1 \|u\|_{2,\mathbb{R}^d} \leq \|\omega \mathcal{F}\{u\}\|_{\mathbb{R}^d} = \|u - \Delta u\|_{\mathbb{R}^d} \leq c_2 \|u\|_{2,\mathbb{R}^d} \qquad \forall u \in H^2(\mathbb{R}^d).$$

Proof. If $u \in H^2(\mathbb{R}^d)$, then u and the partial derivatives $\partial_{x_i} u$ are elements of $H^1(\mathbb{R}^d)$. By Proposition 13.8 we have

$$\omega^{1/2} \mathcal{F}\{u\} \in L^2(\mathbb{R}^d),$$
$$\omega^{1/2} \mathcal{F}\{\partial_{x_i} u\} \in L^2(\mathbb{R}^d) \qquad i = 1, \dots, d,$$

and

$$\mathcal{F}\{\partial_{x_j} \partial_{x_i} u\} = -m_j \mathcal{F}\{\partial_{x_i} u\} = m_i m_j \mathcal{F}\{u\} \qquad \forall i, j.$$

This implies that (note that $|m_i|^2 = -m_i^2$)

$$\omega \mathcal{F}\{u\} = \mathcal{F}\{u\} - \sum_{i=1}^{d} m_i^2 \mathcal{F}\{u\} = \mathcal{F}\{u - \Delta u\}$$

is in $L^2(\mathbb{R}^d)$.

If $\omega \mathcal{F}\{u\} \in L^2(\mathbb{R}^d)$, then $\omega^{1/2} \mathcal{F}\{u\} \in L^2(\mathbb{R}^d)$ and therefore $u \in H^1(\mathbb{R}^d)$ by Proposition 13.8. Furthermore,

$$|\omega^{1/2} \mathcal{F}\{\partial_{x_i} u\}| = |m_i \omega^{1/2} \mathcal{F}\{u\}| \leq |\omega \mathcal{F}\{u\}|,$$

and therefore $\partial_{x_i} u \in H^1(\mathbb{R}^d)$ for $i = 1, \dots, d$, once again by Proposition 13.8. The rightmost inequality in (b) is a direct consequence of the triangle inequality. For the reciprocal estimate, note that

$$\|u\|_{\mathbb{R}^d}^2 + \sum_{i=1}^{d} \|\partial_{x_i} u\|_{1,\mathbb{R}^d}^2 = \|\mathcal{F}\{u\}\|_{\mathbb{R}^d}^2 + \sum_{i=1}^{d} \|\omega^{1/2} \mathcal{F}\{\partial_{x_i} u\}\|_{\mathbb{R}^d}^2$$

$$\leq \|\omega^{1/2} \mathcal{F}\{u\}\|_{\mathbb{R}^d}^2 + \sum_{i=1}^{d} \|\omega^{1/2} m_i \mathcal{F}\{u\}\|_{\mathbb{R}^d}^2$$

$$= \|\omega \mathcal{F}\{u\}\|_{\mathbb{R}^d}^2,$$

which finishes the proof, given the fact that the expression in the left-hand side of the above inequality is equivalent to the square of the norm of $H^2(\mathbb{R}^d)$. □

A surprising consequence of Proposition 13.9 states that elements of $L^2(\mathbb{R}^d)$ with Laplacian in $L^2(\mathbb{R}^d)$ are automatically in $H^2(\mathbb{R}^d)$. In particular the solutions of

$$u \in H^1(\mathbb{R}^d) \qquad -\Delta u + u = f \in L^2(\mathbb{R}^d)$$

are all in the space $H^2(\mathbb{R}^d)$. We prove this result next.

Proposition 13.10. *The bounded linear map*

$$H^2(\mathbb{R}^d) \ni u \longmapsto \mathcal{L}u := u - \Delta u \in L^2(\mathbb{R}^d)$$

admits a bounded inverse.

Proof. The proof is a simple consequence of results that appear in the proof of Proposition 13.9. We can write

$$\mathcal{F}\{u - \Delta u\} = \omega \mathcal{F}\{u\} \qquad \forall u \in H^2(\mathbb{R}^d),$$

and therefore

$$\mathcal{L}u = \mathcal{F}^*\{\omega \mathcal{F}\{u\}\} \qquad \forall u \in H^2(\mathbb{R}^d).$$

However, the alternative expression of $\mathcal{L}$ allows us to define a bounded inverse $\mathcal{L}^{-1}u = \mathcal{F}^*\{\omega^{-1}\mathcal{F}\{u\}\}$, which maps $L^2(\mathbb{R}^d)$ into $H^2(\mathbb{R}^d)$ thanks to Proposition 13.9. □

Corollary 13.1. *We have*

$$H^2(\mathbb{R}^d) = \{u \in H^1(\mathbb{R}^d) : \Delta u \in L^2(\mathbb{R}^d)\}.$$

Proof. Clearly $H^2(\mathbb{R}^d) \subset \{u \in H^1(\mathbb{R}^d) : \Delta u \in L^2(\mathbb{R}^d)\}$. If $u \in H^1(\mathbb{R}^d)$ and $\Delta u \in L^2(\mathbb{R}^d)$, then there exists $v \in H^2(\mathbb{R}^d)$ such that $\Delta v - v = \Delta u - u$. Therefore $w := u - v \in H^1(\mathbb{R}^d)$ satisfies $-\Delta w + w = 0$, i.e.,

$$(w, \varphi)_{\mathbb{R}^d} + (\nabla w, \nabla \varphi)_{\mathbb{R}^d} = 0 \qquad \forall \varphi \in \mathcal{D}(\mathbb{R}^d).$$

Since $\mathcal{D}(\mathbb{R}^d)$ is dense in $H^1(\mathbb{R}^d)$, this proves that $w = 0$ and thus $u = v \in H^2(\mathbb{R}^d)$. □

Note that it follows from Exercise 13.7 below that

$$H^2(\mathbb{R}^d) = \{u \in L^2(\mathbb{R}^d) : \Delta u \in L^2(\mathbb{R}^d)\}.$$

The proof of this equality requires working with tempered distributions, which we will introduce in Section 13.5.

It is an easy consequence of Corollary 13.1 that if $f \in L^2(\mathbb{R}^d)$ and we solve the problem

$$u \in H^1(\mathbb{R}^d) \qquad -\Delta u + u = f \in L^2(\mathbb{R}^d),$$

i.e.,

$$u \in H^1(\mathbb{R}^d), \qquad (\nabla u, \nabla v)_{\mathbb{R}^d} + (u, v)_{\mathbb{R}^d} = (f, v)_{\mathbb{R}^d} \qquad \forall v \in H^1(\mathbb{R}^d),$$

then $u \in H^2(\mathbb{R}^d)$. Moreover, because of Proposition 13.9, we have the bound

$$\|u\|_{2,\mathbb{R}^d} \le C\|f\|_{\mathbb{R}^d}.$$

We will now explore similar situations in the half space $\mathbb{R}^d_+ = \mathbb{R}^{d-1} \times (0, \infty)$ with homogeneous Dirichlet or Neumann boundary conditions.

Proposition 13.11. *Let* $u \in H^1(\mathbb{R}^d_+)$ *satisfy* $\Delta u \in L^2(\mathbb{R}^d_+)$. *If*

   (a) $\gamma u = 0$, *or*

   (b) $\partial_n u = 0$,

then $u \in H^2(\mathbb{R}^d_+)$.

Proof. We will give full details for case (a) and sketch the proof for (b). Let $f := -\Delta u + u \in L^2(\mathbb{R}^d_+)$ and consider the extension by reflection (see the analysis of the kernel of the trace operator in Chapter 4)

$$U(\mathbf{x}) = U(\tilde{\mathbf{x}}, x_d) := \begin{cases} u(\mathbf{x}), & \text{if } x_d > 0, \\ -u(\tilde{\mathbf{x}}, -x_d), & \text{if } x_d < 0. \end{cases}$$

Since $u \in H^1_0(\mathbb{R}^d_+)$, it is easy to prove that $U \in H^1_0(\mathbb{R}^d \setminus \Xi)$, where $\Xi := \mathbb{R}^{d-1} \times \{0\}$ is the interface between the upper half-space and the lower one. Therefore $U \in H^1(\mathbb{R}^d)$. (Why?) Now take $\varphi \in \mathcal{D}(\mathbb{R}^d)$ and define

$$\varphi_+(\mathbf{x}) = \varphi_+(\tilde{\mathbf{x}}, x_d) := \varphi(\mathbf{x}) - \varphi(\tilde{\mathbf{x}}, -x_d).$$

Since $\varphi_+ \in C^\infty(\overline{\mathbb{R}^d_+})$ and $\varphi = 0$ on $\partial \mathbb{R}^d_+ = \Xi$, then $\varphi_+ \in H^1_0(\mathbb{R}^d_+)$. Finally,

$$\begin{aligned} (\nabla U, \nabla \varphi)_{\mathbb{R}^d} + (U, \varphi)_{\mathbb{R}^d} &= (\nabla u, \nabla \varphi_+)_{\mathbb{R}^d_+} + (u, \varphi_+)_{\mathbb{R}^d_+} \quad \text{(change of vars.)} \\ &= (f, \varphi_+)_{\mathbb{R}^d_+} \quad\quad\quad\quad\quad (\varphi_+ \in H^1_0(\mathbb{R}^d_+)) \\ &= (F, \varphi)_{\mathbb{R}^d} \quad \forall \varphi \in \mathcal{D}(\mathbb{R}^d), \end{aligned}$$

where

$$F(\mathbf{x}) = \begin{cases} f(\mathbf{x}), & \text{if } x_d > 0, \\ -f(\tilde{\mathbf{x}}, -x_d), & \text{if } x_d < 0. \end{cases}$$

Therefore $-\Delta U + U = F \in L^2(\mathbb{R}^d)$ and by Corollary 13.1, $U \in H^2(\mathbb{R}^d)$, which implies that $u = U|_{\mathbb{R}^d_+} \in H^2(\mathbb{R}^d_+)$.

Note that condition (b) (with $f := -\Delta u + u \in L^2(\mathbb{R}^d_+)$) is equivalent to

$$(\nabla u, \nabla v)_{\mathbb{R}^d_+} + (u, v)_{\mathbb{R}^d_+} = (f, v)_{\mathbb{R}^d_+} \qquad \forall v \in H^1(\mathbb{R}^d).$$

Now consider the extension by symmetry (see the proof of the extension property in Chapter 4)

$$U(\mathbf{x}) = \begin{cases} u(\mathbf{x}), & \text{if } x_d > 0, \\ u(\widetilde{\mathbf{x}}, -x_d), & \text{if } x_d < 0, \end{cases}$$

and note that for all $\varphi \in \mathcal{D}(\mathbb{R}^d)$, the symmetrization of φ

$$\varphi_-(\mathbf{x}) = \varphi(\mathbf{x}) + \varphi(\widetilde{\mathbf{x}}, -x_d)$$

is an element of $H^1(\mathbb{R}^d)$ and we can write

$$
\begin{aligned}
(\nabla U, \nabla \varphi)_{\mathbb{R}^d} + (U, \varphi)_{\mathbb{R}^d} &= (\nabla u, \nabla \varphi_-)_{\mathbb{R}^d_+} + (u, \varphi_-)_{\mathbb{R}^d_+} && \text{(change of vars.)} \\
&= (f, \varphi_-)_{\mathbb{R}^d_+} && (\varphi_- \in H^1(\mathbb{R}^d)) \\
&= (F, \varphi)_{\mathbb{R}^d} && \forall \varphi \in \mathcal{D}(\mathbb{R}^d),
\end{aligned}
$$

where now F is defined by symmetry as well. Therefore $-\Delta U + U \in L^2(\mathbb{R}^d)$ and the proof is easily finished. $\qquad\square$

On the need for a boundary condition. We now show an example of functions $u \in H^1(\mathbb{R}^d_+)$ with $\Delta u \in L^2(\mathbb{R}^d_+)$, but $u \notin H^2(\mathbb{R}^d_+)$, thus demonstrating that the homogeneous boundary conditions of Proposition 13.11 cannot easily be removed. We first show an example for $d = 2$. (The detailed computations for this 'counterexample' are left to the reader.) To do this, consider the function

$$v(\mathbf{x}) := |\mathbf{x}|^\alpha \sin\left(\alpha \arccos \frac{x_1}{|\mathbf{x}|}\right) = r^\alpha \sin(\alpha\theta) \qquad 0 < \alpha < 1.$$

We note that $v \in \mathcal{C}^\infty(\mathbb{R}^2_+)$, $\Delta v = 0$ in $\mathbb{R}^2_+$ and that

$$\partial_{x_1}\partial_{x_2} v(\mathbf{x}) = \alpha(\alpha - 1)|\mathbf{x}|^{\alpha - 2} \cos\left((\alpha - 2) \arccos \frac{x_1}{|\mathbf{x}|}\right).$$

Now let $\varphi \in \mathcal{D}(\mathbb{R}^2)$ be such that $\varphi \equiv 1$ in a neighborhood of $\mathbf{x} = \mathbf{0}$, so that we can focus on the behavior at the origin and cut-off from infinity. The function $u = \varphi v$ is in $H^1(\mathbb{R}^2_+)$ (we only need to check that v is locally in H^1) and satisfies $\Delta v \in L^2(\mathbb{R}^2_+)$, but

$$\partial_{x_1}\partial_{x_2} u = \underbrace{v\partial_{x_1}\partial_{x_2}\varphi + \partial_{x_1}v\,\partial_{x_2}\varphi + \partial_{x_2}v\,\partial_{x_1}\varphi}_{\in L^2(\mathbb{R}^d_+)} + \varphi\,\partial_{x_1}\partial_{x_2}v \notin L^2(\mathbb{R}^2_+),$$

since $\partial_{x_1}\partial_{x_2} v$ fails to be square integrable near the origin. To find an example in d variables, consider $\varphi \in \mathcal{D}(\mathbb{R}^d)$ in the same conditions and $u(\mathbf{x}) = \varphi(\mathbf{x})v(x_1, x_2)$.

13.4   Topology of the Schwartz class

Locally convex spaces. For readers who are not acquainted with it, we now consider a standard construction in analysis, based on a vector space endowed with a countable family of seminorms, which we will assume to be growing. This construction of metric spaces gives 'natural' topologies to vector spaces like the one giving uniform convergence on compact sets in $\mathcal{C}(\Omega)$ for an open set Ω, or convergence for all derivatives in $\mathcal{C}^\infty(K)$, where K is compact. Exercise 13.12 will be used to examine $\mathcal{C}^\infty(\mathbb{R}^d)$ and its dual space.

Theorem 13.1 (Construction of locally convex spaces). *Let X be a vector space, and $|\cdot|_k : X \to [0,\infty)$ for $k \geq 1$ be a sequence of seminorms in X satisfying*

$$|\cdot|_1 \leq |\cdot|_2 \leq \cdots \leq |\cdot|_k \leq \cdots,$$

and

$$|x|_k = 0 \quad \forall k \quad \Longrightarrow \quad x = 0.$$

The binary function $d : X \times X \to [0,1]$

$$d(x,y) := \sum_{k=1}^{\infty} \frac{1}{2^k} \frac{|x-y|_k}{1+|x-y|_k}$$

defines a metric in X such that $x_n \to x$ if and only if $|x_n - x|_k \to 0$ for all k. Moreover, given a linear map $\ell : X \to \mathbb{R}$, ℓ is continuous if and only if there exist $k \geq 1$ and $C > 0$ such that

$$|\ell(x)| \leq C|x|_k \qquad \forall x \in X. \tag{13.8}$$

Proof. The only nontrivial part of the proof that d is a metric is the triangle inequality. Note that the function $t \mapsto t/(1+t)$ is increasing and therefore, using the triangle inequality for $|\cdot|_k$, it follows that

$$\frac{|x-y|_k}{1+|x-y|_k} \leq \frac{|x-z|_k}{1+|x-z|_k} + \frac{|z-y|_k}{1+|z-y|_k} \qquad \forall k, \qquad \forall x,y,z \in X.$$

The characterization of convergence is also simple. It is clear that convergence in X implies convergence for each of the seminorms. Reciprocally, given $\varepsilon > 0$ we can choose $K \geq 1$ such that

$$\sum_{k=K+1}^{\infty} \frac{1}{2^k} = \frac{1}{2^K} < \frac{\varepsilon}{2},$$

and then $N \geq 0$ such that

$$|x_n - x|_k < \frac{\varepsilon}{2} \qquad \forall k \in \{1,\dots,K\}, \qquad \forall n \geq N.$$

Therefore (note that $t/(1+t) \leq \min\{1, t\}$ for $t \geq 0$)

$$d(x_n, x) \leq \sum_{k=1}^{K} \frac{1}{2^k} |x_n - x|_k + \sum_{k=K+1}^{\infty} \frac{1}{2^k} < \varepsilon \qquad \forall n \geq N,$$

which proves the result.

If $\ell : X \to \mathbb{R}$ satisfies (13.8) and $x_n \to x$, then $|x_n - x|_k \to 0$ and therefore $\ell(x_n) \to \ell(x)$. Assume (13.8) does not hold for any k and C. Therefore, for all $k \geq 1$ we can find $x_k \in X$ such that

$$|x_k|_k = 1 \qquad |\ell(x_k)| \geq k.$$

Consider the sequence $y_n := (1/n)x_n$. Note that for any k

$$|y_n|_k \leq |y_n|_n \leq \frac{1}{n} \qquad \forall n \geq k,$$

and therefore $y_n \to 0$. However $|\ell(y_n)| \geq 1$ for every n, thus ℓ is not continuous at zero. $\qquad\square$

A Fréchet space topology for the Schwartz class. We define the following norms in $\mathcal{S}(\mathbb{R}^d)$:

$$\|\varphi\|_k := \sum_{|\alpha|, |\beta| \leq k} \|m_\beta \partial^\alpha \varphi\|_{L^\infty(\mathbb{R}^d)} = \sum_{|\alpha|, |\beta| \leq k} \max_{\mathbf{x} \in \mathbb{R}^d} |m_\beta(\mathbf{x})(\partial^\alpha \varphi)(\mathbf{x})|.$$

The fact that the $L^\infty(\mathbb{R}^d)$ norm can be given with a maximum is due to the fact that the functions we are dealing with are continuous and decaying to zero at infinity. We use these norms to define the metric

$$d(\varphi, \psi) := \sum_{k=1}^{\infty} \frac{1}{2^k} \frac{\|\varphi - \psi\|_k}{1 + \|\varphi - \psi\|_k}. \tag{13.9}$$

A metric space constructed with the technique of Theorem 13.1 that is also complete is called a **Fréchet space**.

Proposition 13.12. *Endowed with the metric (13.9), the Schwartz class is a complete metric space where convergence $\varphi_n \to \varphi$ is equivalent to*

$$p\partial^\alpha \varphi_n \longrightarrow p\partial^\alpha \varphi \quad \text{uniformly}, \qquad \forall \alpha \in \mathbb{N}^d, \qquad \forall p \in \mathcal{P}(\mathbb{R}^d).$$

Therefore, differentiation of any order and multiplication by polynomials define continuous maps in $\mathcal{S}(\mathbb{R}^d)$. Finally, convergence in $\mathcal{D}(\mathbb{R}^d)$ implies convergence in $\mathcal{S}(\mathbb{R}^d)$.

Proof. For every p and α

$$\|p\partial^\alpha \varphi - p\partial^\alpha \psi\|_{L^\infty(\mathbb{R}^d)} \leq C_p \|\varphi - \psi\|_k, \qquad k := \max\{|\alpha|, \deg p\},$$

which (using Theorem 13.1) proves the first result. It is actually easy to see that convergence in $\mathcal{S}(\mathbb{R}^d)$ is equivalent to uniform convergence

$$(1 + |\cdot|^{2\ell})\partial^\alpha \varphi_n \to (1 + |\cdot|^{2\ell})\partial^\alpha \varphi \qquad \forall \ell \geq 0, \qquad \alpha \in \mathbb{N}^d.$$

If we have a Cauchy sequence $\{\varphi_n\}$ in $\mathcal{S}(\mathbb{R}^d)$, the same idea shows that the sequence $\psi_n^{[\ell,\alpha]} := (1+|\cdot|^{2\ell})\partial^\alpha \varphi_n$ is Cauchy in $L^\infty(\mathbb{R}^d)$ and therefore uniformly convergent to a function $\psi^{[\ell,\alpha]}$. Since

$$\psi^{[0,\alpha]} \longleftarrow \psi_n^{[0,\alpha]} = (1+|\cdot|^{2\ell})^{-1}\psi_n^{[\ell,\alpha]} \longrightarrow (1+|\cdot|^{2\ell})^{-1}\psi^{[\ell,\alpha]},$$

we have that

$$\psi^{[\ell,\alpha]} = (1 + |\cdot|^{2\ell})\psi^{[0,\alpha]} \in L^\infty(\mathbb{R}^d) \cap \mathcal{C}(\mathbb{R}^d). \qquad (13.10)$$

Taking $\varphi := \psi^{[0,0]}$ and comparing convergence in the sense of distributions (this is one of many possible ways) it is simple to see that $\partial^\alpha \varphi = \psi^{[0,\alpha]}$ and therefore $\varphi \in \mathcal{C}^\infty(\mathbb{R}^d)$, but (13.10) shows that $\varphi \in \mathcal{S}(\mathbb{R}^d)$ and finally that $\varphi_n \to \varphi$.

The last assertions in the statement of the proposition are simple consequences of the characterization of convergence and of the definition of convergence in $\mathcal{D}(\mathbb{R}^d)$. $\qquad \square$

We finish this section with a density result of $\mathcal{D}(\mathbb{R}^d)$ in the Schwartz class.

Proposition 13.13. *The space $\mathcal{D}(\mathbb{R}^d)$ is dense in $\mathcal{S}(\mathbb{R}^d)$.*

Proof. This density result is proved by a simple cutoff process, based on rescaling a fixed shape function $\eta \in \mathcal{D}(\mathbb{R}^d)$ satisfying

$$0 \leq \eta \leq 1, \qquad \operatorname{supp} \eta \subset \overline{B(0;2)}, \qquad \eta \equiv 1 \quad \text{in } B(0;1).$$

We then define $\eta_n := \eta(\cdot/n)$. Our goal is to prove that for all $\varphi \in \mathcal{S}(\mathbb{R}^d)$, we have

$$\eta_n \varphi \to \varphi \quad \text{in } \mathcal{S}(\mathbb{R}^d). \qquad (13.11)$$

To prove (13.11), we will show two simple technical results. First of all, given $\psi \in \mathcal{S}(\mathbb{R}^d)$, $p \in \mathcal{P}(\mathbb{R}^d)$, and $\varepsilon > 0$, we can find n such that

$$|p(\mathbf{x})\psi(\mathbf{x})| \leq \varepsilon \qquad |\mathbf{x}| \geq n.$$

Therefore, since

$$|p\eta_n\psi - p\psi| \equiv 0 \quad \text{in } B(0;n),$$
$$|p\eta_n\psi - p\psi| \leq |p\psi| \leq \varepsilon \quad \text{in } \mathbb{R}^d \setminus B(0;n),$$

it follows that

$$\|p\eta_n\psi - p\psi\|_{L^\infty(\mathbb{R}^d)} \to 0 \qquad \forall \psi \in \mathcal{S}(\mathbb{R}^d), \qquad p \in \mathcal{P}(\mathbb{R}^d). \qquad (13.12)$$

Note also that

$$\partial_{x_i}\eta_n = n^{-1}(\partial_{x_i})\eta(\cdot/n) \qquad i = 1, \ldots, d,$$

are supported in the annulus $\overline{B(0; 2n)} \setminus B(0; n)$ and that

$$\|p\psi\partial_{x_i}\eta_n\|_{L^\infty(\mathbb{R}^d)} \le n^{-1}\|\partial_{x_i}\eta\|_{L^\infty(\mathbb{R}^d)}\|p\psi\|_{L^\infty(\mathbb{R}^d)}.$$

Therefore, using (13.12), we have

$$\|p\partial_{x_i}(\eta_n\psi) - p\partial_{x_i}\psi\|_{L^\infty(\mathbb{R}^d)} \le \|p\psi\partial_{x_i}\eta_n\|_{L^\infty(\mathbb{R}^d)} + \|p\eta_n\partial_{x_i}\psi - p\partial_{x_i}\psi\|_{L^\infty(\mathbb{R}^d)}$$
$$\to 0 \qquad \forall \psi \in \mathcal{S}(\mathbb{R}^d), \qquad p \in \mathcal{P}(\mathbb{R}^d). \quad (13.13)$$

An induction argument using (13.12) and (13.13) proves (13.11) and hence the result. □

Proposition 13.14. *For any $s \in \mathbb{R}$, multiplication by the function $\omega^s = (1 + |2\pi \cdot |^2)^s$ defines a continuous invertible operator in $\mathcal{S}(\mathbb{R}^d)$.*

Proof. The following properties of ω^s are straightforward:

$$\omega^s \in \mathcal{C}^\infty(\mathbb{R}^d), \qquad\qquad\qquad\qquad (13.14a)$$
$$\omega^s \le 1 \qquad \text{if } s \le 0, \qquad\qquad\qquad (13.14b)$$
$$\omega^s \le \omega^m \in \mathcal{P}(\mathbb{R}^d) \qquad s \le m \in \mathbb{N}, \qquad (13.14c)$$
$$\partial_{x_i}\omega^s = 4\pi\imath m_i\, s\omega^{s-1}. \qquad\qquad\qquad (13.14d)$$

Using induction and (13.14d), it is clear that for all α and $\varphi \in \mathcal{S}(\mathbb{R}^d)$, the derivative $\partial^\alpha(\omega^s\varphi)$ is a linear combination of terms of the form $q\omega^r\partial^\beta\varphi$ with $\beta \le \alpha, r \le s$, and $q \in \mathcal{P}(\mathbb{R}^d)$. Therefore, from (13.14a)-(13.14c), we have that $\omega^s\varphi \in \mathcal{S}(\mathbb{R}^d)$. Due to the characterization of convergence given in Proposition 13.12, it follows that $\varphi_n \to \varphi$ in $\mathcal{S}(\mathbb{R}^d)$ implies that $\omega^s\varphi_n \to \omega^s\varphi$ in $\mathcal{S}(\mathbb{R}^d)$. Finally, note that multiplication by ω^{-s} is the inverse operator to multiplication by ω^s. □

Proposition 13.15. *The Fourier transform and its inverse define continuous operators in $\mathcal{S}(\mathbb{R}^d)$.*

Proof. Given all the results we have on the Schwartz class and on $\mathcal{F}$, the proof of this result is easy. First, we recall that

$$\|\mathcal{F}\{\varphi\}\|_{L^\infty(\mathbb{R}^d)} \le \|\varphi\|_{L^1(\mathbb{R}^d)} \qquad \forall \varphi \in \mathcal{S}(\mathbb{R}^d). \qquad (13.15)$$

It is also simple to see that

$$\|\varphi\|_{L^1(\mathbb{R}^d)} \le \|\omega^d\varphi\|_{L^\infty(\mathbb{R}^d)} \int_{\mathbb{R}^d} \frac{1}{(1 + |2\pi\mathbf{x}|^2)^d}d\mathbf{x}, \qquad (13.16)$$

and therefore convergence in $\mathcal{S}(\mathbb{R}^d)$ implies convergence in $L^1(\mathbb{R}^d)$. Putting (13.15) and (13.16) together, we thus have proved that if $\varphi_n \to \varphi$ in $\mathcal{S}(\mathbb{R}^d)$, then $\mathcal{F}\{\varphi_n\} \to \mathcal{F}\{\varphi\}$, uniformly in $\mathbb{R}^d$. However, the formulas (see the proof of Proposition 13.3)

$$\partial^\alpha \mathcal{F}\{\varphi\} = \mathcal{F}\{m_\alpha \varphi\}, \quad m_\beta \mathcal{F}\{\varphi\} = (-1)^{|\beta|} \mathcal{F}\{\partial^\beta \varphi\} \qquad \forall \varphi \in \mathcal{S}(\mathbb{R}^d), \ (13.17)$$

and the fact that convergence in $\mathcal{S}(\mathbb{R}^d)$ is preserved by differentiation and multiplication by polynomials (this follows easily by the characterization given in Proposition 13.12), can then be used to prove that for any α and any polynomial $p \in \mathcal{P}(\mathbb{R}^d)$, we have $p\partial^\alpha \mathcal{F}\{\varphi_n\} \to p\partial^\alpha \mathcal{F}\{\varphi\}$ with uniform convergence. $\qquad\square$

13.5 Tempered distributions

Tempered distributions. We have seen that $\mathcal{S}(\mathbb{R}^d)$ can be endowed with a metric that makes it complete, and where convergence is given by Proposition 13.12. We have also seen that convergence in $\mathcal{D}(\mathbb{R}^d)$ implies convergence in $\mathcal{S}(\mathbb{R}^d)$ (Proposition 13.12) and that $\mathcal{D}(\mathbb{R}^d)$ is dense in $\mathcal{S}(\mathbb{R}^d)$. A linear continuous map $T : \mathcal{S}(\mathbb{R}^d) \to \mathbb{C}$ is called a tempered distribution, and we will write $T \in \mathcal{S}'(\mathbb{R}^d)$. Note that by Theorem 13.1, if $T : \mathcal{S}(\mathbb{R}^d) \to \mathbb{C}$ is linear, it is continuous if and only if there exists an integer $m \geq 0$ and $C > 0$ such that

$$|T(\varphi)| \leq C\|\varphi\|_m \qquad \forall \varphi \in \mathcal{S}(\mathbb{R}^d).$$

Also, since $\mathcal{S}(\mathbb{R}^d)$ is a metric space, continuity can be described by preservation of sequential convergence, i.e., T is continuous if and only if $\varphi_n \to \varphi$ in $\mathcal{S}(\mathbb{R}^d)$ implies that $T(\varphi_n) \to T(\varphi)$ (in $\mathbb{C}$). The action of $T \in \mathcal{S}'(\mathbb{R}^d)$ on $\varphi \in \mathcal{S}(\mathbb{R}^d)$ will be denoted

$$[T, \varphi]_{\mathcal{S}' \times \mathcal{S}}.$$

As we did with the space of distributions, convergence in $\mathcal{S}'(\mathbb{R}^d)$ will be defined in a weak way: $T_n \to T$ in $\mathcal{S}'(\mathbb{R}^d)$ when

$$[T_n, \varphi]_{\mathcal{S}' \times \mathcal{S}} \to [T, \varphi] \qquad \forall \varphi \in \mathcal{S}(\mathbb{R}^d).$$

We first justify the name given to the elements of the dual space $\mathcal{S}'(\mathbb{R}^d)$ and the injection

$$\mathcal{S}'(\mathbb{R}^d) \subset \mathcal{D}'(\mathbb{R}^d).$$

Proposition 13.16. *Given* $T \in \mathcal{S}'(\mathbb{R}^d)$, *the map* $T|_{\mathcal{D}(\mathbb{R}^d)}$ *given by*

$$\langle T, \varphi \rangle := [T, \varphi]_{\mathcal{S}' \times \mathcal{S}} \qquad \varphi \in \mathcal{D}(\mathbb{R}^d),$$

defines a distribution. The above restriction process defines an injective non-surjective map $S'(\mathbb{R}^d) \to D'(\mathbb{R}^d)$ that preserves convergence.

Proof. Since $\varphi_n \to \varphi$ in $D(\mathbb{R}^d)$ implies that $\varphi_n \to \varphi$ in $S(\mathbb{R}^d)$, it is clear that $T|_{D(\mathbb{R}^d)}$ is sequentially continuous and therefore defines a distribution. Moreover, due to the density of $D(\mathbb{R}^d)$ in $S(\mathbb{R}^d)$, it follows that if $T|_{D(\mathbb{R}^d)} = 0$, then $T = 0$. To prove that the embedding map $S'(\mathbb{R}^d) \to D'(\mathbb{R}^d)$ is not surjective, we just need to find a distribution that is not tempered. For instance, the map

$$\varphi \longmapsto \int_{\mathbb{R}^d} e^{|\mathbf{x}|^2} \varphi(\mathbf{x}) d\mathbf{x}$$

defines a distribution but cannot be extended to act on general elements of the Schwartz class. To see this, recall the functions $\eta \in D(\mathbb{R}^d)$ where

$$0 \leq \eta \leq 1, \qquad \operatorname{supp} \eta \subset \overline{B(\mathbf{0}; 2)}, \qquad \eta \equiv 1 \quad \text{in } B(\mathbf{0}; 1),$$

and $\eta_n := \eta(\cdot/n)$. We use these functions to define a sequence $\{\varphi_n\}$ in $D(\mathbb{R}^d)$ given by $\varphi_n := \eta_n \exp(-|\cdot|^2)$, and note that this sequence converges in $S(\mathbb{R}^d)$ to $\exp(-|\cdot|^2) \in S(\mathbb{R}^d)$. Furthermore,

$$\int_{\mathbb{R}^d} \varphi_n(\mathbf{x}) e^{|\mathbf{x}|^2} \, d\mathbf{x} \geq \int_{B(\mathbf{0}; n)} 1 \, d\mathbf{x} = \omega_d n^d,$$

where we are using ω_d to be the volume of the unit ball in $\mathbb{R}^d$. Now we need only let n go to infinity to see that this map does not define a tempered distribution, which was our goal. The last assertion of the statement is straightforward. $\qquad\square$

The counter-example given in the proof shows that not all regular distributions are tempered. However, all regular distributions in the spaces $L^p(\mathbb{R}^d)$ for any $1 \leq p \leq \infty$ are tempered. (See Exercise 13.6.) Of particular interest are functions $u \in L^2(\mathbb{R}^d)$, which define tempered distributions via its inner product

$$[u, \varphi]_{S' \times S} = \int_{\mathbb{R}^d} u(\mathbf{x}) \varphi(\mathbf{x}) d\mathbf{x} = (u, \overline{\varphi})_{\mathbb{R}^d}, \qquad (13.18)$$

which we write to emphasize that the bracket used for duality is bilinear, whereas the one for the inner product is sesquilinear.

Proposition 13.17 (Operations with tempered distributions). *The following operations define tempered distributions:*

$$[\partial^\alpha T, \varphi]_{S' \times S} := (-1)^{|\alpha|} [T, \partial^\alpha \varphi]_{S' \times S} \qquad \alpha \in \mathbb{N}^d,$$
$$[p T, \varphi]_{S' \times S} := [T, p\varphi]_{S' \times S} \qquad p \in P(\mathbb{R}^d),$$
$$[\omega^s T, \varphi]_{S' \times S} := [T, \omega^s \varphi]_{S' \times S} \qquad s \in \mathbb{R}.$$

Proof. It is easy to see that $\varphi_n \to \varphi$ in $\mathcal{S}(\mathbb{R}^d)$ implies that $p\varphi_n \to p\varphi$ and $\partial^\alpha \varphi_n \to \partial^\alpha \varphi$ (see Proposition 13.12). This justifies the first two definitions. The last one follows from Proposition 13.14. $\qquad\square$

All three are restrictions of operations that are well-defined in $\mathcal{D}'(\mathbb{R}^d)$. In other words, if $T \in \mathcal{S}'(\mathbb{R}^d) \subset \mathcal{D}'(\mathbb{R}^d)$, then we have shown that the distributions $\partial^\alpha T$, pT, and $\omega^s T$ (the last two are particular cases of multiplication by $\mathcal{C}^\infty(\mathbb{R}^d)$ functions) are tempered.

Proposition 13.18 (Fourier transform). *Given $T \in \mathcal{S}'(\mathbb{R}^d)$, we can define tempered distributions*

$$[\mathcal{F}\{T\}, \varphi]_{\mathcal{S}' \times \mathcal{S}} := [T, \mathcal{F}\{\varphi\}]_{\mathcal{S}' \times \mathcal{S}},$$
$$[\mathcal{F}^*\{T\}, \varphi]_{\mathcal{S}' \times \mathcal{S}} := [T, \mathcal{F}^*\{\varphi\}]_{\mathcal{S}' \times \mathcal{S}}.$$

The linear maps $\mathcal{F}, \mathcal{F}^* : \mathcal{S}'(\mathbb{R}^d) \to \mathcal{S}'(\mathbb{R}^d)$ are sequentially continuous and reciprocally inverse. Furthermore, given $u \in L^2(\mathbb{R}^d)$, the Fourier transform of the associated regular tempered distribution is the same as the tempered distribution associated to the Fourier transform of u, i.e., the Fourier transform in $\mathcal{S}'(\mathbb{R}^d)$ and its inverse extend those defined in $L^2(\mathbb{R}^d)$.

Proof. Proposition 13.15 proves that $\mathcal{F}\{T\}$ and $\mathcal{F}^*\{T\}$ are correctly defined as tempered distributions. It is clear that

$$[\mathcal{F}^*\{\mathcal{F}\{T\}\}, \varphi]_{\mathcal{S}' \times \mathcal{S}} = [T, \mathcal{F}\{\mathcal{F}^*\{\varphi\}\}]_{\mathcal{S}' \times \mathcal{S}} = [T, \varphi]_{\mathcal{S}' \times \mathcal{S}} \qquad \forall \varphi \in \mathcal{S}(\mathbb{R}^d)$$

(recall Proposition 13.5), and therefore $\mathcal{F}^*\{\mathcal{F}\{T\}\} = T$ for all $T \in \mathcal{S}'(\mathbb{R}^d)$. With a similar argument we can prove that $\mathcal{F}^*$ is the inverse of $\mathcal{F}$. The sequential continuity of these maps is a direct consequence of the definition of convergence for tempered distributions: if $T_n \to T$ in $\mathcal{S}'(\mathbb{R}^d)$, then

$$[\mathcal{F}\{T_n\}, \varphi]_{\mathcal{S}' \times \mathcal{S}} = [T_n, \mathcal{F}\{\varphi\}]_{\mathcal{S}' \times \mathcal{S}} \to [T, \mathcal{F}\{\varphi\}]_{\mathcal{S}' \times \mathcal{S}} = [\mathcal{F}\{T\}, \varphi]_{\mathcal{S}' \times \mathcal{S}},$$

and thus $\mathcal{F}\{T_n\} \to \mathcal{F}\{T\}$.

Given $u \in L^2(\mathbb{R}^d)$, let $\mathcal{F}_{L^2}\{u\} \in L^2(\mathbb{R}^d) \subset \mathcal{S}'(\mathbb{R}^d)$ momentarily denote its Fourier transform as defined by density (Proposition 13.7), while $\mathcal{F}\{u\} \in \mathcal{S}'(\mathbb{R}^d)$ is the one defined by transposition. We then have for all $\varphi \in \mathcal{S}(\mathbb{R}^d)$:

$$
\begin{aligned}
[\mathcal{F}\{u\}, \overline{\varphi}]_{\mathcal{S}' \times \mathcal{S}} &= [u, \mathcal{F}\{\overline{\varphi}\}]_{\mathcal{S}' \times \mathcal{S}} \\
&= [u, \overline{\mathcal{F}^*\{\varphi\}}]_{\mathcal{S}' \times \mathcal{S}} && \text{(easy computation)} \\
&= (u, \mathcal{F}^*\{\varphi\})_{\mathbb{R}^d} && \text{(recall (13.18))} \\
&= (\mathcal{F}_{L^2}\{u\}, \varphi)_{\mathbb{R}^d} && \text{(by Proposition 13.7)} \\
&= [\mathcal{F}_{L^2}\{u\}, \overline{\varphi}]_{\mathcal{S}' \times \mathcal{S}}. && \text{(by (13.18) with } \mathcal{F}_{L^2}\{u\} \in L^2(\mathbb{R}^d))
\end{aligned}
$$

The same argument can be repeated for $\mathcal{F}^*$. $\qquad\square$

We thus have 'three' Fourier transforms: (a) the integral one, defined for elements in the Schwartz class; (b) the one in L^2 (that can be written in integral form for elements of $L^2 \cap L^1$) defined by density; (c) the one defined for tempered distributions by duality. As a consequence of Proposition 13.18, we have three invertible operators, which are progressive extensions of the same operator

$$\mathcal{S}(\mathbb{R}^d) \longrightarrow \mathcal{S}(\mathbb{R}^d),$$
$$L^2(\mathbb{R}^d) \longrightarrow L^2(\mathbb{R}^d),$$
$$\mathcal{S}'(\mathbb{R}^d) \longrightarrow \mathcal{S}'(\mathbb{R}^d).$$

The first operator is continuous, the second one is an isometry, and the third one is sequentially continuous (we will not be dealing with any stronger concept of continuity in $\mathcal{S}'(\mathbb{R}^d)$). Finally, by transposition of the formulas (13.17) (using Proposition 13.17), we have

$$\partial^\alpha \mathcal{F}\{T\} = \mathcal{F}\{m_\alpha T\}, \qquad m_\beta \mathcal{F}\{T\} = (-1)^{|\beta|} \mathcal{F}\{\partial^\beta T\} \qquad \forall T \in \mathcal{S}'(\mathbb{R}^d),$$

for arbitrary $\alpha, \beta \in \mathbb{N}^d$.

13.6   Sobolev spaces by Fourier transforms

A full family of Sobolev spaces. For $s \in \mathbb{R}$ we consider the spaces

$$\mathsf{H}^s(\mathbb{R}^d) := \{u \in \mathcal{S}'(\mathbb{R}^d) : \omega^{s/2} \mathcal{F}\{u\} \in L^2(\mathbb{R}^d)\}.$$

Note that $\mathcal{F}\{u\} \in \mathcal{S}'(\mathbb{R}^d)$ can be multiplied by $\omega^{s/2}$ (see Proposition 13.17) yielding another tempered distribution. Our requirement for u to be in $\mathsf{H}^s(\mathbb{R}^d)$ is that this new distribution is equal to an element of $L^2(\mathbb{R}^d)$. In particular, if $u \in \mathsf{H}^s(\mathbb{R}^d)$ for any $s \in \mathbb{R}$, it follows that $\mathcal{F}\{u\}$ is a regular distribution, that is, $\mathcal{F}\{u\}$ is a locally integrable function. It is easy to prove that

$$\|u\|_{\mathsf{H}^s(\mathbb{R}^d)} := \|\omega^{s/2} \mathcal{F}\{u\}\|_{\mathbb{R}^d} = \left(\int_{\mathbb{R}^d} (1 + |2\pi\boldsymbol{\xi}|^2)^{|s|} |\mathcal{F}\{u\}(\boldsymbol{\xi})|^2 d\boldsymbol{\xi} \right)^{1/2}$$

defines a norm in $\mathsf{H}^s(\mathbb{R}^d)$ (the integral expression is justified by the fact that $\mathcal{F}\{u\}$ is actually a function!), associated to the inner product

$$(u, v)_{\mathsf{H}^s(\mathbb{R}^d)} := (\omega^{s/2} \mathcal{F}\{u\}, \omega^{s/2} \mathcal{F}\{v\})_{\mathbb{R}^d}.$$

Note that even while we will shortly recognize some of these spaces as Sobolev spaces, we are going to make a slight notational distinction (using different fonts) to clarify that we are using this new definition. We next list properties of this family of spaces.

Proposition 13.19. *The following properties hold:*

(a) $H^s(\mathbb{R}^d)$ *is a Hilbert space for all $s \in \mathbb{R}$.*

(b) $\mathcal{S}(\mathbb{R}^d)$ *is a dense subset of $H^s(\mathbb{R}^d)$ for all $s \in \mathbb{R}$.*

(c) *For all $s > r$, $H^s(\mathbb{R}^d)$ is continuously and densely embedded into $H^r(\mathbb{R}^d)$.*

(d) *For all $i \in \{1, \ldots, d\}$, the operator $u \mapsto \partial_{x_i} u$ is bounded from $H^s(\mathbb{R}^d)$ to $H^{s-1}(\mathbb{R}^d)$ for all $s \in \mathbb{R}$.*

Proof. To prove (a), let us consider a Cauchy sequence $\{u_n\}$ in $H^s(\mathbb{R}^d)$. The sequence $\{\omega^{s/2} \mathcal{F}\{u_n\}\}$ is then Cauchy in $L^2(\mathbb{R}^d)$ and therefore there exists $v \in L^2(\mathbb{R}^d) \subset \mathcal{S}'(\mathbb{R}^d)$ such that

$$\omega^{s/2} \mathcal{F}\{u_n\} \longrightarrow v \qquad \text{in } L^2(\mathbb{R}^d).$$

Let $u := \mathcal{F}^*\{\omega^{-s/2} v\} \in \mathcal{S}'(\mathbb{R}^d)$ (see Propositions 13.17 and 13.18) and note that $\omega^{s/2} \mathcal{F}\{u\} = v \in L^2(\mathbb{R}^d)$, which implies that $u \in H^s(\mathbb{R}^d)$ and

$$\|u_n - u\|_{H^s(\mathbb{R}^d)} = \|\omega^{s/2} \mathcal{F}\{u_n - u\}\|_{\mathbb{R}^d} = \|\omega^{s/2} \mathcal{F}\{u_n\} - v\|_{\mathbb{R}^d} \to 0,$$

which proves that $H^s(\mathbb{R}^d)$ is complete.

Since $\omega^{s/2} \mathcal{F}\{u\} \in \mathcal{S}(\mathbb{R}^d) \subset L^2(\mathbb{R}^d)$ for every $u \in \mathcal{S}(\mathbb{R}^d)$ (recall Propositions 13.14 and 13.15 concerning multiplication by weights and Fourier transforms in the Schwartz class), it is clear that $\mathcal{S}(\mathbb{R}^d) \subset H^s(\mathbb{R}^d)$ for all $s \in \mathbb{R}$. To prove density, note that

$$\begin{aligned}
(u, \varphi)_{H^s(\mathbb{R}^d)} &= (\omega^{s/2} \mathcal{F}\{u\}, \omega^{s/2} \mathcal{F}\{\varphi\})_{\mathbb{R}^d} \\
&= [\omega^{s/2} \mathcal{F}\{u\}, \omega^{s/2} \overline{\mathcal{F}\{\varphi\}}]_{\mathcal{S}' \times \mathcal{S}} \\
&= [u, \mathcal{F}\{\omega^s \mathcal{F}^*\{\overline{\varphi}\}\}]_{\mathcal{S}' \times \mathcal{S}} \qquad \forall \varphi \in \mathcal{S}(\mathbb{R}^d),
\end{aligned}$$

where we have used that $\mathcal{F}^*\{\overline{\varphi}\} = \overline{\mathcal{F}\{\varphi\}}$, as can be easily verified from the definition of $\mathcal{F}$ and $\mathcal{F}^*$. This proves that $u \in H^s(\mathbb{R}^d)$ is orthogonal to all elements of $\mathcal{S}(\mathbb{R}^d)$ if and only if

$$[u, \mathcal{F}\{\omega^s \mathcal{F}^*\{\overline{\varphi}\}\}]_{\mathcal{S}' \times \mathcal{S}} = 0 \qquad \forall \varphi \in \mathcal{S}(\mathbb{R}^d),$$

but since the map $\varphi \to \mathcal{F}\{\omega^s \mathcal{F}^*\{\overline{\varphi}\}\}$ is invertible in $\mathcal{S}(\mathbb{R}^d)$ (see Propositions 13.14 and 13.15 again), this implies that $u = 0$. This argument shows that the orthogonal complement of $\mathcal{S}(\mathbb{R}^d)$ in $H^s(\mathbb{R}^d)$ is the zero set and, therefore, $\mathcal{S}(\mathbb{R}^d)$ is dense in $H^s(\mathbb{R}^d)$.

Since $\omega \geq 1$, the proof of (c) is straightforward from the definitions and from (b). The proof of (d) is simple since

$$|\omega^{(s-1)/2} \mathcal{F}\{\partial_{x_i} u\}| = |m_i \omega^{(s-1)/2} \mathcal{F}\{u\}| \leq |\omega^{s/2} \mathcal{F}\{u\}|$$

for all s and $u \in H^s(\mathbb{R}^d)$. $\qquad\qquad\qquad\qquad\qquad\qquad\qquad\qquad\qquad \square$

We next recognize some of the Sobolev spaces $H^s(\mathbb{R}^d)$ as spaces we have already studied.

Proposition 13.20. *For $m \in \mathbb{N}$, we have*

$$\mathsf{H}^m(\mathbb{R}^d) = \{u \in L^2(\mathbb{R}^d) : \partial^\alpha u \in L^2(\mathbb{R}^d), \quad |\alpha| \le m\} = H^m(\mathbb{R}^d),$$

with equivalent norms, where in $H^m(\mathbb{R}^d)$ we consider the norm

$$\|u\|_{m,\mathbb{R}^d}^2 = \sum_{|\alpha| \le m} \|\partial^\alpha u\|_{\mathbb{R}^d}^2.$$

Proof. First of all, let us show that $L^2(\mathbb{R}^d) = \mathsf{H}^0(\mathbb{R}^d)$. The key fact is in the statement of Proposition 13.18: the Fourier transform for tempered distributions, restricted to elements of $L^2(\mathbb{R}^d)$ is the Fourier transform in $L^2(\mathbb{R}^d)$. Therefore $\mathcal{F}\{u\} \in L^2(\mathbb{R}^d)$ if and only if $u = \mathcal{F}^*\{\mathcal{F}\{u\}\} \in L^2(\mathbb{R}^d)$. Since $\mathcal{F}$ is an isometry in $L^2(\mathbb{R}^d)$ (Proposition 13.7) we have the equality of the norms of $L^2(\mathbb{R}^d)$ and $\mathsf{H}^0(\mathbb{R}^d)$. Next, note that since $\mathsf{H}^s(\mathbb{R}^d) \subset \mathsf{H}^0(\mathbb{R}^d)$ for $s > 0$ (Proposition 13.19(c)), we can equally define

$$\mathsf{H}^s(\mathbb{R}^d) = \{u \in L^2(\mathbb{R}^d) : \omega^{s/2}\mathcal{F}\{u\} \in L^2(\mathbb{R}^d)\},$$

(the Fourier transform now is the one of $L^2(\mathbb{R}^d)$) and then we have the result for $m = 1$ (Proposition 13.8) and for $m = 2$ (Proposition 13.9). To prove the result for any integer note that

$$H^m(\mathbb{R}^d) = \{u \in H^{m-1}(\mathbb{R}^d) : \partial_{x_i} u \in H^{m-1}(\mathbb{R}^d), i = 1, \dots, d\},$$

as can easily be seen. However, an argument based on how we proved Proposition 13.19(d) shows that

$$u \in \mathsf{H}^s(\mathbb{R}^d) \quad \Longleftrightarrow \quad \begin{cases} u \in \mathsf{H}^{s-1}(\mathbb{R}^d), \\ \partial_{x_i} u \in \mathsf{H}^{s-1}(\mathbb{R}^d) \end{cases} \quad i = 1, \dots, d,$$

(recall that $\omega = 1 + \sum_{i=1}^d |m_i|^2$) and therefore the equality of $\mathsf{H}^m(\mathbb{R}^d)$ and $H^m(\mathbb{R}^d)$ follows by induction. The equivalence of norms follows from the fact that

$$\|u\|_{\mathsf{H}^{s-1}(\mathbb{R}^d)}^2 + \sum_{i=1}^d \|\partial_{x_i} u\|_{\mathsf{H}^{s-1}(\mathbb{R}^d)}^2$$

is an equivalent norm in $\mathsf{H}^s(\mathbb{R}^d)$. $\qquad\square$

Some pseudo-differential operators. Let us now look at some important operators in the space of tempered distributions and their mapping properties among Fourier-Sobolev spaces. Given $r \in \mathbb{R}$, we define the operator

$$\Lambda_r u := \mathcal{F}^*\{\omega^{r/2}\mathcal{F}\{u\}\}.$$

This is an invertible operator in $\mathcal{S}'(\mathbb{R}^d)$, the inverse of Λ_r being Λ_{-r}. We also have that $\Lambda_r : H^s(\mathbb{R}^d) \to H^{s-r}(\mathbb{R}^d)$ is bounded for all $s \in \mathbb{R}$. (It is actually an isometry.) This is a very easy example of what is called a pseudo-differential operator of order r. The case $r = 2$ can easily be recognized as the differential operator $\Lambda_2 u = u - \Delta u$, while its inverse is the operator such that given $u \in \mathcal{S}'(\mathbb{R}^d)$, finds the unique solution of

$$v \in \mathcal{S}'(\mathbb{R}^d), \qquad -\Delta v + v = u.$$

Proposition 13.21 (Duality of Sobolev spaces). *For any $s > 0$,*

$$\mathsf{H}^s(\mathbb{R}^d) \subset L^2(\mathbb{R}^d) \subset \mathsf{H}^{-s}(\mathbb{R}^d)$$

is a well-defined Gelfand triple and, under this representation, the norm of $\mathsf{H}^{-s}(\mathbb{R}^d)$ is equal to the norm of the dual of $\mathsf{H}^s(\mathbb{R}^d)$.

Proof. First of all, by Propositions 13.19(c) (continuous dense embeddings) and 13.20 ($\mathsf{H}^0(\mathbb{R}^d) = L^2(\mathbb{R}^d)$ with the same norm), we can build a Gelfand triple

$$\mathsf{H}^s(\mathbb{R}^d) \subset L^2(\mathbb{R}^d) \subset \mathsf{H}^s(\mathbb{R}^d)'$$

for any $s > 0$. Since $\mathcal{S}(\mathbb{R}^d)$ is dense in $\mathsf{H}^s(\mathbb{R}^d)$, then this representation of the dual of $\mathsf{H}^s(\mathbb{R}^d)$ can be identified with a subspace of the space of tempered distributions (we did the same when we dealt with the Gelfand triple $H_0^1(\Omega) \subset L^2(\Omega) \subset H^{-1}(\Omega)$). We then just need to prove that this representation of the dual of $\mathsf{H}^s(\mathbb{R}^d)$ is exactly $\mathsf{H}^{-s}(\mathbb{R}^d)$ with identical norm.

The following computation is valid for any $s \in \mathbb{R}$ and $u \in \mathsf{H}^{-s}(\mathbb{R}^d)$:

$$\|u\|_{\mathsf{H}^{-s}(\mathbb{R}^d)} = \|\omega^{-s/2}\mathcal{F}\{u\}\|_{\mathbb{R}^d}$$

$$= \sup_{0 \neq \varphi \in \mathcal{S}(\mathbb{R}^d)} \frac{|(\omega^{-s/2}\mathcal{F}\{u\}, \overline{\varphi})_{\mathbb{R}^d}|}{\|\varphi\|_{\mathbb{R}^d}} \qquad (\mathcal{S}(\mathbb{R}^d) \text{ is dense in } L^2)$$

$$= \sup_{0 \neq \varphi \in \mathcal{S}(\mathbb{R}^d)} \frac{|[u, \mathcal{F}\{\omega^{-s/2}\varphi\}]_{\mathcal{S}' \times \mathcal{S}}|}{\|\varphi\|_{\mathbb{R}^d}}$$

$$= \sup_{0 \neq \psi \in \mathcal{S}(\mathbb{R}^d)} \frac{|[u, \psi]_{\mathcal{S}' \times \mathcal{S}}|}{\|\omega^{s/2}\mathcal{F}^*\{\psi\}\|_{\mathbb{R}^d}} \qquad (\text{Propositions 13.14-13.15})$$

$$= \sup_{0 \neq \psi \in \mathcal{S}(\mathbb{R}^d)} \frac{|[u, \psi]_{\mathcal{S}' \times \mathcal{S}}|}{\|\psi\|_{\mathsf{H}^s(\mathbb{R}^d)}}. \qquad (\text{easy to check})$$

The above shows that for $u \in \mathsf{H}^{-s}(\mathbb{R}^d)$, the map $[u, \cdot]_{\mathcal{S}' \times \mathcal{S}} : \mathcal{S}(\mathbb{R}^d) \to \mathbb{C}$ can be extended to a bounded linear map $\ell_u : \mathsf{H}^s(\mathbb{R}^d) \to \mathbb{C}$ and

$$\|u\|_{\mathsf{H}^{-s}(\mathbb{R}^d)} = \|\ell_u\|_{\mathsf{H}^s(\mathbb{R}^d)'},$$

which identifies $\mathsf{H}^{-s}(\mathbb{R}^d)$ with a subset of the dual of $\mathsf{H}^s(\mathbb{R}^d)$. Reciprocally, if $T \in \mathsf{H}^s(\mathbb{R}^d)'$, then

$$T|_{\mathcal{S}(\mathbb{R}^d)} = [T, \cdot]_{\mathcal{S}' \times \mathcal{S}} := \langle T, \cdot \rangle_{\mathsf{H}^s(\mathbb{R}^d)' \times \mathsf{H}^s(\mathbb{R}^d)}|_{\mathcal{S}(\mathbb{R}^d)}$$

defines a tempered distribution such that

$$\sup_{0\neq\varphi\in\mathcal{S}(\mathbb{R}^d)} \frac{|[\omega^{-s/2}\mathcal{F}\{T\},\varphi]_{\mathcal{S}'\times\mathcal{S}}|}{\|\varphi\|_{\mathbb{R}^d}} = \sup_{0\neq\varphi\in\mathcal{S}(\mathbb{R}^d)} \frac{|[T,\mathcal{F}\{\omega^{-s/2}\varphi\}]_{\mathcal{S}'\times\mathcal{S}}|}{\|\varphi\|_{\mathbb{R}^d}}$$

$$= \sup_{0\neq\psi\in\mathcal{S}(\mathbb{R}^d)} \frac{|[T,\psi]_{\mathcal{S}'\times\mathcal{S}}|}{\|\psi\|_{\mathsf{H}^s(\mathbb{R}^d)}}$$

$$= \sup_{0\neq\psi\in\mathcal{S}(\mathbb{R}^d)} \frac{|\langle T,\psi\rangle_{\mathsf{H}^s(\mathbb{R}^d)'\times\mathsf{H}^s(\mathbb{R}^d)}|}{\|\psi\|_{\mathsf{H}^s(\mathbb{R}^d)}}$$

$$= \|T\|_{\mathsf{H}^{-s}(\mathbb{R}^d)}.$$

(We have used that $\mathcal{S}(\mathbb{R}^d)$ is dense in $\mathsf{H}^s(\mathbb{R}^d)$ in the last equality.) This proves that $\omega^{-s/2}\mathcal{F}\{T\} \in L^2(\mathbb{R}^d)$ and therefore $T \in \mathsf{H}^{-s}(\mathbb{R}^d)$. Thus the proof is finished.   □

We finish this section with a result that provides an equivalent norm for $\mathsf{H}^s(\mathbb{R}^d)$ for non-integer $s > 0$, using Solobodetskij-type seminorms for the higher derivatives. We briefly mentioned this result when discussing the intrinsic norms for $H^{1/2}(\Gamma)$ at the end of Chapter 4.

[Proof not provided]

Theorem 13.2. *For $\theta \in (0,1)$ and nonnegative integer $m > 0$,*

$$\|u\|_{m,\mathbb{R}^d}^2 + \sum_{|\alpha|=m} \int_{\mathbb{R}^d\times\mathbb{R}^d} \frac{|\partial^\alpha u(\mathbf{x}) - \partial^\alpha u(\mathbf{y})|^2}{|\mathbf{x}-\mathbf{y}|^{d+2\theta}}\mathrm{dxdy}$$

defines an equivalent norm in $\mathsf{H}^{m+\theta}(\mathbb{R}^d)$.

13.7   The trace space revisited

The goal of this section is the proof of the following result, which partially justifies the fractional Sobolev notation used for the trace space.

Theorem 13.3. *The trace operator $H^1(\mathbb{R}_+^d) \to H^{1/2}(\partial\mathbb{R}_+^d)$ can be identified with the extension of a bounded linear surjective operator*

$$H^1(\mathbb{R}_+^d) \to \mathsf{H}^{1/2}(\mathbb{R}^{d-1}).$$

The proof of this result is made up of several pieces. We will start with two technical lemmas, which contain all the purely mechanical difficulties of the proof. As usual when we work in the half-space we will group variables in pairs

$$\mathbb{R}^d \ni \mathbf{x} = (\widetilde{\mathbf{x}}, x_d) \in \mathbb{R}^{d-1} \times \mathbb{R}.$$

Lemma 13.2. *The linear map*

$$\mathcal{S}(\mathbb{R}^d) \ni u \longrightarrow u(\cdot, 0) \in \mathcal{S}(\mathbb{R}^{d-1})$$

is continuous with respect to the following norms

$$\|u(\cdot, 0)\|_{\mathsf{H}^{1/2}(\mathbb{R}^{d-1})} \leq \frac{1}{\sqrt{2}} \|u\|_{\mathsf{H}^1(\mathbb{R}^d)} \qquad \forall u \in \mathsf{H}^1(\mathbb{R}^d).$$

Proof. If $\varphi \in \mathcal{S}(\mathbb{R}^d)$, then for all $\widetilde{\mathbf{x}} \in \mathbb{R}^{d-1}$, we have $\varphi(\widetilde{\mathbf{x}}, \cdot) \in \mathcal{S}(\mathbb{R})$ and therefore, by the Fourier inversion theorem

$$\varphi(\widetilde{\mathbf{x}}, 0) = \int_{-\infty}^{\infty} \left(\int_{-\infty}^{\infty} e^{-2\pi i x_d \xi_d} \varphi(\mathbf{x}) \mathrm{d}x_d \right) \mathrm{d}\xi_d,$$

which implies that

$$\mathcal{F}_{d-1}\{\varphi(\cdot, 0)\}(\widetilde{\boldsymbol{\xi}}) = \int_{\mathbb{R}^{d-1}} e^{-2\pi i \widetilde{\boldsymbol{\xi}} \cdot \widetilde{\mathbf{x}}} \varphi(\widetilde{\mathbf{x}}, 0) \mathrm{d}\widetilde{\mathbf{x}} = \int_{-\infty}^{\infty} \mathcal{F}\{\varphi\}(\boldsymbol{\xi}) \mathrm{d}\xi_d.$$

Using the identity

$$(1 + a^2)^{1/2} \int_{-\infty}^{\infty} \frac{\mathrm{d}x}{1 + a^2 + (2\pi x)^2} = \frac{1}{2},$$

and the Cauchy-Schwarz inequality, we can estimate

$$\begin{aligned}
\|\varphi(\cdot, 0)\|_{\mathsf{H}^{1/2}(\mathbb{R}^{d-1})}^2 &= \int_{\mathbb{R}^{d-1}} (1 + |2\pi\widetilde{\boldsymbol{\xi}}|^2)^{1/2} |\mathcal{F}_{d-1}\{\varphi(\cdot, 0)\}(\widetilde{\boldsymbol{\xi}})|^2 \mathrm{d}\widetilde{\boldsymbol{\xi}} \\
&= \int_{\mathbb{R}^{d-1}} (1 + |2\pi\widetilde{\boldsymbol{\xi}}|^2)^{1/2} \left| \int_{-\infty}^{\infty} \mathcal{F}\{\varphi\}(\boldsymbol{\xi}) \mathrm{d}\xi_d \right|^2 \mathrm{d}\widetilde{\boldsymbol{\xi}} \\
&\leq \frac{1}{2} \int_{\mathbb{R}^{d-1}} \left(\int_{-\infty}^{\infty} (1 + |2\pi\widetilde{\boldsymbol{\xi}}|^2 + |2\pi\xi_d|^2) |\mathcal{F}\{\varphi\}(\boldsymbol{\xi})|^2 \mathrm{d}\xi_d \right) \mathrm{d}\widetilde{\boldsymbol{\xi}} \\
&= \frac{1}{2} \|\varphi\|_{\mathsf{H}^1(\mathbb{R}^d)}^2,
\end{aligned}$$

as we wanted to prove. $\qquad\square$

Lemma 13.3. *There is a linear map* $L : \mathcal{S}(\mathbb{R}^{d-1}) \to \mathcal{C}^{\infty}(\overline{\mathbb{R}_+^d}) \cap H^1(\mathbb{R}_+^d)$ *such that*

$$\|L\psi\|_{1, \mathbb{R}_+^d} = \|\psi\|_{\mathsf{H}^{1/2}(\mathbb{R}^{d-1})},$$

and $(L\psi)(\cdot, 0) = \psi$, *i.e.,* L *is a lifting of the trace for smooth functions.*

Proof. Given $\psi \in \mathcal{S}(\mathbb{R}^{d-1})$, we define

$$\varphi(\mathbf{x}) = \varphi(\widetilde{\mathbf{x}}, x_d) := \int_{\mathbb{R}^{d-1}} e^{2\pi i \widetilde{\mathbf{x}} \cdot \widetilde{\boldsymbol{\xi}}} e^{-x_d(1+|2\pi\widetilde{\boldsymbol{\xi}}|^2)^{1/2}} \mathcal{F}_{d-1}\{\psi\}(\widetilde{\boldsymbol{\xi}}) d\widetilde{\boldsymbol{\xi}}.$$

We can easily check that $\varphi \in C^\infty(\overline{\mathbb{R}_+^d})$, and $\varphi(\widetilde{\mathbf{x}}, 0) = \psi(\widetilde{\mathbf{x}})$ for all $\widetilde{\mathbf{x}} \in \mathbb{R}^{d-1}$. Moreover,

$$(\partial_{x_d}\varphi)(\mathbf{x}) = -\int_{\mathbb{R}^{d-1}} e^{2\pi i \widetilde{\mathbf{x}} \cdot \widetilde{\boldsymbol{\xi}}} e^{-x_d(1+|2\pi\widetilde{\boldsymbol{\xi}}|^2)^{1/2}}(1 + |2\pi\widetilde{\boldsymbol{\xi}}|^2)^{1/2} \mathcal{F}_{d-1}\{\psi\}(\widetilde{\boldsymbol{\xi}}) d\widetilde{\boldsymbol{\xi}}.$$

For fixed x_d, the expressions for φ and $\partial_{x_d}\varphi$ can be understood as (inverse) Fourier transforms, i.e.,

$$\varphi(\widetilde{\mathbf{x}}, x_d) = \mathcal{F}_{d-1}^*\{e^{-x_d(1+|2\pi\cdot|^2)^{1/2}} \mathcal{F}_{d-1}\{\psi\}\}(\widetilde{\mathbf{x}}),$$

with a similar formulation for $\partial_{x_d}\varphi$. Therefore, using the Fourier space representations of $H^1(\mathbb{R}^{d-1})$ and $L^2(\mathbb{R}^{d-1})$ (recall Propositions 13.7 and 13.8), we have

$$\|\varphi(\cdot, x_d)\|_{1,\mathbb{R}^{d-1}}^2 = \int_{\mathbb{R}^{d-1}} \left(|\varphi(\mathbf{x})|^2 + |\nabla_{\widetilde{\mathbf{x}}}\varphi(\mathbf{x})|^2\right) d\widetilde{\mathbf{x}}$$

$$= \int_{\mathbb{R}^{d-1}} e^{-2x_d(1+|2\pi\widetilde{\boldsymbol{\xi}}|^2)^{1/2}}(1 + |2\pi\widetilde{\boldsymbol{\xi}}|^2)|\mathcal{F}_{d-1}\{\psi\}(\widetilde{\boldsymbol{\xi}})|^2 d\widetilde{\boldsymbol{\xi}}$$

$$\leq e^{-2x_d}\|\psi\|_{1,\mathbb{R}^{d-1}}^2,$$

and

$$\|(\partial_{x_d}\varphi)(\cdot, x_d)\|_{\mathbb{R}^{d-1}}^2 = \int_{\mathbb{R}^{d-1}} e^{-2x_d(1+|2\pi\widetilde{\boldsymbol{\xi}}|^2)^{1/2}}(1 + |2\pi\widetilde{\boldsymbol{\xi}}|^2)|\mathcal{F}_{d-1}\{\psi\}(\widetilde{\boldsymbol{\xi}})|^2 d\widetilde{\boldsymbol{\xi}}$$

$$\leq e^{-2x_d}\|\psi\|_{1,\mathbb{R}^{d-1}}^2.$$

We can now integrate in $x_d \in (0, \infty)$ to estimate

$$\|\varphi\|_{1,\mathbb{R}_+^d}^2 = \int_0^\infty \left(\int_{\mathbb{R}^{d-1}} \left(|\varphi(\mathbf{x})|^2 + |\nabla_{\widetilde{\mathbf{x}}}\varphi(\mathbf{x})|^2 + |(\partial_{x_d}\varphi)(\mathbf{x})|^2\right) d\widetilde{\mathbf{x}}\right) dx_d$$

$$= \int_{\mathbb{R}^{d-1}} \left(2\int_0^\infty e^{-2x_d(1+|2\pi\widetilde{\boldsymbol{\xi}}|^2)^{1/2}} dx_d\right)(1 + |2\pi\widetilde{\boldsymbol{\xi}}|^2)|\mathcal{F}_{d-1}\{\psi\}(\widetilde{\boldsymbol{\xi}})|^2 d\widetilde{\boldsymbol{\xi}}$$

$$= \int_{\mathbb{R}^{d-1}} (1 + |2\pi\widetilde{\boldsymbol{\xi}}|^2)^{1/2}|\mathcal{F}_{d-1}\{\psi\}(\widetilde{\boldsymbol{\xi}})|^2 d\widetilde{\boldsymbol{\xi}} = \|\psi\|_{H^{1/2}(\mathbb{R}^{d-1})}^2.$$

This finishes the proof. It is interesting to note that we have built the optimal lifting of the trace. This can be seen by directly differentiating the function φ and observing that φ satisfies

$$-\Delta\varphi + \varphi = 0 \quad \text{in } \mathbb{R}_+^d, \qquad \varphi(\cdot, 0) = \psi,$$

corresponding to the minimization problem

$$\tfrac{1}{2}\|\varphi\|_{1,\mathbb{R}_+^d}^2 = \text{min!} \qquad \varphi(\cdot, 0) = \psi.$$

$\square$

Proof of Theorem 13.3. The map of Lemma 13.2 can be uniquely extended by density to a bounded linear map $H^1(\mathbb{R}^d) = \mathsf{H}^1(\mathbb{R}^d) \to \mathsf{H}^{1/2}(\mathbb{R}^{d-1})$, which can be understood as a double-sided trace upon the identification $\partial \mathbb{R}^d_+ \equiv \mathbb{R}^{d-1}$. We can then understand the trace operator γ_{Γ_0} in the upper half-space as a composition

$$H^1(\mathbb{R}^d_+) \longrightarrow H^1(\mathbb{R}^d) \longrightarrow \mathsf{H}^{1/2}(\mathbb{R}^{d-1}),$$

where the first map is the extension by symmetry map (see Chapter 4). Moreover,

$$\|\gamma_{\Gamma_0} u\|^2_{\mathsf{H}^{1/2}(\mathbb{R}^{d-1})} \leq \tfrac{1}{2}\|Eu\|^2_{1,\mathbb{R}^d} = \|u\|^2_{1,\mathbb{R}^d_+},$$

which bounds the norm of the (thus reunderstood) trace operator by one. Similarly, the map of Lemma 13.3 can be extended to a bounded linear map $L : \mathsf{H}^{1/2}(\mathbb{R}^{d-1}) \to H^1(\mathbb{R}^d_+)$, which satisfies $\gamma L \psi = \psi$ for all $\psi \in \mathsf{H}^{1/2}(\mathbb{R}^{d-1})$, and therefore is a right inverse of the trace. $\square$

13.8 Interior regularity

In this section, we will use the following local Sobolev spaces

$$H^m_{\mathrm{loc}}(\Omega) := \{u : \Omega \to \mathbb{R} \ : \ u\varphi \in H^m(\Omega) \quad \forall \varphi \in \mathcal{D}(\Omega)\}$$
$$= \{u : \Omega \to \mathbb{R} \ : \ u|_B \in H^m(B) \quad \text{for every ball } B, \text{ s.t. } \overline{B} \subset \Omega\}.$$

Note that these functions are allowed to behave 'badly' near the boundary of Ω, but have otherwise m-th order Sobolev regularity.

Proposition 13.22. *If $u \in H^1(\Omega)$ and $\Delta u \in L^2(\Omega)$, then $u \in H^2_{\mathrm{loc}}(\Omega)$.*

Proof. Let B be a ball with closure contained in Ω and let B_{ext} be a larger concentric ball. We build $\varphi \in \mathcal{D}(B_{\mathrm{ext}})$ such that $\varphi \equiv 1$ in B and note that $v := \widetilde{u\varphi}$ (the tilde subscript is the extension by zero operator) satisfies

$$v \in H^1(\mathbb{R}^d), \qquad -\Delta v + v \in L^2(\mathbb{R}^d).$$

Therefore $v \in H^2(\mathbb{R}^d)$ and $u = v|_B \in H^2(B)$. This proves the result. $\square$

H^2-regular domains. In principle, Proposition 13.22 does not allow us to say anything about the regularity of u all the way to the boundary. There are domains for which $u \in H^1(\Omega)$, $\Delta u \in L^2(\Omega)$, together with a homogeneous Dirichlet or Neumann boundary condition ($\gamma u = 0$ or $\partial_n u = 0$) is enough to guarantee that $u \in H^2(\Omega)$. We have seen that the half plane satisfies this property (Proposition 13.11). It is also satisfied by convex polygons in $\mathbb{R}^2$ and convex polyhedra in $\mathbb{R}^3$, as well as by domains with smooth boundaries.

(See the final comments section for more information about this.) However, this property is not universal to all domains. To illustrate this, we consider the L-shaped domain (see Figure 13.1) $\Omega := (-1,1)^2 \setminus ([0,1] \times [-1,0])$, the function

$$v(\mathbf{x}) := |\mathbf{x}|^{2/3} \sin\left(\tfrac{2}{3} \arccos \frac{x_1}{|\mathbf{x}|}\right) = r^{2/3} \sin(\tfrac{2}{3}\theta) \qquad 0 < \alpha < 1,$$

and a cutoff function $\varphi \in \mathcal{D}(\mathbb{R}^2)$ with $\varphi \equiv 1$ in a neighborhood of the origin and $\operatorname{supp}\varphi \subset B(\mathbf{0};1)$, so that $\varphi \equiv 0$ in a neighborhood of the part of the boundary that lies on the boundary of $(-1,1)^2$. The function $u := \varphi v$ satisfies $u \in H_0^1(\Omega)$, $\Delta u \in L^2(\Omega)$, but $u \notin H^2(\Omega)$, due to a singularity of $\partial_{x_1}\partial_{x_2} u$ at the origin. (See the computations at the end of Section 13.3.)

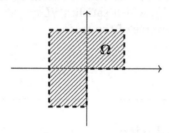

Figure 13.1: The L-shaped domain is not H^2-regular for the Dirichlet problem.

The following results are even stronger, showing that solutions of the Laplace equation are automatically smooth in the interior of the domain. (More results of this kind are proposed in Exercise 13.14.) These results will also be a good excuse to introduce one of the so-called Sobolev embedding theorems. (Some easy extensions are proposed as Exercises 13.15 and 13.16.)

Proposition 13.23. *If $u \in H^1(\Omega)$ satisfies $\Delta u = 0$, then $u \in H_{\mathrm{loc}}^m(\Omega)$ for all m.*

Proof. Note that as a consequence of Proposition 13.22, we have $u \in H_{\mathrm{loc}}^2(\Omega)$. Now let B be any ball with closure contained in Ω and consider a bigger ball B_{ext} whose closure is also contained in Ω. We have $u \in H^2(B_{\mathrm{ext}})$ and therefore $\partial_{x_i} u \in H^1(B_{\mathrm{ext}})$ and $\Delta \partial_{x_i} u = \partial_{x_i} \Delta u = 0$, which implies that $\partial_{x_i} u \in H_{\mathrm{loc}}^2(B_{\mathrm{ext}})$ and, therefore $\partial_{x_i} u \in H^2(B)$. Using this argument for every ball B and every i, it follows that $u \in H_{\mathrm{loc}}^3(\Omega)$. The general result can then be easily proved by induction. $\square$

Proposition 13.24 (Sobolev embedding). *If $s > d/2$, then $\mathsf{H}^s(\mathbb{R}^d)$ is continuously embedded into $L^\infty(\mathbb{R}^d) \cap \mathcal{C}(\mathbb{R}^d)$.*

Proof. For $u \in \mathcal{S}(\mathbb{R}^d)$, we can write

$$|u(\mathbf{x})| = \left| \int_{\mathbb{R}^d} e^{2\pi \imath \mathbf{x} \cdot \boldsymbol{\xi}} \mathcal{F}\{u\}(\boldsymbol{\xi}) \mathrm{d}\boldsymbol{\xi} \right| \leq \left(\int_{\mathbb{R}^d} (1 + |2\pi\boldsymbol{\xi}|^2)^{-s} \mathrm{d}\boldsymbol{\xi} \right)^{1/2} \|u\|_{\mathsf{H}^s(\mathbb{R}^d)},$$

where the integral in the right-hand side is bounded if and only if $2s > d$. This gives a bound

$$\|u\|_{L^\infty(\mathbb{R}^d)} \leq C_{s,d}\|u\|_{\mathsf{H}^s(\mathbb{R}^d)} \qquad \forall u \in \mathcal{S}(\mathbb{R}^d),$$

which can be extended by density to all elements of $\mathsf{H}^s(\mathbb{R}^d)$ (Proposition 13.19(b)), proving the continuous embedding of $\mathsf{H}^s(\mathbb{R}^d)$ into $L^\infty(\mathbb{R}^d)$. Note that in taking limiting sequences, we have uniform limits of continuous functions and therefore the limits are continuous functions as well. □

Corollary 13.2. *If $u \in H^1(\Omega)$ satisfies $\Delta u = 0$, then $u \in C^\infty(\Omega)$.*

Proof. Using cutoff functions, compactly supported in Ω, we can prove that $v := \widetilde{u\varphi} \in H^m(\mathbb{R}^d)$ for all m. By the Sobolev embedding theorem (Proposition 13.24), applied to any partial derivative of v, it follows that $v \in C^\infty(\mathbb{R}^d)$. Making $\varphi \equiv 1$ in arbitrary balls strictly contained in Ω, the result follows. □

Final comments and literature

It is very common to start books on elliptic PDE and Sobolev spaces with the introduction of tempered distributions, Fourier transforms, and the spaces $\mathsf{H}^s(\mathbb{R}^d)$. Sobolev spaces in bounded domains can then be defined through the restriction and density of compactly supported functions. This approach can be found in William McLean's monograph [79], a very complete resource containing a full presentation of the Sobolev spaces of Hilbert type that are used in elliptic PDE. To learn more about regularity of solutions to elliptic PDE, and in particular to see why both smooth and convex domains have additional regularity properties, the reader is directed to the work of Pierre Grisvard [58] and Monique Dauge [43].

The proof of Theorem 13.2 (equivalence of the Slobodetskij norm) is actually not very complicated although it requires some computational work. A nicely presented proof can be found in [3, Theorem 1.7.1]. See also [62, Section 3.2].

Exercises

13.1. Prove Proposition 13.1.

13.2. Prove Proposition 13.3.

13.3. The following steps show that $\mathcal{D}(\mathbb{R}^d)$ is dense in $L^2_\omega(\mathbb{R}^d)$.

(a) Prove that if $u \in L^2_\omega(\mathbb{R}^d)$, then the sequence $\chi_{B(0;n)}u$ converges to u.

(b) Use mollification to approximate $\chi_{B(0;n)}u$ by a sequence in $\mathcal{D}(B(0; n + 1))$ in $L^2_\omega(B(0; n + 1)) \equiv L^2(B(0; n + 1))$.

13.4. Let $u \in L^2(\mathbb{R}^d)$. Prove that $\partial_{x_i} u \in L^2(\mathbb{R}^d)$ if and only if $m_i \mathcal{F}\{u\} \in L^2(\mathbb{R}^d)$.

13.5. Integral form of $\mathcal{F}$. The goal of this exercise is the proof that if $u \in L^1(\mathbb{R}^d) \cap L^2(\mathbb{R}^d)$, then

$$\mathcal{F}\{u\}(\boldsymbol{\xi}) = \int_{\mathbb{R}^d} u(\mathbf{x})e^{-2\pi i \boldsymbol{\xi} \cdot \mathbf{x}} d\mathbf{x}, \qquad (13.19)$$

with equality almost everywhere in $\mathbb{R}^d$, i.e., the Fourier transform in $L^2(\mathbb{R}^d)$ can be given in integral form for those functions that are also integrable.

(a) Prove that $X := L^1(\mathbb{R}^d) \cap L^2(\mathbb{R}^d)$ is dense in $L^2(\mathbb{R}^d)$.

(b) Prove that the following norm makes X a Banach space:

$$\|u\|_X := \|u\|_{L^1(\mathbb{R}^d)} + \|u\|_{L^2(\mathbb{R}^d)}.$$

(c) Prove that $\mathcal{D}(\mathbb{R}^d)$ is dense in X. (**Hint.** Use cutoff and mollification, noting that in a bounded set L^2 convergence implies L^1 convergence.)

(d) Show that the space $Y := C(\mathbb{R}^d) \cap L^\infty(\mathbb{R}^d) \cap L^2(\mathbb{R}^d)$ is a Banach when equipped with the norm

$$\|u\|_Y := \|u\|_{L^\infty(\mathbb{R}^d)} + \|u\|_{L^2(\mathbb{R}^d)}.$$

(e) Show that $\|\mathcal{F}\{\varphi\}\|_Y \le \|\varphi\|_X$ for all $\varphi \in \mathcal{S}(\mathbb{R}^d)$.

(f) Prove that the right-hand side of (13.19) defines a bounded linear map from X to Y.

13.6. Show that if $u \in L^p(\mathbb{R}^d)$, then the map

$$\varphi \longmapsto \int_{\mathbb{R}^d} u(\mathbf{x})\varphi(\mathbf{x})d\mathbf{x}$$

defines a tempered distribution. (**Hint.** Show that convergence in the Schwartz class implies convergence in $L^p(\mathbb{R}^d)$ for all p and use Hölder's inequality.)

13.7. Consider the differential operator $\mathcal{L}u := u - \Delta u$.

(a) Show that $\mathcal{L} : \mathcal{S}(\mathbb{R}^d) \to \mathcal{S}(\mathbb{R}^d)$ is continuous and invertible with continuous inverse. (**Hint.** Use Fourier transforms to give an equivalent representation of $\mathcal{L}$.)

(b) Show that $\mathcal{L} : \mathcal{S}'(\mathbb{R}^d) \to \mathcal{S}'(\mathbb{R}^d)$ is sequentially continuous, invertible, and that its inverse is sequentially continuous. (**Hint.** Same idea as before.)

(c) Show that if $u \in L^2(\mathbb{R}^d)$ and $\Delta u \in L^2(\mathbb{R}^d)$, then $u \in H^2(\mathbb{R}^d)$. (**Hint.** Use (b).)

(d) Repeat parts (a) and (b) of the exercise for the operator $\mathcal{L}_c u := c^2 u - \Delta u$, for arbitrary $c \in \mathbb{C}$ with $\operatorname{Re} c > 0$. (**Hint.** $c^2 + |2\pi \cdot|^2 = c^2 w(\cdot/c)$.)

13.8. A density result. The goal of this exercise is the proof of the density of the injection $\mathcal{D}(\mathbb{R}^d) \subset H^s(\mathbb{R}^d)$ for all $s \in \mathbb{R}$.

(a) Show that convergence in $\mathcal{S}(\mathbb{R}^d)$ implies convergence in $L^2(\mathbb{R}^d)$.

(b) Show that convergence in $\mathcal{S}(\mathbb{R}^d)$ implies convergence in $H^s(\mathbb{R}^d)$ for all s. (**Hint.** Use Propositions 13.14 and 13.15.)

(c) Show finally that $\mathcal{D}(\mathbb{R}^d)$ is dense in $H^s(\mathbb{R}^d)$.

13.9. Conjugation of tempered distributions. Show that if $T \in \mathcal{S}'(\mathbb{R}^d)$, then

$$[\overline{T}, \varphi]_{\mathcal{S}' \times \mathcal{S}} := \overline{[T, \overline{\varphi}]}_{\mathcal{S}' \times \mathcal{S}},$$

defines another tempered distribution with $\mathcal{F}\{\overline{T}\} = \overline{\mathcal{F}^*\{T\}}$.

13.10. Conjugation in Sobolev spaces.. Show that conjugation defines an isometry in $H^s(\mathbb{R}^d)$ for all $s \in \mathbb{R}$.

13.11. A negative result on compact embeddings. In this exercise we show that the embedding $H^s(\mathbb{R}^d) \subset H^r(\mathbb{R}^d)$ for $s > r$ is not compact.

(a) Let $\varphi \in \mathcal{D}(\mathbb{R}^d)$ and consider the sequence of functions $\varphi_n := \varphi(\cdot - n\mathbf{e}_1) \in \mathcal{D}(\mathbb{R}^d)$. Show that for all s, there exists a number $c_s(\varphi)$ such that $\|\varphi_n\|_{H^s(\mathbb{R}^d)} = c_s$ for all n.

(b) Show that $\varphi_n \rightharpoonup 0$ in $L^2(\mathbb{R}^d)$.

(c) Show that $(\varphi_n, \psi)_{H^s(\mathbb{R}^d)} \to 0$ for all $\psi \in \mathcal{S}(\mathbb{R}^d)$.

(d) Conclude that $\varphi_n \rightharpoonup 0$ in $H^s(\mathbb{R}^d)$ and therefore the injection of $H^s(\mathbb{R}^d)$ into $H^r(\mathbb{R}^d)$ (for any $r < s$) cannot be compact.

13.12. Compactly supported distributions. In the space $\mathcal{E}(\mathbb{R}^d) := C^\infty(\mathbb{R}^d)$, we consider the seminorms

$$\|\varphi\|_n := \sum_{|\alpha| \le n} \|\partial^\alpha \varphi\|_{L^\infty(B(0;n))} = \sum_{|\alpha| \le n} \max_{|\mathbf{x}| \le n} |\partial^\alpha \varphi(\mathbf{x})|.$$

(a) Show that the family of seminorms $\|\cdot\|_n$ can be used to define a metric in $C(\mathbb{R}^d)$ such that $\varphi_n \to \varphi$ in $\mathcal{E}(\mathbb{R}^d)$ if and only if $\partial^\alpha \varphi_n \to \partial^\alpha \varphi$ uniformly on compact sets for all α.

(b) Show that $\mathcal{E}(\mathbb{R}^d)$ is complete with respect to that metric.

(c) Show that $\mathcal{D}(\mathbb{R}^d)$ is dense in $\mathcal{E}(\mathbb{R}^d)$. (**Hint.** Use a cutoff sequence.)

(d) Show that if $\varphi_n \to \varphi$ in $\mathcal{E}(\mathbb{R}^d)$ and $\psi \in \mathcal{D}(\mathbb{R}^d)$, then $\psi \varphi_n \to \psi \varphi$ in $\mathcal{D}(\mathbb{R}^d)$.

(e) Show that if $T \in \mathcal{E}'(\mathbb{R}^d)$, then $T|_{\mathcal{D}(\mathbb{R}^d)}$, given by

$$\langle T, \varphi \rangle_{\mathcal{D}'(\mathbb{R}^d) \times \mathcal{D}(\mathbb{R}^d)} := [T, \varphi]_{\mathcal{E}' \times \mathcal{E}},$$

defines a distribution and that this restriction process allows us to consider the inclusions $\mathcal{E}'(\mathbb{R}^d) \subset \mathcal{S}'(\mathbb{R}^d) \subset \mathcal{D}'(\mathbb{R}^d)$. (**Hint.** This is the same process that allowed us to understand tempered distributions as distributions.)

(f) Show that if $T \in \mathcal{E}'(\mathbb{R}^d) \subset \mathcal{D}'(\mathbb{R}^d)$, then there exists $R > 0$ such that

$$\langle T, \varphi \rangle_{\mathcal{D}'(\mathbb{R}^d) \times \mathcal{D}(\mathbb{R}^d)} = 0 \qquad \forall \varphi \in \mathcal{D}(\mathbb{R}^d) \quad \text{s.t.} \quad \operatorname{supp} \varphi \cap B(0; R) = \emptyset.$$

(**Hint.** Use Theorem 13.1.)

(g) Show that if $T \in \mathcal{D}'(\mathbb{R}^d)$ satisfies the equality in (f) for some $R > 0$, and we choose $\psi \in \mathcal{D}(\mathbb{R}^d)$ such that $\psi \equiv 1$ in $B(0; R + \varepsilon)$, then

$$[T, \varphi]_{\mathcal{E}' \times \mathcal{E}} := \langle T, \psi \varphi \rangle_{\mathcal{D}'(\mathbb{R}^d) \times \mathcal{D}(\mathbb{R}^d)}$$

is the only possible extension of T to $\mathcal{E}'(\mathbb{R}^d)$. (**Hint.** Use (d).)

Because of (f) and (g), the elements of $\mathcal{E}'(\mathbb{R}^d)$ are often called compactly supported distributions.

13.13. A density result for solenoidal fields. Consider the spaces

$$V := \{\mathbf{u} \in \mathbf{H}^1(\mathbb{R}^d) \,:\, \nabla \cdot \mathbf{u} = 0\}, \quad H := \{\mathbf{u} \in \mathbf{L}^2(\mathbb{R}^d) \,:\, \nabla \cdot \mathbf{u} = 0\},$$

and the polynomial vector field $\mathbf{m}(\boldsymbol{\xi}) := (m_1(\boldsymbol{\xi}), \ldots, m_d(\boldsymbol{\xi}))$.

(a) Show that $V = \{\mathbf{u} \in L^2(\mathbb{R}^d) \,:\, \mathbf{m} \cdot \mathcal{F}\{\mathbf{u}\} = 0 \quad \text{a.e.}\}$, where the Fourier transform is applied componentwise.

(b) Let $\varphi \in \mathcal{D}(\mathbb{R}^d)$ satisfy $\varphi \equiv 1$ in $B(0; 1)$, and $\varphi \equiv 0$ in $\mathbb{R}^d \setminus B(0; 2)$, and define $\varphi_n := \varphi(\,\cdot\,/n)$. Show that if $\mathbf{u} \in V$, then $\mathbf{u}_n := \mathcal{F}^*\{\varphi_n \mathcal{F}\{\mathbf{u}\}\} \in H$ and $\mathbf{u}_n \to \mathbf{u}$ in $\mathbf{L}^2(\mathbb{R}^d)$. This proves that V is dense in H.

13.14. More interior regularity results. Show that if $u \in H^1(\Omega)$ and $\Delta u \in \mathcal{C}^\infty(\Omega)$, then $u \in \mathcal{C}^\infty(\Omega)$. Extend the result to the operator $u \mapsto \Delta u - \lambda u$ for $\lambda \in \mathbb{C} \setminus (-\infty, 0]$.

13.15. A local Sobolev embedding theorem. Show that if $u \in H^m(\Omega)$ with $m > d/2$, then $u \in \mathcal{C}(\Omega)$.(**Hint.** Multiply by a smooth cutoff function and extend by zero.)

13.16. More Sobolev embeddings. Let Ω be a bounded domain such that the restriction operator $H^m(\mathbb{R}^d) \to H^m(\Omega)$ is surjective. (All Lipschitz domains satisfy this property, although this is far from trivial to prove.) Show that if $m > d/2$, then $H^m(\Omega) \subset \mathcal{C}(\overline{\Omega})$ with continuous embedding.

14

Layer potentials

14.1   Green's functions in free space 327
14.2   Single and double layer Yukawa potentials 330
14.3   Properties of the boundary integral operators 333
14.4   The Calderón calculus 338
14.5   Integral form of the layer potentials 341
14.6   A weighted Sobolev space 343
14.7   Coulomb potentials 347
14.8   Boundary-field formulations 351
Final comments and literature 357
Exercises 358

This chapter is devoted to the introduction of single and double layer potentials for some simple elliptic operators. Potential theory is a very powerful tool to prove important results on existence of solutions to boundary value problems. We will present this theory, first variationally and then in 'integral form' for the Yukawa operators $u \mapsto -\Delta u + c^2 u$ (with $c > 0$), since this avoids bringing in new Sobolev spaces. The goal of these sections will be to understand the variational properties of the layer potentials and their associated boundary integral operators and use them to provide equivalent formulations for homogeneous boundary value problems in the exterior of a Lipschitz domain. We then generalize the results for the Laplacian in three dimensions, which we will use as an excuse to introduce weighted Sobolev spaces. Finally, we will give a taste of the coupling of boundary integral formulations in an exterior domain with variational formulations inside the domain.

14.1   Green's functions in free space

We start with an operator that we have already met in Chapter 13. We fix $c > 0$ and consider the operator $G_c : \mathcal{S}'(\mathbb{R}^d) \to \mathcal{S}'(\mathbb{R}^d)$ given by

$$\mathcal{F}\{G_c f\} := \frac{1}{c^2 + |2\pi \cdot|^2} \mathcal{F}\{f\}.$$

This linear and sequentially continuous operator is well-defined thanks to the possibility of inverting the Fourier transform in the space of tempered distributions, as well as the possibility of multiplying by powers (positive and negative) of the weight function ω. (Note that we are dividing by $c^2\omega(\cdot/c)$.) We will see that this is a convolution operator, even if we will avoid defining general convolutions in the sense of distributions. We can easily prove the following results (see Exercise 13.7):

(a) G_c defines a bounded invertible operator from $H^s(\mathbb{R}^d)$ to $H^{s+2}(\mathbb{R}^d)$ for all $s \in \mathbb{R}$.

(b) G_c defines a continuous invertible operator from $\mathcal{S}(\mathbb{R}^d)$ to itself. (Just follow convergent sequences, applying results in the Schwartz class.)

(c) If $u = G_c f$, then u is the only solution of

$$u \in \mathcal{S}'(\mathbb{R}^d), \qquad -\Delta u + c^2 u = f.$$

This assertion does not exclude the existence of other nontempered solutions to the same equation. For instance, the function $u(\mathbf{x}) := \exp(c\mathbf{x}\cdot\mathbf{d})$, where $\mathbf{d}$ is a constant unit vector is a solution of the equation $-\Delta u + c^2 u = 0$ in a strong sense, and therefore, in the sense of distributions. However, this function does not define a tempered distribution, which can be seen from its exponential growth or, even in a simpler way, from the fact that $u \mapsto -\Delta u + c^2 u$ is an invertible operator in $\mathcal{S}'(\mathbb{R}^d)$, and therefore has no kernel.

Since $(c^2 + |2\pi \cdot |^2)^{-1}) \in L^\infty(\mathbb{R}^d) \subset \mathcal{S}'(\mathbb{R}^d)$, we can then define

$$E_c := \mathcal{F}^{-1}\{(c^2 + |2\pi \cdot |^2)^{-1}\} \in \mathcal{S}'(\mathbb{R}^d).$$

We can easily identify the function E_c, which is the tempered fundamental solution of the Yukawa operator $u \mapsto -\Delta u + c^2 u$, in low dimensions. Higher dimensions involve further use of Bessel-Hankel functions.

Proposition 14.1. *We have*

$$E_c(\mathbf{x}) = \begin{cases} \dfrac{1}{2c}e^{-c|\mathbf{x}|}, & d = 1, \\[2ex] \dfrac{i}{4}H_0^{(1)}(ic\,|\mathbf{x}|) = \dfrac{1}{2\pi}K_0(c|\mathbf{x}|) & d = 2, \\[2ex] \dfrac{e^{-c|\mathbf{x}|}}{4\pi|\mathbf{x}|}, & d = 3. \end{cases} \tag{14.1}$$

Here $H_0^{(1)}$ is the Hankel function of the first kind and order zero, and K_0 is the modified Bessel function of the second kind (Macdonald function) and order zero. Both expressions for E_c yield real values.

Proof. It is simple to see that the functions of the statement are in $L^2(\mathbb{R}^d)$. In the case of $d = 2$, we need to specify that K_0 has a logarithmic singularity at the origin and decays exponentially at infinity. Let Φ be the function defined in the right-hand side of (14.1). The first step is proving that

$$-\Delta\Phi + c^2\Phi = \delta_0 \quad \text{in } \mathcal{D}'(\mathbb{R}^d). \tag{14.2}$$

This is very simple for the case $d = 1$, as the derivatives can be computed without even using test functions:

$$\frac{\mathrm{d}}{\mathrm{d}x}(e^{-c|x|}) = -c\,\mathrm{sign}(x)\,e^{-c|x|}, \qquad \frac{\mathrm{d}^2}{\mathrm{d}x^2}(e^{-c|x|}) = -2c\delta_0 + c^2 e^{-c|x|}.$$

The three-dimensional case was proposed as Exercise 1.13. The proof for the two-dimensional case requires using properties of the Bessel functions, which we ask the reader to do. Once (14.2) has been established, we take Fourier transforms and apply Proposition 13.3(d) (derivatives and Fourier transforms) to obtain

$$(c^2 + |2\pi \cdot |^2)\mathcal{F}\{\Phi\} = \mathcal{F}\{\delta_0\} = 1,$$

from which the result follows. $\qquad\square$

Proposition 14.2. *For $\psi \in \mathcal{S}(\mathbb{R}^d)$, the function*

$$u(\mathbf{x}) := \int_{\mathbb{R}^d} \psi(\mathbf{x} - \mathbf{y})E_c(\mathbf{y})\mathrm{d}\mathbf{y} = \int_{\mathbb{R}^d} \psi(\mathbf{y})E_c(\mathbf{x} - \mathbf{y})\mathrm{d}\mathbf{y}$$

is in the Schwartz class and satisfies $-\Delta u + c^2 u = \psi$. Therefore $u = G_c\psi$.

Proof. It is clear that u is well-defined and continuity follows from the dominated convergence theorem. We can also differentiate under the integral sign as many times as we want, due to the local integrability of E_c and the fast decay of ψ. (Note that we are just doing a convolution product and the basic ideas exposed in Section 2.2 apply to this situation as well.) This proves that $u \in \mathcal{C}^\infty(\mathbb{R}^d)$. Now fix now a point $\mathbf{x} \in \mathbb{R}^d$ and consider the function $\psi_{\mathbf{x}} := \psi(\mathbf{x} - \cdot)$, so that we can write

$$u(\mathbf{x}) = [E_c, \psi_{\mathbf{x}}]_{\mathcal{S}' \times \mathcal{S}}.$$

Differentiating under the integral sign, we have

$$(\Delta u)(\mathbf{x}) = \int_{\mathbb{R}^d} \Delta_{\mathbf{x}}(\psi(\mathbf{x} - \mathbf{y}))E_c(\mathbf{y})\mathrm{d}\mathbf{y} = \int_{\mathbb{R}^d} (\Delta\psi_{\mathbf{x}})(\mathbf{y})E_c(\mathbf{y})\mathrm{d}\mathbf{y}$$

$$= [E_c, \Delta\psi_{\mathbf{x}}]_{\mathcal{S}' \times \mathcal{S}} = [\Delta E_c, \psi_{\mathbf{x}}]_{\mathcal{S}' \times \mathcal{S}}.$$

Therefore,

$$-\Delta u(\mathbf{x}) + c^2 u(\mathbf{x}) = [-\Delta E_c + c^2 E_c, \psi_{\mathbf{x}}]_{\mathcal{S}' \times \mathcal{S}} = [\delta_0, \psi_{\mathbf{x}}]_{\mathcal{S}' \times \mathcal{S}} = \psi(\mathbf{x}).$$

Finally, the fast decay of u can be proved directly or we can just show that $u \in \mathcal{S}'(\mathbb{R}^d)$ and, since $-\Delta u + c^2 u = \psi \in \mathcal{S}'(\mathbb{R}^d)$, as tempered distributions, then

$$\mathcal{F}\{u\} = \frac{1}{c^2 + 4\pi^2 |\cdot|^2} \mathcal{F}\{\psi\} \in \mathcal{S}(\mathbb{R}^d),$$

which proves that $u \in \mathcal{S}(\mathbb{R}^d)$.   □

Proposition 14.3. *We have the identity*

$$[G_c f, \psi]_{\mathcal{S}' \times \mathcal{S}} = [f, G_c \psi]_{\mathcal{S}' \times \mathcal{S}} \qquad \forall f \in \mathcal{S}'(\mathbb{R}^d), \qquad \psi \in \mathcal{S}(\mathbb{R}^d).$$

Proof. Let $\phi_c(\boldsymbol{\xi}) := (c^2 + |2\pi\boldsymbol{\xi}|^2)^{-1}$. Using that $\mathcal{F}\{\psi\} = \mathcal{F}^*\{\psi(-\cdot)\}$ for $\psi \in \mathcal{S}(\mathbb{R}^d)$ (this is easy to prove), we can show that

$$\mathcal{F}\{\phi_c \mathcal{F}^*\{\psi\}\} = \mathcal{F}^*\{\phi_c \mathcal{F}^*\{\psi\}(-\cdot)\} = \mathcal{F}^*\{\phi_c \mathcal{F}\{\psi\}\} = G_c \psi \qquad \forall \psi \in \mathcal{S}(\mathbb{R}^d).$$

Therefore, by definition of the Fourier transform in the sense of tempered distributions, we have

$$\begin{aligned}
[G_c f, \psi]_{\mathcal{S}' \times \mathcal{S}} &= [\mathcal{F}^*\{\phi_c \mathcal{F}\{f\}\}, \psi]_{\mathcal{S}' \times \mathcal{S}} = [\phi_c \mathcal{F}\{f\}, \mathcal{F}^*\{\psi\}]_{\mathcal{S}' \times \mathcal{S}} \\
&= [f, \mathcal{F}\{\phi_c \mathcal{F}^*\{\psi\}\}]_{\mathcal{S}' \times \mathcal{S}} = [f, G_c \psi]_{\mathcal{S}' \times \mathcal{S}},
\end{aligned}$$

which finishes the proof.   □

14.2   Single and double layer Yukawa potentials

In this and the next section, we derive a variational theory of the single and double layer potentials for the Yukawa operator $u \mapsto -\Delta u + c^2 u$. We will loosely refer to the associated boundary operators as boundary integral operators, although we will take some time to actually show that the definitions we provide correspond to the well-known potentials and operators in the literature.

The setting in the coming sections is depicted in Figure 14.1. We will consider a bounded connected domain Ω_- with connected boundary Γ and exterior Ω_+. The normal vector field on Γ will be taken with the orientation that makes it exterior to Ω_-. We then have two trace operators and the corresponding jump

$$\gamma^\pm : H^1(\mathbb{R}^d \setminus \Gamma) \to H^{1/2}(\Gamma), \qquad [\gamma u] := \gamma^- u - \gamma^+ u.$$

When $[\gamma u] = 0$, we have that $u \in H^1(\mathbb{R}^d)$ and we will denote $\gamma u = \gamma^+ u = \gamma^- u$. Similarly, we have two possible normal derivatives and a jump

$$\partial_n^\pm : \{u \in H^1(\mathbb{R}^d \setminus \Gamma) : \Delta u \in L^2(\mathbb{R}^d \setminus \Gamma)\} \to H^{-1/2}(\Gamma), \qquad [\partial_n u] := \partial_n^- u - \partial_n^+ u.$$

When $[\partial_n u] = 0$, we will use $\partial_n u$ to represent both normal derivatives. In particular, note that

$$\langle[\partial_n u], \gamma v\rangle_\Gamma = (\nabla u, \nabla v)_{\mathbb{R}^d\setminus\Gamma} + (\Delta u, v)_{\mathbb{R}^d\setminus\Gamma} \qquad \forall v \in H^1(\mathbb{R}^d). \qquad (14.3)$$

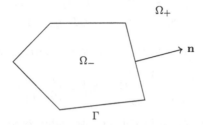

Figure 14.1: The cartoonish representation of the set-up of this chapter: a bounded Lipschitz domain Ω_-, with connected boundary Γ, and the exterior domain Ω_+, with the normal vector pointing from Ω_- to Ω_+.

Layer potentials from a transmission problem. The input for the single and double layer potentials will be transmission data $\lambda \in H^{-1/2}(\Gamma)$ and $\eta \in H^{1/2}(\Gamma)$, for the problem in $\Omega_- \cup \Omega_+ = \mathbb{R}^d \setminus \Gamma$:

$$u \in H^1(\mathbb{R}^d \setminus \Gamma) \qquad -\Delta u + c^2 u = 0, \qquad (14.4a)$$
$$[\gamma u] = \eta, \qquad [\partial_n u] = \lambda. \qquad (14.4b)$$

It is a simple exercise to show that this problem is equivalent to the uniquely solvable variational problem

$$u \in H^1(\mathbb{R}^d \setminus \Gamma), \qquad [\gamma u] = \eta, \qquad (14.5a)$$
$$(\nabla u, \nabla v)_{\mathbb{R}^d\setminus\Gamma} + c^2(u, v)_{\mathbb{R}^d\setminus\Gamma} = \langle\lambda, \gamma v\rangle_\Gamma \qquad \forall v \in H^1(\mathbb{R}^d). \qquad (14.5b)$$

To prove well-posedness we just need a lifting of the side condition $[\gamma u] = \eta$ (take $u_\eta \in H^1(\Omega_-)$ with $\gamma^- u_\eta = \eta$ and extend by zero to Ω_+) and to recognize that $H^1(\mathbb{R}^d) = \{v \in H^1(\mathbb{R}^d \setminus \Gamma) : [\gamma v] = 0\}$. The variational problem (14.5) is equivalent to the minimization problem

$$\tfrac{1}{2}(\|\nabla u\|^2_{\mathbb{R}^d\setminus\Gamma} + c^2\|u\|^2_{\mathbb{R}^d}) - \langle\lambda, \gamma^- u\rangle = \min! \qquad u \in H^1(\mathbb{R}^d \setminus \Gamma), \qquad [\gamma u] = \eta.$$

The unique solution to (14.4) and (14.5) will be denoted by

$$u := S_c\lambda - D_c\eta,$$

thus separating the influence of the two input data. With this process, we have defined two bounded linear operators

$$S_c : H^{-1/2}(\Gamma) \to H^1(\mathbb{R}^d), \qquad D_c : H^{1/2}(\Gamma) \to H^1(\mathbb{R}^d \setminus \Gamma),$$

that will be respectively called **single and double layer potentials.** Both potentials are also bounded when we consider any of $\{u \in H^1(\mathbb{R}^d \setminus \Gamma) : \Delta u = c^2 u\} \subset \{u \in H^1(\mathbb{R}^d \setminus \Gamma) : \Delta u \in L^2(\mathbb{R}^d \setminus \Gamma)\}$ as target spaces. By definition of the potentials, we have the **jump relations**

$$[\gamma S_c \lambda] = 0, \qquad [\gamma D_c \eta] = -\eta, \tag{14.6a}$$

$$[\partial_n S_c \lambda] = \lambda, \qquad [\partial_n D_c \eta] = 0, \tag{14.6b}$$

which can also be presented in this condensed matrix form

$$\begin{bmatrix} [\gamma \cdot] \\ [\partial_n \cdot] \end{bmatrix} \begin{bmatrix} -D_c & S_c \end{bmatrix} = \begin{bmatrix} I & 0 \\ 0 & I \end{bmatrix}. \tag{14.7}$$

The reader will wonder often in this chapter why some signs are 'arbitrarily' changed in some definitions. The main reason comes from looking at the integral expressions of S_c and D_c, where the 'natural signs' are more apparent. An 'intuitive' explanation can be given by looking at the kernel function of the Green's operator, which is a function of the variable $\mathbf{x} - \mathbf{y}$. That means that when we differentiate once, we see different signs if we differentiate with respect to $\mathbf{x}$ or $\mathbf{y}$. In any case, we ask the reader to bear with us and accept this common sign conventions as a given, as a definition if you will.

Four boundary (integral) operators. We can take two-sided traces and normal derivatives of $S_c \lambda$ and $D_c \eta$ (recall that these functions satisfy $\Delta u = c^2 u$ in $\mathbb{R}^d \setminus \Gamma$) and define four operators:

$$V_c \lambda := \gamma^{\pm} S_c \lambda = \tfrac{1}{2}(\gamma^+ + \gamma^-) S_c \lambda,$$

$$K_c \eta := \tfrac{1}{2}(\gamma^+ + \gamma^-) D_c \eta,$$

$$K_c^t \lambda := \tfrac{1}{2}(\partial_n^+ + \partial_n^-) S_c \lambda,$$

$$W_c \eta := -\partial_n^{\pm} D_c \eta = -\tfrac{1}{2}(\partial_n^+ + \partial_n^-) D_c \eta.$$

Once again, the reader might find it odd to have one single negative sign in the definition of W_c. This is done to make W_c positive definite (see Proposition 14.7 below), although its natural tendency is to be negative definite. Considering the averaging operators for two-sided traces and normal derivatives

$$\{\gamma u\} := \tfrac{1}{2}(\gamma^+ u + \gamma^- u), \qquad \{\partial_n u\} := \tfrac{1}{2}(\partial_n^+ u + \partial_n^- u),$$

we can give a joint definition of the four operators using a matrix expression

$$\begin{bmatrix} \{\gamma \cdot\} \\ \{\partial_n \cdot\} \end{bmatrix} \begin{bmatrix} -D_c & S_c \end{bmatrix} = \begin{bmatrix} -K_c & V_c \\ W_c & K_c^t \end{bmatrix}, \tag{14.8}$$

that mimics (14.7). If we swap the two columns, we obtain an expression

$$\begin{bmatrix} \{\gamma \cdot\} \\ \{\partial_n \cdot\} \end{bmatrix} \begin{bmatrix} S_c & -D_c \end{bmatrix} = \begin{bmatrix} V_c & -K_c \\ K_c^t & W_c \end{bmatrix},$$

where we will recognize later (Propositions 14.6 to 14.8) two symmetric positive definite operators in the diagonal and a skew-symmetric off diagonal. We can also write

$$\begin{bmatrix} \{\gamma \cdot\} \\ \{\partial_n \cdot\} \end{bmatrix} [S_c \ \ D_c] = \begin{bmatrix} V_c & K_c \\ K_c^t & -W_c \end{bmatrix},$$

where the matrix in the right-hand side is now a symmetric matrix of operators with some sort of global indefinite sign (V_c is positive and $-W_c$ is negative). In any case, because of the continuity of the potentials and the fact that they map to the set of solutions of the homogeneous Yukawa equation, we have the following result. In the jargon of boundary integral operators, V_c is called the single layer operator, K_c the double layer operator, K_c^t the adjoint double layer operator, and W_c the hypersingular operator.

Proposition 14.4 (Mapping properties of BIO). *The following operators are bounded:*

$$V_c : H^{-1/2}(\Gamma) \longrightarrow H^{1/2}(\Gamma), \qquad K_c : H^{1/2}(\Gamma) \longrightarrow H^{1/2}(\Gamma),$$
$$K_c^t : H^{-1/2}(\Gamma) \longrightarrow H^{-1/2}(\Gamma), \qquad W_c : H^{1/2}(\Gamma) \longrightarrow H^{-1/2}(\Gamma).$$

The trace formulas

$$\gamma^{\pm} D_c = \pm \tfrac{1}{2} I + K_c, \qquad \partial_n^{\pm} S_c = \mp \tfrac{1}{2} I + K_c^t, \tag{14.9}$$

are easy consequences of the jump relations (14.6)-(14.7) and of the definition of the integral operators (14.8). We also have the following representation formula for the solutions of the Yukawa equation.

Proposition 14.5 (Representation formula). *If $u \in H^1(\mathbb{R}^d \setminus \Gamma)$ satisfies $-\Delta u + c^2 u = 0$ in $\mathbb{R}^d \setminus \Gamma$, then*

$$u = S_c[\partial_n u] - D_c[\gamma u].$$

Proof. This proof is very simple and follows from the unique solvability of (14.4). If we define $v := S_c[\partial_n u] - D_c[\gamma u]$ and $w := u - v \in H^1(\mathbb{R}^d \setminus \Gamma)$, then

$$-\Delta w + c^2 w = 0 \quad \text{in } \mathbb{R}^d \setminus \Gamma, \qquad [\gamma w] = 0, \qquad [\partial_n w] = 0,$$

and therefore $w = 0$. $\qquad\qquad\qquad\qquad\qquad\qquad\qquad\qquad\qquad\qquad\square$

14.3 Properties of the boundary integral operators

In this section we prove several important properties of the boundary integral operators, namely the symmetry and coercivity of V_c and W_c, the fact that K_c and K_c^t are reciprocally adjoint, and the invertibility of the four operators 'of the second kind'

$$\pm \tfrac{1}{2}I + K_c, \qquad \pm \tfrac{1}{2}I + K_c^t.$$

We will discuss later on why we use quotation marks in the above sentence, but let it be said right here: in general the operators K_c and K_c^t are not compact. They do happen to be compact, however, when the boundary Γ is smooth enough.

Proposition 14.6 (Symmetry and coercivity of V_c). *There exists a constant c_V, depending on c and Γ such that*

$$\langle \lambda, V_c \mu \rangle_\Gamma = \langle \mu, V_c \lambda \rangle_\Gamma \qquad \forall \lambda, \mu \in H^{-1/2}(\Gamma),$$
$$\langle \lambda, V_c \lambda \rangle_\Gamma \geq c_V \|\lambda\|^2_{-1/2,\Gamma} \qquad \forall \lambda \in H^{-1/2}(\Gamma).$$

Therefore V_c is invertible and $\langle \cdot, V_c \cdot \rangle_\Gamma$ defines an equivalent inner product in $H^{-1/2}(\Gamma)$.

Proof. Let $u := S_c \lambda$ and $v := S_c \mu$. By definition of S_c, we have

$$\langle \lambda, V_c \mu \rangle_\Gamma = \langle [\partial_n u], \gamma v \rangle_\Gamma = (\nabla u, \nabla v)_{\mathbb{R}^d} + c^2 (u, v)_{\mathbb{R}^d},$$

which proves symmetry. Also

$$\begin{aligned}
\langle \lambda, V_c \lambda \rangle_\Gamma &= \|\nabla u\|^2_{\mathbb{R}^d} + c^2 \|u\|^2_{\mathbb{R}^d} = \|\nabla u\|^2_{\mathbb{R}^d} + c^{-2} \|\Delta u\|^2_{\mathbb{R}^d \setminus \Gamma} \\
&\geq C(c, \Gamma)(\|\partial_n^- u\|^2_{-1/2,\Gamma} + \|\partial_n^+ u\|^2_{-1/2,\Gamma}) \\
&\geq \tfrac{1}{2} C(c, \Gamma) \|[\partial_n u]\|^2_{-1/2,\Gamma} = \tfrac{1}{2} C(c, \Gamma) \|\lambda\|^2_{-1/2,\Gamma},
\end{aligned}$$

which proves positivity. The problem $V_c \lambda = g$ is equivalent to the coercive variational problem

$$\lambda \in H^{-1/2}(\Gamma), \qquad \langle \mu, V_c \lambda \rangle_\Gamma = \langle \mu, g \rangle_\Gamma \qquad \forall \mu \in H^{-1/2}(\Gamma),$$

which proves invertibility.   □

Proposition 14.7 (Symmetry and coercivity of W_c). *There exists a constant $c_W > 0$, depending on c and Γ such that*

$$\langle W_c \eta, \psi \rangle_\Gamma = \langle W_c \psi, \eta \rangle_\Gamma \qquad \forall \eta, \psi \in H^{1/2}(\Gamma),$$
$$\langle W_c \eta, \eta \rangle_\Gamma \geq c_W \|\eta\|^2_{1/2,\Gamma} \qquad \forall \eta \in H^{1/2}(\Gamma).$$

Therefore W_c is invertible and $\langle W_c \cdot, \cdot \rangle_\Gamma$ defines an equivalent inner product in $H^{1/2}(\Gamma)$.

Proof. Let $u := D_c\eta$, $v := D_c\psi$, and note that by definition of D_c we have

$$
\begin{aligned}
\langle W_c\eta, \psi\rangle_\Gamma &= \langle \partial_n u, [\gamma v]\rangle_\Gamma = \langle \partial_n^- u, \gamma^- v\rangle_\Gamma - \langle \partial_n^+ u, \gamma^+ v\rangle_\Gamma \\
&= (\nabla u, \nabla v)_{\mathbb{R}^d \backslash \Gamma} + c^2 (u, v)_{\mathbb{R}^d \backslash \Gamma} \\
&= (\nabla u, \nabla v)_{\mathbb{R}^d \backslash \Gamma} + c^{-2} (\Delta u, \Delta v)_{\mathbb{R}^d \backslash \Gamma},
\end{aligned}
$$

and

$$
\begin{aligned}
\langle W_c\eta, \eta\rangle_\Gamma &= \|\nabla u\|_{\mathbb{R}^d \backslash \Gamma}^2 + c^2 \|u\|_{\mathbb{R}^d \backslash \Gamma}^2 \\
&\geq C(\Gamma, c)(\|\gamma^- u\|_{1/2,\Gamma}^2 + \|\gamma^+ u\|_{1/2,\Gamma}^2) \\
&\geq \tfrac{1}{2} C(\Gamma, c)\|[\gamma u]\|_{1/2,\Gamma}^2 = \tfrac{1}{2} C(\Gamma, c)\|\eta\|_{1/2,\Gamma}^2.
\end{aligned}
$$

This proves symmetry and strong positivity of W_c, and therefore invertibility. $\square$

Proposition 14.8. *The following identity holds*

$$
\langle K_c^t \lambda, \eta\rangle_\Gamma = \langle \lambda, K_c\eta\rangle_\Gamma \qquad \forall \lambda \in H^{-1/2}(\Gamma), \qquad \eta \in H^{1/2}(\Gamma),
$$

and therefore $K_c^t = K_c'$.

Proof. Let $u := D_c\eta$ and $v := S_c\lambda \in H^1(\mathbb{R}^d)$. By the variational equations defining $D_c\eta$ we have

$$
\begin{aligned}
0 &= (\nabla u, \nabla v)_{\mathbb{R}^d \backslash \Gamma} + c^2 (u, v)_{\mathbb{R}^d} = (\nabla u, \nabla v)_{\mathbb{R}^d \backslash \Gamma} + (u, \Delta v)_{\mathbb{R}^d \backslash \Gamma} \\
&= \langle \partial_n^- v, \gamma^- u\rangle_\Gamma - \langle \partial_n^+ v, \gamma^+ u\rangle_\Gamma.
\end{aligned}
$$

The latter equation is equivalent to

$$
\langle \tfrac{1}{2}\lambda + K_c^t \lambda, -\tfrac{1}{2}\eta + K_c\eta\rangle_\Gamma = \langle -\tfrac{1}{2}\lambda + K_c^t \lambda, \tfrac{1}{2}\eta + K_c\eta\rangle_\Gamma,
$$

which can be simplified to

$$
-\tfrac{1}{2}\langle K_c^t \lambda, \eta\rangle_\Gamma + \tfrac{1}{2}\langle \lambda, K_c\eta\rangle_\Gamma = \tfrac{1}{2}\langle K_c^t \lambda, \eta\rangle_\Gamma - \tfrac{1}{2}\langle \lambda, K_c\eta\rangle_\Gamma,
$$

and proves the result. Note that this result is equivalent to the fact that single and double layer potentials are orthogonal with respect to the c-weighted $H^1(\mathbb{R}^d \backslash \Gamma)$ inner product, which was the starting point of our proof. $\square$

Proposition 14.9 (Equations of the second kind, I). *The operators*

$$
-\tfrac{1}{2}I + K_c : H^{1/2}(\Gamma) \to H^{1/2}(\Gamma), \qquad -\tfrac{1}{2}I + K_c^t : H^{-1/2}(\Gamma) \to H^{-1/2}(\Gamma)
$$

are invertible.

Proof. This has to be proved for only one of them. Recall that $\gamma^- D_c = -\frac{1}{2}I + K_c$. If $-\frac{1}{2}\eta + K_c\eta = 0$ and $u := D_c\eta$, then

$$-\Delta u + c^2 u = 0 \quad \text{in } \Omega^-, \qquad \gamma^- u = 0,$$

and therefore $u \equiv 0$ in Ω^-. Since $[\partial_n u] = 0$, we have $\partial_n^+ u = \partial_n^- u = 0$ and

$$-\Delta u + c^2 u = 0 \quad \text{in } \Omega^+, \qquad \partial_n^+ u = 0,$$

which implies that $u \equiv 0$ in Ω_+. Finally $\eta = [\gamma u] = 0$ and $-\frac{1}{2}I + K_c$ is injective. Now let $g \in H^{1/2}(\Gamma)$ and progressively solve an interior Dirichlet problem

$$-\Delta u + c^2 u = 0 \quad \text{in } \Omega^-, \qquad \gamma^- u = g,$$

and an exterior Neumann problem

$$-\Delta u + c^2 u = 0 \quad \text{in } \Omega^+, \qquad \partial_n^+ u = \partial_n^- u,$$

so that

$$u = S_c[\partial_n u] - D_c[\gamma u] = D_c\eta, \qquad \eta := -[\gamma u],$$

and $-\frac{1}{2}\eta + K_c\eta = \gamma^- D_c\eta = \gamma^- u = g$. This finishes the proof of the invertibility of $-\frac{1}{2}I + K_c$. Since the second operator in the statement is the adjoint of the first one (recall that $K_c^t = K_c'$), it is also invertible. Also, note that solving $-\frac{1}{2}\lambda + K_c^t\lambda = h$ is equivalent to writing

$$\langle \lambda, \psi \rangle_\Gamma = \langle h, (-\tfrac{1}{2}I + K_c)^{-1}\psi \rangle_\Gamma \qquad \forall \psi \in H^{1/2}(\Gamma),$$

which gives a direct expression of the inverse. $\qquad\square$

Proposition 14.10 (Equations of the second kind, II). *The operators*

$$\tfrac{1}{2}I + K_c : H^{1/2}(\Gamma) \to H^{1/2}(\Gamma), \qquad \tfrac{1}{2}I + K_c^t : H^{-1/2}(\Gamma) \to H^{-1/2}(\Gamma)$$

are invertible.

Proof. The first part is due to the formula $\gamma^+ D_c = \frac{1}{2}I + K_c$. The proof is like the one of Proposition 14.9, swapping the roles of the interior and exterior domains. $\qquad\square$

Once again, we refer to these as second kind equations, although K_c and K_c^t are not compact operators. We will see in the next section that the difference between operators for different c is compact though.

Examples of boundary integral formulations. Before we move on to showing more important properties of the matrix of operators defined by V_c, W_c, K_c, and K_c^t, let us briefly look at what boundary integral formulations are. Details for what follows are proposed as Exercises 14.3 and 14.4. Let us consider the exterior Dirichlet problem

$$-\Delta u + c^2 u = 0 \quad \text{in } \Omega_+, \qquad \gamma^+ u = g \in H^{1/2}(\Gamma).$$

We can use a **potential ansatz** of the solution of this problem for an unknown density $\lambda \in H^{-1/2}(\Gamma)$,

$$u := S_c\lambda.$$

This naturally extends u to a solution of the two-sided Dirichlet problem

$$-\Delta u + c^2 u = 0 \quad \text{in } \mathbb{R}^d \setminus \Gamma, \qquad \gamma^\pm u = g.$$

Using the definition of V_c, we have the boundary integral equation

$$\lambda \in H^{-1/2}(\Gamma), \qquad V_c\lambda = g, \qquad (14.10)$$

which can be used to determine the unknown density in terms of the data (Proposition 14.6 shows that this equation is uniquely solvable). We can actually write an equivalent (coercive and symmetric) variational formulation of (14.10)

$$\lambda \in H^{-1/2}(\Gamma), \qquad \langle \mu, V_c\lambda \rangle_\Gamma = \langle \mu, g \rangle_\Gamma \qquad \forall \mu \in H^{-1/2}(\Gamma). \qquad (14.11)$$

Contrary to the additional work that had to be carried out for variational formulations of boundary value problems, here the variational formulation of (14.10) consists of testing with a general $\mu \in H^{-1/2}(\Gamma) = H^{1/2}(\Gamma)'$. The interest of a formulation like (14.11) stems from the fact that we will provide an integral form of V_c, which will show that (14.11) involves only integration on Γ. This is used in practice as the starting point of Galerkin methods commonly referred to as **boundary element methods**. What we have shown is called an **indirect formulation** (because we used a potential representation), leading to an **equation of the first kind**.

We can proceed differently. Instead of gluing an interior Dirichlet problem with the same data (that was done by proposing a solution of the form $u = S_c\lambda$), we could extend u by zero to Ω_-. By the representation theorem (Proposition 14.5), we have

$$u = S_c[\partial_n u] - D_c[\gamma u] = -S_c\partial_n^+ u + D_c g, \qquad (14.12)$$

after eliminating all interior traces (which vanish, since we have assumed $u \equiv 0$ in Ω_-) and substituting the boundary condition $\gamma^+ u = g$. If we take the exterior trace of (14.12) using (14.9) (specifically, we use that $\gamma^+ D_c = \frac{1}{2}I + K_c$) we have

$$g = \gamma^+ u = -V_c\partial_n^+ u + \tfrac{1}{2}g + K_c g,$$

or equivalently

$$V_c\partial_n^+ u = -\tfrac{1}{2}g + K_c g. \qquad (14.13)$$

We can now understand (14.13) as an equation (which is uniquely solvable), whose unknown is the missing Neumann data for u. Once this equation is solved, (14.12) acts as an explicit integral representation of the solution. This process is called a **direct method** because it deals with a 'physically meaningful' unknown $\partial_n^+ u$ as opposed to a density λ. The equation that we reached is associated to the same operator V_c as the one in (14.10). More formulations, some of them leading to operators of the 'second kind' (those of Propositions 14.9 and 14.10), can also be obtained using indirect or direct methods.

14.4   The Calderón calculus

We consider an interior solution of the Yukawa equation, extended by zero to Ω_+,

$$-\Delta u + c^2 u = 0 \quad \text{in } \Omega_-, \qquad u \equiv 0 \quad \text{in } \Omega_+,$$

so that $[\gamma u] = \gamma^- u$, $[\partial_n u] = \partial_n^- u$, and

$$u = S_c \partial_n^- u - D_c \gamma^- u \tag{14.14}$$

is a representation formula for interior solutions. Taking interior traces in (14.14), we obtain the identities

$$\tfrac{1}{2}\gamma^- u = V_c \partial_n^- u - K_c \gamma^- u, \qquad \tfrac{1}{2}\partial_n^- u = K_c^t \partial_n^- u + W_c \gamma^- u.$$

A matrix form of the above is given by

$$\begin{bmatrix} \tfrac{1}{2}I - K_c & V_c \\ W_c & \tfrac{1}{2}I + K_c^t \end{bmatrix} \begin{bmatrix} \gamma^- u \\ \partial_n^- u \end{bmatrix} = \begin{bmatrix} \gamma^- u \\ \partial_n^- u \end{bmatrix}. \tag{14.15}$$

Now consider the bounded linear operator

$$\mathcal{C}_c^- : H^{1/2}(\Gamma) \times H^{-1/2}(\Gamma) \longrightarrow H^{1/2}(\Gamma) \times H^{-1/2}(\Gamma)$$

that appears in the left-hand side of (14.15), namely

$$\mathcal{C}_c^- := \begin{bmatrix} \tfrac{1}{2}I - K_c & V_c \\ W_c & \tfrac{1}{2}I + K_c^t \end{bmatrix} = \begin{bmatrix} \gamma^- \\ \partial_n^- \end{bmatrix} \begin{bmatrix} -D_c & S_c \end{bmatrix},$$

and the space of Cauchy data for homogeneous solutions of the interior Yukawa equation

$$\mathbb{D}_c^- := \{(\gamma^- u, \partial_n^- u) : u \in H^1(\Omega_-), \quad -\Delta u + c^2 u = 0 \quad \text{in } \Omega_-\}.$$

We have shown that

$$\mathcal{C}_c^- \psi = \psi \qquad \forall \psi \in \mathbb{D}_c^-.$$

Proposition 14.11 (The Calderón projection). *The operator $\mathcal{C}_c^-$ is a projection with range $\mathbb{D}_c^-$.*

Proof. If $u \in H^1(\Omega_-)$ satisfies $-\Delta u + c^2 u = 0$, then $u = S_c \partial_n^- u - D_c \gamma^- u$, or equivalently

$$\begin{bmatrix} -D_c & S_c \end{bmatrix} \begin{bmatrix} \gamma^- \\ \partial_n^- \end{bmatrix} u = u.$$

If $\psi = (\eta, \lambda) \in H^{1/2}(\Gamma) \times H^{-1/2}(\Gamma)$ and $u = -D_c \eta + S_c \lambda$ (Proposition 14.5), then

$$(\mathcal{C}_c^-)^2 \psi = \begin{bmatrix} \gamma^- \\ \partial_n^- \end{bmatrix} \begin{bmatrix} -D_c & S_c \end{bmatrix} \begin{bmatrix} \gamma^- \\ \partial_n^- \end{bmatrix} u = \begin{bmatrix} \gamma^- \\ \partial_n^- \end{bmatrix} u = \mathcal{C}_c^- \psi,$$

which proves that $\mathcal{C}_c^-$ is a projection. On the other hand,

$$\begin{bmatrix} \gamma^- \\ \partial_n^- \end{bmatrix} \begin{bmatrix} -D_c & S_c \end{bmatrix} \psi = \begin{bmatrix} \gamma^- \\ \partial_n^- \end{bmatrix} u \in \mathbb{D}_c^-,$$

and since $\mathcal{C}_c^- \psi = \psi$ for all $\psi \in \mathbb{D}_c^-$, the proof is finished. $\qquad\square$

The exterior Calderón projector. The range of the complementary projection

$$\mathcal{C}_c^+ := \mathcal{I} - \mathcal{C}_c^- = \begin{bmatrix} \frac{1}{2}I + K_c & -V_c \\ -W_c & \frac{1}{2}I - K_c^t \end{bmatrix} = \begin{bmatrix} \gamma^+ \\ \partial_n^+ \end{bmatrix} \begin{bmatrix} D_c & -S_c \end{bmatrix},$$

is the set of Cauchy data of homogeneous solutions to the Yukawa equation

$$\mathbb{D}_c^+ = \{(\gamma^+ u, \partial_n^+ u) : u \in H^1(\Omega_+), \quad -\Delta u + c^2 u = 0 \text{ in } \Omega_+\}.$$

(This is left to prove as Exercise 14.2.)

Corollary 14.1. *The following four identities hold*

$$V_c W_c = \tfrac{1}{4}I - K_c^2, \qquad V_c K_c^t = K_c V_c,$$
$$W_c V_c = \tfrac{1}{4}I - (K_c^t)^2, \qquad W_c K_c = K_c^t W_c.$$

Proof. We just need to expand the matrix form of the identity $\mathcal{C}_c^- \mathcal{C}_c^- = \mathcal{C}_c^-$ and simplify. $\qquad\square$

The following simple lemma (a consequence of elementary distribution theory and H^2 regularity in free space, as seen in Chapter 13) will be instrumental for the results that come next.

Lemma 14.1. *If $f \in L^2(\mathbb{R}^d \setminus \Gamma)$ and*

$$u \in H^1(\mathbb{R}^d \setminus \Gamma), \qquad -\Delta u + c^2 u = f \text{ in } \mathbb{R}^d \setminus \Gamma, \qquad [\gamma u] = 0, \qquad [\partial_n u] = 0,$$

then $u \in H^2(\mathbb{R}^d)$ and $\|u\|_{2,\mathbb{R}^d} \leq C\|f\|_{\mathbb{R}^d}$.

Proof. First, note that $u \in H^1(\mathbb{R}^d)$ and that by the definition of the jump of the normal derivative (see (14.3)), we have

$$(\nabla u, \nabla v)_{\mathbb{R}^d} + c^2(u, v)_{\mathbb{R}^d} = (f, v)_{\mathbb{R}^d} \qquad \forall v \in H^1(\mathbb{R}^d).$$

This proves that $-\Delta u + c^2 u = f$ in $\mathcal{D}'(\mathbb{R}^d)$. On the other hand $u \in \mathcal{S}'(\mathbb{R}^d)$ and therefore, by global regularity in free space (see, for instance Corollary 13.1), it follows that $u \in H^2(\mathbb{R}^d)$. Proposition 13.10 can then be used to prove the bound given in the statement. $\qquad\square$

Proposition 14.12. *The operators*

$$V_c - V_b : H^{-1/2}(\Gamma) \to H^{1/2}(\Gamma), \qquad K_c^t - K_b^t : H^{-1/2}(\Gamma) \to H^{-1/2}(\Gamma)$$

are compact.

Proof. Let $u_c := S_c \lambda$, $u_b := S_b \lambda$, and $u := u_c - u_b \in H^1(\mathbb{R}^d)$. We have

$$[\gamma u] = 0, \qquad [\partial_n u] = 0, \qquad \gamma u = (V_c - V_b)\lambda, \qquad \partial_n u = (K_c^t - K_b^t)\lambda.$$

We also have

$$-\Delta u + c^2 u = (b^2 - c^2)u_b \qquad \text{in } \mathbb{R}^d \setminus \Gamma,$$

but since $[\gamma u] = 0$ and $[\partial_n u] = 0$, we have (by Lemma 14.1)

$$-\Delta u + c^2 u = (b^2 - c^2)u_b \in H^1(\mathbb{R}^d)$$

as distributions in $\mathbb{R}^d$. This implies that $u \in H^3(\mathbb{R}^d)$ and

$$\|u\|_{3,\mathbb{R}^d} \le C\|u_b\|_{1,\mathbb{R}^d} \le C\|\lambda\|_{-1/2,\Gamma}. \tag{14.16}$$

We are going to prove both results now. If $\lambda_n \rightharpoonup \lambda$ in $H^{-1/2}(\Gamma)$, (14.16) shows that $u_n := (S_c - S_b)\lambda_n \rightharpoonup u := (S_c - S_b)\lambda$ in $H^3(\mathbb{R}^d)$ and therefore $u_n|_B \rightharpoonup u|_B$ in $H^3(B)$, where B is any ball containing Γ. By the Rellich-Kondrachov compactness theorem (see Exercise 7.6(b)), we have that $u_n|_B \to u|_B$ in $H^2(B)$. From here we have both results: (a) on the one hand, we have $u_n|_B \to u|_B$ in $H^1(B)$ and therefore $\gamma u_n|_B = \gamma u_n = (V_c - V_b)\lambda_n \to (V_c - V_b)\lambda$ in $H^{1/2}(\Gamma)$; (b) on the other hand, $\partial_n u_n|_B = \partial_n u_n = (K_c^t - K_b^t)\lambda_n \to (K_c^t - K_b^t)\lambda$ in $L^2(\Gamma)$ and therefore in $H^{-1/2}(\Gamma)$.

Note that we have in a way wasted two additional compact injections, $H^2(B) \subset H^1(B)$ for (a) and $L^2(\Gamma) \subset H^{-1/2}(\Gamma)$ for (b). We can be more precise with this result and state that $V_c - V_b : H^{-1/2}(\Gamma) \to \gamma H^3(\mathbb{R}^d)$ is bounded (and the injection $\gamma H^3(\mathbb{R}^d) \subset H^{1/2}(\Gamma)$ is compact) and $K_c^t - K_b^t : H^{-1/2}(\Gamma) \to L^2(\Gamma)$ is compact. $\qquad \square$

Proposition 14.13. *The operators*

$$K_c - K_b : H^{1/2}(\Gamma) \to H^{1/2}(\Gamma), \qquad W_c - W_b : H^{1/2}(\Gamma) \to H^{-1/2}(\Gamma)$$

are compact.

Proof. The assertion about $K_c - K_b$ follows from the compactness of $K_c^t - K_b^t = (K_c - K_b)'$ and the fact that taking adjoints does not affect compactness. The result will also follow from the argument we use to prove the compactness of $W_c - W_b$. Let $\eta \in H^{1/2}(\Gamma)$ and consider $u_c := D_c \eta$, $u_b := D_b \eta$, and $u := u_c - u_b = (D_c - D_b)\eta$. We have

$$[\gamma u] = 0, \qquad [\partial_n u] = 0, \qquad -\Delta u + c^2 u = (b^2 - c^2)u_b \in L^2(\mathbb{R}^d),$$

and therefore $u \in H^2(\mathbb{R}^d)$ and

$$\|u\|_{2,\mathbb{R}^d} \le C\|u_b\|_{\mathbb{R}^d} \le \|\eta\|_{1/2,\Gamma}.$$

If $\eta_n \rightharpoonup \eta$ in $H^{1/2}(\Gamma)$, then $u_n := (D_c - D_b)\eta_n \rightharpoonup u := (D_c - D_b)\eta$ in $H^2(\mathbb{R}^d)$ and $u_n|_B \rightharpoonup u|_B$ in $H^2(B)$, where B is a ball containing Γ. Therefore $u_n|_B \to u|_B$ in $H^1(B)$ and $\gamma u_n|_B = \gamma u_n = (K_c - K_b)\eta_n \to (K_c - K_b)\eta$ in $H^{1/2}(\Gamma)$. Also $(W_c - W_b)\eta_n = -\partial_n u_n \rightharpoonup (W_c - W_b)\eta$ in $L^2(\Gamma)$ and therefore $(W_c - W_b)\eta_n \to (W_c - W_b)\eta$ in $H^{-1/2}(\Gamma)$. $\qquad\square$

In other words, Propositions 14.12 and 14.13 state that the difference $C_c^- - C_b^-$ is compact.

14.5   Integral form of the layer potentials

In this section we present integral forms for the Yukawa layer potentials. We start by deriving two elegant equivalent weak definitions of the layer potentials. To 'redefine' the single layer potential we will use the adjoint

$$\gamma' : H^{-1/2}(\Gamma) \to H^{-1}(\Gamma)$$

of the trace operator $\gamma : H^1(\mathbb{R}^d) \to H^{1/2}(\Gamma)$, given by

$$\langle \gamma'\lambda, v \rangle_{H^{-1}(\mathbb{R}^d) \times H^1(\mathbb{R}^d)} = \langle \lambda, \gamma v \rangle_\Gamma \qquad \forall v \in H^1(\mathbb{R}^d).$$

Equivalently $\gamma'\lambda$ is the distribution

$$\langle \gamma'\lambda, \varphi \rangle_{\mathcal{D}'(\mathbb{R}^d) \times \mathcal{D}(\mathbb{R}^d)} = \langle \lambda, \gamma\varphi \rangle_\Gamma \qquad \forall \varphi \in \mathcal{D}(\mathbb{R}^d).$$

We will now use a restricted version of the Green's operator $G_c : H^{-1}(\mathbb{R}^d) \to H^1(\mathbb{R}^d)$,

$$G_c f = u \in H^1(\mathbb{R}^d), \qquad -\Delta u + c^2 u = f \quad \text{in } \mathcal{D}'(\mathbb{R}^d).$$

Proposition 14.14. *We have $S_c = G_c \circ \gamma'$.*

Proof. If $f := \gamma'\lambda \in H^{-1}(\mathbb{R}^d)$ and $u = G_c f$, then for all $v \in H^1(\mathbb{R}^d)$, we have

$$(\nabla u, \nabla v)_{\mathbb{R}^d} + c^2(u, v)_{\mathbb{R}^d} = \langle f, v \rangle_{H^{-1}(\mathbb{R}^d) \times H^1(\mathbb{R}^d)} = \langle \lambda, \gamma v \rangle_\Gamma,$$

that is, $u = S_c\lambda$. $\qquad\square$

The corresponding representation of the double layer potential is more involved. We start with the operators

$$\partial_n : H^2(\mathbb{R}^d) \to H^{-1/2}(\Gamma), \qquad \partial_n' : H^{1/2}(\Gamma) \to H^{-2}(\mathbb{R}^d),$$

noting that for $u \in H^2(\mathbb{R}^d)$, we have $\partial_n u = (\gamma \nabla u) \cdot \mathbf{n} \in L^2(\Gamma) \subset H^{-1/2}(\Gamma)$, but we still choose $H^{-1/2}(\Gamma)$ as target space. The adjoint operator ∂_n' satisfies

$$\langle \partial_n' \eta, v \rangle_{H^{-2}(\mathbb{R}^d) \times H^2(\mathbb{R}^d)} = \langle \partial_n v, \eta \rangle_\Gamma$$
$$= (\Delta v, w)_{\mathbb{R}^d} + (\nabla v, \nabla w)_{\mathbb{R}^d} \quad w \in H^1(\mathbb{R}^d), \quad \gamma w = \eta.$$

Proposition 14.15. *We have $D_c = G_c \circ \partial_n'$.*

Proof. By definition $u := G_c \partial_n' \eta \in L^2(\mathbb{R}^d)$ satisfies

$$-\Delta u + c^2 u = \partial_n' \eta \quad \text{in } \mathcal{D}'(\mathbb{R}^d).$$

Now let w be the solution of the uncoupled interior and exterior Dirichlet problems

$$w \in H^1(\mathbb{R}^d \setminus \Gamma), \quad -\Delta w + c^2 w = 0, \quad \gamma^- w = \eta, \quad \gamma^+ w = 0,$$

so that $w \equiv 0$ in Ω_+. Therefore

$$\langle -\Delta w + c^2 w, \varphi \rangle_{\mathcal{D}'(\mathbb{R}^d) \times \mathcal{D}(\mathbb{R}^d)} = \langle w, -\Delta \varphi + c^2 \varphi \rangle_{\mathcal{D}'(\mathbb{R}^d) \times \mathcal{D}(\mathbb{R}^d)}$$
$$= (w, -\Delta \varphi + c^2 \varphi)_{\Omega_-}$$
$$= (\nabla w, \nabla \varphi)_{\Omega_-} + c^2 (w, \varphi)_{\Omega_-} - \langle \partial_n \varphi, \gamma^- w \rangle_\Gamma$$
$$= \langle \partial_n^- w, \gamma \varphi \rangle_\Gamma - \langle \partial_n \varphi, \eta \rangle_\Gamma \quad \forall \varphi \in \mathcal{D}(\mathbb{R}^d),$$

or equivalently

$$-\Delta w + c^2 w = \gamma' \partial_n^- w - \partial_n' \eta.$$

Next, we define $v := u + w \in L^2(\mathbb{R}^d)$ and note that

$$-\Delta v + c^2 v = \gamma' \partial_n^- w \in H^{-1}(\mathbb{R}^d),$$

that is,

$$v = G_c \gamma' \partial_n^- w = S_c \partial_n^- w \in H^1(\mathbb{R}^d),$$

which implies that $u = v - w \in H^1(\mathbb{R}^d \setminus \Gamma)$. Recalling that $-\Delta u + c^2 u = 0$ in $\mathbb{R}^d \setminus \Gamma$ (prove it), we have

$$[\partial_n u] = -\partial_n^- w + [\partial_n v] = 0, \quad [\gamma u] = [\gamma v] - [\gamma w] = -\eta,$$

which implies that $u = D_c \eta$. $\qquad\square$

Integral forms. Thanks to Propositions 14.14 and 14.15 we can give integral forms of the layer potentials. By Propositions 14.14 and 14.3 (symmetry of the operator G_c in duality brackets), we can write

$$\langle S_c \lambda, \varphi \rangle_{\mathcal{D}'(\mathbb{R}^d) \times \mathcal{D}(\mathbb{R}^d)} = \langle G_c \gamma' \lambda, \varphi \rangle_{\mathcal{D}'(\mathbb{R}^d) \times \mathcal{D}(\mathbb{R}^d)}$$
$$= [\gamma' \lambda, G_c \varphi]_{\mathcal{S}' \times \mathcal{S}} = \langle \lambda, \gamma G_c \varphi \rangle_\Gamma \quad \forall \varphi \in \mathcal{D}(\mathbb{R}^d).$$

Therefore, if $\lambda \in L^\infty(\Gamma)$, we can apply Fubini's theorem to prove that

$$\langle S_c\lambda, \varphi \rangle_{\mathcal{D}'(\mathbb{R}^d) \times \mathcal{D}(\mathbb{R}^d)} = \int_\Gamma \lambda(\mathbf{y}) \left(\int_{\mathbb{R}^d} E_c(\mathbf{y} - \mathbf{x})\varphi(\mathbf{x})d\mathbf{x} \right) d\Gamma(\mathbf{y})$$

$$= \int_{\mathbb{R}^d} \left(\int_\Gamma E_c(\mathbf{x} - \mathbf{y})\lambda(\mathbf{y})d\Gamma(\mathbf{y}) \right) \varphi(\mathbf{x})d\mathbf{x},$$

and therefore

$$S_c\lambda = \int_\Gamma E_c(\cdot - \mathbf{y})\lambda(\mathbf{y})d\Gamma(\mathbf{y}).$$

Similarly, for $\eta \in H^{1/2}(\Gamma)$ and $\varphi \in \mathcal{D}(\mathbb{R}^d)$, we have

$$\langle D_c\eta, \varphi \rangle_{\mathcal{D}'(\mathbb{R}^d) \times \mathcal{D}(\mathbb{R}^d)} = \langle G_c\partial'_n\eta, \varphi \rangle_{\mathcal{D}'(\mathbb{R}^d) \times \mathcal{D}(\mathbb{R}^d)} = [\partial'_n\eta, G_c\varphi]_{\mathcal{S}' \times \mathcal{S}}$$

$$= \langle \eta, (\nabla G_c\varphi) \cdot \mathbf{n} \rangle_\Gamma$$

$$= \int_\Gamma \eta(\mathbf{y}) \left(\int_{\mathbb{R}^d} \nabla_\mathbf{y} E_c(\mathbf{y} - \mathbf{x}) \cdot \mathbf{n}(\mathbf{y})\varphi(\mathbf{x})d\mathbf{x} \right) d\Gamma(\mathbf{y})$$

$$= \int_{\mathbb{R}^d} \left(\int_\Gamma \nabla_\mathbf{y} E_c(\mathbf{x} - \mathbf{y}) \cdot \mathbf{n}(\mathbf{y})\eta(\mathbf{y})d\Gamma(\mathbf{y}) \right) \varphi(\mathbf{x})d\mathbf{x},$$

and therefore

$$D_c\eta = \int_\Gamma \nabla_\mathbf{y} E_c(\cdot - \mathbf{y}) \cdot \mathbf{n}(\mathbf{y})\eta(\mathbf{y})d\Gamma(\mathbf{y}).$$

14.6 A weighted Sobolev space

The theory of Yukawa potentials can be translated to the formal limit $c = 0$ (the Laplace equation), but this adds some difficulties. We are going to give details for the three-dimensional case and ask the reader to provide details for the two-dimensional case in Exercise 14.9. The main issue can already be seen in the variational theory: the loss of the 'mass-reaction' term when $c = 0$ makes it considerably more difficult to prove coercivity. When dealing with the integral forms, this reflects the poor behavior at infinity of the fundamental solution

$$E_0(\mathbf{x}) = \frac{1}{4\pi|\mathbf{x}|} \qquad \text{when } d = 3.$$

We spend this section with preparatory work on a weighted Sobolev space designed to handle exterior problems and layer potentials for the three-dimensional Laplace equation.

A new space. For an open set $\mathcal{O} \subset \mathbb{R}^3$ we define the space

$$W(\mathcal{O}) := \{u \in \mathcal{D}'(\mathcal{O}) : \rho u \in L^2(\mathcal{O}), \ \nabla u \in \mathbf{L}^2(\Omega)\},$$

where

$$\rho(\mathbf{x}) := \frac{1}{(1 + |\mathbf{x}|^2)^{1/2}}.$$

We consider the norm

$$\|u\|_{W(\mathcal{O})}^2 := \|\rho u\|_{\mathcal{O}}^2 + \|\nabla u\|_{\mathcal{O}}^2,$$

and the inner product associated to it. We note that $\rho \in \mathcal{C}^\infty(\mathbb{R}^d)$ and that ρ behaves like E_0 as $|\mathbf{x}| \to \infty$. The proof of the following collection of properties is proposed as an exercise (Exercise 14.6). Property (e) below is highly relevant in eliminating the possible kernel of the Neumann problem in unbounded domains.

Proposition 14.16. *The following properties hold for any open set $\mathcal{O}$:*

(a) *$W(\mathcal{O})$ is a Hilbert space.*

(b) *If $\mathcal{O}$ is bounded, $H^1(\mathcal{O}) = W(\mathcal{O})$ with equivalent norms.*

(c) *$H^1(\mathcal{O})$ is a dense subset of $W(\mathcal{O})$, and the embedding is continuous.*

(d) *The function ρ is in $W(\mathbb{R}^3)$, but not in $H^1(\mathbb{R}^3)$.*

(e) *If $\mathcal{O}$ is unbounded, then constant functions are not in $W(\mathcal{O})$.*

Traces. The spaces $W(\mathcal{O})$ are locally indistiguishable from the spaces $H^1(\mathcal{O})$: if $u \in W(\mathcal{O})$ and $\varphi \in \mathcal{D}(\mathbb{R}^3)$, then $\varphi u \in H^1(\mathcal{O})$. Therefore, the traces of $W(\mathcal{O})$ are the same as the traces of $H^1(\mathcal{O})$. We thus have the two-sided surjective traces

$$\gamma^\pm : W(\mathbb{R}^3 \setminus \Gamma) \equiv W(\Omega_+) \times H^1(\Omega_-) \longrightarrow H^{1/2}(\Gamma),$$

and we can note that if $u \in W(\mathbb{R}^3 \setminus \Gamma)$, then $[\gamma u] = 0$ is equivalent to $u \in W(\mathbb{R}^3)$. Finally,

$$\ker \gamma^+ := \{u \in W(\Omega_+) : \gamma^+ u = 0\}$$
$$= \{u \in W(\Omega_+) : \exists \{\varphi_n\} \subset \mathcal{D}(\Omega_+), \|\varphi_n - u\|_{W(\Omega_+)} \to 0\} =: W_0(\Omega_+),$$

a set equality that again followss from the fact that we can always find approximating elements in H^1.

Proposition 14.17 (Poincaré inequality). *We have*

$$\|u\|_{W(\mathbb{R}^3)} \le 2\|\nabla u\|_{\mathbb{R}^3} \qquad \forall u \in W(\mathbb{R}^3).$$

Proof. Clearly we only need to prove that

$$\|\rho u\|_{\mathbb{R}^3} \le 2\|\nabla u\|_{\mathbb{R}^3} \qquad \forall u \in \mathcal{D}(\mathbb{R}^3), \tag{14.17}$$

as $\mathcal{D}(\mathbb{R}^3)$ is dense in $H^1(\mathbb{R}^3)$ and $H^1(\mathbb{R}^3)$ is densely and continuously embedded into $W(\mathbb{R}^3)$. If $u \in \mathcal{D}(\mathbb{R}^3)$, we can prove the identity

$$\int_{\mathbb{R}^3} \frac{u(\mathbf{x})^2}{|\mathbf{x}|^2} d\mathbf{x} = \lim_{\varepsilon \to 0} \int_{\mathbb{R}^3 \setminus B(0;\varepsilon)} \frac{u(\mathbf{x})^2}{|\mathbf{x}|^2} d\mathbf{x}$$

$$= -\frac{1}{2} \lim_{\varepsilon \to 0} \int_{\mathbb{R}^3 \setminus B(0;\varepsilon)} u(\mathbf{x})^2 \mathbf{x} \cdot \nabla \frac{1}{|\mathbf{x}|^2} d\mathbf{x}$$

$$= \frac{1}{2} \lim_{\varepsilon \to 0} \left(\int_{\mathbb{R}^3 \setminus B(0;\varepsilon)} \nabla \cdot (u(\mathbf{x})^2 \mathbf{x}) \frac{1}{|\mathbf{x}|^2} d\mathbf{x} - \int_{\partial B(0;\varepsilon)} \frac{u(\mathbf{x})^2 \mathbf{x} \cdot \mathbf{n}(\mathbf{x})}{|\mathbf{x}|^2} d\Gamma(\mathbf{x}) \right)$$

$$= \int_{\mathbb{R}^3} \frac{u(\mathbf{x}) \nabla u(\mathbf{x}) \cdot \mathbf{x}}{|\mathbf{x}|^2} d\mathbf{x} + \frac{3}{2} \int_{\mathbb{R}^3} \frac{u(\mathbf{x})^2}{|\mathbf{x}|^2} d\mathbf{x},$$

using integration by parts and the dominated convergence theorem. Simplifying, we have

$$\int_{\mathbb{R}^3} \frac{u(\mathbf{x})^2}{|\mathbf{x}|^2} d\mathbf{x} = -2 \int_{\mathbb{R}^3} \frac{u(\mathbf{x}) \nabla u(\mathbf{x}) \cdot \mathbf{x}}{|\mathbf{x}|^2} d\mathbf{x}$$

$$\leq 2 \left(\int_{\mathbb{R}^3} \frac{u(\mathbf{x})^2}{|\mathbf{x}|^2} d\mathbf{x} \right)^{1/2} \left(\int_{\mathbb{R}^3} |\nabla u(\mathbf{x})|^2 d\mathbf{x} \right)^{1/2},$$

that is,

$$\int_{\mathbb{R}^3} \frac{u(\mathbf{x})^2}{|\mathbf{x}|^2} d\mathbf{x} \leq 4 \int_{\mathbb{R}^3} |\nabla u(\mathbf{x})|^2 d\mathbf{x} \qquad \forall u \in \mathcal{D}(\mathbb{R}^3).$$

This proves (14.17) since $\rho(\mathbf{x}) \leq |\mathbf{x}|^{-1}$. $\qquad\square$

Corollary 14.2. *We have*

$$\|u\|_{W(\Omega_+)} \leq 2\|\nabla u\|_{\Omega_+} \qquad \forall u \in W_0(\Omega_+).$$

Proof. Functions in $W_0(\Omega_+)$ can be extended by zero to functions in $W(\mathbb{R}^3)$. $\qquad\square$

For the next result (a Poincaré inequality in $W(\Omega_+)$), we will need a small abstract result related to Fredholm theory (Chapter 8).

Lemma 14.2. *If $a : H \times H \to \mathbb{R}$ is a symmetric positive definite bilinear form such that there exists $K : H \to H$ compact and $\alpha > 0$ satisfying*

$$a(u, u) + (Ku, u)_H \geq \alpha \|u\|_H^2 \qquad \forall u \in H, \tag{14.18}$$

then a is coercive.

Proof. Consider $A : H \to H$ given by $(Au, v)_H = a(u, v)$. The operator $A + K$ is invertible by the coercivity condition (14.18). We can then write $A = (A + K) - K$, which proves that A is Fredholm of index zero. Therefore, A is invertible if and only if A is injective, but positive definiteness of the bilinear form a implies injectivity of A. Finally, we can bound

$$\|u\| \leq \|A^{-1}\| \|Au\|_H = \|A^{-1}\| \sup_{0 \neq v \in H} \frac{(Au, v)_H}{\|v\|_H}$$

$$\leq \|A^{-1}\| (Au, u)_H^{1/2} \sup_{0 \neq v \in H} \frac{(Av, v)_H^{1/2}}{\|v\|_H} = \|A^{-1}\| \|A\|^{1/2} a(u, u)^{1/2},$$

which proves the coercivity of a. $\qquad\square$

Proposition 14.18. *There exists a constant $C > 0$ such that*

$$\|u\|_{W(\Omega_+)} \leq C \|\nabla u\|_{\Omega_+} \qquad \forall u \in W(\Omega_+).$$

Proof. First, take $R > 0$ such that $\overline{\Omega_-} \subset B(\mathbf{0}; R)$, and $\varphi \in \mathcal{D}(B(\mathbf{0}; R))$ such that $\varphi \equiv 1$ in a neighborhood of Γ. Since $(1 - \varphi)u \in W_0(\Omega_+)$ and $\varphi u \in \{v \in H^1(\Omega_+ \cap B(\mathbf{0}; R)) : \gamma v = 0 \text{ on } \partial B(\mathbf{0}; R)\}$, we can use Poincaré's inequality in the latter set and Corollary 14.2 to bound

$$\|\varphi u\|_{\Omega_+} \leq \|\varphi u\|_{\Omega_+ \cap B(\mathbf{0}; R)} + \|(1 - \varphi)u\|_{W(\Omega_+)}$$
$$\leq C \|\nabla(\varphi u)\|_{\Omega_+ \cap B(\mathbf{0}; R)} + 2\|\nabla((1 - \varphi)u)\|_{\Omega_+}.$$

Therefore, we can estimate

$$\|u\|_{W(\Omega_+)} \leq C(\|\nabla u\|_{\Omega_+} + \|u\|_{\Omega_+ \cap B(\mathbf{0}; R)}), \qquad (14.19)$$

using the fact that $\nabla \varphi \equiv \mathbf{0}$ outside $B(\mathbf{0}; R)$. Now consider the bounded linear operator $K : W(\Omega_+) \to W(\Omega_+)$ given by

$$(Ku, v)_{W(\Omega_+)} = (u, v)_{\Omega_+ \cap B(\mathbf{0}; R)} \qquad \forall u, v \in W(\Omega_+).$$

Since

$$\|Ku\|_{W(\Omega_+)} \leq C \|u\|_{\Omega_+ \cap B(\mathbf{0}; R)}, \qquad (14.20)$$

we can consider the following diagram

u	$W(\Omega_+)$	$W(\Omega_+)$
$\downarrow$	$\downarrow$ restriction	
$u\|_{\Omega_+ \cap B(\mathbf{0}; R)}$	$H^1(\Omega_+ \cap B(\mathbf{0}; R))$	
$\downarrow$	$\downarrow$ comp. injection	K
$u\|_{\Omega_+ \cap B(\mathbf{0}; R)}$	$L^2(\Omega_+ \cap B(\mathbf{0}; R))$	
$\downarrow$	$\downarrow$ (14.20)	
Ku	$W(\Omega_+)$	$W(\Omega_+)$

using the Rellich-Kondrachov compactness theorem in the bounded domain $\Omega_+ \cap B(\mathbf{0}; R)$, where H^1 and W are indistinguishable, to prove that K is compact. We now have (14.19), the compactness of K, and the fact that constant functions are not elements of $W(\Omega_+)$ (see Proposition 14.16(e)), that is, all the hypotheses of Lemma 14.2 hold for the bilinear form $a(u, v) := (\nabla u, \nabla v)_{\Omega_+}$, which is therefore coercive in $W(\Omega_+)$. This proves the result. $\square$

Exterior Dirichlet and Neumannn problems. We now briefly exploit the Poincaré inequality of Proposition 14.18 (and the simpler one of Corollary 14.2) to study exterior boundary value problems for the Laplacian. We will just focus on boundary conditions, but at the same time we want to warn the reader that right-hand sides in $L^2(\Omega_+)$ are not valid ones, since $W(\Omega_+)$ is not a subset of $L^2(\Omega_+)$. The Dirichlet problem with data $g \in H^{1/2}(\Gamma)$

$$u \in W(\Omega_+), \qquad \Delta u = 0, \qquad \gamma^+ u = g$$

is equivalent to the well posed problem (the trace operator is the same one and coercivity is given by Corollary 14.2)

$$u \in W(\Omega_+), \qquad \gamma^+ u = g, \qquad (\nabla u, \nabla v)_{\Omega_+} = 0 \qquad \forall v \in W_0(\Omega_+),$$

and to the minimization problem

$$\tfrac{1}{2}\|\nabla u\|_{\Omega_+}^2 = \min! \qquad u \in W(\Omega_+), \qquad \gamma^+ u = g.$$

The Neumann problem with data $h \in H^{-1/2}(\Gamma)$

$$u \in W(\Omega_+), \qquad \Delta u = 0, \qquad \partial_n^+ u = h,$$

is equivalent to

$$u \in W(\Omega_+), \qquad (\nabla u, \nabla v)_{\Omega_+} = \langle h, \gamma v \rangle_\Gamma \qquad \forall v \in W(\Omega_+),$$

which is well posed because of Proposition 14.18, and to the minimization problem

$$\tfrac{1}{2}\|\nabla u\|_{\Omega_+}^2 - \langle h, \gamma u \rangle_\Gamma = \min! \qquad u \in W(\Omega_+).$$

14.7 Coulomb potentials

We now adapt some of the results of Sections 14.2-14.5 to the case of the three-dimensional Laplacian. At the variational level, we will substitute the spaces H^1 by the spaces W defined in the previous section. Since the interior Neumann problem for the Laplace equation is not uniquely solvable, we will need to take some care when addressing the invertibility of some of the boundary integral operators. For $\eta \in H^{1/2}(\Gamma)$ and $\lambda \in H^{-1/2}(\Gamma)$, the problem

$$\tfrac{1}{2}\|\nabla u\|_{\mathbb{R}^3 \setminus \Gamma} - \langle \lambda, \gamma^+ u \rangle_\Gamma = \min! \qquad u \in W(\mathbb{R}^3 \setminus \Gamma), \qquad [\gamma u] = \eta,$$

is equivalent to the variational problem

$$u \in W(\mathbb{R}^3 \setminus \Gamma), \qquad [\gamma u] = \eta, \tag{14.21a}$$

$$(\nabla u, \nabla v)_{\mathbb{R}^3 \setminus \Gamma} = \langle \lambda, \gamma v \rangle_\Gamma \qquad \forall v \in W(\mathbb{R}^3), \tag{14.21b}$$

and to the transmission problem

$$u \in W(\mathbb{R}^3 \setminus \Gamma), \qquad \Delta u = 0, \quad \text{in } \mathbb{R}^3 \setminus \Gamma, \tag{14.22a}$$

$$[\gamma u] = \eta, \qquad [\partial_n u] = \lambda. \tag{14.22b}$$

The variational problem is well posed and defines two operators

$$u = S_0 \lambda - D_0 \eta.$$

It is simple to see in the variational formulation (14.21) or in the transmission problem form (14.22) that if $\lambda = 0$ and $\eta \equiv 1$, then the function $u := \chi_{\Omega_-}$ solves the equations and therefore

$$D_0 1 = -\chi_{\Omega_-}. \tag{14.23}$$

We then define the bounded boundary operators

$$V_0 := \gamma^{\pm} S_0 : H^{-1/2}(\Gamma) \longrightarrow H^{1/2}(\Gamma),$$

$$K_0 := \tfrac{1}{2}(\gamma^+ + \gamma^-) D_0 : H^{1/2}(\Gamma) \longrightarrow H^{1/2}(\Gamma),$$

$$K_0^t := \tfrac{1}{2}(\partial_n^+ + \partial_n^-) S_0 : H^{-1/2}(\Gamma) \longrightarrow H^{-1/2}(\Gamma),$$

$$W_0 := -\partial_n D_0 : H^{1/2}(\Gamma) \longrightarrow H^{-1/2}(\Gamma).$$

Note that as a consequence of (14.23), we have

$$K_0 1 \equiv -\tfrac{1}{2}, \qquad W_0 1 = 0,$$

and therefore Propositions 14.7 (coercivity of W_c) and 14.10 (invertibility of $\tfrac{1}{2}I + K_c$) cannot hold for $c = 0$. The changes in the statements just deal with this particular pathological case.

Proposition 14.19. *We have the following symmetry properties*

$$\langle \lambda, V_0 \mu \rangle_\Gamma = \langle \mu, V_0 \lambda \rangle_\Gamma \qquad \forall \lambda, \mu \in H^{-1/2}(\Gamma),$$

$$\langle W_0 \eta, \psi \rangle_\Gamma = \langle W_0 \psi, \eta \rangle_\Gamma \qquad \forall \eta, \psi \in H^{1/2}(\Gamma),$$

$$\langle K_0^t \lambda, \eta \rangle_\Gamma = \langle \lambda, K_0 \eta \rangle_\Gamma \qquad \forall \lambda \in H^{-1/2}(\Gamma), \quad \eta \in H^{1/2}(\Gamma).$$

We also have positivity

$$\langle \lambda, V_0 \lambda \rangle_\Gamma \geq c_V \|\lambda\|_{-1/2,\Gamma}^2 \qquad \forall \lambda \in H^{-1/2}(\Gamma),$$

$$\langle W_0 \eta, \eta \rangle_\Gamma \geq c_W \|\eta\|_{1/2,\Gamma}^2 \qquad \forall \eta \in H_\star^{1/2}(\Gamma),$$

where $H_\star^{1/2}(\Gamma) := \{\eta \in H^{1/2}(\Gamma) : \langle 1, \eta \rangle_\Gamma = 0\}$. In particular V_0 is invertible and the bilinear form

$$\langle W_0 \eta, \psi \rangle_\Gamma + \langle 1, \eta \rangle_\Gamma \langle 1, \psi \rangle_\Gamma$$

defines an invertible operator $H^{1/2}(\Gamma) \to H^{-1/2}(\Gamma)$.

Proof. The proofs of Propositions 14.6, 14.7, and 14.8 just need some slight adjustments to work in the case $c = 0$. A hint is given in Exercise 14.7. □

An example of use. Integral equations of the first kind associated to the operator V_0 can be used in a similar form to those of the Yukawa operator. However, the lack of full positivity of W_0 creates some minor difficulties. Say that you want to solve

$$u \in W(\Omega_+), \qquad \Delta u = 0, \qquad \partial_n u = h,$$

and you want to write $u = D_0 \eta$. You therefore need to solve $W_0 \eta = -h$, but then you need $h \in \text{range } W_0$. We can use the reduced coercivity of W_0 in Proposition 14.19 and write

$$\langle W_0 \eta, \eta \rangle_\Gamma \geq c_W \|\eta - c_\eta\|_{1/2,\Gamma}^2 \qquad c_\eta := \frac{1}{|\Gamma|} \int_\Gamma \eta,$$

from which it follows that the range of W_0 is closed and equal to

$$\{\mu \in H^{-1/2}(\Gamma) : \langle \mu, 1 \rangle_\Gamma = 0\} =: (\ker W_0)^\circ.$$

(This is the first occurrence in this text of a **polar set or annihilator**, which is the duality product equivalent to an orthogonal complement. We will meet this concept again at the end of this section.) A remedy that covers all possible h is easy. We fix a point $\mathbf{x}_0 \in \Omega_-$ and consider the function $\Phi := E_0(\cdot - \mathbf{x}_0) \in W(\Omega_+)$, which satisfies $\Delta \Phi = 0$ in Ω_+. We then represent

$$u = D_0 \eta + \alpha \, \Phi, \qquad \eta \in H_\star^{1/2}(\Gamma), \qquad \alpha \in \mathbb{R},$$

and solve

$$\langle W_0 \eta, \psi \rangle_\Gamma - \alpha \langle \partial_n \Phi, \psi \rangle_\Gamma = -\langle h, \psi \rangle_\Gamma \qquad \forall \psi \in H^{1/2}(\Gamma).$$

This equation can be solved in two rounds. First we use

$$\langle \partial_n \Phi, 1 \rangle_\Gamma = -1$$

(see Exercise 14.8) and $\langle W_0 \eta, 1 \rangle_\Gamma = 0$ to compute α. We then use the coercivity of W in $H_\star^{1/2}(\Gamma)$ (Proposition 14.19) to compute η.

Proposition 14.20. *The operators*

$$V_c - V_0 : H^{-1/2}(\Gamma) \to H^{1/2}(\Gamma), \qquad K_c^t - K_0^t : H^{-1/2}(\Gamma) \to H^{-1/2}(\Gamma),$$
$$K_c - K_0 : H^{1/2}(\Gamma) \to H^{1/2}(\Gamma), \qquad W_c - W_0 : H^{1/2}(\Gamma) \to H^{-1/2}(\Gamma),$$

are compact.

Proof. Let us start with the first row. Because of the slow decay of the layer potentials related to the Laplacian, we need to introduce $\varphi \in \mathcal{D}(\mathbb{R}^3)$ such that $\varphi \equiv 1$ in a neighborhood of Γ, so that traces are not affected. Given $\lambda \in H^{-1/2}(\Gamma)$, we define

$$u_c := S_c\lambda, \qquad u_0 := S_0\lambda, \qquad u := \varphi(S_c - S_0)\lambda \in H^1(\mathbb{R}^3).$$

Multiplying by the cutoff functions allows us to move to $H^1(\mathbb{R}^3)$, since, as we have already mentioned, functions in $W^1(\mathbb{R}^3)$ are locally in $H^1(\mathbb{R}^3)$. Now note that

$$-\Delta u + c^2 u = -2\nabla\varphi \cdot (\nabla u_c - \nabla u_0) - (\Delta\varphi)(u_c - u_0) - c^2\varphi u_0 \quad \text{in } \mathbb{R}^3 \setminus \Gamma.$$

The right-hand side of the above is in $L^2(\mathbb{R}^3)$ and we have $[\gamma u] = 0$ and $[\partial_n u] = 0$, and therefore $-\Delta u + c^2 u \in L^2(\mathbb{R}^3)$ as tempered distributions (recall Lemma 14.1). This shows that $u \in H^2(\mathbb{R}^3)$ and

$$\|u\|_{2,\mathbb{R}^3} \leq C(\|u_c\|_{1,\mathbb{R}^3} + \|u_0\|_{W(\mathbb{R}^3)}) \leq C'\|\lambda\|_{-1/2,\Gamma}.$$

Noting that $\gamma u = (V_c - V_0)\lambda$ and $\partial_n u = (K_c - K_0)\lambda$ (this is where we need $\varphi \equiv 1$ in a neighborhood of Γ), the result can be proved as in Proposition 14.12. We use that the inclusion of $H^2(B)$ into $H^1(B)$ is compact, where B is a ball containing Γ. The compactness of $K_c^t - K_0^t$ follows by transposition.

For the missing result, we define $u := \varphi(D_c - D_0)\eta$ and proceed with similar arguments. The compactness of $W_c - W_0$ follows from the compactness of the inclusion of $L^2(\Gamma)$ into $H^{-1/2}(\Gamma)$. $\square$

Proposition 14.21. *We have*

$$\ker(-\tfrac{1}{2}I + K_0) = \{0\}, \qquad \ker(\tfrac{1}{2}I + K_0) = \mathcal{P}_0(\Gamma).$$

Proof. If $\eta \in \ker(-\tfrac{1}{2}I + K_0)$, then $u = D_0\eta$ satisfies $\gamma^- u = -\tfrac{1}{2}\eta + K_0\eta = 0$. Since $\Delta u = 0$ in Ω_-, then $u \equiv 0$ in Ω_-. Now, $\partial^+ u = \partial^- u = 0$ ($[\partial_n u] = 0$ because u a double layer potential) and then $u \equiv 0$ in Ω_+, and finally $\eta = -[\gamma u] = 0$.

We already know that $K_0 1 = -1/2$, which implies that $\mathcal{P}_0(\Gamma) \subset \ker(\tfrac{1}{2}I + K_0)$. If $\eta \in \ker(\tfrac{1}{2}I + K_0)$ and we define $u := D_0\eta$, we have $\gamma^+ u = 0$, and therefore $u \equiv 0$ in Ω_+ and $\partial_n^- u = \partial_n^+ u = 0$. This implies that $u \in \mathcal{P}_0(\Omega_-)$ and therefore $\eta = -\gamma^- u \in \mathcal{P}_0(\Gamma)$. $\square$

The equilibrium distribution. First of all, since $K_c - K_0$ is compact and $-\frac{1}{2}I + K_c$ is invertible (Proposition 14.9), then $-\frac{1}{2}I + K_0$ is Fredholm of index zero. However, Proposition 14.21 shows that it is injective, and therefore it is invertible and so is its adjoint $-\frac{1}{2}I + K_0^t$. Now consider

$$\lambda_{eq} \in H^{-1/2}(\Gamma) \quad \text{such that} \quad V_0 \lambda_{eq} = 1.$$

Note that $u_{eq} := S_0 \lambda_{eq}$ satisfies $\Delta u_{eq} = 0$ in Ω_- and $\gamma^{\pm} u_{eq} = V_0 \lambda_{eq} = 1$, which implies that $u_{eq} \equiv 1$ in Ω_-. Therefore,

$$\tfrac{1}{2} \lambda_{eq} + K_0^t \lambda_{eq} = \partial_n^- u_{eq} = 0.$$

Since $\dim \ker(\frac{1}{2}I + K_0) = \dim \ker(\frac{1}{2}I + K_0^t)$ (again by Fredholm theory, comparing now with the invertible operator $\frac{1}{2}I + K_c$), we have that

$$\ker(\tfrac{1}{2}I + K_0^t) = \mathrm{span}\{\lambda_{eq}\}.$$

Finally, we can recognize the ranges

$$\mathrm{range}\,(\tfrac{1}{2}I + K_0^t) = \mathcal{P}_0(\Gamma)^\circ = \{\lambda \in H^{-1/2}(\Gamma) : \langle \lambda, 1 \rangle_\Gamma = 0\},$$

$$\mathrm{range}\,(\tfrac{1}{2}I + K_0) = \mathrm{span}\{\lambda_{eq}\}^\circ = \{\eta \in H^{1/2}(\Gamma) : \langle \lambda_{eq}, \eta \rangle_\Gamma = 0\}.$$

Reconciling this variational theory with an integral theory (in the spirit of Section 14.5) requires some additional work, due to the fact that there is no simple extension of the operator G_c to the case $c = 0$. We can define a convolution operator with the fundamental solution, but this operator cannot act on every tempered distribution and its mapping properties between Sobolev spaces are more complicated. We will stop here with this introduction to layer potentials for the Laplacian, reminding the reader that everything we have done is for the three-dimensional case and noting that the two-dimensional case needs even more adjustments.

14.8 Boundary-field formulations

We finish this chapter, which serves as an introduction to layer potentials, by showing how to use them to build nonlocal boundary conditions that are used to rewrite some transmission problems as problems on a bounded domain with integral equations taking care of the exterior domain. Our model problem will be a diffusion equation with variable coefficients in an interior domain $\Omega_- \subset \mathbb{R}^d$ and the Yukawa equation in its exterior. For traces and normal derivatives we will use the following convention: for quantities defined in the interior domain traces will not be tagged with any additional script, but $\pm$ will be used to distinguish traces of potentials. Note that even if the potential representations will be used to reformulate the exterior solution, once the potentials have been introduced, they are automatically extended to the interior domain.

The coefficients of the equation are a strongly positive function $\kappa \in L^\infty(\Omega_-)$ and a constant $c > 0$. Data are $f \in L^2(\Omega_-)$, $g \in H^{1/2}(\Gamma)$, and $h \in H^{-1/2}(\Gamma)$. We look for

$$w \in H^1(\Omega_-), \qquad u \in H^1(\Omega_+) \tag{14.24a}$$

satisfying a diffusion equation in the interior domain and a Yukawa equation in the exterior domain

$$-\nabla \cdot (\kappa \nabla w) = f \quad \text{in } \Omega_-, \qquad -\Delta u + c^2 u = 0 \quad \text{in } \Omega_+, \tag{14.24b}$$

and transmission conditions

$$\gamma w - \gamma^+ u = g, \qquad (\kappa \nabla w) \cdot \mathbf{n} - \partial_n^+ u = h. \tag{14.24c}$$

Using Lemma 14.2, it is simple to prove that the bilinear form

$$(\kappa \nabla w, \nabla v)_{\Omega_-} + (\nabla u, \nabla \omega)_{\Omega_+} + c^2 (u, \omega)_{\Omega_+}$$

is coercive in the space

$$\{(w, u) \in H^1(\Omega_-) \times H^1(\Omega_+) : \gamma w = \gamma^+ u\} \equiv H^1(\mathbb{R}^d).$$

This statement is equivalent to the inequality

$$\|v\|_{1,\mathbb{R}^d}^2 \le C(\|\nabla v\|_{\mathbb{R}^d}^2 + \|v\|_{\Omega_+}^2) \qquad \forall v \in H^1(\mathbb{R}^d).$$

Thanks to this coercivity property, it is easy to prove that the variational problem

$$(w, u) \in H^1(\Omega_-) \times H^1(\Omega_+), \qquad \gamma w = \gamma_+ u + g, \tag{14.25a}$$

$$(\kappa \nabla w, \nabla v)_{\Omega_-} + (\nabla u, \nabla v)_{\Omega_+} + c^2 (u, v)_{\Omega_+} = (f, v)_{\Omega_-} + \langle h, \gamma v \rangle_\Gamma \tag{14.25b}$$
$$\forall v \in H^1(\mathbb{R}^d),$$

equivalent to (14.24), is well posed. We now introduce the bilinear form for the interior domain

$$a(w, v) := (\kappa \nabla w, \nabla v)_{\Omega_-},$$

so that we have

$$a(w, v) - \langle (\kappa \nabla w) \cdot \mathbf{n}, \gamma v \rangle_\Gamma = (f, v)_{\Omega_-} \qquad \forall v \in H^1(\Omega_-), \tag{14.26}$$

or, after substituting the second transmission condition,

$$a(w, v) - \langle \partial_n^+ u, \gamma v \rangle_\Gamma = (f, v)_{\Omega_-} + \langle h, \gamma v \rangle_\Gamma \qquad \forall v \in H^1(\Omega_-). \tag{14.27}$$

Note how (14.27) is (14.25b) susbtituting $c^2 u = \Delta u$ and using the definition of $\partial_n^+ u$. For the exterior part of the solution, we introduce a potential representation in terms of its Cauchy data

$$u = D_c \gamma^+ u - S_c \partial_n^+ u. \tag{14.28}$$

This representation brings along two integral identities (recall Section 14.4)

$$V_c \partial_n^+ u + \tfrac{1}{2}\gamma^+ u - K_c \gamma^+ u = 0, \qquad \tfrac{1}{2}\partial_n^+ u + K_c^t \partial_n^+ u + W_c \gamma^+ u = 0. \tag{14.29}$$

The different ways in which we combine the information of the integral identities (14.29) with the incomplete interior formulation (14.27) will bring three different integral formulations based on the representation (14.28). In the jargon of boundary integral equations, these are direct formulations. There are more possible representations of u, using potential ansatz (indirect formulations).

Two fields, one integral equation. We use $\lambda := \partial_n^+ u$ as an additional unknown, substitute it into the variational equation (14.27) and substitute $\gamma^+ u = \gamma w - g$ in the first of the integral identities (14.29). This leads to a variational formulation in the interior domain, coupled with an integral equation on Γ:

$$w \in H^1(\Omega_-), \ \lambda \in H^{-1/2}(\Gamma), \tag{14.30a}$$

$$a(w, v) - \langle \lambda, \gamma v \rangle_\Gamma = (f, v)_{\Omega_-} + \langle h, \gamma v \rangle_\Gamma \qquad \forall v \in H^1(\Omega_-), \tag{14.30b}$$

$$\tfrac{1}{2}\gamma w - K_c \gamma w + V_c \lambda = \tfrac{1}{2}g - K_c g. \tag{14.30c}$$

Note that (14.30) can also be seen as: (a) an interior Neumann boundary problem with data λ expressed in (14.30b); (b) a uniquely solvable integral equation (14.30c) representing the exterior Dirichlet problem with data $\gamma w - g$. If we invert V_c in (14.30c), we obtain

$$\lambda = V_c^{-1}(\tfrac{1}{2}I - K_c)(g - \gamma w),$$

which we can substitute in (14.30b), thus producing a variational formulation for an interior problem with nonlocal boundary condition

$$a(w, v) + \langle V_c^{-1}(\tfrac{1}{2}I - K_c)\gamma w, \gamma v \rangle_\Gamma \tag{14.31}$$
$$= (f, v)_{\Omega_-} + \langle h, \gamma v \rangle_\Gamma + \langle V_c^{-1}(\tfrac{1}{2}I - K_c)g, \gamma v \rangle_\Gamma \qquad \forall v \in H^1(\Omega_-).$$

The bilinear form in the left-hand side of (14.31) is symmetric due to Exercise 14.3(c) and it is coercive thanks to Exercise 14.3(e). In fact, the operator $V_c^{-1}(\tfrac{1}{2}I - K_c)$ is the negative exterior Dirichlet-to-Neumann operator, which is self-adjoint and strongly positive definite. An equivalent variational formulation of (14.30) can easily be obtained by substituting (14.30c) by the equivalent (tested) equation

$$\langle \mu, \tfrac{1}{2}\gamma w - K_c \gamma w \rangle_\Gamma + \langle \mu, V_c \lambda \rangle_\Gamma = \langle \mu, \tfrac{1}{2}g - K_c g \rangle_\Gamma \qquad \forall \mu \in H^{-1/2}(\Gamma).$$

We will postpone the study of the well-posedness of (14.30) until after we have introduced all of the formulations. We will understand well-posedness of (14.30) as the study of unique solvability of a system like (14.30) with an arbitrary right-hand side in the dual space of $H^1(\Omega_-) \times H^{-1/2}(\Gamma)$. This is equivalent to proving that the bilinear form

$$A_{\mathrm{ns}}((w,\lambda),(v,\mu)) := a(v,w) - \langle \lambda, \gamma v \rangle_\Gamma + \langle \mu, \tfrac{1}{2}\gamma w - K_c \gamma w \rangle_\Gamma + \langle \mu, V_c \lambda \rangle_\Gamma,$$

defines an invertible operator.

Three fields, two integral equations. We now use both exterior Cauchy data as unknowns, introducing $\lambda := \partial_n^+ u$ and $\eta := \gamma^+ u$ and writing

$$u = D_c \eta - S_c \lambda.$$

The first integral identity in (14.29) is rewritten in the following form

$$V_c \lambda - \tfrac{1}{2}\eta - K_c \eta = -\gamma^+ u = g - \gamma w,$$

and the second one is kept as is, after substituting λ and η in it. This leads to the following problem

$$w \in H^1(\Omega_-), \ \lambda \in H^{-1/2}(\Gamma), \ \eta \in H^{1/2}(\Gamma), \tag{14.32a}$$

$$a(w,v) - \langle \lambda, \gamma v \rangle_\Gamma \quad = (f,v)_{\Omega_-} + \langle h, \gamma v \rangle_\Gamma \quad \forall v \in H^1(\Omega_-), \tag{14.32b}$$

$$\gamma w + V_c \lambda - \tfrac{1}{2}\eta - K_c \eta = g, \tag{14.32c}$$

$$\tfrac{1}{2}\lambda + K_c^t \lambda + W_c \eta \quad = 0, \tag{14.32d}$$

coupling an interior variational formulation with two nonlocal boundary conditions, involving two fields on Γ. Testing the two integral equations, we obtain a variational formulation in a product space

$$w \in H^1(\Omega_-), \ \lambda \in H^{-1/2}(\Gamma), \ \eta \in H^{1/2}(\Gamma), \tag{14.33a}$$

$$a(w,v) - \langle \lambda, \gamma v \rangle_\Gamma \quad\quad\quad = (f,v)_{\Omega_-} + \langle h, \gamma v \rangle_\Gamma, \tag{14.33b}$$

$$\langle \mu, \gamma w \rangle_\Gamma + \langle \mu, V_c \lambda \rangle_\Gamma - \langle \mu, \tfrac{1}{2}\eta + K_c \eta \rangle_\Gamma = \langle \mu, g \rangle_\Gamma, \tag{14.33c}$$

$$\langle \tfrac{1}{2}\lambda + K_c^t \lambda, \psi \rangle_\Gamma + \langle W_c \eta, \psi \rangle_\Gamma \quad\quad = 0, \tag{14.33d}$$

for all $v \in H^1(\Omega_-)$, $\mu \in H^{-1/2}(\Gamma)$, and $\psi \in H^{1/2}(\Gamma)$. The relevant bilinear form is

$$
\begin{aligned}
A_{\mathrm{sym}+}((w,\lambda,\eta),(v,\mu,\psi)) :=\ & a(w,v) - \langle \lambda, \gamma v \rangle_\Gamma \\
& + \langle \mu, \gamma w \rangle_\Gamma + \langle \mu, V_c \lambda \rangle_\Gamma - \langle \mu, \tfrac{1}{2}\eta + K_c \eta \rangle_\Gamma \\
& + \langle \tfrac{1}{2}\lambda + K_c^t \lambda, \psi \rangle_\Gamma + \langle W_c \eta, \psi \rangle_\Gamma,
\end{aligned}
$$

and we will have to prove that it defines an invertible operator. Note that $A_{\text{sym}+}$ can be made symmetric by changing the sign of the second equation, that is, the bilinear form

$$A_{\text{sym}+}((w, \lambda, \eta), (v, -\mu, \psi))$$

is symmetric. We also note that

$$A_{\text{sym}+}((w, \lambda, \eta), (w, \lambda, \eta)) = a(w, w) + \langle \lambda, V_c\lambda \rangle_\Gamma + \langle W_c\eta, \eta \rangle_\Gamma. \qquad (14.34)$$

Recall that V_c and W_c define coercive variational forms as we have proved in Propositions 14.6 and 14.7. The right-hand side of (14.34) misses being an equivalent norm in $H^1(\Omega_-) \times H^{-1/2}(\Gamma) \times H^{1/2}(\Gamma)$, since the bilinear form a vanishes on constant inputs. We will use Fredholm theory to prove well-posedness. First of all, the modified bilinear form

$$A_{\text{sym}+}((w, \lambda, \eta), (v, \mu, \psi)) + (w, v)_{\Omega_-}$$

is coercive (we have added an $L^2(\Omega_-)$ term to (14.34)) and its difference with $A_{\text{sym}+}$ is associated to a compact operator. Therefore, well-posedness of (14.33) is equivalent to uniqueness of solution. Let thus (w, λ, η) be a solution to (14.32) with a homogeneous right-hand side, and consider $u := D_c\eta - S_c\lambda$, which, as usual, solves $\Delta u - c^2 u = 0$ in $\mathbb{R}^d \setminus \Gamma$. Equation (14.32d) is equivalent to $\partial_n^- u = 0$ and therefore $u \equiv 0$ in Ω_- and $\lambda = -[\partial_n u] = \partial_n^+ u$ and $\eta = -[\gamma u] = \gamma^+ u$. We then go to (14.32c) (with $g = 0$) and notice that it is equivalent to continuity of the traces $\gamma w - \gamma^+ u = 0$. Finally, we note that

$$a(w, v) - \langle \partial_n^+ u, \gamma v \rangle_\Gamma = 0 \qquad \forall v \in H^1(\Omega_-) \qquad (14.35)$$

is equivalent to the diffusion equation $\nabla \cdot (\kappa \nabla w) = 0$ in Ω_- and continuity of the flux $(\kappa \nabla w) \cdot \mathbf{n} = \partial_n^+ u$. We thus have a homogeneous solution to (14.24) and therefore $w = 0$ and $u = 0$. Since (η, λ) are the exterior Cauchy data of w, this finishes the proof of uniqueness and, by the Fredholm alternative, of well-posedness of (14.33), or equivalently, invertibility of the operator associated to the bilinear form $A_{\text{sym}+}$.

Two fields, two integral equations. We go back to having $\lambda := \partial_n^+ u$ as the only additional boundary unknown. The integral equation (14.30c) is kept, but we now we write

$$(\kappa \nabla w) \cdot \mathbf{n} = \partial_n^+ u + h = \tfrac{1}{2}\lambda - K_c^t\lambda - W_c\gamma^+ u + h$$
$$= \tfrac{1}{2}\lambda - K_c^t\lambda - W_c\gamma w + W_c g + h,$$

using the second integral identity in (14.29). We substitute this in (14.26) and obtain a new formulation of the coupled problem:

$$w \in H^1(\Omega_-), \quad \lambda \in H^{-1/2}(\Gamma), \qquad (14.36a)$$

$$a(w, v) + \langle W_c\gamma w, \gamma v \rangle_\Gamma - \langle \tfrac{1}{2}\lambda - K_c^t\lambda, \gamma v \rangle_\Gamma$$
$$= (f, v)_{\Omega_-} + \langle h, \gamma v \rangle_\Gamma + \langle W_c g, \gamma v \rangle_\Gamma, \quad (14.36b)$$

$$\langle \mu, \tfrac{1}{2}\gamma w - K_c\gamma w \rangle_\Gamma + \langle \mu, V_c\lambda \rangle_\Gamma = \langle \mu, \tfrac{1}{2}g - K_c g \rangle_\Gamma, \qquad (14.36c)$$

for all $v \in H^1(\Omega_-)$ and $\mu \in H^{-1/2}(\Gamma)$. The associated bilinear form is

$$A_{\text{sym}}((w, \lambda), (v, \mu)) := a(w, v) + \langle W_c \gamma w, \gamma v \rangle_\Gamma - \langle \tfrac{1}{2}\lambda - K_c^t \lambda, \gamma v \rangle_\Gamma$$
$$+ \langle \mu, \tfrac{1}{2}\gamma w - K_c \gamma w \rangle_\Gamma + \langle \mu, V_c \lambda \rangle_\Gamma.$$

Up to a sign change in the second equation, we have symmetry, that is, the bilinear form

$$A_{\text{sym}}((w, \lambda), (v, -\mu))$$

is symmetric. This formulation has the easiest possible analysis of the three, since

$$A_{\text{sym}}((w, \lambda), (w, \lambda)) = a(w, w) + \langle W_c \gamma w, \gamma w \rangle_\Gamma + \langle \lambda, V_c \lambda \rangle_\Gamma$$

is an equivalent inner product in the product space. This is due to the coercivity of V_c (Proposition 14.6), the coercivity of W_c (Proposition 14.7), and using a generalized Poincaré inequality in $H^1(\Omega_-)$. Therefore A_{sym} is coercive. This shows the well-posedness of (14.36).

Coercivity analysis for the two field, one integral equation case. First of all, let us prove that the only homogeneous solution to the equations (14.30) is the trivial one. Let thus (w, λ) be a solution to (14.30) with zero right-hand sides and consider the function $u = D_c \gamma w - S_c \lambda$, which satisfies $\Delta u - c^2 u = 0$ in $\mathbb{R}^d \setminus \Gamma$. Equation (14.30c) with $g = 0$ is equivalent to $\gamma^- u = 0$ and therefore $u \equiv 0$ in Ω_-. The cancellation of u in the interior domain implies

$$\gamma w = -[\gamma u] = \gamma^+ u, \qquad \lambda = [\partial_n u] = \partial_n^+ u.$$

Once we have this, we recover (14.35) and the rest of the proof of uniqueness for the formulation with two fields and two integral equations can be used. Existence of a solution to (14.30) will follow from a coercivity argument after adding a compact term to the equation. It will bring along some restrictions on the size of κ_0, the lower bound for the diffusion coefficient.

Proposition 14.22. *If $\kappa_0 > 1/4$, then* (14.30) *is uniquely solvable.*

Proof. We are going to show that if $\kappa_0 > 1/4$, the bilinear form

$$B((w, \lambda), (v, \mu)) := A_{\text{ns}}((w, \lambda), (v, \mu)) + c^2 (w, v)_{\Omega_-}$$

is coercive. Since this corresponds to adding a compact term to the equations, the operator associated to A_{ns} is Fredholm of index zero and, by the Fredholm alternative, it is invertible if and only if it is injective, which we have already verified.

Let $u := S_c \lambda$, so that

$$\langle \lambda, V_c \lambda \rangle_\Gamma = \|\nabla u\|_{\mathbb{R}^d}^2 + c^2 \|u\|_{\mathbb{R}^d}^2, \tag{14.37}$$

and

$$-\langle \lambda, \gamma w \rangle_\Gamma + \langle \lambda, \tfrac{1}{2}\gamma w - K_c \gamma w \rangle_\Gamma = -\langle \tfrac{1}{2}\lambda + K_c^t \lambda, \gamma w \rangle_\Gamma = -\langle \partial_n^- u, \gamma w \rangle_\Gamma$$
$$= -(\nabla u, \nabla w)_{\Omega_-} - c^2 (u, w)_{\Omega_-}.$$

We thus have

$$
\begin{aligned}
B((w,\lambda),(w,\lambda)) =&(\kappa\nabla w,\nabla w)_{\Omega_-} - (\nabla u,\nabla w)_{\Omega_-} + \|\nabla u\|^2_{\Omega_-} + \|\nabla u\|^2_{\Omega_+}\\
&+ c^2\|w\|^2_{\Omega_-} - c^2(u,w)_{\Omega_-} + c^2\|u\|^2_{\Omega_-} + c^2\|u\|^2_{\Omega_+}\\
\geq&\kappa_0\|\nabla w\|^2_{\Omega_-} - \|\nabla u\|_{\Omega_-}\|\nabla w\|_{\Omega_-} + \|\nabla u\|^2_{\Omega_-}\\
&+ \tfrac{1}{2}c^2\|w\|^2_{\Omega_-} + \tfrac{1}{2}c^2\|u\|^2_{\Omega_-} + c^2\|u\|^2_{\Omega_+} + \|\nabla u\|^2_{\Omega_+}.
\end{aligned}
$$

If we focus on the interior terms involving gradients, we can observe the quadratic form

$$
[\|\nabla w\|_{\Omega_-} \quad \|\nabla u\|_{\Omega_-}]
\begin{bmatrix} \kappa_0 & -\tfrac{1}{2}\\ -\tfrac{1}{2} & 1 \end{bmatrix}
\begin{bmatrix} \|\nabla w\|_{\Omega_-}\\ \|\nabla u\|_{\Omega_-} \end{bmatrix}.
$$

The associated matrix is positive definite if and only if $\kappa_0 > 1/4$. In this case, there exists $C > 0$ (the smallest eigenvalue of the matrix, depending on κ_0) such that

$$
\kappa_0\|\nabla w\|^2_{\Omega_-} - \|\nabla u\|_{\Omega_-}\|\nabla w\|_{\Omega_-} + \|\nabla u\|^2_{\Omega_-} \geq C(\|\nabla w\|^2_{\Omega_-} + \|\nabla u\|^2_{\Omega_-}). \tag{14.38}
$$

Therefore

$$
\begin{aligned}
B((w,\lambda),(w,\lambda)) \geq&C\|\nabla w\|^2_{\Omega_-} + \tfrac{1}{2}c^2\|w\|^2_{\Omega_-} + \min\{C,1\}\|\nabla u\|^2_{\mathbb{R}^d} + \tfrac{1}{2}c^2\|u\|^2_{\mathbb{R}^d}\\
\geq&C'(\|w\|^2_{1,\Omega_-} + \langle\lambda,V_c\lambda\rangle_\Gamma),
\end{aligned}
$$

where we have applied (14.37). Since V_c is coercive, this shows the coercivity of the compactly perturbed bilinear form B. Note that $\kappa_0 > 1/4$ is a necessary and sufficient condition for the bound (14.38) to hold, but that this estimate is already a lower bound of the quadratic form associated to B. $\qquad\square$

Final comments and literature

The variational presentation of layer potentials and integral operators follows the seminal work of Jean-Claude Nédélec (for the Laplacian) and the classnotes of his graduate course on boundary integral equations, which have moved around the globe in photocopies and scans. The reconciliation of the variational theory with the integral theory of potentials (Section 14.5) is due to Martin Costabel [34]. A celebrated modern presentation of the theory of boundary integral operators for general elliptic problem on Lipschitz domains is due to William McLean [79]. The massive monograph of George Hsiao and Wolfgang Wendland [64] deals with different approaches for this theory, including a 'parallel' theory based on pseudo-differential operators on smooth manifolds, which requires that the boundary is locally the graph of a smooth function.

As we can see in Sections 14.6 and 14.7, moving from the Yukawa operator to the Laplace operator changes the behavior of the fundamental solution in a dramatic way and H^1 spaces cannot be used. The weighted Sobolev space of Section 14.6 appears in several collections of generalized Sobolev spaces that can be reinterpreted as Beppo-Levi spaces [44, Chapter XI]. In this context, they were used by Nédélec and Jacques Planchard in [84], which sets up the format for the variational introduction of layer potentials that we have given here. The two-dimensional equivalents appear in the doctoral work of Marie-Nöelle LeRoux [74, 75] and bring in yet more difficulties due to the unboundedness of the logarithmic fundamental solution and the occurrence of constant functions in the associated weighted Sobolev space (see Exercise 14.9). To learn about the additional difficulties of trying to use the completion of $\mathcal{D}(\mathbb{R}^d)$ with respect to the norm $\|\nabla \cdot \|_{\mathbb{R}^2}$ as the associated Sobolev space (this would be a Beppo-Levi space), the reader is directed to [50].

The final section on boundary-field formulations (we borrow this name from Gatica and Hsiao's monograph [53]) introduces three possible (there are more) couplings of boundary integral equations and variational formulations. The formulation with one integral equation had appeared in the literature but its first rigorous mathematical treatment can be attributed to Claes Johnson and Jean-Claude Nédélec [65] (two more papers on the topic were written at the time by these authors and Franco Brezzi), with an additional hypothesis on the smoothness of the interface that was shown to be unnecessary much later [92]. The symmetric formulation with two fields and two integral equations is due to Martin Costabel [33] and Houde Han [61], while the correct attribution of the three field formulation is unclear.

Exercises

14.1. Prove Proposition 14.10.

14.2. Prove that

$$\mathbb{D}_c^+ \{(\gamma^+ u, \partial_n^+ u) : u \in H^1(\Omega_+), \quad -\Delta u + c^2 u = 0 \quad \text{in } \Omega_+\}$$

is the range of the exterior Calderón projector $\mathcal{C}_c^+ := \mathcal{I} - \mathcal{C}_c^-$.

14.3. **First kind integral equations – Dirichlet problem.** We consider the problem

$$u \in H^1(\Omega_+), \quad -\Delta u + c^2 u = 0, \quad \gamma^+ u = g.$$

  (a) (Indirect formulation.) Show that the following formulation is well posed and solves the exterior Dirichlet problem

$$V_c \lambda = g, \quad u = S_c \lambda.$$

(b) (Direct formulation.) Show that the following formulation is well posed and solves the exterior Dirichlet problem

$$V_c\lambda = -\tfrac{1}{2}g + K_c g, \qquad u = D_c g - S_c\lambda.$$

Moreover, the function u defined like this is identically zero in Ω_- and $\lambda = \partial_n^+ u$.

(c) Use (a) and (b) to prove that

$$V_c^{-1}(-\tfrac{1}{2}I + K_c) = (-\tfrac{1}{2}I + K_c^t)V_c^{-1}.$$

(d) Show that $V_c K_c^t = K_c V_c$ as a consequence of (c). (This result can also be proved using the Calderón projector.)

(e) Show that

$$\langle V_c^{-1}(\tfrac{1}{2}I - K_c)g, g\rangle_\Gamma \geq C\|g\|_{1/2,\Gamma}^2 \qquad \forall g \in H^{1/2}(\Gamma).$$

(**Hint.** Use (b) to interpret the operator in the left-hand side in terms of the exterior Dirichlet-to-Neumann operator.)

14.4. Second kind integral equations – Dirichlet problem. We consider the problem

$$u \in H^1(\Omega_+), \qquad -\Delta u + c^2 u = 0, \qquad \gamma^+ u = g.$$

(a) (Indirect formulation.) Show that the following formulation is well posed and solves the exterior Dirichlet problem

$$\tfrac{1}{2}\eta + K_c \eta = g, \qquad u = D_c \eta.$$

(b) (Direct formulation.) Show that the following formulation is well posed and solves the exterior Dirichlet problem

$$\tfrac{1}{2}\lambda + K_c^t \lambda = -W_c g, \qquad u = D_c g - S_c\lambda.$$

Moreover, the function u defined like this is identically zero in Ω_- and $\lambda = \partial_n^+ u$.

(c) Use (a) and (b) to prove

$$(\tfrac{1}{2}I + K_c^t)^{-1}W_c = W_c(\tfrac{1}{2}I + K_c)^{-1}.$$

(d) Show that $K_c^t W_c = W_c K_c$ as a consequence of (c). (This result can also be proved using the Calderón projector.)

14.5. Potential formulations for the Neumann problem. We consider the exterior Neumann problem

$$u \in H^1(\Omega_+), \qquad -\Delta u + c^2 u = 0, \qquad \partial_n^+ u = h.$$

Write formulations and associated uniquely solvable integral equations for this problem for each of the following integral ansatz:

(a) $u = S_c \lambda$ (second kind equation)

(b) $u = D_c \eta$ (first kind equation)

(c) $u = D_c \eta - S_c h$ (where $\eta = \gamma^+ u$), using the first and the second integral identities. One of them will lead to a second kind equation and the other one to a first kind equation.

14.6. Prove Proposition 14.16. (**Hint.** If $\{\rho u_n\}$ converges in L^2, then $\{u_n\}$ converges in the sense of distributions, as multiplying by $\rho^{-1} \in \mathcal{C}^\infty(\mathcal{O})$ does not affect convergence in $\mathcal{D}'(\mathcal{O})$. For the density result, use a cutoff argument and note that functions in $W(\mathcal{O})$ that vanish outside a ball are in $H^1(\mathcal{O})$, i.e., the weight only intervenes in unbounded parts of the domain.)

14.7. The boundary operators for the Laplacian. Prove Proposition 14.19. The only part that is different from the case for Yukawa operators is related to the positivity of W_0. In this case, use a Poincaré inequality of the following form

$$\|u\|_{\Omega_-} \leq c\|\nabla u\|_{\Omega_-} \qquad \forall u \in H^1(\Omega_-), \qquad \gamma u \in H_*^{1/2}(\Gamma).$$

14.8. Let $\mathbf{x}_0 \in \mathbb{R}^3$, Γ be the boundary of a Lipschitz domain containing $\mathbf{x}_0$, and

$$\Phi(\mathbf{x}) := \frac{1}{4\pi|\mathbf{x} - \mathbf{x}_0|}.$$

(a) Show that

$$\langle \partial_n \Phi, 1 \rangle_\Gamma = \langle \partial_n \Phi, 1 \rangle_{\partial B(\mathbf{x}_0;\varepsilon)} \qquad \forall \varepsilon > 0.$$

(**Hint.** First prove it for small enough ε, using the annular domain bounded by Γ and $\partial B(\mathbf{x}_0;\varepsilon)$.)

(b) Use a direct computation to show that

$$\langle \partial_n \Phi, 1 \rangle_{\partial B(\mathbf{x}_0;\varepsilon)} = -1.$$

14.9. A weighted space for the two-dimensional Laplacian. In this exercise we consider spaces

$$W(\mathcal{O}) := \{u : \mathcal{O} \to \mathbb{R} : \rho u \in L^2(\mathcal{O}), \quad \nabla u \in \mathbf{L}^2(\mathcal{O})\}$$

for an open set $\mathcal{O} \subset \mathbb{R}^2$ where

$$\rho(\mathbf{x}) := \frac{1}{(1 + |\mathbf{x}|^2)^{1/2}} \frac{1}{1 + \frac{1}{2} \log(1 + |\mathbf{x}|^2)}.$$

In this exercise, we prove all the results that are needed to build a variational definition of the single and double layer potentials for the two-dimensional Laplacian.

(a) Show that $W(\mathcal{O})$ is a Hilbert space, that $H^1(\mathcal{O}) \subset W(\mathcal{O})$ with continuous embedding, and that $\mathcal{P}_0(\mathcal{O}) \subset W(\mathcal{O})$.

(b) Show that functions of $W(\mathcal{O})$ are locally H^1 and that $W(\mathcal{O}) = H^1(\mathcal{O})$ if $\mathcal{O}$ is bounded.

(c) Consider the sequence of functions

$$\varphi_n(\mathbf{x}) := \begin{cases} 1, & |\mathbf{x}| \leq 1, \\ h\left(\frac{n}{\log |\mathbf{x}|} - 1\right), & |\mathbf{x}| > 1, \end{cases}$$

where h is a smooth version of the Heaviside function. Show that $\varphi_n \in \mathcal{D}(\mathbb{R}^2)$, $\varphi_n \equiv 1$ in $B(0; e^{n/2})$, $\operatorname{supp} \varphi_n \subset \overline{B(0; e^n)}$, and that there exists $C > 0$ such that $|\nabla \varphi_n| \leq C\rho$ for all n.

(d) Use the sequence φ_n as a cutoff sequence to prove that $H^1(\mathbb{R}^2)$ is dense in $W(\mathbb{R}^2)$, and therefore so is $\mathcal{D}(\mathbb{R}^2)$.

(e) If Ω_+ is the exterior of a Lipschitz domain, show that $\mathcal{D}(\Omega_+)$ is dense in $\{u \in W(\Omega_+) : \gamma^+ u = 0\}$, where γ^+ is the trace operator.

(f) Prove that if $\mathcal{O} := \mathbb{R}^2 \setminus \overline{B(0; R)}$ for some R, then

$$\|\rho u\|_{\mathcal{O}} \leq C_R \|\nabla u\|_{\mathcal{O}} \qquad \forall u \in \mathcal{D}(\mathcal{O}).$$

(**Hint.** Use polar coordinates and estimate $u(\mathbf{x})$ by a radial integral. The result can be proved for a particular value of R and then a scaling argument can be used.)

(g) Prove that there exists $D_R > 0$ such that.

$$\|\rho u\|_{\mathbb{R}^2} \leq D_R(\|\nabla u\|_{\mathbb{R}^2} + \|u\|_{B(0;R)}) \qquad \forall u \in W(\mathbb{R}^2).$$

(h) **Poincaré inequalities.** Show that if $j : W(\mathcal{O}) \to \mathbb{R}$ is a bounded linear functional such that $j(1) \neq 0$, then

$$\|u\|_{W(\mathcal{O})}^2 \leq C_j(\|\nabla u\|_{\mathcal{O}}^2 + |j(u)|^2) \qquad \forall u \in W(\mathcal{O}).$$

(i) **Exterior Dirichlet and Neumann problems.** Show that the problem

$$u \in W(\Omega_+), \qquad \Delta u = 0, \qquad \gamma^+ u = g \in H^{1/2}(\Gamma)$$

is well posed. Show that if $h \in H^{-1/2}(\Gamma)$ and $\langle h, 1 \rangle_\Gamma = 0$, then the problem

$$u \in W(\Omega_+), \qquad \Delta u = 0, \qquad \partial_n u = h$$

is uniquely solvable up to additive constants.

(j) **Double layer potential.** Show that for $\eta \in H^{1/2}(\Gamma)$, the problem

$$u \in W(\mathbb{R}^2), \qquad \Delta u = 0 \quad \text{in } \mathbb{R}^2 \setminus \Gamma, \qquad [\gamma u] = -\eta,$$

is uniquely solvable. Its solution is denoted $u = D_0 \eta$.

(k) **Single layer potential.** Let $R > 0$ be such that $\overline{\Omega_-} \subset B(0; R)$ and let $\lambda \in H^{-1/2}(\Gamma)$ satisfy $\langle \lambda, 1 \rangle_\Gamma = 0$. Show that the problem

$$u \in W(\mathbb{R}^2 \backslash \Gamma), \quad \Delta u = 0 \quad \text{in } \mathbb{R}^2 \setminus \Gamma, \quad [\partial_n u] = \lambda, \quad \langle 1, \gamma_B u \rangle_{\partial B(0;R)} = 0,$$

is uniquely solvable. Its solution is denoted $S_0 \lambda$.

15

A collection of elliptic problems

15.1   T-coercivity in a dual Helmholtz equation 364
15.2   Diffusion with sign changing coefficient 370
15.3   Dependence with respect to coefficients 374
15.4   Obstacle problems 379
15.5   The Signorini contact problem 385
15.6   An optimal control problem 387
15.7   Friction boundary conditions 391
15.8   The Lions-Stampacchia theorem 395
15.9   Maximal dissipative operators 396
15.10 The evolution of elliptic operators 399
Final comments and literature 404
Exercises 405

This is a peculiar chapter that covers examples, extensions, and applications that do not seem to fit anywhere else in the layout of this book. We start with a detailed example of what has become known as T-coercivity. This is a sort of hidden ellipticity, which we show in a dual formulation of the Helmholtz equation, and in a diffusion problem with sign changing coefficient, loosely related to models of metamaterials. Next, we introduce some tools from differential calculus on normed spaces with the excuse of studying the dependence of diffusion problems with respect to the diffusion coefficient. From here, we move to studying elliptic problems and their interaction with convex optimization tools in Hilbert spaces, by looking for solutions on convex subsets of Sobolev spaces (the obstacle problem and the Signorini contact problem), by minimizing a functional that looks for the best data to match a given solution (a control problem) or by minimizing a nonquadratic convex function (friction boundary conditions). Finally, we finally leave the world of bounded operators and focus on a view of the Laplacian (and related operators) as an unbounded operator. Some of the natural properties that follow from this point of view lead into a black-box use of the theory of evolutionary equations in Hilbert spaces, which will give very simple while general results on the heat and wave equations.

15.1   T-coercivity in a dual Helmholtz equation

We will study an example of a variational problem where the coercivity of the principal part of the operator is hidden and needs to be revealed by using a decomposition of the space. The idea has received the name of T-coercivity in the literature, and it is extremely useful for problems like Maxwell's equations in the time-harmonic regime and for 'diffusion' problems with varying signs.

A dual form of the Helmholtz equation. In a Lipschitz domain Ω, with boundary Γ, we consider the Dirichlet problem for the Helmholtz equation

$$u \in H^1(\Omega), \qquad \Delta u + k^2 u = f, \qquad \gamma u = g, \qquad (15.1)$$

for given wave number $k > 0$, and data $f \in L^2(\Omega)$, $g \in H^{1/2}(\Gamma)$. We will reformulate the problem in the new variable

$$\mathbf{q} := \nabla u, \qquad (15.2)$$

with the possibility of recovering the field u from the equation

$$u = k^{-2}(f - \nabla \cdot \mathbf{q}). \qquad (15.3)$$

The variational formulation is obtained from the boundary condition, by testing it with a general $\mathbf{p} \cdot \mathbf{n}$ for $\mathbf{p} \in \mathbf{H}(\mathrm{div}, \Omega)$ (recall that the operator $\mathbf{p} \mapsto \mathbf{p} \cdot \mathbf{n}$ is surjective), and using (15.2) (the definition of $\mathbf{q}$) and (15.3) (the recovery of u from $\mathbf{q}$ using the Helmholtz equation):

$$\langle \mathbf{p} \cdot \mathbf{n}, g \rangle_\Gamma = \langle \mathbf{p} \cdot \mathbf{n}, \gamma u \rangle_\Gamma = (u, \nabla \cdot \mathbf{p})_\Omega + (\nabla u, \mathbf{p})_\Omega$$
$$= k^{-2}(f, \nabla \cdot \mathbf{p})_\Omega - k^{-2}(\nabla \cdot \mathbf{q}, \nabla \cdot \mathbf{p})_\Omega + (\mathbf{q}, \mathbf{p})_\Omega.$$

We thus reach the dual variational formulation

$$\mathbf{q} \in \mathbf{H}(\mathrm{div}, \Omega), \qquad (15.4\mathrm{a})$$

$$(\nabla \cdot \mathbf{q}, \nabla \cdot \mathbf{p})_\Omega - k^2 (\mathbf{q}, \mathbf{p})_\Omega = (f, \nabla \cdot \mathbf{p})_\Omega - k^2 \langle \mathbf{p} \cdot \mathbf{n}, g \rangle_\Gamma$$
$$\forall \mathbf{p} \in \mathbf{H}(\mathrm{div}, \Omega). \quad (15.4\mathrm{b})$$

Note how, similarly to what happens in mixed formulations for second order problems, the Dirichlet condition has become natural in this formulation. Generally and loosely speaking, the variational formulation in the variable u would be called a primal formulation, a formulation involving only the gradient of u as variable would be called a dual formulation, whereas a mixed formulation (Chapters 10 and 11) would involve both variables.

Preliminary work. Before we address the solvability of (15.4) (which will have the same issues as the problem formulated in the variable u), let us briefly examine this problem. First of all,

$$a(\mathbf{q}, \mathbf{p}) := (\nabla \cdot \mathbf{q}, \nabla \cdot \mathbf{p})_\Omega - k^2 (\mathbf{q}, \mathbf{p})_\Omega,$$
$$\ell(\mathbf{p}) := (f, \nabla \cdot \mathbf{p})_\Omega - k^2 \langle \mathbf{p} \cdot \mathbf{n}, g \rangle_\Gamma,$$

are, respectively, bounded bilinear and linear forms in the space $\mathbf{H}(\mathrm{div}, \Omega)$. Also, if $\mathbf{q}$ is a solution to (15.4) and we define $u := k^{-2}(f - \nabla \cdot \mathbf{q}) \in L^2(\Omega)$, it follows that

$$(u, \nabla \cdot \mathbf{p})_\Omega + (\mathbf{q}, \mathbf{p})_\Omega = \langle \mathbf{p} \cdot \mathbf{n}, g \rangle_\Gamma \qquad \forall \mathbf{p} \in \mathbf{H}(\mathrm{div}, \Omega). \tag{15.5}$$

Testing with $\mathbf{p} \in \mathcal{D}(\Omega)^d$, we obtain the relation $\nabla u = \mathbf{q}$ and, therefore, $u \in H^1(\Omega)$. Substituting this in the definition of u, we prove that u satisfies the Helmholtz equation

$$\Delta u + k^2 u = f.$$

Finally, plugging the equality $\mathbf{q} = \nabla u$ in (15.5), we obtain

$$(u, \nabla \cdot \mathbf{p})_\Omega + (\nabla u, \mathbf{p})_\Omega = \langle \mathbf{p} \cdot \mathbf{n}, g \rangle_\Gamma \qquad \forall \mathbf{p} \in \mathbf{H}(\mathrm{div}, \Omega),$$

and therefore

$$\langle \mathbf{p} \cdot \mathbf{n}, \gamma u - g \rangle_\Gamma = 0 \qquad \forall \mathbf{p} \in \mathbf{H}(\mathrm{div}, \Omega).$$

Since $\mathbf{p} \mapsto \mathbf{p} \cdot \mathbf{n}$ is surjective from $\mathbf{H}(\mathrm{div}, \Omega)$ to $H^{-1/2}(\Gamma) = H^{1/2}(\Gamma)'$, we have proved that $\gamma u = g$. This shows the equivalence of (15.1) and the dual formulation (15.4) via the introduction of $\mathbf{q}$ in (15.2) or the definition of u from $\mathbf{q}$ in (15.3). Moreover, uniqueness of a solution (or lack thereof) to (15.1) and (15.4) is also equivalent. (See Exercise 15.1.)

There is a slight difference between (15.1) and (15.4) and this comes from the regularity of f. In (15.1) we can take $f \in H^{-1}(\Omega) = H_0^1(\Omega)'$, since f will be tested via the $L^2(\Omega)$ inner product with elements of $H_0^1(\Omega)$. However, the argument showing the equivalence needs $f \in L^2(\Omega)$, which means that the dual formulation is meaningful only with stronger data.

The dual formulation (15.4) has the misleading look of a bilinear form for a Fredholm operator (in the sense of coercive plus compact), with a leading term based on the divergence operator and a lower order term with the wrong sign. However, the following result shows that this is not the case.

Proposition 15.1. *The injection of* $\mathbf{H}(\mathrm{div}, \Omega)$ *into* $\mathbf{L}^2(\Omega)$ *is not compact.*

Proof. Consider the operator $K : \mathbf{H}(\mathrm{div}, \Omega) \longrightarrow \mathbf{H}(\mathrm{div}, \Omega)$ given by

$$(K\mathbf{q}, \mathbf{p})_{\mathrm{div}, \Omega} = (\mathbf{q}, \mathbf{p})_\Omega \qquad \forall \mathbf{q}, \mathbf{p} \in \mathbf{H}(\mathrm{div}, \Omega),$$

via the Riesz-Fréchet representation theorem. Since, clearly,

$$\|K\mathbf{q}\|_{\mathrm{div}, \Omega} \leq \|\mathbf{q}\|_\Omega \qquad \forall \mathbf{q} \in \mathbf{H}(\mathrm{div}, \Omega),$$

if the injection $\mathbf{H}(\mathrm{div}, \Omega) \subset \mathbf{L}^2(\Omega)$ were compact, the operator K would be as well. However,

$$K\mathbf{q} = \mathbf{q} \quad \Longleftrightarrow \quad (\nabla \cdot \mathbf{q}, \nabla \cdot \mathbf{p})_\Omega = 0 \quad \forall \mathbf{p} \in \mathbf{H}(\mathrm{div}, \Omega)$$
$$\Longleftrightarrow \quad \nabla \cdot \mathbf{q} = 0,$$

(recall that div : $\mathbf{H}(\mathrm{div}, \Omega) \to L^2(\Omega)$ is surjective, or just take $\mathbf{p} = \mathbf{q}$), but the set of divergence-free vector fields in $\mathbf{H}(\mathrm{div}, \Omega)$ is infinite-dimensional (wait for it), which would produce an infinite-dimensional eigenspace for a nonzero eigenvalue of the compact operator K, thus reaching a contradiction. Let us examine the unproved claim: if $u \in H^1(\Omega)$, then

$$\mathbf{p} = (\partial_{x_2} u, -\partial_{x_1} u, 0, \ldots, 0) \in \mathbf{L}^2(\Omega)$$

has zero divergence. Thus, just using polynomial functions, we can build a sequence of linearly independent divergence-free vector fields.   □

A decomposition comes to the rescue. The previous result basically tells us that rewriting the bilinear form as the sum

$$a(\mathbf{q}, \mathbf{p}) = \underbrace{(\nabla \cdot \mathbf{q}, \nabla \cdot \mathbf{p})_\Omega + (\mathbf{q}, \mathbf{p})_\Omega}_{(\mathbf{q}, \mathbf{p})_{\mathrm{div}, \Omega}} - (k^2 + 1)(\mathbf{q}, \mathbf{p})_\Omega$$

is not the way to go in order to analyze (15.4) using Fredholm theory. It would be premature to assert though, that the operator associated to the bilinear form a is not Fredholm (of index zero). To make a long story short, what will happen is that we will find an isomorphism T in $\mathbf{H}(\mathrm{div}, \Omega)$, such that the 'rotated' bilinear form $a(\mathbf{q}, T\mathbf{p})$ can be written as a coercive bilinear form plus a bilinear form related to a compact operator. Without further ado, we introduce a projection that will save the day. Consider the operator

$$P : \mathbf{H}(\mathrm{div}, \Omega) \longrightarrow \mathbf{H}(\mathrm{div}, \Omega)$$

given by

$$P\mathbf{p} := \nabla u \quad \text{where} \quad \begin{array}{l} u \in H_0^1(\Omega), \\ \Delta u = \nabla \cdot \mathbf{p}. \end{array}$$

If we consider the constant for the Poincaré-Friedrichs inequality

$$\|u\|_\Omega \leq C_{\mathrm{PF}} \|\nabla u\|_\Omega \quad \forall u \in H_0^1(\Omega),$$

it follows from our earliest efforts in solving the homogeneous Dirichlet problem back in Chapter 2 that

$$\|P\mathbf{p}\|_\Omega \leq C_{\mathrm{PF}} \|\nabla \cdot \mathbf{p}\|_\Omega \quad \forall \mathbf{p} \in \mathbf{H}(\mathrm{div}, \Omega).$$

Let us now list several properties of P.

(a) P is linear and bounded, since

$$\|P\mathbf{p}\|^2_{\text{div},\Omega} = \|\nabla u\|^2_\Omega + \|\Delta u\|^2_\Omega \leq (1 + C^2_{\text{PF}})\|\nabla \cdot \mathbf{p}\|_\Omega.$$

(b) P leaves the divergence unchanged, as

$$\nabla \cdot (P\mathbf{p}) = \nabla \cdot \nabla u = \Delta u = \nabla \cdot \mathbf{p} \qquad \forall \mathbf{p} \in \mathbf{H}(\text{div}, \Omega).$$

(c) P is a projection, that is, $P^2 = P$. To see this, note that $P^2\mathbf{p} = P(P\mathbf{p}) = \nabla w$, where

$$w \in H^1_0(\Omega), \qquad \Delta w = \nabla \cdot (P\mathbf{p}) = \nabla \cdot \mathbf{p},$$

and therefore $P\mathbf{p} = \nabla w$ as well. Note that the range of a bounded projection is a closed subspace, as can be proved from a very simple argument.

(d) Finally, and most importantly, the range of P is compactly embedded into $\mathbf{L}^2(\Omega)$, as we show in the next result.

Proposition 15.2. *The range of P is the set*

$$\{\nabla u \in \mathbf{L}^2(\Omega) : u \in H^1_0(\Omega), \quad \Delta u \in L^2(\Omega)\},$$

which is compactly embedded into $\mathbf{L}^2(\Omega)$.

Proof. Consider the set (the domain of the Dirichlet Laplacian in the language that we will introduce in Section 15.10) $D^{\text{dir}}_\Delta := \{u \in H^1_0(\Omega) : \Delta u \in L^2(\Omega)\}$, which can be endowed the norm $\|\Delta \cdot \|_\Omega$, which satisfies

$$\|\Delta u\|^2_\Omega \leq \|\Delta u\|^2_\Omega + \|\nabla u\|^2_\Omega \leq (1 + C^2_{\text{PF}})\|\Delta u\|^2_\Omega.$$

Our argument is based upon two facts: (1) the gradient operator defines an isomorphism between D^{dir}_Δ and range P; (2) the injection of D^{dir}_Δ into $H^1_0(\Omega)$ is compact. The first assertion is easy to prove, since $H^1_0(\Omega)$ does not contain constant functions and the Dirichlet problem is uniquely solvable. For the second one, we recall the spectral decomposition for the Dirichlet problem (see Section 9.5) and how we can identify the spaces

$$H^1_0(\Omega) = \text{range } G^{1/2} = \{u \in L^2(\Omega) : \sum_{n=1}^\infty \lambda_n |(u, \phi_n)_\Omega|^2 < \infty\},$$

$$D^{\text{dir}}_\Delta = \text{range } G = \{u \in L^2(\Omega) : \sum_{n=1}^\infty \lambda^2_n |(u, \phi_n)_\Omega|^2 < \infty\},$$

in terms of the compact Green's operator $G : L^2(\Omega) \to L^2(\Omega)$ or, equivalently, of the Dirichlet eigensystem $\{(\lambda_n, \phi_n)\}$. Since

$$\|\Delta u\|^2_\Omega = \sum_{n=1}^\infty \lambda^2_n |(u, \phi_n)_\Omega|^2, \qquad \|\nabla u\|^2_\Omega = \sum_{n=1}^\infty \lambda_n |(u, \phi_n)_\Omega|^2,$$

the compactness of the injection $D_\Delta^{\mathrm{dir}} \subset H_0^1(\Omega)$ is an easy exercise (see Exercise 9.8(f)). The diagram

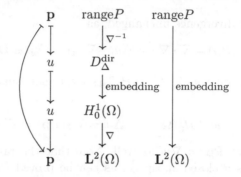

gives a factorization of the embedding as the composition of two isomorphisms with a compact embedding. This finishes the proof. □

The sign flipping operator. The projection P produces a natural decomposition of $\mathbf{H}(\mathrm{div}, \Omega)$ as the direct sum

$$\mathbf{H}(\mathrm{div}, \Omega) = \mathrm{range}\, P \oplus \mathrm{range}\, (I - P).$$

(This decomposition is actually orthogonal, since it is possible to show that P is the orthogonal projection onto its range, as we propose in Exercise 15.2. This does not play any role in our arguments, although it can be used to simplify some expressions below. We will show how when the time comes.) The bounded operator

$$T := 2P - I : \mathbf{H}(\mathrm{div}, \Omega) \to \mathbf{H}(\mathrm{div}, \Omega),$$

picks $\mathbf{p}$, decomposes it in the form $P\mathbf{p} + (I - P)\mathbf{p}$ (note that the first term of the sum is a gradient), and then changes the sign of the second term, $T\mathbf{p} := P\mathbf{p} - (I - P)\mathbf{p}$. Actually, it is easy to see that

$$T^2 = (2P - I)(2P - I) = 4P^2 - 4P + I = I,$$

and T is thus its own inverse (T is an involution). Note that

$$\nabla \cdot (T\mathbf{p}) = 2\nabla \cdot (P\mathbf{p}) - \nabla \cdot \mathbf{p} = \nabla \cdot \mathbf{p}. \tag{15.6}$$

A little bit of algebra based on (15.6) allows us to write

$$\begin{aligned}
a(\mathbf{q}, T\mathbf{p}) =& (\nabla \cdot \mathbf{q}, \nabla \cdot \mathbf{p})_\Omega - k^2(P\mathbf{q} + (I - P)\mathbf{q}, P\mathbf{p} - (I - P)\mathbf{p})_\Omega \\
=& (\nabla \cdot \mathbf{q}, \nabla \cdot \mathbf{p})_\Omega + k^2(P\mathbf{q}, P\mathbf{p})_\Omega + k^2((I - P)\mathbf{q}, (I - P)\mathbf{p})_\Omega \\
& - 2k^2(P\mathbf{q}, P\mathbf{p})_\Omega - k^2((I - P)\mathbf{q}, P\mathbf{p})_\Omega - k^2(P\mathbf{q}, (P - I)\mathbf{p})_\Omega.
\end{aligned}$$

We are almost ready to apply Fredholm's theory for the transformed problem, equivalent to (15.4),

$$\mathbf{q} \in \mathbf{H}(\text{div}, \Omega), \qquad a(\mathbf{q}, T\mathbf{p}) = \ell(T\mathbf{p}) \qquad \forall \mathbf{p} \in \mathbf{H}(\text{div}, \Omega). \tag{15.7}$$

We first note that the bilinear form

$$b(\mathbf{q}, \mathbf{p}) := (\nabla \cdot \mathbf{q}, \nabla \cdot \mathbf{p})_\Omega + k^2 (P\mathbf{q}, P\mathbf{p})_\Omega + k^2 ((I - P)\mathbf{q}, (I - P)\mathbf{p})_\Omega$$

is coercive, since

$$b(\mathbf{q}, \mathbf{q}) = \|\nabla \cdot \mathbf{q}\|_\Omega^2 + k^2 \|P\mathbf{q}\|_\Omega^2 + k^2 \|(I - P)\mathbf{q}\|_\Omega^2$$
$$\geq \|\nabla \cdot \mathbf{q}\|_\Omega^2 + \tfrac{1}{2} k^2 \|\mathbf{q}\|_\Omega^2.$$

(The last inequality can be improved using the fact that $\|\mathbf{q}\|_\Omega^2 = \|P\mathbf{q}\|_\Omega^2 + \|(I - P)\mathbf{q}\|_\Omega^2$ as can be derived from Exercise 15.2.) We condense the missing simple technical work into the following proposition.

Proposition 15.3. *The operator* $K : \mathbf{H}(\text{div}, \Omega) \to \mathbf{H}(\text{div}, \Omega)$ *defined by*

$$(K\mathbf{q}, \mathbf{p})_{\text{div},\Omega} = (P\mathbf{q}, \mathbf{p})_\Omega \qquad \forall \mathbf{q}, \mathbf{p} \in \mathbf{H}(\text{div}, \Omega)$$

is compact.

Proof. This is a direct consequence of Proposition 15.2, which can be restated in the following way: if we have a weakly convergent sequence $\{\mathbf{q}_n\}$ in $\mathbf{H}(\text{div}, \Omega)$, then $\{P\mathbf{q}_n\}$ is strongly convergent in $\mathbf{L}^2(\Omega)$. Noting that by definition of the operator K, we have the bound

$$\|K\mathbf{q}\|_{\text{div},\Omega} \leq \|P\mathbf{q}\|_\Omega \qquad \forall \mathbf{q} \in \mathbf{H}(\text{div}, \Omega),$$

the result follows. □

Operational form. Let $B, C : \mathbf{H}(\text{div}, \Omega) \to \mathbf{H}(\text{div}, \Omega)$ be the operators associated to the bilinear forms of the decomposition of $a(\mathbf{q}, T\mathbf{p})$:

$$(B\mathbf{q}, \mathbf{p})_{\text{div},\Omega} = b(\mathbf{q}, \mathbf{p})$$
$$= (\nabla \cdot \mathbf{q}, \nabla \cdot \mathbf{p})_\Omega + k^2 (P\mathbf{q}, P\mathbf{p})_\Omega + k^2 ((I - P)\mathbf{q}, (I - P)\mathbf{p})_\Omega,$$
$$(C\mathbf{q}, \mathbf{p})_{\text{div},\Omega} = 2(P\mathbf{q}, P\mathbf{p})_\Omega + (P\mathbf{q}, (P - I)\mathbf{p})_\Omega + ((I - P)\mathbf{q}, P\mathbf{p})_\Omega$$
$$= (K\mathbf{q}, 3P\mathbf{p} - \mathbf{p})_{\text{div},\Omega} + ((I - P)\mathbf{q}, K\mathbf{p})_{\text{div},\Omega}.$$

It is clear from the coercivity of b that B is invertible. We can write $C = (3P - I)^* K + K^* (I - P)$, using adjoints in $\mathbf{H}(\text{div}, \Omega)$ and therefore C is compact. (Using a property of P discussed in Exercise 15.2(a), we can prove the simpler formula $C = 2K$.) We can thus write (15.4) or (15.7) in the operator form

$$B\mathbf{q} - k^2 C\mathbf{q} = \mathbf{r}, \tag{15.8}$$

where

$$\mathbf{r} \in \mathbf{H}(\mathrm{div}, \Omega), \qquad (\mathbf{r}, \mathbf{p})_{\mathrm{div}, \Omega} = \ell(T\mathbf{p}) \quad \forall \mathbf{p} \in \mathbf{H}(\mathrm{div}, \Omega).$$

Conclusions. The operator $B - k^2 C$ is invertible if and only if it is injective, that is, if and only if we have uniqueness of solution to (15.7). However, uniqueness for (15.7) is equivalent to uniqueness for (15.4) and for (15.1). Therefore, the dual formulation (15.4) is well posed if and only if $-k^2$ is not a Dirichlet eigenvalue for the Laplacian in Ω. We close this section by going slowly over the compatibility conditions needed for the data so that we have a solution when $-k^2$ is a Dirichlet eigenvalue and we have a finite collection of orthonormal eigenfunctions:

$$(\phi_i, \phi_j)_\Omega = \delta_{ij}, \qquad \begin{array}{l} \phi_i \in H_0^1(\Omega), \\ \Delta\phi_i + k^2\phi_i = 0, \end{array} \qquad i, j, = 1, \ldots, n.$$

By the discussion on uniqueness of solution, the functions $\mathbf{q}_j := \nabla\phi_j$ are a basis of the set of homogeneous solutions to $B\mathbf{q} - k^2 C\mathbf{q} = \mathbf{0}$. Moreover,

$$\begin{aligned} (\nabla\phi_i, \nabla\phi_j)_{\mathrm{div}, \Omega} &= (\Delta\phi_i, \Delta\phi_i)_\Omega + (\nabla\phi_i, \nabla\phi_j)_\Omega \\ &= k^4(\phi_i, \phi_j)_\Omega + k^2(\phi_i, \phi_j)_\Omega = (k^4 + k^2)\delta_{ij}, \end{aligned}$$

so they are orthogonal in $\mathbf{H}(\mathrm{div}, \Omega)$. We also have that $P\nabla\phi_i = \nabla\phi_i$, since $\phi_i \in H_0^1(\Omega)$ has Laplacian in $L^2(\Omega)$ (recall Proposition 15.2). The compatibility conditions for (15.8) are then

$$(\mathbf{r}, \nabla\phi_i)_{\mathrm{div}, \Omega} = 0 \qquad i = 1, \ldots, n,$$

or, in other terms,

$$\ell(T\nabla\phi_i) = 0 \qquad i = 1, \ldots, n.$$

However, since $P\nabla\phi_i = \nabla\phi_i$, we have $T\nabla\phi_i = \nabla\phi_i$ and

$$\begin{aligned} \ell(T\nabla\phi_i) &= \ell(\nabla\phi_i) = (f, \Delta\phi_i)_\Omega - k^2\langle\nabla\phi_i \cdot \mathbf{n}, g\rangle_\Gamma \\ &= -k^2\left((f, \phi_i)_\Omega + \langle\partial_n\phi_i, g\rangle_\Gamma\right), \end{aligned}$$

and the compatibility conditions for the data can then be written in terms of the Dirichlet eigenfunctions

$$(f, \phi_i)_\Omega + \langle\partial_n\phi_i, g\rangle_\Gamma = 0 \qquad i = 1, \ldots, n.$$

The reader is invited to go back to Exercise 8.2 and check that the conditions that are obtained are the same.

15.2   Diffusion with sign changing coefficient

For what follows, we adopt a similar geometric layout to the problem of Stokes-Darcy flow (Section 10.9), but we identify subdomains with a $\pm$

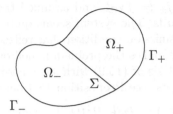

Figure 15.1: The set up for the sign changing coefficient diffusion problem. The domains $\Omega_\pm$ and the global domain are assumed to be Lipschitz.

subscript instead of with the S-D markers. A Lipschitz domain Ω is divided into two Lipschitz subdomains $\Omega_\pm$ by an interface $\Sigma := \partial\Omega_+ \cap \partial\Omega_-$, as in Figure 15.1. The boundary of $\Omega_\pm$ will be thus partitioned into two relatively open subsets, Σ and $\Gamma_\pm$. It will simplify our analysis to assume that both $\Gamma_\pm$ are not empty. We are given two strongly positive functions $\kappa_\pm \in L^\infty(\Omega_\pm)$ so that there are positive constants

$$\kappa_{max}^\pm = \|\kappa_\pm\|_{L^\infty(\Omega_\pm)} \geq \kappa_\pm \geq \kappa_{min}^\pm > 0 \quad \text{a.e. in } \Omega_\pm,$$

and we define

$$\kappa := \kappa_+ \chi_{\Omega_+} - \kappa_- \chi_{\Omega_-}.$$

We look for a solution of the 'diffusion' problem

$$u \in H_0^1(\Omega), \qquad -\nabla \cdot (\kappa \nabla u) = f, \tag{15.9}$$

for given $f \in L^2(\Omega)$. This problem is equivalent to the noncoercive (the bilinear form is indefinite) problem

$$u \in H_0^1(\Omega), \qquad (\kappa \nabla u, \nabla v)_\Omega = (f, v)_\Omega \quad \forall v \in H_0^1(\Omega).$$

We can also write (15.9) as a system of elliptic equations by identifying $u \equiv (u_+, u_-) \in H^1(\Omega_+) \times H^1(\Omega_-)$ and looking for a pair of functions satisfying separate elliptic equations (note the change of sign in Ω_- to compensate for the negative sign in κ)

$$\begin{aligned}
u_+ &\in H^1(\Omega_+), & u_- &\in H^1(\Omega_-), \\
\gamma u_+|_{\Gamma_+} &= 0, & \gamma u_-|_{\Gamma_-} &= 0, \\
-\nabla \cdot (\kappa_+ \nabla u_+) &= f_+, & -\nabla \cdot (\kappa_- \nabla u_-) &= -f_-,
\end{aligned}$$

with coupling conditions

$$\gamma u_+|_\Sigma = \gamma u_-|_\Sigma, \qquad (\kappa_+ \nabla u_+) \cdot \mathbf{n}_+|_\Sigma = -(\kappa_- \nabla u_-) \cdot \mathbf{n}_-|_\Sigma.$$

Here we have denoted $f_\pm := f\chi_{\Omega_\pm}$ and assumed that the normal vectors in $\Omega_\pm$ always point outwards. The system seems quite innocuous, but it is the sign of the second transmission condition that reflects the sign change of the diffusion parameter and makes the problem nonelliptic.

We now give an analysis of (15.9) with some additional hypotheses on the contrast of the coefficients (see Proposition 15.4 below). In the spaces

$$H^1_+(\Omega_+) := \{u \in H^1(\Omega_+) : \gamma u|_{\Gamma_+} = 0\},$$
$$H^1_-(\Omega_-) := \{u_- \in H^1(\Omega_-) : \gamma u|_{\Gamma_-} = 0\},$$

we will use the L^2 norm of the gradient as the standard norm (thanks to the generalized Poincaré inequality), and will also do so in $H^1_0(\Omega)$. For the arguments to come, we will consider a given extension operator

$$E : H^1_+(\Omega_+) \longrightarrow H^1_0(\Omega).$$

This extension can be given using only $\gamma u_+|_\Sigma$ (see the end of this section) or more information of u. The operator $P : H^1_0(\Omega) \to H^1_0(\Omega)$ given by

$$Pu := Eu_+ = Eu|_{\Omega_+}$$

is clearly a projection and therefore $T := 2P - I : H^1_0(\Omega) \to H^1_0(\Omega)$,

$$Tu = \begin{cases} u_+, & \text{in } \Omega_+, \\ 2Eu_+ - u_-, & \text{in } \Omega_-, \end{cases}$$

is an isomorphism saisfying $T^2 = I$.

Proposition 15.4. *If*

$$\left(\frac{\kappa^+_{\min}}{\kappa^-_{\max}}\right)^{1/2} > \|E\| := \sup_{0 \neq u_+ \in H^1_+(\Omega_+)} \frac{\|\nabla Eu_+\|_{\Omega_-}}{\|\nabla u_+\|_{\Omega_+}}, \qquad (15.10)$$

then the bilinear form $a(u, v) := (\kappa \nabla u, \nabla v)_\Omega$ *satisfies*

$$a(u, Tu) \geq \alpha \|\nabla u\|^2_\Omega \qquad \forall u \in H^1_0(\Omega)$$

for a constant $\alpha > 0$, *and the problem* (15.9) *is uniquely solvable.*

Proof. Let us first emphasize that the operator norm in (15.10) is not the one of E as an extension operator, but the norm of an extension and restriction operator from $H^1_+(\Omega_+)$ to $H^1_-(\Omega_-)$. The following computation is simple to follow:

$$\begin{aligned} a(u, Tu) =&(\kappa_+\nabla u_+, \nabla u_+)_{\Omega_+} + (\kappa_-\nabla u_-, \nabla u_-)_{\Omega_-} - 2(\kappa_-\nabla u_-, \nabla Eu_+)_{\Omega_-} \\ \geq& \kappa^+_{\min}\|\nabla u_+\|^2_{\Omega_+} + (\kappa_-\nabla u_-, \nabla u_-)_{\Omega_-} \\ &- 2(\kappa_-\nabla u_-, \nabla u_-)^{1/2}_{\Omega_-}(\kappa_-\nabla Eu_+, \nabla Eu_+)^{1/2}_{\Omega_-} \\ \geq& \kappa^+_{\min}\|\nabla u_+\|^2_{\Omega_+} + \|\kappa^{1/2}_-\nabla u_-\|^2_{\Omega_-} \\ &- 2(\kappa^-_{\max})^{1/2}\|E\|\|\kappa^{1/2}_-\nabla u_-\|_{\Omega_-}\|\nabla u_+\|_{\Omega_+}. \end{aligned}$$

The right-hand side is a quadratic form in the variables

$$(\|\nabla u_+\|_{\Omega_+}, \|\kappa_-^{1/2}\nabla u_-\|_{\Omega_-}) \in \mathbb{R}^2,$$

associated to the matrix

$$\begin{bmatrix} \kappa_{\min}^+ & -(\kappa_{\max}^-)^{1/2}\|E\| \\ -(\kappa_{\max}^-)^{1/2}\|E\| & 1 \end{bmatrix},$$

which is positive definite if and only if (15.10) holds. If c is the smallest of the two positive eigenvalues of the above matrix, we have the estimate

$$a(u, Tu) \geq c\left(\|\nabla u_+\|_{\Omega_+}^2 + \|\kappa_-^{1/2}\nabla u_-\|_{\Omega_-}^2\right) \geq c\,\min\{1, \kappa_{\min}^-\}\|\nabla u\|_{\Omega}^2,$$

and therefore a is coercive with respect to T, i.e., T-coercive. Therefore, (15.9) is equivalent to the coercive (and uniquely solvable) variational problem

$$u \in H_0^1(\Omega), \qquad (\kappa\nabla u, \nabla Tv)_\Omega = (f, Tv)_\Omega \qquad \forall v \in H_0^1(\Omega).$$

This finishes the proof. We note in passing that the roles of Ω_+ and Ω_- can easily be reversed by changing the sign of the entire equation, but then the extension operator E has to be redefined. $\qquad\square$

A trace-based extension. We now explore an extension operator E based on reading $\gamma u_+|_\Sigma$ and extending to Ω_- while trying to minimize $\|E\|$ defined in Proposition 15.4. For clarity, we will use the different symbol γ_+ for the trace operator $H^1(\Omega_+) \to H^{1/2}(\partial\Omega_+)$, and will keep γ untagged in Ω_-. We thus consider the minimization problem

$$\|\nabla w\|_{\Omega_-} = \min! \qquad w \in H_-^1(\Omega_-), \qquad \gamma w|_\Sigma = \gamma_+ u_+|_\Sigma, \qquad (15.11)$$

or equivalently (note that we are minimizing the quadratic positive definite functional $\frac{1}{2}\|\nabla w\|_{\Omega_-}^2$)

$$w \in H_-^1(\Omega_-), \qquad -\Delta w = 0, \qquad \gamma w|_\Sigma = \gamma_+ u_+|_\Sigma. \qquad (15.12)$$

In this case, we can easily prove that

$$\ker P = \{u \in H_0^1(\Omega) : u \equiv 0 \text{ in } \Omega_+\} \equiv \{0\} \times H_0^1(\Omega_-),$$
$$\operatorname{range} P = \{u \in H_0^1(\Omega) : \Delta u = 0 \text{ in } \Omega_-\} = (\ker P)^\perp.$$

Since

$$Pu - u \in \ker P = (\operatorname{range} P)^\perp,$$

we have

$$(\nabla(Pu - u), \nabla v)_\Omega = 0 \qquad \forall v \in \operatorname{range} P,$$

and P is the orthogonal projection onto its range. The fact that P is an orthogonal projection gives

$$\|P\|_{H_0^1(\Omega) \to H_0^1(\Omega)} = 1,$$

but no information about $\|E\|$. However, this is the best option to reduce $\|E\|$ since for given $u \in H_0^1(\Omega)$, the solution of the problem

$$\|\nabla w\|_{\Omega_-} = \min! \qquad w \in H_0^1(\Omega), \qquad w \equiv u \text{ in } \Omega_+,$$

is the same as the solution to (15.11)-(15.12). Note that this does not mean that the operator P (and T) defined through E is the best one to extend the chances of proving the well-posedness of (15.9), but that this is the best option if we want to use the arguments of the proof of Proposition 15.4.

15.3   Dependence with respect to coefficients

In this section we want to understand how the solution of the homogeneous Dirichlet problem

$$u \in H_0^1(\Omega), \qquad (\kappa \nabla u, \nabla v)_\Omega = (f, v)_\Omega \qquad \forall v \in H_0^1(\Omega), \tag{15.13}$$

depends of the diffusion parameter, which is an element of the set

$$
\begin{aligned}
\mathcal{U} :=& \{\kappa \in L^\infty(\Omega) \ : \ \kappa \geq \kappa_0 \text{ a.e., for some } \kappa_0 > 0\} \tag{15.14} \\
=& \{\kappa \in L^\infty(\Omega) \ : \ \kappa^{-1} \in L^\infty(\Omega), \kappa > 0\} \\
=& \{\exp(\rho) \ : \ \rho \in L^\infty(\Omega)\}.
\end{aligned}
$$

The equality of the sets above follows from very simple arguments, but we invite the reader to prove it. The set $\mathcal{U}$ is open in $L^\infty(\Omega)$: given $\kappa \in \mathcal{U}$, you can add any $\delta\kappa \in L^\infty(\Omega)$ such that $\|\delta\kappa\|_{L^\infty(\Omega)} \leq \kappa_0/2$ without abandoning $\mathcal{U}$. We then consider the operator $A : L^\infty(\Omega) \to \mathcal{B}(H_0^1(\Omega), H^{-1}(\Omega))$ given by

$$A(\kappa)u := (\kappa \nabla u, \nabla \cdot)_\Omega \in H^{-1}(\Omega).$$

The map A is linear and bounded as

$$\|A(\kappa)u\|_{-1,\Omega} \leq \|\kappa\|_{L^\infty(\Omega)} \|\nabla u\|_\Omega,$$

where (to simplify some coming formulas) we have preferred to use $\|\nabla \cdot\|_\Omega$ as the norm in $H_0^1(\Omega)$ and have defined the dual norm in $H^{-1}(\Omega)$ accordingly. We thus have

$$\|A(\kappa)\|_{H_0^1(\Omega) \to H^{-1}(\Omega)} \leq \|\kappa\|_{L^\infty(\Omega)}.$$

We care about the map

$$\mathcal{U} \ni \kappa \longmapsto A(\kappa)^{-1} \in \mathcal{B}(H^{-1}(\Omega), H_0^1(\Omega)).$$

We will first discuss continuity and then differentiability. Continuity is a consequence of a general result on the inversion of bounded operators.

Proposition 15.5. *If X and Y are Banach spaces, then the set*

$$\mathcal{B}_{\mathrm{inv}}(X, Y) := \{T \in \mathcal{B}(X, Y) : T \text{ invertible}\}$$

is open, and the map $\mathrm{Inv} : \mathcal{B}_{\mathrm{inv}}(X, Y) \to \mathcal{B}(Y, X)$ *given by* $\mathrm{Inv}(T) := T^{-1}$ *is continuous.*

Proof. If $T \in \mathcal{B}(X, Y)$ is invertible, we can decompose $T + E = T(I + T^{-1}E)$. Therefore, if $\|E\|_{X \to Y} < 1/\|T^{-1}\|_{Y \to X}$, then $(I + T^{-1}E)$ is invertible and so is $T + E$. We also have a bound

$$
\begin{aligned}
\|(T + E)^{-1}\|_{Y \to X} &\leq \|(I + T^{-1}E)^{-1}\|_{X \to X} \|T^{-1}\|_{Y \to X} \\
&\leq \frac{\|T^{-1}\|_{Y \to X}}{1 - \|T^{-1}\|_{Y \to X} \|E\|_{X \to Y}},
\end{aligned}
$$

which is derived from the Neumann series expansion:

$$(I + T^{-1}E)^{-1} = \sum_{n=0}^{\infty} (-1)^n (T^{-1}E)^n.$$

Also

$$(T + E)^{-1} - T^{-1} = (T + E)^{-1}(T - (T + E))T^{-1} = -(T + E)^{-1}ET^{-1}.$$

Assuming that $\|E\|_{X \to Y} \leq \frac{1}{2}\|T^{-1}\|_{Y \to X}$, we have

$$\|(T + E)^{-1} - T^{-1}\|_{Y \to X} \leq 2\|T^{-1}\|_{Y \to X}^2 \|E\|_{X \to Y}, \tag{15.15}$$

which shows continuity of Inv as follows: if $T_n \to T$, we can write $E_n := T_n - T \to 0$ and $\mathrm{Inv}(T_n) \to \mathrm{Inv}(T)$ by (15.15). $\qquad\square$

Note that we are composing a bounded (and therefore continuous) linear map $L^\infty(\Omega) \to \mathcal{B}(H_0^1(\Omega), H^{-1}(\Omega))$, such that $\mathcal{U}$ is mapped into $\mathcal{B}_{\mathrm{inv}}(H_0^1(\Omega), H^{-1}(\Omega))$, with the continuous inversion map of Proposition 15.5. This shows that the map $\kappa \mapsto A(\kappa)^{-1}$ is continuous. Our next step is the study of the 'smoothness' of that map, which we will restrict to continuous differentiability, although it is possible to go much further. That would require some additional work on differential calculus on normed spaces that we prefer not to deal with now. For some of the definitions below, the generic spaces will be called U and V, and we will use these to represent operator spaces, thus creating a second layer of abstraction that we believe will keep the reader on edge.

Fréchet differentiability. Let U and V be Banach spaces and $\mathcal{U} \subset U$ be open. We say that a map $F : \mathcal{U} \subset U \to V$ is Fréchet differentiable at $u_0 \in \mathcal{U}$ if there exists a bounded linear map, which we call $DF(u_0) \in \mathcal{B}(U, V)$ such that

$$\frac{\|F(u_0 + h) - F(u_0) - DF(u_0)h\|_V}{\|h\|_U} \to 0 \quad \text{as } \|h\|_U \to 0, \tag{15.16}$$

that is,

$$h_n \to 0 \text{ in } U \quad \Longrightarrow \quad \frac{1}{\|h_n\|_U} \left(F(u_0 + h_n) - F(u_0) - DF(u_0)h_n \right) \to 0 \text{ in } V.$$

(Obviously, we have to move along the sequence h_n in order to guarantee that $u_0 + h_n \in \mathcal{U}$, but this is a notational precaution, very much in the spirit of basic real analysis, that we will not mention explicitly any more.) We say that F is differentiable in $\mathcal{U}$, if it is differentiable at every point of $\mathcal{U}$. We say that $F \in \mathcal{C}^1(\mathcal{U}; V)$ if the map

$$DF : \mathcal{U} \subset U \to \mathcal{B}(U, V)$$

is continuous. We collect some easy results in the next proposition. We want to emphasize that this theory is just a slight generalization of the theory of differentiability in several variables.

Proposition 15.6. *Let $F : \mathcal{U} \subset U \to V$, where $\mathcal{U}$ is open in U and U and V are Banach spaces.*

(a) *If F is differentiable at u_0, then F is continuous at u_0.*

(b) *The differential $DF(u_0)$ is unique, that is, there exists only one bounded linear map satisfying the definition.*

(c) *If $F : U \to V$ is linear and bounded, then F is differentiable in U and $DF(u_0) = F$ for all u_0, that is, $DF : U \to \mathcal{B}(U, V)$ is the constant map $U \mapsto F$.*

Proof. The proof of (a) is straightforward. To prove (b), assume that D_1 and D_2 are bounded operators from U to V satisfying (15.16). For any $h \in U$ such that $\|h\|_U = 1$, we note that

$$(D_1 - D_2)h = \frac{1}{\varepsilon} \left(F(u_0 + \varepsilon h) - F(u_0) - \varepsilon D_2 h \right)$$

$$- \frac{1}{\varepsilon} \left(F(u_0 + \varepsilon h) - F(u_0) - \varepsilon D_1 h \right)$$

$$= \frac{1}{\|\varepsilon h\|_U} \left(F(u_0 + \varepsilon h) - F(u_0) - \varepsilon D_2 h \right)$$

$$- \frac{1}{\|\varepsilon h\|_U} \left(F(u_0 + \varepsilon h) - F(u_0) - \varepsilon D_1 h \right).$$

The right-hand side of the above converges to zero as $\varepsilon \to 0$ and therefore $D_1 h = D_2 h$ for all h with unit norm. Finally, (c) is a simple consequence of the definition. $\qquad\square$

Differentiability of the inversion map. An important example of a Fréchet differentiable function is the inversion map. We consider Banach spaces X and Y and take $U = \mathcal{B}(X, Y)$, $V = \mathcal{B}(Y, X)$, and $\mathcal{U} := \mathcal{B}_{\text{inv}}(X, Y)$. If it exists, the differential $D\text{Inv}(T) : \mathcal{B}(X, Y) \to \mathcal{B}(Y, X)$ will be a bounded linear map.

Proposition 15.7. *The inversion map* $\text{Inv} : \mathcal{B}_{\text{inv}}(X, Y) \to \mathcal{B}(Y, X)$ *is differentiable and for* $T \in \mathcal{B}_{\text{inv}}(X, Y)$, *we have*

$$D\text{Inv}(T)\delta T = -T^{-1}\delta T\, T^{-1} \qquad \forall \delta T \in \mathcal{B}(X, Y). \tag{15.17}$$

Furthermore, $D\text{Inv} : \mathcal{B}_{\text{inv}}(X, Y) \to \mathcal{B}(\mathcal{B}(X, Y), \mathcal{B}(Y, X))$ *is continuous.*

Proof. The proof of differentiability of the inversion map is simple if the Fréchet derivative is known. Since

$$(T + E)^{-1} - T^{-1} + T^{-1}ET^{-1} = (T^{-1} - (T + E)^{-1})ET^{-1},$$

we can estimate

$$\frac{1}{\|E\|_{X \to Y}}\|(T + E)^{-1} - T^{-1} + T^{-1}ET^{-1}\|_{Y \to X}$$
$$\leq \|T^{-1} - (T + E)^{-1}\|_{Y \to X}\|T^{-1}\|_{Y \to X},$$

and the right-hand side of the inequality goes to zero as $\|E\|_{X \to Y} \to 0$, which proves (15.17). The proof of continuity of the differential is also simple, the only challenge being the handling of two layers of spaces of bounded operators. We leave it to the reader (Exercise 15.7) to derive the remaining details. □

The chain rule can be stated and proved for the Fréchet derivative. Since for our problems we will only need two very simple uses of the rule (composing a differentiable map with a linear map on the left or on the right), we only introduce this restricted result.

Proposition 15.8 (Simplified chain rule). *Let* $F : \mathcal{U} \subset U \to V$ *be differentiable and* $G : V \to W$, $H : W \to U$ *be linear and bounded. The compositions* $G \circ F : \mathcal{U} \to W$ *and* $F \circ H : H^{-1}(\mathcal{U}) \subset W \to V$ *are differentiable and*

$$D(G \circ F) = G \circ DF, \qquad D(F \circ H) = (DF \circ H)H.$$

Here, $H^{-1}(\mathcal{U}) = \{w \in W : Hw \in \mathcal{U}\}$.

Proof. The proof of differentiability of $G \circ F$ is very simple as

$$\|G(F(u_0 + h)) - G(F(u_0)) - G\,DF(u_0)\,h\|_W$$
$$\leq \|G\|_{V \to W}\|F(u_0 + h) - F(u_0) - DF(u_0)h\|_V.$$

Note now that $H^{-1}(\mathcal{U})$ is open because H is continuous. Given $w_0 \in H^{-1}(\mathcal{U})$, $Hw_0 = u_0 \in \mathcal{U}$, and a sequence $\{h_n\}$ in W such that $\|h_n\|_W \to 0$, we consider

$$v_n := \frac{1}{\|h_n\|_W}(F(H(w_0 + h_n)) - F(Hw_0) - DF(Hw_0)Hh_n)$$
$$= \frac{1}{\|h_n\|_W}(F(u_0 + Hh_n) - F(u_0) - DF(u_0)Hh_n).$$

If $Hh_n = 0$, then $v_n = 0$, otherwise

$$\|v_n\|_V \leq \|H\|_{W \to U} \frac{1}{\|Hh_n\|_U} \|F(u_0 + Hh_n) - F(u_0) - DF(u_0)Hh_n\|_V \to 0,$$

which finishes the proof. □

Differentiability with respect to the diffusion coefficient. We again consider $U := L^\infty(\Omega)$, $V := \mathcal{B}(H^{-1}(\Omega), H_0^1(\Omega))$, $\mathcal{U}$ as in (15.14), and the map $\kappa \mapsto A(\kappa)^{-1}$, i.e., $F := \mathrm{Inv} \circ A$ (where the inversion map is done between $X = H_0^1(\Omega)$ and $Y = H^{-1}(\Omega)$). We then have

$$DF(\kappa)\delta\kappa = -A(\kappa)^{-1}A(\delta\kappa)A(\kappa)^{-1}. \qquad (15.18)$$

Since this has become somewhat of a puzzle and we are juggling with many operators, we will look at a slightly simpler map to see what (15.18) means in practice. We fix the right-hand side in the diffusion problem $f \in H^{-1}(\Omega)$ and consider the solution map

$$G_f : \mathcal{U} \subset L^\infty(\Omega) \longrightarrow H_0^1(\Omega)$$

for this particular right-hand side in (15.13). Therefore, $G_f(\kappa) := F(\kappa)f = A(\kappa)^{-1}f$. We can consider this as $G_f = H_f \circ F = H_f \circ \mathrm{Inv} \circ A$, where

$$\mathcal{B}(H^{-1}(\Omega), H_0^1(\Omega)) \ni T \longmapsto H_f T = Tf \in H_0^1(\Omega)$$

is clearly linear and bounded. Therefore, by Proposition 15.8

$$DG_f(\kappa)\delta\kappa = -A(\kappa)^{-1}A(\delta\kappa)A(\kappa)^{-1}f.$$

Now $DG_f(\kappa) \in \mathcal{B}(L^\infty(\Omega), H_0^1(\Omega))$ and $DG_f(\kappa)\delta\kappa \in H_0^1(\Omega)$ is computed as follows:

(a) first solve

$$u_\kappa \in H_0^1(\Omega), \qquad (\kappa\nabla u_\kappa, \nabla v)_\Omega = (f, v)_\Omega \qquad \forall v \in H_0^1(\Omega),$$

(b) for a moment, look at

$$\cdot A(\delta\kappa)A(\kappa)^{-1}f = A(\delta\kappa)u_\kappa = (\delta\kappa\nabla u_\kappa, \nabla \cdot)_\Omega \in H^{-1}(\Omega),$$

(c) and then put it in the right-hand side of the same type of problem

$$u_\delta \in H_0^1(\Omega), \qquad (\kappa\nabla u_\delta, \nabla v)_\Omega = -(\delta\kappa\nabla u_\kappa, \nabla v)_\Omega \qquad \forall v \in H_0^1(\Omega).$$

We have thus computed $u_\delta = DG_f(\kappa)\delta\kappa$. If we go back to F, $DF(\kappa)\delta\kappa$ is the operator that given $f \in H^{-1}(\Omega)$ carries out the entire process (a)-(c) and ends with u_δ.

15.4   Obstacle problems

This section starts the treatment of problems where convexity and optimization play a wider role. Let us go back for a while to one of the first problems we have seen in this book. If we are given $f \in L^2(\Omega)$, and solve the minimization problem

$$\tfrac{1}{2}\|\nabla u\|_\Omega^2 - (f, u)_\Omega = \min! \qquad u \in H_0^1(\Omega), \tag{15.19}$$

we are equivalently solving the variational equations

$$u \in H_0^1(\Omega), \qquad (\nabla u, \nabla v)_\Omega = (f, v)_\Omega \qquad \forall v \in H_0^1(\Omega), \tag{15.20}$$

or the boundary value problem

$$u \in H_0^1(\Omega), \qquad -\Delta u = f.$$

Say now that we have found a function

$$u_0 \in H^1(\Omega), \qquad -\Delta u_0 = f,$$

where we have not worried about the boundary conditions. However, since $(\nabla u_0, \nabla v)_\Omega = (f, v)_\Omega$ for all $v \in H_0^1(\Omega)$, we have

$$\tfrac{1}{2}\|\nabla u - \nabla u_0\|_\Omega^2 = \tfrac{1}{2}\|\nabla u\|_\Omega^2 - (f, u)_\Omega + \tfrac{1}{2}\|\nabla u_0\|_\Omega^2,$$

which means that (15.19) is equivalent to a best approximation problem

$$\|\nabla u - \nabla u_0\|_\Omega = \min! \qquad u \in H_0^1(\Omega). \tag{15.21}$$

Changing the (homogeneous) boundary conditions in (15.20) (to mixed or Neumann) is equivalent to changing the subspace of $H^1(\Omega)$ where we look for the best approximation of u_0 in (15.21). This will be our point of departure: we will change the set where we search for a best approximation. Instead of closed subspaces, we will look for best approximations in closed convex subsets.

Best approximation. A convex set of a Hilbert space, $K \subset H$, is a set such that

$$u, v \in K \quad \Longrightarrow \quad (1 - \theta)u + \theta v \in K \qquad \forall \theta \in (0, 1).$$

Given $u \in H$, we look for its best approximation in K:

$$\|u - w\| = \min! \qquad w \in K. \tag{15.22}$$

Proposition 15.9 (Existence of a best approximation). *If $\emptyset \neq K \subset H$ is convex and closed, then, for every $u \in H$, there exists a unique solution to (15.22).*

Proof. Let

$$d := \inf_{v \in K} \|u - v\| \geq 0.$$

By definition, there exists a sequence $\{v_n\}$ in K such that $\|v_n - u\| \to d$. Note that since K is convex, it follows that

$$\tfrac{1}{2}v_n + \tfrac{1}{2}v_m \in K \qquad \forall n, m,$$

and therefore by the parallelogram law

$$\begin{aligned}
\|v_n - v_m\|^2 &= 2\|v_n - u\|^2 + 2\|v_m - u\|^2 - \|(v_n + v_m) - 2u\|^2 \\
&= 2\|v_n - u\|^2 + 2\|v_m - u\|^2 - 4\left\|\tfrac{1}{2}(v_n + v_m) - u\right\|^2 \\
&\leq 2\left(\|v_n - u\|^2 + \|v_m - u\|^2 - 2d^2\right).
\end{aligned}$$

This implies that $\{v_n\}$ is a Cauchy sequence and hence convergent, $v_n \to w$. Since K is closed, $w \in K$ and $v_n \to w$ implies $\|v_n - u\| \to \|w - u\|$, and therefore $\|w - u\| = d$, which proves the existence of the best approximation.

To prove uniqueness, note that if

$$\|u - w_1\| = \|u - w_2\| = d \qquad w_1, w_2 \in K,$$

then, by the same computation involving the parallelogram law and the convexity of K, we have

$$\begin{aligned}
\|w_1 - w_2\|^2 &= 2\|w_1 - u\|^2 + 2\|w_2 - u\|^2 - 4\left\|\tfrac{1}{2}(w_1 + w_2) - u\right\|^2 \\
&\leq 2d^2 + 2d^2 - 4d^2 = 0,
\end{aligned}$$

and $w_1 = w_2$. □

Projection onto a closed convex set. The unique solution (Proposition 15.9) of

$$\|u - w\| = \min! \qquad w \in K,$$

is called the projection of u onto K and it will be denoted $P_K u$. The next result gives two 'variational inequalities' that are equivalent to the definition of P_K and will be used to prove that P_K is a contraction.

Proposition 15.10 (Equivalent definitions of best approximation). *Let $K \subset H$ be convex and let $u \in H$. The following three problems are equivalent:*

$$\begin{aligned}
w \in K, \qquad \|u - w\| \leq \|u - v\| \qquad &\forall v \in K, &\text{(15.23a)} \\
w \in K, \qquad (u - w, v - w) \leq 0 \qquad &\forall v \in K, &\text{(15.23b)} \\
w \in K, \qquad (u - v, w - v) \geq 0 \qquad &\forall v \in K. &\text{(15.23c)}
\end{aligned}$$

Proof. Let w solve (15.23a). Since

$$w + \theta(v - w) = (1 - \theta)w + \theta v \in K \qquad \forall v \in K, \qquad \theta \in [0, 1],$$

it follows that

$$\|u - w\|^2 \leq \|u - w - \theta(v - w)\|^2$$
$$= \|u - w\|^2 - 2\theta(u - w, v - w) + \theta^2\|v - w\|^2$$

and therefore

$$2(u - w, v - w) \leq \theta\|v - w\|^2 \qquad \forall \theta \in (0, 1), \qquad v \in K,$$

which implies that

$$(u - w, v - w) \leq 0 \qquad \forall v \in K,$$

and w solves (15.23b). Noting that

$$\|u - v\|^2 = \|u - w - (v - w)\|^2$$
$$= \|u - w\|^2 - 2(u - w, v - w) + \|v - w\|^2$$
$$\geq \|u - w\|^2 - 2(u - w, v - w),$$

it is clear that a solution of (15.23b) is a solution of (15.23a).

Next, note that for any v, w

$$(u - v, w - v) = (u - w, w - v) + \|w - v\|^2$$
$$\geq (u - w, w - v) = -(u - w, v - w),$$

which implies that any solution of (15.23b) is a solution of (15.23c). On the other hand if $v, w \in K$, for $\theta \in (0, 1)$ we have $w + \theta(v - w) \in K$ and

$$\theta^{-1}(u - (w + \theta(v - w)), w - (w + \theta(v - w))) = (u - w, w - v) + \theta\|w - v\|^2.$$
$$(15.24)$$

If w solves (15.23c), then the left-hand side of (15.24) is nonnegative for arbitrary $v \in K$ and $\theta \in (0, 1)$ and therefore

$$(u - w, v - w) \leq \theta\|w - v\|^2 \qquad \forall v \in K, \qquad \theta \in (0, 1).$$

Taking the limit as $\theta \to 0$, it follows that w solves (15.23b). $\square$

Corollary 15.1 (Contractivity of the projection)**.** *If $\emptyset \neq K \subset H$ is convex and closed, then*

$$\|P_K u - P_K v\| \leq \|u - v\| \qquad \forall u, v \in H.$$

Proof. Note that the equivalent characterization of the best approximation (15.23b) implies

$$(u - P_K u, P_K v - P_K u) \leq 0.$$

Similarly

$$(v - P_K v, P_K u - P_K v) \leq 0.$$

Adding the above inequalities, it follows that

$$\|P_K v - P_K u\|^2 \leq (u - v, P_K u - P_K v) \leq \|u - v\| \, \|P_K u - P_K v\|,$$

which proves the result.   □

The following result rewrites a certain quadratic minimization problem on K as a projection and derives the equivalent variational inequalities.

Proposition 15.11. *Let H be a Hilbert space, $\emptyset \neq K \subset H$ be closed and convex, $a : H \times H \to \mathbb{R}$ be symmetric, bounded and coercive, and $\ell \in H'$. The minimization problem*

$$\tfrac{1}{2}a(u, u) - \ell(u) = \min! \qquad u \in K, \tag{15.25}$$

is uniquely solvable and equivalent to

$$u \in K, \qquad a(u, v - u) \geq \ell(v - u) \qquad \forall v \in K. \tag{15.26}$$

Proof. We just need to rewrite (15.25) as an equivalent best approximation problem and relate (15.25) to the variational inequalities of Proposition 15.10. Let thus

$$u_0 \in H, \qquad a(u_0, v) = \ell(v) \qquad \forall v \in H,$$

and note that (15.25) is equivalent to

$$\tfrac{1}{2}a(u - u_0, u - u_0) = \min! \qquad u \in K,$$

i.e., u is the projection of u_0 on K when we use the 'energy norm' $\|u\|_a := a(u, u)^{1/2}$. This is characterized by

$$u \in K, \qquad a(u_0 - u, v - u) \leq 0 \qquad \forall v \in K,$$

but this is equivalent to the variational inequality (15.26).   □

A bound. The solution operator $\ell \mapsto u$, associated to the minimization problem (15.25) (or the equivalent variational inequality (15.26)) is not a linear operator unless K is a closed affine subspace of H. However, following the proof of Proposition 15.11 and using the contractivity property of the projection (Corollary 15.1), we can bound

$$\begin{aligned}
\|u\|_H &\leq \alpha^{-1/2}\|u\|_a \leq \alpha^{-1/2}(\|P_K u_0 - P_K 0\|_a + \|P_K 0\|_a) \\
&\leq \alpha^{-1/2}(\|u_0\|_a + \|P_K 0\|_a) \leq C_1 \|\ell\|_{H'} + C_2,
\end{aligned}$$

where α is the coercivity constant for the bilinear form, C_1 depends on the bilinear form, and C_2 depends on both the bilinear form and on the minimization set K. This shows that the map $\ell \mapsto u$ is bounded. With a similar argument, it is simple to show that it is also Lipschitz continuous.

Obstacle problem. Now consider $f \in L^2(\Omega)$ and an 'obstacle function' $u_{\text{obs}} : \Omega \to \mathbb{R}$. We look for a solution to the problem

$$\tfrac{1}{2}\|\nabla u\|_\Omega^2 - (f, u)_\Omega = \min! \qquad u \in H_0^1(\Omega), \qquad u \geq u_{\text{obs}}. \tag{15.27}$$

If

$$u_0 \in H_0^1(\Omega), \qquad -\Delta u_0 = f,$$

we are solving the equivalent best approximation problem

$$\|\nabla(u - u_0)\|_\Omega = \min! \qquad u \in H_0^1(\Omega), \qquad u \geq u_{\text{obs}}.$$

We need to assume that the admissible set

$$K_{\text{obs}} := \{u \in H_0^1(\Omega) : u \geq u_{\text{obs}} \quad \text{a.e.}\}$$

is not empty. (We will give some hypotheses on u_{obs} guaranteeing that this is so later on.) The set K_{obs} is clearly convex. Note that if $u_n \to u$ in $H^1(\Omega)$, then $u_n \to u$ in $L^2(\Omega)$ and there exists a subsequence $\{u_{n_k}\}$ converging to u almost everywhere (by the Riesz-Fischer theorem). This shows that K_{obs} is closed. The variational inequality associated to (15.27) (see Proposition 15.11) is

$$u \in K_{\text{obs}}, \qquad (\nabla u, \nabla(v - u))_\Omega \geq (f, v - u)_\Omega \qquad \forall v \in K_{\text{obs}}. \tag{15.28}$$

Problems (15.27) and (15.28) are uniquely solvable if $K_{\text{obs}} \neq \emptyset$. There are clearly situations where this is not the case. For instance, if $u_{\text{obs}} \equiv 1$, there are no functions in K_{obs} (prove it). On the other hand, if $u_{\text{obs}} \equiv 0$ in a neighborhood of Γ, and $u_{\text{obs}} \in L^\infty(\Omega)$, we can easily find elements of $H_0^1(\Omega)$ which are larger that u_{obs} almost everywhere: to do that, just build a $\mathcal{D}(\Omega)$ function which is constant, with the value $\|u_{\text{obs}}\|_{L^\infty(\Omega)}$, wherever $u_{\text{obs}} \neq 0$.

A partial differential inequality. We now bring back some concepts about the signs of distributions that we briefly looked at when discussing convection-diffusion equations back in Section 5.2. We consider the set

$$\mathcal{D}_+(\Omega) := \{\varphi \in \mathcal{D}(\Omega) : \varphi \geq 0\},$$

and say that $T \in \mathcal{D}'(\Omega)$ satisfies $T \geq 0$, when

$$\langle T, \varphi \rangle \geq 0 \qquad \forall \varphi \in \mathcal{D}_+(\Omega).$$

The same use of the Lebesgue differentiation theorem that we needed to prove the variational lemma in Section 1.7 can be used to prove that if $f \in L^1_{\text{loc}}(\Omega)$, then $f \geq 0$ as a distribution if and only if $f \geq 0$ almost everywhere. (This was proposed as Exercise 5.1.)

If u is the solution to the obstacle problem, then $u \in \mathcal{D}_+(\Omega) \subset K_{\text{obs}}$, and (15.28) implies that

$$(\nabla u, \nabla \varphi)_\Omega \geq (f, \psi)_\Omega \qquad \forall \varphi \in \mathcal{D}_+(\Omega)$$

that is

$$\langle -\Delta u - f, \varphi \rangle_{\mathcal{D}' \times \mathcal{D}} \geq 0 \qquad \forall \varphi \in \mathcal{D}_+(\Omega),$$

which we write as

$$-\Delta u \geq f. \tag{15.29}$$

Note that Δu might not be a function, and (15.29) has to be understood as a distributional inequality. However, if $\Delta u \in L^1_{\text{loc}}(\Omega)$, (15.29) holds almost everywhere. Finally if there exists an open set B and a positive constant ε such that $u > u_{\text{obs}} + \varepsilon$ almost everywhere in B, then $u \pm \eta \varphi \in K_{\text{obs}}$ for every $\varphi \in \mathcal{D}(B)$ and $\eta \leq \varepsilon / \|\varphi\|_{L^\infty(\Omega)}$. Therefore, (15.28) implies that

$$(\nabla u, \nabla \varphi)_B = (f, \varphi)_B \qquad \forall \varphi \in \mathcal{D}(B),$$

that is,

$$-\Delta u = f \quad \text{in } B.$$

This argument shows that a Poisson equation is satisfied in regions where the solution to the obstacle problem stays away from the obstacle. For any more results about this problem we need to assume much more on the regularity of the obstacle function. We will comment on this in the literature review at the end of the chapter. We finish with some more sufficient conditions on u_{obs} which will guarantee the nonemptiness of the set K_{obs}. This result is also used as an excuse to show that if $u \in H^1(\Omega)$, then $|u| \in H^1(\Omega)$.

Proposition 15.12. *If $u_{\text{obs}} \in \mathcal{C}(\overline{\Omega}) \cap H^1(\Omega)$ and $u_{\text{obs}}|_\Gamma \leq 0$, then $K_{\text{obs}} \neq \emptyset$.*

Proof. We first prove this result of independent interest: if $u \in H^1(\Omega)$, then $|u| \in H^1(\Omega)$ and

$$\nabla |u| = \text{sign}(u)\, \nabla u, \qquad \text{sign}(u) := \begin{cases} 1 & \text{in } \{\mathbf{x} : u(\mathbf{x}) > 0\}, \\ 0 & \text{in } \{\mathbf{x} : u(\mathbf{x}) = 0\}, \\ -1 & \text{in } \{\mathbf{x} : u(\mathbf{x}) < 0\}. \end{cases} \tag{15.30}$$

Note that $\text{sign}(u)$ can be defined using a single element of the class of functions grouped in u. The level sets used in the definition of $\text{sign}(u)$ are measurable, and therefore $\text{sign}(u)$ is measurable (characteristic functions of measurable sets are measurable) and is in $L^\infty(\Omega)$. If we take another representative, the variation is on sets of measure zero, and $\text{sign}(u)$ is well-defined almost everywhere as we would expect. To show (15.30), consider the functions

$$u_\varepsilon := (\varepsilon^2 + u^2)^{1/2}.$$

Note that $|u_\varepsilon| \le \varepsilon + |u|$ and $u_\varepsilon \to u$ almost everywhere, and therefore $u_\varepsilon \to u$ in $L^2(\Omega)$. A simple computation (do it!) shows that

$$\nabla u_\varepsilon = \frac{u}{(\varepsilon^2 + u^2)^{1/2}} \nabla u,$$

and then the dominated convergence theorem shows that as $\varepsilon \to 0$, $\nabla u_\varepsilon \to \mathrm{sign}(u)\,\nabla u$ in $L^2(\Omega)$. Equating limits in the sense of distributions we have (15.30). We then have that

$$\max\{u_{\mathrm{obs}}, 0\} = \tfrac{1}{2}(u_{\mathrm{obs}} + |u_{\mathrm{obs}}|) \in \mathcal{C}(\overline{\Omega}) \cap H^1(\Omega),$$

and since the trace is the restriction for continuous functions, we have

$$\gamma \max\{u_{\mathrm{obs}}, 0\} = 0.$$

We have thus found an element of K_{obs}. $\qquad\square$

15.5 The Signorini contact problem

We now study a problem that fits in the same theoretical framework (variational inequalities, minimization of quadratic functionals in closed convex sets) but leads to a partial differential equation with boundary conditions given by inequalities. We consider an open bounded Lipschitz domain Ω and a nonempty relatively open subset of its boundary $\Sigma \subset \Gamma$. We will be working on the elasticity system: Section 4.6 contains the basic concepts and Section 7.7 includes a presentation of the Navier-Lamé equations for nonhomogeneous anisotropic materials, using Korn's inequality (Section 7.6) to handle ellipticity issues. In the Sobolev space

$$\mathbf{H}_\Sigma^1(\Omega) := \{\mathbf{u} \in \mathbf{H}^1(\Omega) : \gamma\mathbf{u} = \mathbf{0} \text{ on } \Sigma\},$$

the expression $\|\boldsymbol{\varepsilon}(\mathbf{u})\|_\Omega$ (recall that $\boldsymbol{\varepsilon}(\mathbf{u}) = \tfrac{1}{2}(\nabla\mathbf{u} + (\nabla\mathbf{u})^\top)$) defines an equivalent norm by Korn's second inequality (see Proposition 7.14). The material coefficients are collected in a four index tensor $\mathbf{C} : \Omega \to \mathbb{R}^{d\times d\times d\times d}$ with components in $L^\infty(\Omega)$, satisfying the symmetry conditions

$$C_{ijkl} = C_{jikl} = C_{klij} = C_{ijlk} \qquad \text{a.e.} \qquad i,j,k,l = 1,\dots,d,$$

and the positivity condition

$$\sum_{i,j,k,l=1}^{d} C_{ijkl}\xi_{ij}\xi_{kl} \ge c_0 \sum_{i,j=1}^{d} \xi_{ij}^2 \qquad \text{a.e.} \qquad \forall \xi_{ij} = \xi_{ji} \in \mathbb{R}.$$

The above conditions prove that

$$\|\mathbf{u}\|_C^2 := (C\varepsilon(\mathbf{u}), \varepsilon(\mathbf{u}))_\Omega$$

defines an equivalent norm in $\mathbf{H}_\Sigma^1(\Omega)$. The stress tensor is defined as

$$\boldsymbol{\sigma} := C\varepsilon(\mathbf{u}) : \Omega \to \mathbb{R}_{\text{sym}}^{d\times d}.$$

The Signorini problem. As data, we consider functions $\mathbf{f} \in \mathbf{L}^2(\Omega)$, $\mathbf{g} \in \mathbf{L}^2(\Gamma \setminus \Sigma)$, and we will look for solutions in the set

$$K := \{\mathbf{u} \in \mathbf{H}_\Sigma^1(\Omega) : \gamma\mathbf{u} \geq 0\}.$$

This set is clearly nonempty (it contains $\mathcal{D}(\Omega)^d$ for instance). Convexity follows from the linearity of the trace operator. Finally, to see that it is closed, note that convergence in $H^{1/2}(\Gamma)$ implies convergence in $L^2(\Gamma)$ and, therefore, almost everywhere convergence of a subsequence. We then look for the unique solution to

$$\tfrac{1}{2}(C\varepsilon(\mathbf{u}), \varepsilon(\mathbf{u}))_\Omega - (\mathbf{f}, \mathbf{u})_\Omega - \langle\mathbf{g}, \gamma\mathbf{u}\rangle_\Gamma = \min! \qquad \mathbf{u} \in K. \qquad (15.31)$$

(Note that the hypotheses of Proposition 15.11 are satisfied.) If we take the unique solution of the mixed boundary value problem

$$\mathbf{u}_0 \in \mathbf{H}_\Sigma^1(\Omega), \qquad -\operatorname{div} C\varepsilon(\mathbf{u}_0) = \mathbf{f}, \qquad (C\varepsilon(\mathbf{u}_0))\mathbf{n} = \mathbf{g} \quad \text{on } \Gamma \setminus \Sigma,$$

that is, the unique solution of the variational problem

$$\mathbf{u}_0 \in \mathbf{H}_\Sigma^1(\Omega), \qquad (C\varepsilon(\mathbf{u}_0), \varepsilon(\mathbf{v}))_\Omega = (\mathbf{f}, \mathbf{v})_\Omega + \langle\mathbf{g}, \gamma\mathbf{v}\rangle_\Gamma \qquad \forall\mathbf{v} \in \mathbf{H}_\Sigma^1(\Omega),$$

the Signorini problem (15.31) is equivalent to the best approximation problem

$$\|\mathbf{u} - \mathbf{u}_0\|_C = \min! \qquad \mathbf{u} \in K.$$

(This is the process explained in the proof of Proposition 15.11.) The associated variational inequality is

$$\mathbf{u} \in K, \qquad (C\varepsilon(\mathbf{u}), \varepsilon(\mathbf{v} - \mathbf{u}))_\Omega \geq (\mathbf{f}, \mathbf{v} - \mathbf{u})_\Omega + \langle\mathbf{g}, \gamma(\mathbf{v} - \mathbf{u})\rangle_\Gamma \qquad \forall\mathbf{v} \in K.$$

Since $2\mathbf{u} \in K$ and $\mathbf{0} \in K$ we can also write the equivalent form

$$\mathbf{u} \in K, \qquad (C\varepsilon(\mathbf{u}), \varepsilon(\mathbf{u}))_\Omega = (\mathbf{f}, \mathbf{u})_\Omega + \langle\mathbf{g}, \gamma\mathbf{u}\rangle_\Gamma, \qquad\qquad (15.32\text{a})$$
$$(C\varepsilon(\mathbf{u}), \varepsilon(\mathbf{v}))_\Omega \geq (\mathbf{f}, \mathbf{v})_\Omega + \langle\mathbf{g}, \gamma\mathbf{v}\rangle_\Gamma \qquad \forall\mathbf{v} \in K. \qquad (15.32\text{b})$$

If $\mathbf{v} \in \mathcal{D}(\Omega)^d$, then $\pm\mathbf{v} \in K$ and (15.32b) implies that

$$-\operatorname{div}\boldsymbol{\sigma} = \mathbf{f}, \qquad \boldsymbol{\sigma} := C\varepsilon(\mathbf{u}). \qquad (15.33)$$

The weak integration by parts formula for elasticity (Betti's formula) or, properly speaking, the definition of the normal traction $\sigma \mathbf{n} \in \mathbf{H}^{-1/2}(\partial\Omega)$ (see Section 7.7), means that (15.33) translates into

$$\langle \sigma\mathbf{n}, \gamma\mathbf{v}\rangle_\Gamma = (C\varepsilon(\mathbf{u}), \varepsilon(\mathbf{v}))_\Omega - (\mathbf{f}, \mathbf{v})_\Omega \qquad \forall \mathbf{v} \in \mathbf{H}^1(\Omega).$$

Therefore, (15.32a) implies

$$\langle \sigma\mathbf{n} - \mathbf{g}, \gamma\mathbf{u}\rangle_{\Gamma\setminus\Sigma} = 0, \tag{15.34}$$

whereas (15.32b) implies

$$\langle \sigma\mathbf{n} - \mathbf{g}, \boldsymbol{\xi}\rangle_{\Gamma\setminus\Sigma} \geq 0 \qquad \forall \boldsymbol{\xi} \in \widetilde{\mathbf{H}}^{1/2}(\Gamma\setminus\Sigma) \qquad \boldsymbol{\xi} \geq \mathbf{0}, \tag{15.35}$$

where $\widetilde{\mathbf{H}}^{1/2}(\Gamma\setminus\Sigma)$ is the range of the trace operator acting on $\mathbf{H}^1_\Sigma(\Omega)$. We can formally write two inequalities

$$\gamma\mathbf{u} \geq \mathbf{0}, \qquad \sigma\mathbf{n} \geq \mathbf{g} \qquad \text{on } \Gamma\setminus\Sigma. \tag{15.36}$$

The first of these takes place in $\widetilde{\mathbf{H}}^{1/2}(\Gamma\setminus\Sigma)$ and just follows from the fact that $\mathbf{u} \in K$. The second one can be understood as an inequality in the dual space $\mathbf{H}^{-1/2}(\Gamma\setminus\Sigma)$, with the precise meaning of (15.35). Finally (15.34) can be formally understood as a complementary condition for the inequalities (15.36): if one of the inequalities is strict in a part of $\Gamma\setminus\Sigma$, then the other inequality has to become an equality. We will avoid entering into more details, since they require studying regularity properties of the sets where the inequalities (15.36) are strict.

15.6 An optimal control problem

In this section we introduce and analyze a simple optimal control problem. In this kind of problem we try to find data for a boundary value problem (state equation) so that the solution is as close as possible to a given desired solution (desired state). To make the problem uniquely solvable, we typically balance the functional measuring closeness to the desired state with another one measuring the size of data. We start with some easy concepts of convexity.

Convex functionals. Let $j : H \to \mathbb{R}$ be a functional. We say that j is convex when

$$j(\theta u + (1-\theta)v) \leq \theta j(u) + (1-\theta)j(v) \qquad \forall u, v \in H, \qquad \theta \in (0,1),$$

while we say that it is strictly convex when

$$j(\theta u + (1-\theta)v) < \theta j(u) + (1-\theta)j(v) \qquad \forall u \neq v \in H, \qquad \theta \in (0,1).$$

In any inner product space H, the function $u \mapsto \|u\|_H^2$ is strictly convex. To see that, note that for any $u \neq v$, the map

$$\phi(t) := \|v + t(u - v)\|_H^2 = \|v\|_H^2 + t^2\|u - v\|_H^2 + 2t(v, u - v)_H$$

is strictly convex (it is a parabola) and

$$\|\theta u + (1 - \theta)v\|_H^2 = \phi(\theta) = \phi((1 - \theta)\,0 + \theta\,1)$$
$$< (1 - \theta)\phi(0) + \theta\phi(1) = \theta\|u\|_H^2 + (1 - \theta)\|v\|_H^2.$$

If $F(u) := Au - b$, where $A : H_1 \to H_2$ is linear, then

$$u \longmapsto \|F(u)\|_{H_2}^2 = \|Au - b\|_{H_2}^2$$

is convex. This is easy, as

$$F(\theta u + (1 - \theta)v) = \theta F(u) + (1 - \theta)F(v).$$

(Note that this one is not necessarily strictly convex.) Therefore, functionals of the form

$$\tfrac{1}{2}\|Au - b\|_{H_2}^2 + \tfrac{\alpha}{2}\|u\|_{H_1}^2$$

are strictly convex.

A distributed control problem. Let $\kappa \in L^\infty(\Omega)$ be strongly positive, $c \in L^\infty(\Omega)$, $c \geq 0$, and $B \subset \Omega$ be an open subset of Ω. We consider the problem

$$u \in H_0^1(\Omega), \qquad -\nabla \cdot (\kappa \nabla u) + cu = f\chi_B. \tag{15.37}$$

In principle $f \in L^2(B)$ and $f\chi_B$ is what we would have called $\widetilde{f}$ elsewhere. Let $S : L^2(B) \to L^2(\Omega)$ be the solution operator corresponding to the above problem. (Note that S is compact.) We will refer to (15.37) as the state equation. Our optimal control problem is: given a desired state $u_d \in L^2(\Omega)$ and a regularization parameter $\alpha > 0$,

$$\tfrac{1}{2}\|u - u_d\|_\Omega^2 + \tfrac{\alpha}{2}\|f\|_B^2 = \min! \qquad f \in L^2(B), \qquad Sf = u. \tag{15.38}$$

The reduced functional eliminates the state variable u and is defined as

$$j(f) := \tfrac{1}{2}\|Sf - u_d\|_\Omega^2 + \tfrac{\alpha}{2}\|f\|_B^2. \tag{15.39}$$

It is clear that $j : L^2(B) \to \mathbb{R}$ is continuous and by the arguments given after the definition of convex functionals, j is strictly convex. The next abstract result is needed for the proof of the existence of a minimizer. In the language of convex analysis it can be phrased as: every continuous convex functional is weakly lower semicontinuous.

Proposition 15.13. *If $j : H \to \mathbb{R}$ is continuous and convex, then*

$$u_n \rightharpoonup u \quad \Longrightarrow \quad \liminf_{n \to \infty} j(u_n) \geq j(u).$$

Proof. For any $a \in \mathbb{R}$, if the level set

$$H_a := \{u \in H : j(u) \leq a\}$$

is not empty, it is closed (j is continuous) and convex (j is convex). Now assume that $j(u_n) \leq a$ for all n, so that H_a is not empty, and let $P_a : H \to H_a$ be the projection on H_a. It then follows that

$$(u - P_a u, u_n - P_a u) \leq 0 \qquad \forall n \qquad (u_n \in H_a),$$

and by the weak limit $\|u - P_a u\| = 0$ so $P_a u = u$ and $j(u) \leq a$ as well. If $\liminf_n j(u_n) < j(u)$ we can take a subsequence and $\varepsilon > 0$ such that

$$j(u_{n_k}) \leq j(u) - \varepsilon,$$

but this contradicts the previous argument. □

Proposition 15.14. *If $j : H \to \mathbb{R}$ is continuous, strictly convex, bounded below, and such that $j(u) \to \infty$ as $\|u\| \to \infty$, then the minimization problem*

$$j(u) = \min! \qquad u \in H$$

has a unique solution.

Proof. Let $\{u_n\}$ be a sequence such that

$$j(u_n) \to c := \inf\{j(u) : u \in H\} > -\infty.$$

Note that $\{u_n\}$ has to be bounded, otherwise it would have a subsequence $\{u_{n_k}\}$ such that $\|u_{n_k}\| \to \infty$ and therefore $j(u_{n_k}) \to \infty$. Now, take a weakly convergent subsequence $u_{n_k} \rightharpoonup u$ and note that by Proposition 15.13

$$c = \lim_k j(u_{n_k}) = \liminf_k j(u_{n_k}) \geq j(u) \geq c,$$

which shows that u is a minimizer. If $v \neq u$ is another one, then by strict convexity

$$c \leq j(\tfrac{1}{2}u + \tfrac{1}{2}v) < \tfrac{1}{2}j(u) + \tfrac{1}{2}j(v) = c,$$

which is a contradiction. □

Proposition 15.14 proves that we have a unique solution for the optimal control problem (15.38). We will now write some variational equations that are satisfied by the unique solution to (15.38). The functional $j : L^2(B) \to \mathbb{R}$ is quadratic and continuous, hence differentiable (see Exercise 15.6). We can then deal with its Fréchet derivative $Dj : H \to H'$ (we will denote the action of $Dj(f) \in H'$ on $\delta f \in H$ by simple juxtaposition $Dj(f)\delta f$. In the literature it is also common to deal with the gradient map $j' : H \to H$ obtained by composing with the Riesz-Fréchet representation

$$(j'(f), \delta f)_H = Dj(f)\delta f.$$

The next result is the infinite-dimensional equivalent of the well-known result that the gradient of a differentiable function vanishes at extrema.

Proposition 15.15 (Euler equations). *If $j : H \to \mathbb{R}$ is Fréchet differentiable, and $u \in H$ is a local minimum of j, then $Dj(u) = 0$.*

Proof. Let $h \in H$ and $\phi(t) := j(u + th)$. We have $\phi'(0) = Dj(u)h$, but $t = 0$ is a local minimum of ϕ.   □

Computation of the derivative. We are now going to derive a formula for the Fréchet derivative Dj and for the Euler equations associated to minimizing the functional j in (15.39). It is clear that

$$Dj(f)\delta f = (Sf - u_d, S\delta f)_\Omega + \alpha(f, \delta f)_B = (S^*(Sf - u_d) + \alpha f, \delta f)_B,$$

where $S^* : L^2(\Omega) \to L^2(B)$ is the adjoint of the solution operator S. The Euler equations, written in terms of the gradient, are clearly

$$S^*(Sf - u_d) + \alpha f = 0. \tag{15.40}$$

Let us now relate the operator S^* to the solution of an **adjoint problem** to the state equation. Note that $S\delta f \in L^2(\Omega)$ is found by solving

$$S\delta f \in H_0^1(\Omega), \qquad a(S\delta f, \omega) = (\delta f, \omega)_B \qquad \forall \omega \in H_0^1(\Omega),$$

where $a(u, v) = (\kappa \nabla u, \nabla v)_\Omega + (cu, v)_\Omega$. Given $v \in L^2(\Omega)$, consider the problem

$$w \in H_0^1(\Omega), \qquad -\nabla \cdot (\kappa \nabla w) + cw = v.$$

It is then easy to see that

$$(v, S\delta f)_\Omega = a(w, S\delta f) = a(S\delta f, w) = (\delta f, w)_B,$$

and therefore $S^* v = w|_B$.

Solvability of the Euler equations. We know that the Euler equations (15.40) have at least one solution, namely the global minimum of the functional j. Convexity of j precludes the existence of other local minima and of any local maximum. In fact, in this simple case, with a quadratic strictly convex functional, it is possible to prove that the only possible solution of the Euler equations has to be the minimum. An even simpler argument can be invoked by noticing that $\alpha I + S^*S$ is a self-adjoint bounded strongly positive definite operator and it is therefore invertible. We are now going to examine what (15.40) looks like in our example problem. To compute $S^*(Sf - u_d)$, we first compute $u := Sf$ by solving

$$u \in H_0^1(\Omega), \qquad -\nabla \cdot (\kappa \nabla u) + cu = f\chi_B, \tag{15.41a}$$

and then we compute $S^*(u - u_d)$ by solving

$$w \in H_0^1(\Omega), \qquad -\nabla \cdot (\kappa \nabla w) + cw = u - u_d, \tag{15.41b}$$

and then restricting $w|_B$. Therefore, (15.40) is equivalent to

$$w|_B + \alpha f = 0. \tag{15.41c}$$

The Euler equations are then the linear system of algebraic-differential equations (15.41). Using the last equation, we can eliminate $f = -\alpha^{-1}w|_B$ from the system and reduce the equations to

$$(u, w) \in H_0^1(\Omega)^2, \qquad -\nabla \cdot (\kappa \nabla u) + cu + \alpha^{-1}w|_B = 0, \tag{15.42a}$$
$$-\nabla \cdot (\kappa \nabla w) + cw - u \quad = -u_d, \tag{15.42b}$$

and then take $f = -\alpha^{-1}w|_B$ as control variable. Once again, we have already seen that the Euler equations are uniquely solvable, and therefore, (15.41) are uniquely solvable. This proves that the system (15.42) –which, at first sight, can be seen to be Fredholm of index zero (eliminate the coupling terms to obtain two uncoupled elliptic problems)– is uniquely solvable.

15.7 Friction boundary conditions

In this section we study another minimization problem, now related to a nonlinear perturbation of a quadratic functional and associated to a simplified version of the Tresca friction problem. On a Lipschitz domain Ω and with data $f \in L^2(\Omega)$ and a strongly positive coefficient $\kappa \in L^\infty(\Omega)$, we look for the solution to

$$\tfrac{1}{2}\|\kappa^{1/2}\nabla u\|_\Omega^2 + \tfrac{1}{2}\|u\|_\Omega^2 - (f, u)_\Omega + \langle 1, |\gamma u|\rangle_\Gamma = \min! \qquad u \in H^1(\Omega). \tag{15.43}$$

This problem will fit in the following penalized best approximation framework. We consider the functional

$$j(u) := \tfrac{1}{2}\|u - u_0\|^2 + s(u), \tag{15.44}$$

where $u_0 \in H$ is given and $s : H \to \mathbb{R}$ is continuous and convex. We look for a global minimum

$$j(u) = \min! \qquad u \in H. \tag{15.45}$$

The next result shows a variational inequality (typically called of the second kind) associated to the minimization problem.

Proposition 15.16 (Equivalent variational inequality). *If j is of the form (15.44) with s continuous and convex, the minimization problem (15.45) is equivalent to*

$$(u - u_0, v - u) + s(v) - s(u) \geq 0 \qquad \forall v \in H. \tag{15.46}$$

Proof. We start the proof with an easy computation:

$$j(v) - j(u) = \tfrac{1}{2}\|v - u + u - u_0\|^2 - \tfrac{1}{2}\|u - u_0\|^2 + s(v) - s(u)$$
$$= \tfrac{1}{2}\|v - u\|^2 + (u - u_0, v - u) + s(v) - s(u).$$

Because of this, if the inequality (15.46) holds, then

$$j(v) - j(u) \geq \tfrac{1}{2}\|v - u\|^2 \geq 0 \qquad \forall v \in H,$$

and we have a minimum of j at u. By convexity

$$s(u + \theta\,w) - s(u) = s((1 - \theta)u + \theta(w + u)) - s(u)$$
$$\leq \theta(s(u + w) - s(u)) \qquad \forall u, w \in H, \quad \theta \in [0, 1].$$

Therefore, if we have a minimum of j at u, we have

$$0 \leq j(u + \theta w) - j(u)$$
$$= \tfrac{\theta^2}{2}\|w\|^2 + \theta(u - u_0, w) + s(u + \theta w) - s(u)$$
$$\leq \tfrac{\theta^2}{2}\|w\|^2 + \theta\left((u - u_0, w) + s(u + w) - s(u)\right) \qquad \forall w \in H, \quad \theta \in [0, 1].$$

Dividing by θ and taking the limit as $\theta \to 0$, it follows that

$$(u - u_0, w) + s(u + w) - s(u) \geq 0 \qquad \forall w \in H,$$

which is equivalent to (15.46). $\qquad\square$

Proposition 15.17. *If j is of the form (15.44) with s continuous and convex, then*

$$j(v) \to \infty \quad as \quad \|v\| \to \infty. \tag{15.47}$$

Therefore (15.45) and the equivalent (15.46) have a unique solution.

Proof. Let $r(v) := s(v) - s(0) + 1$ and note that $r : H \to \mathbb{R}$ is continuous and convex. It also satisfies $r(0) = 1$. Now consider the set

$$K := \{(v, c) \in H \times \mathbb{R} : r(v) \leq c\},$$

which is closed (r is continuous), convex (r is convex) and nonempty (it contains the graph of r). Note also that $(0, 0) \notin K$ since $r(0) = 1$. We can then consider the best approximation of $(0, 0)$ on K (Proposition 15.9) and thus let $(0, 0) \neq (u, d) := P_K(0, 0)$ and note that

$$(-(u, d), (v, r(v)) - (u, d))_{H \times \mathbb{R}} \leq 0 \qquad \forall v \in H.$$

Therefore,

$$(u, v)_H + d\,r(v) \geq \|u\|^2 + d^2 > 0. \tag{15.48}$$

(We have tagged the inner product, to distinguish it from the pairing of elements in $H \times \mathbb{R}$.) Taking $v = 0$ in (15.48) and using that $r(0) = 1$, we have that $d > 0$ and we thus can define $w := -(1/d)u$. What we have proved is the existence of $w \in H$ such that

$$r(v) > (w, v)_H \qquad \forall v \in H,$$

and therefore

$$s(v) > (w, v)_H - s(0) + 1 \qquad \forall v \in H.$$

This proves that a continuous convex functional can be bounded below by an affine functional. Therefore

$$j(u) \geq \tfrac{1}{2}\|u\|^2 - (u, u_0)_H + (u, w)_H + \tfrac{1}{2}\|u_0\|^2 - s(0) + 1$$
$$\geq \tfrac{1}{2}\|u\|^2 - \|u\|\|w - u_0\| + \tfrac{1}{2}\|u_0\|^2 - s(0) + 1,$$

which proves (15.47). Since j is continuous, strictly convex (it is the sum of a strictly convex and a convex functional), and satisfies (15.47), Proposition 15.14 shows existence and uniqueness of a global minimum. □

Application to the model problem. Using the same tools as in Sections 15.4 and 15.5, we can rewrite the minimization problem (15.43) as the minimization of a functional j of the form (15.44), where $s(u) := \langle 1, |\gamma u|\rangle_\Gamma$ is continuous (the absolute value defines a continuous function in $L^2(\Gamma)$ and all other operations in the definition of s are bounded linear operators) and convex (this is an easy exercise). Proposition 15.16 then gives the equivalent variational inequality

$$a(u, v - u) - (f, v - u)_\Omega + s(v) - s(u) \geq 0 \qquad \forall v \in H^1(\Omega),$$

where

$$a(u, v) := (\kappa\nabla u, \nabla v)_\Omega + (cu, v)_\Omega.$$

We will work with this inequality in the following equivalent form

$$a(u, w) - (f, w)_\Omega + s(u + w) - s(u) \geq 0 \qquad \forall w \in H^1(\Omega). \qquad (15.49)$$

Given the special form of s (which vanishes on elements of $H_0^1(\Omega)$) we have, taking $w = \pm\varphi \in \mathcal{D}(\Omega)$,

$$a(u, \varphi) = (f, \varphi)_\Omega \qquad \forall \varphi \in \mathcal{D}(\Omega),$$

and therefore

$$-\nabla \cdot (\kappa\nabla u) + cu = f. \qquad (15.50)$$

Substituting this into (15.49) we have that the minimization problem and the variational inequality are equivalent to the elliptic equation (15.50), together with the inequality

$$\langle(\kappa\nabla u) \cdot \mathbf{n}, \gamma w\rangle_\Gamma + s(u + w) - s(u) \geq 0 \qquad \forall w \in H^1(\Omega), \qquad (15.51)$$

which will act as a boundary condition. To emphasize the boundary condition character of (15.51), we can express it in terms of a different functional $\tau : L^2(\Gamma) \to \mathbb{R}$

$$\tau(\xi) := \langle 1, |\xi| \rangle_\Gamma, \qquad \tau(\gamma u) = s(u),$$

so that we can equivalently write

$$\langle (\kappa \nabla u) \cdot \mathbf{n}, \xi \rangle_\Gamma + \tau(\gamma u + \xi) - \tau(\gamma u) \geq 0 \qquad \forall \xi \in H^{1/2}(\Gamma). \qquad (15.52)$$

Taking $\xi = -\gamma u$ in (15.52) and using that $\tau(0) = 0$, we have

$$-\langle (\kappa \nabla u) \cdot \mathbf{n}, \gamma u \rangle_\Gamma - \tau(\gamma u) \geq 0.$$

Taking now $\xi = \gamma u$ in (15.52) and using that $\tau(2\xi) = 2\tau(\xi)$, we obtain

$$\langle (\kappa \nabla u) \cdot \mathbf{n}, \gamma u \rangle_\Gamma + \tau(\gamma u) \geq 0.$$

and therefore

$$\langle (\kappa \nabla u) \cdot \mathbf{n}, \gamma u \rangle_\Gamma + \langle 1, |\gamma u| \rangle_\Gamma = 0.$$

We can then substitute this into (15.52) to obtain

$$\langle (\kappa \nabla u) \cdot \mathbf{n}, \xi + \gamma u \rangle_\Gamma + \langle 1, |\gamma u + \xi| \rangle_\Gamma \geq 0 \qquad \forall \xi \in H^{1/2}(\Gamma),$$

or equivalently

$$\langle (\kappa \nabla u) \cdot \mathbf{n}, \xi \rangle_\Gamma + \langle 1, |\xi| \rangle_\Gamma \geq 0 \qquad \forall \xi \in H^{1/2}(\Gamma). \qquad (15.53)$$

Some further conclusions. While we will not go all the way to the end and write (15.53) as a set of boundary conditions, we are going to extract an easy consequence from (15.53). To do this, we need to introduce inequalities in the dual space $H^{-1/2}(\Gamma)$. Given $h \in H^{-1/2}(\Gamma)$, we say that $h \geq 0$, when

$$\langle h, \xi \rangle_\Gamma \geq 0 \qquad \forall \xi \in H^{1/2}(\Gamma), \qquad \xi \geq 0.$$

The reader is invited to check that when $h \in L^2(\Gamma)$ this condition is equivalent to $h \geq 0$ (use elements of $\mathcal{D}(\Gamma)$ as tests). Testing (15.53) with $\pm \xi$, where $\xi \geq 0$, we can easily prove that

$$-1 \leq (\kappa \nabla u) \cdot \mathbf{n} \leq 1$$

as elements of $H^{-1/2}(\Gamma)$. This condition is typically shortened to

$$|(\kappa \nabla u) \cdot \mathbf{n}| \leq 1. \qquad (15.54)$$

We stop here, but not before warning that (15.53) contains more information than (15.54).

15.8   The Lions-Stampacchia theorem

Even if we will not explore further applications needing variational inequalities, we include here the statement and proof of the Lions-Stampacchia theorem. In short, this relates to the results on variational inequalities for symmetric problems (Proposition 15.11) as the Lax-Milgram lemma relates to the Riesz-Fréchet theorem. Its proof, converting the variational inequality to a fixed point problem, is reminiscent of the proof we have used for the existence of solutions to nonlinear strongly monotone Lipschitz operator equations (Theorem 12.2). It also provides an alternative proof to the Lax-Milgram lemma, which is the particular case $K = H$.

Proposition 15.18 (Lions-Stampacchia). *Let H be a Hilbert space, $\emptyset \neq K \subset H$ be closed and convex, $a : H \times H \to \mathbb{R}$ be bilinear bounded, and such that*

$$a(u - v, u - v) \geq \alpha \|u - v\|^2 \qquad \forall u, v \in K.$$

For every $\ell \in H'$, the problem

$$u \in K, \qquad a(u, v - u) \geq \ell(v - u) \qquad \forall v \in K \tag{15.55}$$

is uniquely solvable.

Proof. Defining the bounded linear operator $A : H \to H$ such that $(Au, v) = a(u, v)$ for all u, v in H and letting f be the Riesz-Fréchet representative of the functional ℓ, we can write (15.55) as

$$u \in K, \qquad (Au - f, v - u) \geq 0 \qquad \forall v \in K. \tag{15.56}$$

We now define

$$\Phi(u) := P_K(u - \alpha M^{-2}(Au - f)), \qquad M := \|A\|.$$

Since P_K is a nonstrict contraction (Corollary 15.1), we have

$$\begin{aligned}
\|\Phi(u) - \Phi(v)\|^2 &\leq \left\| (u - \alpha M^{-2}(Au - f)) - (v - \alpha M^{-2}(Av - f)) \right\|^2 \\
&= \|u - v\|^2 + \alpha^2 M^{-4}\|A(u - v)\|^2 - 2\alpha M^{-2}a(u - v, u - v) \\
&\leq (1 - \alpha^2 M^{-2})\|u - v\|^2,
\end{aligned}$$

and Φ is a contraction. Therefore, by the Banach fixed point theorem (Theorem 12.1), Φ has a unique fixed point, i.e., there exists a unique u such that $\Phi(u) = u \in K$. It follows that

$$(u - \alpha M^{-2}(Au - f) - u, v - u) \leq 0 \qquad \forall v \in K,$$

and we have a solution of (15.56) and (15.55). If u and w are solutions to (15.55), then

$$a(u, w - u) \geq \ell(w - u) = -\ell(u - w) \geq -a(w, u - w) = a(w, w - u).$$

Therefore $a(u - w, w - u) \geq 0$, $a(u - w, u - w) \leq 0$, and $u = w$ by the coercivity hypotheses. □

A particular case. In addition to the hypotheses of Proposition 15.18, let us further assume that K is a cone (this means that if $u \in K$, then $\lambda u \in K$ for all $\lambda \in (0, \infty)$). Therefore $0 \in K$ (take the limit as $\lambda \to 0$) and $2u \in K$, which can easily be used to show that the solution of (15.55) satisfies

$$a(u, u) = \ell(u).$$

Therefore, the problem (15.55) is equivalent to

$$u \in K, \qquad a(u, u) = \ell(u), \qquad a(u, v) \geq \ell(v) \qquad \forall v \in K.$$

15.9   Maximal dissipative operators

In the next two sections we explore a simple concept associated to the differential operators that we have seen in past chapters. The main change in our point of view will consist of thinking of the differential operators as unbounded operators, defined in dense subspaces of the space where the data live. We will recover the inverses of the Green's operators that we used in Chapter 9 to describe Sobolev spaces in terms of Fourier series. The main application of the concept of maximal dissipative operators is related to Theorem 15.1 below, proving existence of solutions for evolutionary equations that are associated to them. Even if this is a textbook on elliptic equations, we hope the reader will enjoy such an easy and direct application of the theory to a large class of nonelliptic equations. The proof of Theorem 15.1 (which opens a door to the shining world of semigroups of operators and the Hille-Yosida theorem) is not difficult and we will give a precise reference for an elegant proof.

Maximal dissipative operators. A linear operator defined on a subspace of a Hilbert space $A : D(A) \subset H \to H$ is said to be maximal dissipative when

(a) $(Au, u) \leq 0$ for all $u \in D(A)$,

(b) for all $f \in H$, there exists $u \in D(A)$ such that $u = Au + f$.

Property (a) is referred to as A being dissipative. The name comes from the fact (which we will not prove) that if an operator is maximal dissipative, it cannot be extended to another dissipative operator, i.e., if $B : D(B) \subset H \to H$ is dissipative and $D(A) \subset D(B)$ and $B|_{D(A)} = A$, then $D(A) = D(B)$. In part of the literature the signs are changed and maximal monotone operators are introduced: an operator A is maximal monotone when $-A$ is maximal dissipative. In a nutshell, Δ will be dissipative and $-\Delta$ will be monotone.

Proposition 15.19. *If A is maximal dissipative, then*

(a) $I - A : D(A) \to H$ *is invertible and its inverse* $(I - A)^{-1} : H \to H$ *is bounded*

$$\|(I - A)^{-1}v\| \leq \|v\| \qquad \forall v \in H.$$

(b) A *is closed, i.e., the graph of* A

$$G(A) := \{(u, Au) : u \in D(A)\} \subset H \times H$$

is closed.

(c) $D(A)$ *is dense in* H.

(d) $D(A)$ *is a Hilbert space when endowed with the norm*

$$\|u\|^2_{D(A)} := \|u\|^2 + \|Au\|^2.$$

Proof. If $Au = u$, then
$$\|u\|^2 = (Au, u) \leq 0,$$

which shows that for any dissipative operator A, the operator $I - A$ is injective. Therefore, every maximal dissipative operator defines an unbounded bijection $I - A : D(A) \to H$. If $v \in H$ and $u - Au = v$ (such u exists and is unique) then

$$\|u\|^2 = (Au, u) + (u, v) \leq (u, v) \leq \|u\| \|v\|,$$

which proves that $(I - A)^{-1} : H \to H$ is bounded and $\|(I - A)^{-1}\|_{H \to H} \leq 1$. This finishes the proof of (a). Now let $\{(u_n, Au_n)\}$ converge to $(u, v) \in H$. It follows that

$$u_n - Au_n \to u - v,$$

and therefore

$$u \leftarrow u_n \to (I - A)^{-1}(u - v) \in D(A).$$

This proves that $u \in D(A)$ and $u - Au = u - v$, which simplifies to $Au = v$. This finishes the proof that $G(A)$ is closed. To prove that $D(A)$ is dense in H we just need to show that $D(A)^\perp = \{0\}$. Let $v \in H$ satisfy $(v, u) = 0$ for all $u \in D(A)$ and take $w \in D(A)$ such that $w - Aw = v$, so

$$0 = (v, w) = (w, w) - (Aw, w) \geq \|w\|^2,$$

which proves that $w = 0$, hence $v = 0$. The final property is simple. $\square$

Functions of the 'time' variable. In what follows we will consider functions $u : [0, \infty) \to X$ that are differentiable (Fréchet differentiable). The derivative of u at $t > 0$ will be identified (via the Riesz-Fréchet theorem) with an element of X and denoted $\dot{u}(t)$. We will also handle right differentiability at the origin for functions such that there exists $\dot{u}(0) \in X$ satisfying

$$\tfrac{1}{h} \| u(h) - u(0) - \dot{u}(0)\, h \|_X \to 0 \quad \text{as } h \to 0^+.$$

Initial value problems. We will first consider **strong solutions** of

$$\dot{u}(t) = Au(t) \qquad t \geq 0, \qquad u(0) = u_0 \in D(A). \tag{15.57a}$$

For this to make sense at $t = 0$, we need $u_0 \in D(A)$, and we will require

$$u \in \mathcal{C}^1([0, \infty); H) \cap \mathcal{C}([0, \infty); D(A)). \tag{15.57b}$$

Note that if such a solution exists, then the function $t \mapsto \|\dot{u}(t)\|^2$ is differentiable $[0, \infty) \to \mathbb{R}$ (prove it) and

$$\frac{\mathrm{d}}{\mathrm{d}t}\left(\tfrac{1}{2}\|u(t)\|^2\right) = (\dot{u}(t), u(t)) = (Au(t), u(t)) \leq 0 \qquad \forall t \geq 0,$$

and thus

$$\|u(t)\| \leq \|u(0)\| = \|u_0\| \qquad \forall t \geq 0.$$

This proves uniqueness of solution. The difficulty in dealing with (15.57) is proving existence of solutions. In some cases, we will also be able to deal with problems of the form

$$\dot{u}(t) = Au(t) \qquad t > 0, \qquad u(0) = u_0 \in H, \tag{15.58a}$$

with

$$u \in \mathcal{C}^1((0, \infty); H) \cap \mathcal{C}((0, \infty); D(A)) \cap \mathcal{C}([0, \infty); H). \tag{15.58b}$$

In this case, we still have that $\|u(t)\|$ is nonincreasing. Note that this function is continuous in the closed interval, so we can take the limit and still prove that $\|u(t)\| \leq \|u_0\|$ which proves uniqueness and, once again, proving existence of solutions is the challenging part. Finally, we will also consider operators $A : D(A) \subset H \to H$ satisfying

$$(Au, u) = 0 \qquad \forall u \in D(A), \tag{15.59}$$

that is, both A and $-A$ are dissipative. If $I \pm A : D(A) \to H$ are surjective, then $\pm A$ are maximal dissipative. Note that in this case the solution to (15.57) satisfies $\|\dot{u}(t)\|^2 = 0$ and

$$\|u(t)\| = \|u_0\| \qquad \forall t.$$

In this case, we can also define problems for negative t, that is, we can consider the two-sided initial value problem

$$\dot{u}(t) = Au(t) \qquad t \in \mathbb{R}, \qquad u(0) = u_0 \in D(A), \qquad (15.60a)$$

with regularity

$$u \in \mathcal{C}^1(\mathbb{R}; H) \cap \mathcal{C}(\mathbb{R}; D(A)). \qquad (15.60b)$$

[Proof not provided]

Theorem 15.1. *If $A : D(A) \subset H \to H$ is maximal dissipative, problem (15.57) is uniquely solvable. If $\pm A$ are maximal dissipative, then problem (15.60) is uniquely solvable. Finally if A is maximal dissipative and symmetric*

$$(Au, v) = (u, Av) \qquad \forall u, v \in D(A),$$

then problem (15.58) is uniquely solvable.

15.10   The evolution of elliptic operators

We now present some examples of maximal dissipative operators and use Theorem 15.1 to derive some easy consequences involving evolutionary equations.

The Dirichlet Laplacian. We first take

$$H := L^2(\Omega), \qquad D(A) := \{u \in H_0^1(\Omega) : \Delta u \in L^2(\Omega)\}, \qquad Au := \Delta u.$$

The operator A is dissipative since

$$(Au, u)_\Omega = -(\nabla u, \nabla u)_\Omega \le 0 \qquad \forall u \in D(A).$$

Also, given $f \in L^2(\Omega)$, the solution to

$$u \in H_0^1(\Omega), \qquad -\Delta u + u = f$$

is in $D(A)$, which shows that A is maximal dissipative. It is simple to prove that A is symmetric. The associated heat equation

$$u : [0, \infty) \to H_0^1(\Omega), \qquad \dot{u}(t) = \Delta u(t) \qquad t \ge 0, \qquad u(0) = u_0 \in D(A)$$

has a unique solution in the sense of (15.57), but we can also have initial data in $L^2(\Omega)$

$$u(t) \in H_0^1(\Omega), \qquad \dot{u}(t) = \Delta u(t) \qquad t > 0, \qquad u(0) = u_0 \in L^2(\Omega),$$

understanding this instance of the heat equation as (15.58).

The Neumann Laplacian. We use the same space H, and define

$$D(A) := \{u \in H^1(\Omega) : \Delta u \in L^2(\Omega), \quad \partial_n u = 0\}, \qquad Au := \Delta u.$$

Note that even if the operator A seems to be the same (it is the Laplacian in the sense of distributions), A has a different domain of definition. The boundary condition $\partial_n u = 0$ is equivalent to

$$(\nabla u, \nabla v)_\Omega + (\Delta u, v)_\Omega = 0 \qquad \forall v \in H^1(\Omega), \qquad u \in D(A),$$

which proves that A is dissipative and symmetric. Maximal dissipativity follows from the solvability of

$$u \in H^1(\Omega), \quad -\Delta u + u = f, \qquad \partial_n u = 0.$$

In this case we can consider problems in the form of (15.57) and (15.58), namely heat equations with homogeneous Neumann boundary conditions and smooth or general initial values.

Convection-diffusion. If we take the same H, and $D(A)$ as in the Dirichlet Laplacian, but change the operator

$$Au := \Delta u - \mathbf{b} \cdot \nabla u,$$

where $\mathbf{b} \in \mathbf{L}^\infty(\Omega)$ satisfies $\nabla \cdot \mathbf{b} = 0$ (see Section 5.2). Since $(\mathbf{b} \cdot \nabla u, u)_\Omega = 0$ for all $u \in H_0^1(\Omega)$, it is easy to prove that A is dissipative. Maximal dissipativity follows from an easy argument about the coercivity of the problem

$$u \in H_0^1(\Omega), \qquad -\Delta u + \mathbf{b} \cdot \nabla u + u = f.$$

Note, however, that the operator A is not symmetric, and therefore, in the associated time-dependent convection-diffusion equation

$$\dot{u}(t) = \Delta u(t) - \mathbf{b} \cdot \nabla u(t),$$

we can take initial data in $D(A)$, but not in $L^2(\Omega)$.

Lemma 15.1. *If* $A : D(A) \subset H \to H$ *satisfies (15.59) (that is, $\pm A$ are dissipative), A is maximal dissipative and there exists an isomorphism $T : H \to H$ satisfying*

$$Tu \in D(A) \qquad \forall u \in D(A), \qquad ATu = -TAu \qquad \forall u \in D(A),$$

then $-A$ *is maximal dissipative.*

Proof. To solve $u + Au = f$, we just need to realize that $Tu - ATu = Tf$. $\square$

A wave equation. Let us now consider the spaces

$$H := L^2(\Omega) \times \mathbf{L}^2(\Omega), \qquad D(A) := H_0^1(\Omega) \times \mathbf{H}(\mathrm{div}, \Omega),$$

and the operator $A(u, \mathbf{p}) := (\nabla \cdot \mathbf{p}, \nabla u)$. With these definitions, it readily follows that

$$\begin{aligned}(A(u, \mathbf{p}), (u, \mathbf{p})) &= (\nabla \cdot \mathbf{p}, u)_\Omega + (\mathbf{p}, \nabla u)_\Omega \\ &= \langle \mathbf{p} \cdot \mathbf{n}, \gamma u \rangle_\Gamma = 0 \qquad \forall (u, \mathbf{p}) \in D(A),\end{aligned}$$

and therefore $\pm A$ are dissipative. Given $(f, \mathbf{f}) \in H$, we want to find a solution

$$(u, \mathbf{p}) \in H_0^1(\Omega) \times \mathbf{H}(\mathrm{div}, \Omega), \qquad u = \nabla \cdot \mathbf{p} + f, \qquad \mathbf{p} = \nabla u + \mathbf{f}.$$

This can be done by eliminating $\mathbf{p}$, and instead solving

$$u \in H_0^1(\Omega), \qquad u = \Delta u + f + \nabla \cdot \mathbf{f}, \qquad \mathbf{p} := \nabla u + \mathbf{f},$$

or, in variational form,

$$u \in H_0^1(\Omega), \qquad (\nabla u, \nabla v)_\Omega + (u, v)_\Omega = (f, v)_\Omega - (\mathbf{f}, \nabla v)_\Omega \qquad \forall v \in H_0^1(\Omega).$$

The reader is invited to prove that the unique solution of the latter, followed by the definition of $\mathbf{p} = \nabla u + \mathbf{f}$, provides the pair $(u, \mathbf{p}) \in D(A)$ such that $(I - A)(u, \mathbf{p}) = (f, \mathbf{f}) \in H$. To prove that $-A$ is also maximal dissipative, we can use the trick provided by Lemma 15.1. The sign flipping operator $T(u, \mathbf{p}) := (u, -\mathbf{p})$, clearly maps H to H and $D(A)$ to $D(A)$ isometrically and isomorphically, and it satisfies $AT(u, \mathbf{p}) = A(u, -\mathbf{p}) = (-\nabla \cdot \mathbf{p}, \nabla u) = -TA(u, \mathbf{p})$. This proves that $\pm A$ are maximal dissipative. The associated initial value problem is (15.60). With given initial data $u_0 \in H_0^1(\Omega)$ and $\mathbf{p}_0 \in \mathbf{H}(\mathrm{div}, \Omega)$, we have a solution

$$u \in \mathcal{C}^1([0, \infty); L^2(\Omega)) \cap \mathcal{C}([0, \infty); H_0^1(\Omega)),$$

$$\mathbf{p} \in \mathcal{C}^1([0, \infty); \mathbf{L}^2(\Omega)) \cap \mathcal{C}([0, \infty); \mathbf{H}(\mathrm{div}, \Omega)),$$

of

$$\dot{u}(t) = \nabla \cdot \mathbf{p}(t), \qquad \dot{\mathbf{p}}(t) = \nabla u(t), \qquad t \in \mathbb{R}.$$

Note that $\nabla \cdot \dot{\mathbf{p}} \in \mathcal{C}([0, \infty); H^{-1}(\Omega))$ and therefore $u \in \mathcal{C}^2([0, \infty); H^{-1}(\Omega))$ satisfies

$$\ddot{u}(t) = \Delta u(t) \qquad t \in \mathbb{R},$$

with equality as functions of the variable t with values in $H^{-1}(\Omega)$. The initial conditions can also be written in terms of u alone

$$u(0) = u_0, \qquad \dot{u}(0) = \nabla \cdot \mathbf{p}_0.$$

Another formulation for the wave equation. We now consider the space and inner product

$$H := H_0^1(\Omega) \times L^2(\Omega), \qquad [(u,v),(w,z)] := (\nabla u, \nabla w)_\Omega + (v,z)_\Omega,$$

and the operator

$$D(A) := \{u \in H_0^1(\Omega) : \Delta u \in L^2(\Omega)\} \times H_0^1(\Omega), \qquad A(u,v) := (v, \Delta u).$$

For further use, we will identify the domain of the Dirichlet Laplacian as $D_\Delta^{\mathrm{dir}} := \{u \in H_0^1(\Omega) : \Delta u \in L^2(\Omega)\}$. A simple computation shows that for all $(u,v) \in D(A)$, we have

$$[A(u,v),(u,v)] = [(v,\Delta u),(u,v)] = (\nabla v, \nabla u)_\Omega + (\Delta u, v)_\Omega = 0,$$

and therefore $\pm A$ are dissipative. To prove that A is maximal dissipative, we proceed as follows. Given $f \in H_0^1(\Omega)$, $g \in L^2(\Omega)$, we look for

$$u,v \in H_0^1(\Omega), \qquad u = v + f, \qquad v = \Delta u + g. \qquad (15.61)$$

In order to do that, we eliminate v and look for a solution of

$$u \in H_0^1(\Omega), \qquad -\Delta u + u = f + g,$$

and then define $v := u - f \in H_0^1(\Omega)$. This is enough to have a solution (the unique solution) of (15.61). The sign-flipping operator $T(u,v) = (u,-v)$ satisfies the hypotheses of Lemma 15.1 and can thus be used to prove that $-A$ is maximal dissipative as well. It is also clear that A is not symmetric. If we choose initial data $u_0 \in D_\Delta^{\mathrm{dir}}$ and $v_0 \in H_0^1(\Omega)$, we have a unique solution to

$$\dot{u}(t) = v(t), \qquad \dot{v}(t) = \Delta u(t) \qquad t \in \mathbb{R}, \qquad u(0) = u_0, \qquad v(0) = v_0,$$

with regularity

$$u \in \mathcal{C}^1(\mathbb{R}; H_0^1(\Omega)) \cap \mathcal{C}^0(\mathbb{R}; D_\Delta^{\mathrm{dir}}),$$
$$v \in \mathcal{C}^1(\mathbb{R}; L^2(\Omega)) \cap \mathcal{C}^0(\mathbb{R}; H_0^1(\Omega)),$$

but, since $\dot{u} = v$, we also have $u \in \mathcal{C}^2(\mathbb{R}; L^2(\Omega))$ and the second order equation

$$\ddot{u}(t) = \Delta u(t) \qquad t \in \mathbb{R}, \qquad u(0) = u_0, \qquad \dot{u}(0) = v_0. \qquad (15.62)$$

If we compare this with the previous description of the wave equation (in the variables u and $\mathbf{p}$), we now have a stronger solution, with the equation taking place in $L^2(\Omega)$ for all t instead of in $H^{-1}(\Omega)$, but we pay the price of having initial data in a more restrictive space.

Parabolic evolution of Fourier series. All of the maximal dissipative symmetric operators that we have seen also fit in the framework of the spectral analysis of Chapter 9, namely, it happens that A is the inverse of a self-adjoint compact operator $G : L^2(\Omega) \to L^2(\Omega)$ with range $G = D(A)$. We can then describe A using the spectral series associated to G. We thus start with a Hilbert basis $\{\phi_n\}$ of H and take a nondecreasing nonnegative divergent sequence $\{\lambda_n\}$. Consider the space

$$D(A) := \{u \in H : \sum_{n=1}^{\infty} \lambda_n^2 |(u, \phi_n)_H|^2 < \infty\},$$

and

$$Au = - \sum_{n=1}^{\infty} \lambda_n (u, \phi_n)_H \phi_n.$$

The operator A is symmetric and dissipative. Moreover, if $f \in H$, then

$$u := \sum_{n=1}^{\infty} \frac{1}{1 + \lambda_n} (f, \phi_n)_H \phi_n \in D(A), \qquad u - Au = f,$$

i.e., A is maximal dissipative. In this case, the solution to the evolutionary equations (15.57) and (15.58) can be computed by hand

$$u(t) = \sum_{n=1}^{\infty} e^{-\lambda_n t} (u_0, \phi_n)_H \phi_n.$$

(Note that this is the solution by separation of variables.) This gives two kinds of behavior depending on whether $u_0 \in D(A)$ or $u_0 \in H$. The reader is invited to check that the function u thus defined has the correct regularity. Note that if $\lambda_1 = \ldots = \lambda_K = 0$, then the space

$$R := \mathrm{span}\{\phi_1, \ldots, \phi_K\}$$

is the kernel of A and if $u_0 \in R$, then $u(t) \equiv u_0$ for all t. Also, note that

$$u(t) \longrightarrow P_R u_0 = \sum_{n=1}^{K} (u_0, \phi_n)_H \phi_n \qquad \text{as } t \to \infty,$$

that is, all solutions converge to the space of steady-states R. Keep in mind that R could be the zero space if $\lambda_n > 0$ for all n.

Hyperbolic evolution of Fourier series. We continue with the spaces H and $D(A)$ defined as above. We now introduce an intermediate space

$$V := \{u \in H : \sum_{n=1}^{\infty} \lambda_n |(u, \phi_n)_H|^2 < \infty\},$$

endowed with the norm

$$\|u\|_V^2 := \sum_{n=K+1}^{\infty} \lambda_n |(u, \phi_n)_H|^2 + \sum_{n=1}^{K} |(u, \phi_n)_H|^2.$$

We consider $\mathcal{H} := V \times H$ and $D(\mathcal{A}) := D(A) \times V$ and the operator

$$\mathcal{A}(u, v) := (v, Au).$$

The reader can easily prove that $\pm\mathcal{A}$ are maximal dissipative. The series expansion

$$u(t) = \sum_{n=1}^{K} (u_0, \phi_n)_H \phi_n + t \sum_{n=1}^{K} (u_0, \phi_n)_H \phi_n$$

$$+ \sum_{n=K+1}^{\infty} \cos(\lambda_n^{1/2} t)(u_0, \phi_n)_H \phi_n + \sum_{n=K+1}^{\infty} \lambda_n^{-1/2} \sin(\lambda_n^{1/2} t)(v_0, \phi_n)_H \phi_n$$

provides the solution of (15.62). Note that the finite sums in the first line of the formula (corresponding to $\lambda_n = 0$) provide terms whose second derivative with respect to time vanishes (rigid motions).

Final comments and literature

The concept of T-coercivity seems to have been coined by Anne Sophie Bonnet-Ben Dhia, Patrick Ciarlet and their then student Carlo Maria Zwölf [17], to refer to bilinear forms such that

$$|a(u, Tu)| \geq \alpha \|u\|^2,$$

and then the concept was extended to allow for compact perturbations in the more abstract treatment of the idea by Annalisa Buffa [29]. The concept had already appeared (with no associated name) in previous work on integral equations for electromagnetism [30] and even in abstract theory of Galerkin methods [102]. Properly speaking, a is T-coercive for a particular operator T, but the name T-coercivity (forgetting where the T is coming from) has caught on [18]. It is tempting the say that the concept is moot in some way, but the idea behind it is not the concept, but its particular use (with well-chosen operators T, typically associated to a stable decomposition of the space V) to discover the hidden coercivity of some bilinear forms. In fact, if $A : V \to V$ is the operator associated to the bilinear form a

$$(Au, v)_V = a(u, v) \qquad \forall u, v \in V,$$

then the invertibility of A implies that, with $T := (A^*)^{-1} = (A^{-1})^*$, we have

$$a(u, Tu) = (Au, (A^{-1})^* u)_V = (u, u)_V = \|u\|_V^2,$$

so every bilinear form for a well posed problem is automatically T-coercive. The example of sign changing coefficients is taken from the literature on diffraction [17], while the dual formulation of the Helmholtz equation is borrowed from Manuel Solano's master's thesis [96].

The functional language of optimal control of distributed systems (problems described with partial differential equations with functional control variables) is not that different from the language of functional optimization and includes the study of the dependence of PDE with respect to its coefficients. We have just explored some elementary examples of this, but it would be unfair not to mention that the very prolific Jacques-Louis Lions set the basis for much of what is now considered basic in the area [76].

The theoretical and practical literature on variational inequalities is quite vast, especially with respect to their applications in contact mechanics. Let us acknowledge here some of the earlier creators and expositors of this theory (Georges Duvaut, Lions once again, David Kinderlehrer, Guido Stampacchia, Tinsley Oden, and Noboru Kikuchi) and some of the seminal books on the topic [48, 68, 88, 67]. Without getting into all the details of the problems, the textbook [6] contains an approachable introduction to this topic.

A nice self-contained proof of Theorem 15.1 can be found in [23, Theorem 7.4] and [23, Theorem 7.7] (symmetric case). In the language of that text we refer to maximal monotone operators, where A is maximal monotone whenever $-A$ is maximal dissipative. The classical reference for evolutionary PDE is Amnon Pazy's celebrated monograph on semigroups [89]. A simple introduction to the topic can also be found in [22, Chapter 7].

Exercises

15.1. Show that Equation (15.1) admits a nontrivial homogeneous solution (k^2 is a Dirichlet eigenvalue of the negative Laplacian) if and only if (15.4) admits a nontrivial homogeneous solution.

15.2. More on the Helmholtz decomposition. Consider the operator $P :$ $\mathbf{H}(\mathrm{div}, \Omega) \to \mathbf{H}(\mathrm{div}, \Omega)$ given in Section 15.1, i.e., $P\mathbf{p} = \nabla u$, where

$$u \in H_0^1(\Omega), \qquad \Delta u = \nabla \cdot \mathbf{p}.$$

(a) Show that

$$(P\mathbf{p}, \mathbf{q})_\Omega = (\mathbf{p}, P\mathbf{q})_\Omega \qquad \forall \mathbf{p}, \mathbf{q} \in \mathbf{H}(\mathrm{div}, \Omega).$$

(b) Show that P is self-adjoint, and therefore P is the orthogonal projection onto its range.

(c) Show that $\mathrm{range}(I - P) = \{\mathbf{p} \in \mathbf{H}(\mathrm{div}, \Omega) : \nabla \cdot \mathbf{p} = 0\}$.

15.3. **Dual formulation with Neumann conditions.** In this exercise we explore the dual formulation for the problem

$$u \in H^1(\Omega), \qquad \Delta u + k^2 u = f, \qquad \partial_n u = h, \qquad (15.63)$$

for given $f \in L^2(\Omega)$ and $h \in H^{-1/2}(\Gamma)$. This will require deriving a new Helmholtz decomposition, since the key space involved now is $\mathbf{H}_0(\mathrm{div}, \Omega) = \{\mathbf{p} \in \mathbf{H}(\mathrm{div}, \Omega) : \mathbf{p} \cdot \mathbf{n} = 0\}$.

(a) Show that if $\mathbf{q} = \nabla u$, then an equivalent variational formulation of (15.63) is

$$\mathbf{q} \in \mathbf{H}(\mathrm{div}, \Omega), \qquad \mathbf{q} \cdot \mathbf{n} = h, \qquad (15.64a)$$
$$(\nabla \cdot \mathbf{q}, \nabla \cdot \mathbf{p})_\Omega - k^2(\mathbf{q}, \mathbf{p})_\Omega = (f, \nabla \cdot \mathbf{p})_\Omega \qquad \forall \mathbf{p} \in \mathbf{H}_0(\mathrm{div}, \Omega), \qquad (15.64b)$$

providing the formula for the reconstruction of u from $\mathbf{q}$.

(b) Consider the operator $P : \mathbf{H}_0(\mathrm{div}, \Omega) \to \mathbf{H}_0(\mathrm{div}, \Omega)$ given by $P\mathbf{p} = \nabla u$, where

$$u \in H^1(\Omega), \qquad \Delta u = \nabla \cdot \mathbf{p}, \qquad \partial_n u = 0.$$

Show that P is well-defined (note that we are dealing with the Neumann problem for the Laplacian), and that P is a bounded projection in the space $\mathbf{H}_0(\mathrm{div}, \Omega)$.

(c) If $a(\mathbf{q}, \mathbf{p}) := (\nabla \cdot \mathbf{q}, \nabla \cdot \mathbf{p})_\Omega - k^2(\mathbf{q}, \mathbf{p})_\Omega$, and $T := 2P - I$ show that

$$\mathbf{H}_0(\mathrm{div}, \Omega) \ni \mathbf{q}, \mathbf{p} \longmapsto a(\mathbf{q}, T\mathbf{p})$$

can be decomposed as the sum of a coercive bilinear form plus a bilinear form associated to a compact operator.

(d) Using a lifting of the normal component, transform (15.64) into a problem where Fredholm's alternative can be applied. Show well-posedness assuming the unique solvability of (15.63). Show that when k^2 is a Neumann eigenvalue for the negative Laplacian in Ω, the compatibility conditions for (15.63) and (15.64) are the same.

15.4. Consider the problem

$$u \in H^1(\Omega), \qquad \nabla \cdot (\kappa \nabla u) + k^2 c u = f, \qquad \gamma u = g,$$

where $\kappa, c \in L^\infty(\Omega)$ are strongly positive, and the data are $f \in L^2(\Omega)$, $k > 0$, and $g \in H^{1/2}(\Gamma)$.

(a) Show that it can be given an equivalent variational formulation in the dual variable $\mathbf{q} = \kappa \nabla u$, associated to the bilinear form

$$(c^{-1}(\nabla \cdot \mathbf{q}), \nabla \cdot \mathbf{p})_\Omega - k^2(\kappa^{-1}\mathbf{q}, \mathbf{p})_\Omega.$$

(b) Use the operator T of Section 15.1 to show that the Fredholm alternative applies to this formulation.

15.5. The Gâteaux derivative. Let $F : \mathcal{U} \subset U \to V$ be differentiable at u_0. Show that for all $h \in U$, the function $\phi_h(t) := F(u_0 + t\,h)$ is differentiable and that $\phi_h'(0) = DF(u_0)h$.

15.6. Quadratic functionals. Let H be a Hilbert space, $a : H \times H \to \mathbb{R}$ be symmetric bilinear bounded and $\ell \in H'$. Show that $F : H \to \mathbb{R}$ given by

$$F(u) := \tfrac{1}{2}a(u, u) - \ell(u)$$

is Fréchet differentiable everywhere and

$$DF(u) = a(u, \cdot) - \ell \in \mathcal{B}(H, \mathbb{R}) = H'.$$

15.7. Show that the inversion map $\mathrm{Inv} : \mathcal{B}_{\mathrm{inv}}(X, Y) \to \mathcal{B}(Y, X)$ of Section 15.3 is $\mathcal{C}^1$.

15.8. Study the well-posedness of the problem

$$\tfrac{1}{2}\|\nabla u\|_\Omega^2 - (f, u)_\Omega = \min! \qquad u \in H^1(\Omega), \qquad \gamma u = g, \qquad u \geq u_{\mathrm{obs}}.$$

15.9. Show that there is $c_0 > 0$ such that if $c \in L^\infty(\Omega)$, $c \geq -c_0$, then the problem

$$u \in H_0^1(\Omega), \qquad -\Delta u + c\,u = f \in H^{-1}(\Omega)$$

is uniquely solvable. Consider the set $\mathcal{U} := \{c \in L^\infty(\Omega) : c > c_0 \text{ a.e.}\}$ and the map $F : \mathcal{U} \subset L^\infty(\Omega) \to H_0^1(\Omega)$ given by $c \mapsto F(c) := u$ for fixed f. Prove that F is differentiable and compute $DF(0)$. (**Hint.** You will need to extend the chain rule to affine maps.)

15.10. A nonsymmetric control problem. Consider the solution map $L^2(\Omega) \ni f \mapsto u \in L^2(\Omega)$,

$$u \in H_0^1(\Omega) \qquad -\nabla \cdot (\kappa \nabla u) + \mathbf{b} \cdot \nabla u = f\chi_B,$$

where $\nabla \cdot \mathbf{b} = 0$ and $\mathbf{b} \in \mathbf{L}^\infty(\Omega)$. Compute S^*. Write down the Euler equations for the minimization problem

$$\tfrac{1}{2}\|Sf - u_d\|_\Omega^2 + \tfrac{\alpha}{2}\|f\|_B^2 = \min! \qquad f \in L^2(\Omega).$$

15.11. Control on a stronger norm. Let $S : L^2(B) \to H_0^1(\Omega)$ be given by

$$u \in H_0^1(\Omega), \qquad -\nabla \cdot (\kappa \nabla u) = f\chi_B,$$

and consider the control problem

$$\tfrac{1}{2}\|\nabla u - \nabla u_d\|_\Omega^2 + \tfrac{1}{2}\|f\|_B^2 = \min! \qquad f \in L^2(B), \qquad Sf = u,$$

for given $u_d \in H_0^1(\Omega)$. Show unique solvability and write down the Euler equations. (**Hint.** They are less trivial and involve some divergence operators.)

15.12. Let $\rho \in L^\infty(\Omega)$ be strongly positive and consider $H = L^2(\Omega)$ with inner product $(u, v)_H := (\rho u, v)_\Omega$. Let $\kappa, c \in L^\infty(\Omega)$, with $\kappa \geq \kappa_0 > 0$ and $c \geq 0$. Show that the operator

$$Au := \rho^{-1}(\nabla \cdot (\kappa \nabla u) - cu), \quad D(A) := \{u \in H_0^1(\Omega) : \nabla \cdot (\kappa \nabla u) \in L^2(\Omega)\},$$

is maximal dissipative and symmetric.

15.13. Consider $H := L^2(\Omega)$ and $Au := \Delta u$ defined on

$$D(A) := \{u \in H^1(\Omega) : \Delta u \in L^2(\Omega), \quad \partial_n u + k\gamma u = 0\},$$

where $k \in L^\infty(\Gamma)$, $k \geq 0$. Show that A is maximal dissipative and symmetric.

15.14. If $H := L^2(\Omega) \times \mathbf{L}^2(\Omega)$, $D(A) := H^1(\Omega) \times \mathbf{H}_0(\text{div}, \Omega)$, and $A(u, \mathbf{p}) =: (\nabla \cdot \mathbf{p}, \nabla u)$, show that $\pm A$ are maximal dissipative. (**Hint.** In this case you need to use the solution of

$$u \in H^1(\Omega), \quad (\nabla u, \nabla v)_\Omega + (u, v)_\Omega = (f, v)_\Omega - (\mathbf{f}, \nabla v)_\Omega \quad \forall v \in H^1(\Omega),$$

to define $\mathbf{p} = \nabla u + \mathbf{f}$ and then show that $\mathbf{p} \cdot \mathbf{n} = 0$.) Write the associated initial value problems in terms of u only.

15.15. Let $H := \mathbf{L}^2(\Omega) \times L^2(\Omega; \mathbb{R}_{\text{sym}}^{d \times d})$ with norm $(\rho \mathbf{u}, \mathbf{u})_\Omega + (\mathrm{C}^{-1}\boldsymbol{\sigma}, \boldsymbol{\sigma})_\Omega$, where C is defined as in Section 15.5. If $A(\mathbf{u}, \boldsymbol{\sigma}) := (\rho^{-1}\text{div}\,\boldsymbol{\sigma}, \mathrm{C}\boldsymbol{\varepsilon}(\mathbf{u}))$, with domain

$$D(A) := \mathbf{H}_0^1(\Omega) \times \{\boldsymbol{\sigma} \in L^2(\Omega; \mathbb{R}_{\text{sym}}^{d \times d}) : \text{div}\,\boldsymbol{\sigma} \in \mathbf{L}^2(\Omega)\},$$

show that A is maximal dissipative and write down the associated evolutionary equation.

16

Curl spaces and Maxwell's equations

16.1	Sobolev spaces for the curl	409
16.2	A first look at the tangential trace	412
16.3	Curl-curl equations	417
16.4	Time-harmonic Maxwell's equations	423
16.5	Two de Rham sequences	428
16.6	Maxwell eigenvalues	431
16.7	Normally oriented trace fields	432
16.8	Tangential trace spaces and their rotations	435
16.9	Tangential definition of the tangential traces	439
16.10	The curl-curl integration by parts formula	444
Final comments and literature		448
Exercises		449

In this chapter we will study several problems related to Maxwell's equations, phrased as equations where the curl operator is used. The natural space for these equations $\mathbf{H}(\mathrm{curl}, \Omega)$ is considerably more complicated than the classical Sobolev spaces, and this will be particularly visible in two aspects: (a) the natural trace in this space is defined in a weak form (like the normal component in $\mathbf{H}(\mathrm{div}, \Omega)$), but its range will be quite hard to identify; (b) the lack of compactness of the injection of $\mathbf{H}(\mathrm{curl}, \Omega)$ into $\mathbf{L}^2(\Omega)$ will cause trouble when dealing with time-harmonic equations and eigenvalues. The entire chapter will deal with vector fields in domains of three-dimensional space or defined on their boundaries. To simplify notation, we will write

$$\boldsymbol{\mathcal{D}}(\Omega) := \mathcal{D}(\Omega)^3, \qquad \boldsymbol{\mathcal{C}}^\infty(\mathbb{R}^3) := \mathcal{C}^\infty(\mathbb{R}^3)^3, \qquad \boldsymbol{\mathcal{S}}(\mathbb{R}^3) := \mathcal{S}(\mathbb{R}^3)^3,$$

and so on.

16.1 Sobolev spaces for the curl

On an open set $\Omega \subset \mathbb{R}^3$, we consider the space

$$\mathbf{H}(\mathrm{curl}, \Omega) := \{\mathbf{u} \in \mathbf{L}^2(\Omega) \, : \, \nabla \times \mathbf{u} \in \mathbf{L}^2(\Omega)\},$$

where as usual $\mathbf{L}^2(\Omega) = L^2(\Omega; \mathbb{R}^3) \equiv L^2(\Omega)^3$ and, where the curl has the classical definition

$$\nabla \times (u_1, u_2, u_3) := (\partial_{x_2} u_3 - \partial_{x_3} u_2, \partial_{x_3} u_1 - \partial_{x_1} u_3, \partial_{x_1} u_2 - \partial_{x_2} u_1),$$

now understood in the sense of distributions. We define the associated 'natural' inner product

$$(\mathbf{u}, \mathbf{v})_{\mathrm{curl}, \Omega} := (\nabla \times \mathbf{u}, \nabla \times \mathbf{v})_\Omega + (\mathbf{u}, \mathbf{v})_\Omega,$$

and the norm

$$\|\mathbf{u}\|_{\mathrm{curl}, \Omega}^2 = \|\nabla \times \mathbf{u}\|_\Omega^2 + \|\mathbf{u}\|_\Omega^2.$$

Endowed with this inner product and associated norm, it is simple to see that $\mathbf{H}(\mathrm{curl}, \Omega)$ is a Hilbert space. We also consider the closed subspace

$$\mathbf{H}_0(\mathrm{curl}, \Omega) := \{\mathbf{u} \in \mathbf{H}(\mathrm{curl}, \Omega) : \exists \{\boldsymbol{\varphi}_n\} \text{ in } \mathcal{D}(\Omega) \quad \|\boldsymbol{\varphi}_n - \mathbf{u}\|_{\mathrm{curl}, \Omega} \to 0\},$$

that is, $\mathbf{H}_0(\mathrm{curl}, \Omega)$ is the closure of $\mathcal{D}(\Omega)$ with respect to the norm of $\mathbf{H}(\mathrm{curl}, \Omega)$. In the following section we will recognize $\mathbf{H}_0(\mathrm{curl}, \Omega)$ as the kernel of a certain tangential trace operator when Ω is a Lipschitz domain. For the moment being, our arguments will be valid for a general open domain Ω, not necessarily bounded or Lipschitz.

A first boundary value problem. We consider a coefficient $c \in L^\infty(\Omega)$ such that $c \geq c_0 > 0$ almost everywhere (for constant c_0) and $\mathbf{f} \in \mathbf{L}^2(\Omega)$. We look for a solution to

$$\mathbf{u} \in \mathbf{H}_0(\mathrm{curl}, \Omega), \qquad \nabla \times \nabla \times \mathbf{u} + c\,\mathbf{u} = \mathbf{f}, \tag{16.1}$$

with the equation holding, as usual, in the sense of distributions. The differential operator on the left-hand side of the equation is simply called the curl-curl operator. It is a simple exercise using the definition of distributional derivatives to show that

$$\langle \nabla \times \nabla \times \mathbf{u}, \boldsymbol{\varphi} \rangle_{\mathcal{D}' \times \mathcal{D}} = \langle \nabla \times \mathbf{u}, \nabla \times \boldsymbol{\varphi} \rangle_{\mathcal{D}' \times \mathcal{D}} \qquad \forall \boldsymbol{\varphi} \in \mathcal{D}(\Omega),$$

and therefore (16.1) is equivalent to the variational formulation

$$\mathbf{u} \in \mathbf{H}_0(\mathrm{curl}, \Omega), \tag{16.2a}$$

$$(\nabla \times \mathbf{u}, \nabla \times \mathbf{v})_\Omega + (c\mathbf{u}, \mathbf{v})_\Omega = (\mathbf{f}, \mathbf{v})_\Omega \qquad \forall \mathbf{v} \in \mathbf{H}_0(\mathrm{curl}, \Omega), \tag{16.2b}$$

which is clearly equivalent to the minimization problem

$$\tfrac{1}{2}\|\nabla \times \mathbf{u}\|_\Omega^2 + \tfrac{1}{2}\|c^{1/2}\mathbf{u}\|_\Omega^2 - (\mathbf{f}, \mathbf{u})_\Omega = \min! \qquad \mathbf{u} \in \mathbf{H}_0(\mathrm{curl}, \Omega). \tag{16.3}$$

The bilinear form in (16.2) is clearly equivalent to the inner product (and thus bounded and coercive), which shows that (16.2) is uniquely solvable (and thus so are (16.1) and (16.3)) and we can bound the solution in terms of the data using the coercivity constant

$$\|\mathbf{u}\|_{\mathrm{curl},\Omega} \leq \max\{1, 1/c_0\}\|\mathbf{f}\|_\Omega.$$

Note that the solution of (16.1) satisfies $\nabla \times \mathbf{u} \in \mathbf{H}(\mathrm{curl},\Omega)$. The above argument can be easily extended to the more general problem

$$\mathbf{u} \in \mathbf{H}_0(\mathrm{curl},\Omega), \qquad \nabla \times \kappa(\nabla \times \mathbf{u}) + c\,\mathbf{u} = \mathbf{f},$$

where $\kappa : \Omega \to \mathbb{R}_{\mathrm{sym}}^{3\times3}$ has L^∞ components and satisfies the positivity condition

$$\boldsymbol{\xi} \cdot (\kappa\boldsymbol{\xi}) \geq \kappa_0 |\boldsymbol{\xi}|^2 \qquad \forall \boldsymbol{\xi} \in \mathbb{R}^3, \quad \text{a.e.}$$

A second variational problem. Now let us start with a minimization problem, where we relax the 'boundary condition' associated to the space:

$$\tfrac{1}{2}\|\nabla \times \mathbf{u}\|_\Omega^2 + \tfrac{1}{2}\|c^{1/2}\mathbf{u}\|_\Omega^2 - (\mathbf{f},\mathbf{u})_\Omega = \min! \qquad \mathbf{u} \in \mathbf{H}(\mathrm{curl},\Omega). \qquad (16.4)$$

This problem is equivalent to the coercive problem

$$\mathbf{u} \in \mathbf{H}(\mathrm{curl},\Omega), \hspace{6.5cm} (16.5\mathrm{a})$$
$$(\nabla \times \mathbf{u}, \nabla \times \mathbf{v})_\Omega + (c\mathbf{u},\mathbf{v})_\Omega = (\mathbf{f},\mathbf{v})_\Omega \qquad \forall \mathbf{v} \in \mathbf{H}(\mathrm{curl},\Omega), \hspace{0.5cm} (16.5\mathrm{b})$$

and its unique solution clearly satisfies

$$\nabla \times \nabla \times \mathbf{u} + c\,\mathbf{u} = \mathbf{f}.$$

Substituting back into (16.5b), we obtain the condition

$$(\nabla \times \mathbf{u}, \nabla \times \mathbf{v})_\Omega - (\nabla \times \nabla \times \mathbf{u}, \mathbf{v})_\Omega = 0 \qquad \forall \mathbf{v} \in \mathbf{H}(\mathrm{curl},\Omega),$$

which we will interpret as a homogeneous boundary condition in the next section. We finish this section by showing that $\mathbf{H}_0(\mathrm{curl},\mathbb{R}^3) = \mathbf{H}(\mathrm{curl},\mathbb{R}^3)$ and by revealing an important subspace of $\mathbf{H}_0(\mathrm{curl},\Omega)$.

Proposition 16.1. *The space $\mathcal{D}(\mathbb{R}^3)$ is dense in $\mathbf{H}(\mathrm{curl},\mathbb{R}^3)$. Therefore, $\mathbf{H}^1(\mathbb{R}^3)$ is dense in $\mathbf{H}(\mathrm{curl},\mathbb{R}^3)$.*

Proof. This result can be proved by cutoff and mollification using techniques that we learned in Chapter 2. Note that for a smooth function $\varphi \in \mathcal{C}^\infty(\mathbb{R}^3)$, we have

$$\nabla \times (\varphi\,\mathbf{u}) = \varphi\,\nabla \times \mathbf{u} + \nabla\varphi \times \mathbf{u},$$

which shows that multiplication by cutoff functions defines bounded operators in $\mathbf{H}(\mathrm{curl},\mathbb{R}^3)$. Instead of pursuing this proof (which we recommend to the reader as a review exercise), we will use the more advanced tools provided by tempered distributions to find a simpler argument. If we take $\psi_n := \chi_{B(0;n)}$, then $\psi_n \mathbf{v} \longrightarrow \mathbf{v}$ in $\mathbf{L}^2(\mathbb{R}^3)$ for every $\mathbf{v} \in \mathbf{L}^2(\mathbb{R}^3)$. Next, note that

$$|\omega^{s/2}\psi_n \mathbf{v}| \le (1 + 4\pi^2 n^2)^{s/2}|\mathbf{v}| \in \mathbf{L}^2(\mathbb{R}^3)$$

for every $s > 0$, and therefore

$$\mathbf{u}_n := \mathcal{F}^{-1}\{\psi_n \mathbf{v}\} \in \mathbf{H}^s(\mathbb{R}^3) \qquad \forall n, \quad \forall s \in \mathbb{R}.$$

We then take $\mathbf{u} \in \mathbf{H}(\mathrm{curl}, \mathbb{R}^3)$ and $\mathbf{v} := \mathcal{F}\{\mathbf{u}\} \in \mathbf{L}^2(\mathbb{R}^3)$ (recall that the Fourier transform is an isometry in $L^2(\mathbb{R}^3)$ and here we are just applying it componentwise). We have

$$\|\mathbf{u}_n - \mathbf{u}\|_{\mathbb{R}^3} = \|\psi_n \mathcal{F}\{\mathbf{u}\} - \mathcal{F}\{\mathbf{u}\}\|_{\mathbb{R}^3} \longrightarrow 0.$$

Furthermore (recall the monomials $m_j(\mathbf{x}) = 2\pi \imath x_j$)

$$\mathcal{F}\{\nabla \times \mathbf{u}\} = \boldsymbol{m} \times \mathcal{F}\{\mathbf{u}\} \in \mathbf{L}^2(\mathbb{R}^3) \qquad \boldsymbol{m} = (m_1, m_2, m_3),$$

and therefore

$$\|\nabla \times \mathbf{u}_n - \nabla \times \mathbf{u}\|_{\mathbb{R}^3} = \|(\psi_n - 1)\boldsymbol{m} \times \mathcal{F}\{\mathbf{u}\}\|_{\mathbb{R}^3} \longrightarrow 0.$$

We have thus proved that $\mathbf{u}_n \to \mathbf{u}$ in $\mathbf{H}(\mathrm{curl}, \mathbb{R}^3)$ for a sequence $\{\mathbf{u}_n\}$ in $\mathbf{H}^1(\mathbb{R}^3)$. This proves that $\mathbf{H}^1(\mathbb{R}^3)$ is dense in $\mathbf{H}(\mathrm{curl}, \mathbb{R}^3)$. (It also proves that $\mathbf{H}^s(\mathbb{R}^3)$ is dense in the same space for $s \ge 1$.) Since $\mathcal{D}(\mathbb{R}^3)$ is dense in $\mathbf{H}^1(\mathbb{R}^3)$ and convergence in $\mathbf{H}^1(\mathbb{R}^3)$ implies convergence in $\mathbf{H}(\mathrm{curl}, \mathbb{R}^3)$, the proof is finished. $\qquad \square$

Proposition 16.2. *The space* $\nabla H_0^1(\Omega) = \{\nabla u : u \in H_0^1(\Omega)\}$ *is a subset of* $\mathbf{H}_0(\mathrm{curl}, \Omega)$.

Proof. First, note that, for any distribution $u \in \mathcal{D}'(\Omega)$ we have $\nabla \times \nabla u = \mathbf{0}$. Now let $u \in H_0^1(\Omega)$ and consider a sequence $\{\varphi_n\}$ in $\mathcal{D}(\Omega)$ such that $\varphi_n \to u$ in $H^1(\Omega)$. Therefore $\mathcal{D}(\Omega) \ni \nabla \varphi_n \to \nabla u$ in $\mathbf{L}^2(\Omega)$, and $\mathbf{0} = \nabla \times \nabla \varphi_n \to \nabla \times \nabla u = \mathbf{0}$, which proves that $\nabla u \in \mathbf{H}_0(\mathrm{curl}, \Omega)$. $\qquad \square$

16.2   A first look at the tangential trace

Let us start with a technical lemma that will be used for the proof of our first characterization of $\mathbf{H}_0(\mathrm{curl}, \Omega)$.

Lemma 16.1. *If* Ω *is a bounded domain that is star-shaped with respect to* $B(\mathbf{x}_0; \varepsilon)$ *and*

$$\Omega_\theta := \{\mathbf{x}_0 + \theta(\mathbf{x} - \mathbf{x}_0) : \mathbf{x} \in \Omega\} \qquad \theta \in (0, 1),$$

then the distance between $\overline{\Omega_\theta}$ *and* $\partial\Omega$ *is positive.*

Proof. We just need to prove that $\overline{\Omega_\theta} \subset \Omega$ (see Figure 16.1), which would then prove that $\overline{\Omega_\theta} \cap \Gamma = \emptyset$. Since distances are translation-invariant, we can prove the result assuming that $\mathbf{x}_0 = \mathbf{0}$ and then $\Omega_\theta = \theta \Omega$. Note that, if $\mathbf{x} \in \Omega$, then $(1 - \theta)\mathbf{y} + \theta\mathbf{x} \in \Omega$ for all $\mathbf{y} \in B(\mathbf{0}; \varepsilon)$ and $\theta \in (0, 1)$ and therefore

$$\Omega_\theta \subset \bigcup_{\mathbf{x} \in \Omega_\theta} B(\mathbf{x}; (1 - \theta)\varepsilon) \subset \Omega.$$

If $\mathbf{x} \in \overline{\Omega_\theta}$, then $\theta^{-1}\mathbf{x} \in \overline{\Omega}$, and we can find $\mathbf{y} \in \Omega$ such that $|\mathbf{y} - \theta^{-1}\mathbf{x}| < (1 - \theta)\varepsilon$, which implies that

$$|\theta\mathbf{y} - \mathbf{x}| = \theta|\mathbf{y} - \theta^{-1}\mathbf{x}| < \theta(1 - \theta)\varepsilon < (1 - \theta)\varepsilon.$$

This shows that $\mathbf{x} \in B(\theta\mathbf{y}; (1 - \theta)\varepsilon) \subset \Omega$ since $\theta\mathbf{y} \in \Omega_\theta$, and the proof is thus finished. $\qquad\square$

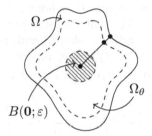

Figure 16.1: An illustration of the situation in Lemma 16.1, that a contraction of a star-shaped domain separates points from the boundary.

From here on out, we will use the extension-by-zero operator very often, frequently starting from different domains: if $\mathbf{u} : \Omega \to \mathbb{R}^3$ is given, then we will denote

$$\tilde{\mathbf{u}} := \begin{cases} \mathbf{u}, & \text{in } \Omega, \\ \mathbf{0}, & \text{in } \mathbb{R}^3 \setminus \Omega. \end{cases}$$

Proposition 16.3. *If Ω is a bounded Lipschitz domain, then*

$$\mathbf{u} \in \mathbf{H}_0(\mathrm{curl}, \Omega) \qquad \Longleftrightarrow \qquad \tilde{\mathbf{u}} \in \mathbf{H}(\mathrm{curl}, \mathbb{R}^3).$$

Proof. One implication is simple, given the fact that

$$\mathbf{H}_0(\mathrm{curl}, \Omega) \supset \mathcal{D}(\Omega) \ni \mathbf{u} \longmapsto \tilde{\mathbf{u}} \in \mathcal{D}(\mathbb{R}^3) \subset \mathbf{H}(\mathrm{curl}, \mathbb{R}^3)$$

is an isometry. This map can then be uniquely extended to the closure of $\mathcal{D}(\Omega)$, namely $\mathbf{H}_0(\mathrm{curl}, \Omega)$. The extended map has to be the extension-by-zero. (Prove this!)

To prove the reciprocal statement we start with a simpler class of domains. Let, for the moment being, Ω be a bounded domain which is *star-shaped* with respect to $B(\mathbf{x}_0; \varepsilon)$, take $\mathbf{u} \in \mathbf{H}(\mathrm{curl}, \Omega)$ such that $\widetilde{\mathbf{u}} \in \mathbf{H}(\mathrm{curl}, \mathbb{R}^3)$, and define

$$\mathbf{u}_n := \widetilde{\mathbf{u}}(\mathbf{x}_0 + \delta_n(\,\cdot\, - \mathbf{x}_0)) \qquad \delta_n < 1, \qquad \delta_n \to 1.$$

It is very simple to show that $\mathbf{u}_n \to \widetilde{\mathbf{u}}$ in $\mathbf{H}(\mathrm{curl}, \mathbb{R}^3)$, and that $\mathbf{u}_n \equiv \mathbf{0}$ in a neighborhood of $\partial\Omega$ (see Lemma 16.1). Using an approximation of the identity $\{\varphi_\varepsilon\}$ as in Proposition 2.3, we have that

$$\mathbf{u}_n * \varphi_\varepsilon \in \mathcal{D}(\mathbb{R}^3), \qquad \mathbf{u}_n * \varphi_\varepsilon \to \mathbf{u}_n \quad \text{in } \mathbf{H}(\mathrm{curl}, \Omega).$$

(The convolution of the scalar function and the vector field are defined, as one would expect, by convoluting each component of the vector field with the scalar field.) However, for each n and $\varepsilon = \varepsilon(n)$ small enough, we have that $\mathbf{u}_n * \varphi_\varepsilon \equiv \mathbf{0}$ in a neighborhood of $\partial\Omega$ and outside Ω. Therefore, $\mathbf{u}_n \in \mathbf{H}_0(\mathrm{curl}, \Omega)$ for all n and finally $\mathbf{u} \in \mathbf{H}_0(\mathrm{curl}, \Omega)$ since the latter space is closed.

We now go for the *general case* of a bounded Lipschitz domain Ω. We use the fact (Proposition 10.11) that we can cover $\overline{\Omega} \subset \cup_{j=1}^{J}\mathcal{O}_j$, where each $\Omega_j := \Omega \cap \mathcal{O}_j$ is (Lipschitz and) star-shaped with respect to a ball. We then build a smooth partition of unity, using functions $\varphi_j \in \mathcal{D}(\mathcal{O}_j)$ such that $\sum_{j=1}^{J} \varphi_j \equiv 1$ in $\overline{\Omega}$, define $\mathbf{u}_j := \varphi_j\mathbf{u}$, and note that

$$\widetilde{\mathbf{u}} = \left(\sum_{j=1}^{J} \varphi_j \right) \widetilde{\mathbf{u}} = \sum_{j=1}^{J} \widetilde{\varphi_j\mathbf{u}} = \sum_{j=1}^{J} \widetilde{\mathbf{u}}_j.$$

We can then consider $\mathbf{u}_j|_{\Omega_j} \in \mathbf{H}(\mathrm{curl}, \Omega_j)$ (we have just multiplied $\mathbf{u}$ by a smooth function) and note that its extension by zero to the rest of Ω is $\mathbf{u}_j$ and, therefore, its extension by zero from Ω_j to free space is $\widetilde{\mathbf{u}}_j = \varphi_j\widetilde{\mathbf{u}} \in \mathbf{H}(\mathrm{curl}, \mathbb{R}^3)$. Since Ω_j is star-shaped, it follows from the first part of the proof that $\mathbf{u}_j|_{\Omega_j} \in \mathbf{H}_0(\mathrm{curl}, \Omega_j)$. The same sequence of smooth compactly supported functions that approximates $\mathbf{u}_j|_{\Omega_j}$ in Ω_j approximates $\mathbf{u}_j$ in Ω, that is, $\mathbf{u}_j \in \mathbf{H}_0(\mathrm{curl}, \Omega)$. Since $\mathbf{u} = \sum_{j=1}^{J} \mathbf{u}_j$, the result follows. $\qquad\square$

Proposition 16.4. *Let Ω be a bounded Lipschitz domain and $\mathbf{u} \in \mathbf{H}(\mathrm{curl}, \Omega)$. The following statements are equivalent:*

(a) $\mathbf{u} \in \mathbf{H}_0(\mathrm{curl}, \Omega)$,

(b) $(\mathbf{u}, \nabla \times \mathbf{v})_\Omega - (\nabla \times \mathbf{u}, \mathbf{v})_\Omega = 0 \qquad \forall \mathbf{v} \in \mathbf{H}(\mathrm{curl}, \Omega)$.

(c) $(\mathbf{u}, \nabla \times \mathbf{v})_\Omega - (\nabla \times \mathbf{u}, \mathbf{v})_\Omega = 0 \qquad \forall \mathbf{v} \in \mathbf{H}^1(\Omega)$.

Proof. We can easily prove the identity

$$(\varphi, \nabla \times \mathbf{v})_\Omega - (\nabla \times \varphi, \mathbf{v})_\Omega = 0 \qquad \forall \varphi \in \mathcal{D}(\Omega), \qquad \mathbf{v} \in \mathbf{H}(\mathrm{curl}, \Omega),$$

and therefore

$$(\mathbf{u}, \nabla \times \mathbf{v})_\Omega - (\nabla \times \mathbf{u}, \mathbf{v})_\Omega = 0 \qquad \forall \mathbf{u} \in \mathbf{H}_0(\mathrm{curl}, \Omega), \qquad \mathbf{v} \in \mathbf{H}(\mathrm{curl}, \Omega).$$

This proves that (a) implies (b). Note that (b) implies (c). Now assume that (c) holds. Our goal is proving that $\tilde{\mathbf{u}} \in \mathbf{H}(\mathrm{curl}, \mathbb{R}^3)$. If $\mathbf{u} \in \mathbf{H}(\mathrm{curl}, \Omega)$ is such that condition (c) holds, and we consider $\mathbf{v} = \nabla \times \mathbf{u} \in \mathbf{L}^2(\Omega)$, we have the equality

$$(\mathbf{u}, \nabla \times \boldsymbol{\varphi})_\Omega - (\mathbf{v}, \boldsymbol{\varphi})_\Omega = 0 \qquad \forall \boldsymbol{\varphi} \in \mathcal{D}(\mathbb{R}^3),$$

since $\mathcal{C}^\infty(\overline{\Omega}) \subset \mathbf{H}^1(\Omega)$. Therefore

$$(\tilde{\mathbf{u}}, \nabla \times \boldsymbol{\varphi})_{\mathbb{R}^3} - (\tilde{\mathbf{v}}, \boldsymbol{\varphi})_{\mathbb{R}^3} = 0 \qquad \forall \boldsymbol{\varphi} \in \mathcal{D}(\mathbb{R}^3),$$

which is equivalent to

$$\nabla \times \tilde{\mathbf{u}} = \tilde{\mathbf{v}} \in \mathbf{L}^2(\mathbb{R}^3).$$

This proves that $\tilde{\mathbf{u}} \in \mathbf{H}(\mathrm{curl}, \mathbb{R}^3)$ and by Proposition 16.3, $\mathbf{u} \in \mathbf{H}_0(\mathrm{curl}, \Omega)$. $\square$

A tangential trace. Now consider $\mathbf{u} \in \mathbf{H}(\mathrm{curl}, \Omega)$. We define the map $\gamma_T \mathbf{u} \in \mathbf{H}^{-1/2}(\Gamma) = \mathbf{H}^{1/2}(\Gamma)'$ in a weak way (like we did for the normal component in $\mathbf{H}(\mathrm{div}, \Omega)$ in Chapter 6) with the formula

$$\langle \gamma_T \mathbf{u}, \gamma \mathbf{v} \rangle_\Gamma := (\mathbf{u}, \nabla \times \mathbf{v})_\Omega - (\nabla \times \mathbf{u}, \mathbf{v})_\Omega \qquad \forall \mathbf{v} \in \mathbf{H}^1(\Omega).$$

To see that $\gamma_T \mathbf{u}$ is actually a function of the trace of $\mathbf{v}$, note that

$$(\mathbf{u}, \nabla \times \mathbf{v})_\Omega - (\nabla \times \mathbf{u}, \mathbf{v})_\Omega = 0 \qquad \mathbf{v} \in \mathbf{H}_0^1(\Omega),$$

as we can easily prove by density by first using test functions in $\mathcal{D}(\Omega)$. It is also clear that

$$|\langle \gamma_T \mathbf{u}, \gamma \mathbf{v} \rangle_\Gamma| \le \|\mathbf{u}\|_{\mathrm{curl},\Omega} \|\mathbf{v}\|_{\mathrm{curl},\Omega} \le C \|\mathbf{u}\|_{\mathrm{curl},\Omega} \|\mathbf{v}\|_{1,\Omega},$$

and taking the infimum over all $\mathbf{v} \in \mathbf{H}^1(\Omega)$ sharing the same trace, it follows that

$$|\langle \gamma_T \mathbf{u}, \boldsymbol{\xi} \rangle_\Gamma| \le C \|\mathbf{u}\|_{\mathrm{curl},\Omega} \|\boldsymbol{\xi}\|_{1/2,\Gamma} \qquad \forall \boldsymbol{\xi} \in \mathbf{H}^{1/2}(\Gamma),$$

or equivalently

$$\|\gamma_T \mathbf{u}\|_{-1/2,\Gamma} \le C \|\mathbf{u}\|_{\mathrm{curl},\Omega}.$$

This proves that $\gamma_T : \mathbf{H}(\mathrm{curl}, \Omega) \to \mathbf{H}^{-1/2}(\Gamma)$ is a bounded linear operator. Using classical integration by parts (see Proposition 6.3) we have

$$(\mathbf{u}, \nabla \times \mathbf{v})_\Omega - (\nabla \times \mathbf{u}, \mathbf{v})_\Omega = \langle (\gamma \mathbf{u}) \times \mathbf{n}, \gamma \mathbf{v} \rangle_\Gamma \qquad \forall \mathbf{u}, \mathbf{v} \in \mathcal{C}^\infty(\overline{\Omega}).$$

By density this identity can be extended to $\mathbf{u}, \mathbf{v} \in \mathbf{H}^1(\Omega)$ and, therefore, if $\mathbf{u} \in \mathbf{H}^1(\Omega)$, then

$$\gamma_T \mathbf{u} = (\gamma \mathbf{u}) \times \mathbf{n} \in \mathbf{L}^2(\Gamma).$$

Actually, this shows that for $\mathbf{u} \in \mathbf{H}^1(\Omega)$, the tangential trace $\gamma_T \mathbf{u}$ is orthogonal to the normal vector field and is thus a tangential vector field on Γ. The kernel of γ_T is quite simple to recognize but this will not be the case for its range, which will require much more additional work, starting in Section 16.7 and culminating with its definition in Section 16.9.

Proposition 16.5. *On any bounded Lipschitz domain we have*

$$\mathbf{H}_0(\mathrm{curl}, \Omega) = \ker \gamma_T.$$

Proof. This is a simple rephrasing of the equivalence of (a) and (c) in Proposition 16.4. □

The two problems of Section 16.1. On a Lipschitz domain, the problem (16.1) (or equivalently (16.2) and (16.3)) can be written as a boundary value problem

$$\mathbf{u} \in \mathbf{H}(\mathrm{curl}, \Omega), \quad \nabla \times \nabla \times \mathbf{u} + c\mathbf{u} = \mathbf{f}, \quad \gamma_T \mathbf{u} = \mathbf{0},$$

while (16.4) and (16.5) are equivalent to

$$\mathbf{u} \in \mathbf{H}(\mathrm{curl}, \Omega), \quad \nabla \times \nabla \times \mathbf{u} + c\mathbf{u} = \mathbf{f}, \quad \gamma_T(\nabla \times \mathbf{u}) = \mathbf{0}.$$

Note that in contrast to the Dirichlet and Neumann problems for the Laplace equation (or similar ones), the boundary condition for these two problems is defined in an equally weak form. However, the first one can be understood in more general domains by using the more general form $\mathbf{u} \in \mathbf{H}_0(\mathrm{curl}, \Omega)$, which does not need any hypothesis on the domain.

We finish this section with a density result, whose proof uses arguments that are very similar to those developed in this section.

Proposition 16.6. *If Ω is a Lipschitz domain, then $\mathbf{H}^1(\Omega)$ is dense in $\mathbf{H}(\mathrm{curl}, \Omega)$.*

Proof. We will prove that $\mathcal{C}^\infty(\overline{\Omega})$ is dense in $\mathbf{H}(\mathrm{curl}, \Omega)$, which implies the result. We will prove this by showing that any $\mathbf{u} \in \mathbf{H}(\mathrm{curl}, \Omega)$ which is orthogonal to all elements of $\mathcal{C}^\infty(\overline{\Omega})$, i.e., such that

$$(\nabla \times \mathbf{u}, \nabla \times \boldsymbol{\varphi})_\Omega + (\mathbf{u}, \boldsymbol{\varphi})_\Omega = 0 \quad \forall \boldsymbol{\varphi} \in \mathcal{D}(\mathbb{R}^3), \tag{16.6}$$

must vanish. Condition (16.6) implies that

$$\nabla \times \nabla \times \mathbf{u} + \mathbf{u} = \mathbf{0}, \tag{16.7}$$

and therefore $\mathbf{v} := \nabla \times \mathbf{u} \in \mathbf{H}(\mathrm{curl}, \Omega)$. Moreover, (16.6) can be written using extensions-by-zero of both $\mathbf{u}$ and $\mathbf{v}$ in the equivalent form

$$(\widetilde{\mathbf{v}}, \nabla \times \boldsymbol{\varphi})_{\mathbb{R}^3} + (\widetilde{\mathbf{u}}, \boldsymbol{\varphi})_{\mathbb{R}^3} = 0 \quad \forall \boldsymbol{\varphi} \in \mathcal{D}(\mathbb{R}^3),$$

or equivalently

$$\nabla \times \widetilde{\mathbf{v}} = -\widetilde{\mathbf{u}} \in \mathbf{L}^2(\mathbb{R}^3).$$

Therefore, by Proposition 16.3, we have that $\mathbf{v} \in \mathbf{H}_0(\mathrm{curl}, \Omega)$. Taking the curl of (16.7) (recall that we have defined $\mathbf{v} = \nabla \times \mathbf{u}$), we have

$$\mathbf{v} \in \mathbf{H}_0(\mathrm{curl}, \Omega), \quad \nabla \times \nabla \times \mathbf{v} + \mathbf{v} = \mathbf{0},$$

and therefore $\mathbf{v} = \mathbf{0}$ by the well-posedness of (16.1). Going back to (16.7) we have that $\mathbf{u} = -\nabla \times \mathbf{v} = \mathbf{0}$, which proves that the space orthogonal to $\mathcal{C}^\infty(\overline{\Omega})$ in $\mathbf{H}(\text{curl}, \Omega)$ is the trivial subspace, which is equivalent to the desired density property. $\qquad\qquad\qquad\qquad\qquad\qquad\qquad\qquad\qquad\qquad\qquad\qquad\quad$ □

16.3 Curl-curl equations

Our goals. In this section we will explore the solvability of the following problems: given $\mathbf{f} \in \mathbf{L}^2(\Omega)$, we look for a solution to

$$\mathbf{u} \in \mathbf{H}_0(\text{curl}, \Omega), \qquad \nabla \times \nabla \times \mathbf{u} = \mathbf{f}, \qquad\qquad (16.8)$$

and to

$$\mathbf{u} \in \mathbf{H}(\text{curl}, \Omega), \qquad \nabla \times \nabla \times \mathbf{u} = \mathbf{f}, \qquad \gamma_T(\nabla \times \mathbf{u}) = \mathbf{0}, \qquad (16.9)$$

First, note that we need $\nabla \cdot \mathbf{f} = 0$ since

$$\nabla \cdot (\nabla \times \mathbf{u}) = 0 \qquad \forall \mathbf{u} \in \mathcal{D}(\Omega)'.$$

Also, note that if $\mathbf{u} \in \nabla H_0^1(\Omega) \subset \mathbf{H}_0(\text{curl}, \Omega)$ (recall Proposition 16.2), then $\mathbf{u}$ is a homogeneous solution to (16.8). Similarly, any $\mathbf{u} \in \nabla H^1(\Omega)$ defines a homogeneous solution to (16.9). This means that problems (16.8) and (16.9) need some additional side conditions to make them uniquely solvable. A simple way to get rid of gradients will be to enforce the value of $\nabla \cdot \mathbf{u}$. This will motivate the study of some key subspaces of the space

$$\mathbf{H}(\text{curl}, \Omega) \cap \mathbf{H}(\text{div}, \Omega) = \{\mathbf{u} \in \mathbf{H}(\text{curl}, \Omega) : \nabla \cdot \mathbf{u} \in L^2(\Omega)\}.$$

This space will be endowed with its natural norm

$$\left(\|\mathbf{u}\|_\Omega^2 + \|\nabla \times \mathbf{u}\|_\Omega^2 + \|\nabla \cdot \mathbf{u}\|_\Omega^2 \right)^{1/2},$$

and it will be an important goal of what is coming to see if we can eliminate the L^2 term in the norm with some kind of Poincaré inequality. Now recall that the Poincaré inequality in $H^1(\Omega)$ was proved based on a compactness embedding property (the Rellich-Kondrachov theorem), which will motivate a similar study about compact embeddings of $\mathbf{H}(\text{curl}, \Omega) \cap \mathbf{H}(\text{div}, \Omega)$ into $\mathbf{L}^2(\Omega)$. The first property below is very promising, but that will be compensated by the bad news that in general $\mathbf{H}^1(\Omega)$ is a proper subspace of $\mathbf{H}(\text{curl}, \Omega) \cap \mathbf{H}(\text{div}, \Omega)$, and it will not be the case that the latter is compactly embedded into $\mathbf{L}^2(\Omega)$.

Proposition 16.7. *If Ω is a Lipschitz domain, then*

$$\mathbf{H}_0(\mathrm{curl}, \Omega) \cap \mathbf{H}_0(\mathrm{div}, \Omega) = \mathbf{H}_0^1(\Omega).$$

Proof. If $\mathbf{u} \in \mathbf{H}_0^1(\Omega)$, there exists a sequence $\{\boldsymbol{\varphi}_n\}$ in $\mathcal{D}(\Omega)$ converging to $\mathbf{u}$ in $\mathbf{H}^1(\Omega)$ and, therefore, also in $\mathbf{H}(\mathrm{curl}, \Omega)$ and $\mathbf{H}(\mathrm{div}, \Omega)$. This proves that $\mathbf{u} \in \mathbf{H}_0(\mathrm{curl}, \Omega) \cap \mathbf{H}_0(\mathrm{div}, \Omega)$.

Now if $\mathbf{u} \in \mathbf{H}_0(\mathrm{curl}, \Omega) \cap \mathbf{H}_0(\mathrm{div}, \Omega)$, then $\widetilde{\mathbf{u}} \in \mathbf{H}(\mathrm{curl}, \mathbb{R}^3)$ (by Proposition 16.3) and

$$\begin{aligned}
\langle \nabla \cdot \widetilde{\mathbf{u}}, \varphi \rangle_{\mathcal{D}'(\mathbb{R}^3) \times \mathcal{D}(\mathbb{R}^3)} &= - \langle \widetilde{\mathbf{u}}, \nabla \varphi \rangle_{\mathcal{D}'(\mathbb{R}^3) \times \mathcal{D}(\mathbb{R}^3)} \\
&= - (\mathbf{u}, \nabla \varphi)_\Omega \\
&= (\nabla \cdot \mathbf{u}, \varphi)_\Omega - \langle \mathbf{u} \cdot \mathbf{n}, \gamma \varphi \rangle_\Gamma \qquad \forall \varphi \in \mathcal{D}(\mathbb{R}^3).
\end{aligned}$$

Since $\mathbf{H}_0(\mathrm{div}, \Omega)$ is the kernel of the normal component operator, this shows that

$$\langle \nabla \cdot \widetilde{\mathbf{u}}, \varphi \rangle_{\mathcal{D}'(\mathbb{R}^3) \times \mathcal{D}(\mathbb{R}^3)} = (\widetilde{\nabla \cdot \mathbf{u}}, \varphi)_{\mathbb{R}^3} \qquad \forall \varphi \in \mathcal{D}(\mathbb{R}^3),$$

that is, $\nabla \cdot \widetilde{\mathbf{u}} = \widetilde{\nabla \cdot \mathbf{u}}$. Therefore

$$\widetilde{\mathbf{u}} \in \mathbf{H}(\mathrm{curl}, \mathbb{R}^3) \cap \mathbf{H}(\mathrm{div}, \mathbb{R}^3).$$

If we can show

$$\mathbf{H}(\mathrm{curl}, \mathbb{R}^3) \cap \mathbf{H}(\mathrm{div}, \mathbb{R}^3) = \mathbf{H}^1(\mathbb{R}^3), \tag{16.10}$$

then it follows that $\widetilde{\mathbf{u}} \in \mathbf{H}^1(\mathbb{R}^3)$ and therefore $\mathbf{u} \in \mathbf{H}_0^1(\Omega)$.

The set equality (16.10) can be proved with the following argument. First of all, we have

$$\Delta \mathbf{w} = \nabla(\nabla \cdot \mathbf{w}) - \nabla \times \nabla \times \mathbf{w}$$

for every vector-valued distribution. Therefore, if $\mathbf{w} \in \mathbf{H}(\mathrm{curl}, \mathbb{R}^3) \cap \mathbf{H}(\mathrm{div}, \mathbb{R}^3)$, we have

$$-\Delta \mathbf{w} + \mathbf{w} = \mathbf{w} - \nabla(\nabla \cdot \mathbf{w}) + \nabla \times (\nabla \times \mathbf{w}) \in \mathbf{H}^{-1}(\mathbb{R}^3).$$

(Recall that first order derivatives map $L^2(\mathbb{R}^3)$ to $H^{-1}(\mathbb{R}^3)$.) Therefore $\mathbf{w} \in \boldsymbol{\mathcal{S}}'(\mathbb{R}^3)$ satisfies $-\Delta \mathbf{w} + \mathbf{w} \in \mathbf{H}^{-1}(\mathbb{R}^3)$, which implies that $\mathbf{w} \in \mathbf{H}^1(\mathbb{R}^3)$ (see the arguments after Proposition 13.20). $\qquad \square$

We are now ready to introduce two extremely important spaces for our theory

$$\begin{aligned}
\mathbf{X} &:= \mathbf{H}_0(\mathrm{curl}, \Omega) \cap \mathbf{H}(\mathrm{div}, \Omega), \\
\mathbf{Y} &:= \mathbf{H}(\mathrm{curl}, \Omega) \cap \mathbf{H}_0(\mathrm{div}, \Omega),
\end{aligned}$$

as well as a key theorem about them, which will be offered without proof. We will however extract many consequences from it.

> [**Proof not provided**]
>
> **Theorem 16.1** (Weber compactness theorem). *If Ω is a Lipschitz domain, the spaces $\mathbf{X}$ and $\mathbf{Y}$ are compactly embedded into $\mathbf{L}^2(\Omega)$.*

Harmonic fields. We will generically refer to the elements of the two following subspaces

$$\mathcal{H}_X := \{\mathbf{u} \in \mathbf{X} : \nabla \times \mathbf{u} = \mathbf{0}, \quad \nabla \cdot \mathbf{u} = 0\},$$
$$\mathcal{H}_Y := \{\mathbf{u} \in \mathbf{Y} : \nabla \times \mathbf{u} = \mathbf{0}, \quad \nabla \cdot \mathbf{u} = 0\},$$

as harmonic fields. Next, we will identify these spaces as the respective spaces of homogeneous divergence-free solutions to (16.8) and (16.9). We will also show that they are finite-dimensional. A very important issue we will not touch upon is what the dimensions of $\mathcal{H}_X$ and $\mathcal{H}_Y$ are. They happen to be topological invariants: the dimension of $\mathcal{H}_Y$ is the number of 'handles' of the domain (the first Betti number), and the dimension of $\mathcal{H}_X$ is the number of 'cavities' (second Betti number). This is just the tip of the iceberg of the deep and difficult theory of cohomology and de Rham complexes, which we will very briefly introduce in Section 16.5.

Proposition 16.8 (Harmonic fields). *The spaces $\mathcal{H}_X$ and $\mathcal{H}_Y$ are respectively the sets of solutions to*

$$\nabla \times \nabla \times \mathbf{u} = \mathbf{0}, \qquad \gamma_T \mathbf{u} = \mathbf{0}, \qquad \nabla \cdot \mathbf{u} = 0, \tag{16.11a}$$
$$\nabla \times \nabla \times \mathbf{u} = \mathbf{0}, \qquad \gamma_T(\nabla \times \mathbf{u}) = \mathbf{0}, \qquad \nabla \cdot \mathbf{u} = 0, \qquad \mathbf{u} \cdot \mathbf{n} = 0. \tag{16.11b}$$

Both spaces are finite-dimensional.

Proof. First of all, if $\mathbf{u}$ satisfies (16.11a), then

$$(\nabla \times \mathbf{u}, \nabla \times \mathbf{v})_\Omega = 0 \qquad \forall \mathbf{v} \in \mathbf{H}_0(\mathrm{curl}, \Omega),$$

and therefore $\nabla \times \mathbf{u} = \mathbf{0}$, so $\mathbf{u} \in \mathcal{H}_X$. If the set of solutions of (16.11) were infinite-dimensional, we would be able to construct an orthonormal sequence $\{\mathbf{u}_n\}$ in $\mathbf{X}$ satisfying $\nabla \times \mathbf{u}_n = \mathbf{0}$ and $\nabla \cdot \mathbf{u}_n = 0$. Therefore $\mathbf{u}_n \rightharpoonup \mathbf{0}$ in $\mathbf{X}$ (every orthonormal sequence converges weakly to zero) and by compactness (Theorem 16.1) $\mathbf{u}_n \to \mathbf{0}$ in $\mathbf{L}^2(\Omega)$. However, $\|\mathbf{u}_n\|_\Omega = \|\mathbf{u}_n\|_\mathbf{X} = 1$, which contradicts our hypothesis.

Similarly, if $\mathbf{u}$ satisfies (16.11b), then

$$(\nabla \times \mathbf{u}, \nabla \times \mathbf{v})_\Omega = 0 \qquad \forall \mathbf{v} \in \mathbf{H}(\mathrm{curl}, \Omega)$$

(recall the definition of tangential trace), and therefore $\mathbf{u} \in \mathcal{H}_Y$. The proof of the finite dimensionality of $\mathcal{H}_Y$ is very similar to the one given for $\mathcal{H}_X$ and is left as Exercise 16.7(b). □

Proposition 16.9. *On a bounded Lipschitz domain Ω, we have that $\mathcal{H}_X = \{\mathbf{0}\}$, that is, the following problem only admits the trivial solution*

$$\mathbf{u} \in \mathbf{X}, \qquad \nabla \times \nabla \times \mathbf{u} = \mathbf{0}, \qquad \nabla \cdot \mathbf{u} = 0, \qquad (16.12)$$

if and only if there exists $C > 0$ such that

$$\|\mathbf{u}\|_\Omega \le C(\|\nabla \times \mathbf{u}\|_\Omega + \|\nabla \cdot \mathbf{u}\|_\Omega) \qquad \forall \mathbf{u} \in \mathbf{X}. \qquad (16.13)$$

Proof. Due to Proposition 16.8, we only need to prove that $\mathcal{H}_X = \{\mathbf{0}\}$ implies the norm estimate. Assume thus that (16.13) does not hold. We can therefore find a sequence $\{\mathbf{u}_n\}$ in $\mathbf{X}$ such that

$$\nabla \times \mathbf{u}_n \to \mathbf{0} \quad \text{in } \mathbf{L}^2(\Omega), \qquad \nabla \cdot \mathbf{u}_n \to 0 \quad \text{in } L^2(\Omega), \qquad (16.14)$$

while $\|\mathbf{u}_n\|_\Omega = 1$ for all n. Since $\{\mathbf{u}_n\}$ is bounded in $\mathbf{X}$ (this follows from (16.14)), we can find a subsequence that is weakly convergent in $\mathbf{L}^2(\Omega)$. Denoting this subsequence with the same name we have $\mathbf{u}_n \rightharpoonup \mathbf{u}$ in $\mathbf{L}^2(\Omega)$ and therefore (first order differentiation is bounded from L^2 to H^{-1})

$$\nabla \times \mathbf{u}_n \to \mathbf{0} \quad \text{in } \mathbf{H}^{-1}(\Omega), \qquad \nabla \cdot \mathbf{u}_n \to 0 \quad \text{in } H^{-1}(\Omega).$$

Therefore $\mathbf{u}_n \rightharpoonup \mathbf{u}$ in $\mathbf{H}(\mathrm{curl}, \Omega) \cap \mathbf{H}(\mathrm{div}, \Omega)$ with $\nabla \times \mathbf{u} = \mathbf{0}$ and $\nabla \cdot \mathbf{u} = 0$. We also have that $\mathbf{0} = \gamma_T \mathbf{u}_n \rightharpoonup \gamma_T \mathbf{u}$ and therefore $\mathbf{u} \in \mathbf{X}$. We have thus proved that $\mathbf{u}_n \rightharpoonup \mathbf{u}$ in $\mathbf{X}$, where $\mathbf{u}$ is a solution to (16.12). By the Weber compactness theorem we have that $\mathbf{u}_n \to \mathbf{u}$ in $\mathbf{L}^2(\Omega)$ and therefore $\|\mathbf{u}\|_\Omega = \lim_n \|\mathbf{u}_n\|_\Omega = 1$, which implies that we have found a nonzero element of $\mathcal{H}_X$. $\qquad \square$

A regularized formulation. Given $\mathbf{f} \in \mathbf{L}^2(\Omega)$ with $\nabla \cdot \mathbf{f} = 0$, we consider the problem

$$\mathbf{u} \in \mathbf{X}, \qquad \nabla \times \nabla \times \mathbf{u} = \mathbf{f}, \qquad \nabla \cdot \mathbf{u} = 0. \qquad (16.15)$$

If $\mathbf{u}$ solves the above, then

$$(\nabla \times \mathbf{u}, \nabla \times \mathbf{v})_\Omega = (\mathbf{f}, \mathbf{v})_\Omega \qquad \forall \mathbf{v} \in \mathbf{X} \subset \mathbf{H}_0(\mathrm{curl}, \Omega),$$

and therefore we have a solution to

$$\mathbf{u} \in \mathbf{X}, \qquad (16.16a)$$

$$(\nabla \times \mathbf{u}, \nabla \times \mathbf{v})_\Omega + (\nabla \cdot \mathbf{u}, \nabla \cdot \mathbf{v})_\Omega = (\mathbf{f}, \mathbf{v})_\Omega \qquad \forall \mathbf{v} \in \mathbf{X}. \qquad (16.16b)$$

Reciprocally, if $\mathbf{u}$ solves (16.16) and we take

$$\mathbf{v} = \nabla w, \quad \text{where} \quad \begin{cases} w \in H_0^1(\Omega), \\ \Delta w = \nabla \cdot \mathbf{u}, \end{cases}$$

then $\nabla \times \mathbf{v} = \mathbf{0}$ and $\mathbf{v} \in \mathbf{X}$ (recall Proposition 16.2 which showed that $\nabla H_0^1(\Omega)$ was a subset of $\mathbf{H}_0(\mathrm{curl}, \Omega)$). We also have

$$\|\nabla \cdot \mathbf{u}\|_\Omega^2 = (\mathbf{f}, \nabla w)_\Omega = 0$$

given the fact that $\nabla \cdot \mathbf{f} = 0$ and $w \in H_0^1(\Omega)$. Therefore $\nabla \cdot \mathbf{u} = 0$ and (16.16) implies that

$$(\nabla \times \mathbf{u}, \nabla \times \varphi)_\Omega = (\mathbf{f}, \varphi)_\Omega \qquad \forall \varphi \in \mathcal{D}(\Omega),$$

or equivalently $\nabla \times \nabla \times \mathbf{u} = \mathbf{f}$. We have thus proved that problems (16.15) and (16.16) are equivalent. Proposition 16.9 proves that having a unique solution to (16.15) is equivalent to having coercivity of the bilinear form in (16.16). Therefore, uniqueness of solution to (16.16) implies well-posedness of this problem. This can also be seen by noticing that the Weber compactness theorem implies that the operator $K : \mathbf{X} \to \mathbf{X}$ defined by

$$(K\mathbf{u}, \mathbf{v})_\mathbf{X} = (\mathbf{u}, \mathbf{v})_\Omega \qquad \forall \mathbf{u}, \mathbf{v} \in \mathbf{X}$$

is compact. This allows us to rewrite (16.16) as an operator equation associated to an operator of the form invertible plus compact. This also gives an alternative proof of the finite dimensionality of the set of harmonic fields $\mathcal{H}_X$ (Proposition 16.8).

A mixed formulation. Now let us briefly go back to the minimization problem associated to (16.8), where we have added a restriction to enforce uniqueness:

$$\tfrac{1}{2}\|\nabla \times \mathbf{u}\|_\Omega^2 - (\mathbf{f}, \mathbf{u})_\Omega = \min! \qquad \mathbf{u} \in \mathbf{H}_0(\mathrm{curl}, \Omega), \qquad \nabla \cdot \mathbf{u} = 0.$$

Following Section 10.4 (which related quadratic minimization problems with linear restrictions to saddle point problems and mixed formulations), we are going to insert a Lagrange multiplier and then use Proposition 10.6 to obtain a mixed formulation. To prepare for that, we need to write the side restriction through a bilinear form which is bounded in the space $V = \mathbf{H}_0(\mathrm{curl}, \Omega)$ and a corresponding space for a Lagrange multiplier. Since $\nabla \cdot \mathbf{u} = 0$ if and only if

$$(\mathbf{u}, \nabla q)_\Omega = 0 \qquad \forall q \in H_0^1(\Omega),$$

this leads to the introduction of a Lagrange multiplier $p \in M = H_0^1(\Omega)$ and to the mixed problem

$$\begin{aligned}
&(\mathbf{u}, p) \in \mathbf{H}_0(\mathrm{curl}, \Omega) \times H_0^1(\Omega), &&&&\text{(16.17a)}\\
&(\nabla \times \mathbf{u}, \nabla \times \mathbf{v})_\Omega + (\mathbf{v}, \nabla p)_\Omega = (\mathbf{f}, \mathbf{v})_\Omega &&\forall \mathbf{v} \in \mathbf{H}_0(\mathrm{curl}, \Omega), &&\text{(16.17b)}\\
&(\mathbf{u}, \nabla q)_\Omega \qquad\qquad\qquad\quad = 0 &&\forall q \in H_0^1(\Omega). &&\text{(16.17c)}
\end{aligned}$$

Due to the fact that $\mathcal{D}(\Omega) \times \mathcal{D}(\Omega)$ is dense in $V \times M$, and making the boundary conditions explicit, it follows that equations (16.17) are equivalent to

$$\begin{aligned}
\nabla \times \nabla \times \mathbf{u} + \nabla p = \mathbf{f}, \qquad & \nabla \cdot \mathbf{u} = 0, && \text{(16.18a)}\\
\gamma_T \mathbf{u} = \mathbf{0}, \qquad & \gamma p = 0. && \text{(16.18b)}
\end{aligned}$$

If we take the divergence in the first equation of (16.18a) (recall that $\nabla \cdot \mathbf{f} = 0$ is a necessary condition for existence of solutions to the original problem (16.8)), it follows that

$$p \in H_0^1(\Omega), \qquad \Delta p = 0,$$

and therefore $p = 0$. (The problem (16.18) makes sense even if $\nabla \cdot \mathbf{f} \neq 0$, but then it is not a reformulation of (16.8).) We have already chosen the spaces V and M in the abstract mixed framework, and just need to verify the hypotheses on the bilinear forms

$$a(\mathbf{u}, \mathbf{v}) := (\nabla \times \mathbf{u}, \nabla \times \mathbf{v})_\Omega, \qquad b(\mathbf{u}, q) := (\mathbf{u}, \nabla p)_\Omega.$$

Since $\nabla H_0^1(\Omega) \subset \mathbf{H}_0(\mathrm{curl}, \Omega)$, the inf-sup condition is straightforward: using that $\nabla \times \nabla q = \mathbf{0}$ and the Cauchy-Schwarz inequality, we have

$$\sup_{0 \neq \mathbf{u} \in \mathbf{H}_0(\mathrm{curl}, \Omega)} \frac{(\mathbf{u}, \nabla q)_\Omega}{\|\mathbf{u}\|_{\mathrm{curl}, \Omega}} = \sup_{0 \neq \mathbf{u} \in \mathbf{H}_0(\mathrm{curl}, \Omega)} \frac{(\mathbf{u}, \nabla q)_{\mathrm{curl}, \Omega}}{\|\mathbf{u}\|_{\mathrm{curl}, \Omega}}$$

$$= \|\nabla q\|_{\mathrm{curl}, \Omega} = \|\nabla q\|_\Omega \qquad \forall q \in H_0^1(\Omega).$$

We can easily identify the kernel associated to the bilinear form b:

$$\{\mathbf{u} \in V : b(\mathbf{u}, q) = 0 \quad \forall q \in M\} = \{\mathbf{u} \in \mathbf{H}_0(\mathrm{curl}, \Omega) : \nabla \cdot \mathbf{u} = 0\} =: \mathbf{X}_0 \subset \mathbf{X}.$$

Finally, all that remains is the difficult work of showing the invertibility of the operator associated to the bilinear form a when restricted to $\mathbf{X}_0$. Note that Proposition 16.9 can be rephrased to say that uniqueness of solution of

$$\mathbf{u} \in \mathbf{X}_0, \qquad \nabla \times \nabla \times \mathbf{u} = \mathbf{0}$$

is equivalent to $\|\nabla \times \mathbf{u}\|_\Omega$ being an equivalent norm in $\mathbf{X}_0$. In other words, the bilinear form a is coercive in $\mathbf{X}_0$ if and only if the original problem (16.15) is well posed. This completes the checklist of mixed formulations (the Babuska-Brezzi conditions of Theorem 10.1) and shows that (16.17) is a well posed problem if and only if (16.15) only admits trivial homogeneous solutions. The case where there are harmonic fields is examined in Exercise 16.5.

The second curl-curl problem. We give just a brief overview of what to do with our second boundary value problem for the curl-curl operator (16.9). The reader is asked to provide details in Exercise 16.7. We now have that all elements of $\nabla H^1(\Omega)$ are homogeneous solutions to (16.9), but imposing zero divergence of an element of $\nabla H^1(\Omega)$ is not enough to ensure that it vanishes. Instead we use two side conditions and write the formulation

$$\mathbf{u} \in \mathbf{H}(\mathrm{curl}, \Omega) \cap \mathbf{H}(\mathrm{div}, \Omega),$$

$$\nabla \times \nabla \times \mathbf{u} = \mathbf{f}, \qquad\qquad \nabla \cdot \mathbf{u} = 0,$$

$$\gamma_T(\nabla \times \mathbf{u}) = \mathbf{0}, \qquad\qquad \mathbf{u} \cdot \mathbf{n} = 0.$$

This can be rephrased using the space $\mathbf{Y} = \mathbf{H}(\mathrm{curl}, \Omega) \cap \mathbf{H}_0(\mathrm{div}, \Omega)$ as

$$\mathbf{u} \in \mathbf{Y}, \qquad \nabla \times \nabla \times \mathbf{u} = \mathbf{f}, \qquad \gamma_T(\nabla \times \mathbf{u}) = \mathbf{0}, \qquad \nabla \cdot \mathbf{u} = 0. \qquad (16.19)$$

The set of homogeneous solutions of (16.19) is finite-dimensional and constitutes the second set of harmonic fields $\mathcal{H}_Y$, which is the zero subspace if and only if

$$\left(\|\nabla \times \mathbf{u}\|_\Omega^2 + \|\nabla \cdot \mathbf{u}\|_\Omega\right)^{1/2}$$

defines an equivalent norm in $\mathbf{Y}$. This justifies uniqueness of solution to the regularized formulation

$$\mathbf{u} \in \mathbf{Y}, \qquad (\nabla \times \mathbf{u}, \nabla \times \mathbf{v})_\Omega + (\nabla \cdot \mathbf{u}, \nabla \cdot \mathbf{v})_\Omega = (\mathbf{f}, \mathbf{v})_\Omega \qquad \forall \mathbf{v} \in \mathbf{Y},$$

which is equivalent to (16.19). Now consider a mixed formulation of the form

$$\begin{aligned}
&(\mathbf{u}, p) \in \mathbf{H}(\text{curl}, \Omega) \times H_\star^1(\Omega), \\
&(\nabla \times \mathbf{u}, \nabla \times \mathbf{v})_\Omega + (\mathbf{v}, \nabla p)_\Omega = (\mathbf{f}, \mathbf{v})_\Omega \qquad \forall \mathbf{v} \in \mathbf{H}(\text{curl}, \Omega), \\
&(\mathbf{u}, \nabla q)_\Omega \qquad\qquad\qquad\quad = 0 \qquad\qquad \forall q \in H_\star^1(\Omega),
\end{aligned}$$

using the space $H_\star^1(\Omega) := \{p \in H^1(\Omega) : (p, 1)_\Omega = 0\}$ (this is done to eliminate constants, which constitute the kernel of the gradient restricted to $H^1(\Omega)$). This mixed problem is the variational formulation of the problem

$$\begin{aligned}
\nabla \times \nabla \times \mathbf{u} + \nabla p &= \mathbf{f}, & \nabla \cdot \mathbf{u} &= 0, \\
\gamma_T(\nabla \times \mathbf{u}) &= \mathbf{0}, & \mathbf{u} \cdot \mathbf{n} &= 0, \\
(p, 1)_\Omega &= 0, & \partial_n p &= 0,
\end{aligned}$$

which combines the solution of (16.19) with a vanishing Lagrange multiplier $p = 0$ when $\nabla \cdot \mathbf{f} = 0$.

16.4 Time-harmonic Maxwell's equations

We now move on to study the simplest boundary value problem for the Maxwell equations in a time-harmonic regime with wave number $\omega \neq 0$:

$$\nabla \times \nabla \times \mathbf{u} - \omega^2 \mathbf{u} = \mathbf{f}, \qquad \gamma_T \mathbf{u} = \mathbf{0}. \tag{16.20}$$

In principle we will only assume that $\mathbf{f} \in \mathbf{L}^2(\Omega)$ (the regularized formulation below will require that $\mathbf{f} \in \mathbf{H}(\text{div}, \Omega)$) and note that

$$\nabla \cdot \mathbf{u} = -\omega^{-2} \nabla \cdot \mathbf{f} \in H^{-1}(\Omega).$$

First, we will give formulations based on what we used in Section 16.3 for the curl-curl operator and then we will pick up ideas from T-coercivity (See Section 15.1) to give an analysis of the simplest variational formulation. Uniqueness will be an issue in this section too. Since we will look at eigenvalues in the next section, we will just assume here that ω^2 is not a Dirichlet eigenvalue of the curl-curl operator in Ω, that is, we will assume that

$$\mathbf{u} \in \mathbf{H}_0(\text{curl}, \Omega), \qquad \nabla \times \nabla \times \mathbf{u} = \omega^2 \mathbf{u} \qquad \Longrightarrow \qquad \mathbf{u} = \mathbf{0}. \qquad (16.21)$$

Mixed formulation. Considering the spaces $V := \mathbf{H}_0(\text{curl}, \Omega)$ and $M := H_0^1(\Omega)$, and the bilinear forms

$$a_\omega(\mathbf{u}, \mathbf{v}) := (\nabla \times \mathbf{u}, \nabla \times \mathbf{v})_\Omega - \omega^2(\mathbf{u}, \mathbf{v})_\Omega, \qquad b(\mathbf{u}, q) := (\mathbf{u}, \nabla q)_\Omega,$$

we can give the following equivalent variational formulation of (16.20):

$$\begin{align}
&(\mathbf{u}, p) \in V \times M, &&\text{(16.22a)}\\
&a_\omega(\mathbf{u}, \mathbf{v}) + b(\mathbf{v}, p) = (\mathbf{f}, \mathbf{v})_\Omega &&\forall \mathbf{v} \in V, &&\text{(16.22b)}\\
&b(\mathbf{u}, q) \qquad\quad = -\omega^{-2}(\mathbf{f}, \nabla q)_\Omega &&\forall q \in M. &&\text{(16.22c)}
\end{align}$$

This formulation is easily shown to be equivalent to the problem

$$\begin{align}
&(\mathbf{u}, p) \in V \times M,\\
&\nabla \times \nabla \times \mathbf{u} - \omega^2 \mathbf{u} + \nabla p = \mathbf{f},\\
&\nabla \cdot \mathbf{u} = -\omega^{-2} \nabla \cdot \mathbf{f},
\end{align}$$

and therefore implies that $p = 0$ (take the divergence in the first equation, substitute the information given by the second one, and note that $p \in H_0^1(\Omega)$), which is why (16.22) is an equivalent variational formulation for (16.20). The inf-sup condition for (16.22) is the same as the one that we proved in Section 16.3 for the curl-curl operator and the solvability analysis is related to the invertibility of the operator $A_\omega : \mathbf{X}_0 \to \mathbf{X}_0$ (here, as then, $\mathbf{X}_0 = \{\mathbf{u} \in \mathbf{X} : \nabla \cdot \mathbf{u} = 0\}$) given by the expression

$$(A_\omega \mathbf{u}, \mathbf{v})_{\mathbf{X}} = a_\omega(\mathbf{u}, \mathbf{v}) = (\mathbf{u}, \mathbf{v})_{\mathbf{X}} - (\omega^2 + 1)(\mathbf{u}, \mathbf{v})_\Omega \qquad \forall \mathbf{u}, \mathbf{v} \in \mathbf{X}_0.$$

We can apply the Fredholm alternative to A_ω (we use the Weber compactness theorem to show Fredholmness) and all that remains is the realization that the condition (16.21) is equivalent to the injectivity of A_ω. Let us perform a small computation to prepare the argument: if $\mathbf{v} \in \mathbf{X}$, we can define $\mathbf{w} := \mathbf{v} - \nabla w \in \mathbf{X}_0$ by taking $w \in H_0^1(\Omega)$ such that $\Delta w = \nabla \cdot \mathbf{v}$ (note that $\nabla w \in \mathbf{H}_0(\text{curl}, \Omega)$) and then write for all $\mathbf{u} \in \mathbf{X}_0$

$$\begin{align}
a_\omega(\mathbf{u}, \mathbf{v}) &= a_\omega(\mathbf{u}, \mathbf{w}) + \omega^2(\mathbf{u}, \nabla w)_\Omega &&(\nabla \times \nabla w = \mathbf{0})\\
&= a_\omega(\mathbf{u}, \mathbf{w}) - \omega^2(\nabla \cdot \mathbf{u}, w)_\Omega &&(w \in H_0^1(\Omega))\\
&= a_\omega(\mathbf{u}, \mathbf{w}) &&(\mathbf{u} \in \mathbf{X}_0).
\end{align}$$

This computation shows that

$$\begin{align}
\ker A_\omega &= \{\mathbf{u} \in \mathbf{X}_0 : a_\omega(\mathbf{u}, \mathbf{v}) = 0 \quad \forall \mathbf{v} \in \mathbf{X}\}\\
&= \{\mathbf{u} \in \mathbf{X}_0 : \nabla \times \nabla \times \mathbf{u} - \omega^2 \mathbf{u} = \mathbf{0}\}.
\end{align}$$

(The reader is invited to verify the last equality.) This proves that injectivity of A_ω (and with it the well-posedness of (16.22)) is equivalent to (16.21).

Regularized formulation. If $\mathbf{f} \in \mathbf{H}(\text{div}, \Omega)$, we can look for $\mathbf{u} \in \mathbf{X}$ (we now have that $\omega^2 \nabla \cdot \mathbf{u} = -\nabla \cdot \mathbf{f} \in L^2(\Omega)$) and try the regularized formulation

$$\mathbf{u} \in \mathbf{X}, \tag{16.23a}$$

$$c_{\omega,r}(\mathbf{u}, \mathbf{v}) = (\mathbf{f}, \mathbf{v})_\Omega - r\,(\nabla \cdot \mathbf{f}, \nabla \cdot \mathbf{v})_\Omega \qquad \forall \mathbf{v} \in \mathbf{X}, \tag{16.23b}$$

where

$$c_{\omega,r}(\mathbf{u}, \mathbf{v}) := (\nabla \times \mathbf{u}, \nabla \times \mathbf{v})_\Omega + \omega^2 r\,(\nabla \cdot \mathbf{u}, \nabla \cdot \mathbf{v})_\Omega - \omega^2(\mathbf{u}, \mathbf{v})_\Omega, \tag{16.24}$$

and we have chosen a parameter $r > 0$. This parameter can be chosen with some (but not complete) freedom. The argument showing that (16.23) is a variational formulation for (16.20) is a little more involved than usual, and we include the key step for the proof in Proposition 16.10 below, leaving the reader with all missing (and, after Proposition 16.10, quite trivial) details. An already familiar compactness argument can now be invoked to show that (16.23) is well posed if and only if

$$\left. \begin{array}{c} \mathbf{u} \in \mathbf{X} \\ c_{\omega,r}(\mathbf{u}, \mathbf{v}) = 0 \qquad \forall \mathbf{v} \in \mathbf{X} \end{array} \right\} \quad \Longrightarrow \quad \mathbf{u} = \mathbf{0}.$$

However, using Proposition 16.10, we prove that any $\mathbf{u}$ satisfying $c_{\omega,r}(\mathbf{u}, \cdot) = 0$ as a functional in $\mathbf{X}$ is divergence-free and thus satisfies $\nabla \times \nabla \times \mathbf{u} - \omega^2 \mathbf{u} = \mathbf{0}$. Therefore, uniqueness of solution is equivalent to the eigenvalue condition (16.21). This closes the analysis of this weighted regularized formulation.

Proposition 16.10. *If $-r^{-1}$ is not a Dirichlet eigenvalue of the Laplacian, then any solution to (16.23) satisfies $\omega^2 \nabla \cdot \mathbf{u} = -\nabla \cdot \mathbf{f}$.*

Proof. Let $\mathbf{w} := \omega^2 \mathbf{u} + \mathbf{f} \in \mathbf{H}(\text{div}, \Omega)$, consider the unique solution to

$$w \in H_0^1(\Omega), \qquad \Delta w + r^{-1}w = \nabla \cdot \mathbf{w},$$

and take $\mathbf{v} := \nabla w \in \mathbf{X}$ as a test function in (16.23). Reordering terms, and applying that $w \in H_0^1(\Omega)$ at the time of integrating by parts, we have

$$r(\nabla \cdot \mathbf{w}, \nabla \cdot \mathbf{w} - r^{-1}w)_\Omega = (\mathbf{w}, \nabla w)_\Omega = -(\nabla \cdot \mathbf{w}, w)_\Omega.$$

Therefore $\nabla \cdot \mathbf{w} = 0$ as we wanted to prove. $\qquad \square$

A direct formulation. Probably the simplest attempt to reach a variational formulation of (16.20) would be testing with elements of $\mathbf{H}_0(\text{curl}, \Omega)$ and dealing with the obviously equivalent variational formulation

$$\mathbf{u} \in \mathbf{H}_0(\text{curl}, \Omega), \quad (\nabla \times \mathbf{u}, \nabla \times \mathbf{v})_\Omega - \omega^2(\mathbf{u}, \mathbf{v})_\Omega = (\mathbf{f}, \mathbf{v})_\Omega \quad \forall \mathbf{v} \in \mathbf{H}_0(\text{curl}, \Omega).$$

The lack of compactness of the injection of $\mathbf{H}_0(\mathrm{curl}, \Omega)$ into $\mathbf{L}^2(\Omega)$ (see Exercise 16.8) makes any naive attempt to use Fredholm theory useless. However, the situation is very similar to when we were dealing with the bilinear form $(\nabla \cdot \mathbf{u}, \nabla \cdot \mathbf{v})_\Omega - k^2(\mathbf{u}, \mathbf{v})_\Omega$ in $\mathbf{H}(\mathrm{div}, \Omega)$ in Section 15.1. Since the analysis will easily allow for variable coefficients (this is not the case for the previous formulations, which would require a variant of Weber's compactness theorem, where $\mathbf{H}(\mathrm{div}, \Omega)$ is substituted by a weighted version), we will approach a slightly more general problem. We are thus given two matrix-valued functions $\boldsymbol{\kappa}, \boldsymbol{\rho} : \Omega \to \mathbb{R}^{3\times3}_{\mathrm{sym}}$, with $L^\infty(\Omega)$ components and such that

$$\boldsymbol{\xi} \cdot (\boldsymbol{\kappa}\boldsymbol{\xi}) \geq \kappa_0|\boldsymbol{\xi}|^2, \qquad \boldsymbol{\xi} \cdot (\boldsymbol{\rho}\boldsymbol{\xi}) \geq \rho_0|\boldsymbol{\xi}|^2 \qquad \forall \boldsymbol{\xi} \in \mathbb{R}^3, \quad \text{a.e.},$$

for given positive constants κ_0, ρ_0. The problem is then

$$\mathbf{u} \in \mathbf{H}_0(\mathrm{curl}, \Omega), \qquad \nabla \times \boldsymbol{\kappa}(\nabla \times \mathbf{u}) - \omega^2 \boldsymbol{\rho} \mathbf{u} = \mathbf{f}, \qquad (16.25)$$

for a positive frequency (wave-number) ω and for $\mathbf{f} \in \mathbf{L}^2(\Omega)$. This problem is clearly equivalent to

$$\mathbf{u} \in \mathbf{H}_0(\mathrm{curl}, \Omega), \qquad\qquad\qquad\qquad\qquad\qquad (16.26a)$$

$$(\boldsymbol{\kappa}(\nabla \times \mathbf{u}), \nabla \times \mathbf{v})_\Omega - \omega^2(\boldsymbol{\rho}\mathbf{u}, \mathbf{v})_\Omega = (\mathbf{f}, \mathbf{v})_\Omega \qquad \forall \mathbf{v} \in \mathbf{H}_0(\mathrm{curl}, \Omega). \quad (16.26b)$$

We consider the *self-adjoint* operator $A : \mathbf{H}_0(\mathrm{curl}, \Omega) \to \mathbf{H}_0(\mathrm{curl}, \Omega)$ given by the bilinear form in (16.26) and the Riesz-Fréchet theorem

$$(A\mathbf{u}, \mathbf{v})_{\mathrm{curl},\Omega} = (\boldsymbol{\kappa}(\nabla \times \mathbf{u}), \nabla \times \mathbf{v})_\Omega - \omega^2(\boldsymbol{\rho}\mathbf{u}, \mathbf{v})_\Omega \qquad \forall \mathbf{u}, \mathbf{v} \in \mathbf{H}_0(\mathrm{curl}, \Omega).$$

Our ultimate goal can be narrated as follows. We want to prove that $A = B_0 + C_0$, where B_0 is invertible and C_0 is compact (so we can use the Fredholm alternative). To do that we will find an invertible operator T such that $AT = B + C$, where $(B\cdot, \cdot)_{\mathrm{curl},\Omega}$ is coercive and C is compact. The operator T will be built by flipping a decomposition of the space provided by a bounded projection P in the following form: $T = 2P - I$, where P is a bounded projection such that range P is compactly embedded into $\mathbf{L}^2(\Omega)$ and $\nabla \times P\mathbf{u} = \nabla \times \mathbf{u}$. (Note that in the literature the operator T is typically applied to the test function, as in Section 15.1, and we end up with a decomposition of T^*A instead of AT, but this does not change our conclusions.) First, we will show how the process works assuming that the projection P has been found, and then we will construct the projection. This is done in the next two propositions. Note that a projection P satisfying the requirements of Proposition 16.11 decomposes $\mathbf{u} \in \mathbf{H}_0(\mathrm{curl}, \Omega)$ as the sum of a divergence-free field $P\mathbf{u}$ plus a curl-free field $\mathbf{u} - P\mathbf{u}$. This is an example of a **Helmholtz decomposition.**

Proposition 16.11. *Let* $P : \mathbf{H}_0(\mathrm{curl}, \Omega) \to \mathbf{H}_0(\mathrm{curl}, \Omega)$ *be a bounded projection satisfying*

$$\nabla \times P\mathbf{u} = \nabla \times \mathbf{u}, \qquad \nabla \cdot (P\mathbf{u}) = 0 \qquad \forall \mathbf{u} \in \mathbf{H}_0(\mathrm{curl}, \Omega). \qquad (16.27)$$

The operator $T := 2P - I = P + (P - I)$ *is an isomorphism and there exists an invertible operator B such that $AT - B$ is compact. Therefore,* $\ker A = \{0\}$ *if and only if A is invertible.*

Proof. First of all, from the fact that $P^2 = P$, we show that $T^2 = I$. The first condition in (16.27) implies that $\nabla \times T\mathbf{u} = \nabla \times \mathbf{u}$ for all $\mathbf{u}$. The second condition in (16.27) implies that $P : \mathbf{H}_0(\mathrm{curl}, \Omega) \to \mathbf{X}$ is bounded and therefore $P : \mathbf{H}_0(\mathrm{curl}, \Omega) \to \mathbf{L}^2(\Omega)$ is compact. We then decompose (the algebra of the following computation is really easy and left to the reader): for all $\mathbf{u}, \mathbf{v}$,

$$(AT\mathbf{u}, \mathbf{v})_{\mathrm{curl}, \Omega} = (B_\omega \mathbf{u}, \mathbf{v})_{\mathrm{curl}, \Omega} - \omega^2 (D\mathbf{u}, \mathbf{v})_{\mathrm{curl}, \Omega},$$

where

$$(B_\omega \mathbf{u}, \mathbf{v})_{\mathrm{curl}, \Omega} = (\kappa(\nabla \times \mathbf{u}), \nabla \times \mathbf{v})_\Omega$$
$$+ \omega^2 \left((\rho P \mathbf{u}, P\mathbf{v})_\Omega + (\rho(I - P)\mathbf{u}, (I - P)\mathbf{v})_\Omega \right),$$

and

$$(D\mathbf{u}, \mathbf{v})_{\mathrm{curl}, \Omega} = 2(\rho P \mathbf{u}, P\mathbf{v})_\Omega + (\rho(P - I)\mathbf{u}, P\mathbf{v})_\Omega + (\rho P \mathbf{u}, (I - P)\mathbf{v})_\Omega.$$

The symmetric bilinear form $(B_\omega \mathbf{u}, \mathbf{v})_{\mathrm{curl}, \Omega}$ defines an equivalent inner product in $\mathbf{H}_0(\mathrm{curl}, \Omega)$ and therefore B_ω is invertible. At the same time, the operator D can easily be seen to be compact. This proves the desired decomposition. Moreover, $A = B_\omega T^{-1} - \omega^2 DT^{-1}$, where $B_\omega T^{-1}$ is invertible and DT^{-1} is compact, which shows that the Fredholm alternative holds for A. $\square$

Proposition 16.12. *The operator $P : \mathbf{H}_0(\mathrm{curl}, \Omega) \to \mathbf{H}_0(\mathrm{curl}, \Omega)$ given by*

$$P\mathbf{u} := \mathbf{u} - \nabla w, \qquad \text{where} \quad \begin{cases} w \in H_0^1(\Omega), \\ \Delta w = \nabla \cdot \mathbf{u} \in H^{-1}(\Omega), \end{cases}$$

satisfies the conditions of Proposition 16.11.

Proof. It is a simple verification, recalling that $\nabla H_0^1(\Omega) \subset \mathbf{H}_0(\mathrm{curl}, \Omega)$. $\square$

Conclusions. Propositions 16.11 and 16.12 and the Fredholm alternative allow us to draw the following easy conclusions:

(a) The set of solutions of

$$\mathbf{v} \in \mathbf{H}_0(\mathrm{curl}, \Omega), \qquad \nabla \times \kappa(\nabla \times \mathbf{v}) - \omega^2 \rho \mathbf{v} = \mathbf{0}, \qquad (16.28)$$

is finite-dimensional.

(b) If (16.28) has no solutions apart from the trivial one, problems (16.26) and (16.25) are well posed.

(c) The compatibility conditions for the general case are just $(\mathbf{f}, \mathbf{v})_\Omega = 0$ where $\mathbf{v}$ is any nontrivial solution to (16.28).

16.5   Two de Rham sequences

Four spaces and three operators. Next, we will see that we have developed two collections of spaces connected by operators and satisfying some common properties, which we will denote in the following way

$$A \xrightarrow{d_A} B \xrightarrow{d_B} C \xrightarrow{d_C} D,$$

where the arrows are operators and

(a) d_A is injective,

(b) range $d_A \subset \ker d_B$,

(c) range $d_B \subset \ker d_C$,

(d) d_C is surjective.

We will also have the following additional (topological) properties:

(e) range d_A and range d_B are closed,

(f) the spaces

$$h_B := \ker d_B \cap (\text{range } d_A)^\perp \equiv \ker d_B / \text{range } d_A,$$
$$h_C := \ker d_C \cap (\text{range } d_B)^\perp \equiv \ker d_C / \text{range } d_B$$

are finite-dimensional. (Note that the identification of the quotient space with the definition of the spaces h_B and h_C holds because the ranges of d_A and d_B are closed.)

When h_B and h_C are the zero subspace, that is, when range $d_A = \ker d_B$ and range $d_B = \ker d_C$, this structure is called an **exact sequence**. We will not plunge any deeper into theoretical aspects of these kinds of 'complexes,' leaving the reader to satisfy his or her curiosity by investigating the literature. The following result holds as a consequence of the Weber compactness theorem.

Proposition 16.13. *Let $P_X : \mathbf{X} \to \mathcal{H}_X$ and $P_Y : \mathbf{Y} \to \mathcal{H}_Y$ be the orthogonal projections onto the corresponding spaces of harmonic fields. There exists a constant such that*

$$C\|\mathbf{u}\|_\Omega \le \|\nabla \times \mathbf{u}\|_\Omega + \|\nabla \cdot \mathbf{u}\|_\Omega + \|P_X \mathbf{u}\|_\Omega \qquad \forall \mathbf{u} \in \mathbf{X},$$
$$C\|\mathbf{u}\|_\Omega \le \|\nabla \times \mathbf{u}\|_\Omega + \|\nabla \cdot \mathbf{u}\|_\Omega + \|P_Y \mathbf{u}\|_\Omega \qquad \forall \mathbf{u} \in \mathbf{Y}.$$

Proof. Follow the proof of Proposition 16.9, assuming that the inequalities do not hold, and reach a contradiction. □

A de Rham sequence with zero boundary conditions. On a connected (we assume this for simplicity of the exposition) bounded Lipschitz domain Ω, we consider the spaces

$$A := H_0^1(\Omega), \qquad B := \mathbf{H}_0(\mathrm{curl}, \Omega), \qquad C := \mathbf{H}_0(\mathrm{div}, \Omega), \qquad D := L_0^2(\Omega),$$

where $L_0^2(\Omega) = \{u \in L^2(\Omega) : (u, 1)_\Omega = 0\}$ is the space of L^2 functions with vanishing integral. The operators connecting the spaces are

$$d_A := \mathrm{grad} = \nabla, \qquad d_B := \mathrm{curl} = \nabla \times, \qquad d_C := \mathrm{div} = \nabla \cdot.$$

We have already seen that $\nabla H_0^1(\Omega) \subset \mathbf{H}_0(\mathrm{curl}, \Omega)$ and it is clear (recall that $\mathbf{H}_0(\mathrm{div}, \Omega)$ is the kernel of the normal trace) that $\nabla \cdot \mathbf{H}_0(\mathrm{div}, \Omega) \subset L_0^2(\Omega)$. To see that $\nabla \times \mathbf{H}_0(\mathrm{curl}, \Omega) \subset \mathbf{H}_0(\mathrm{div}, \Omega)$, take a sequence in $\mathcal{D}(\Omega)$ converging in $\mathbf{H}(\mathrm{curl}, \Omega)$ and note that the curl of this sequence converges in $\mathbf{H}(\mathrm{div}, \Omega)$. Since $H_0^1(\Omega)$ does not contain constants, ∇ is injective, and the surjectivity of the divergence can be proved by solving a Neumann problem. This finishes the proof of (a)-(d).

The range of the gradient of $H_0^1(\Omega)$ is closed in $\mathbf{L}^2(\Omega)$ because of the Poincaré-Friedrichs inequality, which can be restated as

$$\|u\|_\Omega \leq C\|\nabla u\|_{\mathrm{curl}, \Omega} \qquad \forall u \in H_0^1(\Omega).$$

If $\mathbf{u} \in \mathbf{H}_0(\mathrm{curl}, \Omega)$, and we choose

$$w \in H_0^1(\Omega), \qquad \Delta w = \nabla \cdot \mathbf{u} \in H^{-1}(\Omega),$$

then $\mathbf{u} - \nabla w \in \mathbf{X}$ and

$$\mathbf{v} := \mathbf{u} - \nabla w - P_X(\mathbf{u} - \nabla w) \in \mathbf{X}$$

satisfies $\nabla \cdot \mathbf{v} = 0$ and $P_X \mathbf{v} = \mathbf{0}$. By Proposition 16.13, we have

$$\|\mathbf{v}\|_\Omega \leq C\|\nabla \times \mathbf{v}\|_\Omega \qquad \mathbf{v} \in \mathbf{X}, \qquad \nabla \cdot \mathbf{v} = 0, \qquad P_X \mathbf{v} = \mathbf{0},$$

and this can be used to show that the range of the curl is closed as follows. If $\nabla \times \mathbf{u}_n$ converges, then $\nabla \times \mathbf{u}_n = \nabla \times \mathbf{v}_n$ (with $\mathbf{v}_n \in \mathbf{X}$ satisfying $\nabla \cdot \mathbf{v}_n = 0$, $P_X \mathbf{v}_n = \mathbf{0}$) is Cauchy and therefore $\mathbf{v}_n \to \mathbf{v} \in \mathbf{L}^2(\Omega)$. Comparing curls in the sense of distributions, it follows that $\nabla \times \mathbf{u}_n = \nabla \times \mathbf{v}_n \to \nabla \times \mathbf{v}$ in $\mathbf{L}^2(\Omega)$; since $\mathbf{H}_0(\mathrm{curl}, \Omega)$ is closed, this proves that $\mathbf{v} \in \mathbf{H}_0(\mathrm{curl}, \Omega)$ and therefore the range is closed. Finally, we need to identify the spaces h_B and h_C. A simple computation shows that

$$\begin{aligned}
h_B &= \{\mathbf{u} \in \mathbf{H}_0(\mathrm{curl}, \Omega) : \nabla \times \mathbf{u} = \mathbf{0}, \quad (\mathbf{u}, \nabla p)_{\mathrm{curl}, \Omega} = 0 \quad \forall p \in H_0^1(\Omega)\} \\
&= \{\mathbf{u} \in \mathbf{H}_0(\mathrm{curl}, \Omega) : \nabla \times \mathbf{u} = \mathbf{0}, \quad \nabla \cdot \mathbf{u} = 0\} = \mathcal{H}_X, \\
h_C &= \{\mathbf{u} \in \mathbf{H}_0(\mathrm{div}, \Omega) : \nabla \cdot \mathbf{u} = 0, \quad (\mathbf{u}, \nabla \times \mathbf{w})_{\mathrm{div}, \Omega} = 0 \quad \forall \mathbf{w} \in \mathbf{H}_0(\mathrm{curl}, \Omega)\} \\
&= \{\mathbf{u} \in \mathbf{H}_0(\mathrm{div}, \Omega) : \nabla \cdot \mathbf{u} = 0, \quad \nabla \times \mathbf{u} = \mathbf{0}\} = \mathcal{H}_Y,
\end{aligned}$$

which proves that the two quotient spaces are isomorphic to the spaces of harmonic fields. This finishes the proof of the topological conditions (e) and (f).

A de Rham sequence without boundary conditions. We now consider the spaces

$$A := H^1_\star(\Omega), \qquad B := \mathbf{H}(\mathrm{curl}, \Omega), \qquad C := \mathbf{H}(\mathrm{div}, \Omega), \qquad D := L^2(\Omega),$$

where $H^1_\star(\Omega) = \{u \in H^1(\Omega) : (u, 1)_\Omega = 0\}$. The operators between these spaces are the same as above. Conditions (a) to (d) follow readily. The range of d_A is closed, since Poincaré's inequality can be written in the equivalent form

$$\|u\|_\Omega \leq C\|\nabla u\|_{\mathrm{curl}, \Omega} \qquad \forall u \in H^1_\star(\Omega).$$

To prove that the range of the curl is closed we use similar arguments to those employed in the sequence with boundary conditions, using the other inequality in Proposition 16.13. Noting that for a vector field $\mathbf{u} \in \mathbf{L}^2(\Omega)$ we have

$$\left.\begin{array}{l} \nabla \cdot \mathbf{u} = 0 \\ \mathbf{u} \cdot \mathbf{n} = 0 \end{array}\right\} \quad \Longleftrightarrow \quad (\mathbf{u}, \nabla p)_\Omega = 0 \qquad \forall p \in H^1(\Omega),$$

and

$$\left.\begin{array}{l} \nabla \times \mathbf{u} = \mathbf{0} \\ \gamma_T \mathbf{u} = \mathbf{0} \end{array}\right\} \quad \Longleftrightarrow \quad (\mathbf{u}, \nabla \times \mathbf{v})_\Omega = 0 \qquad \forall \mathbf{v} \in \mathbf{H}(\mathrm{curl}, \Omega),$$

and this time we obtain $h_B = \mathcal{H}_Y$ and $h_C = \mathcal{H}_X$, that is, the spaces of harmonic fields have swapped places as representations of the quotient spaces in condition (f).

Exact sequences. The following result (which we offer without proof) states that in any ball, we have equality of the subspaces $\mathrm{range}\, d_A = \ker d_B$ and $\mathrm{range}\, d_B = \ker d_C$ for both of the de Rham sequences we have defined. This is equivalent to showing that the spaces of harmonic fields are trivial. The result can be proved in a single ball, since by translation and dilation the result would follow for any other ball. In fact, the result can be extended to domains that are connected, simply connected, and with connected boundary, but this is not an easy proof. We will only use the result for a single domain.

[**Proof not provided**]

Theorem 16.2. *In any ball, only trivial harmonic fields exist, that is* $\mathcal{H}_X = \mathcal{H}_Y = \{\mathbf{0}\}$.

16.6 Maxwell eigenvalues

Eigenvalue problems. In this section we study the two basic eigenvalue problems for the Maxwell equations, with zero tangential condition

$$\mathbf{u} \in \mathbf{H}(\text{curl}, \Omega), \qquad \nabla \times \nabla \times \mathbf{u} = \lambda \mathbf{u}, \qquad \gamma_T \mathbf{u} = \mathbf{0}, \qquad (16.29)$$

and with zero tangential condition on the curl

$$\mathbf{v} \in \mathbf{H}(\text{curl}, \Omega), \qquad \nabla \times \nabla \times \mathbf{v} = \lambda \mathbf{v}, \qquad \gamma_T(\nabla \times \mathbf{v}) = \mathbf{0}. \qquad (16.30)$$

A simple argument (integration by parts) shows that we only need to worry about $\lambda \in \mathbb{R}$, $\lambda \geq 0$. The zero eigenvalue has infinitely many eigenfunctions. The solutions of (16.29) with $\lambda = 0$ are the elements of $\nabla H_0^1(\Omega) \oplus \mathcal{H}_X$, while those of (16.30) with $\lambda = 0$ are the elements of $\nabla H^1(\Omega) \oplus \mathcal{H}_Y$. (See Exercises 16.13 and 16.14.) Furthermore, if $\mathbf{v}$ is a solution of (16.30) with $\lambda \neq 0$, then $\mathbf{u} := \nabla \times \mathbf{v}$ is obviously a solution to (16.29) for the same eigenvalue. The reciprocal holds as well, given the fact that if $\mathbf{u}$ is a solution of (16.29) and $\mathbf{v} := \nabla \times \mathbf{u}$, then $\gamma_T(\nabla \times \mathbf{v}) = \lambda \gamma_T \mathbf{u} = \mathbf{0}$. This means that we only need to worry about (16.29). The way we will handle this problem will be similar to the Neumann eigenvalues for the Laplacian (by shifting the operator, recall Section 9.5), imposing a condition on the divergence to eliminate the infinitely many gradient eigenfunctions associated to $\lambda = 0$, while recovering harmonic fields as part of the theory.

The source operator and the Maxwell eigenvalues. We will work on the space

$$\mathbf{H} := \{\mathbf{f} \in \mathbf{L}^2(\Omega) \; : \; \nabla \cdot \mathbf{f} = 0\},$$

where we define the bounded linear operator $\mathbf{G} : \mathbf{H} \to \mathbf{H}$ given by

$$\mathbf{Gf} = \mathbf{u}, \qquad \text{where} \qquad \begin{cases} \mathbf{u} \in \mathbf{H}_0(\text{curl}, \Omega), \\ \nabla \times \nabla \times \mathbf{u} + \mathbf{u} = \mathbf{f}. \end{cases}$$

Note that the fact that $\mathbf{Gf} \in \mathbf{H}$ follows from the fact that $\mathbf{f} \in \mathbf{H}$. We could extend $\mathbf{G}$ to $\mathbf{L}^2(\Omega)$, but that extension will fail to be compact. By the variational formulation of the problem defining $\mathbf{G}$, we have

$$(\mathbf{Gf}, \mathbf{v})_{\text{curl}, \Omega} = (\mathbf{f}, \mathbf{v})_\Omega \qquad \forall \mathbf{v} \in \mathbf{H}_0(\text{curl}, \Omega), \qquad (16.31)$$

and therefore

$$(\mathbf{Gf}, \mathbf{Gg})_{\text{curl}, \Omega} = (\mathbf{f}, \mathbf{Gg})_\Omega \qquad \forall \mathbf{f}, \mathbf{g} \in \mathbf{H}.$$

This proves that $\mathbf{G}$ is self-adjoint and positive semidefinite, as well as the bound

$$\|\mathbf{Gf}\|_{\mathbf{X}} = \|\mathbf{Gf}\|_{\text{curl}, \Omega} \leq \|\mathbf{f}\|_\Omega \qquad \forall \mathbf{f} \in \mathbf{H}. \qquad (16.32)$$

This proves that $\mathbf{G} : \mathbf{H} \to \mathbf{H}$ is compact (the injection of $\mathbf{X}$ into $\mathbf{L}^2(\Omega)$ is compact). Since $\mathcal{D}(\Omega)$ is dense in $\mathbf{L}^2(\Omega)$, (16.31) implies that $\mathbf{G}$ is injective and all its eigenvalues are positive. From (16.32), it follows that all the eigenvalues of $\mathbf{G}$ are less than or equal to one. We thus have a nonincreasing sequence of positive numbers $0 < \sigma_n \le 1$, such that $\sigma_n \to 0$, and a Hilbert basis of $\mathbf{H}$, $\{\phi_n\}$, such that

$$\mathbf{G} = \sum_{n=1}^{\infty} \sigma_n(\,\cdot\,, \phi_n)_\Omega \phi_n.$$

It is then clear that these are solutions to (16.29), as

$$\phi_n \in \mathbf{H}_0(\mathrm{curl}, \Omega), \qquad \nabla \times \nabla \times \phi_n = (\sigma_n^{-1} - 1)\phi_n, \qquad \nabla \cdot \phi_n = 0.$$

The Maxwell eigenvalues $\lambda_n = \sigma_n^{-1} - 1 \ge 0$ diverge to infinity in a nondecreasing sequence. The eigenvalue $\lambda = 0$ can have a finite number of linearly independent eigenfunctions, those in $\mathcal{H}_X$. Note that we have orthogonality of the curls of the eigenfunctions, as we have

$$(\nabla \times \phi_n, \nabla \times \phi_m)_\Omega = \lambda_n(\phi_n, \phi_m) = \lambda_n \delta_{nm} \qquad \forall n, m \ge 1.$$

Using the tools from Section 9.5, we can identify the range of $\mathbf{G}$ and the intermediate space $\mathbf{H} \cap \mathbf{H}_0(\mathrm{curl}, \Omega)$ with the associated Fourier series:

$$\mathbf{X}_0 = \{\mathbf{u} \in \mathbf{X} : \nabla \cdot \mathbf{u} = 0\} = \{\mathbf{u} \in \mathbf{H} : \sum_{n=1}^{\infty} \lambda_n |(\mathbf{u}, \phi_n)_\Omega|^2 < \infty\},$$

$$\{\mathbf{u} \in \mathbf{X}_0 : \nabla \times \nabla \times \mathbf{u} \in \mathbf{L}^2(\Omega)\} = \{\mathbf{u} \in \mathbf{H} : \sum_{n=1}^{\infty} \lambda_n^2 |(\mathbf{u}, \phi_n)_\Omega|^2 < \infty\}.$$

16.7   Normally oriented trace fields

The goal of the coming four sections is the precise understanding of the trace space of the tangential trace and, even more, the extension of the integration by parts formula

$$(\mathbf{u}, \nabla \times \mathbf{v})_\Omega - (\nabla \times \mathbf{u}, \mathbf{v})_\Omega = \langle \gamma_T \mathbf{u}, \gamma \mathbf{v} \rangle_\Gamma,$$

which so far we accept for $\mathbf{u} \in \mathbf{H}(\mathrm{curl}, \Omega)$ and $\mathbf{v} \in \mathbf{H}^1(\Omega)$. While it is clear that the left-hand side is a bounded skew-symmetric bilinear form in $\mathbf{H}(\mathrm{curl}, \Omega)$, it is much less obvious how to understand the right-hand side. We will do this by redefining the tangential trace as a purely tangential vector field (as opposed to a vector field that is tangential; see later) and then working carefully through rotations of tangential vector fields. We will borrow a result from the literature to be able to fully characterize the range of the new tangential trace and finally the curl-curl integration by parts formula will be ready to be proved. This will be essential to handle both the Dirichlet and Neumann problems for the Maxwell equation.

A space of normally oriented traces. In this section we study the subspace of $\mathbf{H}^{1/2}(\Gamma)$ consisting of vector fields that are purely normal:

$$
\begin{aligned}
\mathbf{H}_n^{1/2}(\Gamma) :&=\{\boldsymbol{\phi} \in \mathbf{H}^{1/2}(\Gamma) \; : \; \boldsymbol{\phi} \times \mathbf{n} = \mathbf{0}\} \\
&=\{\boldsymbol{\phi} \in \mathbf{H}^{1/2}(\Gamma) \; : \; \boldsymbol{\phi} = (\boldsymbol{\phi} \cdot \mathbf{n})\mathbf{n}\} \\
&=\{\gamma\mathbf{v} \; : \; \mathbf{v} \in \mathbf{H}^1(\Omega), \gamma_T\mathbf{v} = \mathbf{0}\}.
\end{aligned}
$$

(Recall that for elements of $\mathbf{H}^1(\Omega)$, we have $\gamma_T\mathbf{v} = (\gamma\mathbf{v}) \times \mathbf{n} = \mathbf{0}$, which proves the last equality.) This closed subspace of $\mathbf{H}^{1/2}(\Gamma)$ is related to the operator γ_T in yet another way. Recall that we have defined $\gamma_T : \mathbf{H}(\mathrm{curl}, \Omega) \to \mathbf{H}^{-1/2}(\Gamma)$ with the integration by parts formula

$$
\langle \gamma_T\mathbf{u}, \gamma\mathbf{v} \rangle_\Gamma = (\mathbf{u}, \nabla \times \mathbf{v})_\Omega - (\nabla \times \mathbf{u}, \mathbf{v})_\Omega \qquad \forall \mathbf{v} \in \mathbf{H}^1(\Omega).
$$

We have also seen (in Proposition 16.4) that

$$
(\mathbf{u}, \nabla \times \mathbf{v})_\Omega - (\nabla \times \mathbf{u}, \mathbf{v})_\Omega = 0 \qquad \forall \mathbf{v} \in \mathbf{H}(\mathrm{curl}, \Omega)
$$

is equivalent to $\gamma_T\mathbf{u} = \mathbf{0}$. Therefore,

$$
\begin{aligned}
\mathbf{H}_n^{1/2}(\Gamma) =&\{\boldsymbol{\phi} \in \mathbf{H}^{1/2}(\Gamma) \; : \; \langle \gamma_T\mathbf{u}, \boldsymbol{\phi} \rangle_\Gamma = 0 \quad \forall \mathbf{u} \in \mathbf{H}(\mathrm{curl}, \Omega)\} \\
=&(\mathrm{range}\, \gamma_T)^\circ,
\end{aligned}
$$

and we have

$$
\begin{aligned}
\mathrm{range}\, \gamma_T \subset &((\mathrm{range}\, \gamma_T)^\circ)^\circ = \mathbf{H}_n^{1/2}(\Gamma)^\circ \\
=&\{\boldsymbol{\lambda} \in \mathbf{H}^{-1/2}(\Gamma) \; : \; \langle \boldsymbol{\lambda}, \boldsymbol{\phi} \rangle_\Gamma = 0, \quad \forall \boldsymbol{\phi} \in \mathbf{H}_n^{1/2}(\Gamma)\}.
\end{aligned}
$$

Before proceeding in this line (which will require basically admitting yet another result that is not trivial to prove), we are going to give another characterization of $\mathbf{H}_n^{1/2}(\Gamma)$ that will clarify from where the next result is coming. We need two technical results to pave the way for the proof of the next characterization of $\mathbf{H}_n^{1/2}(\Gamma)$.

Lemma 16.2. *For any* $\mathbf{u} \in \mathbf{H}_0(\mathrm{curl}, \Omega)$, *there exists* $\mathbf{v} \in \mathbf{H}^1(\Omega)$ *such that* $\nabla \times \mathbf{u} = \nabla \times \mathbf{v}$.

Proof. Consider a ball B such that $\overline{\Omega} \subset B$, and let $\mathbf{w}$ be the unique solution of

$$
\mathbf{w} \in \mathbf{H}_0(\mathrm{curl}, B), \qquad \nabla \times \nabla \times \mathbf{w} = \nabla \times \widetilde{\mathbf{u}}, \qquad \nabla \cdot \mathbf{w} = 0,
$$

which can be found since $\nabla \times \tilde{\mathbf{u}} \in \mathbf{L}^2(\Omega)$ is divergence-free and we do not have harmonic fields in B. Now note that

$$\Delta \mathbf{w} = \nabla(\nabla \cdot \mathbf{w}) - \nabla \times \nabla \times \mathbf{w} = -\nabla \times \tilde{\mathbf{u}} \in \mathbf{L}^2(\Omega),$$

and, therefore (recall the results of Section 13.8, stating that functions with Laplacian in L^2 have interior H^2 regularity), we have that $\mathbf{w} \in \mathbf{H}^2(\Omega)$. The field $\mathbf{v} := \nabla \times \mathbf{w}$ satisfies the requirements of the statement. $\quad\square$

Lemma 16.3. *If $\Theta \subset \mathbb{R}^3$ is a domain where $\mathcal{H}_Y = \{\mathbf{0}\}$, then*

$$\{\mathbf{u} \in \mathbf{H}(\mathrm{curl}, \Theta) : \nabla \times \mathbf{u} = \mathbf{0}\} = \nabla H^1(\Theta).$$

Proof. This is another way of phrasing the results of Section 16.5 about the de Rham sequence with no boundary conditions. $\quad\square$

We will admit the following restriction/extension theorem.

[Proof not provided]

Theorem 16.3 (Lipschitz domains and H^2-extension property). *For Lipschitz domains in d dimensions, the restriction operator $H^2(\mathbb{R}^d) \to H^2(\Omega)$ is surjective. Therefore, every function in $H^2(\Omega)$ can be extended to a function in $H^2(\mathbb{R}^d)$ using a bounded linear operator.*

Proposition 16.14. *On any Lipschitz domain Ω with boundary Γ, we have*

$$\mathbf{H}_n^{1/2}(\Gamma) = \gamma \nabla(H^2(\Omega) \cap H_0^1(\Omega)) = \{\gamma \nabla u : u \in H^2(\Omega), \gamma u = 0\}.$$

Proof. If $u \in H_0^1(\Omega)$, then $\nabla u \in \mathbf{H}_0(\mathrm{curl}, \Omega)$ and therefore $\gamma_T \nabla u = \mathbf{0}$. Therefore, if $u \in H^2(\Omega) \cap H_0^1(\Omega)$ we have $\mathbf{v} := \nabla u \in \mathbf{H}^1(\Omega)$ satisfies $\gamma_T \mathbf{v} = \mathbf{0}$, that is, $\gamma \mathbf{v} \in \mathbf{H}_n^{1/2}(\Gamma)$.

The reciprocal inclusion requires a three step construction. We first take an open ball Θ such that $\overline{\Omega} \subset \Theta$ and define $\Omega_+ := \Theta \setminus \overline{\Omega}$. Now let $\mathbf{u} \in \mathbf{H}^1(\Omega) \cap \mathbf{H}_0(\mathrm{curl}, \Omega)$, that is, $\gamma \mathbf{u} \in \mathbf{H}_n^{1/2}(\Gamma)$. We apply Lemma 16.2 in Θ to find $\mathbf{v} \in \mathbf{H}^1(\Theta)$ with $\nabla \times \mathbf{v} = \nabla \times \tilde{\mathbf{u}}$, which we can do since $\tilde{\mathbf{u}} \in \mathbf{H}_0(\mathrm{curl}, \Theta)$. We then apply Lemma 16.3 to the curl-free field $\tilde{\mathbf{u}} - \mathbf{v}$, which can therefore be written as a gradient. We have thus decomposed

$$\tilde{\mathbf{u}} = \mathbf{v} + \nabla p, \qquad \mathbf{v} \in \mathbf{H}^1(\Theta), \qquad p \in H^1(\Theta).$$

Note that

$$\nabla p|_{\Theta \setminus \Gamma} = \begin{cases} \mathbf{u} - \mathbf{v}, & \text{in } \Omega, \\ -\mathbf{v}, & \text{in } \Omega_+, \end{cases}$$

and therefore $p \in H^2(\Theta \setminus \Gamma)$. Now we use Theorem 16.3 to find $q \in H^2(\Theta)$ such that $p = q$ in Ω_+. Finally, we define

$$u := (p - q)|_\Omega : \Omega \to \mathbb{R}, \qquad u \in H^2(\Omega) \cap H_0^1(\Omega).$$

(The fact that $\gamma p = \gamma q$ is due to the equality of p and q in Ω_+ and the fact that both functions are in $H^1(\Theta)$.) We can write

$$\tilde{\mathbf{u}} = (\mathbf{v} + \nabla q) + \nabla(p - q).$$

The first part of this decomposition is in $\mathbf{H}^1(\Theta)$ and is identically zero in Ω_+, since $(\mathbf{v} + \nabla q)|_{\Omega_+} = (\mathbf{v} + \nabla p)|_{\Omega_+} = \tilde{\mathbf{u}}|_{\Omega_+} = \mathbf{0}$. Taking the trace from inside, we have

$$\gamma \mathbf{u} = \gamma \nabla(p - q)|_\Omega = \gamma \nabla u,$$

and the result is proved. $\qquad \square$

16.8   Tangential trace spaces and their rotations

A new space of vector fields on Γ. Let us consider the space

$$\mathbf{L}_t(\Gamma) := \{\boldsymbol{\phi} : \Gamma \to \mathbb{R}^3 \ : \ \boldsymbol{\phi} \cdot \mathbf{n} = 0, \quad |\boldsymbol{\phi}| \in L^2(\Gamma)\},$$

endowed with the inner product

$$\langle \boldsymbol{\phi}, \boldsymbol{\psi} \rangle_t := \int_\Gamma \boldsymbol{\phi}(\mathbf{x}) \cdot \boldsymbol{\psi}(\mathbf{x}) \mathrm{d}\Gamma(\mathbf{x}).$$

We will not consider $\mathbf{L}_t^2(\Gamma)$ as a closed subspace of $\mathbf{L}^2(\Gamma)$, even if it clearly is so, but as a Hilbert space on its own. The reason to do this is that we will use it as a pivot space for two new Gelfand triples. In this way, we will distinguish between tangential vector fields on Γ (this would be the elements of $\mathbf{L}_t(\Gamma)$) and vector fields on Γ that happen to be tangential (these are elements of $\mathbf{L}^2(\Gamma)$ that are normal to the normal vector field). The distinction is apparently minor, but it will play an important role in the constructions that follow.

Tangential $\mathbf{H}^{1/2}$ spaces. We consider the spaces (recall the identity $\mathbf{a} \times (\mathbf{b} \times$

c) $= (\mathbf{a} \cdot \mathbf{c})\mathbf{b} - (\mathbf{a} \cdot \mathbf{b})\mathbf{c}$ to see one of the equalities)

$$\mathbf{H}_{||}^{1/2}(\Gamma) := \{\phi \in \mathbf{L}_t^2(\Gamma) : \phi = \mathbf{n} \times (\gamma\mathbf{u} \times \mathbf{n}), \quad \mathbf{u} \in \mathbf{H}^1(\Omega)\}$$
$$= \{\phi \in \mathbf{L}_t^2(\Gamma) : \phi = \mathbf{n} \times (\boldsymbol{\psi} \times \mathbf{n}), \quad \boldsymbol{\psi} \in \mathbf{H}^{1/2}(\Gamma)\}$$
$$= \{\phi \in \mathbf{L}_t^2(\Gamma) : \phi = \boldsymbol{\psi} - (\boldsymbol{\psi} \cdot \mathbf{n})\mathbf{n}, \quad \boldsymbol{\psi} \in \mathbf{H}^{1/2}(\Gamma)\},$$
$$\mathbf{H}_{\perp}^{1/2}(\Gamma) := \{\boldsymbol{\xi} \in \mathbf{L}_t^2(\Gamma) : \boldsymbol{\xi} = \gamma\mathbf{u} \times \mathbf{n}, \quad \mathbf{u} \in \mathbf{H}^1(\Omega)\}$$
$$= \{\boldsymbol{\xi} \in \mathbf{L}_t^2(\Gamma) : \boldsymbol{\xi} = \boldsymbol{\psi} \times \mathbf{n}, \quad \boldsymbol{\psi} \in \mathbf{H}^{1/2}(\Gamma)\}$$
$$= \{\boldsymbol{\xi} \in \mathbf{L}_t^2(\Gamma) : \boldsymbol{\xi} = \mathbf{n} \times \phi, \quad \phi \in \mathbf{H}_{||}^{1/2}(\Gamma)\},$$

and endow them with image norms

$$\|\phi\|_{1/2,||,\Gamma} := \inf\{\|\boldsymbol{\psi}\|_{1/2,\Gamma} : \phi = \mathbf{n} \times (\boldsymbol{\psi} \times \mathbf{n}), \quad \boldsymbol{\psi} \in \mathbf{H}^{1/2}(\Gamma)\}$$
$$= \inf\{\|\gamma\mathbf{u}\|_{1/2,\Gamma} : \phi = \mathbf{n} \times (\gamma\mathbf{u} \times \mathbf{n}), \quad \mathbf{u} \in \mathbf{H}^1(\Omega)\},$$
$$\|\boldsymbol{\xi}\|_{1/2,\perp,\Gamma} := \inf\{\|\boldsymbol{\psi}\|_{1/2,\Gamma} : \boldsymbol{\xi} = \boldsymbol{\psi} \times \mathbf{n}, \quad \boldsymbol{\psi} \in \mathbf{H}^{1/2}(\Gamma)\}$$
$$= \inf\{\|\gamma\mathbf{u}\|_{1/2,\Gamma} : \boldsymbol{\xi} = \gamma\mathbf{u} \times \mathbf{n}, \quad \mathbf{u} \in \mathbf{H}^1(\Omega)\}.$$

Since the space

$$\mathbf{H}_n^{1/2}(\Gamma) = \{\boldsymbol{\psi} \in \mathbf{H}^{1/2}(\Gamma) : \boldsymbol{\psi} \times \mathbf{n} = \mathbf{0}\}$$
$$= \{\boldsymbol{\psi} \in \mathbf{H}^{1/2}(\Gamma) : \mathbf{n} \times (\boldsymbol{\psi} \times \mathbf{n}) = \mathbf{0}\}$$

is closed in $\mathbf{H}^{1/2}(\Gamma)$ (this is very easy to verify), the above norms are well-defined and make $\mathbf{H}_{||}^{1/2}(\Gamma)$ and $\mathbf{H}_{\perp}^{1/2}(\Gamma)$ Hilbert spaces. The operators

$$\Pi_\tau : \mathbf{H}^{1/2}(\Gamma) \longrightarrow \mathbf{H}_{||}^{1/2}(\Gamma), \qquad \Upsilon_\tau : \mathbf{H}^{1/2}(\Gamma) \longrightarrow \mathbf{H}_{\perp}^{1/2}(\Gamma), \qquad (16.33)$$
$$\boldsymbol{\psi} \longmapsto \mathbf{n} \times (\boldsymbol{\psi} \times \mathbf{n}), \qquad\qquad \boldsymbol{\psi} \longmapsto \boldsymbol{\psi} \times \mathbf{n},$$

are surjective by definition. Their common kernel is the space $\mathbf{H}_n^{1/2}(\Gamma)$. The norms defined for $\mathbf{H}_{||}^{1/2}(\Gamma)$ and $\mathbf{H}_{\perp}^{1/2}(\Gamma)$ make Π_τ and Υ_τ bounded with

$$\|\Pi_\tau\boldsymbol{\psi}\|_{1/2,||,\Gamma} \leq \|\boldsymbol{\psi}\|_{1/2,\Gamma}$$
$$\|\Upsilon_\tau\boldsymbol{\psi}\|_{1/2,\perp,\Gamma} \leq \|\boldsymbol{\psi}\|_{1/2,\Gamma} \qquad \forall\boldsymbol{\psi} \in \mathbf{H}^{1/2}(\Gamma).$$

Note that $\phi \in \mathbf{H}_{||}^{1/2}(\Gamma)$ if and only if there exists $\mathbf{u} \in \mathbf{H}^1(\Omega)$ such that $\phi = \Pi_\tau\gamma\mathbf{u} = \mathbf{n} \times (\gamma\mathbf{u} \times \mathbf{n})$, which implies that

$$\mathbf{n} \times \phi = \mathbf{n} \times (\gamma\mathbf{u} - (\gamma\mathbf{u} \cdot \mathbf{n})\mathbf{n}) = \mathbf{n} \times (\gamma\mathbf{u}) = -\Upsilon_\tau\gamma\mathbf{u},$$

and therefore $\mathbf{n} \times \phi \in \mathbf{H}_{\perp}^{1/2}(\Gamma)$ (we have already seen this in the definition). We can argue backwards and note that if $\boldsymbol{\xi} \in \mathbf{H}_{\perp}^{1/2}(\Gamma)$, then $\mathbf{n} \times \boldsymbol{\xi} \in \mathbf{H}_{||}^{1/2}(\Gamma)$. This proves that the rotation of tangential vector fields

$$\mathbf{H}_{||}^{1/2}(\Gamma) \ni \phi \longmapsto \mathbf{n} \times \phi \in \mathbf{H}_{\perp}^{1/2}(\Gamma)$$

defines an algebraic isomorphism between the two tangential trace spaces. At this time, however, it is not clear whether this operator is bounded. We will see that this is the case by taking a long detour through the dual spaces.

Proposition 16.15. *The spaces* $\mathbf{H}_{||}^{1/2}(\Gamma)$ *and* $\mathbf{H}_{\perp}^{1/2}(\Gamma)$ *are continuously and densely embedded into* $\mathbf{L}_t(\Gamma)$.

Proof. Let us first prove the continuity of the embedding for the first space. (The proof for the second one is almost identical.) Given $\phi \in \mathbf{H}_{||}^{1/2}(\Gamma)$ and any $\psi \in \mathbf{H}^{1/2}(\Gamma)$ such that $\phi = \Pi_\tau \psi$, we have

$$\|\phi\|_t = \|\Pi_\tau \psi\|_t \leq \|\psi\|_\Gamma \leq C_\Gamma \|\psi\|_{1/2,\Gamma},$$

and therefore

$$\|\phi\|_t \leq C_\Gamma \|\phi\|_{1/2,||,\Gamma},$$

where C_Γ is the continuity constant for the trace operator $\mathbf{H}^1(\Omega) \to \mathbf{L}^2(\Omega)$, and where we have used that $|\Pi_\tau \psi| \leq |\psi|$ almost everywhere. To prove density, let $\phi \in \mathbf{L}_t^2(\Gamma)$ satisfy

$$\langle \phi, \Pi_\tau \psi \rangle_t = 0 \qquad \forall \psi \in \mathbf{H}^{1/2}(\Gamma).$$

Therefore, since $\phi \cdot \mathbf{n} = 0$, we have

$$\int_\Gamma \phi(\mathbf{x}) \cdot \psi(\mathbf{x}) d\Gamma(\mathbf{x}) = 0 \qquad \forall \psi \in \mathbf{H}^{1/2}(\Gamma), \tag{16.34}$$

but this implies that $\phi = \mathbf{0}$ because $\mathbf{H}^{1/2}(\Gamma)$ is dense in $\mathbf{L}^2(\Gamma)$. Note that in (16.34), we have embedded ϕ in the set of all L^2 vector fields on Γ. This shows that the orthogonal complement of $\mathbf{H}_{||}^{1/2}(\Gamma)$ in $\mathbf{L}_t^2(\Gamma)$ is the zero space, which is equivalent to the density of $\mathbf{H}_{||}^{1/2}(\Gamma)$ in $\mathbf{L}_t^2(\Gamma)$. The proof for $\mathbf{H}_{\perp}^{1/2}(\Gamma)$ is basically the same. $\qquad\square$

Two tangential Gelfand triples. Because of Proposition 16.15, we can define two independent Gelfand triples around $\mathbf{L}_t^2(\Gamma)$,

$$\mathbf{H}_{||}^{1/2}(\Gamma) \subset \mathbf{L}_t^2(\Gamma) \subset \mathbf{H}_{||}^{-1/2}(\Gamma) := \mathbf{H}_{||}^{1/2}(\Gamma)',$$
$$\mathbf{H}_{\perp}^{1/2}(\Gamma) \subset \mathbf{L}_t^2(\Gamma) \subset \mathbf{H}_{\perp}^{-1/2}(\Gamma) := \mathbf{H}_{\perp}^{1/2}(\Gamma)'.$$

Note that since we are using $\mathbf{L}_t^2(\Omega)$ as the pivot space in both triples, we also have the dense embeddings $\mathbf{H}_{||}^{1/2}(\Gamma) \subset \mathbf{H}_{\perp}^{-1/2}(\Gamma)$ and $\mathbf{H}_{\perp}^{1/2}(\Gamma) \subset \mathbf{H}_{||}^{-1/2}(\Gamma)$. We will make use of these inclusions several times in the sequel. With these triples defined, we can take adjoints of the operators in (16.33) to define two operators

$$\imath_\pi := \Pi_\tau' : \mathbf{H}_{||}^{-1/2}(\Gamma) \to \mathbf{H}^{-1/2}(\Gamma), \qquad \imath_\upsilon := \Upsilon_\tau' : \mathbf{H}_{\perp}^{-1/2}(\Gamma) \to \mathbf{H}^{-1/2}(\Gamma),$$

by the relations

$$\langle \imath_\pi \mu, \psi \rangle_\Gamma = \langle \mu, \Pi_\tau \psi \rangle_{t,||}, \qquad \forall \mu \in \mathbf{H}_{||}^{-1/2}(\Gamma), \qquad \psi \in \mathbf{H}^{1/2}(\Gamma),$$

$$\langle \imath_\upsilon \lambda, \psi \rangle_\Gamma = \langle \lambda, \Upsilon_\tau \psi \rangle_{t,\perp}, \qquad \forall \lambda \in \mathbf{H}_\perp^{-1/2}(\Gamma), \qquad \psi \in \mathbf{H}^{1/2}(\Gamma).$$

As a measure of precaution, we have used a different symbol for the two possible extensions of the inner product in $\mathbf{L}_t^2(\Gamma)$ to duality pairings. In the way that we 'project' the space $\mathbf{H}^{1/2}(\Gamma)$ into $\mathbf{H}_\square^{1/2}(\Gamma)$, we can understand that these adjoints embed $\mathbf{H}_\square^{-1/2}(\Gamma)$ into $\mathbf{H}^{-1/2}(\Gamma)$ for $\square \in \{||,\perp\}$. Since Π_τ and Υ_τ are surjective, $\imath_\pi$ and $\imath_\upsilon$ are injective and have closed range, and

$$\text{range } \imath_\pi = (\ker \Pi_\tau)^\circ = \mathbf{H}_n^{1/2}(\Gamma)^\circ = (\ker \Upsilon_\tau)^\circ = \text{range } \imath_\upsilon. \qquad (16.35)$$

This argument justifies the following definition of a rotation of weak vector fields.

Proposition 16.16 (A weak rotation operator). *The bounded invertible operator*

$$r := -\imath_\upsilon^{-1} \circ \imath_\pi : \mathbf{H}_{||}^{-1/2}(\Gamma) \to \mathbf{H}_\perp^{-1/2}(\Gamma),$$

is the only bounded extension of the rotation operator $\boldsymbol{\eta} \mapsto \mathbf{n} \times \boldsymbol{\eta}$ *defined in* $\mathbf{L}_t^2(\Gamma)$.

Proof. Because of the coincidence of ranges of $\imath_\pi$ and $\imath_\upsilon$ (see (16.35)) and the fact that the range is a closed subspace, we can define a bounded inverse $\imath_\upsilon^{-1} : \text{range } \imath_\pi \to \mathbf{H}_\perp^{-1/2}(\Gamma)$ and the composition defining r provides a bounded linear operator, whose inverse is clearly $-\imath_\pi^{-1} \circ \imath_\upsilon$. If $\boldsymbol{\eta} \in \mathbf{L}_t^2(\Gamma)$ and $\psi \in \mathbf{H}^{1/2}(\Gamma)$, then

$$\langle r\boldsymbol{\eta}, \Upsilon_\tau \psi \rangle_t = -\langle \imath_\upsilon \imath_\upsilon^{-1} \imath_\pi \boldsymbol{\eta}, \psi \rangle_\Gamma = -\langle \boldsymbol{\eta}, \Pi_\tau \psi \rangle_t$$

$$= -\int_\Gamma \boldsymbol{\eta}(\mathbf{x}) \cdot \psi(\mathbf{x}) \, d\Gamma(\mathbf{x}) \qquad (\boldsymbol{\eta} \cdot \mathbf{n} = 0)$$

$$= \int_\Gamma (\mathbf{n}(\mathbf{x}) \times \boldsymbol{\eta}(\mathbf{x})) \cdot (\psi(\mathbf{x}) \times \mathbf{n}(\mathbf{x})) d\Gamma(\mathbf{x}) \qquad (\mathbf{n} \cdot \mathbf{n} = 1)$$

$$= \int_\Gamma (\mathbf{n}(\mathbf{x}) \times \boldsymbol{\eta}(\mathbf{x})) \cdot (\Upsilon_\tau \psi)(\mathbf{x}) d\Gamma(\mathbf{x})$$

$$= \langle \mathbf{n} \times \boldsymbol{\eta}, \Upsilon_\tau \psi \rangle_t,$$

which proves that $r\boldsymbol{\eta} = \mathbf{n} \times \boldsymbol{\eta}$ for all $\boldsymbol{\eta} \in \mathbf{L}_t^2(\Gamma)$. The fact that r is the only extension of the strong rotation operator follows from the density of $\mathbf{L}_t^2(\Gamma)$ in both dual spaces. $\qquad \square$

Proposition 16.17 (Rotation of trace spaces). *The invertible bounded operator*

$$r' : \mathbf{H}_\perp^{1/2}(\Gamma) \to \mathbf{H}_{||}^{1/2}(\Gamma),$$

adjoint of $r = -\imath_\upsilon^{-1} \circ \imath_\pi$, *is the rotation of tangential vector fields* $r'\boldsymbol{\xi} = \boldsymbol{\xi} \times \mathbf{n} = -\mathbf{n} \times \boldsymbol{\xi}$ *for all* $\boldsymbol{\xi} \in \mathbf{H}_\perp^{1/2}(\Gamma)$.

Proof. Let $\boldsymbol{\xi} = \Upsilon_\tau \psi \in \mathbf{H}_\perp^{1/2}(\Gamma)$ for $\psi \in \mathbf{H}^{1/2}(\Gamma)$. For all $\boldsymbol{\mu} \in \mathbf{H}_{||}^{-1/2}(\Gamma)$, we have

$$\langle \boldsymbol{\mu}, r'\Upsilon_\tau \psi \rangle_{t,||} = \langle r\,\boldsymbol{\mu}, \Upsilon_\tau \psi \rangle_{t,\perp} = -\langle \imath_\pi \boldsymbol{\mu}, \psi \rangle_\Gamma = -\langle \boldsymbol{\mu}, \Pi_\tau \psi \rangle_{t,||},$$

and therefore $r'\boldsymbol{\xi} = -\Pi_\tau \psi = \Upsilon_\tau \psi \times \mathbf{n} = \boldsymbol{\xi} \times \mathbf{n}$, which finishes the proof. $\square$

16.9 Tangential definition of the tangential traces

The tangential trace as a tangential field. Recall that we defined $\gamma_T : \mathbf{H}(\mathrm{curl}, \Omega) \to \mathbf{H}^{-1/2}(\Gamma)$ by

$$\langle \gamma_T \mathbf{u}, \gamma \mathbf{v} \rangle_\Gamma = (\mathbf{u}, \nabla \times \mathbf{v})_\Omega - (\nabla \times \mathbf{u}, \mathbf{v})_\Omega \qquad \forall \mathbf{v} \in \mathbf{H}^1(\Omega).$$

The goal of the next two sections is related to extending the left-hand side of the above identity to $\mathbf{v} \in \mathbf{H}(\mathrm{curl}, \Omega)$, given the fact that the right-hand side can be extended to arbitrary elements of the same space. Also, recall that

$$\mathrm{range}\,\gamma_T \subset \mathbf{H}_n^{1/2}(\Gamma)^\circ = \mathrm{range}\,\imath_\pi = \mathrm{range}\,\imath_v.$$

We can thus define the bounded linear operator

$$\gamma_\tau := \imath_\pi^{-1} \circ \gamma_T : \mathbf{H}(\mathrm{curl}, \Omega) \to \mathbf{H}_{||}^{-1/2}(\Gamma).$$

We then have

$$
\begin{aligned}
\langle \gamma_\tau \mathbf{u}, \Pi_\tau \gamma \mathbf{v} \rangle_{t,||} &= \langle \gamma_T \mathbf{u}, \gamma \mathbf{v} \rangle_\Gamma \\
&= (\mathbf{u}, \nabla \times \mathbf{v})_\Omega - (\nabla \times \mathbf{u}, \mathbf{v})_\Omega \qquad \forall \mathbf{v} \in \mathbf{H}^1(\Omega), \quad (16.36)
\end{aligned}
$$

an expression that could have been used as an alternative and equivalent definition of γ_τ. Obviously

$$\ker \gamma_\tau = \ker \gamma_T = \mathbf{H}_0(\mathrm{curl}, \Omega), \qquad \imath_\pi \circ \gamma_\tau = \gamma_T,$$

and

$$\gamma_\tau \mathbf{u} = \gamma \mathbf{u} \times \mathbf{n} = \Upsilon_\tau \gamma \mathbf{u} \in \mathbf{L}_t^2(\Gamma) \subset \mathbf{H}_{||}^{-1/2}(\Gamma) \qquad \forall \mathbf{u} \in \mathbf{H}^1(\Omega).$$

This identity is due to the fact that $\gamma_T \mathbf{u} = \gamma \mathbf{u} \times \mathbf{n} \in \mathbf{L}^2(\Gamma)$ for $\mathbf{u} \in \mathbf{H}^1(\Omega)$.

A second tangential trace. We define a second rotated tangential trace by means of the bounded linear operator

$$\pi_\tau := r \circ \gamma_\tau = -\imath_v^{-1} \circ \gamma_T : \mathbf{H}(\mathrm{curl}, \Omega) \to \mathbf{H}_\perp^{-1/2}(\Gamma),$$

and note that
$$\ker \pi_\tau = \ker \gamma_\tau = \ker \gamma_T = \mathbf{H}_0(\mathrm{curl}, \Omega),$$

and

$$\langle \pi_\tau \mathbf{u}, \Upsilon_\tau \gamma \mathbf{v} \rangle_{t,\perp} = -\langle \gamma_T \mathbf{u}, \gamma \mathbf{v} \rangle_\Gamma = (\nabla \times \mathbf{u}, \mathbf{v})_\Omega - (\mathbf{u}, \nabla \times \mathbf{v})_\Omega \qquad \forall \mathbf{v} \in \mathbf{H}^1(\Omega).$$

We also have

$$\pi_\tau \mathbf{u} = \Pi_\tau \gamma \mathbf{u} \qquad \forall \mathbf{u} \in \mathbf{H}^1(\Omega).$$

A trace space with smoother functions. We define the space

$$H^{3/2}(\Gamma) := \gamma H^2(\Omega) = \{ \gamma u \, : \, u \in H^2(\Omega) \} \subset H^{1/2}(\Gamma),$$

and endow it with the image norm

$$\|\phi\|_{3/2,\Gamma} = \inf \{ \|u\|_{2,\Omega} \, : \, \gamma u = \phi \},$$

which makes $H^{3/2}(\Gamma)$ a Hilbert space (the kernel of $\gamma : H^2(\Omega) \to L^2(\Gamma)$ is the closed space $H^2(\Omega) \cap H_0^1(\Omega)$) and the trace operator $\gamma : H^2(\Omega) \to H^{3/2}(\Gamma)$ is bounded with

$$\|\gamma u\|_{3/2,\Gamma} \leq \|u\|_{2,\Omega} \qquad \forall u \in H^2(\Omega).$$

We also have that

$$\|\phi\|_{1/2,\Gamma} \leq \|\phi\|_{3/2,\Gamma} \qquad \forall \phi \in H^{3/2}(\Gamma),$$

as $\|u\|_{1,\Omega} \leq \|u\|_{2,\Omega}$ and that $H^{3/2}(\Gamma)$ is dense in $H^{1/2}(\Gamma)$ since $\mathcal{D}(\Gamma) \subset H^{3/2}(\Gamma)$ is dense in $H^{1/2}(\Gamma)$ as follows from the density of $C^\infty(\overline{\Omega})$ in $H^1(\Omega)$. The notation $H^{3/2}(\Gamma)$ is convenient from the point of view of reminding us that it is the trace of a space with $1/2$-higher regularity in the domain. However, it is misleading since for general Lipschitz domains functions in $H^{3/2}(\Gamma)$ are not locally in $H^{3/2}(\mathbb{R}^2)$ (using pullback to the reference configuration), unless the boundary Γ has some additional smoothness properties that we will not explore here. We will define $H^{-3/2}(\Gamma)$ as the representation of the dual of $H^{3/2}(\Gamma)$ in the Gelfand triple

$$H^{3/2}(\Gamma) \subset L^2(\Gamma) \subset H^{-3/2}(\Gamma).$$

We define two independent liftings (bounded right inverses of the trace)

$$\gamma_{3/2}^+ : H^{3/2}(\Gamma) \to H^2(\Omega), \qquad \gamma_{1/2}^+ : H^{1/2}(\Gamma) \to H^1(\Omega),$$

by looking for the element that minimizes the corresponding $H^m(\Omega)$ norm. These two liftings are independent and in general $\gamma_{3/2}^+ \phi \neq \gamma_{1/2}^+ \phi$ for $\phi \in H^{3/2}(\Gamma)$, although both liftings share the same trace. We will use these liftings to define expedited versions of the tangential gradient and then, by transposition, of the tangential divergence.

Proposition 16.18 (Strong and weak tangential gradients). *The bounded linear operators*

$$\nabla_\Gamma^s := \pi_\tau \nabla \gamma_{3/2}^+ = \Pi_\tau \gamma \nabla \gamma_{3/2}^+ : H^{3/2}(\Gamma) \longrightarrow \mathbf{H}_{||}^{1/2}(\Gamma),$$

$$\nabla_\Gamma^w := \pi_\tau \nabla \gamma_{1/2}^+ : H^{1/2}(\Gamma) \longrightarrow \mathbf{H}_\perp^{-1/2}(\Gamma)$$

satisfy

$$\nabla_\Gamma^s \gamma u = \Pi_\tau \gamma \nabla u \qquad \forall u \in H^2(\Omega), \tag{16.37}$$

and

$$\nabla_\Gamma^w \gamma u = \pi_\tau \nabla u \qquad \forall u \in H^1(\Omega), \tag{16.38}$$

and thus coincide in $H^{3/2}(\Gamma)$. Therefore we can define

$$\nabla_\Gamma \phi := \nabla_\Gamma^w \phi \qquad \forall \phi \in H^{1/2}(\Gamma).$$

Proof. We have that $\gamma_{3/2}^+ \gamma u - u \in H^2(\Omega) \cap H_0^1(\Omega)$ for all $u \in H^2(\Omega)$ and therefore (recall the characterization of purely normal vector fields in Proposition 16.14)

$$\gamma \nabla(\gamma_{3/2}^+ \gamma u - u) \in \mathbf{H}_n^{1/2}(\Gamma) \qquad \forall u \in H^2(\Omega),$$

which is equivalent to saying that

$$\Pi_\tau \gamma \nabla(\gamma_{3/2}^+ \gamma u - u) = \mathbf{0} \qquad \forall u \in H^2(\Omega),$$

and thus to (16.37). Similarly $\gamma_{1/2}^+ \gamma u - u \in H_0^1(\Omega)$ for all $u \in H^1(\Omega)$ and therefore

$$\nabla(\gamma_{1/2}^+ \gamma u - u) \in \mathbf{H}_0(\mathrm{curl}, \Omega) \qquad \forall u \in H^1(\Omega),$$

or equivalently (we recently saw that $\ker \pi_\tau = \mathbf{H}_0(\mathrm{curl}, \Omega)$)

$$\pi_\tau \nabla(\gamma_{1/2}^+ \gamma u - u) = \mathbf{0} \qquad \forall u \in H^1(\Omega),$$

that is, (16.38) holds. $\qquad\square$

Proposition 16.19 (Weak and strong tangential divergence). *The bounded linear operators*

$$\mathrm{div}_\Gamma^w := -(\nabla_\Gamma^s)' : \mathbf{H}_{||}^{-1/2}(\Gamma) \longrightarrow H^{-3/2}(\Gamma),$$

$$\mathrm{div}_\Gamma^s := -(\nabla_\Gamma^w)' : \mathbf{H}_\perp^{1/2}(\Gamma) \longrightarrow H^{-1/2}(\Gamma)$$

coincide in $\mathbf{H}_\perp^{1/2}(\Gamma)$ and allow us to define

$$\nabla_\Gamma \cdot \boldsymbol{\mu} := -(\nabla_\Gamma^s)' \boldsymbol{\mu} \qquad \forall \boldsymbol{\mu} \in \mathbf{H}_{||}^{-1/2}(\Gamma).$$

Proof. This is an easy exercise, but let us write down all the details to clarify the definition. We have

$$\langle \mathrm{div}^s_\Gamma \boldsymbol{\phi}, \phi \rangle_\Gamma = -\langle \nabla_\Gamma \phi, \boldsymbol{\phi} \rangle_{t,\perp} \qquad \boldsymbol{\phi} \in \mathbf{H}^{1/2}_\perp(\Gamma), \qquad \phi \in H^{1/2}(\Gamma),$$

and

$$\langle \mathrm{div}^w_\Gamma \boldsymbol{\mu}, \phi \rangle_\Gamma = -\langle \boldsymbol{\mu}, \nabla_\Gamma \phi \rangle_{t,\|} \qquad \boldsymbol{\mu} \in \mathbf{H}^{-1/2}_\|(\Gamma), \qquad \phi \in H^{3/2}(\Gamma),$$

admitting the same symbol for the duality $H^{-1/2}(\Gamma) \times H^{1/2}(\Gamma)$ that extends to the duality $H^{-3/2}(\Gamma) \times H^{3/2}(\Gamma)$. However,

$$\langle \nabla_\Gamma \phi, \boldsymbol{\phi} \rangle_{t,\|} = \langle \nabla_\Gamma \phi, \boldsymbol{\phi} \rangle_t = \langle \boldsymbol{\phi}, \nabla_\Gamma \phi \rangle_{t,\perp} \qquad \forall \boldsymbol{\phi} \in \mathbf{H}^{1/2}_\perp(\Gamma), \qquad \phi \in H^{3/2}(\Gamma),$$

since both extensions of the $\mathbf{L}^2_t(\Gamma)$ inner product coincide 'in the middle.' This proves the result. $\qquad\qquad\square$

A space for the tangential divergence. We next consider the space which will end up being the range of γ_τ:

$$\mathbf{H}^{-1/2}(\mathrm{div}_\Gamma, \Gamma) := \{ \boldsymbol{\mu} \in \mathbf{H}^{-1/2}_\|(\Gamma) : \nabla_\Gamma \cdot \boldsymbol{\mu} \in H^{-1/2}(\Gamma) \},$$

endowed with the norm (clearly associated to an inner product)

$$\|\boldsymbol{\mu}\|^2_{-1/2,\mathrm{div},\Gamma} := \|\boldsymbol{\mu}\|^2_{-1/2,\|,\Gamma} + \|\nabla_\Gamma \cdot \boldsymbol{\mu}\|^2_{-1/2,\Gamma},$$

which makes $\mathbf{H}^{-1/2}(\mathrm{div}_\Gamma, \Gamma)$ a Hilbert space. The first important result is an easy consequence of Proposition 16.19, and it is just the realization that

$$\mathbf{H}^{1/2}_\perp(\Gamma) \subset \mathbf{H}^{-1/2}(\mathrm{div}_\Gamma, \Gamma),$$

with continuous injection. The next two key results are actually not complicated, but due to the large diversity of Gelfand triples and spaces that we are handling, we will look at them carefully.

Proposition 16.20. *For all* $\mathbf{u} \in \mathbf{H}(\mathrm{curl}, \Omega)$, *we have the identity*

$$\nabla_\Gamma \cdot \gamma_\tau \mathbf{u} = (\nabla \times \mathbf{u}) \cdot \mathbf{n}. \tag{16.39}$$

Therefore, the map $\gamma_\tau : \mathbf{H}(\mathrm{curl}, \Omega) \longrightarrow \mathbf{H}^{-1/2}(\mathrm{div}_\Gamma, \Gamma)$ *is bounded.*

Proof. If $\mathbf{u} \in \mathbf{H}(\mathrm{curl}, \Omega)$ and $v \in H^2(\Omega)$, we have

$$
\begin{aligned}
\langle \nabla_\Gamma \cdot \gamma_\tau \mathbf{u}, \gamma v \rangle_\Gamma &= -\langle \gamma_\tau \mathbf{u}, \nabla_\Gamma \gamma v \rangle_{t,\|} \\
&= -\langle \gamma_\tau \mathbf{u}, \Pi_\tau \gamma \nabla v \rangle_{t,\|} && \text{(Proposition 16.18)} \\
&= (\nabla \times \mathbf{u}, \nabla v)_\Omega - (\mathbf{u}, \nabla \times \nabla v)_\Omega && \text{(by (16.36))} \\
&= (\nabla \times \mathbf{u}, \nabla v)_\Omega \\
&= (\nabla \times \mathbf{u}, \nabla v)_\Omega + (\nabla \cdot (\nabla \times \mathbf{u}), v)_\Omega \\
&= \langle (\nabla \times \mathbf{u}) \cdot \mathbf{n}, \gamma v \rangle_\Gamma.
\end{aligned}
$$

Note that we have started with an $H^{-3/2}(\Gamma) \times H^{3/2}(\Gamma)$ duality and ended with the duality pair $H^{-1/2}(\Gamma) \times H^{1/2}(\Gamma)$. The continuity of the map γ_τ then follows from (16.39) since

$$\|\gamma_\tau \mathbf{u}\|^2_{-1/2,\mathrm{div},\Gamma} = \|\gamma_\tau \mathbf{u}\|^2_{-1/2,\|,\Gamma} + \|(\nabla \times \mathbf{u}) \cdot \mathbf{n}\|^2_{-1/2,\Gamma}$$
$$\leq C\|\mathbf{u}\|^2_{\mathrm{curl},\Omega} + \|\nabla \times \mathbf{u}\|^2_{\mathrm{div},\Omega} \leq (C+1)\|\mathbf{u}\|^2_{\mathrm{curl},\Omega},$$

and the proof is thus finished. □

Proposition 16.21. *Given $\boldsymbol{\mu} \in \mathbf{H}_{\|}^{-1/2}(\Gamma)$, we have that $\boldsymbol{\mu} \in \mathbf{H}^{-1/2}(\mathrm{div}_\Gamma, \Gamma)$ if and only if there exists $C > 0$ such that*

$$|\langle \boldsymbol{\mu}, \Pi_\tau \gamma \nabla u \rangle_{t,\|}| \leq C\|\gamma u\|_{1/2,\Gamma} \qquad \forall u \in H^2(\Omega). \qquad (16.40)$$

Proof. If $\boldsymbol{\mu} \in \mathbf{H}^{-1/2}(\mathrm{div}_\Gamma, \Gamma)$, then

$$\langle \boldsymbol{\mu}, \Pi_\tau \gamma \nabla u \rangle_{t,\|} = \langle \boldsymbol{\mu}, \nabla_\Gamma \gamma u \rangle_{t,\|} = -\langle \nabla_\Gamma \cdot \boldsymbol{\mu}, \gamma u \rangle_\Gamma \qquad \forall u \in H^2(\Omega),$$

which implies (16.40) since $\nabla_\Gamma \cdot \boldsymbol{\mu} \in H^{-1/2}(\Gamma)$. If (16.40) holds, we can find (by density) a unique $\eta \in H^{-1/2}(\Gamma)$ such that

$$\langle \eta, \gamma u \rangle_\Gamma = \langle \boldsymbol{\mu}, \Pi_\tau \gamma \nabla u \rangle_\Gamma \qquad \forall u \in H^2(\Omega).$$

Therefore

$$\langle \eta, \gamma u \rangle_\Gamma = \langle \boldsymbol{\mu}, \Pi_\tau \gamma \nabla u \rangle_\Gamma = \langle \boldsymbol{\mu}, \nabla_\Gamma \gamma u \rangle_{t,\|} = -\langle \nabla_\Gamma \cdot \boldsymbol{\mu}, \gamma u \rangle_\Gamma \qquad \forall u \in H^2(\Omega),$$

and $\nabla_\Gamma \cdot \boldsymbol{\mu} = -\eta$, when acting on elements of $H^{3/2}(\Gamma)$, which is dense in $H^{1/2}(\Gamma)$. In other words, $\nabla_\Gamma \cdot \boldsymbol{\mu} \in H^{-1/2}(\Gamma)$. □

We now bring a result from the literature characterizing the range of γ_T. The proof of this was never published by its discoverer, but appears in a reference that we give in our literature review at the end of the chapter. It does not require tools that we do not know so far.

[Proof not provided]

Theorem 16.4. *The range of γ_T is the set*

$$\left\{ \boldsymbol{\theta} \in \mathbf{H}^{-1/2}(\Gamma) : \sup_{0 \neq u \in H^2(\Omega)} \frac{\langle \boldsymbol{\theta}, \gamma \nabla u \rangle_\Gamma}{\|\gamma u\|_{1/2,\Gamma}} < \infty \right\}.$$

Corollary 16.1. *The range of γ_τ is $\mathbf{H}^{-1/2}(\mathrm{div}_\Gamma, \Gamma)$. In particular, there exists $C > 0$ such that*

$$\inf\{\|\mathbf{u}\|_{\mathrm{curl},\Omega} : \gamma_\tau \mathbf{u} = \boldsymbol{\mu}\} \leq C\|\boldsymbol{\mu}\|_{-1/2,\mathrm{div},\Gamma}, \qquad (16.41)$$

and the expression of the left-hand side defines an equivalent norm in $\mathbf{H}^{-1/2}(\mathrm{div}_\Gamma, \Gamma)$.

Proof. Since $\iota_\pi \circ \gamma_\tau = \gamma_T$, we have that $\boldsymbol{\mu} \in \operatorname{range} \gamma_\tau$ if and only if $\iota_\pi \boldsymbol{\mu} \in$ range γ_T. Comparing Theorem 16.4 and Proposition 16.21, we have that $\boldsymbol{\mu} \in$ $\mathbf{H}^{-1/2}(\operatorname{div}_\Gamma, \Gamma)$ if and only if $\iota_\pi \boldsymbol{\mu} \in \operatorname{range} \gamma_T$, which fully characterizes the range of γ_τ. Note finally that, given $\boldsymbol{\mu} \in \mathbf{H}^{-1/2}(\operatorname{div}_\Gamma, \Gamma)$, the minimization problem

$$\tfrac{1}{2}\|\mathbf{u}\|^2_{\operatorname{curl},\Omega} = \min! \quad \mathbf{u} \in \mathbf{H}(\operatorname{curl}, \Omega), \quad \gamma_\tau \mathbf{u} = \boldsymbol{\mu}$$

is equivalent to the well posed (coercive) variational problem

$$\mathbf{u} \in \mathbf{H}(\operatorname{curl}, \Omega), \quad \gamma_\tau \mathbf{u} = \boldsymbol{\mu},$$
$$(\nabla \times \mathbf{u}, \nabla \times \mathbf{v})_\Omega + (\mathbf{u}, \mathbf{v})_\Omega = 0 \quad \forall \mathbf{v} \in \mathbf{H}_0(\operatorname{curl}, \Omega) = \ker \gamma_\tau,$$

and to the nonhomogeneous Dirichlet problem

$$\mathbf{u} \in \mathbf{H}(\operatorname{curl}, \Omega), \quad \nabla \times \nabla \times \mathbf{u} + \mathbf{u} = \mathbf{0}, \quad \gamma_\tau \mathbf{u} = \boldsymbol{\mu}.$$

This proves the estimate (16.41). Note that this argument is looking for the Moore-Penrose pseudoinverse of γ_τ. $\qquad\square$

16.10 The curl-curl integration by parts formula

Vector and scalar tangential curls. The tangential curls (in vector and scalar forms) arise from applying a rotation to the tangential gradient and divergence. We thus first define the weak form of the tangential curl

$$\nabla^\perp_\Gamma := r^{-1}\nabla_\Gamma = r^{-1}\pi_\tau \nabla \gamma^+_{1/2} = \gamma_\tau \nabla \gamma^+_{1/2} : H^{1/2}(\Gamma) \to \mathbf{H}^{-1/2}_{\|}(\Gamma).$$

It is clear that

$$\nabla^\perp_\Gamma \phi = r\nabla_\Gamma \phi = -(r')^{-1}\nabla_\Gamma \phi \in \mathbf{H}^{1/2}_\perp(\Gamma) \quad \forall \phi \in H^{3/2}(\Gamma),$$

and that $\nabla^\perp_\Gamma : H^{3/2}(\Gamma) \to \mathbf{H}^{1/2}_\perp(\Gamma)$ is a bounded operator. We take the adjoint of this stronger version of the vector curl to define

$$\operatorname{curl}_\Gamma : \mathbf{H}^{-1/2}_\perp(\Gamma) \to H^{-3/2}(\Gamma)$$

as

$$\langle \operatorname{curl}_\Gamma \boldsymbol{\lambda}, \phi \rangle_\Gamma = \langle \nabla^\perp_\Gamma \cdot \boldsymbol{\lambda}, \phi \rangle_\Gamma = \langle \boldsymbol{\lambda}, \nabla^\perp_\Gamma \phi \rangle_{t,\perp} \quad \boldsymbol{\lambda} \in \mathbf{H}^{-1/2}_\perp(\Gamma), \quad \phi \in H^{3/2}(\Gamma).$$

Since for all $\boldsymbol{\lambda} \in \mathbf{H}^{-1/2}_\perp(\Gamma)$ and $\phi \in H^{3/2}(\Gamma)$ we have

$$\langle \nabla^\perp_\Gamma \cdot \boldsymbol{\lambda}, \phi \rangle_\Gamma = -\langle \boldsymbol{\lambda}, (r^{-1})'\nabla_\Gamma \phi \rangle_{t,\perp} = \langle \nabla_\Gamma \cdot r^{-1}\boldsymbol{\lambda}, \phi \rangle_\Gamma,$$

we have the formula

$$\nabla_\Gamma^\perp \cdot \boldsymbol{\lambda} = \nabla_\Gamma \cdot (r^{-1}\boldsymbol{\lambda}) \qquad \forall \boldsymbol{\lambda} \in \mathbf{H}_\perp^{-1/2}(\Gamma), \qquad (16.42)$$

which shows that

$$\nabla_\Gamma^\perp \cdot \boldsymbol{\lambda} \in H^{-1/2}(\Gamma) \qquad \forall \boldsymbol{\lambda} \in \mathbf{H}_{||}^{1/2}(\Gamma), \qquad (16.43)$$

and gives an alternative definition for a stronger form of the scalar curl. The formula (16.42) also shows that (recall Proposition 16.20)

$$\nabla_\Gamma^\perp \cdot \pi_\tau \mathbf{u} = \nabla_\Gamma^\perp \cdot (r\gamma_\tau \mathbf{u}) = \nabla_\Gamma \cdot \gamma_\tau \mathbf{u} = (\nabla \times \mathbf{u}) \cdot \mathbf{n} \qquad \forall \mathbf{u} \in \mathbf{H}(\mathrm{curl}, \Omega). \quad (16.44)$$

In particular, if $\mathbf{u} \in \mathbf{H}(\mathrm{curl}, \Omega)$, then $\pi_\tau \mathbf{u}$ is an element of the space

$$\mathbf{H}^{-1/2}(\mathrm{curl}_\Gamma, \Gamma) := \{\boldsymbol{\lambda} \in \mathbf{H}_\perp^{-1/2}(\Gamma) : \nabla_\Gamma^\perp \cdot \boldsymbol{\lambda} \in H^{-1/2}(\Gamma)\}.$$

We endow this space with its natural norm

$$\|\boldsymbol{\lambda}\|_{-1/2,\mathrm{curl},\Gamma}^2 := \|\boldsymbol{\lambda}\|_{-1/2,\perp,\Gamma}^2 + \|\nabla_\Gamma^\perp \cdot \boldsymbol{\lambda}\|_{-1/2,\Gamma}^2.$$

The following result collects several key properties of this space.

Proposition 16.22. *We have*

$$\mathbf{H}_{||}^{1/2}(\Gamma) \subset \mathbf{H}^{-1/2}(\mathrm{curl}_\Gamma, \Gamma) = \mathrm{range}\,\pi_\tau, \qquad (16.45)$$

and $r : \mathbf{H}^{-1/2}(\mathrm{div}_\Gamma, \Gamma) \to \mathbf{H}^{-1/2}(\mathrm{curl}_\Gamma, \Gamma)$ *is an isometric isomorphism. Finally, there exists C such that*

$$\inf\{\|\mathbf{u}\|_{\mathrm{curl},\Omega} : \pi_\tau \mathbf{u} = \boldsymbol{\lambda}\} \leq C\|\boldsymbol{\lambda}\|_{-1/2,\mathrm{curl},\Gamma} \qquad \forall \boldsymbol{\lambda} \in \mathbf{H}^{-1/2}(\mathrm{curl}_\Gamma, \Gamma).$$

Proof. The inclusion in (16.45) is a direct consequence of (16.43). The inclusion of range π_τ in $\mathbf{H}^{-1/2}(\mathrm{curl}_\Gamma, \Gamma)$ is similarly a direct consequence of (16.44). The fact that $r : \mathbf{H}_{||}^{-1/2}(\Gamma) \to \mathbf{H}_\perp^{-1/2}(\Gamma)$ is an isomorphism and (16.42) prove that r defines an isomorphism between $\mathbf{H}^{-1/2}(\mathrm{div}_\Gamma, \Gamma)$ and $\mathbf{H}^{-1/2}(\mathrm{curl}_\Gamma, \Gamma)$. Note that r is also an isometry (the reader is asked to show this in Exercise 16.17). Finally, if $\boldsymbol{\lambda} \in \mathbf{H}^{-1/2}(\mathrm{curl}_\Gamma, \Gamma)$, then there exists $\mathbf{u} \in \mathbf{H}(\mathrm{curl}, \Omega)$ such that $\gamma_\tau \mathbf{u} = r^{-1}\boldsymbol{\lambda} \in \mathbf{H}^{-1/2}(\mathrm{div}_\Gamma, \Gamma) = \mathrm{range}\,\gamma_\tau$ (Corollary 16.1), and therefore $\boldsymbol{\lambda} = r\gamma_\tau \mathbf{u} = \pi_\tau \mathbf{u} \in \mathrm{range}\,\pi_\tau$, which finishes the proof of (16.45). The final inequality is a strightforward consequence of Corollary 16.1 and the already proved assertions of this proposition. $\square$

Proposition 16.23 (Curl commutator formula). *There exists a bounded bilinear form*

$$\langle \cdot, \cdot \rangle_{t,o} : \mathbf{H}^{-1/2}(\mathrm{div}_\Gamma, \Gamma) \times \mathbf{H}^{-1/2}(\mathrm{curl}_\Gamma, \Gamma) \longrightarrow \mathbb{R}$$

such that the following integration by parts formula holds

$$\langle \gamma_\tau \mathbf{u}, \pi_\tau \mathbf{v} \rangle_{t,\circ} = (\mathbf{u}, \nabla \times \mathbf{v})_\Omega - (\nabla \times \mathbf{u}, \mathbf{v})_\Omega \qquad \forall \mathbf{u}, \mathbf{v} \in \mathbf{H}(\mathrm{curl}, \Omega).$$

The bilinear form extends the tangential $\mathbf{L}_t^2(\Gamma)$ *inner product in the following way*

$$\langle \boldsymbol{\mu}, \boldsymbol{\lambda} \rangle_{t,\circ} = \langle \boldsymbol{\mu}, \boldsymbol{\lambda} \rangle_t \qquad \forall (\boldsymbol{\mu}, \boldsymbol{\lambda}) \in \mathbf{H}_\perp^{1/2}(\Gamma) \times \mathbf{H}_{||}^{1/2}(\Gamma).$$

Proof. We define

$$\langle \boldsymbol{\mu}, \boldsymbol{\lambda} \rangle_{t,\circ} := (\mathbf{u}, \nabla \times \mathbf{v})_\Omega - (\nabla \times \mathbf{u}, \mathbf{v})_\Omega,$$

where $\gamma_\tau \mathbf{u} = \boldsymbol{\mu}$ and $\pi_\tau \mathbf{v} = \boldsymbol{\lambda}$. The definition is correct since $\ker \gamma_\tau = \ker \pi_\tau = \mathbf{H}_0(\mathrm{curl}, \Omega)$ and $\mathbf{w} \in \mathbf{H}_0(\mathrm{curl}, \Omega)$ if and only if (Proposition 16.4)

$$(\mathbf{w}, \nabla \times \mathbf{r})_\Omega - (\nabla \times \mathbf{v}, \mathbf{r})_\Omega = 0 \qquad \forall \mathbf{r} \in \mathbf{H}(\mathrm{curl}, \Omega).$$

By Corollary 16.1 and Proposition 16.22, we can estimate

$$|\langle \boldsymbol{\mu}, \boldsymbol{\lambda} \rangle_{t,\circ}| \leq \inf\{\|\mathbf{u}\|_{\mathrm{curl},\Omega} : \gamma_\tau \mathbf{u} = \boldsymbol{\mu}\} \inf\{\|\mathbf{v}\|_{\mathrm{curl},\Omega} : \pi_\tau \mathbf{v} = \boldsymbol{\lambda}\}$$
$$\leq C\|\boldsymbol{\mu}\|_{-1/2,\mathrm{div},\Gamma}\|\boldsymbol{\lambda}\|_{-1/2,\mathrm{curl},\Gamma},$$

which proves the continuity of the bilinear form. When $\mathbf{u}, \mathbf{v} \in \mathbf{H}^1(\Omega)$ we have

$$\langle \gamma_\tau \mathbf{u}, \pi_\tau \mathbf{v} \rangle_{t,\circ} = (\mathbf{u}, \nabla \times \mathbf{v})_\Omega - (\nabla \times \mathbf{u}, \mathbf{v})_\Omega$$
$$= \langle \gamma_T \mathbf{u}, \gamma \mathbf{v} \rangle_\Gamma = \langle \imath_\pi \gamma_\tau \mathbf{u}, \gamma \mathbf{v} \rangle_\Gamma$$
$$= \langle \gamma_\tau \mathbf{u}, \Pi_\tau \gamma \mathbf{v} \rangle_{t,||} = \langle \gamma_\tau \mathbf{u}, \pi_\tau \mathbf{v} \rangle_t,$$

and this finishes the proof. $\qquad\qquad\square$

Proposition 16.24 (Reciprocal duality estimates). *There exist constants such that for all* $\boldsymbol{\mu} \in \mathbf{H}^{-1/2}(\mathrm{div}_\Gamma, \Gamma)$

$$c_1\|\boldsymbol{\mu}\|_{-1/2,\mathrm{div},\Gamma} \leq \sup_{0 \neq \boldsymbol{\lambda} \in \mathbf{H}(\mathrm{curl}_\Gamma,\Gamma)} \frac{\langle \boldsymbol{\mu}, \boldsymbol{\lambda} \rangle_{t,\circ}}{\|\boldsymbol{\lambda}\|_{-1/2,\mathrm{curl},\Gamma}} \leq c_2\|\boldsymbol{\mu}\|_{-1/2,\mathrm{div},\Gamma}, \quad (16.46)$$

and for all $\boldsymbol{\lambda} \in \mathbf{H}^{-1/2}(\mathrm{curl}_\Gamma, \Gamma)$

$$c_1\|\boldsymbol{\lambda}\|_{-1/2,\mathrm{curl},\Gamma} \leq \sup_{0 \neq \boldsymbol{\mu} \in \mathbf{H}(\mathrm{div}_\Gamma,\Gamma)} \frac{\langle \boldsymbol{\mu}, \boldsymbol{\lambda} \rangle_{t,\circ}}{\|\boldsymbol{\mu}\|_{-1/2,\mathrm{div},\Gamma}} \leq c_2\|\boldsymbol{\lambda}\|_{-1/2,\mathrm{curl},\Gamma}. \quad (16.47)$$

Therefore, the maps

$$\mathbf{H}^{-1/2}(\mathrm{div}_\Gamma, \Gamma) \ni \boldsymbol{\mu} \longmapsto \langle \boldsymbol{\mu}, \cdot \rangle_{t,\circ} \in \mathbf{H}^{-1/2}(\mathrm{curl}_\Gamma, \Gamma)', \qquad (16.48\mathrm{a})$$
$$\mathbf{H}^{-1/2}(\mathrm{curl}_\Gamma, \Gamma) \ni \boldsymbol{\lambda} \longmapsto \langle \cdot, \boldsymbol{\lambda} \rangle_{t,\circ} \in \mathbf{H}^{-1/2}(\mathrm{div}_\Gamma, \Gamma)' \qquad (16.48\mathrm{b})$$

are isomorphisms and the bracket $\langle \cdot, \cdot \rangle_{t,\circ}$ *can be considered as the duality pairing of* $\mathbf{H}^{-1/2}(\mathrm{div}_\Gamma, \Gamma) \times \mathbf{H}^{-1/2}(\mathrm{curl}_\Gamma, \Gamma)$.

Proof. The rightmost inequalities in (16.46) and (16.47) are due to the continuity of the bracket $\langle \cdot, \cdot \rangle_{t,o}$. Now let $\boldsymbol{\mu} \in \mathbf{H}^{-1/2}(\mathrm{div}_\Gamma, \Gamma)$ and consider the problem (recall the proof of Corollary 16.1)

$$\mathbf{u} \in \mathbf{H}(\mathrm{curl}, \Omega), \qquad \nabla \times \nabla \times \mathbf{u} + \mathbf{u} = 0, \qquad \gamma_\tau \mathbf{u} = \boldsymbol{\mu}.$$

Note that $\nabla \times \mathbf{u} \in \mathbf{H}(\mathrm{curl}, \Omega)$ and $\|\nabla \times \mathbf{u}\|_{\mathrm{curl}, \Omega} = \|\mathbf{u}\|_{\mathrm{curl}, \Omega}$. Since

$$\begin{aligned}
\langle \boldsymbol{\mu}, -\pi_\tau(\nabla \times \mathbf{u}) \rangle_{t,o} &= -\langle \gamma_\tau \mathbf{u}, \pi_\tau(\nabla \times \mathbf{u}) \rangle_{t,o} \\
&= (\nabla \times \mathbf{u}, \nabla \times \mathbf{u})_\Omega - (\mathbf{u}, \nabla \times \nabla \times \mathbf{u})_\Omega \\
&= \|\mathbf{u}\|_{\mathrm{curl}, \Omega}^2 = \|\mathbf{u}\|_{\mathrm{curl}, \Omega} \|\nabla \times \mathbf{u}\|_{\mathrm{curl}, \Omega} \\
&\geq C \|\boldsymbol{\mu}\|_{-1/2, \mathrm{div}, \Gamma} \|\pi_\tau(\nabla \times \mathbf{u})\|_{-1/2, \mathrm{curl}, \Gamma},
\end{aligned}$$

the leftmost inequality in (16.46) follows. To prove the missing inequality in (16.47), we solve the problem

$$\mathbf{u} \in \mathbf{H}(\mathrm{curl}, \Omega), \qquad \nabla \times \nabla \times \mathbf{u} + \mathbf{u} = 0, \qquad \pi_\tau \mathbf{u} = \boldsymbol{\lambda},$$

and proceed with the same argument.

The operator defined in (16.48a) is bounded, injective and has closed range because of (16.46). The operator defined in (16.48b) is the adjoint of the one defined in (16.48a) and is, therefore, surjective. The same argument can be reversed using (16.47). $\qquad\qquad\square$

Nonhomogeneous boundary value problems. Given $\mathbf{f} \in \mathbf{L}^2(\Omega)$ and $\boldsymbol{\mu} \in \mathbf{H}^{-1/2}(\mathrm{div}_\Gamma, \Gamma)$, the boundary value problem

$$\mathbf{u} \in \mathbf{H}(\mathrm{curl}, \Omega), \qquad \nabla \times \nabla \times \mathbf{u} + \mathbf{u} = \mathbf{f}, \qquad \gamma_\tau \mathbf{u} = \boldsymbol{\mu},$$

is equivalent to

$$\begin{aligned}
&\mathbf{u} \in \mathbf{H}(\mathrm{curl}, \Omega), \qquad \gamma_\tau \mathbf{u} = \boldsymbol{\mu}, \\
&(\nabla \times \mathbf{u}, \nabla \times \mathbf{v})_\Omega + (\mathbf{u}, \mathbf{v})_\Omega = (\mathbf{f}, \mathbf{v})_\Omega \qquad \forall \mathbf{v} \in \mathbf{H}_0(\mathrm{curl}, \Omega).
\end{aligned}$$

Note that this problem could have been approached right at the beginning of this chapter by imposing a nonhomogeneous condition $\gamma_T \mathbf{u} = \boldsymbol{\xi}$ and assuming that $\boldsymbol{\xi} \in \mathrm{range}\, \gamma_T$, even if we were not able to characterize this space. The situation is not the same for the second boundary problem though. We now look for the solution of

$$\mathbf{u} \in \mathbf{H}(\mathrm{curl}, \Omega), \qquad \nabla \times \nabla \times \mathbf{u} + \mathbf{u} = \mathbf{f}, \qquad \gamma_\tau(\nabla \times \mathbf{u}) = \boldsymbol{\mu}. \qquad (16.49)$$

Because of Propositions 16.24 (which identifies $\mathbf{H}^{-1/2}(\mathrm{curl}_\Gamma, \Gamma)$ as the dual of $\mathbf{H}^{-1/2}(\mathrm{div}_\Gamma)$) and 16.22 (which identifies the range of π_τ as $\mathbf{H}^{-1/2}(\mathrm{curl}_\Gamma, \Gamma)$), the boundary condition is equivalent to

$$\langle \boldsymbol{\mu}, \pi_\tau \mathbf{v} \rangle_{t,o} = \langle \gamma_\tau(\nabla \times \mathbf{u}), \pi_\tau \mathbf{v} \rangle_{t,o} \qquad \forall \mathbf{v} \in \mathbf{H}(\mathrm{curl}, \Omega).$$

This places us in a situation very similar to the Neumann problem for the Laplace equation when trying to find the equivalent variational formulation for (16.49). We start with the boundary condition and unfold the variational formulation from there, obtaining that (16.49) is equivalent to

$$\mathbf{u} \in \mathbf{H}(\operatorname{curl}, \Omega),$$
$$(\nabla \times \mathbf{u}, \nabla \times \mathbf{v})_\Omega + (\mathbf{u}, \mathbf{v})_\Omega = (\mathbf{f}, \mathbf{v})_\Omega + \langle \boldsymbol{\mu}, \pi_\tau \mathbf{v} \rangle_{t,\circ} \qquad \forall \mathbf{v} \in \mathbf{H}(\operatorname{curl}, \Omega).$$

The reader is invited to finish up the details, including the study of well-posedness, and extensions to problems for the equation $\nabla \times \nabla \times \mathbf{u} - \omega^2 \mathbf{u} = \mathbf{f}$ with $\omega \geq 0$.

Before we finish this chapter, let us make a short summary of rotations and Gelfand triples. The following pairs of spaces are isometrically isomorphic under rotations.

$\boldsymbol{\xi} \mapsto \mathbf{n} \times \boldsymbol{\xi} = r\boldsymbol{\xi} = -(r')^{-1}\boldsymbol{\xi}$	$\mathbf{H}_{\parallel}^{1/2}(\Gamma)$	$\mathbf{H}_{\perp}^{1/2}(\Gamma)$
$\boldsymbol{\eta} \mapsto \mathbf{n} \times \boldsymbol{\eta}$	$\mathbf{L}_t^2(\Gamma)$	$\mathbf{L}_t^2(\Gamma)$
$r := -\imath_\upsilon^{-1} \circ \imath_\pi$	$\mathbf{H}_{\parallel}^{-1/2}(\Gamma)$	$\mathbf{H}_{\perp}^{-1/2}(\Gamma)$
$r\vert_{\mathbf{H}^{-1/2}(\operatorname{div}_\Gamma, \Gamma)}$	$\mathbf{H}^{-1/2}(\operatorname{div}_\Gamma, \Gamma)$	$\mathbf{H}^{-1/2}(\operatorname{curl}_\Gamma, \Gamma)$

In the next diagram, we list seven of the spaces that we have used in the preceding sections, the arrows denoting inclusions, all of which are continuous and dense. The central cross of arrows corresponds to the pair of Gelfand triples pivotal to $\mathbf{L}_t^2(\Gamma)$. The limiting spaces in the second row are dual to each other with respect to an extension of the inner product of the central space in that row, although that space is not of a subspace of either of them and, therefore, this row fails to define a Gelfand triple.

$$
\begin{array}{ccccc}
\mathbf{H}_{\parallel}^{1/2}(\Gamma) & & & & \mathbf{H}_{\perp}^{1/2}(\Gamma) \\
\swarrow \qquad \searrow & & \swarrow \qquad \searrow \\
\mathbf{H}^{-1/2}(\operatorname{curl}_\Gamma, \Gamma) & & \mathbf{L}_t^2(\Gamma) & & \mathbf{H}^{-1/2}(\operatorname{div}_\Gamma, \Gamma) \\
\searrow \qquad \swarrow & & \searrow \qquad \swarrow \\
& \mathbf{H}_{\perp}^{-1/2}(\Gamma) & & \mathbf{H}_{\parallel}^{-1/2}(\Gamma)
\end{array}
$$

Final comments and literature

We first give hints at where to find the missing proofs in this chapter.

(a) The H^2 extension property (Theorem 16.3) for strong Lipschitz domains is part of what appears in [79, Appendix A]. It uses a nontrivial construction called the Sobolev representation formula, which brings us all the way to the realm of harmonic analysis.

(b) The proof of the Weber compactness theorem (Theorem 16.1) can be found in [101]. (See also Picard [90].) Some additional properties of this kind (about vector fields in three dimensions) can be found in [11, 4].

(c) The proof of Theorem 16.2 follows from the use of regularized Poincaré and Bogovskiĭ operators in [41]: on strongly star-shaped domains with trivial cohomology the de Rham sequences of section 16.5 are exact, i.e., the ranges of the operators are the kernels of the following ones.

(d) The proof of Theorem 16.4, a result due to Luc Tartar [97], is technical (using local charts) but not particularly difficult. It can be found in full detail in [28, Theorem 7.1].

The characterization of the trace spaces and the curl-curl integration by parts formula (curl commutator formula) on non-smooth domains is a very recent result. When the domains are smooth so that the normal vector field is a smooth function defined on the boundary, there is no problem in handling the rotation operators and all the spaces are easy to understand. The difficulties for Lipschitz polyhedra were first tackled by Annalisa Buffa and Patrick Ciarlet [26, 27]. Shortly thereafter, they were extended to Lipschitz domains by Buffa, Martin Costabel and Dongwoo Sheen [28].

For more on the de Rham sequences and extensions of the same ideas to any dimensions, one has to change the language to differential forms. A modern, recent introduction is due to Douglas Arnold, Richard Falk, and Ragnar Winther [5], who also create a whole theory for discretization of differential complexes. On the topic of regularized formulations for Maxwell, [37, 39, 35] are standard references, while [38] is a good source for knowledge about Maxwell eigenvalues. For more on general mathematical techniques in electromagnetism, the lecture notes [60] contain careful explanations of spaces and formulations. Peter Monk's book on numerical electromagnetism [81], which is due a second edition shortly, is another excellent resource for Maxwell's equations. Finally, the treatment of the time harmonic Maxwell equations given here is inspired by [54], where a problem with impedance boundary conditions is studied.

Exercises

16.1. Show that $\mathbf{H}(\mathrm{curl}, \Omega)$ is a Hilbert space. Prove that

$$\|\mathbf{u}\|_{\mathrm{curl},\Omega} \leq \|\mathbf{u}\|_{1,\Omega} \qquad \forall \mathbf{u} \in \mathbf{H}^1(\Omega).$$

(**Remark.** The only difficulty of the last part of the exercise is showing that there is no multiplicative constant in the inequality.)

16.2. Show that $\nabla \times \mathbf{H}_0(\mathrm{curl}, \Omega) \subset \mathbf{H}_0(\mathrm{div}, \Omega)$.

16.3. **The eddy current problem.** Understanding that we allow for complex-valued functions in the definition of $\mathbf{H}(\mathrm{curl}, \Omega)$, show that for every $\omega \neq 0$, the problem

$$\mathbf{u} \in \mathbf{H}_0(\mathrm{curl}, \Omega), \qquad \nabla \times \nabla \times \mathbf{u} + \imath \omega\, \mathbf{u} = \mathbf{f},$$

is well posed. (Here $\mathbf{f} \in \mathbf{L}^2(\Omega)$.)

16.4. **A local regularity theorem.** Let $\mathbf{u} \in \mathbf{H}(\mathrm{curl}, \Omega) \cap \mathbf{H}(\mathrm{div}, \Omega)$. Show that $\varphi\mathbf{u} \in \mathbf{H}^1(\Omega)$ for every $\varphi \in \mathcal{D}(\Omega)$ and therefore $\mathbf{u} \in \mathbf{H}^1(B)$ for every open ball B such that $\overline{B} \subset \Omega$. (**Hint.** For the first question, extend by zero.)

16.5. **More on mixed formulations for the curl-curl operator.** Consider the problem

$$
\begin{aligned}
&(\mathbf{u}, p) \in \mathbf{H}_0(\mathrm{curl}, \Omega) \times H_0^1(\Omega), \\
&(\nabla \times \mathbf{u}, \nabla \times \mathbf{v})_\Omega + (\mathbf{v}, \nabla p)_\Omega = (\mathbf{f}, \mathbf{v})_\Omega && \forall \mathbf{v} \in \mathbf{H}_0(\mathrm{curl}, \Omega), \\
&(\mathbf{u}, \nabla q)_\Omega && = 0 && \forall q \in H_0^1(\Omega),
\end{aligned}
$$

where $\mathbf{f} \in \mathbf{L}^2(\Omega)$ (we do not need $\nabla \cdot \mathbf{f} = 0$ for this argument). Study this problem using the Fredholm alternative in the case where there are harmonic fields. (**Hint.** It is very easy to relate all homogeneous solutions of the above problem to harmonic fields.)

16.6. Show that $\mathrm{div} : \mathbf{H}_0(\mathrm{curl}, \Omega) \to H^{-1}(\Omega)$ is surjective. To do that, consider the mixed problem

$$
\begin{aligned}
&(\mathbf{u}, p) \in \mathbf{H}_0(\mathrm{curl}, \Omega) \times H_0^1(\Omega), \\
&(\mathbf{u}, \mathbf{v})_{\mathrm{curl},\Omega} + (\nabla p, \mathbf{v})_\Omega = 0 && \forall \mathbf{v} \in \mathbf{H}_0(\mathrm{curl}, \Omega), \\
&(\mathbf{u}, \nabla q)_\Omega = -\langle f, q\rangle_{H^{-1}(\Omega) \times H_0^1(\Omega)} && \forall q \in H_0^1(\Omega).
\end{aligned}
$$

16.7. **The second boundary value problem for the curl-curl operator.** Prove the following statements:

(a) If $\mathbf{u} \in \mathbf{H}(\mathrm{div}, \Omega)$, then

$$\left.\begin{array}{l} \nabla \cdot \mathbf{u} = 0 \\ \mathbf{u} \cdot \mathbf{n} = 0 \end{array}\right\} \iff (\mathbf{u}, \nabla q)_\Omega = 0 \qquad \forall q \in H^1(\Omega).$$

(b) The set of solutions to

$$\mathbf{u} \in \mathbf{Y}, \qquad \nabla \times \mathbf{u} = \mathbf{0}, \qquad \nabla \cdot \mathbf{u} = 0, \tag{16.50}$$

is finite-dimensional.

(c) Uniqueness of solution to (16.50) is equivalent to uniqueness of solution to

$$\mathbf{u} \in \mathbf{Y}, \qquad \nabla \times \nabla \times \mathbf{u} = \mathbf{f}, \qquad \gamma_T(\nabla \times \mathbf{u}) = \mathbf{0}, \qquad \nabla \cdot \mathbf{u} = 0, \tag{16.51}$$

and to the existence of $C > 0$ such that

$$\|\nabla \times \mathbf{u}\|_\Omega + \|\nabla \cdot \mathbf{u}\|_\Omega \geq C \|\mathbf{u}\|_\Omega \qquad \forall \mathbf{u} \in \mathbf{Y}.$$

(d) For $\mathbf{f} \in \mathbf{L}^2(\Omega)$ with $\nabla \cdot \mathbf{f} = 0$, problem (16.51) is equivalent to

$$\mathbf{u} \in \mathbf{Y}, \qquad\qquad\qquad\qquad\qquad\qquad\qquad\qquad (16.52\text{a})$$
$$(\nabla \times \mathbf{u}, \nabla \times \mathbf{v})_\Omega + (\nabla \cdot \mathbf{u}, \nabla \cdot \mathbf{v})_\Omega = (\mathbf{f}, \mathbf{v})_\Omega \qquad \forall \mathbf{v} \in \mathbf{Y}. \qquad (16.52\text{b})$$

(e) With the same hypotheses on $\mathbf{f}$, problem (16.52) is equivalent to

$$(\mathbf{u}, p) \in \mathbf{H}(\text{curl}, \Omega) \times H^1_*(\Omega),$$
$$(\nabla \times \mathbf{u}, \nabla \times \mathbf{v})_\Omega + (\mathbf{v}, \nabla p)_\Omega = (\mathbf{f}, \mathbf{v})_\Omega \qquad \forall \mathbf{v} \in \mathbf{H}(\text{curl}, \Omega),$$
$$(\mathbf{u}, \nabla q)_\Omega = 0 \qquad \forall q \in H^1_*(\Omega),$$

where $H^1_*(\Omega) := \{p \in H^1(\Omega) : (p, 1)_\Omega = 0\}$.

Finally, study the well-posedness of the problems in (d) and (e), considering the nonuniquely solvable cases too, and write down minimization problems associated to all the above.

16.8. Show that $\mathbf{H}_0(\text{curl}, \Omega)$ and $\mathbf{H}(\text{curl}, \Omega)$ are not compactly embedded into $\mathbf{L}^2(\Omega)$. (**Hint.** Use gradients of functions in $H^1_0(\Omega)$ and $H^1(\Omega)$ to build counterexamples.)

16.9. **Neumann time-harmonic Maxwell problems.** We consider the problems

$$\mathbf{u} \in \mathbf{H}(\text{curl}, \Omega), \qquad \nabla \times \nabla \times \mathbf{u} - \omega^2 \mathbf{u} = \mathbf{f}, \qquad \gamma_T(\nabla \times \mathbf{u}) = \mathbf{0}, \qquad (16.53)$$

and

$$\mathbf{u} \in \mathbf{Y}, \qquad c_{\omega,r}(\mathbf{u}, \mathbf{v}) = (\mathbf{f}, \mathbf{v})_\Omega - r(\nabla \cdot \mathbf{f}, \nabla \cdot \mathbf{v})_\Omega \qquad \forall \mathbf{v} \in \mathbf{Y}, \qquad (16.54)$$

where $c_{\omega,r}$ is given by (16.24) and $\mathbf{f} \in \mathbf{H}_0(\text{div}, \Omega)$.

(a) Show that (16.53) and (16.54) are equivalent if $-r^{-1}$ is not a Neumann eigenvalue for the Laplacian in Ω. (Hint. Follow the arguments of Proposition 16.10, using Neumann boundary conditions.)

(b) Derive conditions for the well-posedness of (16.53), including the case where nontrivial homogeneous solutions exist.

16.10. **Neumann time-harmonic Maxwell problems.** Consider the problem

$$\mathbf{u} \in \mathbf{H}(\text{curl}, \Omega), \qquad \nabla \times \boldsymbol{\kappa}(\nabla \times \mathbf{u}) - \omega^2 \boldsymbol{\rho} \mathbf{u} = \mathbf{f}, \qquad \gamma_T(\boldsymbol{\kappa}(\nabla \times \mathbf{u})) = \mathbf{0},$$

where $\mathbf{f} \in \mathbf{L}^2(\Omega)$ and the matrix-valued coefficients $\boldsymbol{\kappa}$ and $\boldsymbol{\rho}$ are L^∞ and strongly positive definite as in Section 16.4. Study this problem with a direct formulation using the operator $P : \mathbf{H}(\text{curl}, \Omega) \to \mathbf{H}(\text{curl}, \Omega)$ given by

$$P\mathbf{u} = \mathbf{u} + \nabla w, \quad \text{where} \quad \begin{cases} w \in H^1(\Omega), \\ (\nabla w, \nabla v)_\Omega = -(\mathbf{u}, \nabla v)_\Omega \qquad \forall v \in H^1(\Omega). \end{cases}$$

16.11. Prove Proposition 16.13.

16.12. Show that $\{\nabla \times \mathbf{u} : \mathbf{u} \in \mathbf{H}(\mathrm{curl}, \Omega)\}$ is closed in $\mathbf{L}^2(\Omega)$ and $\mathbf{H}(\mathrm{div}, \Omega)$.

16.13. Show that if $\mathbf{u} \in \mathbf{H}_0(\mathrm{curl}, \Omega)$ satisfies $\nabla \times \nabla \times \mathbf{u} = \mathbf{0}$, then there exists $w \in H_0^1(\Omega)$ such that $\mathbf{u} - \nabla w \in \mathcal{H}_X$.

16.14. Show that if $\mathbf{u} \in \mathbf{H}(\mathrm{curl}, \Omega)$ satisfies $\nabla \times \nabla \times \mathbf{u} = \mathbf{0}$ and $\gamma_T(\nabla \times \mathbf{u}) = \mathbf{0}$, then there exists $w \in H^1(\Omega)$ such that $\mathbf{u} - \nabla w \in \mathcal{H}_Y$.

16.15. **A lifting of the trace in $H^{3/2}(\Gamma)$.** Given $\phi \in \gamma H^2(\Omega)$, we define $u \in H^2(\Omega)$ as the solution of

$$\|u\|_{2,\Omega} = \min! \qquad u \in H^2(\Omega), \qquad \gamma u = \phi. \tag{16.55}$$

If we have defined the $H^2(\Omega)$ norm to include double occurrences of the crossed derivatives, i.e.,

$$\|u\|_{2,\Omega}^2 := \|u\|_{\Omega}^2 + \|\nabla u\|_{\Omega}^2 + \sum_{i,j=1}^{d} \|\partial_{x_i}\partial_{x_j}u\|_{\Omega}^2,$$

show that the solution to (16.55) satisfies

$$\Delta^2 u - \Delta u + u = 0, \qquad \gamma u = \phi,$$

and the weak boundary condition

$$\sum_{i,j=1}^{d}(\partial_{x_i}\partial_{x_j}u, \partial_{x_i}\partial_{x_j}v)_\Omega - (\Delta^2 u, v)_\Omega = 0 \qquad \forall v \in H^2(\Omega) \cap H_0^1(\Omega).$$

16.16. Prove that $\mathbf{H}^{-1/2}(\mathrm{div}_\Gamma, \Gamma)$ is a Hilbert space. (**Hint.** If $\{\boldsymbol{\mu}_n\}$ converges in $\mathbf{H}_{\|}^{-1/2}(\Gamma)$, then $\{\nabla_\Gamma \cdot \boldsymbol{\mu}_n\}$ converges in $H^{-3/2}(\Gamma)$.)

16.17. **The rotation operator is an isometry.**

(a) Prove the following lemma: If $\zeta : X \to Y$ is a linear surjective operator between a Banach space X and a vector space Y, $\ker \zeta$ is closed, and we endow Y with the image norm, then $\zeta' : Y' \to X'$ is an isometry with range $(\ker \zeta)^\circ$.

(b) Use part (a) to prove that $r : \mathbf{H}_{\|}^{-1/2}(\Gamma) \to \mathbf{H}_{\perp}^{-1/2}(\Gamma)$ is an isometry.

(c) Use part (b) to show that

$$\|r\phi\|_{1/2,\perp,\Gamma} = \|\mathbf{n} \times \phi\|_{1/2,\perp,\Gamma} = \|\phi\|_{1/2,\|,\Gamma} \qquad \forall \phi \in \mathbf{H}_{\|}^{1/2}(\Gamma).$$

17

Elliptic equations on boundaries

17.1   Surface gradient and Laplace-Beltrami operator 453
17.2   The Poincaré inequality on a surface 456
17.3   More on boundary spaces 459
Final comments and literature 461
Exercises 462

In this chapter we give the fundamentals for a theory of elliptic partial differential equations defined on the boundary of a Lipschitz domain. The key problem will be the Laplace-Beltrami equation

$$-\Delta_\Gamma g = f,$$

associated to a Dirichlet form $\langle \nabla_\Gamma g, \nabla_\Gamma h \rangle_\Gamma$ on a suitably defined Sobolev space $H^1(\Gamma)$. Apart from the difficulties of defining the surface differential operators, we will be challenged with the property (not easy to prove) that if the surface gradient of a field vanishes, then the field needs to be constant.

17.1   Surface gradient and Laplace-Beltrami operator

We start by recalling that, as part of the definition of Lispchitz domain (see Section 3.5), we have a collection of (Lipschitz continuous) local charts $\boldsymbol{\Phi}_\ell : B_{d-1}(\mathbf{0}; 1) \to \Gamma$ that parametrize patches Γ_ℓ for $\ell = 1, \ldots, L$ (we will use the distinctly recognizable index ℓ to count the charts). We also recall that, at a point $\mathbf{y} = \boldsymbol{\Phi}_\ell(\widetilde{\mathbf{y}}) \in \Gamma_\ell$, such that $\boldsymbol{\Phi}_\ell$ is differentiable at $\widetilde{\mathbf{y}}$, the normal vector $\mathbf{n}(\mathbf{y})$ is orthogonal to the linearly independent vectors $\partial_{x_i} \boldsymbol{\Phi}_\ell(\widetilde{\mathbf{y}})$ for $i \in \{1, \ldots, d-1\}$. As a first step toward a definition of a tangential gradient we show that smooth functions that vanish on Γ have outward oriented normal vectors.

Lemma 17.1. *If $\varphi \in \mathcal{D}(\mathbb{R}^d)$ and $\varphi \equiv 0$ on Γ, then*

$$\nabla \varphi = (\nabla \varphi \cdot \mathbf{n}) \mathbf{n}.$$

Proof. By the above hypothesis, we have $\varphi \circ \boldsymbol{\Phi}_\ell \equiv 0$ and therefore

$$\mathbf{0} = \nabla(\varphi \circ \boldsymbol{\Phi}_\ell) = (\nabla \varphi \circ \boldsymbol{\Phi}_\ell) D\boldsymbol{\Phi}_\ell \qquad \ell = 1, \ldots, L.$$

Therefore $\nabla\varphi \circ \Phi_\ell$ is orthogonal to $\partial_{x_i}\Phi_\ell$ for $i = 1, \ldots, d-1$, and thus $\nabla\varphi \circ \Phi_\ell$ is parallel to $\mathbf{n}$. $\qquad\qquad\square$

The surface gradient. Given $g \in \mathcal{D}(\Gamma)$, that is, when $g = \varphi|_\Gamma$ where $\varphi \in C^\infty(\mathbb{R}^d)$ or equivalently, where $\varphi \in \mathcal{D}(\mathbb{R}^d)$, we define

$$\nabla_\Gamma g := (\nabla\varphi - (\nabla\varphi \cdot \mathbf{n})\mathbf{n}) : \Gamma \to \mathbb{R}^d.$$

Lemma 17.1 shows that $\|\nabla_\Gamma\varphi\|_\Gamma$ is well-defined for all $\varphi \in \mathcal{D}(\Gamma)$. We then define $H^1(\Gamma)$ to be the closure of $\mathcal{D}(\Gamma)$ with respect to the norm

$$\|g\|_{1,\Gamma}^2 := \|g\|_\Gamma^2 + \|\nabla_\Gamma g\|_\Gamma^2.$$

Proposition 17.1. *The following properties hold:*

(a) $H^1(\Gamma)$ *is continuously and densely embedded into* $L^2(\Gamma)$.

(b) *The surface gradient can be uniquely extended to a bounded linear operator*

$$\nabla_\Gamma : H^1(\Gamma) \to \mathbf{L}^2(\Gamma)$$

satisfying

$$\nabla_\Gamma g \cdot \mathbf{n} = 0 \qquad \forall g \in H^1(\Gamma).$$

(c) *Constant functions have vanishing surface gradient, that is,* $\nabla_\Gamma 1 = \mathbf{0}$.

Proof. To prove (a), note that $\mathcal{D}(\Gamma)$ is dense in $L^2(\Gamma)$. To prove (b), note that by definition $\nabla_\Gamma : \mathcal{D}(\Gamma) \to \mathbf{L}^2(\Gamma)$ is bounded when we use the $H^1(\Gamma)$ norm in $\mathcal{D}(\Gamma)$ and thus the operator has a unique extension. Since the map $\mathcal{D}(\Gamma) \ni g \mapsto \nabla_\Gamma g \cdot \mathbf{n} \in L^2(\Gamma)$ is the zero map, so is its extension to $H^1(\Gamma)$. Finally, $g \equiv 1$ is the restriction to Γ of a smooth function which is constant in a neighborhood of Γ and therefore $\nabla_\Gamma 1 = \mathbf{0}$. $\qquad\square$

Part (a) of Proposition 17.1 allows us to consider the Gelfand triple

$$H^1(\Gamma) \subset L^2(\Gamma) \subset H^{-1}(\Gamma) := H^1(\Gamma)'.$$

Since we have already extended the $L^2(\Gamma)$ inner product to the (in principle independent) duality pairing $H^{-1/2}(\Gamma) \times H^{1/2}(\Gamma)$, we will tag the duality $H^{-1}(\Gamma) \times H^1(\Gamma)$ with the product symbol (see below) to avoid confusion. We will see in Section 17.3 that $H^1(\Gamma)$ is a dense subset of $H^{1/2}(\Gamma)$, which implies that the newly defined extension of the $L^2(\Gamma)$ inner product to the duality product $H^{-1}(\Gamma) \times H^1(\Gamma)$ is also an extension of the $H^{-1/2}(\Gamma) \times H^{1/2}(\Gamma)$ duality product.

Some elliptic problems on Γ. Given $f \in H^{-1}(\Gamma)$, the problem

$$g \in H^1(\Gamma), \tag{17.1a}$$

$$\langle \nabla_\Gamma g, \nabla_\Gamma h \rangle_\Gamma + \langle g, h \rangle_\Gamma = \langle f, h \rangle_{H^{-1}(\Gamma) \times H^1(\Gamma)} \qquad \forall h \in H^1(\Gamma) \tag{17.1b}$$

is well posed, since it just provides the Riesz-Fréchet representation of f in $H^1(\Gamma)$. We can define $\Delta_\Gamma : H^1(\Gamma) \to H^{-1}(\Gamma)$ in the form

$$\langle \Delta_\Gamma g, h \rangle_{H^{-1}(\Gamma) \times H^1(\Gamma)} := -\langle \nabla_\Gamma g, \nabla_\Gamma h \rangle_\Gamma \qquad \forall h \in H^1(\Gamma).$$

This is a bounded linear operator, called the **Laplace-Beltrami operator.** Note that $\Delta_\Gamma 1 = 0$. The problem (17.1) is obviously equivalent to

$$g \in H^1(\Gamma), \qquad -\Delta_\Gamma g + g = f.$$

The reader might be wondering where the boundary conditions are. The answer lies in the fact that Γ itself does not have a boundary and, therefore, there is no place to impose side conditions. A simple variant of (17.1) can be obtained using a strongly positive function $c \in L^\infty(\Gamma)$, $c \geq c_0 > 0$ almost everywhere,

$$g \in H^1(\Gamma), \qquad -\Delta_\Gamma g + c\,g = f,$$

since

$$\langle \nabla_\Gamma g, \nabla_\Gamma h \rangle_\Gamma + \langle c\,g, h \rangle_\Gamma$$

is coercive in $H^1(\Gamma)$. We can also consider the variational problem

$$g \in H^1(\Gamma), \tag{17.2a}$$

$$\langle \kappa \nabla_\Gamma g, \nabla_\Gamma h \rangle_\Gamma + \langle c\,g, h \rangle_\Gamma = \langle f, h \rangle_{H^{-1}(\Gamma) \times H^1(\Gamma)} \qquad \forall h \in H^1(\Gamma) \tag{17.2b}$$

where both κ and c are stricly positive functions in $L^\infty(\Gamma)$. This can be seen as a weak formulation of a surface partial differential equation. To do that, we define the **surface divergence** operator

$$\mathrm{div}_\Gamma := -\nabla'_\Gamma : \mathbf{L}^2(\Gamma) \to H^{-1}(\Gamma)$$

given by

$$\langle \nabla_\Gamma \cdot \mathbf{p}, g \rangle_{H^{-1}(\Gamma) \times H^1(\Gamma)} := -\langle \mathbf{p}, \nabla_\Gamma g \rangle_\Gamma \qquad \forall g \in H^1(\Gamma).$$

Note that we do not demand $\mathbf{p}$ to be a tangential vector field, but normal components added to $\mathbf{p}$ do not influence the surface divergence (Exercise 17.1). The problem (17.2) is equivalent to

$$-\nabla_\Gamma \cdot (\kappa \nabla_\Gamma g) + c\,g = f.$$

We also have

$$\Delta_\Gamma g = \nabla_\Gamma \cdot \nabla_\Gamma g,$$

that is $\Delta_\Gamma = \mathrm{div}_\Gamma \circ \nabla_\Gamma$. The next goal is to be able to eliminate (at least partially) the reaction coefficient c, i.e., getting some kind of Poincaré inequality. This comes with a little surprise, as the proof that $\nabla_\Gamma g = \mathbf{0}$ implies that g is constant is far from trivial. (We should, however, recall from Chapter 1 that this result for distributions on an open set was not trivial either.)

17.2   The Poincaré inequality on a surface

Our first goal is the proof of the compact embedding of $H^1(\Gamma)$ into $L^2(\Gamma)$. This will show several equivalent characterizations of the same problem (the fact that $\ker \nabla_\Gamma = \mathcal{P}_0(\Gamma)$), which we will finally admit without proof. The compact injection will automatically bring along a spectral decomposition of $L^2(\Gamma)$ in terms of eigenfunctions of the Laplace-Beltrami operator. Before we continue, recall (see Section 3.4) the notation $\{\Omega_\ell\}$ for an open cover of a Lipschitz boundary and $\{\varphi_\ell\}$ for its associated partition of unity.

Lemma 17.2. *The localization and pullback map*

$$\mathcal{D}(\Gamma) \ni g \longmapsto P_\ell g := (\varphi_\ell \, g) \circ \mathbf{\Phi}_\ell$$

is bounded from $H^1(\Gamma)$ to $H_0^1(B_{d-1}(\mathbf{0};1))$. It can therefore be extended in a unique way to a bounded linear map $H^1(\Gamma) \to H_0^1(B_{d-1}(\mathbf{0};1))$.

Proof. First, note that $P_\ell g \in L^2(B_{d-1}(\mathbf{0};1)) \cap \mathcal{C}(B_{d-1}(\mathbf{0};1))$ vanishes in a neighborhood of the boundary of $B_{d-1}(\mathbf{0};1)$. In what follows, we will use the gradient symbols ∇ and ∇_Γ as row vectors, so that ∇ represents the differential. We choose $\varphi \in \mathcal{D}(\mathbb{R}^d)$ such that $\varphi|_\Gamma = g$ and note that we have

$$\nabla(P_\ell g) = ((\varphi \nabla \varphi_\ell + \varphi_\ell \nabla \varphi) \circ \mathbf{\Phi}_\ell) D\mathbf{\Phi}_\ell.$$

Since $\varphi \nabla \varphi_\ell + \varphi_\ell \nabla \varphi$ is continuous and compactly supported in the volumetric patch Ω_ℓ and $D\mathbf{\Phi}_\ell$ is bounded, it follows that $\nabla(P_\ell g) \in \mathbf{L}^2(B_{d-1}(\mathbf{0};1))$. However, since the normal vector is orthogonal to the columns of $D\mathbf{\Phi}_\ell$, we have

$$(\nabla \varphi \circ \mathbf{\Phi}_\ell) D\mathbf{\Phi}_\ell = \left((\nabla \varphi - (\nabla \varphi \cdot \mathbf{n})\mathbf{n}^\top) \circ \mathbf{\Phi}_\ell \right) D\mathbf{\Phi}_\ell = (\nabla_\Gamma g \circ \mathbf{\Phi}_\ell) D\mathbf{\Phi}_\ell,$$

and therefore

$$\nabla(P_\ell g) = (g \circ \mathbf{\Phi}_\ell)\mathbf{b}_\ell + (\nabla_\Gamma g \circ \mathbf{\Phi}_\ell)(\varphi_\ell \circ \mathbf{\Phi}_\ell) D\mathbf{\Phi}_\ell,$$

where $\mathbf{b}_\ell$ and $(\varphi_\ell \circ \mathbf{\Phi}_\ell) D\mathbf{\Phi}_\ell$ are bounded. The rest of the proof is then easy. $\quad\square$

Proposition 17.2. *The space $H^1(\Gamma)$ is compactly embedded into $L^2(\Gamma)$.*

Proof. If $\{g_n\}$ is weakly convergent in $H^1(\Gamma)$, then for all ℓ, $\{P_\ell g_n\}$ is weakly convergent in $H_0^1(B_{d-1}(\mathbf{0};1))$ (by Lemma 17.2) and then $\{P_\ell g_n\}$ is strongly convergent in $L^2(B_{d-1}(\mathbf{0};1))$ by the Rellich-Kondrachov theorem. Therefore $\{\varphi_\ell g_n\}$ is strongly convergent in $L^2(\Gamma)$ and $\{g_n\}$ is strongly convergent in $L^2(\Gamma)$, since $\sum_\ell \varphi_\ell \equiv 1$ on Γ. $\quad\square$

As a consequence of the compactness of the embedding of $H^1(\Gamma)$ into $L^2(\Gamma)$ we have a collection of equivalent statements that can be considered as different forms of Poincaré's inequality. We first prove that they are all equivalent and we will then admit without proof that the first of them holds, and therefore all of them.

Proposition 17.3. *If* Γ *is connected, then the following statements are equiv-alent.*

(a) *If* $g \in H^1(\Gamma)$ *and* $\nabla_\Gamma g = \mathbf{0}$, *then* $g \in \mathcal{P}_0(\Gamma)$.

(b) *If* $g \in H^1(\Gamma)$, $\nabla_\Gamma g = \mathbf{0}$ *and* $\langle g, 1 \rangle_\Gamma = 0$, *then* $g = 0$.

(c) *There exists* C *such that*

$$\|g\|_\Gamma \le C \left(|\langle g, 1 \rangle_\Gamma| + \|\nabla_\Gamma g\|_\Gamma \right) \qquad \forall g \in H^1(\Gamma).$$

(d) *There exists* C *such that*

$$\left\| g - |\Gamma|^{-1} \langle g, 1 \rangle_\Gamma \right\|_\Gamma \le C \|\nabla_\Gamma g\|_\Gamma \qquad \forall g \in H^1(\Gamma).$$

(e) *The quantity*

$$\|\nabla_\Gamma g\|_\Gamma^2 + |\langle g, 1 \rangle_\Gamma|^2$$

defines an equivalent norm in $H^1(\Gamma)$.

Proof. Clearly (a) implies (b). If (b) holds, subtracting the average of g, we can easily prove (a). It is also clear that (c) implies (b). Now assume that (c) does not hold. We can then find a sequence $\{g_n\}$ such that

$$\|g_n\|_\Gamma = 1, \qquad g_n \rightharpoonup g \text{ in } L^2(\Gamma), \qquad \nabla_\Gamma g_n \to \mathbf{0} \text{ in } \mathbf{L}^2(\Gamma), \qquad \langle g_n, 1 \rangle_\Gamma \to 0.$$

Consider the functionals

$$\ell_n(v) := \langle \nabla_\Gamma g_n, \nabla_\Gamma v \rangle_\Gamma + \langle g_n, v \rangle_\Gamma = \langle g_n, v \rangle_{1,\Gamma}.$$

The conditions on the sequence $\{g_n\}$ show that

$$\ell_n(v) \to \ell(v) := \langle g, v \rangle_\Gamma \qquad \forall v \in H^1(\Gamma).$$

We now take $h \in H^1(\Gamma)$ such that $-\Delta_\Gamma h + h = g$, that is,

$$\langle g, v \rangle_\Gamma = \langle \nabla_\Gamma h, \nabla_\Gamma v \rangle_\Gamma + \langle h, v \rangle_\Gamma \qquad \forall v \in H^1(\Gamma), \qquad (17.3)$$

which proves that

$$\langle g_n, v \rangle_{1,\Gamma} = \ell_n(v) \longrightarrow \ell(v) = \langle h, v \rangle_{1,\Gamma} \qquad \forall v \in H^1(\Gamma),$$

and thus $g_n \rightharpoonup h$ in $H^1(\Gamma)$. Since $H^1(\Gamma)$ is continuously embedded into $L^2(\Gamma)$, we have $g_n \rightharpoonup h$ in $L^2(\Gamma)$ and therefore $g = h \in H^1(\Gamma)$ satisfies $\nabla_\Gamma g = \mathbf{0}$ (take $v = g$ in (17.3)) and $\langle g, 1 \rangle_\Gamma = 0$. However, by compactness (Proposition 17.2), we have $g_n \to g$ in $L^2(\Gamma)$ and therefore $\|g\|_\Gamma = 1$. This means that (b) does not hold. So far we have shown the equivalence of (a), (b) and (c), but it is easy to see that (c), (d) and (e) are equivalent, which finishes the proof. $\square$

[Proof not provided]

Theorem 17.1. *If a Lipschitz domain has connected boundary* Γ, *then* $\ker \nabla_\Gamma = \mathcal{P}_0(\Gamma)$.

The Laplace-Beltrami equation. Let $f \in H^{-1}(\Gamma)$ satisfy

$$\langle f, 1 \rangle_{H^{-1}(\Gamma) \times H^1(\Gamma)} = 0. \tag{17.4}$$

The problem

$$g \in H^1(\Gamma) \qquad -\Delta_\Gamma g = f, \qquad \langle g, 1 \rangle_\Gamma = 0,$$

is equivalent to

$$g \in H^1(\Gamma), \qquad \langle g, 1 \rangle_\Gamma = 0, \tag{17.5a}$$

$$\langle \nabla_\Gamma g, \nabla_\Gamma h \rangle_\Gamma = \langle f, h \rangle_{H^{-1}(\Gamma) \times H^1(\Gamma)} \qquad \forall h \in H^1(\Gamma), \tag{17.5b}$$

and to

$$g \in H^1(\Gamma), \tag{17.6a}$$

$$\langle \nabla_\Gamma g, \nabla_\Gamma h \rangle_\Gamma + \langle g, 1 \rangle_\Gamma \langle h, 1 \rangle_\Gamma = \langle f, h \rangle_{H^{-1}(\Gamma) \times H^1(\Gamma)} \qquad \forall h \in H^1(\Gamma). \tag{17.6b}$$

The equivalence of (17.5) and (17.6) follows from the compatibility condition (17.4). Problem (17.6) is coercive due to Proposition 17.3 and Theorem 17.1 (note that the latter shows that all assertions of Proposition 17.3 actually hold). Thus (17.6) is uniquely solvable even without the compatibility condition. The minimization problem

$$\tfrac{1}{2} \|\nabla_\Gamma g\|_\Gamma^2 - \langle f, g \rangle_{H^{-1}(\Gamma) \times H^1(\Gamma)} = \min! \qquad g \in H^1(\Gamma),$$

is equivalent to (17.5) without the side restriction on the integral of g.

Eigenvalues of the Laplace-Beltrami operator. We finish this section with a result on a spectral decomposition based on an eigensystem for the Laplace-Beltrami operator. The proof of the following proposition is left for the reader as Exercise 17.4.

Proposition 17.4. *Assume that* Γ *is the connected boundary of a Lipschitz domain. There exists a Hilbert basis of* $L^2(\Gamma)$, $\{\phi_n\}_{n \geq 0}$ *and a nondecreasing divergent sequence of nonnegative real numbers* $\{\lambda_n\}$, *such that*

$$-\Delta_\Gamma \phi_n = \lambda_n \phi_n \qquad \forall n \geq 0.$$

The first eigenpair is $\lambda_0 = 0$ *with* $\phi_0 \equiv |\Gamma|^{-1/2}$. *Furthermore, the set* $\{\phi_n\}$ *is orthogonal complete in* $H^1(\Gamma)$.

17.3   More on boundary spaces

In this section we show that

$$H^1(\Gamma) \subset H^{1/2}(\Gamma) \subset L^2(\Gamma),$$

with continuous and dense embeddings. (Note that only the inclusion $H^1(\Gamma) \subset H^{1/2}(\Gamma)$ and the continuity of this embedding have to be proved. The denseness of the inclusion follows from the fact that $\mathcal{D}(\Gamma)$ is dense in $H^1(\Gamma)$.)

We redefine the operator P_ℓ of Lemma 17.2 to include an extension by zero and to handle functions in $L^2(\Gamma)$ only. We thus consider two operators on each patch

$$P_\ell : L^2(\Gamma) \to L^2(\mathbb{R}^{d-1}), \qquad Q_\ell : H^1(\mathbb{R}^d) \to H^1(\mathbb{R}^d)$$

given by

$$P_\ell g := \begin{cases} (\varphi_\ell g) \circ \Phi_\ell, & \text{in } B_{d-1}(0;1), \\ 0 & \text{elsewhere,} \end{cases}$$

$$Q_\ell u := \begin{cases} (\varphi_\ell u) \circ F_\ell, & \text{in } B_{d-1}(0;1) \times (-1,1), \\ 0 & \text{elsewhere.} \end{cases}$$

The operators P_ℓ and Q_ℓ are linear and bounded, and we have bounds

$$\|P_\ell g\|_{\mathbb{R}^{d-1}} \le C\|g\|_\Gamma, \qquad \|P_\ell g\|_{1,\mathbb{R}^{d-1}} \le C\|g\|_{1,\Gamma}, \qquad \|Q_\ell u\|_{1,\mathbb{R}^d} \le C\|u\|_{1,\mathbb{R}^d}.$$

Proposition 17.5. *If $g \in H^{1/2}(\Gamma)$, then $P_\ell g \in \mathsf{H}^{1/2}(\mathbb{R}^{d-1})$ and we have*

$$\sum_{\ell=1}^{L} \|P_\ell g\|_{\mathsf{H}^{1/2}(\mathbb{R}^{d-1})} \le C\|g\|_{1/2,\Gamma} \qquad \forall g \in H^{1/2}(\Gamma).$$

Proof. Take $u \in H^1(\mathbb{R}^d)$ such that $\gamma u = g$. We have that $\varphi_\ell u \in H^1(\mathbb{R}^d)$ and

$$\gamma(\varphi_\ell u) = \varphi_\ell|_\Gamma \gamma u = \varphi_\ell|_\Gamma g.$$

We also have $Q_\ell u \in H^1(\mathbb{R}^d)$ and $\gamma_{\Gamma_0} Q_\ell u = P_\ell g$, where γ_{Γ_0} is the trace operator in half-space (see Theorem 13.3 and Section 4.2). Using the estimate of Theorem 13.3, we have

$$\|P_\ell g\|_{\mathsf{H}^{1/2}(\mathbb{R}^{d-1})} \le \|Q_\ell u\|_{1,\mathbb{R}^d} \le C\|u\|_{1,\mathbb{R}^d} \qquad \forall u \in H^1(\mathbb{R}^d) \text{ s.t. } \gamma u = g.$$

However,

$$\inf\{\|u\|_{1,\mathbb{R}^d} : \gamma u = g\}$$

is an equivalent norm in $H^{1/2}(\Gamma)$. (Note that we originally took the norm as an infimum over liftings to the interior domain, while this one lifts both ways.) □

Proposition 17.6. *We have the inequality*

$$\|g\|_{1/2,\Gamma} \leq C \sum_{\ell=1}^{L} \|P_\ell g\|_{\mathsf{H}^{1/2}(\mathbb{R}^{d-1})} \qquad \forall g \in H^{1/2}(\Gamma).$$

Proof. We are going to build a continuous lifting by pullback. We will use the notation for the reference domain $\mho$ as in Chapter 3. We use a lifting (recall Lemma 13.3) $R : \mathsf{H}^{1/2}(\mathbb{R}^{d-1}) \to H^1(\mathbb{R}^d)$ which is cut-off in the 'vertical direction.' Taking $g \in H^{1/2}(\Gamma)$, we have $P_\ell g \in \mathsf{H}^{1/2}(\mathbb{R}^{d-1})$ for all ℓ, and $RP_\ell g \in H^1(\mathbb{R}^d)$. The lifting can be chosen so that $RP_\ell g \equiv 0$ in a neighborhood of $\partial\mho$ and

$$RP_\ell g|_\mho \in H^1_0(\mho).$$

Therefore

$$u_\ell := (RP_\ell g) \circ F_\ell^{-1} \in H^1_0(\Omega_\ell),$$

and extending by zero, we can define

$$u := \sum_{\ell=1}^{L} \widetilde{u}_\ell \in H^1(\mathbb{R}^d).$$

Since

$$\gamma\widetilde{u}_\ell = \widetilde{P_\ell g \circ \Phi_\ell^{-1}} = \widetilde{\varphi_\ell g} = \varphi_\ell g,$$

we have that $\gamma u = g$. This proves the result, since

$$\|g\|_{1/2,\Gamma} \leq C\|u\|_{1,\mathbb{R}^d} \leq C \sum_{\ell=1}^{L} \|u_\ell\|_{1,\Omega_\ell},$$

and R is bounded. $\square$

Proposition 17.7. *If $g \in H^1(\Gamma)$, then $g \in H^{1/2}(\Gamma)$ and the inclusion $H^1(\Gamma) \subset H^{1/2}(\Gamma)$ is bounded and dense.*

Proof. Take $g \in H^1(\Gamma)$ and a sequence $\{\varphi_n\}$ in $\mathcal{D}(\Gamma)$ such that $\varphi_n \to g$ in $H^1(\Gamma)$. We have

$$P_\ell\varphi_n \to P_\ell g \qquad \text{in } H^1(\mathbb{R}^{d-1}),$$

and therefore

$$P_\ell g \in H^1(\mathbb{R}^{d-1}) \subset \mathsf{H}^{1/2}(\mathbb{R}^{d-1}).$$

We can use the same lifting used in the proof of Proposition 17.6 to show that $g \in H^{1/2}(\Gamma)$ and

$$\|g\|_{1/2,\Gamma} \leq C \sum_{\ell=1}^{L} \|P_\ell g\|_{\mathsf{H}^{1/2}(\mathbb{R}^{d-1})} \leq C \sum_{\ell=1}^{L} \|P_\ell g\|_{1,\mathbb{R}^d} \leq C'\|g\|_{1,\Gamma}.$$

As we have already mentioned in the introductory paragraph to this section, the density follows from that of $\mathcal{D}(\Gamma)$ in $H^1(\Gamma)$. $\square$

Ventcel boundary conditions. We now use the embedding of $H^1(\Gamma)$ into $H^{1/2}(\Gamma)$ to justify the well-posedness of a boundary value problem with boundary conditions of the type $\partial_n u - \Delta_\Gamma \gamma u = h$. We first consider the space

$$V := \{u \in H^1(\Omega) : \gamma u \in H^1(\Gamma)\},$$

endowed with the norm

$$\|u\|_V^2 := \|u\|_{1,\Omega}^2 + \|\gamma u\|_{1,\Gamma}^2.$$

This is a Hilbert space due to the fact that if $u_n \to u$ in $H^1(\Omega)$ and $\gamma u_n \to g$ in $H^1(\Gamma)$, then $\gamma u = g$. If $c \geq 0$, $c \in L^\infty(\Omega)$ and $c \neq 0$, then

$$\|\nabla u\|_\Omega^2 + \|c^{1/2} u\|_\Omega^2 + \|\nabla_\Gamma \gamma u\|_\Gamma^2$$

is an equivalent norm (see Exercise 17.7 for a similar result) and the problem

$$u \in V,$$
$$(\nabla u, \nabla v)_\Omega + (c\, u, v)_\Omega + \langle \nabla_\Gamma \gamma u, \nabla_\Gamma \gamma v \rangle_\Gamma = (f, v)_\Omega + \langle h, \gamma v \rangle_\Gamma \qquad \forall v \in V$$

is well posed with data $f \in L^2(\Omega)$ and $h \in H^{-1/2}(\Gamma)$. This variational problem is equivalent to the elliptic problem with Ventcel conditions

$$u \in H^1(\Omega), \qquad -\Delta u + c\, u = f, \qquad \partial_n u - \Delta_\Gamma \gamma u = h. \qquad (17.7)$$

The last expression assumes that $\gamma u \in H^1(\Gamma)$ so that we can apply the Laplace-Beltrami operator. Note that if u solves (17.7), we have $\Delta_\Gamma \gamma u \in H^{-1/2}(\Gamma)$ given the data regularity. Since we have the reverse inclusion of dual spaces

$$L^2(\Gamma) \subset H^{-1/2}(\Gamma) \subset H^{-1}(\Gamma),$$

we can easily extend (17.7) to handle $h \in H^{-1}(\Gamma)$ by changing the duality product in the right-hand side of its variational formulation.

Final comments and literature

The fact that the kernel of the tangential gradient is the set of constant functions seems like a logical statement to accept (and it is typically hidden under the rug in simplified expositions), but with the current definition of tangential gradient, it is not obvious to prove. In the three-dimensional case, there is a proof, using the tools that we developed in the previous chapter, in [28]. This is also related to the problem of proving Hodge decompositions in non-smooth domains [7].

The inclusion $H^1(\Gamma) \subset H^{1/2}(\Gamma)$ is also compact and $H^{1/2}(\Gamma)$ can be characterized as an intermediate interpolation space based on the Laplace-Beltrami eigensystem (see Sections 9.5 and 9.8) although this is far from obvious with our approach and requires going deeper into harmonic analysis.

Finally, the Ventcel boundary conditions are probably the simplest set of conditions that use tangential differential operators on surfaces. For more about them, see [16, 32, 42].

Exercises

17.1. On the surface divergence. Show that if $\mathbf{p} \in \mathbf{L}^2(\Gamma)$ is purely normal, i.e., $\mathbf{p} = (\mathbf{p} \cdot \mathbf{n})\mathbf{n}$, then $\nabla_\Gamma \cdot \mathbf{p} = 0$.

17.2. Show that if $f \in H^{-1}(\Gamma)$ satisfies $\langle f, 1 \rangle_{H^{-1}(\Gamma) \times H^1(\Gamma)} = 0$, and $\kappa \in L^\infty(\Gamma)$ is strongly positive, then

$$-\nabla_\Gamma \cdot (\kappa \nabla_\Gamma g) = f$$

is uniquely solvable up to additive constants.

17.3. Let $\kappa, c \in L^\infty(\Gamma)$ satisfy: κ is strongly positive, $c \geq 0$, and $cg = 0$ implies $g = 0$. Show that the problem

$$-\nabla_\Gamma \cdot (\kappa \nabla_\Gamma g) + cg = f$$

is uniquely solvable for all $f \in H^{-1}(\Gamma)$.

17.4. The Laplace-Beltrami eigensystem. Prove Proposition 17.4. Characterize $H^1(\Gamma)$ and

$$\{g \in H^1(\Gamma) : \Delta_\Gamma g \in L^2(\Gamma)\}$$

in terms of eigenfunction expansion.

17.5. Screens. Let $\Gamma_{\mathrm{scr}} \subset \Gamma$ be such that $\chi_{\Gamma_{\mathrm{scr}}} \in L^\infty(\Gamma)$. We define the space

$$\widetilde{H}^1(\Gamma_{\mathrm{scr}}) := \{g \in H^1(\Gamma) : g\chi_{\Gamma_{\mathrm{scr}}} = g\} = \{g \in H^1(\Gamma) : g = 0 \text{ in } \Gamma \setminus \Gamma_{\mathrm{scr}}\}.$$

(a) Show that $\widetilde{H}^1(\Gamma_{\mathrm{scr}})$ is closed and does not contain constants.

(b) Show that $\|\nabla_\Gamma \cdot \|_\Gamma$ is an equivalent norm in this space.

(c) For data $f \in L^2(\Gamma)$, prove the well-posedness of the problem

$$g \in \widetilde{H}^1(\Gamma_{\mathrm{scr}}), \qquad \langle \nabla_\Gamma g, \nabla_\Gamma h \rangle_\Gamma = \langle f, h \rangle_\Gamma \qquad \forall h \in \widetilde{H}^1(\Gamma_{\mathrm{scr}}).$$

17.6. Prove that

$$\|u\|_{H^{1/2}(\mathbb{R}^{d-1})} \leq \|u\|_{\mathbb{R}^{d-1}}^{1/2} \|u\|_{1, \mathbb{R}^{d-1}}^{1/2} \qquad \forall u \in H^1(\mathbb{R}^d) = H^1(\mathbb{R}^d),$$

and therefore

$$\|g\|_{1/2,\Gamma} \le C\|g\|_{\Gamma}^{1/2}\|g\|_{1,\Gamma}^{1/2} \qquad \forall g \in H^1(\Gamma).$$

17.7. Show that in the space $V := \{u \in H^1(\Omega) : \gamma u \in H^1(\Gamma)\}$, we have the inequality

$$\|\gamma u\|_{\Gamma}^2 \le C(\|\nabla u\|_{\Omega}^2 + \|\nabla_\Gamma \gamma u\|_{\Gamma}^2 + |j(u)|^2),$$

where $j : V \to \mathbb{R}$ is any bounded linear functional such that $j(1) \ne 0$. (**Hint.** Use compactness. The argument can be made very simple using Lemma 14.2.)

17.8. Study the problem (with data $f \in L^2(\Omega)$ and $h \in H^{-1/2}(\Gamma)$)

$$u \in H^1(\Omega), \qquad -\Delta u = f, \qquad \partial_n u - \Delta_\Gamma \gamma u = h, \qquad (u,1)_\Omega = 0,$$

showing that

$$(f,1)_\Omega + \langle h,1 \rangle_\Gamma = 0$$

is a necessary and sufficient condition for existence and uniqueness of solutions.

Appendix A

Review material

A.1   The divergence theorem 465
A.2   Analysis 466
A.3   Banach spaces 469
A.4   Hilbert spaces 471

A.1   The divergence theorem

Let $\Omega = (a_1, b_1) \times (a_2, b_2) \times \ldots \times (a_d, b_d)$ and $f \in \mathcal{C}^1(\overline{\Omega})$. We have

$$\int_\Omega (\partial_{x_d} f)(\mathbf{x}) d\mathbf{x} = \int_{\Omega_{d-1}} f(\tilde{\mathbf{x}}, b_d) d\tilde{\mathbf{x}} - \int_{\Omega_{d-1}} f(\tilde{\mathbf{x}}, a_d) d\tilde{\mathbf{x}}, s$$

where $\Omega_{d-1} = (a_1, b_1) \times \ldots \times (a_{d-1}, b_{d-1})$. If $\mathbf{n} = (n_1, \ldots, n_d) : \partial\Omega \to \mathbb{R}^d$ is the unit outward oriented normal vector field, the above can be rewritten as

$$\int_\Omega (\partial_{x_d} f)(\mathbf{x}) d\mathbf{x} = \int_{\partial\Omega} f(\mathbf{x}) n_d(\mathbf{x}) d\Gamma(\mathbf{x}).$$

The result can clearly be proved for the other variables, leading to

$$\int_\Omega (\partial_{x_i} f)(\mathbf{x}) d\mathbf{x} = \int_{\partial\Omega} f(\mathbf{x}) n_i(\mathbf{x}) d\Gamma(\mathbf{x}). \tag{A.1}$$

Note that, once (A.1) has been proved in a domain Ω, we have the divergence theorem in the same domain, by applying (A.1) to the components of a vector field $\mathbf{f} \in \mathcal{C}^1(\overline{\Omega}; \mathbb{R}^d)$:

$$\int_\Omega (\nabla \cdot \mathbf{f})(\mathbf{x}) d\mathbf{x} = \int_{\partial\Omega} \mathbf{f}(\mathbf{x}) \cdot \mathbf{n}(\mathbf{x}) d\Gamma(\mathbf{x}). \tag{A.2}$$

The d-dimensional divergence theorem implies the 'integration by parts formula'

$$\int_\Omega (\nabla \cdot \mathbf{f})(\mathbf{x}) g(\mathbf{x}) d\mathbf{x} + \int_\Omega \mathbf{f}(\mathbf{x}) \cdot (\nabla g)(\mathbf{x}) d\mathbf{x} = \int_{\partial\Omega} g(\mathbf{x}) \mathbf{f}(\mathbf{x}) \cdot \mathbf{n}(\mathbf{x}) d\Gamma(\mathbf{x}), \tag{A.3}$$

for $\mathcal{C}^1$ scalar and vector fields g and $\mathbf{f}$. Note that (A.3) implies (A.2) by taking $g \equiv 1$. Finally, (A.3) implies Green's formula

$$\int_\Omega (\Delta f)(\mathbf{x}) g(\mathbf{x}) d\mathbf{x} + \int_\Omega (\nabla f)(\mathbf{x}) \cdot (\nabla g)(\mathbf{x}) d\mathbf{x} = \int_{\partial\Omega} g(\mathbf{x}) \nabla f(\mathbf{x}) \cdot \mathbf{n}(\mathbf{x}) d\Gamma(\mathbf{x}).$$

This shows that the only key theorem to be proved (in order to have the kind of integration by parts formulas that we handle in this book) is (A.1). We have seen (A.1) for a d-cell domain, that is, for a parallelepiped aligned along the coordinate axes. It is then very easy to prove (A.1) for domains Ω such that $\overline{\Omega} = \overline{\Omega_1} \cup \ldots \cup \overline{\Omega_L}$, where Ω_ℓ are pairwise disjoint d-cells. The proof follows from the realization that all integrals in internal boundaries (interfaces) vanish due to the opposing signs of the normal vectors thereon. This is the integration by parts formula that we use in Chapter 1.

The basic form of the divergence theorem (A.1) can easily be proved for domains of the form

$$\Omega = \{(\widetilde{\mathbf{x}}, x_d) : \widetilde{\mathbf{x}} \in \Omega_{d-1}, \quad a_d < x_d < \psi(\widetilde{\mathbf{x}})\}$$

where Ω_{d-1} is a $(d-1)$-cell and $\psi \in \mathcal{C}^1(\overline{\Omega}_{d-1})$ is such that $\psi(\widetilde{\mathbf{x}}) > a_d$ for all $\widetilde{\mathbf{x}}$. In this case, the gradient of ψ plays an important role in defining the exterior normal vector field to Ω. With simple transformations (swapping and flipping variables), we can also prove (A.1) for domains that are finite unions of domains like the above, and with this we can deal with many piecewise smooth domains.

A.2   Analysis

We assume that the reader is acquainted with the basic notions of Lebesgue measure and integration. Given two functions f and g defined on the same domain of $\mathbb{R}^d$, we will say that $f = g$ almost everywhere (typically shortened to a.e.) when the set

$$\{\mathbf{x} \in \Omega : f(\mathbf{x}) \neq g(\mathbf{x})\}$$

has zero measure. As usual in analysis, we will identify functions that coincide almost everywhere in a single equivalence class. Therefore, when we talk about measurable or integrable functions, we actually refer to classes of functions coinciding up to sets of zero measure, and any representative of the class can be used for the argument where it is used. However, when a continuous function is mentioned in a context of integrable functions, it is the continuous representative of the class (and not the other functions that are equal to it almost everywhere) that is chosen for the argument and de facto the class is reduced to a single element, with all other almost equal functions being ignored.

The space on a measurable set Ω (in this book Ω will typically be an open set) $L^p(\Omega)$, with $1 \leq p < \infty$ will be the space of functions f (classes of functions) such that $|f|^p$ is integrable, and we will denote

$$\|f\|_{L^p(\Omega)} := \left(\int_\Omega |f(\mathbf{x})|^p \mathrm{d}\mathbf{x} \right)^{1/p}.$$

We will say that a function $f : \Omega \to \mathbb{R}$ is essentially bounded, and we will write $f \in L^\infty(\Omega)$, when there exists $C \in \mathbb{R}$ such that $|f| \leq C$ almost everywhere. We will write

$$\|f\|_{L^\infty(\Omega)} := \inf\{C \in \mathbb{R} : |f| \leq C \quad \text{a.e.}\}.$$

We will also need to consider a class of weighted Lebesgue spaces. If $\omega \in L^\infty(\Omega)$ satisfies $\omega > 0$ almost everywhere, we will denote

$$\|f\|_{L^p_\omega(\Omega)} := \left(\int_\Omega |f(\mathbf{x})|^p \omega(\mathbf{x}) d\mathbf{x} \right)^{1/p}, \tag{A.4}$$

and consider the space $L^p_\omega(\Omega)$ of Lebesgue measurable functions such that the integral (A.4) is finite. In the language of functional analysis that we will introduce in the next section, the Riesz-Fischer theorem says that $L^p_\omega(\Omega)$ are Banach spaces. Note that the theorem also contains an interesting *furthermore* statement, about almost everywhere convergence.

Theorem A.1 (Riesz-Fischer). *The spaces $L^p_\omega(\Omega)$ with $1 \leq p < \infty$ are complete as metric spaces, where $d(f,g) := \|f - g\|_{L^p_\omega(\Omega)}$. Moreover, every convergent sequence in $L^p_\omega(\Omega)$ contains a subsequence that converges almost everywhere. Finally, $L^\infty(\Omega)$ is complete with respect to the metric $d(f,g) := \|f - g\|_{L^\infty(\Omega)}$.*

Theorem A.2 (Lebesgue's dominated convergence theorem). *Let $\{f_n\}$ be a sequence of integrable functions in Ω and $f : \Omega \to \mathbb{R}$, such that:*

(a) *$f_n \to f$ almost everywhere,*

(b) *there exists an integrable function g such that $|f_n| \leq g$ almost everywhere for all n.*

The above hypotheses imply that $f \in L^1(\Omega)$, that the sequence $\{f_n\}$ converges to f in $L^1(\Omega)$, and

$$\lim_{n\to\infty} \int_\Omega f_n(\mathbf{x}) d\mathbf{x} = \int_\Omega f(\mathbf{x}) d\mathbf{x}.$$

Theorem A.3 (Differentiation under the integral sign). *Let $f : (a,b) \times \Omega \to \mathbb{R}$ satisfy:*

(a) *$f(t, \cdot) \in L^1(\Omega)$ for all $t \in (a, b)$,*

(b) *for almost every $\mathbf{x} \in \Omega$, the function $t \mapsto f(t, \mathbf{x})$ is differentiable in (a, b),*

(c) *there exists $g \in L^1(\Omega)$ such that*

$$|\partial_t f(t, \mathbf{x})| \leq g(\mathbf{x}) \quad \text{a.e. in } \Omega \quad \forall t \in (a, b).$$

The above hypotheses imply that

$$\frac{d}{dt} \int_\Omega f(t, \mathbf{x})d\mathbf{x} = \int_\Omega \partial_t f(t, \mathbf{x})d\mathbf{x}.$$

Theorem A.4 (Fubini-Tonelli). *Let* $f : \mathbb{R}^{d_1+d_2} \equiv \mathbb{R}^{d_1} \times \mathbb{R}^{d_2} \to \mathbb{R}$ *be a measurable function.*

(a) *If* $f \in L^1(\mathbb{R}^{d_1+d_2})$, *then*

$$\int_{\mathbb{R}^{d_1+d_2}} f(\mathbf{x}, \mathbf{y})d\mathbf{x}d\mathbf{y} = \int_{\mathbb{R}^{d_1}} \left(\int_{\mathbb{R}^{d_2}} f(\mathbf{x}, \mathbf{y})d\mathbf{y} \right) d\mathbf{x}.$$

(b) *If*

$$\int_{\mathbb{R}^{d_1}} \left(\int_{\mathbb{R}^{d_2}} |f(\mathbf{x}, \mathbf{y})|d\mathbf{y} \right) d\mathbf{x} < \infty,$$

then $f \in L^1(\mathbb{R}^{d_1+d_2})$.

Theorem A.5 (Lebesgue's differentiation theorem). *If* $f \in L^1(\mathbb{R}^d)$, *then*

$$\lim_{\varepsilon \to 0^+} \frac{1}{|B(\mathbf{x}; \varepsilon)|} \int_{B(\mathbf{x};\varepsilon)} |f(\mathbf{y}) - f(\mathbf{x})|d\mathbf{y} = 0$$

for almost every $\mathbf{x}$.

Theorem A.6 (Rademacher). *If* $f : \Omega \to \mathbb{R}$ *is Lipschitz, then* f *is differentiable almost everywhere in* Ω.

Note that as a consequence of Rademacher's theorem, the partial derivatives of a Lipschitz function are measurable (they can be written as almost everywhere limits of sequences of measurable functions) and essentially bounded by the Lipschitz constant.

Theorem A.7 (Stone-Weierstrass). *If* K *is a compact subset of* $\mathbb{R}^d$ *and* $\mathcal{A} \subset \mathcal{C}(K; \mathbb{C})$ *satisfies:*

(a) $\mathcal{A}$ *is closed under conjugation and multiplication,*

(b) *for every* $\mathbf{x} \in K$, *there is an* $f \in \mathcal{A}$ *such that* $f(\mathbf{x}) \neq \mathbf{0}$,

(c) *for all* $\mathbf{x}, \mathbf{y} \in K$, *there exists an* $f \in \mathcal{A}$ *such that* $f(\mathbf{x}) \neq f(\mathbf{y})$,

then $\mathcal{A}$ *is dense in* $\mathcal{C}(K; \mathbb{C})$. *In the real case, the condition that* $\mathcal{A}$ *is closed under conjugation can be dropped.*

A.3 Banach spaces

We include here some basic concepts and results on normed and Banach spaces. The reader who is not acquainted with this language should spend some quality time with any of the many excellent textbooks on functional analysis that can be found in the market or with online materials.

Norms. A vector space X over the field $\mathbb{K} \in \{\mathbb{R}, \mathbb{C}\}$, endowed with a function $\| \cdot \| : X \to [0, \infty)$ satisfying

$$\|x\| = 0 \qquad \Longleftrightarrow \qquad x = 0,$$

positive homogeneity

$$\|\lambda x\| = |\lambda|\, \|x\| \qquad \forall \lambda \in \mathbb{K}, \quad x \in X,$$

and the triangle inequality

$$\|x + y\| \le \|x\| + \|y\| \qquad \forall x, y \in X$$

is called a normed space and the function $\| \cdot \|$ is called a norm in X. Every normed space includes a concept of distance $d(x, y) := \|x - y\|$, which makes it a metric space. Therefore, classical concepts of convergence and Cauchy sequences, open and closed sets, etc. are inherited from the basic theory of metric spaces. A normed space that is complete (every Cauchy sequence converges) is called a Banach space.

Operators. Since normed spaces are metric spaces, continuity of functions between normed spaces is equivalent to sequential continuity, that is, if X and Y are normed spaces over the same field, then $f : X \to Y$ is continuous if and only if it transforms convergent sequences $x_n \to x$ to convergent sequences $f(x_n) \to f(x)$. The space of continuous linear maps from a normed space X to its associated field is called the dual space of X and denoted X'. It is simple to see that for linear maps, continuity (and therefore sequential continuity) is equivalent to boundedness (the image of any bounded set is bounded) and that X' can be endowed with the norm

$$\|\ell\|_{X'} = \sup_{0 \neq x \in X} \frac{|\ell(x)|}{\|x\|},$$

which makes X' a Banach space, even if X is only a normed space. The absolute value in the definition of $\|\ell\|_{X'}$ is only needed if the field is $\mathbb{C}$, and it can be eliminated (this is easy to check) if the space is real. For the action of a bounded linear functional $\ell \in X'$ on an element of the space $x \in X$, it is common to use the (angled) bracket notation $\langle \ell, x \rangle_{X' \times X}$, where the bracket is linear in both components: in the second one because ℓ is linear and in the first one because that is how we define addition of functions and the product of a scalar by a function. Similar to what happens with the dual space, we can consider the set of continuous linear operators between two normed spaces X and Y. Continuous (that is, sequentially continuous) linear operators between normed spaces coincide with bounded linear operators and the vector space that contains all of them, $\mathcal{B}(X, Y)$, can be endowed with the norm

$$\|T\|_{X \to Y} := \sup_{0 \neq x \in X} \frac{\|Tx\|_Y}{\|x\|_X} = \sup\{\|Tx\|_Y : x \in X, \quad \|x\|_X = 1\}$$
$$= \inf\{C \geq 0 : \|Tx\|_Y \leq C\|x\|_X \quad \forall x \in X\}.$$

When Y is a Banach space, $\mathcal{B}(X, Y)$ is a Banach space too. A bounded linear operator between normed spaces, $T : X \to Y$, defines an adjoint operator

$$T' : Y' \to X', \qquad \langle T'y', x \rangle_{X' \times X} = \langle y', Tx \rangle_{Y' \times Y} \qquad y' \in Y', \qquad x \in X,$$

that is if $\chi \in Y'$, then $T'\chi = \chi \circ T : X \to \mathbb{K}$. The operator T' is also linear and bounded.

Reflexivity. The following concept of Banach space theory is not always easy to grasp on a first attempt. Let us start with a Banach space X (we will soon see why we need to start with a Banach space and a normed space will not do) and consider the dual space X' and the bidual space $X'' = (X')' := \{\rho : X' \to \mathbb{K} : \rho$ linear and bounded$\}$, which is a Banach space. Given $x \in X$ we can fix it in the second component of the duality product and define a linear functional

$$\langle \cdot, x \rangle_{X' \times X} : X' \to \mathbb{K}.$$

This functional is linear and bounded and, moreover

$$\|x\|_X = \|\langle \cdot, x \rangle_{X' \times X}\|_{X''} = \sup_{0 \neq \ell \in X'} \frac{|\langle \ell, x \rangle_{X' \times X}|}{\|\ell\|_{X'}}.$$

This equality is far from trivial and is one of the important consequences of the **Hahn-Banach** theorem. This map defines an isometric linear transformation

$$X \ni x \longmapsto \langle \cdot, x \rangle_{X' \times X} \in X'',$$

which is injective (as it is isometric) and allows us to understand X as a subspace of X''. When the transformation is surjective, that is, when for every $\rho \in X''$ there exists an element $x \in X$ such that $\rho = \langle \cdot, x \rangle_{X' \times X}$, we say that the space X is reflexive. Since X and X'' are isometrically identifiable when X is reflexive, and since X'' is a Banach space even if X is not complete, it is clear that the concept of reflexive space is meaningful only when applied to a Banach space. It is important to remember that reflexivity is not the property of X and X'' being isometrically isomorphic, but of the particular map $x \mapsto \langle \cdot, x \rangle_{X' \times X}$ being an isomorphism.

Bilinear forms and operators. If we have two real normed spaces X and Y, we can consider bilinear forms (functions) $a : X \times Y \to \mathbb{R}$. The space $X \times Y$ can be made a normed space by using the product norm

$$\|(x,y)\|^2_{X \times Y} := \|x\|^2_X + \|y\|^2_Y$$

(we can define many equivalent norms, that is, norms that define the same concept of convergence). A bilinear form $a : X \times Y \to \mathbb{R}$ is continuous if and only if it is bounded, which is easily seen to be equivalent to the existence of $C \geq 0$ such that

$$|a(x,y)| \leq C\|x\|_X \|y\|_Y \qquad \forall x \in X, \qquad y \in Y.$$

A bounded bilinear form defines two continuous linear operators

$$x \in X \longmapsto Ax := a(x,\cdot) \in Y', \qquad y \in Y \longmapsto A'y := a(\cdot,y) \in X'.$$

Here we have to be careful with the notation, since $A : X \to Y'$ would have an adjoint $A' : Y'' \to X'$ and we are considering $A' : Y \to X'$ instead. In the case of a reflexive space Y, both definitions can be identified. When X and Y are complex normed forms, we have the choice to study bilinear or sesquilinear forms. Bilinear forms lead to the same concepts above, where for sesquilinear forms (functions that are linear in the first variable and conjugate linear in the second), we would have to consider duals made up of continuous conjugate linear functionals.

Theorem A.8 (Banach isomorphism theorem). *If a bounded linear map between Banach spaces is bijective, its inverse is bounded.*

Theorem A.9 (Uniform boundedness principle). *If we have a sequence of bounded linear functionals $\{\ell_n\}$ on a Banach space X, and if for every $x \in X$ there exists C_x such that*

$$|\langle \ell_n, x \rangle_{X' \times X}| \leq C_x \qquad \forall n,$$

then there exists $C > 0$ such that

$$\|\ell_n\|_{X'} \leq C \qquad \forall n.$$

A.4   Hilbert spaces

An **inner product** on a vector space X over $\mathbb{R}$ is a bilinear function, which is typically named with a bracket

$$(\cdot,\cdot)_X : X \times X \to \mathbb{R},$$

which is symmetric

$$(x, y)_X = (y, x)_X \qquad \forall x, y \in X,$$

and positive definite

$$(x, x)_X > 0 \qquad \Longleftrightarrow \qquad x \neq 0.$$

(Note that as a bilinear form, $(0, 0)_X = 0$.) If X is a complex vector space, we demand the bracket to be sesquilinear (linear in the first component and conjugate linear in the second) and the symmetry is substituted by the Hermitian property

$$(x, y)_X = \overline{(y, x)_X} \qquad \forall x, y \in X,$$

which implies that $(x, x)_X \in \mathbb{R}$ for all x. Inner products satisfy the **Cauchy-Schwarz inequality**

$$|(x, y)_X| \leq (x, x)_X^{1/2} (y, y)_X^{1/2} \qquad \forall x, y \in X,$$

with equality if and only if x and y are proportional to each other. Due to the Cauchy-Schwarz inequality, it is easy to show that if X is an inner product space, then

$$\|x\|_X := (x, x)_X^{1/2}$$

defines a norm in X. In this sense we consider all inner product spaces to be normed spaces. The Cauchy-Schwarz inequality can be written in operational form as

$$\|x\|_X = \sup_{0 \neq y \in X} \frac{|(x, y)_X|}{\|y\|_X} \qquad \forall x \in X.$$

(The absolute value in the numerator is only needed in the complex case.) An inner product space is said to be a Hilbert space when it is a Banach space as a normed space, or equivalently, when it is complete as a metric space. An interesting property of inner product spaces is that their associated norm satisfies the parallelogram identity

$$\|x + y\|^2 + \|x - y\|^2 = 2(\|x\|^2 + \|y\|^2) \qquad \forall x, y \in X.$$

Just out of curiosity, and since the proof is quite surprising (it deals with 'purely' algebraic concepts and yet it uses some analytical tools) and not often included in textbooks, let us give here a proof of the reciprocal statement.

Proposition A.1. *Every normed space satisfying the parallelogram identity is an inner product space.*

Proof. We will do it for the real case and will leave the reader to fill in the gaps for the complex case. The inner product we propose is defined by

$$\langle u, v \rangle := \tfrac{1}{4}(\|v + u\|^2 - \|v - u\|^2).$$

We first note that (using the property $\|w\| = \|-w\|$)

$$\langle v, u \rangle = \tfrac{1}{4}(\|v + u\|^2 - \|v - u\|^2) = \tfrac{1}{4}(\|u + v\|^2 - \|u - v\|^2) = \langle u, v \rangle,$$

that is we have symmetry. Also

$$\langle u, u \rangle = \tfrac{1}{4}\|2u\|^2 = \|u\|^2,$$

so $\langle u, u \rangle \geq 0$ and $\langle u, u \rangle = 0$ if and only if $u = 0$. (This follows from the axioms defining a norm.)

Linearity in the first variable (the only one we need to check) is quite tricky. We first prove the following:

$$
\begin{aligned}
\langle u_1, v_1 \rangle + \langle u_2, v_2 \rangle =& \tfrac{1}{4}\big(\|u_1 + v_1\|^2 + \|u_2 + v_2\|^2 - \|u_1 - v_1\|^2 - \|u_2 - v_2\|^2\big) \\
=& \tfrac{1}{8}\big(\|u_1 + u_2 + v_1 + v_2\|^2 + \|u_1 - u_2 + v_1 - v_2\|^2 \\
& - \|u_1 + u_2 - v_1 - v_2\|^2 - \|u_1 - u_2 - v_1 + v_2\|^2\big) \\
=& \tfrac{1}{2}\big(\langle u_1 + u_2, v_1 + v_2 \rangle + \langle u_1 - u_2, v_1 - v_2 \rangle\big).
\end{aligned}
$$

We have just used the definition of the inner product to be and the parallelogram law. For easy reference, we repeat the formula:

$$\langle u_1, v_1 \rangle + \langle u_2, v_2 \rangle = \tfrac{1}{2}\big(\langle u_1 + u_2, v_1 + v_2 \rangle + \langle u_1 - u_2, v_1 - v_2 \rangle\big). \tag{A.5}$$

Using (A.5) we obtain

$$\langle u, v \rangle = \langle u, v \rangle + \langle u, 0 \rangle = \tfrac{1}{2}\langle 2u, v \rangle$$

(since $\langle u, 0 \rangle = 0$ as follows from the definition of the bracket), and by symmetry

$$\langle 2u, v \rangle = 2\langle u, v \rangle = \langle u, 2v \rangle. \tag{A.6}$$

Therefore

$$
\begin{aligned}
\langle u_1, v \rangle + \langle u_2, v \rangle =& \tfrac{1}{2}\langle u_1 + u_2, 2v \rangle && \text{(by (A.5) and } \langle w, 0 \rangle = 0) \\
=& \langle u_1 + u_2, v \rangle, && \text{(by (A.6))}
\end{aligned}
$$

which proves additivity in the first variable. Using this property $n - 1$ times, we show that

$$\langle n\,u, v \rangle = \langle u + \ldots + u, v \rangle = n\langle u, v \rangle \qquad \forall n \in \mathbb{Z}, \quad n \geq 0,$$

and also

$$\langle u, v \rangle + \langle -u, v \rangle = 0,$$

so

$$\langle nu, v \rangle = n\langle u, v \rangle \qquad \forall n \in \mathbb{Z}. \tag{A.7}$$

Applying this to $w = nu$, we obtain

$$\tfrac{1}{n}\langle w, v \rangle = \langle \tfrac{1}{n}w, v \rangle \qquad \forall n \in \mathbb{Z} \setminus \{0\}. \tag{A.8}$$

As a simple consequence of (A.7) and (A.8), we can show

$$\langle qu, v\rangle = q\langle u, v\rangle \qquad \forall q \in \mathbb{Q}. \tag{A.9}$$

The extension of (A.9) to real scalars needs a continuity argument. First, note that

$$
\begin{aligned}
\langle u, v\rangle &= \tfrac{1}{2}(\|u+v\|^2 - \|u\|^2 - \|v\|^2) && \text{(parallelogram law)}\\
&\le \tfrac{1}{2}((\|u\|+\|v\|)^2 - \|u\|^2 - \|v\|^2) && \text{(triangle inequality)}\\
&= \|u\|\|v\|,
\end{aligned}
$$

which (after applying this inequality to $-u$ and v, noticing that $\|-u\| = \|u\|$), yields the Cauchy-Schwarz inequality

$$|\langle u, v\rangle| \le \|u\|\|v\|.$$

Finally, if $r \in \mathbb{R}$ and $q \in \mathbb{Q}$, we have

$$
\begin{aligned}
|\langle ru, v\rangle - r\langle u, v\rangle| &= |\langle ru, v\rangle - \langle qu, v\rangle - (r-q)\langle u, v\rangle| && \text{(by (A.9))}\\
&= |\langle (r-q)u, v\rangle - (r-q)\langle u, v\rangle| && \text{(additivity)}\\
&\le |\langle (r-q)u, v\rangle| + |r-q||\langle u, v\rangle|\\
&\le \|(r-q)u\|\|v\| + |r-q|\|u\|\|v\| && \text{(Cauchy-Schwarz)}\\
&= 2|r-q|\|u\|\|v\|.
\end{aligned}
$$

Since we can take a sequence of rational numbers $\{q_n\}$ such that $q_n \to r$, it follows that $\langle ru, v\rangle = r\langle u, v\rangle$, which finishes the proof. $\qquad\square$

Orthogonality. Given a subspace V of a Hilbert space H we consider the set

$$V^\perp := \{u \in H : (u, v)_H = 0 \quad \forall v \in V\}.$$

It is simple to see that $V^\perp$ is a closed subspace of H and that $\overline{V}^\perp = V^\perp$.

Proposition A.2 (Best approximation - orthogonal projection). *Let V be a closed subspace of a Hilbert space H. Given $u \in H$, the problems*

$$\|u - v\| = \min! \qquad v \in V, \tag{A.10}$$

and

$$v \in V, \qquad v - u \in V^\perp \tag{A.11}$$

are uniquely solvable

Proof. We will show: (a) existence of solutions of the approximation problem (A.10), (b) uniqueness of a solution of the orthogonal decomposition problem (A.11); (c) equivalence of the problems.

Existence. Let $\delta := \inf\{\|u - v\| : v \in V\} \geq 0$. By definition there exists a sequence $\{v_n\}$ in V such that

$$\|u - v_n\| \to \delta.$$

By the parallelogram identity,

$$
\begin{aligned}
\|v_n - v_m\|^2 &= \|(u - v_n) - (u - v_m)\|^2 \\
&= 2\|u - v_n\|^2 + 2\|u - v_m\|^2 - 4\left\|u - \tfrac{1}{2}(v_n + v_m)\right\|^2 \\
&\leq 2\|u - v_n\|^2 + 2\|u - v_m\|^2 - 4\delta^2.
\end{aligned}
$$

It is easy to prove from the above inequality that $\{v_n\}$ is a Cauchy sequence. Therefore, there exists $v \in V$ (we use that V is closed in a Hilbert space) such that $v_n \to v$. In particular $v \in V$ satisfies

$$\|u - v\| \leq \|u - w\| \qquad \forall w \in V,$$

that is v solves (A.10).

Uniqueness. Let $v_1, v_2 \in V$ solve (A.11). Since $v_1 - v_2 \in V$, we have

$$(u - v_1, v_1 - v_2) = 0 \qquad (u - v_2, v_2 - v_1) = 0.$$

Subtracting these equalities it follows that $\|v_1 - v_2\|^2 = 0$ and therefore the solution to (A.11) is unique.

Problem (A.10) *implies problem* (A.11). If $w \in V$ satisfies $\|w\| = 1$, then

$$
\begin{aligned}
\|u - v\|^2 &\leq \|u - (v + (u - v, w)w)\|^2 \\
&= \|u - v\|^2 + |(u - v, w)|^2 - 2|(u - v, w)|^2,
\end{aligned}
$$

which, after elementary simplifications implies that $(u - v, w) = 0$.

Problem (A.11) *implies problem* (A.10). If $v \in V$ is a solution to (A.11), then, since $(u - v, v - w) = 0$, we have

$$\|u - w\|^2 = \|u - v + (v - w)\|^2 = \|u - v\|^2 + \|v - w\|^2 \geq \|u - v\|^2 \qquad \forall w \in V,$$

and v solves (A.10). This finishes the proof. $\qquad\square$

Consequences. The following results are simple consequences of Proposition A.2

(a) The operator $P : H \to V$ defined by the bext approximation onto V

$$Pu \in V, \qquad (Pu - u, v) = 0 \qquad \forall v \in V$$

is linear and bounded with $\|P\| = 1$ if V is not the trivial subspace. This is due to the fact that

$$\|u\|^2 = \|u - Pu\|^2 + \|Pu\|^2,$$

since Pu and $u - Pu$ are orthogonal and to the fact that $Pv = v$ for all $v \in V$.

(b) The orthogonal projection P provides an orthogonal decomposition of any space

$$H = V \oplus V^\perp$$

if V is closed, and

$$H = \overline{V} \oplus V^\perp$$

in general. (For the latter assertion, apply the orthogonal projection onto $\overline{V}$ and note that $V^\perp = (\overline{V})^\perp$.)

(c) A subspace $V \subset H$ is dense if and only if $V^\perp = \{0\}$.

Orthonormal sequences. Now let $\{\phi_n\}$ be an orthonormal sequence in H, i.e., a sequence such that $(\phi_n, \phi_m) = \delta_{nm}$, and let $V := \text{span}\{\phi_n : n \geq 1\}$. The convergent orthogonal series

$$Pu := \sum_{n=1}^\infty (u, \phi_n)\phi_n$$

defines the orthogonal projection of u onto $\overline{V}$ and we have **Bessel's inequality**

$$\sum_{n=1}^\infty |(u, \phi_n)|^2 = \|Pu\|^2 \leq \|u\|^2 \qquad \forall u \in H.$$

The completeness condition

$$(u, \phi_n) = 0 \qquad \forall n \qquad \implies \qquad u = 0 \tag{A.12}$$

is therefore equivalent to $V^\perp = \{0\}$, and thus to the denseness of V in H. This is clearly equivalent to $Pu = u$ for all $u \in H$ and implies **Parseval's identity**

$$\|u\|^2 = \sum_{n=1}^\infty |(u, \phi_n)|^2 \qquad \forall u \in H. \tag{A.13}$$

Since Parseval's identity clearly implies that $V^\perp = \{0\}$, we have that all the following statements are equivalent for a given orthonormal sequence:

(a) $\{\phi_n\}$ is complete orthonormal (i.e., (A.12) holds),

(b) $\text{span}\{\phi_n\}$ is dense in H,

(c) Parseval's identity (A.13) holds,

(d) every element can be reconstructed from the orthonormal series

$$u = \sum_{n=1}^\infty (u, \phi_n)\phi_n \qquad \forall u \in H.$$

Appendix B

Glossary

B.1   Commonly used terms 477
B.2   Some key spaces 478

B.1   Commonly used terms

1. **Adjoints in Hilbert spaces.** Given two Hilbert spaces X and Y and a bounded linear operator $A : X \to Y$, its adjoint is the operator $A' : Y' \to X'$ defined by

$$A'y' := \langle y', A \cdot \rangle_{Y' \times Y} \qquad \forall y' \in Y'.$$

The Hilbert space adjoint of A is the operator $A^* : Y \to X$ given by the relation

$$(Ax, y)_Y = (x, A^* y)_X \qquad \forall x \in X, \quad y \in Y.$$

2. **Coercivity.** A bilinear form in a real Hilbert space $a : V \times V \to \mathbb{R}$ is said to be coercive, when there exists $\alpha > 0$ such that

$$a(u, u) \geq \alpha \|u\|^2 \qquad \forall u \in V.$$

In the complex case, coercivity of a sesquilinear form is the condition

$$|a(u, u)| \geq \alpha \|u\|^2 \qquad \forall u \in V.$$

3. **Fredholm operators of index zero.** Given two Hilbert spaces X, Y, a bounded linear operator $A : X \to Y$ is Fredholm of index zero, whenever $A = B + K$, where B is a bounded isomorphism and K is compact. The Fredholm alternative holds for Fredholm operators of index zero. In particular, for those operators injectivity is equivalent to invertibility. Non-invertible Fredholm operators of index zero have finite-dimensional kernel and range $A = (\ker A^*)^\perp$.

4. **Strictly positive functions.** A function $c : \Omega \to \mathbb{R}$ is strictly positive when there exists $c_0 \in \mathbb{R}$, $c_0 > 0$ such that $c \geq c_0$ almost everywhere.

B.2   Some key spaces

The space $\mathcal{D}(\Omega)$ is the set containing $\varphi \in \mathcal{C}^\infty(\Omega)$ such that $\operatorname{supp}\varphi$ is compact and contained in Ω. The space of distributions $\mathcal{D}'(\Omega)$ is the set of sequentially continuous linear functionals defined on $\mathcal{D}(\Omega)$.

$$H^m(\Omega) := \{u \in L^2(\Omega) \,:\, \partial^\alpha u \in L^2(\Omega), \quad |\alpha| \le m\},$$
$$H_0^m(\Omega) := \text{closure of } \mathcal{D}(\Omega) \text{ in } H^m(\Omega),$$
$$H^{-1}(\Omega) := H_0^1(\Omega)',$$
$$H^{1/2}(\Gamma) := \{g \in L^2(\Gamma) \,:\, g = \gamma u, \quad u \in H^1(\Omega)\},$$
$$H^{-1/2}(\Gamma) := H^{1/2}(\Gamma)',$$
$$\mathbf{H}(\operatorname{div}, \Omega) := \{\mathbf{u} \in L^2(\Omega)^d \,:\, \nabla \cdot \mathbf{u} \in L^2(\Omega)\},$$
$$\mathbf{H}_0(\operatorname{div}, \Omega) := \text{closure of } \mathcal{D}(\Omega)^d \text{ in } \mathbf{H}(\operatorname{div}, \Omega),$$

The Schwartz class $\mathcal{S}(\mathbb{R}^d)$ contains all $\psi \in \mathcal{C}^\infty(\mathbb{R}^d)$ such that $p\,\partial^\alpha \psi \in L^\infty(\mathbb{R}^d)$ for all $p \in \mathcal{P}(\mathbb{R}^d)$ (polynomials in d variables) and $\alpha \in \mathbb{N}^d$. The space of tempered distributions $\mathcal{S}'(\mathbb{R}^d)$ is the set of continuous functionals defined on $\mathcal{S}(\mathbb{R}^d)$.

Bibliography

[1] Gabriel Acosta and Ricardo G. Durán. *Divergence operator and related inequalities*. SpringerBriefs in Mathematics. Springer, New York, 2017.

[2] Robert A. Adams and John J. F. Fournier. *Sobolev spaces*, volume 140 of *Pure and Applied Mathematics*. Elsevier/Academic Press, Amsterdam, second edition, 2003.

[3] Mikhail S. Agranovich. *Sobolev spaces, their generalizations and elliptic problems in smooth and Lipschitz domains*. Springer Monographs in Mathematics, 2015. Revised translation of the 2013 Russian original.

[4] C. Amrouche, C. Bernardi, M. Dauge, and V. Girault. Vector potentials in three-dimensional non-smooth domains. *Math. Methods Appl. Sci.*, 21(9):823–864, 1998.

[5] Douglas N. Arnold, Richard S. Falk, and Ragnar Winther. Finite element exterior calculus: from Hodge theory to numerical stability. *Bull. Amer. Math. Soc. (N.S.)*, 47(2):281–354, 2010.

[6] Kendall Atkinson and Weimin Han. *Theoretical numerical analysis*, volume 39 of *Texts in Applied Mathematics*. Springer, Dordrecht, third edition, 2009. A functional analysis framework.

[7] Andreas Axelsson and Alan McIntosh. Hodge decompositions on weakly Lipschitz domains. In *Advances in analysis and geometry*, pages 3–29. Birkhäuser, Basel, 2004.

[8] Ivo Babuška and A. K. Aziz. Survey lectures on the mathematical foundations of the finite element method. pages 1–359, 1972. With the collaboration of G. Fix and R. B. Kellogg.

[9] Ivo Babuška and Gabriel N. Gatica. On the mixed finite element method with Lagrange multipliers. *Numer. Methods Partial Differential Equations*, 19(2):192–210, 2003.

[10] Constantin Bacuta. Sharp stability and approximation estimates for symmetric saddle point systems. *Appl. Anal.*, 95(1):226–237, 2016.

[11] Faker Ben Belgacem, Christine Bernardi, Martin Costabel, and Monique Dauge. Un résultat de densité pour les équations de Maxwell. *C. R. Acad. Sci. Paris Sér. I Math.*, 324(6):731–736, 1997.

[12] Christine Bernardi, Claudio Canuto, and Yvon Maday. Generalized inf-sup conditions for Chebyshev spectral approximation of the Stokes problem. *SIAM J. Numer. Anal.*, 25(6):1237–1271, 1988.

[13] Christine Bernardi, Martin Costabel, Monique Dauge, and Vivette Girault. Continuity properties of the inf-sup constant for the divergence. *SIAM J. Math. Anal.*, 48(2):1250–1271, 2016.

[14] Daniele Boffi, Franco Brezzi, and Michel Fortin. *Mixed finite element methods and applications*, volume 44 of *Springer Series in Computational Mathematics*. Springer-Verlag, Heidelberg, 2013.

[15] M. E. Bogovskiĭ. Solution of the first boundary value problem for an equation of continuity of an incompressible medium. *Dokl. Akad. Nauk SSSR*, 248(5):1037–1040, 1979.

[16] V. Bonnaillie-Noël, M. Dambrine, F. Hérau, and G. Vial. On generalized Ventcel's type boundary conditions for Laplace operator in a bounded domain. *SIAM J. Math. Anal.*, 42(2):931–945, 2010.

[17] A. S. Bonnet-Ben Dhia, P. Ciarlet, Jr., and C. M. Zwölf. Time harmonic wave diffraction problems in materials with sign-shifting coefficients. *J. Comput. Appl. Math.*, 234(6):1912–1919, 2010.

[18] Anne-Sophie Bonnet-Ben Dhia, Lucas Chesnel, and Patrick Ciarlet, Jr. T-coercivity for scalar interface problems between dielectrics and metamaterials. *ESAIM Math. Model. Numer. Anal.*, 46(6):1363–1387, 2012.

[19] Dietrich Braess. *Finite elements*. Cambridge University Press, Cambridge, third edition, 2007. Theory, fast solvers, and applications in elasticity theory, Translated from the German by Larry L. Schumaker.

[20] J. H. Bramble and S. R. Hilbert. Estimation of linear functionals on Sobolev spaces with application to Fourier transforms and spline interpolation. *SIAM J. Numer. Anal.*, 7:112–124, 1970.

[21] James H. Bramble. A proof of the inf-sup condition for the Stokes equations on Lipschitz domains. *Math. Models Methods Appl. Sci.*, 13(3):361–371, 2003. Dedicated to Jim Douglas, Jr. on the occasion of his 75th birthday.

[22] Alberto Bressan. *Lecture notes on functional analysis*, volume 143 of *Graduate Studies in Mathematics*. American Mathematical Society, Providence, RI, 2013. With applications to linear partial differential equations.

[23] Haim Brezis. *Functional analysis, Sobolev spaces and partial differential equations*. Universitext. Springer-Verlag, New York, 2011.

[24] F. Brezzi. On the existence, uniqueness and approximation of saddle-point problems arising from Lagrangian multipliers. *Rev. Française Automat. Informat. Recherche Opérationnelle Sér. Rouge*, 8(R-2):129–151, 1974.

[25] Franco Brezzi and Michel Fortin. *Mixed and hybrid finite element methods*, volume 15 of *Springer Series in Computational Mathematics*. Springer-Verlag, New York, 1991.

[26] A. Buffa and P. Ciarlet, Jr. On traces for functional spaces related to Maxwell's equations. I. An integration by parts formula in Lipschitz polyhedra. *Math. Methods Appl. Sci.*, 24(1):9–30, 2001.

[27] A. Buffa and P. Ciarlet, Jr. On traces for functional spaces related to Maxwell's equations. II. Hodge decompositions on the boundary of Lipschitz polyhedra and applications. *Math. Methods Appl. Sci.*, 24(1):31–48, 2001.

[28] A. Buffa, M. Costabel, and D. Sheen. On traces for $\mathbf{H}(\mathbf{curl}, \Omega)$ in Lipschitz domains. *J. Math. Anal. Appl.*, 276(2):845–867, 2002.

[29] Annalisa Buffa. Remarks on the discretization of some noncoercive operator with applications to heterogeneous Maxwell equations. *SIAM J. Numer. Anal.*, 43(1):1–18, 2005.

[30] Annalisa Buffa and Ralf Hiptmair. Galerkin boundary element methods for electromagnetic scattering. In *Topics in computational wave propagation*, volume 31 of *Lect. Notes Comput. Sci. Eng.*, pages 83–124. Springer-Verlag, Berlin, 2003.

[31] Jean Céa. Approximation variationnelle des problèmes aux limites. *Ann. Inst. Fourier (Grenoble)*, 14(fasc. 2):345–444, 1964.

[32] Ph. Clément and C. A. Timmermans. On C_0-semigroups generated by differential operators satisfying Ventcel's boundary conditions. *Nederl. Akad. Wetensch. Indag. Math.*, 48(4):379–387, 1986.

[33] M. Costabel. Symmetric methods for the coupling of finite elements and boundary elements (invited contribution). In *Boundary elements IX, Vol. 1 (Stuttgart, 1987)*, pages 411–420. Computational Mechanics, Southampton, 1987.

[34] Martin Costabel. Boundary integral operators on Lipschitz domains: elementary results. *SIAM J. Math. Anal.*, 19(3):613–626, 1988.

[35] Martin Costabel. A coercive bilinear form for Maxwell's equations. *J. Math. Anal. Appl.*, 157(2):527–541, 1991.

[36] Martin Costabel, Michel Crouzeix, Monique Dauge, and Yvon Lafranche. The inf-sup constant for the divergence on corner domains. *Numer. Methods Partial Differential Equations*, 31(2):439–458, 2015.

[37] Martin Costabel and Monique Dauge. Un résultat de densité pour les équations de Maxwell régularisées dans un domaine lipschitzien. *C. R. Acad. Sci. Paris Sér. I Math.*, 327(9):849–854, 1998.

[38] Martin Costabel and Monique Dauge. Maxwell and Lamé eigenvalues on polyhedra. *Math. Methods Appl. Sci.*, 22(3):243–258, 1999.

[39] Martin Costabel and Monique Dauge. Weighted regularization of Maxwell equations in polyhedral domains. A rehabilitation of nodal finite elements. *Numer. Math.*, 93(2):239–277, 2002.

[40] Martin Costabel and Monique Dauge. On the inequalities of Babuška-Aziz, Friedrichs and Horgan-Payne. *Arch. Ration. Mech. Anal.*, 217(3):873–898, 2015.

[41] Martin Costabel and Alan McIntosh. On Bogovskiĭ and regularized Poincaré integral operators for de Rham complexes on Lipschitz domains. *Math. Z.*, 265(2):297–320, 2010.

[42] M. Dambrine and D. Kateb. Persistency of wellposedness of Ventcel's boundary value problem under shape deformations. *J. Math. Anal. Appl.*, 394(1):129–138, 2012.

[43] Monique Dauge. *Elliptic boundary value problems on corner domains*, volume 1341 of *Lecture Notes in Mathematics*. Springer-Verlag, Berlin, 1988. Smoothness and asymptotics of solutions.

[44] Robert Dautray and Jacques-Louis Lions. *Mathematical analysis and numerical methods for science and technology. Vol. 4*. Springer-Verlag, Berlin, 1990. Integral equations and numerical methods, With the collaboration of Michel Artola, Philippe Bénilan, Michel Bernadou, Michel Cessenat, Jean-Claude Nédélec, Jacques Planchard and Bruno Scheurer, Translated from the French by John C. Amson.

[45] Françoise Demengel and Gilbert Demengel. *Functional spaces for the theory of elliptic partial differential equations*. Universitext. Springer-Verlag, London; EDP Sciences, Les Ulis, 2012. Translated from the 2007 French original by Reinie Erné.

[46] J. Deny and J. L. Lions. Les espaces du type de Beppo Levi. *Ann. Inst. Fourier, Grenoble*, 5:305–370 (1955), 1953–54.

[47] Ricardo G. Durán and Maria Amelia Muschietti. An explicit right inverse of the divergence operator which is continuous in weighted norms. *Studia Math.*, 148(3):207–219, 2001.

[48] G. Duvaut and J.-L. Lions. *Les inéquations en mécanique et en physique.* Dunod, Paris, 1972. Travaux et Recherches Mathématiques, No. 21.

[49] Alexandre Ern and Jean-Luc Guermond. *Theory and practice of finite elements,* volume 159 of *Applied Mathematical Sciences.* Springer-Verlag, New York, 2004.

[50] G. P. Galdi. *An introduction to the mathematical theory of the Navier-Stokes equations.* Springer Monographs in Mathematics. Springer-Verlag, New York, second edition, 2011. Steady-state problems.

[51] Giovanni P. Galdi. *An introduction to the mathematical theory of the Navier-Stokes equations. Vol. I,* volume 38 of *Springer Tracts in Natural Philosophy.* Springer-Verlag, New York, 1994. Linearized steady problems.

[52] Martin J. Gander and Gerhard Wanner. From Euler, Ritz, and Galerkin to modern computing. *SIAM Rev.*, 54(4):627–666, 2012.

[53] Gabriel N. Gatica and George C. Hsiao. *Boundary-field equation methods for a class of nonlinear problems,* volume 331 of *Pitman Research Notes in Mathematics Series.* Longman, Harlow, 1995.

[54] Gabriel N. Gatica and Salim Meddahi. Finite element analysis of a time harmonic Maxwell problem with an impedance boundary condition. *IMA J. Numer. Anal.*, 32(2):534–552, 2012.

[55] Gabriel N. Gatica and Francisco-Javier Sayas. Characterizing the inf-sup condition on product spaces. *Numer. Math.*, 109(2):209–231, 2008.

[56] David Gilbarg and Neil S. Trudinger. *Elliptic partial differential equations of second order.* Classics in Mathematics. Springer-Verlag, Berlin, 2001. Reprint of the 1998 edition.

[57] Vivette Girault and Pierre-Arnaud Raviart. *Finite element methods for Navier-Stokes equations,* volume 5 of *Springer Series in Computational Mathematics.* Springer-Verlag, Berlin, 1986. Theory and algorithms.

[58] P. Grisvard. *Singularities in boundary value problems,* volume 22 of *Recherches en Mathématiques Appliquées [Research in Applied Mathematics].* Masson, Paris; Springer-Verlag, Berlin, 1992.

[59] Pierre Grisvard. *Elliptic problems in nonsmooth domains,* volume 69 of *Classics in Applied Mathematics.* Society for Industrial and Applied Mathematics (SIAM), Philadelphia, PA, 2011.

[60] Houssem Haddar, Ralf Hiptmair, Peter Monk, and Rodolfo Rodríguez. *Computational electromagnetism,* volume 2148 of *Lecture Notes in Mathematics.* Springer-Verlag, Cham; Fondazione C.I.M.E., Florence,

2015. Notes from the CIME School held in Cetraro, June 9–14, 2014, Edited by Alfredo Bermúdez de Castro and Alberto Valli, Fondazione CIME/CIME Foundation Subseries.

[61] Hou De Han. A new class of variational formulations for the coupling of finite and boundary element methods. *J. Comput. Math.*, 8(3):223–232, 1990.

[62] Dorothee D. Haroske and Hans Triebel. *Distributions, Sobolev spaces, elliptic equations.* EMS Textbooks in Mathematics. European Mathematical Society (EMS), Zürich, 2008.

[63] Stefan Hildebrandt and Ernst Wienholtz. Constructive proofs of representation theorems in separable Hilbert space. *Comm. Pure Appl. Math.*, 17:369–373, 1964.

[64] George C. Hsiao and Wolfgang L. Wendland. *Boundary integral equations*, volume 164 of *Applied Mathematical Sciences*. Springer-Verlag, Berlin, 2008.

[65] Claes Johnson and J.-Claude Nédélec. On the coupling of boundary integral and finite element methods. *Math. Comp.*, 35(152):1063–1079, 1980.

[66] Mika Juntunen and Rolf Stenberg. Analysis of finite element methods for the Brinkman problem. *Calcolo*, 47(3):129–147, Sep 2010.

[67] N. Kikuchi and J. T. Oden. *Contact problems in elasticity: a study of variational inequalities and finite element methods*, volume 8 of *SIAM Studies in Applied Mathematics*. Society for Industrial and Applied Mathematics (SIAM), Philadelphia, PA, 1988.

[68] David Kinderlehrer and Guido Stampacchia. *An introduction to variational inequalities and their applications*, volume 31 of *Classics in Applied Mathematics*. Society for Industrial and Applied Mathematics (SIAM), Philadelphia, PA, 2000. Reprint of the 1980 original.

[69] Rainer Kress. *Linear integral equations*, volume 82 of *Applied Mathematical Sciences*. Springer-Verlag, New York, third edition, 2014.

[70] James R. Kuttler. Remarks on a Stekloff eigenvalue problem. *SIAM J. Numer. Anal.*, 9:1–5, 1972.

[71] O. A. Ladyzhenskaya. *The mathematical theory of viscous incompressible flow.* Second English edition, revised and enlarged. Translated from the Russian by Richard A. Silverman and John Chu. Mathematics and its Applications, Vol. 2. Gordon and Breach, New York, 1969.

[72] P. D. Lax and A. N. Milgram. Parabolic equations. In *Contributions to the theory of partial differential equations*, Annals of Mathematics Studies, no. 33, pages 167–190. Princeton University Press, Princeton, 1954.

[73] William J. Layton, Friedhelm Schieweck, and Ivan Yotov. Coupling fluid flow with porous media flow. *SIAM J. Numer. Anal.*, 40(6):2195–2218 (2003), 2002.

[74] M. N. Le Roux. Méthode d'éléments finis pour la résolution numérique de problèmes extérieurs en dimension 2. *RAIRO Anal. Numér.*, 11(1):27–60, 112, 1977.

[75] Marie-Noëlle Le Roux. équations intégrales pour le problème du potentiel électrique dans le plan. *C. R. Acad. Sci. Paris Sér. A*, 278:541–544, 1974.

[76] J.-L. Lions. *Contrôle optimal de systèmes gouvernés par des équations aux dérivées partielles*. Avant propos de P. Lelong. Dunod, Paris; Gauthier-Villars, Paris, 1968.

[77] Antonio Márquez, Salim Meddahi, and Francisco-Javier Sayas. A decoupled preconditioning technique for a mixed Stokes-Darcy model. *J. Sci. Comput.*, 57(1):174–192, 2013.

[78] Antonio Márquez, Salim Meddahi, and Francisco-Javier Sayas. Strong coupling of finite element methods for the Stokes-Darcy problem. *IMA J. Numer. Anal.*, 35(2):969–988, 2015.

[79] William McLean. *Strongly elliptic systems and boundary integral equations*. Cambridge University Press, Cambridge, 2000.

[80] Norman G. Meyers and James Serrin. $H = W$. *Proc. Nat. Acad. Sci. U.S.A.*, 51:1055–1056, 1964.

[81] Peter Monk. *Finite element methods for Maxwell's equations*. Numerical Mathematics and Scientific Computation. Oxford University Press, New York, 2003.

[82] Gustavo A. Muñoz, Yannis Sarantopoulos, and Andrew Tonge. Complexifications of real Banach spaces, polynomials and multilinear maps. *Studia Math.*, 134(1):1–33, 1999.

[83] J. Naumann. Notes on the prehistory of Sobolev spaces. *Bol. Soc. Port. Mat.*, (63):13–55, 2010.

[84] J.-C. Nédélec and J. Planchard. Une méthode variationnelle d'éléments finis pour la résolution numérique d'un problème extérieur dans R^3. *Rev. Française Automat. Informat. Recherche Opérationnelle Sér. Rouge*, 7(R-3):105–129, 1973.

[85] Jindřich Nečas. Sur une méthode pour résoudre les équations aux dérivées partielles du type elliptique, voisine de la variationnelle. *Ann. Scuola Norm. Sup. Pisa (3)*, 16:305–326, 1962.

[86] Jindřich Nečas. *Equations aux Dérivées Partielles.* Presses de l'Université de Montréal, Montréal, 1965.

[87] Jindřich Nečas. *Direct methods in the theory of elliptic equations.* Springer Monographs in Mathematics. Springer-Verlag, Heidelberg, 2012. Translated from the 1967 French original by Gerard Tronel and Alois Kufner, Editorial coordination and preface by Šárka Nečasová and a contribution by Christian G. Simader.

[88] J.T. Oden. *Qualitative Methods in Nonlinear Mechanics.* Prentice-Hall, 1986.

[89] A. Pazy. *Semigroups of linear operators and applications to partial differential equations*, volume 44 of *Applied Mathematical Sciences.* Springer-Verlag, New York, 1983.

[90] Rainer Picard. On the boundary value problems of electro- and magnetostatics. *Proc. Roy. Soc. Edinburgh Sect. A*, 92(1-2):165–174, 1982.

[91] Francisco Javier Sayas. Infimum-supremum. *Bol. Soc. Esp. Mat. Apl. SeMA*, (41):19–40, 2007.

[92] Francisco-Javier Sayas. The validity of Johnson-Nédélec's BEM-FEM coupling on polygonal interfaces [reprint of mr2551202]. *SIAM Rev.*, 55(1):131–146, 2013.

[93] Laurent Schwartz. Sur l'impossibilité de la multiplication des distributions. *C. R. Acad. Sci. Paris*, 239:847–848, 1954.

[94] Laurent Schwartz. *Mathematics for the physical sciences.* Hermann, Paris; Addison-Wesley, Reading, MA, 1966.

[95] Laurent Schwartz. *Théorie des distributions.* Publications de l'Institut de Mathématique de l'Université de Strasbourg, No. IX-X. Nouvelle édition, entiérement corrigée, refondue et augmentée. Hermann, Paris, 1966.

[96] Manuel Solano. *Metodos de Elementos Finitos Mixtos para Problemas No-coercivos. Aplicaciones en Acustica y Elastodinamica.* 2007. Thesis (M.S.)–Universidad de Concepción, Chile.

[97] Luc Tartar. On the characterization of traces of a Sobolev space used for maxwell's equation. In *Proceedings of a Meeting.* Bordeaux, France, 1997.

[98] Luc Tartar. *An introduction to Sobolev spaces and interpolation spaces*, volume 3 of *Lecture Notes of the Unione Matematica Italiana*. Springer-Verlag, Berlin; UMI, Bologna, 2007.

[99] Michael E. Taylor. *Partial differential equations III. Nonlinear equations*, volume 117 of *Applied Mathematical Sciences*. Springer-Verlag, New York, second edition, 2011.

[100] Roger Temam. *Navier-Stokes equations*. AMS Chelsea Publishing, Providence, RI, 2001. Theory and numerical analysis, Reprint of the 1984 edition.

[101] Ch. Weber and P. Werner. A local compactness theorem for Maxwell's equations. *Mathematical Methods in the Applied Sciences*, 2(1):12–25, 1980.

[102] W.L. Wendland. Boundary element methods for ellipctic problems. In *Mathematical Theory of Finite and Boundary Element Methods*. Birkhäuser-Verlag, Basel, 1990.

[103] Jinchao Xu and Ludmil Zikatanov. Some observations on Babuška and Brezzi theories. *Numer. Math.*, 94(1):195–202, 2003.

Index

$\mathcal{C}^k(\overline{\Omega})$, 28
$\mathcal{D}(\Omega)$, 3
$\mathcal{D}'(\Omega)$, 7
H^2-regularity
    domains, 323
    in $\mathbb{R}^d$, 306
    in $\mathbb{R}^d_+$, 307
    interior, 323
$L^2_\circ(\Omega)$, 181, 229
$L^1_{\mathrm{loc}}(\Omega)$, 5
$\mathbf{L}^2(\Omega)$, 28, 80, 107
$\mathcal{P}(\mathbb{R}^d)$, 298
$\mathcal{P}_0(\Omega)$, 132, 159
$\mathcal{P}_1(\Omega)$, 162
$\mathcal{P}_k(\Omega)$, 148

adjoint, 76
annihilator, 351
approximation of identity, 31

Babuška-Brezzi conditions, 217
Banach fixed point theorem, 279
Beavers-Joseph-Saffman condition, 240
Bessel's inequality, 152, 478
bi-Lipschitz homeomorphism, 48
    transformation of H^1 functions, 49
    transformation of H^1_0 functions, 49
Bogovskiĭ operator, 232
boundary condition
    Dirichlet, 38, 73
    friction, 393
    impedance, 114, 169
    Neumann, 107
    nonlocal, 118, 353

    periodic, 149
    traction, 145
    Ventcel, 462
    weak Dirichlet, 219
boundary integral operators
    boundary integral equations, 338, 351, 360
    compactness properties, 342
    definition, 334
    double layer operator, 337
    equilibrium distribution, 353
    hypersingular operator, 336
    Laplacian in 3D, 350
    single layer operator, 336
Bramble-Hilbert lemma, 149
Browder-Minty theorem, 292

Calderón projection, 340
closed range theorem, 213
coercive
    bilinear form, 43
    functional, 394
compact operator, 128
    adjoint, 164
    sequential characterization, 129
complexification, 95
convection-diffusion, 93, 168, 402
    mixed form, 261
convergence
    $L^p(\Omega)$ implies $\mathcal{D}'(\Omega)$, 14
    in $\mathcal{D}(\Omega)$, 7
    of distributions, 13
convex
    functional, 389
    projection on sets, 381
convolution, 30
    by a test function, 31

490   *Index*

cutoff, 29

density
    of $\mathcal{C}^{\infty}(\overline{\Omega})$ in $H^1(\Omega)$, 67
    of $\mathcal{D}(\mathbb{R}^d)$ in $\mathsf{H}^s(\mathbb{R}^d)$, 327
    of $\mathcal{D}(\mathbb{R}^d)$ in $\mathcal{S}(\mathbb{R}^d)$, 311
    of $\mathcal{D}(\mathbb{R}^d)$ in $H^1(\mathbb{R}^d)$, 31
    of $H^{1/2}(\Gamma)$ in $L^2(\Gamma)$, 108
Deny-Lions theorem, 131, 147
Dirichlet form, 28
distribution, 7
    compactly supported, 327
    complex-valued, 25
    derivative, 10
    Dirac delta, 8
    Leibniz rule, 29
    multiplication by smooth
        functions, 26
    regular, 9
    tempered, 313
dual space, 41

eigenvalues
    Dirichlet, 181
    Laplace-Beltrami, 460
    Neumann, 181
    spectrum of a self-adjoint
        compact operator, 183
    Steklov, 197
embedding
    of $\mathsf{H}^s(\mathbb{R}^d)$ in $\mathcal{C}(\mathbb{R}^d)$, 324
    of $H^1(\Omega)$ in $L^p(\Omega)$, 281, 282
    of $H^m(\Omega)$ in $\mathcal{C}(\overline{\Omega})$, 328
extension
    $H^1(\Omega)$ functions, 66
    H^2 functions, 436
    by zero, 36
    harmonic, 206
    of $H^1(\mathbb{R}^d_+)$ functions, 84
    of $H^2(\mathbb{R}^d_+)$ functions, 88

flow
    Brinkman, 244
    Darcy, 228
    hydrodynamic stress, 238
    Navier-Stokes, 284
    Oseen, 253
    Stokes, 236
    Stokes-Darcy, 238, 268
Fourier transform
    in $L^2(\mathbb{R}^d)$, 302
    in $\mathcal{S}(\mathbb{R}^d)$, 299, 312
    in $\mathcal{S}'(\mathbb{R}^d)$, 315
Fréchet derivative, 378
Fréchet space, 310
Fredholm alternative
    general case, 166
    self-adjoint case, 153
Fredholm operator of index
    zero, 164
fundamental solution, 15, 26
    Yukawa operator, 330

Galerkin
    approximation of coercive
        problems, 102
    compact perturbations, 171
    Lipschitz strongly monotone
        operator equations, 295
    mixed problems, 275
    projection, 174
    Ritz-Galerkin projection, 101
    stability, 172
Gelfand triple, 76, 77
Green's operator, 180

Heaviside function
    derivative, 11
    smoothened, 4
Helmholtz decomposition, 428
Helmholtz equation, 99, 154, 169
    Fourier series expansion, 204
Hilbert bases, 137
Hilbert-Schmidt theorem, 184, 187,
    204

inf-sup condition
    divergence operator, 230
    product spaces, 266
    theory, 214
infinitesimal rigid motions, 141
interpolation theorems, 201, 207

jump relations, 334

Kirchhoff plate, 46, 161
    mixed formulation, 254
Korn
    first inequality, 81
    second inequality, 142–143, 149

Lamé constants, 80
Laplace-Beltrami
    equation, 460
    operator, 457
Lax-Milgram lemma
    complex case, 97
    real case, 90
    with side conditions, 92
layer potentials
    integral forms, 344
    Laplacian in 3D, 350
    operational definition, 343
    variational definition, 333
Lipschitz
    domain, 52
    epigraph, 53
    map, 48
    weak Lipschitz domain, 55
locally convex space, 309
locally integrable function, 5

maximal dissipative operator, 398
minimization problem
    constrained quadratic, 223
    Euler equations, 392
    quadratic, 40, 97
    variational inequality, 393
mixed variational problem
    Darcy, 228
    diffusion, 225
    theory, 215
mollification, 30, 31
Moore-Penrose pseudoinverse, 66

Navier-Lamé equations, 79

Parseval's identity, 137, 192, 478
partition of unity, 17

covering a Lipschitz domain, 56
Picard's criterion, 193
Plancherel's identity, 302
Poincaré-Friedrichs inequality, 35
Poincaré inequality, 132, 133
    in weighted spaces, 346
pullback, 57

reaction-diffusion, 45
    mixed form, 255
Reissner-Mindlin plates, 247
Rellich-Kondrachov theorem
    in $H^1(\Omega)$, 130
    in $H_0^1(\Omega)$, 130
resolvent set, 183
Riesz-Fréchet theorem
    complex case, 97
    real case, 41
    reflexivity, 222
    variational problem form, 42
right inverse, 66

saddle point problem
    mixed problems, 224
    relation to constrained
        minimization, 224
Schwartz class
    definition, 298
    metric, 310
singular value decomposition, 204
Sobolev space
    $H^1(\Omega)$, 27
    $H^1(\Omega; \mathbb{C})$, 96
    $H_0^1(\Omega)$, 34
    $H^2(\Omega)$, 44, 161
    $H^2(\mathbb{R}^d)$, 304
    $H_0^2(\Omega)$, 46, 163, 249
    $H^{-1/2}(\Gamma)$, 108
    $H^{-1}(\Omega)$, 77
    $H^{1/2}(\Gamma)$, 70
    $\mathbf{H}(\mathrm{curl}, \Omega)$, 411
    $\mathbf{H}^1(\Omega)$, 44, 79, 111, 139
    $\mathbf{H}^{-1}(\Omega)$, 230
    $\mathbf{H}_0(\mathrm{div}, \Omega)$, 228
    $\mathbf{H}_0^1(\Omega)$, 81, 139, 211

$\mathbf{H}(\text{div}, \Omega)$, 109
$H^s(\mathbb{R}^d)$, 316
$\widetilde{H}^{-1}(\Omega)$, 77
$W(\Omega)$ weighted, 345, 362
$W^{1,p}(\Omega)$, 46
star-shaped domain
    relation with Lipschitz domains,
        235
    with respect to a ball, 232
strongly monotone map, 280
support of a function, 3
surface gradient, 456
symmetric Jacobian, 80

trace, 68
    compactness of trace operator,
        158

half-space, 320
kernel, 71
Lipschitz domain, 69
normal trace of $\mathbf{H}(\text{div}, \Omega)$
    functions, 109
normal traction, 145
range, 70
reference cylinder, 69
two-sided, 116

variational inequality
    obstacle problem, 385
    Signorini problem, 387
Variational lemma, 6, 22

weak convergence, 128
well-posedness, 42